Springer Texts in Business and Economics

More information about this series at http://www.springer.com/series/10099

Wolfgang Eichhorn • Winfried Gleißner

Mathematics and Methodology for Economics

Applications, Problems and Solutions

Springer

Wolfgang Eichhorn
Karlsruhe Institute of Technology (KIT)
Karlsruhe, Germany

Winfried Gleißner
University of Applied Sciences Landshut
Landshut, Germany

ISSN 2192-4333 ISSN 2192-4341 (electronic)
Springer Texts in Business and Economics
ISBN 978-3-319-23352-9 ISBN 978-3-319-23353-6 (eBook)
DOI 10.1007/978-3-319-23353-6

Library of Congress Control Number: 2016932103

Springer Cham Heidelberg New York Dordrecht London

Printed on acid-free paper

Springer International Publishing AG Switzerland is part of Springer Science+Business Media (www.springer.com)

Preface

This book about mathematics and methodology for economics is the result of the lifelong teaching experience of the authors. It is written for university students as well as for students of a university of applied sciences. It is completely self-contained and does not assume any previous knowledge of high school mathematics. At the end of all chapters and sections, there are exercises such that the reader can test how familiar she or he is with the material of the preceding stuff. After each set of exercises, the answers are given to encourage the reader to tackle the problems.

The idea to write such a book was born in 1990 during an international meeting on functional equations which took place at the University of Graz, Austria. At this meeting a lot of fascinating applications of functional equations to solve mathematically formulated economic problems inspired János Aczél, Distinguished Professor of Mathematics, University of Waterloo, Ontario, Canada: He proposed to one of us (W.E.) to start such an adventure in a form of a textbook for beginners. Since then he supported the tentative steps into this direction by a great wealth of brilliant scientific advices. Later on he became for both of us the lodestar for our endeavour. Dear János, we owe you a great debt of gratitude.

For a basic course Chaps. 1 (sets, vectors, trigonometric functions, complex numbers), 3 (mappings and functions), 4 (vectors, matrices, systems of linear equations), 6 (functions, limits, derivations), 7 (important nonlinear functions), and 10 (integration) are sufficient. If a later course will discuss discrete models of economics, Chap. 12 (difference equations) should be covered, too. For continuous models, Chap. 11 (differential equations) is necessary. (However, we decided not to go very far into details.)

Chapter 2 gives an introduction to linear optimisation and game theory using production systems. These ideas are continued in Chaps. 5 and 9, which discusses the notion of a Nash Equilibrium. Chapter 8 deals with nonlinear optimisation.

Chapter 13, as the conclusion, reflects methodologically most of all that what we optimistically offered in Chaps. 1, 2, 3, 4, 5, 6, 7, 8, 9, 10, 11 and 12.

Many thanks go to Thomas Schlink for typing most of the manuscript in LATEX very conscientiously and to Dr. Roland Peyrer for his inspiring drawings, which were transformed to PSTricks, an additional package for graphics in Latex.

Karlsruhe, Germany
Landshut, Germany
Summer, 2015

Wolfgang Eichhorn
Winfried Gleißner

Contents

List of Figures

List of Tables

1 Sets, Numbers and Vectors

God created the natural numbers,
everything else is the work of mankind.
LEOPOLD KRONECKER (1823–1891)

1.1 Introduction

Notions like sets, numbers and vectors with which this introductory chapter deals, among others, are fundamental both to mathematical (quantitative) representations of relations in economics and to mathematical notions and methods which will be the subject of this book. The belief that mathematics and its applications to economics are just about calculations is mistaken. Mathematics and mathematicians are needed to discover or create and analyse structures in a logically sound way. Chapter 13 at the end of the book will deal, among other things, with the basics of mathematical–logical reasoning.

In the present chapter we not only summarise basic knowledge about *natural numbers, integers, rational* and *real numbers* but define also *complex numbers* as a particular case of vectors. They will make, among others, the derivation of important *trigonometric formulas* easier than usual. *Vectors* and *sets*, to be introduced in this chapter, form the basis of much that will follow.

1.2 Basics

Most of the contents of this section just restates the obvious or the well known. It may, however, be useful to remind the reader of these building stones in what follows.

A *set* is a collection of *distinct* objects (this is really just paraphrasing not defining; we do not define such apparently simple things in this book). The objects, of which it consists, are the *elements* of the set. For instance you are an element of

W. Eichhorn, W. Gleißner, *Mathematics and Methodology for Economics*,
Springer Texts in Business and Economics, DOI 10.1007/978-3-319-23353-6_1

(or: belong to) the set of all people who are reading this sentence ("belongs to" is a synonym of "element of"). A set is usually given by enumerating all its elements (if there are only finitely many of them) or by giving a procedure (often called "algorithm") enabling us to determine all its elements.

For instance, the set S consisting of the elements A, B, C is usually written as

$$S = \{A, B, C\} \quad \text{or} \quad S = \{A, C, B\} \quad \text{or} \quad S = \{B, A, C\} \quad \text{or}$$
$$S = \{B, C, A\} \quad \text{or} \quad S = \{C, A, B\} \quad \text{or} \quad S = \{C, B, A\}.$$

The order of the elements is irrelevant (unless told otherwise; if the order is of partial or total relevance then we speak of partially or totally ordered sets; to the latter belong the sequences with which we will deal in detail in Sect. 5.4; compare also Sects. 1.5 and 3.7).

The set of all positive integers, in other words the set of all *natural numbers* $1, 2, 3, \ldots$ is written as

$$\mathbb{N} = \{1, 2, 3, \ldots\}.$$

We also mention the notation

$$\mathbb{N} = \{n \mid n \text{ is a natural number}\}.$$

After n follows the condition imposed on n separated form n by $|$.

The symbol $\in$ reads "element of", while $\notin$ means "is not among the elements of" (or "does not belong to"). For instance,

$$B \in \{A, B, C\}, \ 126 \in \mathbb{N}, \ 3 \notin \{1, 4, 7\}.$$

In addition to the "natural numbers" we are also familiar with 0 (zero) and the negative integers (like $-5°$ in temperature). The *set of all integers* is denoted by

$$\mathbb{Z} = \{0, 1, -1, 2, -2, 3, -3, \ldots\}.$$

Similarly familiar are (or should be) the *set of all rational numbers*:

$$\mathbb{Q} = \left\{ \frac{m}{n} \mid m \in \mathbb{Z}, n \in \mathbb{N}, gcd(m, n) = 1 \right\},$$

that is, the set of *fractions* with integer *numerator* and positive integer *denominator*, whose greatest common divisor (gcd) is 1. We assume also that the rules for addition, subtraction, multiplication and division of rational numbers are known.

As is also known, *rational numbers can be represented as finite or periodic infinite decimal fractions*. Confining ourselves, for simplicity, to positive rational numbers, a *finite decimal fraction* can be written as

$$a_1a_2\dots a_n.b_1b_2\dots b_m = a_1 10^{n-1} + a_2 10^{n-2} + \dots + a_{n-1}10 + a_n$$
$$+\frac{b_1}{10} + \frac{b_2}{10^2} + \dots + \frac{b_m}{10^m}$$

(where $n \in \mathbb{N}$, $m \in \mathbb{N}$, $a_j \in \{0, 1, \dots, 9\}$ $(j = 1, 2, \dots, n)$, $b_k \in \{0, 1, \dots, 9\}$ $(k = 1, 2, \dots, m)$ and, of course, $10^2 = 100$, $10^3 = 1000, \dots$). For instance,

$$\frac{17}{8} = 2.125.$$

An *infinite decimal fraction* can be written as

$$a_1a_2\dots a_n.b_1b_2b_3\dots = a_1 10^{n-1} + a_2 10^{n-2} + \dots + a_{n-1}10 + a_n$$
$$+\frac{b_1}{10} + \frac{b_2}{10^2} + \frac{b_3}{10^3} + \dots .$$

On the right hand side we really have an infinite series. We will deal with infinite series in detail in Chaps. 5 and 6, here the example

$$\frac{237}{70} = 3.38571428571428\dots$$

should suffice to show what an infinite decimal fraction, for that matter what a periodic infinite decimal fraction is. The latter means that the same segment, here 857142, keeps repeating.

We demonstrate on the simple examples $\frac{101}{8}$ and $\frac{30}{13}$ why *every rational number equals either a finite or a periodic infinite decimal fraction*. In the long division the remainders have to be smaller than the denominator, so they have to be one of the numbers 0, 1, 2, 3, 4, 5, 6, 7 in the first case and one of 0, 1, 2, ..., 12 in the second. So sooner or later either the division ends or we get a previous remainder again and the period restarts. Indeed

$$\frac{101}{8} = 12.625 \quad \text{and} \quad \frac{30}{13} = 2.307692307692307692\dots .$$

We also show on another simple example why, conversely, *all periodic infinite decimal fractions equal rational numbers*. (That *finite decimal fractions are rational numbers*, is obvious: for instance $34.125 = \frac{34125}{1000} = \frac{273}{8}$.) Take

$$x = 5.4181818\dots .$$

Then

$$\begin{aligned} 1000x &= 5418.1818\ldots \\ 10x &= \quad 54.1818\ldots \end{aligned}$$

and, by subtraction (really multiplication and subtraction of infinite decimal fraction have to be justified but they are quite intuitive here),

$$990x = 5364, \quad \text{so} \quad x = \frac{5364}{990} = \frac{298}{55}.$$

There is an obvious way to make a (periodic) infinite decimal fraction out of a finite one:

$$31.46 = 31.460000\ldots$$

but we agree that, if in a decimal fraction (finite or infinite) there are only 0's from a place on (after the decimal point), then we omit them. There is also a less obvious way:

$$31.46 = 31.459999\ldots.$$

Indeed, using the above procedure for

$$x = 31.45999\ldots$$

we get

$$\begin{array}{rl} 1000x = & 31459.999\ldots \\ -\quad 100x = & 3145.999\ldots \\ \hline 900x = & 28314 \\ x = \dfrac{28314}{900} = & \dfrac{3146}{100} = 31.46. \end{array}$$

Actually, *those ending with 999... and those ending with 000... are the only infinite decimal fractions which equal finite ones and they are the only pairs of infinite decimal fractions which are equal without all their digits being equal* (in the same order).

Clearly there are also *non-periodic decimal fractions*; for instance

$$111.1010010001000010\ldots.$$

(While only 1's and 0's figure in it, there is no finite segment which keeps exactly repeating.) These (and their products by (-1)) are the *irrational numbers*. The numbers 2π (the length of the circumference of the unit circle) and $\sqrt{2}$ (the number whose square is 2) are also irrational. Actually, in a certain sense, which can be made precise, there are "many more" irrational than rational numbers. This is quite intuitive: we would be rather surprised if the same numbers in the same order kept repeating as winners in a lottery every fixed (albeit possibly large) number of weeks.

The rational and irrational numbers together form the set $\mathbb{R}$ of *real numbers*. It follows from the above that *every real number can be represented as a finite or infinite decimal fraction*—multiplied by (-1) if the real number was negative.

There is a pretty proof showing that $\sqrt{2}$ is indeed *irrational*, that is, it cannot be a rational number. We prove this *by contradiction*. (see Appendix): Suppose

$$\sqrt{2} = \frac{m}{n}$$

($n \in \mathbb{N}$, $m \in \mathbb{N}$ since $\sqrt{2}$ is positive). *We may choose m and n so that not both are even* (either just one or neither of them is even; an even number is an integer divisible by 2; an integer which is not even, is odd) because, if both the numerator and the denominator were even, then we could cancel the highest power of 2 by which both would be divisible (for instance $\frac{16}{24} = \frac{2}{3}$).

Squaring the above equation, we get

$$2 = \frac{m^2}{n^2}, \quad \text{that is,} \quad m^2 = 2n^2,$$

so m^2 is even. But then *also m would be even* (because the squares of odd numbers are odd):

$$m = 2k.$$

Substituting this into $m^2 = 2n^2$, we obtain

$$4k^2 = 2n^2, \quad \text{that is,} \quad n^2 = 2k^2.$$

So n^2 would be even, thus, by the above argument, *n would be even too.* But at the beginning of this proof we had *excluded* that both m and n are even.

This contradiction shows that our original supposition, that $\sqrt{2}$ is rational, cannot be true. Therefore $\sqrt{2}$ is irrational, as asserted.

The expression "irrational number" (like later "imaginary number") comes from a time in the distant past when only integers and fractions of integers were considered "reasonable". But there is nothing "unreasonable" about irrational numbers. In fact, in their geometric *representation on the straight line* they are quite indistinguishable from the rational numbers: If one chooses (Fig. 1.1) a point 0 and a point 1 on the line then every point represents a real number (either rational or irrational) and, conversely, every real number is represented by a point of that

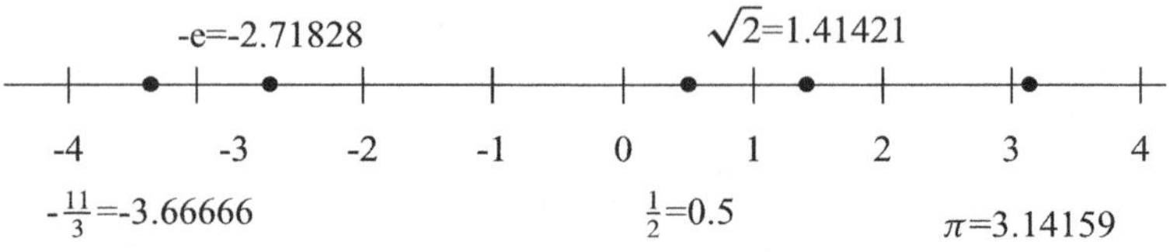

Fig. 1.1 Representation of real numbers on the straight line. The rational numbers $\frac{1}{2} = 0.5$ and $-\frac{11}{3} = -3.66\cdots$ are represented by the points • between 0 and 1, and -3 and -4, respectively. The irrational numbers $\sqrt{2} = 1.41\cdots$, $\pi = 3.14\cdots$, and $-e = -2.71\cdots$ are represented by the points • between 1 and 2, 3 and 4, and -3 and -2, respectively

line. We will identify that point with the real number which it represents (use them interchangeably) and call this line the "*real line*" or the "*number line*". We note that *any real number can be approximated both by rational and by irrational numbers as closely as one wants*, that is, the distance from the real number to an appropriately chosen rational resp. irrational number can be made as small as one wishes. The *distance* of two (real or rational or integer or positive) numbers x and y is defined by

$$d(a, b) = |b - a|,$$

where

$$|x| := \begin{cases} x & \text{if } x > 0 \\ 0 & \text{if } x = 0 \\ -x & \text{if } x < 0 \end{cases}$$

is the *absolute value* of x. (Here and in what follows $A := B$ *or* $B =: A$ *means that* A *is defined by* B.) Note (see also Fig. 1.1) that even for pairs a, b of positive numbers the difference $b - a$ may be negative but $|b - a|$ is always *nonnegative* (that is, either positive or 0).

We denote the set of nonnegative real numbers by $\mathbb{R}_+$, that of positive real numbers by $\mathbb{R}_{++}$:

$$\mathbb{R}_+ := \{x \,|\, x \in \mathbb{R} \text{ and } x \geq 0\}, \ \mathbb{R}_{++} := \{x \,|\, x \in \mathbb{R}, x > 0\}.$$

Similarly

$$\begin{aligned} \mathbb{R}_- &:= \{x \,|\, x \in \mathbb{R}, x \leq 0\}, \ \mathbb{R}_{--} := \{x \,|\, x \in \mathbb{R}, x < 0\}, \\ \mathbb{Q}_+ &:= \{x \,|\, x \in \mathbb{Q}, x \geq 0\}, \ \mathbb{Q}_{++} := \{x \,|\, x \in \mathbb{Q}, x > 0\}, \\ \mathbb{Q}_- &:= \{x \,|\, x \in \mathbb{Q}, x \leq 0\}, \ \mathbb{Q}_{--} := \{x \,|\, x \in \mathbb{Q}, x < 0\}. \end{aligned}$$

1.2.1 Exercises

1. Express the following periodic infinite decimal fractions as rational numbers:
   (a) 2.38888..., (b) 7.074074074...,
   (c) 5.76432143214321..., (d) 28.571428571428571428...,
   (e) 3.59999..., (f) 3.60000....
2. Express the following rational numbers as infinite fractions:
   (a) 61234/3, (b) 98765/6,
   (c) 11/123, (d) 77/666.
3. Write in the notation of this section the sets of numbers which can be described verbally as follows:
   (a) All numbers $x \in \mathbb{R}$ whose distance from $x = 3.50$ is smaller than or equal to 4.18.
   (b) All rational numbers smaller than $x = \sqrt{2}$.
   (c) All irrational numbers greater than or equal to $x = \sqrt{2}$.
4. Which of the following expressions are sets?
   (a) $\{2, 4, 7, 9\}$, (b) $d(3, 8) = |3 - 8|$,
   (c) $\{1, 6, 5, 8, 1\}$, (d) $\{\{5, 7\}, \{2\}, \{1, 4, 3\}\}$,
   (e) $\{\{8, 9\}, \{7, 8\}, \{8\}\}$, (f) $\{x \,|\, x \in \mathbb{R},\ x > 1,\ x < 2\}$.
5. (a) Let a and b be rational numbers. Are $a + b$, $a - b$, and, with $b \neq 0$, a/b rational numbers?
   (b) Let a be a rational number and λ be an irrational number. Are $a + \lambda$, $a - \lambda$, $a\lambda$, a/λ irrational numbers?
   (c) Is for any pair λ, μ of distinct irrational numbers $\lambda + \mu$, $\lambda\mu$, λ/μ irrational?

1.2.2 Answers

1. (a) $\frac{43}{18}$, (b) $\frac{191}{27}$, (c) $\frac{1\,152\,749}{199\,980}$, (d) $\frac{200}{7}$, (e) $\frac{18}{5}$, (f) $\frac{18}{5}$.
2. (a) 20 411.3333..., (b) 16 460.83333...,
   (c) 0.089 430 894 3..., (d) 0.115 615 6....

3. (a) $\{x \mid x \in \mathbb{R},\ d(x, 3.50) = |x - 3.50| \leq 4.18\}$
   $= \{x \mid x \in \mathbb{R},\ -0.68 \leq x \leq 7.68\}$,
   (b) $\left\{x \,\middle|\, x \in \mathbb{R},\ x = \frac{m}{n},\ m \in \mathbb{Z},\ n \in \mathbb{N},\ \frac{m}{n} < \sqrt{2}\right\}$
   $= \left\{x \,\middle|\, x \in \mathbb{Q},\ x < \sqrt{2}\right\}$,
   (c) $\left\{x \,\middle|\, x \in \mathbb{R} x \neq \frac{m}{n},\ m \in \mathbb{Z},\ n \in \mathbb{N},\ x \geq \sqrt{2}\right\}$
   $= \left\{x \,\middle|\, x \in \mathbb{R},\ \backslash\mathbb{Q},\ x \geq \sqrt{2}\right\}$.
4. The expressions (a), (d), (e), (f) are sets. The expression (b) means the distance (number) 5, *not* the set consisting of the single element 5 (that would be $\{5\}$). The expression (c) is no set, since not all numbers (elements) are distinct.

5. (a) Yes, (b) Yes,
   (c) No. For $\lambda = 1 + \sqrt{2}$, $\mu = 1 - \sqrt{2}$ we get $\lambda + \mu = 1$, for $\lambda = \sqrt{2}$, $\mu = \sqrt{1/2}$ we get $\lambda\mu = 1$ and $\lambda/\mu = 2$.

1.3 Subsets, Operations Between Sets

A set T is a subset of a set S if every element of T is also element of S (while elements of *S may or may not* be elements of *T*). This is written as

$$T \subset S \quad \text{or, what is the same,} \quad S \supset T,$$

and is sometimes verbalised as "*S contains T*". For instance,

$$\mathbb{N} \subset \mathbb{Z},\ \mathbb{Z} \subset \mathbb{Q},\ \mathbb{Q} \subset \mathbb{R}$$

(which also can be written as $\mathbb{N} \subset \mathbb{Z} \subset \mathbb{Q} \subset \mathbb{R}$),

$$\mathbb{R} \subset \mathbb{R},$$

$$\{3,5\} \subset \{8,5,3\},\ \{8\} \subset \{3,5,8\}.$$

Note from the last example that there are sets having only one element. It is often convenient to speak also about a *set with no element*, the *empty set* which is denoted by $\emptyset$. This is *not to be confused with the set* $\{0\}$ which has one element: the number 0.

Clearly, *if $T \subset S$ and $S \subset T$ then $S = T$*, that is, *S* and *T* are the same set (because every element of *T* belongs also to *S* and every element of *S* is also element of *T*).

The set *T* needs not be a subset of *S* in order to define

$$S \setminus T = \{x \mid x \in S \quad \text{but} \quad x \notin T\}$$

(which may be empty) as the *complement of T with respect to S*. But $S \setminus T$ is a subset of *S*. *Examples*:

$$\{3,4,6\} \setminus \{3,6\} = \{4\}, \quad \{3,4,6\} \setminus \{1,2,3\} = \{4,6\}, \quad \mathbb{R}_+ \setminus \mathbb{R}_{++} = \{0\}.$$

The union of the sets S and T (neither of which needs to be a subset of the other) *is the set V which contains those elements which belong either to S or to T* (or to both). In symbols:

$$V = S \cup T := \{x \mid x \in S \quad \text{or} \quad x \in T\}.$$

Examples

$$\{1,3,5\} \cup \{2,7,9\} = \{1,2,3,5,7,9\},$$
$$\{1,3,5\} \cup \{1,3,6\} = \{1,3,5,6\},$$
$$\{2,4,6,\ldots\} \cup \mathbb{N} = \mathbb{N},$$
$$\{2,4,6,\ldots\} \cup \{1,3,5,\ldots\} = \mathbb{N}$$

(in the last two examples $\{2,4,6,\ldots\}$ is, of course, the set of all even numbers and, in the last one, $\{1,3,5,\ldots\}$ is the set of all odd numbers).

Also

$$\mathbb{N} \cup \mathbb{Z} = \mathbb{Z},\ \mathbb{N} \cup \mathbb{R} = \mathbb{R},\ \mathbb{R}_{+} \cup \mathbb{R}_{-} = \mathbb{R},\ \mathbb{R}_{++} \cup \mathbb{R}_{-} = \mathbb{R}.$$

The reader can easily check that, for *all* sets S, T, W,

$$S \cup S = S, \qquad S \cup \emptyset = S,\ (S \setminus T) \cup T = S \cup T,$$
$$(S \setminus T) \cup S = S, \quad (S \cup T) \cup W = S \cup (T \cup W).$$

One can also define the union of three sets S, T, W:

$$S \cup T \cup W := \{x \mid x \in S \text{ or } x \in T \text{ or } x \in W\} \quad (= (S \cup T) \cup W),$$

or the union of any (even infinite) number of sets. One may use in this case the more convenient notation

$$\bigcup_{k=1}^{n} S_k = \{x \mid x \in S_1 \text{ or } x \in S_2 \text{ or } \ldots \text{ or } x \in S_n\}.$$

We use this occasion to call attention to a fine point. Let the sets A, B and C consist of the employees ("elements"; of course, a company consists of more than its employees but we will ignore this here) $a_1, a_2, \ldots, a_{10}, b_1, b_2, \ldots, b_{90}, c_1, c_2, \ldots, c_{35}$, respectively:

$$A = \{a_1, a_2, \ldots, a_{10}\},\ B = \{b_1, b_2, \ldots, b_{90}\},\ C = \{c_1, c_2, \ldots, c_{35}\}.$$

Then the set S defined in the next line is a *set of sets*

$$S = \{A, B, C\}$$

(continued)

which has three elements (A, B, and C) while the union

$$A \cup B \cup C = \{a_1, a_2, \ldots, a_{10}, b_1, b_2, \ldots b_{90}, c_1, c_2, \ldots, c_{35}\}$$

has 135 elements assuming that no individual is employed by more than one company.

The intersection of the sets S and T is the set W, the elements of which belong to both S and T. In symbols:

$$W = S \cap T = \{x \,|\, x \in S \quad \text{and} \quad x \in T\}.$$

If S and T have no element in common, then

$$S \cap T = \emptyset.$$

Examples

$$\{1,3,5\} \cap \{1,3,6\} = \{1,3\},\ \mathbb{R}_{--} \cap \mathbb{R}_{+} = \emptyset,\ \mathbb{R}_{-} \cap \mathbb{R}_{+} = \{0\}.$$

Again one can define also

$$S \cap T \cap V = \{x \mid x \in S \text{ and } x \in T \text{ and } x \in V\}$$

and

$$\bigcap_{k=1}^{n} S_k = \{x \mid x \in S_1 \text{ and } x \in S_2 \text{ and } \ldots \text{ and } x \in S_n\}.$$

and verify for any sets S, T, V

$$S \cap T \cap V = (S \cap T) \cap V = S \cap (T \cap V),$$

$$\begin{aligned} S \cap (T \cup V) &= (S \cap T) \cup (S \cap V), \\ S \cup (T \cap V) &= (S \cup T) \cap (S \cup V). \end{aligned} \tag{1.1}$$

We have also the *commutativity* of both $\cap$ and $\cup$:

$$S \cap T = T \cap S \quad \text{and} \quad S \cup T = T \cup S$$

(continued)

(why?), while

$$(S \cap T) \cap V = S \cap (T \cap V) \quad \text{and} \quad (S \cup T) \cup V = S \cup (T \cup V)$$

is called the *associativity* of $\cap$ and $\cup$, respectively, and the first and second part of (1.1) is the *distributivity* of $\cap$ over $\cup$ and of $\cup$ over $\cap$, respectively.

While these "identities" are quite important, one can construct many others.

The symbols $\forall$ ("for all") and $\exists$ ("there exists") help express some mathematical facts. For instance,

$$\forall x \in T : x \in S \quad \text{means} \quad T \subset S$$

and

$$\exists x \in S \quad \text{means} \quad S \neq \emptyset.$$

1.3.1 Exercises

1. Do the following expressions describe sets?

(a) $\{\{8,9\},\emptyset,\{0\}\}$, (b) $\{\{3,4\}\cap\{4,5\},\{4,5\}\}$,
(c) $\{\{3,4\}\cup\{4,5\},\{3,4,5\}\}$, (d) $\{\{3,4,5\}\cap\{4,5,6\},\{4,5\}\}$,
(e) $\emptyset\cap\{0\}$, (f) $\emptyset\cup\{0\}$.

2. Let $S = \{3,4,5\}, T = \{2,3\}$. Which of the following statements are correct?

(a) $T \subset S$, (b) $S \subset T$, (c) $S \neq T$, (d) $5 \subset S$,
(e) $2 \in T$, (f) $\{\{3,4\},5\} \subset S$, (g) $\{5,3\} \subset S$.

3. Write the elements of the following sets in a simpler form:
   (a) $(\{\alpha,\beta,\gamma,\delta\}\cup\{\alpha,\delta,\epsilon\})\cup\{\alpha,\omega\},\{\alpha,\beta,\gamma,\delta\}\cup(\{\alpha,\delta,\epsilon\}\cup\{\alpha,\omega\})$,
   (b) $(\{\alpha,\beta,\gamma,\delta\}\cap\{\alpha,\delta,\epsilon\})\cap\{\alpha,\delta,\omega\},\{\alpha,\beta,\gamma,\delta\}\cap(\{\alpha,\delta,\epsilon\}\cap\{\alpha,\delta,\omega\})$,
   (c) $(\{\alpha,\beta,\gamma,\delta\}\cap\{\alpha,\delta,\epsilon\})\cup\{\alpha,\omega\}$,
   (d) $\{\alpha,\beta,\gamma,\delta\}\cap(\{\alpha,\delta,\epsilon\}\cup\{\alpha,\omega\})$,
   (e) $(\{\alpha,\beta,\gamma,\delta\}\cap\{\alpha,\delta,\epsilon\})\cup\{\alpha,\beta,\gamma,\delta\}\cap\{\alpha,\omega\}$,
   (f) $\{\alpha,\beta,\gamma,\delta\}\setminus\{\gamma,\delta,1,2,3,\ldots\}$.
4. Show that for arbitrary sets S, T, V
   (a) $(S\cap T)\cap V = S\cap(T\cap V)$ (associativity of $\cap$),
   (b) $(S\cup T)\cup V = S\cup(T\cup V)$ (associativity of $\cup$),

(c) $S \cap (T \cup V) = (S \cap T) \cup (S \cap V)$ (distributivity of $\cap$ over $\cup$),
(d) $S \cup (T \cap V) = (S \cup T) \cap (S \cup V)$ (distributivity of $\cup$ over $\cap$),

5. Verify for arbitrary sets S, T, V
   (a) $S \subset T$ and $T \subset V$ imply $S \subset V$,
   (b) $S \setminus T = S$ implies $T \setminus S = T$ and $S \cap T = \emptyset$,
   (c) $S \cap (T \setminus V) = (S \cap T) \setminus (S \cap V)$.
6. Give examples of sets S, T, V such that
   (a) $S \cup (T \setminus V) \neq (S \cup T) \setminus (S \cup V)$,
   (b) $S \cup (T \setminus V) \neq (S \cup T) \setminus V$,
   (c) $S \setminus (T \cup V) \neq (S \setminus T) \cup V$,
   (d) $S \setminus (T \cap V) \neq (S \setminus T) \cap V$,
   (e) $S \setminus (T \setminus V) \neq (S \setminus T) \setminus V$.

1.3.2 Answers

1. (a) and (b) are set, their elements, the sets $\{8, 9\}, \emptyset, \{0\}$ and $\{4\}, \{4, 5\}$, respectively are distinct.
   (c) is not a set, since its elements, the sets $\{3, 4\} \cup \{4, 5\} = \{3, 4, 5\}$ and $\{3, 4, 5\}$, are not distinct.
   (d) is not a set, since its elements, the sets $\{3, 4, 5\} \cap \{4, 5, 6\} = \{4, 5\}$ and $\{4, 5\}$, are not distinct.
   (e) and (f) are the sets $\emptyset$ and $\{0\}$, respectively.
2. The statements (c), (e), (g) are correct.
3. (a) $\{\alpha, \beta, \gamma, \delta, \epsilon, \omega\}$, (b) $\{\alpha, \delta\}$, (c) $\{\alpha, \delta, \omega\}$,
   (d) $\{\alpha, \delta\}$, (e) $\{\alpha, \delta\}$, (f) $\{\alpha, \beta\}$.

1.4 Cartesian Products of Sets, $\mathbb{R}^n$, Vectors

Another important operation between sets is the *Cartesian product*, defined as follows. The *Cartesian product* $S \times T$ of the sets S and T is the set of *ordered pairs* (s, t) where $s \in S$, $t \in T$, in symbols:

$$S \times T := \{(s, t) \mid s \in S, t \in T\}.$$

A few remarks may be useful here: This is a "set of sets" as discussed in the previous section on the example of a "set of companies": The elements of $S \times T$ are the *ordered pairs* (s, t) just as the elements of the *Cartesian product of n sets* (the *notations on the left and in the middle can be used* interchangeably):

$$\begin{aligned} \bigtimes_{k=1}^{n} S_k &:= S_1 \times S_2 \times \ldots \times S_n \\ &:= \{(s_1, s_2, \ldots, s_n) \mid s_1 \in S_1, s_2 \in S_2, \ldots, s_n \in S_n\} \end{aligned}$$

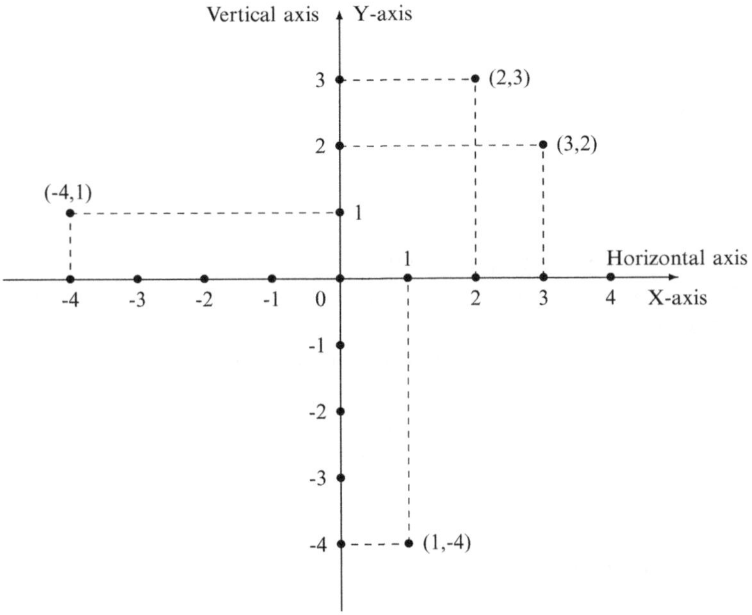

Fig. 1.2 The points in the plane are represented by pairs of real numbers. If the numbers of such a pair are written in different order, we usually get different points

are *ordered n*-tuples $(s_1, s_2, \ldots, s_n)$. "Ordered", because their order is of importance (at the beginning of Sect. 1.2 we have already indicated that later some sets may be ordered or, at least, partially ordered). The importance of ordering is seen on the example in Fig. 1.2: As usual (see also below), a point in the *Cartesian plane* is represented by its "x and y coordinates", that is, its distances from the "vertical axis" $\{(0, y) \mid y \in \mathbb{R}\}$ and from the "horizontal axis" $\{(x, 0) \mid x \in \mathbb{R}\}$, respectively. Both "Cartesian product" and "Cartesian plane" refer to the name of the French mathematician René Descartes (1596–1650). We emphasise that the couples and n-tuples are *ordered*: As we see in the Fig. 1.2, (2,3) and (3,2) are *two different points*.

(Actually (s, t) and (t, s) give the same points *only* in the obvious case $t = s$).

Example The Cartesian product of the sets

$$S_1 = \{a, b, c\}, \quad S_2 = \{x, y\}, \quad S_3 = \{z\} \text{ and } S_4 = \{w\}$$

(continued)

is given by

$$S_1 \times S_2 \times S_3 \times S_4$$
$$= \{(a,x,z,w),\ (a,y,z,w),\ (b,x,z,w),\ (b,y,z,w),\ (c,x,z,w),\ (c,y,z,w)\}.$$

This is a set of six elements $(a,x,z,w), \ldots, (c,y,z,w)$ and *not* of seven elements a,b,c,x,y,z,w: *the ordered sets* (a,x,z,w), $(a,y,z,w), \ldots$ themselves *are the elements of* $S_1 \times S_2 \times S_3 \times S_4$.

By the way, the $S_1 \times S_2 \times \ldots \times S_n$ notation is legitimate because *the Cartesian product is associative*:

$$(S_1 \times S_2) \times S_3 = S_1 \times (S_2 \times S_3) = S_1 \times S_2 \times S_3$$
$$= \{(s_1, s_2, s_3) \mid s_1 \in S_1, s_2 \in S_2, s_3 \in S_3\}.$$

But the *Cartesian product is not commutative*:

$$S_1 \times S_2 = \{(s,t) \mid s \in S_1,\ t \in S_2\} \neq \{(s,t) \mid s \in S_2, t \in S_1\} = S_2 \times S_1,$$

for instance

$$\{a,b,c\} \times \{x,y\} = \{(a,x),(a,y),(b,x),(b,y),(c,x),(c,y)\}$$

and

$$\{x,y\} \times \{a,b,c\} = \{(x,a),(x,b),(x,c),(y,a),(y,b),(y,c)\}.$$

While the latter equals $\{(x,a),(y,a),(x,b),(y,b),(x,c),(y,c)\}$ (compare the introduction of sets at the beginning of Sect. 1.2), this is still not the same as $\{a,b,c\} \times \{x,y\}$ above, because (x,a) is not the same *ordered* pair as (a,x), and (y,a) not the same as (a,y), and so on.

If all sets $S_1, S_2, \ldots, S_n$ are the same

$$S_1 = S_2 = \ldots = S_n = S$$

then their Cartesian product is the n-th *Cartesian power*

$$S^n := \{(s_1, s_2, \ldots, s_n) \mid s_1 \in S, s_2 \in S, \ldots, s_n \in S\}.$$

In particular, for $S = \mathbb{R}$, we get

$$\mathbb{R}^n = \{(x_1, x_2, \ldots, x_n) \mid x_k \in \mathbb{R}\ (k = 1, 2, \ldots, n)\}.$$

In other words, the elements of $\mathbb{R}^n$ are the *vectors* with n real components or "*n-component real vectors*". Similarly, the elements of S^n are "*vectors with n components in S*". For instance, the elements of $\mathbb{R}^n_{++}$ are the vectors with n positive components, those of $\mathbb{N}^n$ are the vectors whose all n components are natural numbers, similarly for $\mathbb{N}^n_0$, where $\mathbb{N}_0 = \mathbb{N} \cup \{0\}$ is the set of nonnegative integers, and so on.

There are many examples of such vectors in economics and other social sciences, for instance the *price vector* $(p_1, \ldots, p_n) \in \mathbb{R}^n_{++}$ of the present prices and the *vector of quantities* $(q_1, q_2, \ldots, q_n) \in \mathbb{R}^n_{++}$ in a "basket of goods". Further, the component of the vector

$$(m_1, m_2, \ldots, m_n) \in \mathbb{N}^n_0$$

could be, say, the number of unemployed in n different job categories or the number of students enrolled in n faculties of a university, and so on.

As mentioned (compare Figs. 1.2 and 1.3), for $n = 2$, every element (x_1, x_2) of $\mathbb{R}^2$ can be identified with the point in the (Cartesian) plane, whose coordinates are x_1 and x_2. We identify $(x_1, x_2) \in \mathbb{R}^2$ also with the *directed segment* of the straight line connecting the origin (the point (0,0)) with the point (x_1, x_2) (Fig. 1.3). That directed segment is the arrow usually associated with the word "fig1.3", in this case a "2-component real vector" (x_1, x_2 are its components). Similarly a 3-component real vector can be identified with a point in the three-dimensional (Euclidean) space and also with a directed segment from the origin (0,0,0) to that point. As a generalisation

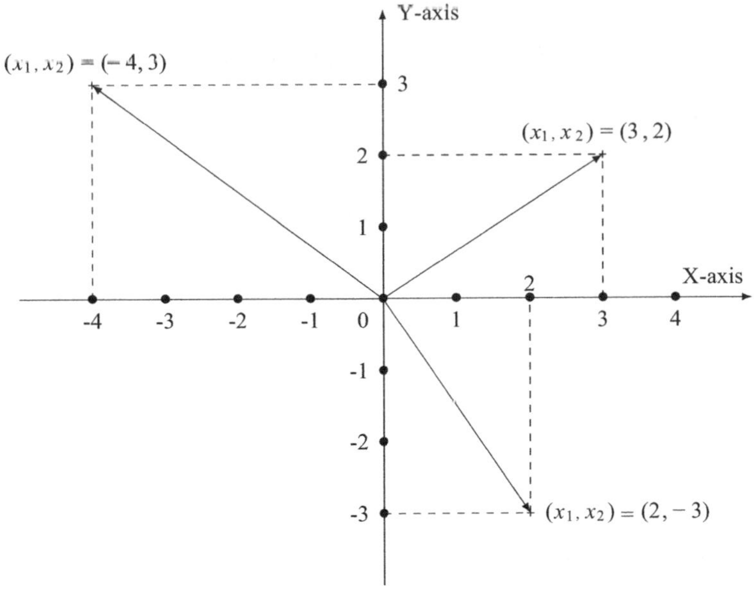

Fig. 1.3 (x_1, x_2) as point and as vector (directed segment) in the plane

we call the *n-component real vector* $(x_1, x_2, \ldots, x_n) \in \mathbb{R}^n$ ($x_1, x_2, \ldots, x_n$ are its *components*) also a *point* in the *n-dimensional (Cartesian) space*.

We will write bold face letters for vectors, in particular for real vectors:

$$\mathbf{x} = (x_1, \ldots, x_n) \in \mathbb{R}^n.$$

This manner of writing really defines "*row vectors*". It is sometimes more convenient to write the components in a column. Then we speak about "*column vectors*":

$$\mathbf{x} = \begin{pmatrix} x_1 \\ \vdots \\ x_n \end{pmatrix}.$$

At present we treat these interchangeably: we will not distinguish them till Chap. 4, where they will turn out to be two different special cases of *matrices*.

For $n = 2$ the *length* of the vector (directed segment) $\mathbf{x} = (x_1, x_2)$ is $||\mathbf{x}|| = (x_1^2 + x_2^2)^{1/2}$ by the *theorem of Pythagoras*. While the reader is surely familiar with this theorem, the simple proof in Fig. 1.4 may not be so well known. Actually, Pythagoras's theorem proves

$$||\mathbf{x}|| = (x_1^2 + x_2^2)^{1/2}$$

only for positive x_1, x_2 but it implies the same expression for the length of all $\mathbf{x} = (x_1, x_2) \in \mathbb{R}^2$ and we accept as definition of $||\mathbf{x}||$ the similar formula

$$||\mathbf{x}|| = (x_1^2 + \ldots + x_n^2)^{1/2} \in \mathbb{R}_+$$

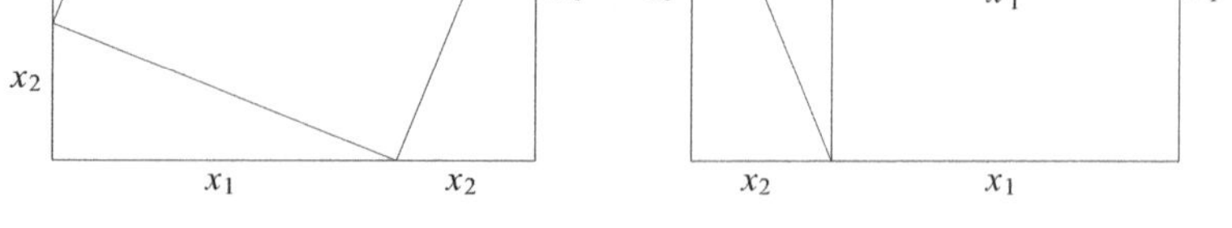

Fig. 1.4 $||\mathbf{x}||^2 = x_1^2 + x_2^2$: Pythagoras's theorem proved by taking away four equal rectilinear triangles each from the two equal (*big*) squares

for all $\mathbf{x} = (x_1, \ldots, x_n) \in \mathbb{R}^n$ ($n = 1, 2, 3, \ldots$; note that, for $n = 1$, $||\mathbf{x}|| = |\mathbf{x}|$) and call it the *Euclidean norm* (though "Pythagorean" may be appropriate). For $n = 3$ it still has the geometric meaning of *length* of $\mathbf{x}$. Vectors $\mathbf{e}$ with norm 1 ($||\mathbf{e}|| = 1$) are called *unit vectors*.

We emphasised that the n-tuples of components are *ordered*. In another sense, the set $\mathbb{R}$ of real numbers is *ordered* ("*totally ordered*", to be exact): for any $a, b \in \mathbb{R}$ either $a < b$ or $a = b$ or $a > b$ (one and only one of these can hold). "Greater" (or "smaller" and, of course, "equal") can be usefully defined also for *n-component real vectors* with $n > 1$, even in two, in general different, ways. One is

$$\mathbf{x} > \mathbf{y} \quad \text{(the same as } \mathbf{y} < \mathbf{x}) \quad \textit{if} \quad x_1 > y_1, x_2 > y_2, \ldots, x_n > y_n;$$

Of course,

$$\mathbf{x} = \mathbf{y} \quad \textit{means} \quad x_1 = y_1, x_2 = y_2, \ldots, x_n = y_n.$$

If this does not hold (that is, $\mathbf{x}$ and $\mathbf{y}$ are not the same vector) *then we write* $\mathbf{x} \neq \mathbf{y}$. Knowing that $x_k \geq y_k$ *for real numbers* means that x_k is either greater or equal y_k, we *define* for n-component real vectors the second "greater" (or "smaller") relation by

$$\mathbf{x} \geq \mathbf{y} \quad \text{(the same as } \mathbf{y} \leq \mathbf{x}) \quad \textit{if} \quad x_1 \geq y_1, x_2 \geq y_2, \ldots, x_n \geq y_n \textit{ but } \mathbf{x} \neq \mathbf{y},$$

that is $x_k \geq y_k$ for all $k(= 1, 2, \ldots, n)$ but, at least for one ℓ, "sharply" $x_\ell > y_\ell$ ($\ell \in \{1, 2, \ldots, n\}$). This is not the same as

$$\mathbf{x} \geqq \mathbf{y} \quad \text{(or } \mathbf{y} \leqq \mathbf{x}) \quad \text{which means that} \quad x_1 \geq y_1, x_2 \geq y_2, \ldots, x_n \geq y_n$$

but no x_ℓ needs to be really greater than y_ℓ. In other words, $\mathbf{x} \geqq \mathbf{y}$ contains $\mathbf{x} = \mathbf{y}$ as particular case, but $\mathbf{x} \geq \mathbf{y}$ does not. Strictly speaking, in $\mathbb{R}^1$ (=$\mathbb{R}$, that is, for reals), we should write $x \geqq y$ if x can be either greater or equal y but it is traditional to use the simpler $x \geq y$ notation in this (exceptional) $n = 1$ case (where the "$\geq$" in the above sense is not needed, because it means *the same* as "$>$" for $n = 1$, which is *not* the case if $n > 1$).

Under either of these "greater" relations (there are also others, these are the most useful ones), $\mathbb{R}^n$ *is not totally ordered*, it is only *partially ordered*, meaning that, *while for some* pairs of vectors $\mathbf{x} \in \mathbb{R}^n$, $\mathbf{y} \in \mathbb{R}^n$ we have $\mathbf{x} > \mathbf{y}$ (or $\mathbf{x} < \mathbf{y}$ or $\mathbf{x} = \mathbf{y}$) or $\mathbf{x} \geq \mathbf{y}$ (or $\mathbf{x} \leq \mathbf{y}$ or $\mathbf{x} = \mathbf{y}$ and *at most one of these three*), *there are* $\mathbf{x} \in \mathbb{R}^n$, $\mathbf{y} \subset \mathbb{R}^n$ *for which neither* $\mathbf{x} > \mathbf{y}$ *nor* $\mathbf{x} < \mathbf{y}$ *nor* $\mathbf{x} = \mathbf{y}$ (neither $\mathbf{x} \geq \mathbf{y}$ nor $\mathbf{x} \leq \mathbf{y}$ nor $\mathbf{x} = \mathbf{y}$) holds. For instance, of the two vectors $(3, 2)$ and $(2, 3)$ in Fig. 1.2 *neither is greater* (either in the sense $>$ or $\geq$) than the other. (Their *norms* happen to be equal, both are $\sqrt{13}$, but they are not equal according to the above definition, since already their first components are different.) Another example is given by the three vectors of goods

$$\mathbf{a} = (3, 2), \ \mathbf{b} = (4, 5), \ \mathbf{c} = (6, 3)$$

(the first components being, say, pounds of butter, the second pounds of honey). Clearly

$$\mathbf{a} < \mathbf{b} \quad (\text{because } 3 < 4,\ 2 < 5) \quad \text{and} \quad \mathbf{a} < \mathbf{c} \quad (\text{since } 3 < 6,\ 2 < 3)$$

but neither $\mathbf{b} < \mathbf{c}$ nor $\mathbf{b} = \mathbf{c}$, not even $\mathbf{b} \leq \mathbf{c}$ or $\mathbf{b} \geq \mathbf{c}$ (since $4 < 6$ but $5 > 3$). This is not only of theoretical importance: because of this it is not clear which of the two vectors of quantities of goods, **b** or **c** is of more economic utility. This is what makes synthesising (merging, index) methods necessary.

We note that there does exist a total order on $\mathbb{R}^n$, the *lexicographical order*. In this order, the point with the greater first component is considered greater; in case of equal first components that with greater second component, and so on. The ordering is called "lexicographic" because that is how "lexicons" (dictionaries, phone directories, etc.) are ordered: in the alphabetical order of the first letter; if that is the same in two words then by the second letter, and so on. The words can consist of differently many letters. Any word W stands in front of every longer word starting with W. Applying this rule accordingly we can establish a complete (lexicographical) order for all vectors of $\mathbb{R}^2$, $\mathbb{R}^3$, $\mathbb{R}^4, \ldots$ The lexicographical order is, however, not practical for most applications in economics.

1.4.1 Exercises

1. For the sets $S_1 = \{a, b\}$, $S_2 = \{c, d, e, f\}$, $S_3 = \{x\}$ determine
   (a) $S_1 \times S_2$,
   (b) $S_2 \times S_1$,
   (c) the Cartesian product $S_1 \times S_2 \times S_3$,
   (d) the fourth Cartesian power of S_1.
2. Calculate the length of the vectors
   (a) (3,4), (5,12), (6,7),
   (b) (3,4,5), (1,2,3), (2,2,2).
3. Calculate the Euclidean norms of the vectors
   $(3, 4, 5, 6)$, (2,2,2,2), (1,2,3,4,5,6,7).
4. Take the vectors

$$\mathbf{u} = (4,7),\ \mathbf{v} = (1,8),\ \mathbf{w} = (2,8),\ \mathbf{x} = (3,9),\ \mathbf{y} = (-5,8),\ \mathbf{z} = (2,4,6).$$

   Which of the following relations are correct?
   (a) $\mathbf{u} < \mathbf{v}$, (b) $\mathbf{v} \leq \mathbf{w}$, (c) $\mathbf{x} > \mathbf{y}$,
   (d) $\mathbf{z} < \mathbf{u}$, (e) $\mathbf{u} > \mathbf{y}$, (f) $\mathbf{w} \geqq \mathbf{v}$.
5. Take the vectors
   $\mathbf{a} = (-3, 4)$, $\mathbf{b} = (5, 2, 1)$, $\mathbf{c} = (4, 5, 6, 7)$, $\mathbf{d} = (-1, 7)$,
   $\mathbf{e} = (4, 5, 6, 8)$, $\mathbf{f} = (6, 1)$, $\mathbf{g} = (6, 1, 2)$.

(a) Which of these vectors are comparable with respect to $<$, $\leq$, $\leqq$?
(b) Order them in the lexicographical order. (Start with **a** which has the smallest first component.)

1.4.2 Answers

1. (a) $\{(a,c),(a,d),(a,e),(a,f),(b,c),(b,d),(b,e),(b,f)\}$,
   (b) $\{(c,a),(c,b),(d,a),(d,b),(e,a),(e,b),(f,a),(f,b)\}$,
   (c) $\{(a,c,x),(a,d,x),(a,e,x),(a,f,x),(b,c,x),(b,d,x),(b,e,x),$ $(b,f,x)\}$.
   (d) $S_1^4 = \{(a,a,a,a),\ (a,a,a,b),\ (a,a,b,a),\ (a,b,a,a),\ (b,a,a,a),$ $(a,a,b,b),\ (a,b,a,b),\ (b,a,a,b),\ (a,b,b,a),\ (b,a,b,a),$ $(b,b,a,a),(a,b,b,b),\ (b,a,b,b),\ (b,b,a,b),\ (b,b,b,a),\ (b,b,b,b)\}$.

2. (a) 5, 13, $\sqrt{85}$, (b) $5\sqrt{2}$, $\sqrt{14}$, $2\sqrt{3}$.
3. $\sqrt{86}$, 4, $2\sqrt{35}$.
4. (b), (c), (f).
5. (a) $\mathbf{a} < \mathbf{d}$, $\mathbf{a} \leq \mathbf{d}$, $\mathbf{a} \leqq \mathbf{d}$, $\mathbf{c} \leq \mathbf{e}$, $\mathbf{c} \leqq \mathbf{d}$.
   (b) **a**, **d**, **c**, **e**, **b**, **f**, **g**.

1.5 Operations for Vectors, Linear Dependence and Independence

While not any two vectors could be compared in the sense of the above "$>$" or "$\geq$" order, any two (n-component real) vectors can be added, subtracted, any vector can be multiplied by a real number ("scalar" in this context) and even any two n-component vectors can be multiplied in a sense (giving a "scalar product", not an n-component vector as product).

1.5.1 Sums, Differences, Linear Combinations of Vectors

If the prices $p_1^0, p_2^0, \ldots, p_n^0$ of n goods in a "basket of goods" in the base year are considered to be the components of a vector

$$\mathbf{p}^0 = (p_1^0, \ldots, p_n^0) \in \mathbb{R}_{++}^n$$

and during a certain time-interval the prices increase by $d_1, \ldots, d_n$, which we collect again into a vector

$$\mathbf{d} = (d_1, \ldots, d_n) \in \mathbb{R}_{++}^n,$$

then the new prices will be $p_1^0 + d_1, p_2^0 + d_2, \ldots, p_n^0 + d_n$, forming the new price vector

$$\mathbf{p} = (p_1^0 + d_1, \ldots, p_n^0 + d_n).$$

It is natural to consider this $\mathbf{p}$ the sum of the two vectors $\mathbf{p}^0$ and $\mathbf{d}$:

$$\mathbf{p} = \mathbf{p^0} + \mathbf{d} \ := (p_1^0 + d_1, \ldots, p_n^0 + d_n).$$

The sum of two vectors $\mathbf{x} = (x_1, \ldots, x_n), \mathbf{y} = (y_1, \ldots, y_n)$ in $\mathbb{R}^n$ (practical since, incredible as it may seem, prices can also go down or remain unchanged) *is* therefore *defined by*

$$\mathbf{x} + \mathbf{y} = (x_1, \ldots, x_n) + (y_1, \ldots, y_n) \ := (x_1 + y_1, \ldots, x_n + y_n).$$

Obviously, the addition of real vectors is *commutative* and *associative*:

$$\mathbf{x} + \mathbf{y} = \mathbf{y} + \mathbf{x} \quad \text{and} \quad (\mathbf{x} + \mathbf{y}) + \mathbf{z} = \mathbf{x} + (\mathbf{y} + \mathbf{z}) \ =: \mathbf{x} + \mathbf{y} + \mathbf{z},$$

because the addition of real numbers has these properties (write the above equation in components). The sum of more than three vectors can be defined similarly.

As motivation for the rule on *multiplication of vectors by scalars*, consider a bank which pays on 90-day term deposit 4 % (nominal) yearly interest, that is 1 % for the 90 day period. Denote the amounts of n term deposits by $t_1^0, t_2^0, \ldots, t_n^0$, again forming a vector

$$\mathbf{t}^0 = (t_1^0, \ldots, t_n^0) \in \mathbb{R}^n_{++}.$$

By the end of the 90-days term, the depositors will be paid the amounts $1.01\, t_1^0,\ 1.01\, t_2^0, \ldots,\ 1.01\, t_n^0$. It is natural to consider the vector

$$\mathbf{t} = (1.01\, t_1^0, \ldots,\ 1.01\, t_n^0)$$

as 1.01 times the original vector $\mathbf{t}^0$:

$$1.01\, \mathbf{t}^0 = 1.01(t_1^0, \ldots, t_n^0) \ := (1.01\, t_1^0, \ldots,\ 1.01\, t_n^0).$$

In general, *multiplication of a vector* $\mathbf{x} = (x_1, \ldots, x_n)$ in $\mathbb{R}^n$ *by a real number ("scalar")* $\lambda \in \mathbb{R}$ *is defined by*

$$\lambda\mathbf{x} = \lambda(x_1, \ldots, x_n) \ := (\lambda x_1, \ldots, \lambda x_n).$$

By definition this is also $\mathbf{x}\lambda$ (multiplication of a scalar by a vector):

$$\mathbf{x}\lambda := \lambda\mathbf{x} \qquad (\mathbf{x} \in \mathbb{R}^n, \lambda \in \mathbb{R})$$

(a kind of "*commutativity*"). Observe also

$$(\lambda\mu)\mathbf{x} = \lambda(\mu\mathbf{x}) \qquad (\lambda \in \mathbb{R}, \mu \in \mathbb{R}, \mathbf{x} \in \mathbb{R}^n)$$

(a kind of "associativity") and the two *distributivity* identities:

$$\lambda(\mathbf{x}+\mathbf{y}) = \lambda\mathbf{x} + \lambda\mathbf{y} \qquad (\lambda \in \mathbb{R},\ \mathbf{x} \in \mathbb{R}^n,\ \mathbf{y} \in \mathbb{R}^n)$$

$$(\lambda + \mu)\mathbf{x} = \lambda\mathbf{x} + \mu\mathbf{x} \qquad (\lambda \in \mathbb{R},\ \mu \in \mathbb{R},\ \mathbf{x} \in \mathbb{R}^n).$$

Notice that the *null-vector* $\mathbf{0} := (0, \ldots, 0) \in \mathbb{R}^n$ satisfies

$$\mathbf{x} + \mathbf{0} = \mathbf{0} + \mathbf{x} = \mathbf{x},\ 0\mathbf{x} = \mathbf{0} \text{ for all } \mathbf{x} \in \mathbb{R}^n \text{ and } r\mathbf{0} = \mathbf{0} \text{ for all } r \in \mathbb{R}.$$

Combining (as in the distributivity identities) the rules for multiplication of a vector by a scalar and for addition of vectors (and the associativity of the latter), we get the definition of the *linear combination*

$$\lambda\mathbf{x} + \mu\mathbf{y} = \lambda(x_1, \ldots, x_n) + \mu(y_1, \ldots, y_n) := (\lambda x_1 + \mu y_1, \ldots, \lambda x_n + \mu y_n),$$

and a similar definition for p vectors (with n components each):

$$\lambda_1\mathbf{x}_1 + \ldots + \lambda_p\mathbf{x}_p = \sum_{j=1}^{p} \lambda_j\mathbf{x}_j$$

(the right hand side is just a short way of writing the left, in the same vein as we had $\cup_{k=1}^n S_k$, $\cap_{k=1}^n S_k$, $\mathsf{X}_{k=1}^n S_k$). A particular case is the *difference of two vectors*:

$$\mathbf{x} - \mathbf{y} = (x_1, \ldots, x_n) - (y_1, \ldots, y_n) := (x_1 - y_1, \ldots, x_n - y_n).$$

For instance, the vector $\mathbf{d}$ considered above is the difference $\mathbf{p} - \mathbf{p}^0$ of the vectors $\mathbf{p}$ and $\mathbf{p}^0$ of the new and the base year prices, respectively.

1.5.2 Linear Dependence, Independence

If a vector $\mathbf{x}_m$ is, as above a linear combination of $p = m - 1$ vectors,

$$\begin{gathered}\mathbf{x}_m = \lambda_1\mathbf{x}_1 + \lambda_2\mathbf{x}_2 + \ldots + \lambda_{m-1}\mathbf{x}_{m-1} \\ (\lambda_1 \in \mathbb{R}, \ldots, \lambda_{m-1} \in \mathbb{R};\ \mathbf{x}_1 \in \mathbb{R}^n,\ \ldots, \mathbf{x}_{m-1} \in \mathbb{R}^n,\ \mathbf{x}_m \in \mathbb{R}^n),\end{gathered} \tag{1.2}$$

then we say that $\mathbf{x}_m$ is *linearly dependent* upon $\mathbf{x}_1, \ldots, \mathbf{x}_{m-1}$. Of course also, for instance, $\mathbf{x}_1$ could be linearly dependent upon $\mathbf{x}_2, \ldots, \mathbf{x}_m$. This is not quite the same as (1.2): if $\lambda_1 = 0$ (which was not excluded) then it does not follow from (1.2) that $\mathbf{x}_1$ is a linear combination of $\mathbf{x}_2, \ldots, \mathbf{x}_{m-1}, \mathbf{x}_m$. A symmetrical definition for all these linear dependencies (also for $\mathbf{x}_2$ or ... or $\mathbf{x}_{m-1}$ to be linearly dependent on the others) is: $\mathbf{x}_1, \ldots, \mathbf{x}_m$ *are linearly dependent if there exist* $\lambda_1 \in \mathbb{R}, \ldots, \lambda_m \in \mathbb{R}$, *not all 0, such that*

$$\lambda_1 \mathbf{x}_1 + \ldots + \lambda_{m-1}\mathbf{x}_{m-1} + \lambda_m \mathbf{x}_m = \mathbf{0}.$$

Equation (1.2) is the special case where $\lambda_m = -1$.

The opposite of linear dependence is the linear independence: *the vectors* $\mathbf{x}_1, \ldots, \mathbf{x}_m$ *are linearly independent* if

$$\lambda_1 \mathbf{x}_1 + \ldots + \lambda_m \mathbf{x}_m = \mathbf{0} \quad \textit{can hold only for} \quad \lambda_1 = \ldots = \lambda_m = 0.$$

As an important example, take the *basis vectors* of $\mathbb{R}^n$

$$\mathbf{e}_1 = (1, 0, 0, \ldots, 0, 0), \quad \mathbf{e}_2 = (0, 1, 0, \ldots, 0, 0), \quad \ldots, \quad \mathbf{e}_n = (0, 0, \ldots, 0, 1).$$

They are clearly unit vectors ($||\mathbf{e}_k|| = 1$; $k = 1, 2, \ldots, n$). *They are* also *linearly independent.* Indeed,

$$\lambda_1 \mathbf{e}_1 + \lambda_2 \mathbf{e}_2 + \ldots + \lambda_n \mathbf{e}_n = (\lambda_1, \lambda_2, \ldots \lambda_n)$$

can be the $\mathbf{0}$ vector only if $\lambda_1 = \lambda_2 = \ldots = \lambda_n = 0$. Of course any number ($< n$) of them are also linearly independent. Note that *a single vector of nonzero* (therefore positive) *length (norm) is always linearly independent* ($\lambda_1 \mathbf{x} = \mathbf{0}$ only if $\lambda_1 = 0$) while *the zero vector is always linearly dependent* ($\lambda_1 \mathbf{0} = \mathbf{0}$ for all, also nonzero λ_1). In particular, each single basis vector $\mathbf{e}_k$ is linearly independent.

The basis vectors (all n of them together) span $\mathbb{R}^n$. By this we mean the following. *The vectors* $\mathbf{x}_1 \in \mathbb{R}^n, \ldots, \mathbf{x}_m \in \mathbb{R}^n$ *span* a subset S of $\mathbb{R}^n$ (possibly $\mathbb{R}^n$ itself) *if every* $\mathbf{x} \in S$ *is a linear combination of* $\mathbf{x}_1, \ldots, \mathbf{x}_m$. This is indeed the case for $\mathbf{e}_1, \ldots, \mathbf{e}_n$ with $S = \mathbb{R}^n$: for every $\mathbf{x} = (x_1, \ldots, x_n) \in \mathbb{R}^n$

$$\begin{aligned} \mathbf{x} = (x_1, \ldots, x_n) &= x_1(1, 0, \ldots, 0, 0) + \ldots + x_n(0, 0, \ldots 0, 1) \\ &= x_1 \mathbf{e}_1 + \ldots + x_n \mathbf{e}_n. \end{aligned}$$

But it is easy to see that fewer than n of the basis vectors cannot span $\mathbb{R}^n$. In general, *if* $m < n$ *then no* m *vectors can span* $\mathbb{R}^n$ *but if the* m *vectors are independent they span a "space* S *of dimension* m*"*. Furthermore, *n vectors span* $\mathbb{R}^n$ *exactly if they are linearly independent (n vectors which span* $\mathbb{R}^n$ *are* often called a *basis* of $\mathbb{R}^n$). On the other hand *more than n nonzero vectors in* $\mathbb{R}^n$ *can never be linearly independent.*

We will not prove these assertions here but the reader may wish to try. However, we give *examples*:

Example 1 The vectors $\mathbf{x}_1 = (1, 3, 1)$, $\mathbf{x}_2 = (4, 2, 1)$, $\mathbf{x}_3 = (2, 0, 1/5) \in \mathbb{R}^3$ are *linearly dependent*, because

$$2\mathbf{x}_1 + (-3)\mathbf{x}_2 + 5\mathbf{x}_3 = (2 - 12 + 10,\ 6 - 6 + 0,\ 2 - 3 + 1) = (0, 0, 0) = \mathbf{0}.$$

Example 2 The vectors $\mathbf{y}_1 = (3, 4)$, $\mathbf{y}_2 = (2, 1) \in \mathbb{R}^2$ are *linearly independent*:

$$\begin{aligned}\lambda_1\mathbf{y}_1 + \lambda_2\mathbf{y}_2 &= \lambda_1(3, 4) + \lambda_2(2, 1) = (3\lambda_1, 4\lambda_1) + (2\lambda_2, \lambda_2)\\ &= (3\lambda_1 + 2\lambda_2,\ 4\lambda_1 + \lambda_2) = (0, 0)\end{aligned}$$

if $3\lambda_1 + 2\lambda_2 = 0$ and $4\lambda_1 + \lambda_2 = 0$. From the second equation $\lambda_2 = -4\lambda_1$ which we put into the first:

$$3\lambda_1 + 2(-4\lambda_1) = -5\lambda_1 = 0, \quad \text{so} \quad \lambda_1 = 0 \quad \text{and} \quad \lambda_2 = -4\lambda_1 = 0,$$

that is $\mathbf{y}_1 = (3, 4)$ and $\mathbf{y}_2 = (2, 1)$ indeed satisfy the definition of linear independence. They also *span* $\mathbb{R}^2$ (they form a *basis* of $\mathbb{R}^2$). Indeed, for any $\mathbf{x} = (x_1, x_2) \in \mathbb{R}^2$ one can find μ_1, μ_2 such that $\mathbf{x} = \mu_1\mathbf{y}_1 + \mu_2\mathbf{y}_2$. These are $\mu_1 = (2x_2 - x_1)/5$, $\mu_2 = (4x_1 - 3x_2)/5$:

$$\begin{aligned}\mu_1(3, 4) + \mu_2(2, 1) &= \frac{2x_2 - x_1}{5}(3, 4) + \frac{4x_1 - 3x_2}{5}(2, 1)\\ &= \left(\frac{6x_2 - 3x_1 + 8x_1 - 6x_2}{5}, \frac{8x_2 - 4x_1 + 4x_1 - 3x_2}{5}\right)\\ &= (x_1, x_2).\end{aligned}$$

Example 3 The vectors $\mathbf{y}_1 = (3, 4)$, $\mathbf{y}_2 = (2, 1)$, $\mathbf{y}_3 = (1, -1) \in \mathbb{R}^2$ are *linearly dependent*, because

$$3(3, 4) + (-7)(2, 1) + 5(1, -1) = (9 - 14 + 5,\ 12 - 7 - 5) = (0, 0).$$

(continued)

Actually, since we showed in Example 2 that *every* $\mathbf{x}$ is linearly dependent upon $\mathbf{y}_1, \mathbf{y}_2$ these $\mathbf{y}_1 = (3,4), \mathbf{y}_2 = (2,1)$ would be linearly dependent not only with this $\mathbf{y}_3 = (1,-1)$ but also with any other $\mathbf{y}_3$. (Moreover, as mentioned above, no three vectors in $\mathbb{R}^2$ are linearly independent.)

1.5.3 Inner Product

If the quantities of goods in a "basket of goods" are $x_1, x_2, \ldots, x_n$, their prices $p_1, p_2, \ldots, p_n$ then this basket of goods costs $x_1p_1 + x_2p_2 + \ldots + x_np_n$. In vectorial terminology we define this scalar as the "inner or scalar product" of $\mathbf{x} = (x_1, \ldots, x_n)$ and $\mathbf{p} = (p_1, \ldots, p_n)$ and write

$$\mathbf{x} \cdot \mathbf{p} = (x_1, \ldots, x_n) \cdot (p_1, \ldots, p_n) := x_1p_1 + \ldots + x_np_n$$

(the notations $(\mathbf{x}, \mathbf{p})$ or $\mathbf{xp}$ without dots are also used in some books). In general, if $\mathbf{x} = (x_1, \ldots, x_n) \in \mathbb{R}^n$, $\mathbf{y} = (y_1, \ldots, y_n) \in \mathbb{R}^n$, then

$$\mathbf{x} \cdot \mathbf{y} = x_1y_1 + \ldots + x_ny_n \in \mathbb{R}$$

is the *inner product* of $\mathbf{x}$ and $\mathbf{y}$ (also called *scalar product* because the result is a scalar; there are also "outer products", "vectorial products" but they are less important for us and we will not deal with them here).

Notice that the inner product is *commutative* and *distributive over vector addition*:

$$\mathbf{x} \cdot \mathbf{y} = x_1y_1 + \ldots + x_ny_n = \mathbf{y} \cdot \mathbf{x} \quad \text{and}$$
$$\mathbf{x} \cdot (\mathbf{y} + \mathbf{z}) = (x_1y_1 + x_2y_2, \ldots, x_ny_n + x_nz_n) = \mathbf{x} \cdot \mathbf{y} + \mathbf{x} \cdot \mathbf{z}.$$

Associativity makes no sense for $n > 1$, because $\mathbf{x} \cdot \mathbf{y}$ is in $\mathbb{R}$, not $\mathbb{R}^n$ and if we would consider once the scalar product and once multiplication of a vector by a scalar on each side, we would get

$$\begin{aligned}
(\mathbf{x} \cdot \mathbf{y}) \cdot \mathbf{z} &= (x_1y_1 + \ldots + x_ny_n)(z_1, \ldots, z_n) \\
&= (x_1y_1z_1 + \ldots + x_ny_nz_1, \ldots, x_1y_1z_n + \ldots + x_ny_nz_n), \\
\mathbf{x} \cdot (\mathbf{y} \cdot \mathbf{z}) &= (x_1, \ldots, x_n)(y_1z_1 + \ldots + y_nz_n) \\
&= (x_1y_1z_1 + \ldots + x_1y_nz_n, \ldots, x_ny_1z_1 + \ldots + x_ny_nz_n)
\end{aligned}$$

which are not the same, as the example (check!)

$$\mathbf{x} = (1,2), \mathbf{y} = (2,1), \mathbf{z} = (1,3)$$

shows. But *we have*

$$\lambda(\mathbf{x} \cdot \mathbf{y}) = (\lambda \mathbf{x}) \cdot \mathbf{y} = \mathbf{x} \cdot (\lambda \mathbf{y}) \quad \textit{for all} \quad \mathbf{x}, \mathbf{y} \in \mathbb{R}^n,\ \lambda \in \mathbb{R}. \tag{1.3}$$

We see that, for $n = 1$, the inner product (just as the multiplication of a vector by a scalar) reduces to ordinary multiplication of real numbers: if $\mathbf{x} = x_1 \in \mathbb{R}$, $\mathbf{y} = y_1 \in \mathbb{R}$ then $\mathbf{x} \cdot \mathbf{y} = x_1 y_1 = x_1 \mathbf{y} = \mathbf{x} y_1$. (And, similarly, for $n = 1$, addition of vectors reduces to addition of real numbers, etc.) Indeed, the value of the quantity $x_1 \in \mathbb{R}$ of *one* good with price p_1 is $x_1 p_1$ (and that is also the value at maturity of *one* 90-day term deposit of x_1 with interest factor $p_1 = 1 + (q/400)$, where q is the bank's interest rate).

There is a remarkable connection between the inner product and the norm: *the scalar product of a vector with itself equals the square of its norm*:

$$\mathbf{x} \cdot \mathbf{x} = (x_1, \ldots, x_n) \cdot (x_1, \ldots, x_n) = x_1^2 + \ldots + x_n^2 = ||\mathbf{x}||^2.$$

1.5.4 Exercises

1. From the vectors $\mathbf{x} = (1, 3, 5), \mathbf{y} = (2, -4, -1), \mathbf{z} = (-2, 8, 7)$ calculate the following vectors:
   (a) $3\mathbf{x} - 6\mathbf{y} + 9\mathbf{z}$,
   (b) $1.04\mathbf{x} + 1.05\mathbf{y} - 1.06\mathbf{z}$,
   (c) $\lambda \mathbf{x} + \eta \mathbf{y} + \nu \mathbf{z}$.
2. Is the vector (2,1,3) linearly dependent upon
   (a) the vectors (1,2,3) and (1,3,4),
   (b) the vectors (1,2,3) and (0,3,4)?
3. (a) Are the vectors (1,2) and (1,-2) linearly independent?
   (b) Are the vectors (1,2,3), (1,2,-3), (7,14,-9) linearly dependent?
4. For the vectors **x**, **y**, **z** in Exercise **1** calculate
   (a) $(\mathbf{x} \cdot \mathbf{y}) \cdot \mathbf{z}$, (b) $\mathbf{x} \cdot (\mathbf{y} \cdot \mathbf{z})$,
   (c) $\mathbf{x} \cdot (\mathbf{y} + \mathbf{z})$, (d) $(\mathbf{x} + \mathbf{y}) \cdot \mathbf{z}$,
   (e) $\mathbf{x}(\mathbf{y} - \mathbf{z})$, (f) $(\mathbf{x} - \mathbf{y}) \cdot \mathbf{z}$.
5. For the vectors **x**, **y**, **z** in Exercise **1** calculate
   (a) $5\mathbf{x} \cdot \mathbf{y} - 3\mathbf{y} \cdot \mathbf{z} + 2\mathbf{z} \cdot \mathbf{x}$,
   (b) $6(\mathbf{x} - \mathbf{z}) \cdot \mathbf{y} - 4(\mathbf{x} + \mathbf{z}) \cdot \mathbf{y}$,
   (c) $7\mathbf{x} \cdot \mathbf{x} - 8\mathbf{y} \cdot \mathbf{y} + \mathbf{z} \cdot \mathbf{z}$.
6. For the vectors

$$\mathbf{a} = (a_1, a_2) \in \mathbb{R}^2,\ \mathbf{u} = (u_1, u_2, u_3) \in \mathbb{R}^3,\ \mathbf{x} = (x_1, x_2, x_3, x_4) \in \mathbb{R}^4$$

   determine vectors $\mathbf{b} \in \mathbb{R}^2$, $\mathbf{v} \in \mathbb{R}^3$, $\mathbf{y} \in \mathbb{R}^4$ such that
   (a) $\mathbf{a} \cdot \mathbf{b} = 0$, (b) $\mathbf{u} \cdot \mathbf{v} = 0$, (c) $\mathbf{x} \cdot \mathbf{y} = 0$.

1.5.5 Answers

1. (a) $(-27,\ 105,\ 84)$, (b) $(5.26,\ -9.56,\ -3.27)$,
   (c) $(\lambda + 2\mu - 2\nu,\ 3\lambda - 4\mu + 8\nu,\ 5\lambda - \mu + 7\nu)$.

2. (a) Yes, $(2,1,3) = 5\,(1,2,3) - 3\,(1,3,4)$.
   (b) No, there do not exist λ_1, λ_2 such that

$$\lambda_1(1,2,3) + \lambda_2(0,3,4) = (2,1,3).$$

3. (a) Yes, $\lambda_1(1,2) + \lambda_2(1,-2) = (0,0)$ if and only if $\lambda_1 = \lambda_2 = 0$.
   (b) Yes, $2(1,2,3) + 5(1,2,-3) + (-1)(7,14,-9) = (0,0,0)$.
4. (a) $(30,\ -120,\ -105)$, (b) $(-43,\ -129,\ -215)$.
   (c) 42, (d) 14, (e) -72, (f) 100.

5. (a) 168, (b) 400, (c) 210.

1.6 Geometric Interpretations. Distance. Orthogonal Vectors

While we have just seen that the operations with one-component real vectors are just the familiar operations with real numbers, we saw also in Sect. 1.4 that two-component real vectors nicely and simply illustrate the general situation (three-component real vectors do this even better, though the drawings are less simple, therefore we stick to $n = 2$).

Figure 1.5 illustrates the addition of two component real vectors, Fig. 1.6 their multiplication by a real number (scalar). We see that, by completing the two vectors

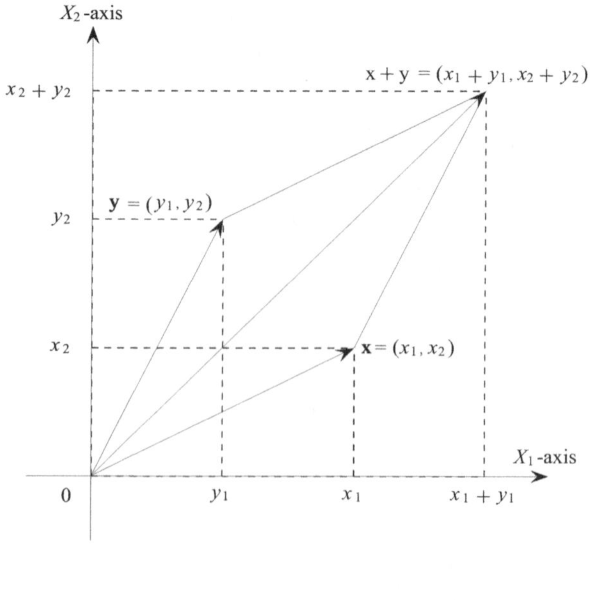

Fig. 1.5 The vector $\mathbf{x} + \mathbf{y}$ is represented by the fourth vertex of the parallelogram, whose other three vertices represent $\mathbf{0}$, $\mathbf{x}$ and $\mathbf{y}$

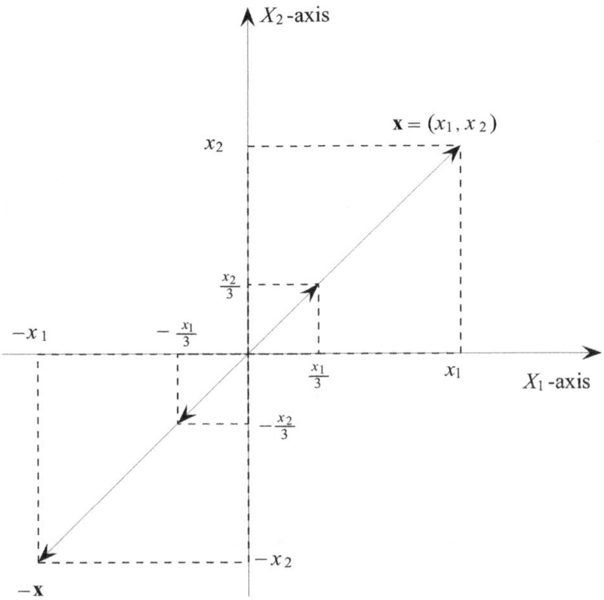

Fig. 1.6 The vector $\mathbf{x}$ has the same direction but three times the length of $\frac{1}{3}\mathbf{x}$, $-\mathbf{x}$ has also three times the length but opposite direction to $\frac{1}{3}\mathbf{x}$, $-\mathbf{x} = (-1)\mathbf{x}$ has the same length as $\mathbf{x}$ but opposite direction

to a *parallelogram* (quadrangle with two pairs of parallel sides), the vertex opposite to the origin $\mathbf{0}$ represents the vector $\mathbf{x} + \mathbf{y}$.

On the other hand, for $\mathbf{x} \in \mathbb{R}^n$, $\lambda \in \mathbb{R}$

$$\lambda\mathbf{x} = (\lambda x_1, \dots, \lambda x_n),$$
$$|\lambda\mathbf{x}| = (\lambda^2 x_1^2 + \dots + \lambda^2 x_n^2)^{1/2} = |\lambda|(x_1^2 + \dots + x_n^2)^{1/2} = |\lambda|\,|\mathbf{x}|\,.$$

(We know that the *absolute value* $|\lambda|$ is the positive square root of λ^2, that is, the nonnegative number whose square is λ^2, still otherwise put: $|\lambda| = \lambda$ if $\lambda \geq 0$ but $|\lambda| = -\lambda$ if $\lambda < 0$. Notice that, on the right end of the above equation, both the absolute value and the norm figure. We mentioned already in Sect. 1.4 that, for $n = 1$, the norm *is* the absolute value, so *the norm is a generalisation of the absolute value from* $n = 1$ *to* $n > 1$.) Thus the length of $\lambda\mathbf{x}$ is $|\lambda|$ times the length of $\mathbf{x}$ and the direction of $\lambda\mathbf{x}$ is that of $\mathbf{x}$ if $\lambda > 0$, but opposite to $\mathbf{x}$ if $\lambda < 0$ (of course, for $\lambda = 0$ we get $0\,\mathbf{x} = \mathbf{0}$, the null-vector).

Figure 1.7, based on Figs. 1.5 and 1.6, illustrates the subtraction of vectors:

$$\mathbf{x} - \mathbf{y} = \mathbf{x} + (-1)\mathbf{y}$$

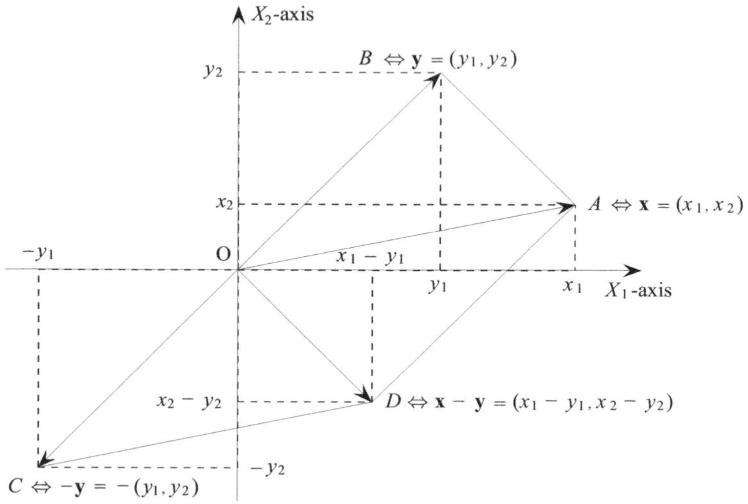

Fig. 1.7 Construction of $\mathbf{x} - \mathbf{y} = \mathbf{x} + (-1)\mathbf{y}$ based on Figs. 1.5 and 1.6. Note that $|\mathbf{x} - \mathbf{y}|$ is the distance between the points $\mathbf{x}$ and $\mathbf{y}$

is the vector connecting O and D (we write also OD). One sees immediately (since not only OCDA but also ODAB is a parallelogram) that BA has the same length as OD and *that length is* $|\mathbf{x} - \mathbf{y}|$. Now, *the length of BA is the distance of the points* $\mathbf{x}$ and $\mathbf{y}$. We accept

$$d(\mathbf{x}, \mathbf{y}) := |\mathbf{x} - \mathbf{y}|$$

as *definition* of the *distance* $d(\mathbf{x}, \mathbf{y}) \in \mathbb{R}_+$ of $\mathbf{x} \in \mathbb{R}^n, \mathbf{y} \in \mathbb{R}^n$ *for all n* (For $n = 3$ we have still a similar geometric picture and proof).

From Fig. 1.5 we see that the "triangle inequality"

$$|\mathbf{x} + \mathbf{y}| \leq |\mathbf{x}| + |\mathbf{y}|$$

holds, because the sum of the lengths of the two sides of a triangle is at least as large as the length of the third side. We see also that *there is* "=" *in place of* "≤" *in the triangle inequality* if, and only if, the two vectors "have the same direction", that is, (compare Fig. 1.6) either there exists $\lambda \in \mathbb{R}$ such that

$$\mathbf{y} = -\mathbf{x} \quad \text{or} \quad \mathbf{x} = \mathbf{0},$$

in other words: *exactly when* $\mathbf{x}$ *and* $\mathbf{y}$ *are linearly dependent* (why are the last two statements equivalent?). These results are true also in n-dimensional spaces with $n > 2$.

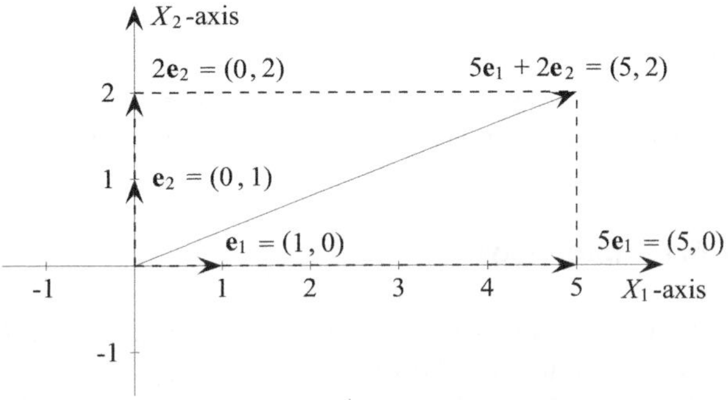

Fig. 1.8
$(5, 2) = 5(1, 0) + 2(0, 1)$

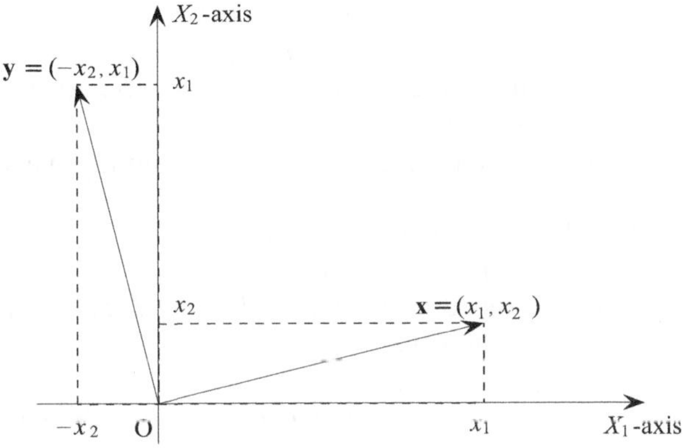

Fig. 1.9 For the orthogonal vectors **x, y** we have $\mathbf{x} \cdot \mathbf{y} = (x_1, x_2) \cdot (-x_2, x_1) = -x_1x_2 + x_1x_2 = 0$

Figure 1.8 gives an example that the base vectors $\mathbf{e}_1 = (1, 0)$ and $\mathbf{e}_2 = (0, 1)$ indeed span $\mathbb{R}^2$, that is, to every $\mathbf{x} \in \mathbb{R}^2$ there exist $\mu_1, \mu_2 \in \mathbb{R}$ such that $\mathbf{x} = \mu_1\mathbf{e}_1 + \mu_2\mathbf{e}_2$. Here $\mathbf{x} = (5, 2)$, $\mu_1 = 5$, $\mu_2 = 2$.

We will return in Sect. 1.7.2 to the geometric interpretation of the inner product. Here we deal only with *the case when* $\mathbf{x} \cdot \mathbf{y} = 0$ (Fig. 1.9).

In Fig. 1.9, $\mathbf{x} = (x_1, x_2)$ and $\mathbf{y} = (-x_2, x_1)$ are orthogonal ("perpendicular": their angle is a *right angle*; look at all angles at **0**) and their inner product is **0**:

$$\mathbf{x} \cdot \mathbf{y} = (x_1, x_2) \cdot (-x_2, x_1) = x_1(-x_2) + x_2x_1 = 0.$$

These **x** and **y** are of equal length $(x_1^2 + x_2^2)^{1/2}$ but it follows now from (1.3) in Sect. 1.4 that the inner product of *any* two perpendicular 2-component real vectors is 0. In general we can *define*, also for $n > 2$, *two* n-component vectors *to be orthogonal* exactly *when their inner product is* 0. (Again it could still be illustrated geometrically for $n = 3$; for $n = 1$ it does not make much sense because for $\mathbf{x} = x_1$, $\mathbf{y} = y_1 \in \mathbb{R}$ we have $\mathbf{x} \cdot \mathbf{y} = x_1y_1 = 0$ exactly if either $x_1 = 0$ or $y_1 = 0$.) Notice that *the null-vector is orthogonal to every vector in* $\mathbb{R}^n$ (including itself):

$$\mathbf{x} \cdot \mathbf{0} = (x_1, \ldots, x_n) \cdot (0, \ldots, 0) = x_10 + \ldots + x_n0 = 0$$

(true also for $\mathbf{x} = \mathbf{0}$). Since the inner product is commutative, the orthogonality relation is "symmetric", that is, if $\mathbf{x}$ is orthogonal to $\mathbf{y}$ then $\mathbf{y}$ is orthogonal to $\mathbf{x}$ ($0 = \mathbf{x} \cdot \mathbf{y} = \mathbf{y} \cdot \mathbf{x}$); in particular $\mathbf{0} \cdot \mathbf{x} = \mathbf{x} \cdot \mathbf{0}$. That is why we chose the "symmetric" expression "$\mathbf{x}$ and $\mathbf{y}$ are orthogonal (to each other)".

1.6.1 Exercises

1. Draw the vectors $\mathbf{x} = (1, 2)$, $\mathbf{y} = (3, 4)$, $\mathbf{z} = (-2, 4)$ in the plane and calculate with their aid also the vectors
   (a) $\mathbf{x} + \mathbf{y}$, (b) $\mathbf{x} - \mathbf{y}$, (c) $\mathbf{x} + \mathbf{z}$, (d) $\mathbf{x} - \mathbf{z}$,
   (e) $\mathbf{y} + \mathbf{z}$, (f) $\mathbf{y} - \mathbf{z}$, (g) $3\mathbf{x}$, (h) $2\mathbf{y}$,
   (i) $\mathbf{x} + \mathbf{y} + \mathbf{z}$, (j) $\mathbf{x} - \mathbf{y} + \mathbf{z}$, (k) $\mathbf{x} + \mathbf{y} - \mathbf{z}$, (l) $3\mathbf{x} + 2\mathbf{y} + \mathbf{z}$,
2. Calculate the distances $d(\mathbf{x}, \mathbf{y})$, $d(\mathbf{x}, \mathbf{z})$, $d(\mathbf{y}, \mathbf{z})$, where $\mathbf{x}$, $\mathbf{y}$, $\mathbf{z}$ are the vectors defined in Exercise 1.
3. Construct vectors of length 1 which are orthogonal to the vectors

   (a) (3,4), (b) (4,−1,−1),
   (c) (−6,2,2,2), (d) (2,2,3,4,5).
4. For the vectors $\mathbf{a} = (1, 2, 3, 4)$, $\mathbf{b} = (-3, 5, -6, 2)$, $\mathbf{c} = (-3, -2, -1, 8)$, $\mathbf{d} = (6, -4, 7, 3)$ show that

$$|\mathbf{a} + \mathbf{b} + \mathbf{c} + \mathbf{d}| < |\mathbf{a}| + |\mathbf{b}| + |\mathbf{c}| + |\mathbf{d}|\,,$$

$$|\mathbf{a} - \mathbf{b} + \mathbf{c} - \mathbf{d}| < |\mathbf{a}| + |\mathbf{b}| + |\mathbf{c}| + |\mathbf{d}|\,.$$

5. (a) For pairs of vectors $\mathbf{x}$, $\mathbf{y}$ taken from Exercise 4 and for arbitrary real numbers $\lambda \neq 0$, $\mu \neq 0$, show that

$$|\lambda\mathbf{x} + \mu\mathbf{y}| \leq |\lambda|\,|\mathbf{x}| + |\mu|\,|\mathbf{y}|\,.$$

   (b) Prove this inequality for all $\mathbf{x} \in \mathbb{R}^n$, $\mathbf{y} \in \mathbb{R}^n$, $\lambda \in \mathbb{R}$, $\mu \in \mathbb{R}$.
   (c) When does = hold for the above inequality?

1.6.2 Answers

1. $d(\mathbf{x}, \mathbf{y}) = 2\sqrt{2}$, $d(\mathbf{x}, \mathbf{z}) = \sqrt{13}$, $d(\mathbf{y}, \mathbf{z}) = 5$.
2. (a) $\frac{1}{5}(4, -3)$, (b) $\frac{1}{3}(1, 2, 2)$,
   (c) $\frac{1}{2}(1, 1, 1, 1)$, (d) $\frac{1}{7}(-5, -4, 0, 2, 2)$.

   Note that these vectors are not unique.

1.7 Complex Numbers; the Cosine, Sine, Tangent and Cotangent

Vectors in $\mathbb{R}^2$ are particularly important because they can be interpreted as "complex numbers". Why do we need another kind of number? Negative numbers were introduced so that an equation like $3 + x = 2$ have a solution ($x = -1$); (noninteger) rational numbers came in so that an equation like $3x = 2$ have a solution ($x = 2/3$). As one reason for introducing irrational numbers we can name the desire that an equation like $3x^2 = 2$ have a solution ($x = \sqrt{2}/\sqrt{3}$); there are also other reasons, for instance that π (the half circumference length of the unit circle) be a number or that the "number line" (Sect. 1.1, compare Fig. 1.1) be "filled". In the sixteenth century the need to solve equations like $3 + x^2 = 2$ or $25 + z^2 = 8z$ arose. But there are *no real numbers* (rational or irrational) x and z satisfying these equations and the one-dimensional number line is already "full". So we have to move into two dimensions. In the two-dimensional space there are the two-component (real) vectors. But we had till now no multiplication of vectors (and so no squaring) which results in a vector (though we have multiplication of a vector by a scalar and multiplication of two vectors, the inner product, which results in a scalar). Fortunately, two is a dimension in which a multiplication of vectors can be defined which results in a vector of the same dimension (two) and which has quite similar properties to the product of real numbers. (There are not many such dimensions: in a sense two is the only one, though there is a multiplication in four dimensions, that of "quaternions" which has many properties of multiplication for real numbers but not commutativity, that is, there in general $\mathbf{xy} \neq \mathbf{yx}$. We will not deal with them here. The multiplication of "*complex numbers*", as we will define it, is commutative.) Actually, this multiplication is what makes "*complex numbers*" out of two-component (real) vectors.

We could define multiplication right away but that definition could seem quite arbitrary. So, rather than hitting the reader over the head with these formulas, we try some gentle persuasion and preparation:

1.7.1 Multiplication of Complex Numbers

As we saw in Sect. 1.5.2, any two 2-component real vectors can be written as

$$\mathbf{a} = a_1\mathbf{e}_1 + a_2\mathbf{e}_2, \quad \mathbf{b} = b_1\mathbf{e}_1 + b_2\mathbf{e}_2, \quad \text{where} \quad \mathbf{e}_1 = (1, 0), \quad \mathbf{e}_2 = (0, 1)$$

(and $a_1\mathbf{e}_1$ is the product of the scalar a_1 and of the vector $\mathbf{e}_1$ as defined in Sect. 1.4, and so on). If we want our new multiplication of complex numbers (2-component real vectors) to be distributive (and satisfy $(\lambda\mathbf{x})(\mu\mathbf{y}) = (\lambda\mu)\mathbf{xy}$) then

$$\begin{aligned}\mathbf{ab} &= (a_1\mathbf{e}_1 + a_2\mathbf{e}_2)(b_1\mathbf{e}_1 + b_2\mathbf{e}_2)\\ &= a_1b_1\mathbf{e}_1\mathbf{e}_1 + a_1b_2\mathbf{e}_1\mathbf{e}_2 + a_2b_1\mathbf{e}_2\mathbf{e}_1 + a_2b_2\mathbf{e}_2\mathbf{e}_2.\end{aligned}$$

This shows that the only products we have to define are $\mathbf{e}_1\mathbf{e}_1$, $\mathbf{e}_2\mathbf{e}_2$ and $\mathbf{e}_1\mathbf{e}_2 = \mathbf{e}_2\mathbf{e}_1$, if we want the multiplication to be also commutative. We want the product to be again a (two-component real) *vector* (not a scalar as in the scalar product) so the above three fundamental products have also to be vectors (with two components, though one of them may be 0). There are still several possibilities left but, since we want the horizontal line to be our old "real line", it is rather pleasing to *define* $\mathbf{e}_1\mathbf{e}_1 = \mathbf{e}_1$ and then, similarly,

$$\mathbf{e}_1\mathbf{x} = \mathbf{x}\mathbf{e}_1 = \mathbf{x} \quad \text{for every } \mathbf{x} \text{ in } \mathbb{R}^2. \tag{1.4}$$

In particular,

$$\mathbf{e}_1\mathbf{e}_2 = \mathbf{e}_2\mathbf{e}_1 = \mathbf{e}_2. \tag{1.5}$$

Only $\mathbf{e}_2\mathbf{e}_2$ remains to be defined. It may be tempting to equate it to $\mathbf{e}_2$ but then we could loose an important property of multiplication, the *cancellativity*: if $\mathbf{xz} = \mathbf{yz}$ and $\mathbf{z} \neq \mathbf{0}$ then $\mathbf{x}=\mathbf{y}$ or, equivalently, $\mathbf{t}\,\mathbf{z} = \mathbf{0}$ *only if* $\mathbf{t} = \mathbf{0}$ *or* $\mathbf{z} = \mathbf{0}$ *or both.* Cancellativity yields $\mathbf{e}_1 = \mathbf{e}_2$ from $\mathbf{e}_2\mathbf{e}_2 = \mathbf{e}_2 = \mathbf{e}_1\mathbf{e}_2$ (see (1.5)) and $\mathbf{e}_2 \neq \mathbf{0}$. The same problem would arise if we chose $\mathbf{e}_2\mathbf{e}_2 = -\mathbf{e}_2$. So we do not want $\mathbf{e}_2\mathbf{e}_2$ to equal $\mathbf{e}_2$ or $-\mathbf{e}_2$. The same reason *excludes* $\mathbf{e}_2\mathbf{e}_2 = \mathbf{e}_1$ (be careful *not to use* later *the equations in the last three sentences and in the next sentence*). Indeed we would have then, by (1.4), $\mathbf{e}_2\mathbf{e}_2 = \mathbf{e}_1 = \mathbf{e}_1\mathbf{e}_1$, that is, $\mathbf{0} = \mathbf{e}_1\mathbf{e}_1 - \mathbf{e}_2\mathbf{e}_2 = (\mathbf{e}_1 + \mathbf{e}_2)(\mathbf{e}_1 - \mathbf{e}_2)$ contradicting cancellativity, because neither $\mathbf{e}_1 + \mathbf{e}_2 = \mathbf{0}$ (since $\mathbf{e}_2 \neq -\mathbf{e}_1$), nor $\mathbf{e}_1 - \mathbf{e}_2 = \mathbf{0}$ (since $\mathbf{e}_2 \neq \mathbf{e}_1$) hold. So we choose the next best thing and define

$$\mathbf{e}_2\mathbf{e}_2 = -\mathbf{e}_1. \tag{1.6}$$

From (1.4), (1.5) and (1.6) we have now the *definition of the product of two complex numbers*:

$$\begin{aligned} \mathbf{ab} &= (a_1\mathbf{e}_1 + a_2\mathbf{e}_2)(b_1\mathbf{e}_1 + b_2\mathbf{e}_2) \\ &= (a_1b_1 - a_2b_2)\mathbf{e}_1 + (a_1b_2 + a_2b_1)\mathbf{e}_2 \end{aligned} \tag{1.7}$$

or, what is the same, the definition with which we could have started but which in our opinion needed some explanation:

$$(a_1, a_2)(b_1, b_2) := (a_1b_1 - a_2b_2, a_1b_2 + a_2b_1).$$

Note that this product satisfies cancellativity (see Exercise 2).

Because of (1.4), $\mathbf{e}_1$ plays the same role in this multiplication of complex numbers as 1 in the multiplication of real numbers. Therefore we will define $\mathbf{1} := \mathbf{e}_1$. Then (1.6) reads

$$\mathbf{e}_2^2 = \mathbf{e}_2\mathbf{e}_2 = -\mathbf{1}$$

(just as for numbers, we define also for complex numbers the product, of a "number" by itself as the square of that number). We write $\mathbf{i} := \mathbf{e}_2$ because of this property, since the "number" $\mathbf{i}$, for which $\mathbf{i}^2 = -\mathbf{1}$ was what mathematicians were looking for. (Of course, there exists no such real number. This $\mathbf{i}$ would be, in particular, a solution of the equation $3 + x^2 = 2$, mentioned at the beginning of this section or, what is the same, a solution of $x^2 + 1 = 0$.) With this notation, every complex number $\mathbf{a}$ can be written as

$$\mathbf{a} = a_1\mathbf{1} + a_2\mathbf{i}.$$

Now, if we only remember the all-important relation $\mathbf{i}^2 = -1$ (and distributivity), we can get rid of the vector notation and write complex numbers just as real ones:

$$a = a_1 + \mathbf{i}a_2.$$

Indeed then, as in (1.7),

$$\begin{aligned} ab = (a_1 + \mathbf{i}a_2)(b_1 + \mathbf{i}b_2) &= a_1b_1 + \mathbf{i}a_1b_2 + \mathbf{i}a_2b_1 + \mathbf{i}^2a_2b_2 \\ &= (a_1b_1 - a_2b_2) + \mathbf{i}(a_1b_2 + a_2b_1). \end{aligned} \tag{1.8}$$

As "non real", $\mathbf{i}$ got the name "imaginary unit" and every $b\,\mathbf{i}$ with $b \in \mathbb{R}$ was called "imaginary number".

Not only does $x = \mathbf{i}$ satisfy $x^2 + 1 = 0$ and $z = 4 + 3\mathbf{i}$ satisfy the equation $25 + z^2 = 8z$ $(25 + (4 + 3\mathbf{i})^2 = 25 + 16 + 24\mathbf{i} - 9 = 32 + 24\mathbf{i} = 8(4 + 3\mathbf{i}))$ but, for every equation of the form

$$a_px^p + a_{p-1}^{p-1} + \ldots + a_1x + a_0 = 0$$

$(a_0, a_1, \ldots, a_p$ (real or) complex, $a_p \neq 0)$, there exists a complex number x which satisfies it ("fundamental theorem of algebra", see also Sect. 6.2).

Note that particular cases of the complex numbers

$$a = a_1 + \mathbf{i}a_2 \quad (a_1, a_2 \in \mathbb{R})$$

are real numbers (for $a_2 = 0$) and the imaginary numbers (for $a_1 = 0$). From now on $\mathbf{i}$ will no longer be represented bold-faced, i.e. we shall write i instead of $\mathbf{i}$.

Let us recapitulate: *Complex numbers are 2-component real vectors with the multiplication (1.7) [or (1.8)] defined for them.* If such a multiplication is defined then, instead of the "real plane" $\mathbb{R}^2$, we speak about the "complex plane" or "Gaussian plane" $\mathbb{C}$ (Carl Friedrich Gauss (1777–1855) was a very famous German mathematician but he did not invent complex numbers, only clarified and enhanced their role). This time the horizontal and vertical axes are called the "real axis" and the "imaginary axis", respectively. *We write the complex numbers as*

$$a = a_1 + ia_2 \quad (a_1, a_2 \in \mathbb{R})$$

and can add, subtract, multiply, divide (see below), *square etc. them as we would do with other two-term sums. We have to remember only that*

$$i^2 = -1.$$

In particular, the multiplication (1.8) is commutative, associative and distributive upon addition:

$$ab = ba, \quad (ab)c = a(bc), \quad a(b + c) = ab + ac \quad (a, b, c \in \mathbb{C}).$$

(We leave the proofs to the reader). Clearly *two complex numbers are equal*, that is $a_1 + ia_2 = b_1 + ib_2$ *exactly if their real parts are equal:* $a_1 = b_1$ *and so are their imaginary parts*: $a_2 = b_2$.

The addition of complex numbers and their multiplication by a real number has, of course, the same meaning as for 2-component real vectors, see Figs. 1.5 and 1.6. To interpret geometrically the multiplication of complex numbers is easier in their trigonometric form.

1.7.2 Trigonometric Form of Complex Numbers; Sine, Cosine

A complex number in the Gaussian plane (or, for that matter, a vector in $\mathbb{R}^2$) can be described not only by its two components, but also by its length (the norm $|\mathbf{z}| = (z_1^2 + z_2^2)^{1/2}$ of the vector $\mathbf{z} = z_1\mathbf{e}_1 + z_2\mathbf{e}_2 = z_1\mathbf{1} + z_2 i = z_1 + iz_2$; for complex numbers, just as for real ones, we use the *absolute value* name and sign for norms: $|z| = |z_1 + iz_2| = |z_1\mathbf{1} + z_2 i| = (z_1^2 + z_2^2)^{1/2}$) and by its "angle", "argument" or "amplitude". The latter is the (directed) angle ϕ between the basis vector $\mathbf{e}_1 = (1, 0)$ and the vector $\mathbf{z} = (z_1, z_2)$ (see Fig. 1.10), it is denoted by arg

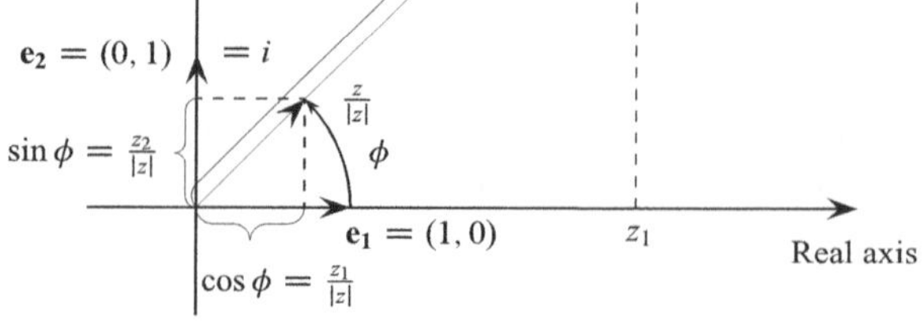

Fig. 1.10 Trigonometric form of a complex number: $z = z_1 + iz_2 = |z|\,(\cos\phi + i\sin\phi)$

$z = \phi$. Two reservations have to be made: *For the complex number* 0 (the null-vector **0**) *the amplitude* (angle) *is arbitrary* (while the absolute value makes sense: $|z| = 0$; conversely, *if* $|z| = 0$ *then* $z = 0$. And, if $|z| = |\mathbf{z}| > 0$ (the absolute value is *never* negative), then *the amplitude of* z *is determined only up to multiples of* 2π. Note that, while for some practical purposes angles are measured by *degrees* or even "*decimal degrees*" (with which the right angle is 90° or 100 decimal degrees, respectively), in mathematics angles are mostly measured in *radians* (length of the arc of the unit circle belonging to that central angle: right angle = $\pi/2$, full angle = full turn = 2π); we will see the advantage of this, among others, in Sects. 5.2 and 5.4.

The *triangle inequality* in Sect. 1.6 can now be written as

$$|z_1 + z_2| \leq |z_1| + |z_2| \quad \textit{for all} \quad z_1, z_2 \in \mathbb{C}$$

with equality if and only if

$$\arg z_1 = \arg z_2 \quad (\text{or } \arg z_1 = \arg z_2 + 2k\pi,\ k \in \mathbb{Z}).$$

This implies (why? prove it by induction) the following generalisation:

$$|z_1 + z_2 + \ldots + z_m| \leq |z_1| + |z_2| + \ldots + |z_m|$$

for all $z_1, z_2, \ldots, z_m$ in $\mathbb{C}$, with equality if and only if

$$\arg z_1 = \arg z_2 = \ldots = \arg z_m.$$

Often this too is called triangle inequality.

We also need to define $\sin\phi$ and $\cos\phi$, the "sine" and the "cosine" of the amplitude ϕ. This can be done in many ways. For our purpose the following seems to be convenient. For every vector $\mathbf{z} \neq \mathbf{0}$, the vector (see Fig. 1.10) $(1/|\mathbf{z}|)\mathbf{z}$ is a *unit vector* with the same amplitude ϕ as that of $\mathbf{z}$. *The components of this unit vector are* $\cos\phi$ *and* $\sin\phi$, *they define the cosine and sine of* ϕ. This holds also if $\phi \geq \pi/2$ or $\phi \leq 0$; *the* $\cos\phi$ *and the* $\sin\phi$ *do not change if we add* $2k\pi$ $(k \in \mathbb{Z})$ *to* ϕ. It follows from the definition that $-1 \leq \cos\phi \leq 1$, $-1 \leq \sin\phi \leq 1$. From the similarity of the two rectangular triangles in Fig. 1.10 (both with the angle ϕ at **0**)

$$\mathbf{z} = |\mathbf{z}| \cos\phi\, \mathbf{e}_1 + |\mathbf{z}| \sin\phi\, \mathbf{e}_2.$$

Now, identifying, as above, the vector $\mathbf{z} \in \mathbb{R}^2$ with the complex number $z \in \mathbb{C}$, its norm $|\mathbf{z}|$ with the absolute value $|z|$, $\mathbf{e}_1$ with 1 and $\mathbf{e}_2$ with i, we get

$$z = |z| (\cos\phi + i \sin\phi) = r(\cos\phi + i \sin\phi),$$

the *trigonometric form of the complex number* z (it is customary to write $r = |z|$) and it is easy to see that this remains valid also for $\phi \geq \pi/2$ and $\phi \leq 0$.

Now we can give a geometrical interpretation for the multiplication of complex numbers. First, we know that multiplication of *any* real (2-component) vector by a real number c means (see Fig. 1.6) stretching (or compressing) it by c and keeping its direction (this is for positive c, for simplicity we will restrict ourselves in this argument to positive c; for negative c, we stretch by $|c| = -c$ and reverse the direction; for $c = 0$ the product is the null-vector) and this is also how we multiply a complex number by a real number. We now look at multiplication of a complex number by i:

$$zi = (z_1 + iz_2)i = -z_2 + z_1 i.$$

But, as we have seen (look at Fig. 1.9), the vector $(-z_2, z_1)$ *is orthogonal (perpendicular) to* $\mathbf{z} = (z_1, z_2)$. This is all we need. Figure 1.11 shows the steps yielding

$$(a_1 + ia_2)(b_1 + ib_2) = ab = a(b_1 + ib_2) = ab_1 + (ai)b_2.$$

Since the rectilinear triangle OC_1C can be obtained by "enlarging" (if $|a| > 1$, otherwise "shrinking") the triangle OB_1B "$|a|$-times" (look at the sides enclosing the rectangle in both triangles), so OC, the length of the product vector ab will also

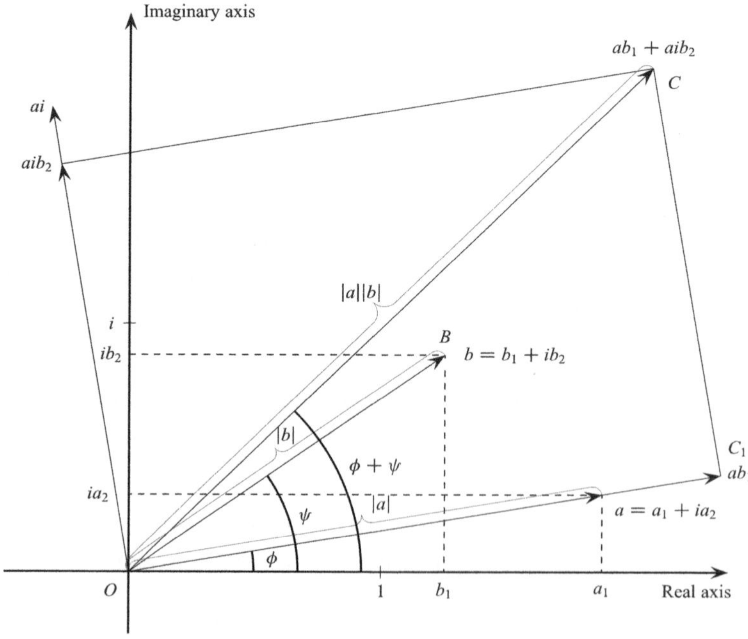

Fig. 1.11 Multiplication of complex numbers: the absolute values are multiplied, the amplitudes are added up

be $|a|$ times the "length" (absolute value) OB of b:

$$|ab| = |a|\ |b| \quad (a \in \mathbb{C},\ b \in \mathbb{C})$$

and the angle $\angle C_1OC$ will equal the angle $\angle B_1OB$. So, we *multiply two nonzero complex numbers by multiplying their absolute values and adding up their amplitudes*:

$$\begin{aligned} ab &= (|a|\,(\cos\phi + i\,\sin\phi))\ \ (|b|\,(\cos\psi + i\,\sin\psi)) \\ &= |a|\ |b|\,(\cos(\phi+\psi) + i\,\sin(\phi+\psi)). \end{aligned} \tag{1.9}$$

As an added bonus we get "free of charge" significant properties of the cosine and sine, which otherwise are usually proved by rather painful and lengthy arguments.

Comparing our last formula to that obtained by straightforward multiplication, as in (1.12 and 1.13):

$$\begin{aligned} ab &= (|a|\cos\phi + \mathrm{i}\,|a|\sin\phi)\ (|b|\cos\psi + \mathrm{i}\,|b|\sin\psi) \\ &= |a|\ |b|\,(\cos\phi\cos\psi - \sin\phi\sin\psi + \mathrm{i}(\sin\phi\cos\psi + \cos\phi\sin\psi))\,, \end{aligned}$$

we get

$$\cos(\phi+\psi) = \cos\phi\cos\psi - \sin\phi\sin\psi, \tag{1.10}$$

$$\sin(\phi+\psi) = \sin\phi\cos\psi + \cos\phi\sin\psi, \tag{1.11}$$

the all–important *addition formulas of the cosine and of the sine, respectively.*

The above definitions of the cosine and sine imply (Fig. 1.12)

$$\begin{aligned} &\cos 0 = 1, && \sin 0 = 0, \\ &\cos(\pi/2) = 0, && \sin(\pi/2) = 1, \\ &\cos\pi = -1, && \sin\pi = 0, \\ &\cos(\pi-\psi) = -\cos\psi, && \sin(\pi-\psi) = \sin\psi, \\ &\cos(-\psi) = \cos\psi, && \sin(-\psi) = -\sin\psi. \end{aligned}$$

Using the last two equations when replacing ψ by $-\psi$ in (1.10), (1.11), we obtain

$$\cos(\phi-\psi) = \cos\phi\cos\psi + \sin\phi\sin\psi, \tag{1.12}$$

$$\sin(\phi-\psi) = \sin\phi\cos\psi - \cos\phi\sin\psi, \tag{1.13}$$

the *subtraction formulas of the cosine and of the sine*. Choosing here $\phi = \pi/2$, the above equations give (see also Fig. 1.12)

$$\cos\left(\frac{\pi}{2} - \psi\right) = \sin\psi, \quad \sin\left(\frac{\pi}{2} - \psi\right) = \cos\psi,$$

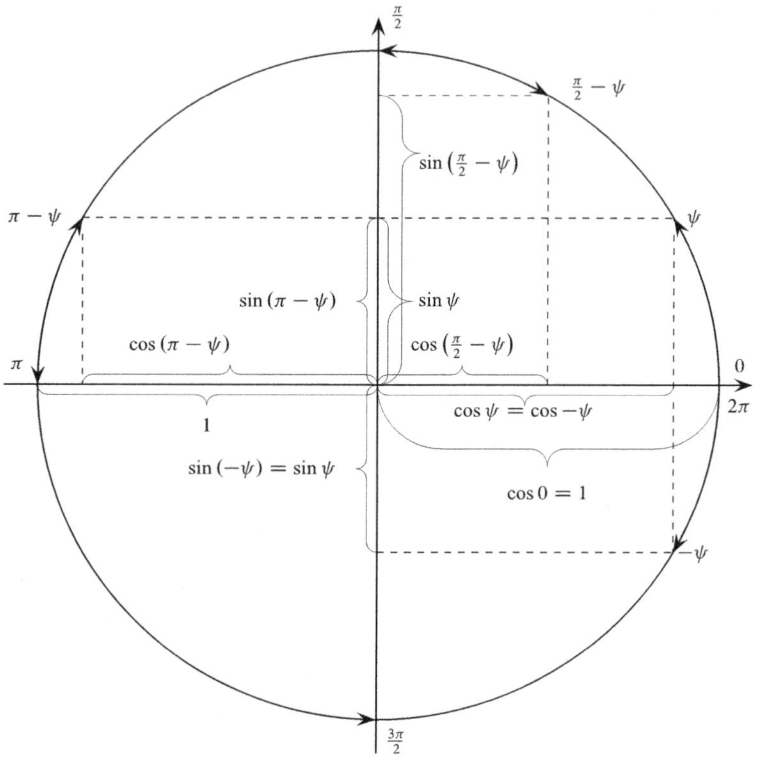

Fig. 1.12 Cosines and sines of 0, $\pi/2$, π, $\pi - \psi$, $-\psi$, $(\pi/2) - \psi$. The circle is the unit circle, i.e., has radius 1

an important link between the cosine and the sine and which is now true for *all* ψ. Also another important link between the cosine and the sine follows from (1.12) and from $\cos 0 = 1$: Putting $\psi = \phi$ into (1.12) we get the *fundamental equation*

$$\cos^2 \phi + \sin^2 \phi = 1 \tag{1.14}$$

($\cos^2\phi$ and $\sin^2\phi$ are frequently used—though not very fortunate—abbreviations for $(\cos\phi)^2$ and for $(\sin\phi)^2$, respectively). This follows also from the definition of cos ϕ and sin ϕ as components of a unit vector with amplitude ϕ (see Fig. 1.10), via the Pythagoras theorem (see Fig. 1.4) at least when $0 < \phi < \pi/2$; but we obtained it now right away for *all* $\phi \in \mathbb{R}$. From (1.11), (1.12), (with $\psi = \phi$) and (1.10) (with $\psi = 0$), (1.14) we get also

$$\begin{aligned} \sin 2\phi &= 2\sin\phi\cos\phi, \\ \cos 2\phi &= \cos^2\phi - \sin^2\phi = 2\cos^2\phi - 1 = 1 - \sin^2\phi. \end{aligned}$$

Another application of the subtraction formula (1.12) for the cosine reaches back to the definition of the inner product. As Fig. 1.10 shows, two–component real vectors can be written as

$$\mathbf{x} = (|\mathbf{x}| \cos\phi,\ |\mathbf{x}| \sin\phi), \qquad \mathbf{y} = (|\mathbf{y}| \cos\psi,\ |\mathbf{y}| \sin\psi).$$

Their inner product is

$$\begin{aligned}\mathbf{x}\cdot\mathbf{y} &= (|\mathbf{x}| \cos\phi)\,(|\mathbf{y}| \cos\psi) + (|\mathbf{x}| \sin\phi)\,(|\mathbf{y}| \sin\psi)\\ &= |\mathbf{x}|\ |\mathbf{y}| \cos(\phi - \psi).\end{aligned}$$

Since $\phi - \psi$ is the angle between $\mathbf{x}$ and $\mathbf{y}$ (see Fig. 1.13) so, at least for two-component real vectors, the *inner product of two vectors is the product of their norms* (lengths) *multiplied by the cosine of the angle between the two vectors.*

The same rule can be proved also for three-component real vectors and, if the angle between vectors is appropriately defined, also for n-component vectors (actually, then the angle is defined just so that this rule should remain valid).

But let us return now to complex numbers.

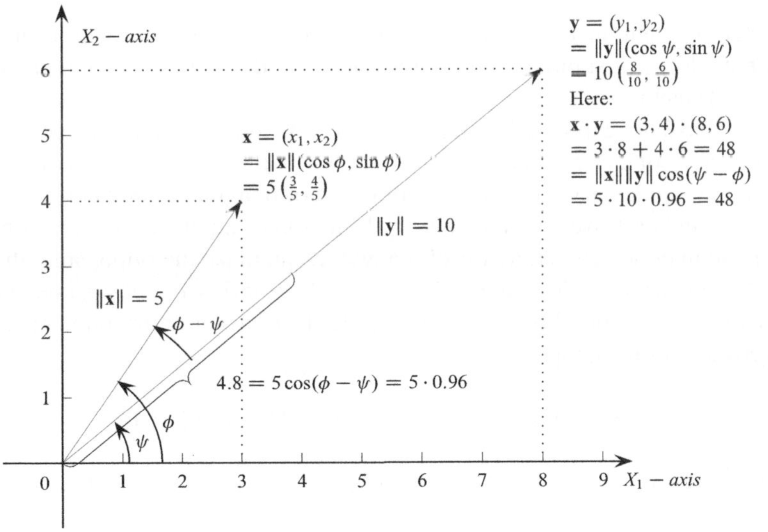

Fig. 1.13 The inner product $\mathbf{x}\cdot\mathbf{y} = |\mathbf{x}|\ |\mathbf{y}| \cos(\phi - \psi)$

1.7.3 Division of Complex Numbers; Equations

As for reals, also for complex numbers $b = c/a$ is by definition the (complex) number for which $ab = c$ $(a \neq 0)$. From (1.9),

$$c = |c|\,(\cos\chi + \mathrm{i}\sin\chi) = ab = |a|\;|b|\,(\cos(\phi + \psi) + \mathrm{i}\sin(\phi + \psi)),$$

we get the rule *for division of two complex numbers in trigonometric form*:

$$b = \frac{c}{a} = \frac{|c|}{|a|}(\cos(\chi - \phi) + \mathrm{i}\sin(\chi - \phi)), \quad \text{if} \quad a \neq 0$$

because $b = |b|\,(\cos\psi + \mathrm{i}\sin\psi)$ so, from the above equation,

$$|c| = |a|\;|b|\,, \quad \text{that is,} \quad |b| = \frac{|c|}{|a|} \quad (\text{since } |a| \neq 0)$$

and

$$\chi = \phi + \psi, \quad \text{that is,} \quad \psi = \chi - \phi.$$

(Note that, as mentioned above, ϕ, ψ, and χ are determined only up to multiples of 2π, but adding or subtracting multiples of 2π to the angles does not change the cosines and sines).

We still have not expressed, however, $b = c/a$ by the *real* and "*imaginary*" parts a_1, a_2 and c_1, c_2 of $a = a_1 + \mathrm{i}a_2 \neq 0$ and of $c = c_1 + \mathrm{i}c_2$ $(a_1, a_2, c_1, c_2$ real numbers; note that the "imaginary parts" a_2, c_2 are real, not imaginary numbers, $\mathrm{i}a_2$, $\mathrm{i}c_2$ are imaginary numbers). We could do this "by brute force" but the concept of conjugate complex number, which is rather useful anyway, can make the riding smoother.

The "conjugate complex number" of $a = a_1 + \mathrm{i}a_2$ is $\bar{a} = a_1 - \mathrm{i}a_2$ (same real part, imaginary part multiplied by (-1), see Fig. 1.14; if written in trigonometric form, $a = r(\cos\phi + \mathrm{i}\sin\phi)$, then

$$\bar{a} = r(\cos\phi - \mathrm{i}\sin\phi) = r(\cos(-\phi) + \mathrm{i}\sin(-\phi)),$$

same absolute value, amplitude multiplied by (-1)). Important is that

$$\bar{a}a = a\bar{a} = (a_1 + \mathrm{i}a_2)(a_1 - \mathrm{i}a_2) = a_1^2 + a_2^2 = |a|^2 \;\; (= r^2),$$

that is, the *product of a complex number with its conjugate is the square of their common absolute value*. (Compare: the inner product of a vector with itself was the square of its norm; the product, in the sense of multiplication of complex numbers, of a complex number with its conjugate is the square of its absolute value, that is, the square of its norm.)

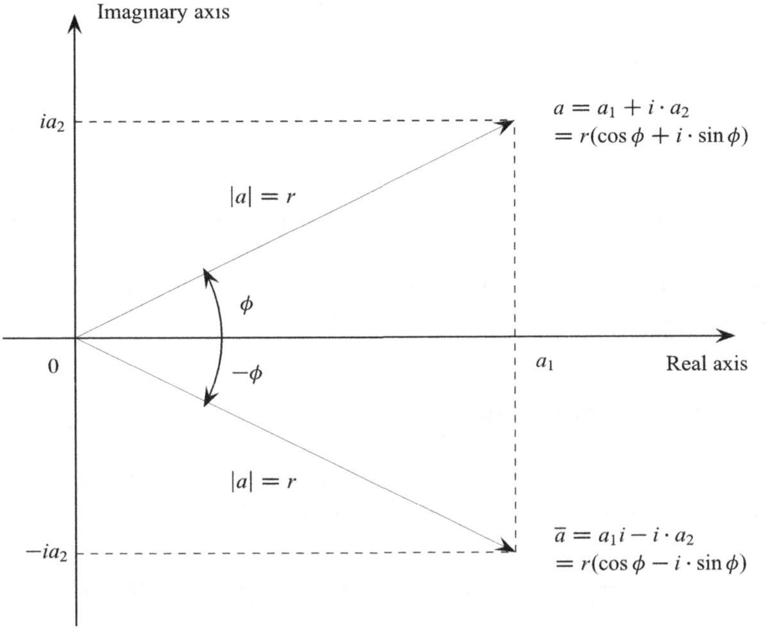

Fig. 1.14 Conjugate complex numbers

Now, multiplying $c = ab$ by $\bar{a}$, we get $\bar{a}c = \bar{a}ab = |a|^2 b$. Since $|a|^2 \neq 0$ is real, we can multiply by $1/|a|^2$, as in Sect. 1.4, and obtain

$$\frac{c}{a} = b = \frac{\bar{a}c}{|a|^2} = \frac{(a_1 - \mathrm{i}a_2)(c_1 + \mathrm{i}c_2)}{a_1^2 + a_2^2} = \frac{a_1c_1 + a_2c_2}{a_1^2 + a_2^2} + \mathrm{i}\frac{a_1c_2 - a_2c_1}{a_1^2 + a_2^2}$$

as *formula for* $c/a = (c_1 + ic_2)/(a_1 + ia_2)$ *in terms of the real and imaginary parts* $a_1 \in \mathbb{R}$, $a_2 \in \mathbb{R}$, $c_1 \in \mathbb{R}$, $c_2 \in \mathbb{R}$, *when* $a \neq 0$.

Not only the equation $az = c$ (with a, c in $\mathbb{C}$ and $a \neq 0$) but, as mentioned before, every equation

$$a_p z^p + a_{p-1}z^{p-1} + \ldots + a_1 z + a_0 = 0 \quad (a_0, a_1, \ldots, a_p \in \mathbb{C};\ a_p \neq 0)$$

has a complex solution (a complex number z which satisfies it).

As is well known the numbers

$$z_1 = b + (b^2 - c)^{1/2} \qquad \text{and} \qquad z_2 = b - (b^2 - c)^{1/2}$$

are the (only) solutions of the equation $z^2 - 2bz + c = 0$. If b and c are real but $c > b^2$ then the conjugate complex numbers $z_1 = b + \mathrm{i}(c - b^2)^{1/2}$, $z_2 = \bar{z_1} = b - \mathrm{i}(c - b^2)^{1/2}$ are the solutions of $z^2 - 2bz + c$ $(c > b^2)$.

Since

$$\begin{aligned}\overline{c+d} &= \overline{c_1 + \mathrm{i}c_2 + d_1 + \mathrm{i}d_2} = (c_1 + d_1) - \mathrm{i}(c_2 + d_2) \\ &= c_1 - \mathrm{i}c_2 + d_2 - \mathrm{i}d_2 = \bar{c} + \bar{d}, \\ \overline{cd} &= \overline{(c_1 + \mathrm{i}c_2)(d_1 + \mathrm{i}d_2)} \\ &= (c_1d_1 + c_2d_2) - \mathrm{i}(c_1d_2 + c_2d_1) = (c_1 - \mathrm{i}c_2)(d_1 - \mathrm{i}d_2) = \bar{c}\bar{d},\end{aligned}$$

if a complex number z is a solution of $a_pz^p + \ldots + a_1, z + a_0 = 0$ with real "coefficients" $a_0, a_1, \ldots, a^p$, then $\bar{z}$ is also a solution.

1.7.4 Tangent, Cotangent

At least two "relatives" of sine and cosine are often important, the *tangent* and the *cotangent* and the *cotangent* abbreviated as tan and cot:

$$\tan\phi := \frac{\sin\phi}{\cos\phi} \text{ if } \cos\phi \neq 0, \quad \cot\phi := \frac{\cos\phi}{\sin\phi} \text{ if } \sin\phi \neq 0.$$

(Sometimes 1/cos ϕ and 1/sin ϕ are also given separate names, sec and cosec, respectively; note, by the way, that cot ϕ=1/tan ϕ if tan $\phi \neq 0$.) For two–component real vectors $\mathbf{z}$, that is for complex numbers $z = r(\cos\phi + \mathrm{i}\sin\phi)$, the real number tan $\phi = \frac{\sin\phi}{\cos\phi}$ (if sin $\phi \neq 0$) is called the "*slope* of the vector" $\mathbf{z}$ (or of the complex number z), and, more generally, if a straight line forms an angle ϕ with the positive horizontal axis, then $\tan\phi$ is the *slope* of that straight line.

From our formulas for the cosine and the sine, and from the above definition of the tangent and cotangent, we get easily that, whenever the denominator is not 0, then

$$\begin{aligned}\tan(\phi+\psi) &= \frac{\tan\phi + \tan\psi}{1 - \tan\phi\tan\psi}, & \cot(\phi+\psi) &= \frac{\tan\phi\cot\psi - 1}{\cot\phi + \cot\psi}, \\ \tan(\phi-\psi) &= \frac{\tan\phi - \tan\psi}{1 + \tan\phi\tan\psi}, & \cot(\phi-\psi) &= \frac{\cot\phi\cot\psi + 1}{\cot\psi - \cot\phi}, \\ \tan 2\phi &= \frac{2\tan\phi}{1 - \tan^2\phi}, & \cot 2\phi &= \frac{\cot^2\phi - 1}{2\cot\phi}, \\ \tan\left(\frac{\pi}{2} - \phi\right) &= \cot\phi, & \cot\left(\frac{\pi}{2} - \phi\right) &= \tan\phi, \\ \tan 0 &= \cot(\pi/2) = 0,\end{aligned}$$

while $\tan(\pi/2)$ *and* $\cot 0$ *are not defined*, since $\cos(\pi/2) = \mathit{sin}\, 0 = 0$, and neither are $\cot(k\pi)$, $\tan((2k+1)\pi/2)$ with $k \in \mathbb{Z}$ defined.

1.7.5 Exercises

1. For the complex numbers $a = 2\mathrm{i}$, $b = -3 + \mathrm{i}$, $c = 4 - 5\mathrm{i}$ determine
   (a) ab, (b) ac (c) bc, (d) $b\bar{c}$,
   (e) $\bar{b}c$, (f) abc, (g) $a\bar{b} + \bar{a}c$, (h) $a(b + c)$,
   (i) $(a + b)c$, (j) a^2, (k) b^3, (l) c^4,
   (m) a/b, (n) b/c, (o) c/a, (p) $a/(b + c)$.
2. Show that complex multiplication satisfies cancellativity.
3. Determine the absolute values of the complex numbers
   (a) $5 + \mathrm{i} \cdot 12$, (b) $(1 + \mathrm{i}7)/(1 + \mathrm{i})$, (c) $(-3 + 2\mathrm{i})^4$.
4. Write the following complex numbers in their trigonometric form:
   (a) $1 + \mathrm{i}$, (b) $1 - \mathrm{i}$, (c) $\sqrt{8} - \mathrm{i}\sqrt{8}$.
5. For the complex numbers $x = 3(\cos \frac{2}{3}\pi + \mathrm{i} \sin \frac{2}{3}\pi)$ and $y = 4(\cos \frac{3}{5}\pi + \mathrm{i} \sin \pi)$ determine
   (a) xy, (b) x/y, (c) y/x, (d) $x\bar{x}$,
   (e) $y\bar{y}$, (f) x^5, (g) $1/x^5$, (h) x/y^2.
6. Solve the equations
   (a) $(3 + 4\mathrm{i})z = -1 + \mathrm{i}$, (b) $z^2 - (8 - 2\mathrm{i})z + 23 - 2\mathrm{i} = 0$.
7. Determine all complex numbers z that solve
   (a) $z^3 - (4 + \mathrm{i})z^2 + (13 + 4\mathrm{i})z - 13\mathrm{i} = 0$,
   (b) $z^4 - 2z^3 + 6z^2 - 2z + 5 = 0$.
   Hint: In both cases, one of the solutions is $z = \mathrm{i}$. Hence the left hand sides in (a) and (b) can be written in the forms $(z-\mathrm{i})(z^2+az+b)$ and $(z-\mathrm{i})(z+\mathrm{i})(z^2+cz+d)$, respectively. For (b), $z = -\mathrm{i}$ is also a solution.
8. Show that the triangle inequality
   $$|a + b + c| < |a| + |b| + |c|$$
   holds for the complex numbers defined in Exercise 1.
9. Draw the complex numbers $a = 5 + 2\mathrm{i}$ and $b = 1 + 3\mathrm{i}$ as vectors in the complex plane. With the aid of these vectors construct the vector that represents the complex number ab. (Hint: see Fig. 1.11.)

1.7.6 Answers

1. (a) $-2 - \mathrm{i}6$, (b) $10 + \mathrm{i}8$, (c) $-7 + \mathrm{i}19$, (d) $-17 - \mathrm{i}11$,
   (e) $-17 + \mathrm{i}11$, (f) $-38 - \mathrm{i}14$, (g) $-8 - \mathrm{i}14$, (h) $8 + \mathrm{i}2$,
   (i) $3 + \mathrm{i}27$, (j) -4, (k) $-18 + \mathrm{i}26$, (l) $-1519 + \mathrm{i}720$,
   (m) $(1 - \mathrm{i}3)/5$, (n) $-(17+\mathrm{i}11)/41$, (o) $-(5 + \mathrm{i}4)/2$, (p) $(-8 + \mathrm{i}2)/17$.
2. Multiplying both sides of the equation $xz = yz$ (x, y, z complex numbers, $z = z_1 + \mathrm{i}z_2 \neq 0$) by $\bar{z} = z_1 - \mathrm{i}z_2$ yields $x(z_1^2 + z_2^2) = y(z_1^2 + z_2^2)$. Both sides of this equation can be divided by the real number $z_1^2 + z_2^2 \neq 0$. Hence $x = y$.
3. (a) 13, (b) 5, (c) 169.

4. (a) $\sqrt{2}(\cos\frac{\pi}{4} + \mathrm{i}\sin\frac{\pi}{4}) = \sqrt{2}(\frac{1}{\sqrt{2}} + \frac{\mathrm{i}}{\sqrt{2}})$,
   (b) $\sqrt{2}(\cos(\frac{-\pi}{4}) + \mathrm{i}\sin(\frac{\pi}{4})) = \sqrt{2}(\cos\frac{\pi}{4} - \mathrm{i}\sin\frac{\pi}{4}) = \sqrt{2}(\frac{1}{\sqrt{2}} - \frac{\mathrm{i}}{\sqrt{2}})$,
   (c) $4(\cos(\frac{-\pi}{4}) + \mathrm{i}\sin(\frac{-\pi}{4})) = 4(\frac{1}{\sqrt{2}} - \frac{\mathrm{i}}{\sqrt{2}}) = 4(\frac{\sqrt{2}}{2} - \frac{\mathrm{i}\sqrt{2}}{2}) = 2\sqrt{2}(1-\mathrm{i})$.
5. (a) $12(\cos\frac{19}{15}\pi) + \mathrm{i}\sin(\frac{19}{15}\pi) = 12(-\cos\frac{4}{15}\pi - \mathrm{i}\sin\frac{4}{15}\pi)$,
   (b) $\frac{3}{4}(\cos\frac{\pi}{15} + \mathrm{i}\sin\frac{\pi}{15})$,
   (c) $\frac{4}{3}(\cos(\frac{\pi}{15}) + \mathrm{i}\sin(\frac{\pi}{15})) = \frac{4}{3}(\cos\frac{\pi}{15} - \mathrm{i}\sin\frac{\pi}{15})$,
   (d) 9, (e) 16,
   (f) $243(\cos\frac{10}{3}\pi + \mathrm{i}\sin\frac{10}{3}\pi) = 243(-\cos\frac{\pi}{3} - \mathrm{i}\sin\frac{\pi}{3})$,
   (g) $\frac{1}{243}(-\cos\frac{\pi}{3} + \mathrm{i}\sin\frac{\pi}{3})$,
   (h) $\frac{-3}{16}(\cos\ + \mathrm{i}\sin\frac{\pi}{5})$.
6. (a) $z = \frac{1}{25}(1+7\mathrm{i})$, (b) $z_1 = 3+2\mathrm{i}$, (c) $z_2 = 5-4\mathrm{i}$.
7. (a) $z_1 = \mathrm{i}$, $z_2 = 2+3\mathrm{i}$, $z_3 = 2-3\mathrm{i}$,
   (b) $z_1 = \mathrm{i}$, $z_2 = -\mathrm{i}$, $z_3 = 1+2\mathrm{i}$, $z_4 = 1-2\mathrm{i}$.

Production Systems Production Processes, Technologies, Efficiency, Optimisation 2

This is a non-profit organisation.
We didn't plan it that way, but it is.
SIGN IN AN OFFICE

2.1 Introduction

In Chap. 1 we introduced, among others, the concepts of sets, numbers and vectors. In this chapter we start to apply them to fundamental definitions, formulation of problems, indicating also elementary and intuitive methods for solving them. We will get acquainted with concepts like *production system, production process, technology, efficiency* and *optimisation*. The knowledge gained from Chap. 1 suffices already for formulating basic problems of efficiency of production of goods and services and some ideas for their (approximative) solution. These are *optimisation problems*, in particular *linear optimisation problems*.

An example of linear optimisation is the following. A company is willing to spend at most \$30 000 for buying machines to produce a certain product. Two types of machines are available, one costs \$1 000 the other \$3 000. But the second machine produces twice as many items as the first during the same time span. Moreover, in experience, the first machine stands still an average 50 hours a year for repairs while the second only 30 hours. On the other hand, these repairs are expected to cost \$300 or \$400 a year per machine of the first or second type, respectively. The company does not want to spend more than \$5 000 per year for repairs and wants the total time-off for all machines be at most 650 hours a year. How many machines of each type would be best for the company to buy?

W. Eichhorn, W. Gleißner, *Mathematics and Methodology for Economics*, Springer Texts in Business and Economics, DOI 10.1007/978-3-319-23353-6_2

If it buys x_1 and x_2 pieces of the first and the second machine, respectively, then the problem is to

maximise $\quad x_1 + 2x_2$

under the conditions

$$\begin{aligned} 1\,000\,x_1 + 3\,000\,x_2 &\leq 30\,000, \\ 50\,x_1 + \quad 30\,x_2 &\leq \quad 650, \\ 300\,x_1 + \quad 400\,x_2 &\leq \; 5\,000. \end{aligned}$$

We will solve this problem in Sect. 2.4 by geometric methods.

2.2 Basics

A real-world economic system is called a *production system* if

(a) it consists of persons and things and *produces goods and services* and
(b) it exists in an *environment from which it can obtain and to which it can deliver* such goods and services.

Products are goods and services produced in a production system by aid of *production factors*, that is, of goods and services which are used in the production.

Vectors $\mathbf{a} = (a_1, \ldots, a_n) \in \mathbb{R}^n_+$ which consist of amounts ("quantities") $a_1, \ldots, a_n$ of n production factors are called *input vectors*. Similarly, a vector $\mathbf{b} = (b_1, \ldots, b_m) \in \mathbb{R}^m_+$, is an *output vector* if its *components* are quantities of products.

A *production process (process*, for short) consists of an input vector $\mathbf{a} \in \mathbb{R}^n_+$ and of an output vector $\mathbf{b} \in \mathbb{R}^m_+$ written as

$$(\mathbf{a}, \mathbf{b}) = ((a_1, \ldots, a_n), (b_1, \ldots, b_m)) \in \mathbb{R}^{n+m}_+. \tag{2.1}$$

It expresses that the output vector $\mathbf{b}$ can be produced from the input vector $\mathbf{a}$

(i) in a given production system,
(ii) at a certain stage of technical progress, and
(iii) during a given time span.

In (2.1) we wrote row vectors; we will write the same later as column vectors. The set T of all production processes possible in a production system, given the state of technical progress and the time at disposal, is called *technology* (of the production system).

If two production processes **(a,b)** and **(c,d)** of the same technology T satisfy the inequalities $\mathbf{a} \leqq \mathbf{c}$, $\mathbf{b} \geq \mathbf{d}$ or $\mathbf{a} \leq \mathbf{c}$, $\mathbf{b} \geqq \mathbf{d}$ (see Sect. 1.4), which can be condensed into

$$(-\mathbf{a}, \mathbf{b}) \geq (-\mathbf{c}, \mathbf{d}), \tag{2.2}$$

then, roughly speaking, the process (**a**,**b**) "produces more from less", So the production process (**c**,**d**) should, as a rule, not be "run" at all, since it is economically "*inefficient*". Accordingly, we say that a process (**c**,**d**) is *efficient* in a technology T if no process $(\mathbf{a}, \mathbf{b}) \in T$ exists for which (2.2) holds.

Since, as we saw in Sect. 1.4, the vectors in $\mathbb{R}^n$ $(n \geq 2)$ are *not totally ordered* under the relation $\geq$, it may happen that *all* production processes in a technology T are efficient.

Example 1 For instance, the five processes of the technology

$$T = \{((1,2),(3,4)),\ ((2,1),(3,4)),\ ((1,2),(4,3)),\ ((2,1),(4,3)),\ ((3,3),(5,5))\}$$

are all efficient.

It is also possible that *none* of the processes in technology is efficient.

Example 2 Neither of the infinitely many processes

$$(\mathbf{a}^0, \mathbf{b}) = ((\mathbf{a}_1^0, \ldots, \mathbf{a}_n^0),\ (\mathbf{b}_1, \ldots, \mathbf{b}_n))$$

in the following technology T is efficient:

$$T = \left\{((a_1^0, \ldots, a_n^0), (b_1, \ldots, b_n)) \;\middle|\; a_k^0 \leq b_k < b_k^0 \quad (k = 1, \ldots, n)\right\},$$

where $a_1^0, \ldots, a_n^0, b_1^0, \ldots, b_n^0$ are given positive numbers $(a_k^0 < b_k^0)$. This T contains indeed infinitely many processes, because the b_k $(k = 1, \ldots, n)$ may take *any* value between a_k^0 and b_k^0.

Looking for efficient processes is only the first step in the search for optimal production processes in the technology T. A process is *optimal* if in that process either the costs are minimal, given the output vector, or the value of the output vector is maximal, given the input vector, or the profit is maximal.

Cost, output and profit are, of course, measured by prices.

Lack of optimality makes some efficient processes uninteresting from the point of view of economics. So the following *optimisation problems* arise:

Optimisation problem 1 Let T be a technology, with processes as in (2.1) and $\mathbf{p} = (p_1, \ldots, p_n) \in \mathbb{R}^n_{++}$ the vector of *input prices* (that is, the prices of the production factors in the input vector; no input factor of 0 price is considered here). If we want to produce at least the quantities (components) in the output vector $\mathbf{b}^* \in \mathbb{R}^m_+$ with greatest cost-efficiency then we have to *find an* $\mathbf{a} = \mathbf{a}^* \in \mathbb{R}^n_+$ *for which* $(\mathbf{a}, \mathbf{b}) \in T$, $\mathbf{b} \geqq \mathbf{b}^*$, *and the inner product* $\mathbf{a} \cdot \mathbf{p} = a_1p_1 + \ldots + a_np_n$ (see Sect. 1.4) *is the smallest possible*. (We wrote $\mathbf{b} \geqq \mathbf{b}^*$ since *at least* $\mathbf{b}^*$ should be produced.)

This is an optimisation problem. Every solution $\mathbf{a}^*$ of this problem is called an *optimal input vector* or a *minimal cost combination for producing at least the output vector* $\mathbf{b}^*$ *in the price situation* $\mathbf{p}$.

Optimisation problem 2 Similarly, *an optimal output vector* or a *maximal revenue combination* $\mathbf{b} \in \mathbb{R}^m_+$ *from the input vector* $\hat{\mathbf{a}} \in \mathbb{R}^n_+$ *in the output price situation* $\mathbf{q} \in \mathbb{R}^m_{++}$ is obtained by *finding some* $\mathbf{b} = \hat{\mathbf{b}} \in \mathbb{R}^m_+$ *for which* $(\mathbf{a}, \mathbf{b}) \in T$, $\mathbf{a} \leqq \hat{\mathbf{a}}$ *and the inner product* $\mathbf{b} \cdot \mathbf{q} = b_1q_1 + \ldots + b_mq_m$ *is the greatest possible*. (Similarly as in optimisation problem 1, here we wrote $\mathbf{a} \leqq \hat{\mathbf{a}}$ because the input should be *at most* $\hat{\mathbf{a}}$.)

If the input *and* output price vectors $\mathbf{p} \in \mathbb{R}^n_{++}$, $\mathbf{q} \in \mathbb{R}^m_{++}$ are given, then we may try to solve the following optimisation problem.

Optimisation problem 3 *Determine those processes* $(\mathbf{a}, \mathbf{b}) \in T$ *for which the profit*

$$\mathbf{b} \cdot \mathbf{q} - \mathbf{a} \cdot \mathbf{p} = b_1q_1 + \ldots + b_mq_m - a_1p_1 - \ldots - a_np_n$$

is maximal.

Take, for instance, the *technology in Example 1 with the price vectors* $\mathbf{p} = (4, 4)$, $\mathbf{q} = (1, 5)$. Then the two processes ((1,2),(3,4)) *and* ((2,1),(3,4)) *are optimal* in the sense of optimisation problem 3, since *the profit*

$$\begin{aligned}(3,4) \cdot (1,5) - (1,2) \cdot (4,4) &= (3,4) \cdot (1,5) - (2,1) \cdot (4,4) \\ &= 3 \cdot 1 + 4 \cdot 5 - 2 \cdot 4 - 1 \cdot 4 = 11\end{aligned}$$

is maximal within T. The remaining processes furnish smaller *profits* 6 and 7. We see that also *more than one process in a technology can be optimal*. On the other hand, *if for the technology in Example 1 the price vectors are* $\mathbf{p} = (1, 1)$ *and* $\mathbf{q} = (1, 5)$, *then the profit of the process* ((3,3),(5,5)),

$$(5,5) \cdot (1,5) - (3,3) \cdot (1,1) = 5 \cdot 1 + 5 \cdot 5 - 3 \cdot 1 - 3 \cdot 1 = 24,$$

is maximal. Indeed, in this case the other processes yield profits 16 and 20.

2.2.1 Exercises

1. Take a technology T given by the production processes

$$((3,2,1),(5,7)),\ ((2,1,1),(3,8)),\ ((4,2,1),(5,6)),$$
$$((2,2,2),(2,8)),\ ((5,5,5),(6,9)),\ ((6,5,5),(5,6)).$$

   Which of these are efficient?
2. In the technology T in Exercise 1 let the input prices $\mathbf{p} = (p_1, p_2, p_3)$ and the output prices $\mathbf{q} = (q_1, q_2)$ be
   (a) $\mathbf{p} = (1,2,3), \mathbf{q} = (4,5)$, (b) $\mathbf{p} = (4,4,4), \mathbf{q} = (7,6)$.
   Which processes are optimal with respect to (a) and which with respect to (b)?
3. For the technology T in Exercise 1 determine an input price vector $\mathbf{p} = (p_1, p_2, p_3)$ and an output price vector $\mathbf{q} = (q_1, q_2)$ such that the process ((5,5,5),(6,9)) is optimal.
4. Define a technology T, different from Example 1, which is given by five production processes such that each of these processes is efficient.
5. Define a technology T given by seven production processes such that exactly one of the processes is efficient.

2.2.2 Answers

1. $((3,2,1),\ (5,7))$, $((2,1,1),\ (3,8))$, $((5,5,5),\ (6,9))$.
2. (a) $((3,2,1),\ (5,7))$, $((2,1,1),\ (3,8))$, profit in both cases is 45,
   (b) $((3,2,1),\ (5,7))$, profit $= 53$.
3. For the price vectors $\mathbf{p} = (1,2,3)$ and $\mathbf{q} = (20,20)$ the profit under process $((5,5,5),\ (6,9))$ is 270, while the profit under the two processes $((3,2,1),\ (5,7))$ and $((2,1,1),\ (3,8))$, efficient processes as $((5,5,5),\ (6,9))$, are 230 and 213, respectively.

2.3 Linear Production Models, Linear Optimisation Problems

Production models are (often simplified) images of relations in a production system S. Their main purpose is to get information about "the" (or one) "best possible" way of production in S. They consist of statements and assumptions about the relations in S, as informative and as easy to check as possible. Production models can refer to technologies, as defined in Sect. 2.2.

A technology T is *additive* if, with any pair of processes, say

$$((a_1,\ldots,a_n),(b_1,\ldots,b_m)) = (\mathbf{a},\mathbf{b}) \quad \text{and}$$
$$((c_1,\ldots,c_n),(d_1,\ldots,d_m)) = (\mathbf{c},\mathbf{d}),$$

also their vector sum (see Sect. 1.5.1)

$$(\mathbf{a}, \mathbf{b}) + (\mathbf{c}, \mathbf{d}) = (\mathbf{a} + \mathbf{c}, \mathbf{b} + \mathbf{d})$$

belongs to T. Explicitly: *A technology T is additive if, with the sum of input vectors of any two processes in T, as input of a new process, the sum of the output vectors of the original processes can be produced as output.* This is the case, in particular, if in a production system the processes can run parallel, independent of each other.

A technology T is (positively) *linearly homogeneous* if it contains with any process $((a_1, \ldots, a_n), (b_1, \ldots, b_m)) = (\mathbf{a}, \mathbf{b})$ also its *multiple* by any $x \in \mathbb{R}_+$, that is, the process

$$x(\mathbf{a}, \mathbf{b}) = (x\mathbf{a}, x\mathbf{b}) = ((xa_1, \ldots, xa_n), (xb_1, \ldots, xb_m)) = (\mathbf{a}, \mathbf{b})x.$$

To put it otherwise: *If the output vector* $\boldsymbol{b}$ *can be produced by use of the input vector* $\boldsymbol{a}$ *then the output* $x\mathbf{b}$ *can be produced by the input* $x\mathbf{a}$.

A technology and thus its production model is *linear* if it is both additive and linearly homogeneous. In what follows, we will be interested in *linear technologies which can be generated by a finite number* (say r) *of its production processes.* Writing this time column vectors, this means that *there exist*

$$\begin{pmatrix} a_{11} \\ \vdots \\ a_{n1} \\ b_11 \\ \vdots \\ b_{m1} \end{pmatrix} = \begin{pmatrix} \mathbf{a}_1 \\ \mathbf{b}_1 \end{pmatrix} \in T, \quad \ldots, \quad \begin{pmatrix} a_{1r} \\ \vdots \\ a_{nr} \\ b_{1r} \\ \vdots \\ b_{mr} \end{pmatrix} = \begin{pmatrix} \mathbf{a}_r \\ \mathbf{b}_r \end{pmatrix} \in T$$

such that every process $\begin{pmatrix} \mathbf{a} \\ \mathbf{b} \end{pmatrix} \in T$ *is a linear combination* (as in Sect. 1.5.1 but here *with nonnegative coefficients*) of $\begin{pmatrix} \mathbf{a}_1 \\ \mathbf{b}_1 \end{pmatrix}, \ldots, \begin{pmatrix} \mathbf{a}_r \\ \mathbf{b}_r \end{pmatrix}$, that is, to *every* $\begin{pmatrix} \mathbf{a} \\ \mathbf{b} \end{pmatrix} \in T$ there exist multiplicators ("intensities") $x_1 \in \mathbb{R}_+, \ldots, x_r \in \mathbb{R}_+$ such that

$$\begin{pmatrix} \mathbf{a} \\ \mathbf{b} \end{pmatrix} = \begin{pmatrix} \mathbf{a}_1 \\ \mathbf{b}_1 \end{pmatrix} x_1 + \ldots + \begin{pmatrix} \mathbf{a}_r \\ \mathbf{b}_r \end{pmatrix} x_r.$$

Suppose that in such a technology we have *at most the inputs* $\mathbf{a}^* = (a_1^*, \ldots, a_n^*)$ (united into the input vector) at our disposal. We want to produce *at least the outputs* $\mathbf{b}^*$ (again united into a vector) with *minimal expenses* for the input items, given the vector $\mathbf{p} = (p_1, \ldots, p_n)$ of the input prices. Then the input $\mathbf{a}_j$ costs $\mathbf{p} \cdot \mathbf{a}_j$ $(j = 1, 2, \ldots, r)$ and the input

$$\mathbf{a} := \mathbf{a}_1 x_1 + \ldots + \mathbf{a}_r x_r$$

costs accordingly

$$\mathbf{p} \cdot \mathbf{a} = (\mathbf{p} \cdot \mathbf{a}_1)x_1 + \ldots + (\mathbf{p} \cdot \mathbf{a}_r)x_r.$$

Since $\mathbf{a}_1 \in \mathbb{R}^n_+, \ldots, \mathbf{a}_r \in \mathbb{R}^n_+$ and $\mathbf{p} \in \mathbb{R}^n_+$ are constant, so *the inner products*

$$c_1 := \mathbf{p} \cdot \mathbf{a}_1, \quad \ldots, \quad c_r := \mathbf{p} \cdot \mathbf{a}_r$$

are nonnegative constants. Therefore, if we wish to *minimise the input expenses* in producing at least the *output* $(b_1^*, \ldots, b_m^*) = \mathbf{b}^*$, we have the following.

Linear optimisation problem 1 *Determine the minimum of*

$$\mathbf{p} \cdot \mathbf{a} = c_1x_1 + \ldots + c_rx_r \qquad (c_1 \geq 0, \ldots, c_r \geq 0) \tag{2.3}$$

by appropriate choice of the intensities

$$x_1 \geq 0, \ldots, x_r \geq 0 \tag{2.4}$$

under the further conditions

$$\mathbf{a}_1x_1 + \ldots + \mathbf{a}_rx_r \leqq \mathbf{a}^*, \quad \mathbf{b}_1x_1 + \ldots + \mathbf{b}_rx_r \geqq \mathbf{b}^* \tag{2.5}$$

or, in components,

$$\begin{aligned} a_{11}x_1 &+ \ldots + a_{1r}x_r &\leq a_1^*, \\ &\vdots & \vdots \\ a_{n1}x_1 &+ \ldots + a_{nr}x_r &\leq a_n^*, \\ b_{11}x_1 &+ \ldots + b_{1r}x_r &\leq b_1^*, \\ &\vdots & \vdots \\ b_{m1}x_1 &+ \ldots + a_{mr}x_r &\leq b_m^*. \end{aligned} \tag{2.6}$$

We say that $\mathbf{p} \cdot \mathbf{a}$ in (2.3), which assigns to the input vector $\mathbf{a}$ its cost, is the *objective function* (see more about *functions* in Chap. 3 and later), $c_1, \ldots, c_r$ are its *coefficients* and the $r + n + m$ conditions (2.4) and (2.6) form a *linear system of inequalities* (linear because the left hand sides in (2.4) and (2.6) are linear in the above sense and in the sense of Sect. 4.2).

If $\mathbf{q} = (q_1, \ldots, q_m)$ is the vector of output prices (here taken as constants) then the *revenue* obtained through the *output vector*

$$\mathbf{b} := \mathbf{b}_1x_1 + \ldots + \mathbf{b}_rx_r$$

is

$$\mathbf{q}\cdot\mathbf{b} = (\mathbf{q}\cdot\mathbf{b}_1)x_1 + \ldots + (\mathbf{q}\cdot\mathbf{b}_r)x_r \;=:\; \gamma_1 x_1 + \ldots + \gamma_r x_r \tag{2.7}$$

where $\gamma_1 := \mathbf{q}\cdot\mathbf{b}_1, \ldots, \gamma_z := \mathbf{q}\cdot b_r$ are nonnegative constants.

If, using *at most input* $(a_1^*, \ldots, a_n^*) = \mathbf{a}^*$, one wishes to *maximise revenue* in the above model, then one has to solve the following.

Linear optimisation problem 2 *Find the maximum of*

$$\mathbf{q}\cdot\mathbf{b} = \gamma_1 x_1 + \ldots + \gamma_r x_r$$

under the conditions

$$x_1 \geq 0, \ldots, x_r \geq 0;\; \mathbf{a}_1 x_1 + \ldots + \mathbf{a}_r x_r \leqq \mathbf{a}^*. \tag{2.8}$$

Finally, if we want to *maximise the profit* in the model while *limiting the input* by $\mathbf{a}^*$, we have the following.

Linear optimisation problem 3 *Determine the maximum of*

$$\mathbf{q}\cdot\mathbf{b} - \mathbf{p}\cdot\mathbf{a} = (\gamma_1 - c_1)x_1 + \ldots + (\gamma_r - c_r)x_r \tag{2.9}$$

under the conditions

$$x_1 \geq 0, \ldots, x_r \geq 0,\; \mathbf{a}_1 x_1 + \ldots + \mathbf{a}_r x_r \leqq \mathbf{a}^*.$$

Note that, different from (2.3) and (2.7) *here* in (2.9) *we may have negative coefficients.*

Notice also the relation of the linear optimisation problems 1, 2, 3 to the optimisation problems 1, 2, 3 in Sect. 2.1.

2.3.1 Exercises

1. Consider a linear technology T that can be generated by the following three of its production processes:

$$(\mathbf{a},\mathbf{b}) = ((2,3,4),(1,7,5,6)),$$
$$(\mathbf{c},\mathbf{d}) = ((8,2,5),(3,9,7,4)),$$
$$(\mathbf{g},\mathbf{h}) = ((3,3,3),(6,6,6,6)).$$

Do the processes
(a) $(\mathbf{r},\mathbf{s}) = ((33,19,32),(15,57,43,38))$,

(b) $(\mathbf{t}, \mathbf{u}) = ((48, 29, 46), (29, 83, 65, 58))$ belong to T?
(c) Does there exist a $y \in \mathbb{R}$ such that

$$(\mathbf{v}, \mathbf{w}) = ((33, 19, 32), (16, y, y, y))$$

belongs to T?

2. For the technology T defined in Exercise 1 let the inputs be limited by (51,70,92). Is it possible to produce, within T, output vectors $\mathbf{b} = (b_1, b_2, b_3, b_4)$ which are greater than or equal to

(a) (102,102,102,102), (b) (97,103,101,102),
(c) (103,100,100,100), (d) (20,161,110,130)?

3. Formulate the linear optimisation problems 1, 2, 3 for the technology T defined in Exercise 1.
4. Define a technology T consisting of infinitely many processes which is neither additive nor linearly homogeneous.
5. Define a technology T consisting of infinitely many processes which is linearly homogeneous but not additive.

2.3.2 Answers

1. (a) Yes, $4(\mathbf{a}, \mathbf{b}) + 3(\mathbf{c}, \mathbf{d}) + \frac{1}{3}(\mathbf{g}, \mathbf{h}) = (\mathbf{r}, \mathbf{s})$.
   (b) Yes, $5(\mathbf{a}, \mathbf{b}) + 4(\mathbf{c}, \mathbf{d}) + 2(\mathbf{g}, \mathbf{h}) = (\mathbf{t}, \mathbf{u})$.
   (c) No, there do not exist any real numbers y, x_1, x_2, x_3 such that $(\mathbf{a}, \mathbf{b})x_1 + (\mathbf{c}, \mathbf{d})x_2 + (\mathbf{g}, \mathbf{h})x_3 = (\mathbf{v}, \mathbf{w})$.
2. (a) $17((3, 3, 3), (6, 6, 6, 6)) = ((51, 51, 51), (102, 102, 102, 102))$.
   (b) Yes, $((2, 3, 4), (1, 7, 5, 6)) + 16((3, 3, 3), (6, 6, 6, 6)) = ((50, 51, 52), (97, 103, 101, 102))$.
   (c) No, there does not exist any vector (x_1, x_2, x_3) such that the fourth component of $(\mathbf{a}, \mathbf{b})x_1 + (\mathbf{c}, \mathbf{d})x_2 + (\mathbf{e}, \mathbf{f})x_3$ is 103 and the first one is smaller than or equal to 51.
   (d) Yes, $23((2, 3, 4), (1, 7, 6, 5)) = ((46, 69, 92), (23, 161, 115, 138))$.
4. See T in Example 2, Sect. 2.2.
5. $T := \{\mu((1, 2), (3, 4)), \mu((5, 6), (7, 8)) \mid \mu \in \mathbb{R}_{++}\}$ is linearly homogeneous, but the sum $((6, 8), (10, 12))$ of the processes $((1, 2), (3, 4)) \in T$ and $((5, 6), (7, 8)) \in T$ does not belong to T.

2.4 Simple Approaches to Linear Optimisation Problems

We use the problem formulated in the introduction to this chapter as an example of linear optimisation problems of type 2 or 3 in the case $r = 2$ (see Sect. 2.3). After cancellations (division of both sides of the inequalities by 1,000, 10 or 100, respectively) our problem is the following. Maximise

$$x_1 + 2x_2 \tag{2.10}$$

under the conditions $x_1 \geq 0,\ x_2 \geq 0$ and

$$\begin{aligned} x_1 + 3\,x_2 &\leq 30, \\ 5\,x_1 + 3\,x_2 &\leq 65, \\ 3\,x_1 + 4\,x_2 &\leq 50. \end{aligned} \tag{2.11}$$

A solution $(x_1, x_2) = (x_1^*, x_2^*) \leqq \mathbf{0}$ of (2.11) is *optimal* for which (2.10) is maximal, that is,

$$x_1 + 2x_2 \leq x_1^* + 2x_2^*$$

for all solutions $(x_1, x_2) \geqq \mathbf{0}$ of (2.11).

In Fig. 2.1, the first straight line segment, starting at the point (0,10) (the mark 10 of the vertical axis), is described by the equation

$$x_1 + 3x_2 = 30,$$

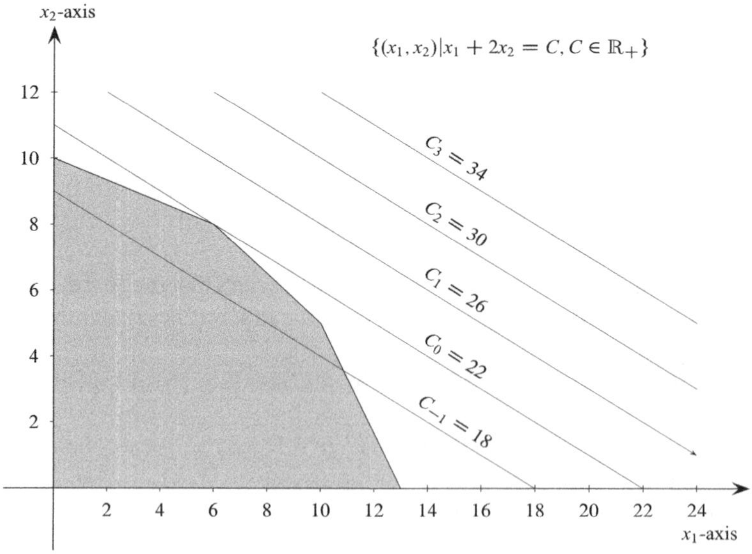

Fig. 2.1 Geometrical representation of the problem of maximizing $x_1 + 2x_2$ by nonnegative x_1, x_2 satisfying (2.11). The shaded region contains the points $(x_1, x_2) \in \mathbb{R}_+^2$ for which (2.11) holds. The parallel line segments marked C_k represent those $(x_1, x_2) \in \mathbb{R}_+^2$ for which $x_1 + 2x_2 = C_k$ $(k = -1, 0, 1, 2, 3)$. Since $C_{-1} < C_0 < C_1 < C_2 < C_3$, we see that for $C_0 = 22$ one has the line segment with maximal C_k which has nonempty intersection with the shaded area. The intersection is a single point, so the linear optimization problem (2.10), (2.10) has exactly one solution

the adjoining straight line segment (from the point (6,8) to the point (10,5)) by the equation

$$3x_1 + 4x_2 = 50$$

and the last segment (connecting the points (10,5) and (13,0)) by the equation

$$5x_1 + 3x_2 = 65.$$

Therefore, the points (x_1, x_2) with $x_1 \geq 0$, $x_2 \geq 0$ satisfying the inequalities (2.11) lie under these straight line segments, in the shaded area bordered by them and the vertical and horizontal axes.

The *parallel lines* with the $C_{-1}, C_0, C_1, C_2, C_3$ marks have the *equations*

$$x_1 + 2x_2 = C_k \qquad (k = -1, 0, 1, 2, 3).$$

As we see from Fig. 2.1, $x_1 + 2x_2$, that is (2.10), is *maximal* on the *set* of *points* satisfying (2.11) and $x_1 \geq 0, x_2 \geq 0$, that is in the shaded area, at the single point $(x_1^*, x_2^*) = (6, 8)$ and the maximal value is

$$x_1^* + 2x_2^* = 22.$$

(Fortunately, we got integers as solutions; the company could hardly buy, for instance, 30/7 pieces of the first and 60/7 pieces of the second kind of machine. But this is another story of approximation).

If, however, we replace (2.10) by $x_1 + 3x_2$ or $3x_1 + 4x_2$ or $5x_1 + 3x_2$ as values to be maximised under the same conditions (2.11) (and $x_1 \geq 0, x_2 \geq 0$) then, as Fig. 2.2 shows, there are infinitely many solutions, those represented by the first, second or third line segment, respectively on the upper border of the shaded area.

As we see, there is a simple geometric way to solve linear optimisation problems in the case $r = 2$. For the case $r = 3$ we would need spatial images and for $r > 3$ the capacity of most of us to "see" in r-dimensional space runs out. In Sect. 5.2 we will introduce an "algorithm", the "simplex method" for solving linear optimisation problems like problems 1, 2, 3.

Here, however, we mention another method which approximates but does not necessarily yield the optimal processes.

We consider the following linear optimisation problem.

Maximise

$$c_1x_1 + \ldots + c_rx_r \qquad (2.12)$$

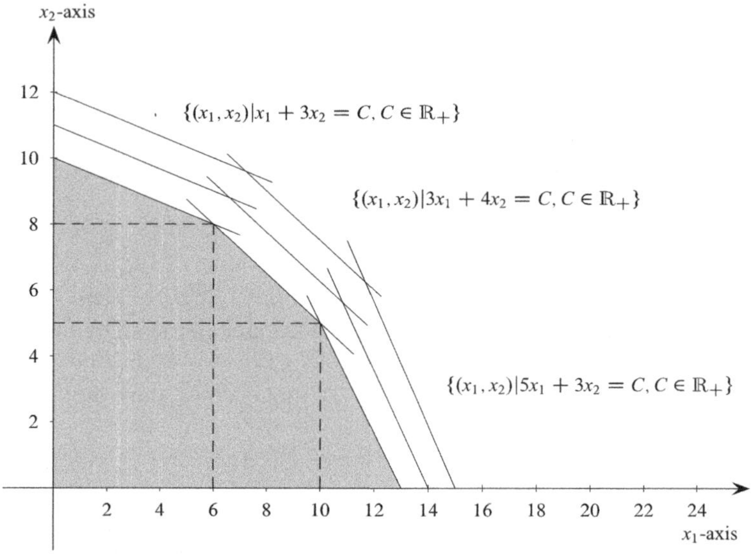

Fig. 2.2 Geometric representation of the problem of maximizing x_1+3x_2 or $3x_1+4x_2$ or $5x_1+3x_2$ by nonnegative x_1, x_2 satisfying (2.11). In each case there are infinitely many solutions, namely those represented by the line segments from (0,10) to (6,8), from (6,8) to (10,5), or from (10,5) to (13,0), respectively

under the conditions

$$\begin{aligned} x_1 &\geq 0, \ldots, x_r \geq 0, \\ a_{11}x_1 + \ldots + a_{1r}x_r &\leq b_1, \\ \vdots \qquad\quad \vdots \\ a_{n_1}x_1 + \ldots + a_{n_r}x_r &\leq b_n, \end{aligned} \tag{2.13}$$

where

$$\begin{aligned} (c_1, \ldots, c_r) &\geq \mathbf{0}, \\ (b_1, \ldots, b_n) &> \mathbf{0}, \\ \mathbf{a}_k := (a_{k_1}, \ldots, a_{k_r}) &\geq \mathbf{0} \qquad (k = 1, \ldots, n). \end{aligned} \tag{2.14}$$

Notice that in (2.13) and (2.14) below we have $\geq$, not $\geqq$, that is, compare Sect. 1.4, we exclude the $\mathbf{0}$ vector. If

$$\tilde{\mathbf{x}} = (\tilde{x}_1, \ldots, \tilde{x}_r) \geq \mathbf{0} \tag{2.15}$$

satisfies the conditions (2.13) and $c_1\tilde{x}_1 + \ldots + c_r\tilde{x}_r$ is the greatest possible solution of (2.12) under these conditions then we have a *solution of the linear optimisation problem*. If not and if

$$\mathbf{a}_k \cdot \tilde{\mathbf{x}} < b_k \qquad (k = 1, \ldots, n) \tag{2.16}$$

then there exists a $\lambda > 1$ such that

$$\mathbf{x} = (x_1, \ldots, x_r) = \lambda(\tilde{x}_1, \ldots, \tilde{x}_r) \quad (\geq \mathbf{0}) \tag{2.17}$$

still satisfies (2.13) but gives a greater $c_1x_1 + \ldots + c_rx_r$. Indeed, then by (2.15) and (2.14),

$$c_1x_1 + \ldots + c_rx_r = \lambda(c_1\tilde{x}_1 + \ldots + c_r\tilde{x}_r) > c_1\tilde{x}_1 + \ldots + c_r\tilde{x}_r$$

(a positive number increases when multiplied by $\lambda > 1$) while, by (2.15), (2.14) and (2.16), there is space for $\lambda > 1$ but sufficiently close to 1 so that

$$\mathbf{a}_k \cdot \mathbf{x} = \lambda(\mathbf{a_k} \cdot \tilde{\mathbf{x}}) \leq b_k \qquad (k = 1, \ldots, n) \tag{2.18}$$

that is, (2.13) still holds. For every $\lambda > 1$ (in particular, *for the largest* $\lambda > 1$) *for which (2.18) is valid,*

$$(x_1, \ldots, x_r) = \lambda(\tilde{x}_1, \ldots, \tilde{x}_r) = \lambda\tilde{\mathbf{x}}$$

is a better approximation than $\tilde{x}$ *to the solution of the linear approximation problem (2.12), (2.13).* Of course, *there may be* $(x_1, \ldots, x_r)$ *with greater value of (2.12)* but we have described how to get the *greatest value for vectors of the form (2.17).*

In order to find a solution (2.15) of (2.19), (2.20) and (2.21) in the case where $b_k > 0,\ a_{kj} \geq 0,\ c_j \geq 0 \quad (k = 1, \ldots, n;\ j = 1, \ldots, r)$, we can proceed as follows. Since

$$(x_1, \ldots, x_r) = \mu(c_1, \ldots, c_r) \tag{2.19}$$

satisfies (2.13) if $\mu = 0$ (though then $(x_1, \ldots, x_r) = \mathbf{0}$ and not $\geq \mathbf{0}$, therefore, *for small enough positive* μ, *(2.19) will be a solution of (2.13)*:

$$a_{k1} + \ldots + a_{kr}x_r = \mu(a_{k1}c_1 + \ldots + a_{kr}c_r) \leq b_k \qquad (k = 1, \ldots, n).$$

(This and the above discussion of (2.18) are called "*continuity arguments*"; we will discuss *continuity* in detail from Sect. 5.3 on). *If we take in (2.19) the largest* μ *for which (2.13) is still satisfied we get the "best solution of (2.13) in the direction"* $(c_1, \ldots, c_r)$.

We apply these considerations to the optimisation problem (2.10), (2.11); compare Fig. 2.3, where the "best solution in the direction (1,2)" gives for $x_1 + 2x_2$, that is (2.10), the value 150/7, while the solution of the optimisation problem (2.10), (2.11) is (6,8), yielding $x_1 + 2x_2 = 22 > 150/7$.

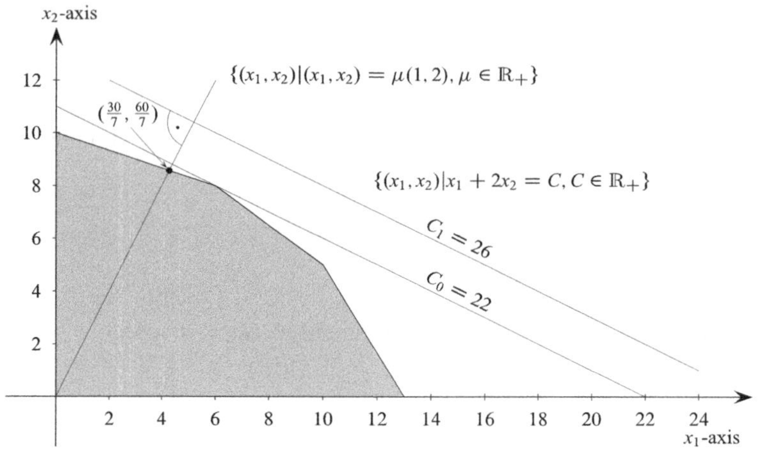

Fig. 2.3 Geometric representation of the inequalities (2.11) and of their "best solution in the direction (1,2)". The vector (1,2) is orthogonal to the straight line $\{(x_1, x_2) \in \mathbb{R}^2 \mid x_1 + 2x_2 = 22\}$

Notice that $c_1x_1 + c_2x_2 = 0$ is the equation of a straight line (compare Fig. 2.3), $c_1x_1 + c_2x_2 + c_3x_3 = 0$ is the equation of a plane and so are

$$c_1x_1 + c_2x_2 = C, \qquad c_1x_1 + c_2x_2 + c_3x_3 = C \tag{2.20}$$

for all real C. The vector (c_1, c_2) is *orthogonal to the line* with equation $c_1x_1 + c_2x_2 = 0$ and (c_1, c_2, c_3) is *orthogonal to the plane* with equation $c_1x_1 + c_2x_2 + c_3x_3 = 0$. More exactly, they are orthogonal to the vectors in the sets

$$\{(x_1, x_2) \in \mathbb{R}^2 \mid c_1x_1 + c_2x_2 = 0\},$$
$$\{(x_1, x_2, x_3) \in \mathbb{R}^3 \mid c_1x_1 + c_2x_2 + c_3x_3 = 0\},$$

by the definition of orthogonality in Sect. 1.5:

$$\begin{aligned} (c_1, c_2) \cdot (x_1, x_2) &= c_1x_1 + c_2x_2 &&= 0, \\ (c_1, c_2, c_3) \cdot (x_1, x_2, x_3) &= c_1x_1 + c_2x_2 + c_3x_3 &&= 0. \end{aligned}$$

Since the lines and planes with Eq. (2.20) are parallel to this line and this plane, the vectors (c_1, c_2), (c_1, c_2, c_3) are orthogonal to the lines and planes with the Eq. (2.20), too. By the same argument, in general the vector $(c_1, \ldots, c_r)$ is orthogonal to the set

$$\{(x_1, \ldots, x_r) \in \mathbb{R}^r \mid c_1x_1 + \ldots + c_rx_r = C\} \tag{2.21}$$

for all $C \in \mathbb{R}$. These sets are called "*hyperplanes*". So, what we described above is the "*best solution of (2.13) in the direction orthogonal to the hyperplane (2.21)*".

Notice also that $c_1x_1 + \ldots + c_rx_r$ in (2.21) is exactly (2.12), the largest value of which under the conditions (2.13) we were looking for. Of course, as also Fig. 2.3 shows, this *"best solution" of (2.13) in the direction orthogonal to (2.20) need not be a solution of the linear optimisation problem* (2.12), (2.13). It *may be a solution*, however: in the situations represented in Fig. 2.2, *it is a solution* (but *not the only one*).

Remark For approximating a solution of (2.13), we could have started in (2.19) with an arbitrary vector $(\tilde{c}_1, \ldots, \tilde{c}_r)$ in place of $(c_1, \ldots, c_r)$. We chose the latter in order to establish a connection with (2.12): In absence of other preferences one may wish to advance on the shortest path to the lines (planes, hyperplanes) on which the expression (2.12) to be maximised is constant. As in the two-dimensional space, so also in three and more dimensional spaces this shortest path is orthogonal to that plane or hyperplane, respectively. One may think of climbing a mountain on the path orthogonal to the lines of equal height, that is, on the steepest path in order to gain height on the shortest way. As in "real life" this may or may not lead to the summit of the mountain—it certainly leads to considerable height. This approach is called the "*method of steepest ascent*".

2.4.1 Exercises

1. Maximise $x_1 + 3x_2$ under the conditions

$$\begin{aligned} x_1 + 2x_2 &\le 110, \\ x_1 + 4x_2 &\le 160, \\ x_1 + x_2 &\le 100, \\ x_1 \ge 0,\ x_2 &\ge 0. \end{aligned}$$

2. Maximise $3x_1 + 4x_2$ under the conditions

$$\begin{aligned} 3x_1 + 2x_2 &\le 21, \\ x_1 + 2x_2 &\le 12, \\ 2x_1 + 3x_2 &\le 19, \\ x_1 \ge 0,\ x_2 &\ge 0. \end{aligned}$$

3. Minimise $5x_1 + 7x_2$ under the conditions

$$\begin{aligned} 2x_1 + 4x_2 &\ge 12, \\ 2x_1 + x_2 &\ge 6, \\ 4x_2 &\ge 4, \\ x_1 \ge 0,\ x_2 &\ge 0. \end{aligned}$$

4. What is the best solution
   (a) of optimisation problem 1 in the direction (1,3),
   (b) of optimisation problem 2 in the direction (3,4),
   (c) of optimisation problem 3 in the direction (5,7)?
5. What is the best solution
   (a) of optimisation problem 1 in the direction (3,1),
   (b) of optimisation problem 2 in the direction (4,3),
   (c) of optimisation problem 3 in the direction (7,5)?

2.4.2 Answers

1. Maximum = 135 at $x_1 = 60, x_2 = 25$.
2. Maximum = 27 at $x_1 = 5, x_2 = 3$.
3. Maximum = 24 at $x_1 = 2, x_2 = 2$.
4. (a) $\frac{1600}{13} = 123.08$ at $x_1 = \frac{160}{13} = 12.31, x_2 = \frac{480}{13} = 36.92$,
   (b) $\frac{475}{18} = 26.39$ at $x_1 = \frac{19}{6} = 3.17, x_2 = \frac{38}{9} = 4.22$,
   (c) $\frac{444}{17} = 26.12$ at $x_1 = \frac{30}{17} = 1.76, x_2 = \frac{42}{17} = 2.27$,
5. (a) 132 at $x_1 = 66, x_2 = 22$,
   (b) $\frac{273}{11} = 24.82$ at $x_1 = \frac{63}{11} = 5.73, x_2 = \frac{21}{11} = 1.91$,
   (c) 24.6 at $x_1 = 2.4, x_2 = 1.8$.

Mappings, Functions 3

That flower of modern mathematical thought—the function.

THOMAS J. MCCORMACK (*1900)

3.1 Introduction

While we have dealt till now with just a few basic notions in mathematics and economics, we have several times encountered, though not emphasised, the fundamental concept of *mapping*. This is the situation, where certain objects are *mapped* to others, that is, the latter are *assigned* to the former, following specific instructions. Rather than an abstract definition, we give *examples*:

1. The *norm* in Sect. 1.4 maps a vector $\mathbf{x} \in \mathbb{R}^n$ to a nonnegative number $||\mathbf{x}||$ (the length of $\mathbf{x}$). In particular, for $n = 1$ (Sect. 1.2) and for $n = 2$ (Sect. 1.7.2) the absolute value maps the real or complex number z, respectively, again to a nonnegative number, namely $|z|$.
2. The *distance* in Sect. 1.6 (for $n = 1$ in Sect. 1.2) maps a *pair of vectors* $\mathbf{x} \in \mathbb{R}^n$, $\mathbf{y} \in \mathbb{R}^n$ to one nonnegative number $d(\mathbf{x}, \mathbf{y})$.
3. *Addition of vectors* in Sect. 1.5.1 maps *a pair of vectors* $\mathbf{x} \in \mathbb{R}^n$, $\mathbf{y} \in \mathbb{R}^n$ to the vector $\mathbf{x} + \mathbf{y} \in \mathbb{R}^n$.
4. *Multiplication of a vector by a scalar* in Sect. 1.5.1 maps *a pair consisting of a vector* $\mathbf{x} \in \mathbb{R}^n$ *and a scalar* $\lambda \in \mathbb{R}$ to the vector $\lambda\mathbf{x} \in \mathbb{R}^n$.
5. The *linear combination of vectors*, also in Sect. 1.5.1, maps a *2p-tuple consisting of p scalars* $\lambda_k \in \mathbb{R}$ *and p vectors* $\mathbf{x}_k \in \mathbb{R}^n$ $(k = 1, 2, \ldots, p)$ to *the vector* $\lambda_1\mathbf{x}_1 + \ldots + \lambda_p\mathbf{x}_p \in \mathbb{R}^n$.
6. The *inner product* in Sect. 1.5.3 assigns to *a pair of vectors* $\mathbf{x} \in \mathbb{R}^n$, $\mathbf{y} \in \mathbb{R}^n$ *the scalar* $\mathbf{x} \cdot \mathbf{y} \in \mathbb{R}$. As mentioned there, the *price level*, important in economic statistics, is the inner product of the quantity vector $\mathbf{x} = (x_1, \ldots, x_n) \in \mathbb{R}^n_+$ and of the price vector $\mathbf{p} = (p_1, \ldots, p_n) \in \mathbb{R}^n_{++}$ for a basket of goods.

W. Eichhorn, W. Gleißner, *Mathematics and Methodology for Economics*,
Springer Texts in Business and Economics, DOI 10.1007/978-3-319-23353-6_3

7. *Multiplication of complex numbers* in Sect. 1.7.1 maps *a pair of complex numbers* $a = a_1 + \mathrm{i}a_2 \in \mathbb{C}$, $b = b_1 + \mathrm{i}b_2 \in \mathbb{C}$ to *the complex number* $ab = (a_1b_1 - a_2b_2) + \mathrm{i}(a_1b_2 + a_2b_1) \in \mathbb{C}$.
8. The *conjugation* of a complex number in Sect. 1.7.3 maps $a = a_1 + \mathrm{i}a_2 \in \mathbb{C}$ to $\bar{a} = a_1 - \mathrm{i}a_2 \in \mathbb{C}$.
9. The *cosine* and *sine* in Sect. 1.7.2 map *an angle* (measured, say, in radians) $\phi \in \mathbb{R}$ to $\cos\phi$ or $\sin\phi$, respectively; both are real numbers between -1 and 1. The *cotangent* in Sect. 1.7.4 maps *the angles* (in radians) $\phi \in \mathbb{R} \setminus \{k\pi \mid k = 0, \pm1, \pm2, \dots\}$ (all angles except those of the form $k\pi$, where $k \in \mathbb{Z}$) to $\cot\phi \in \mathbb{R}$. Similarly, the *tangent* maps $\phi \in \mathbb{R} \setminus \{(2k+1)\pi/2 \mid k = 0, \pm1, \pm2, \dots\}$ to $\tan\phi \in \mathbb{R}$.
10. For the purpose of comparison of the *efficiency of k technologies* $T_1, \dots, T_k$, as in Sect. 2.2, it would be nice to know the mapping which assigns to each T_j $(j = 1, 2, \dots, k)$ the *set* of its efficient production processes.

    This example shows that the instructions defining a mapping may be quite involved. This is even more so in the following example, where an optimisation problem has to be solved before we know what is mapped to what.
11. In the framework of the *linear optimisation problem* 1 in Sect. 2.4 we assign to each vector $\mathbf{p} = (p_1, \dots, p_n) \in \mathbb{R}^n_+$ of *input prices* the *set* of those *intensity vectors* $(x_1, \dots, x_n) \in \mathbb{R}^n_+$ which solve the linear optimisation problem 1, that is, which result in *minimal cost*.

    Here are some further examples of mappings from economics:
12. *Utility*. Here a vector $\mathbf{x} \in \mathbb{R}^n_{++}$ of *quantities of goods and services* is mapped to a real number, its *utility*, say for a person or for an "average household".
13. *Price index*. Take again a vector $\mathbf{x} \in \mathbb{R}^n_{++}$ of *quantities* of goods (and services) and let the components of the price vectors $\mathbf{p}^0 = (p^0_1, \dots, p^0_n) \in \mathbb{R}^n_{++}$ and $\mathbf{p} = (p_1, \dots, p_n) \in \mathbb{R}^n_{++}$ be the *prices* of these goods at a *base time* or at a *comparison time* (usually the present), giving the cost of this "basket of goods" as the inner products $\mathbf{x} \cdot \mathbf{p^0}$ and $\mathbf{x} \cdot \mathbf{p}$ respectively. These three vectors are mapped to the *price index*

$$\frac{\mathbf{x} \cdot \mathbf{p}}{\mathbf{x} \cdot \mathbf{p}^0} \in \mathbb{R}^n_{++}.$$

    This quotient provides information on the change of price levels (for the given basket of goods) between the base time and the *comparison time*. (There are also other price indices, see Sect. 3.7.)
14. The *gross national product* maps all *goods and services produced in a country* to a real number. Of course, one has to state also the method by which the value of the various goods and services are calculated, which are included and which not, and so on.
15. *Purchasing power*. A vector $\mathbf{p} \in \mathbb{R}^n_{++}$ of *prices* in a "basket of goods" and an amount $M \in \mathbb{R}_{++}$ of *money* are given. If the question is what quantities $\mathbf{x} = (x_1, \dots, x_n) \in \mathbb{R}^n_+$ of goods (and services) from this basket of goods the amount of money M can buy, then the mapping assigns to the *pair* $\mathbf{p} \in \mathbb{R}^n_{++}$, $M \in \mathbb{R}_{++}$

the *set*

$$\{\mathbf{x} \mid x \cdot p \le M\}.$$

16. If we are interested in the *production potential* of a *production system* (compare Sect. 2.2), we need the mapping which assigns to each *input vector* $\mathbf{x} \in \mathbb{R}^n_+$ of the system the set of output vectors in $\mathbb{R}^m_+$ which can be produced in the system from $\mathbf{x}$ during a given time interval at the present state of technical progress. Conversely, another mapping (in a sense the "inverse mapping" of the previous one) assigns to each output vector $\mathbf{x} \in \mathbb{R}^m_+$ the set of all input vectors in $\mathbb{R}^n_+$ which can produce $\mathbf{u}$ in the system during that time with the given technology.

We emphasise that in the Examples 10, 11, 15 and 16 the mapping assigns to an object (vector, pair consisting of a vector and a scalar) a whole *set* of objects (in these examples: a set of processes or a set of vectors). This is somewhat different from the mappings in the other examples which assign a single object to the given object. Of course "object" is not defined and we could also consider the *sets* in Examples 10, 11, 15, and 16 as "single objects". But we may recognise the difference if we compare the situation in Example 16 to the special case where the production system happens to be kept so rigid that each input vector $\mathbf{x} \in \mathbb{R}^n_+$ can produce *just one* output vector $\mathbf{u} \in \mathbb{R}^m_+$ during the given time with the given technology. In this case the mapping again assigns to each object (vector) just one object (vector) or, alternatively, a set having only one element (which, as we saw in Sect. 1.3, is not quite the same thing).

Another variation on the theme of Example 16 is the following.

17. *Maximal production potential.* If just *one* good is produced in a production system, that is $m = 1$, then we may be interested in the mapping which assigns to each input vector $\mathbf{x} \in \mathbb{R}^n_+$ of the system the *maximal quantity* $u \in \mathbb{R}_+$ of the only output good, which can be produced using the input vector $\mathbf{x}$ during the given time with the given technology (*if* such a maximal quantity *exists*). Notice that normally it would *not* make sense to ask for a *maximal* output vector $\mathbf{u} \in \mathbb{R}^m_+$ if $m > 1$ because, as we have seen in Sect. 1.4, *vectors of more than one component are not ordered* in general.

We could give many more examples but we hope that already those above show what a central role mappings play in mathematics and economics. In a certain sense this entire book deals with mappings.

3.2 Basics. Domains, Ranges, Images (Codomains). Mappings (Binary Relations), Functions, Injections, Surjections, Bijections. Graphs

In Sect. 3.1 we described what some specific mappings do to individual objects, even though we indicated most of the time from what sets these objects are taken. In what follows we treat these matters somewhat more systematically.

We start with two sets, S the *domain* and T the *range*. A *mapping (or a binary relation) assigns to each element of S an element of T* (*at least one* element of T). (More formally, a *binary relation* is a *set of ordered pairs* (s, t) where $s \in S$, $t \in T$, in this case such that *each element of S has to be used* as s, but *not all elements of T need to come up* as second element in an (s, t).) If the mapping (or *binary relation*) assigns to each element of S *exactly one element* of T (it maps each $s \in S$ into exactly one $t \in T$; still all elements of S but not necessarily all elements of T have to be "used up") then we speak of a *function*. We note here that the usage of these names is not uniform, sometimes what we called "function" is called also "mapping" or "single-valued function" while what we called "mapping" is often called "function" or "multivalued function" or the words "mapping" and "function" are used interchangeably. Be it as it may, if t is *the* element (or, in case of mappings or multivalued functions, *one of the* elements) of T assigned to an element s of S then t is *the* (or *a*) *function value* belonging to s, or, if the function itself is denoted by f (one uses also g, h, ϕ, ψ, F, G, ... and many other letters to denote functions), then the function value belonging to s is $t = f(s)$. Even though this is not always done, it is *preferable to distinguish between the function f and its value $f(s)$* (for one or more $s \in S$) otherwise ambiguities result. (For instance, what does $f(s) = 0$ mean then? does $f(s)$ equal zero for one $s \in S$? for several? for all $s \in S$? $f = 0$ certainly means the latter.) If it is inconvenient to introduce a function symbol as, for instance the square, one can write

$$s \longmapsto s^2$$

or the function defined by $t = s^2$. (On the other hand, it is quite all right to call, for instance, the function $y \mapsto \cos y$, just "cos" or "cosine" or "cosine function".) Note the funny arrow $\mapsto$ in the above notation; the simple arrow $\longrightarrow$ is used to connect the function domain and range, so:

$$f : S \longrightarrow T$$

(while the double arrow $\Rightarrow$ means that one statement *implies* the other, again something completely different). The fact that S is the domain of f is often expressed so: "f is *defined on* S"" (and on every subset of S).

As we emphasised above, not all elements of the range T need to be function values belonging to some elements of the domain. This is convenient, because that way we need not calculate in advance what all possible function values of a given f are. For instance, as we saw in Sect. 1.7.2 the function $\cos : \mathbb{R} \to \mathbb{R}$ can perfectly well be defined on the domain $\mathbb{R}$ before we know that its values have to lie between -1 and 1 (these values included), see also Sect. 3.1, Example 9. (In some cases, as in Examples 10, 11, 12, 13, 14, 15, 16 and 17 of Sect. 3.1 it is not even obvious or simple to determine what all possible function values are.) But we define also *the set of all values of a function f on the domain S*, called *the "image"*, the range, or the *"codomain" of S under f*. The definition and notation is:

$$f(S) := \{t \mid \exists s \in S : f(s) = t\}.$$

Clearly, $f(S) \subset T$. One can define exactly the same way the *image under f of every subset A of S* by $f(A) = \{t \mid \exists s \in A : f(s) = t\}$ (the colon serves in these formulas to avoid the confusion of a second bar |; both mean "such that").

If, however, $T = f(S)$, that is, *if every element of T is a value of f belonging to some $s \in S$, then $f : S \to T$ is a surjection* (or: is "*surjective*" or "the mapping f is *onto T*", while otherwise or if we do not know, then it is "*into*" *T*).

For instance, the square $t \mapsto t^2$, defined on $\mathbb{R}$ is not surjective to $\mathbb{R}$, it is "*into* $\mathbb{R}$ *but not onto* $\mathbb{R}$"; however it is "*onto*" $\mathbb{R}_+$. (Note that previously we wrote that a mapping maps an *individual object* "into" another one, this does not interfere with this "mapping into a *set*".)

If, for $f : S \to T$, given any $t \in T$, *the equation $t = f(s)$ is satisfied by at most one $s \in S$, then f is an "injection"* (or "*injective*"). Equivalently, *for each $t \in f(S)$ there is then exactly one $s \in S$ so that $t = f(s)$*, t is the function value of f at s. For instance, the square $s \mapsto s^2$ *with domain* $\mathbb{R}$ and range, say, $\mathbb{R}_+$ is not injective (for example $4 = 2^2 = (-2)^2$) but the square $s \mapsto s^2$ with *domain* $\mathbb{R}_+$ and range, say, $\mathbb{R}$ is injective (for every real number $t \in \mathbb{R}_+$ there is exactly one number $s \in \mathbb{R}_+$ such that $t = s^2$, namely $s = \sqrt{t}$; obviously, for $t \in \mathbb{R}_{--}$ such an s does *not* exist).

If $f : S \to T$ is both surjective and injective then it is "bijective" or a "bijection" (in older terminology "*one-to-one*" or "1–1" for short; but there was some ambiguity; some people meant by "one-to-one" sometimes "injective" rather than "bijective"; that may be the reason why the new terminology has been introduced). In other words the function $f : S \to T$ is called *bijective if, for every $t \in T$, there exists exactly one $s \in S$ for which $t = f(s)$*. Clearly $s \mapsto s^2 : \mathbb{R}_+ \to \mathbb{R}_+$ and $s \mapsto s^3 : \mathbb{R} \to \mathbb{R}$ are bijections.

For bijective functions $f : S \to T$ there exists an "*inverse function*" which, by definition, assigns to every $t \in T$ that (in case of bijections *unique*) $s \in S$ for which $f(s) = t$. This function is usually denoted by $f^{-1} : T \to S$, so $t = f(s)$ is equivalent to $s = f^{-1}(t)$. The *inverse function* can be defined also for injective functions but then it will not be defined outside $f(S)$, because only for $t \in f(S)$ does exist a (but then unique) $s \in S$ such that $t = f(s)$. This is clear also if we recognise that every injection $f : S \to T$ is a bijection if the range is confined to the image $f(S)$ of S: $f : S \to f(S)$; then $f^{-1} : f(S) \to S$.

By the way, for every function (or, for that matter, for every mapping) there exists an "*inverse mapping*", that is an inverse (possibly) multivalued function: for $f : S \to f(S)$ this inverse mapping $f^{-1} : f(S) \to S$ assigns to each $t \in f(S)$ *all* $s \in S$ for which $f(s) = t$ (that is why possibly f^{-1} is not a "single-valued" function anymore).

Figures 3.1, 3.2, 3.3, 3.4 and 3.5 schematically show mappings, functions, injections, surjections and bijections respectively. The elements of S and T are

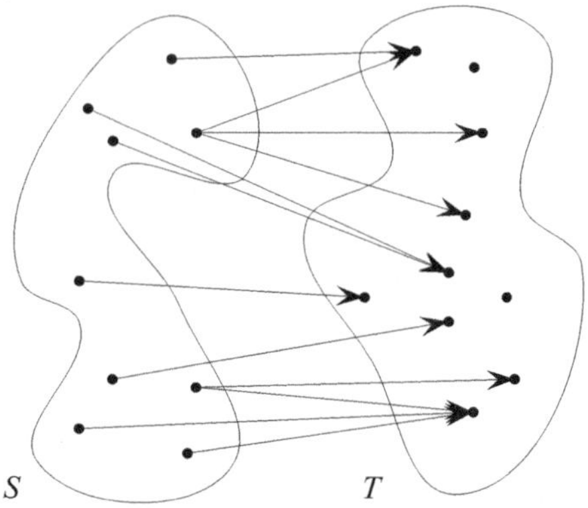

Fig. 3.1 Mapping (multivalued function): To each element of S an element of T is assigned but not all elements of T need to be assigned to any element of S and several elements of T may be assigned to one element of S and an element of T may be assigned to several elements of S

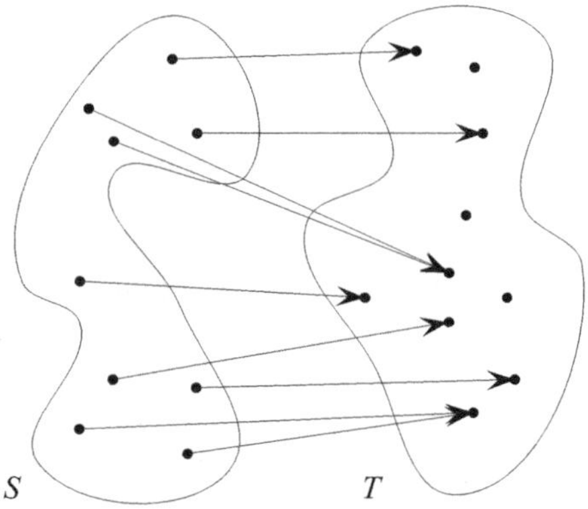

Fig. 3.2 (Single-valued) function: Representation as in Fig. 3.1 but only each element of the set T can be assigned to one element of S. In general inverse (single-valued) functions cannot be defined, but inverse mappings (inverse multivalued functions) can always be defined

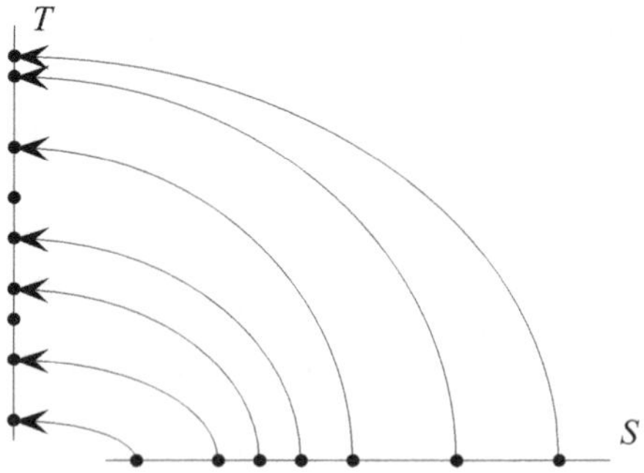

Fig. 3.3 Injection: To each element $s \in S$ exactly one element $t = f(s) \in T$ is assigned, but not all elements of T need to be so assigned. Every $t \in f(S)$ is assigned to exactly one $s \in f(S)$. Inverse function $f^{-1} : f(S) \to S$ exists

represented by the indicated points. The instruction assigning an element (point) of S to an element (point) of T is indicated by an arrow. These arrows thus represent the mapping, that is, multivalued function or (single-valued) function. We get the inverse mapping or function by inverting the direction of the arrows.

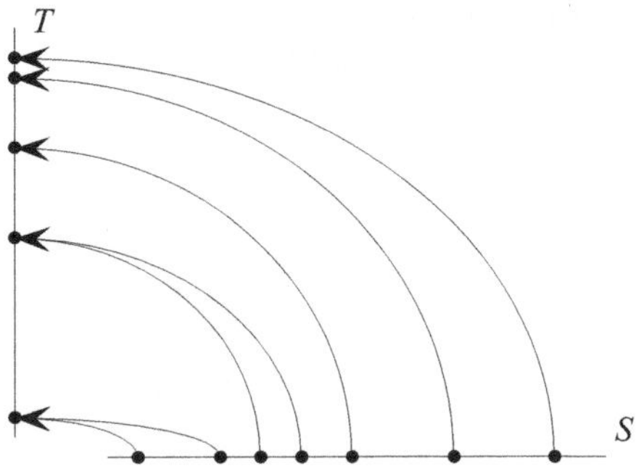

Fig. 3.4 Surjection: Every element of T is assigned to an element of S but it may be assigned to more than one. In general no inverse function exists (only inverse multivalued function)

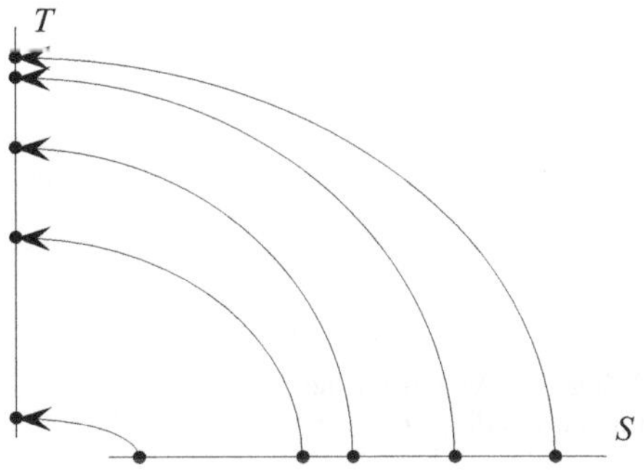

Fig. 3.5 Bijection: To each element of S an element of T is assigned and every element of T is assigned to exactly one of S. Inverse function $f^{-1} : T \to S$ exists

At least if the domain is a subset of $\mathbb{R}$ (or of $\mathbb{R}^2$) and the range is $\mathbb{R}$, the so-called graph is a much more intuitive picture of the function (or of the mapping, that is, "multivalued function") but *graphs* can be defined for *any* function or mapping by

$$G(f) := \{(s, t) \mid s \in S,\ t = f(s) \in f(S)\}.$$

The *graph of the inverse mapping* (whether this inverse is a single-valued or a "multivalued function") is clearly

$$G(f^{-1}) = \{(t, s) \mid s \in S,\ t = f(s) \in f(S)\}.$$

For functions (or "mappings") mapping subsets of $\mathbb{R}$ into $\mathbb{R}$, the graph is a subset of the plane $\mathbb{R}^2$ which we *identify* with the Cartesian plane (see Sect. 1.4). If, for instance, $S = \{-3, -2, -1, 0, 1, 2, 3\}$ and the "instruction" defining the function f is

$$t = f(s) = s^3 - 8s - 1,$$

then the graph of this function is

$$G(f) = \big\{(s, t) \bigm| s \in \{-3, -2, -1, 0, 1, 2, 3\},\ t = s^3 - 8s - 1\big\},$$

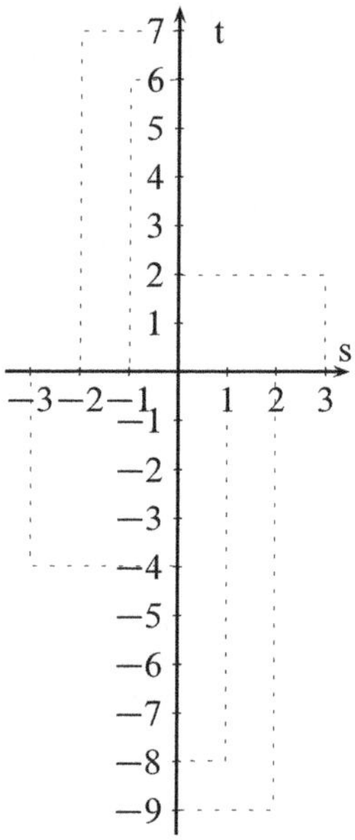

Fig. 3.6 Graph of the function given by Table 3.1

Table 3.1 Values for the function in Fig. 3.6

s	−3	−2	−1	0	1	2	3
t	−4	7	6	−1	−8	−9	2

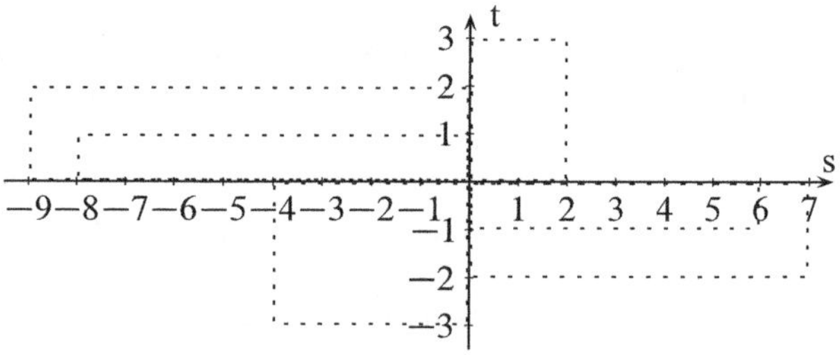

Fig. 3.7 Graph showing the inverse mapping of the function whose graph is shown in Fig. 3.6

or simply

$$G(f) = \{(-3,-4), (-2,7), (-1,6), (0,-1), (1,-8), (2,-9), (3,2)\}.$$

This can be represented either by plotting these *points* on $\mathbb{R}^2$ (Fig. 3.6) or by a Table 3.1. By the definition of $G(f^{-1})$ above we get the graph of the inverse mapping, namely

$$G(f^{-1}) = \{(-4,-3), (7,-2), (6,-1), (-1,0), (-8,1), (-9,2), (2,3)\}.$$

We represent this graph on $\mathbb{R}^2$ in the same coordinate system into which we plotted the points of $G(f)$; see Fig. 3.7. Notice that we get Fig. 3.7 from Fig. 3.6 by *exchanging the roles of s and t*, that is, of the horizontal axis ("X-axis") and the vertical axis ("Y-axis"); see Sect. 1.4.

We present now the graphs of some functions that, in contrast to the examples above, are defined on infinitely many points or real numbers. Figures 3.8, 3.9, 3.10, 3.11 and 3.12 show the graphs of the functions

(i) $x \longmapsto x : \mathbb{R} \longrightarrow \mathbb{R}$ ("*identity function*"),
(ii) $x \longmapsto c : \mathbb{R} \longrightarrow \{c\}$ $c \in \mathbb{R}$, fixed ("*constant function*"),
(iii) $x \longmapsto x^2 : \mathbb{R}_+ \longrightarrow \mathbb{R}_+$ ("*square function*"),
(iv) $x \longmapsto \sqrt{x} : \mathbb{R}_+ \longrightarrow \mathbb{R}_+$ ("*square root function*"),
(v) $\cos : \mathbb{R} \longrightarrow \mathbb{R}$ ("*cosine function*", see Sect. 1.7.2),

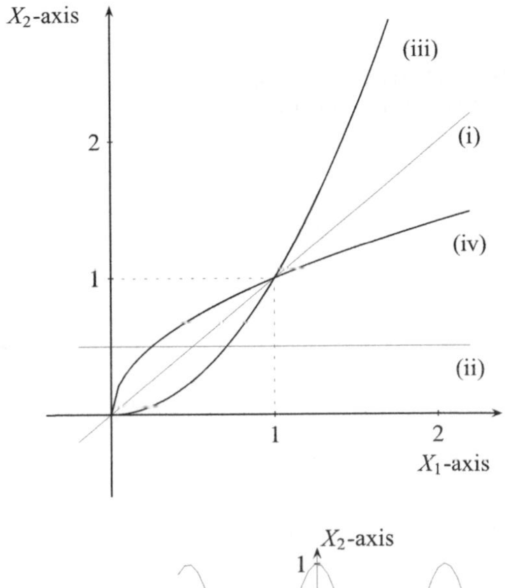

Fig. 3.8 (Parts of the) graphs of the (*i*) identity, (*ii*) constant, (*iii*) square and (*iv*) square root function

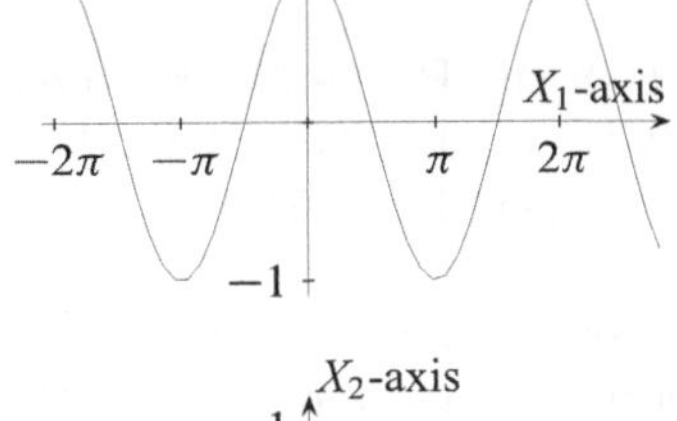

Fig. 3.9 (Part of the) graph of the (v) cosine function

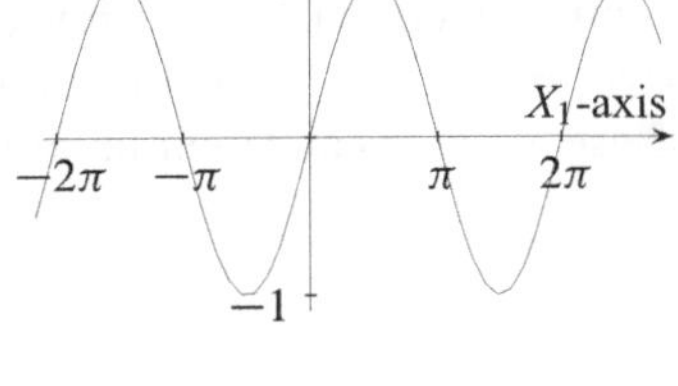

Fig. 3.10 (Part of the) graph of the (vi) sine function

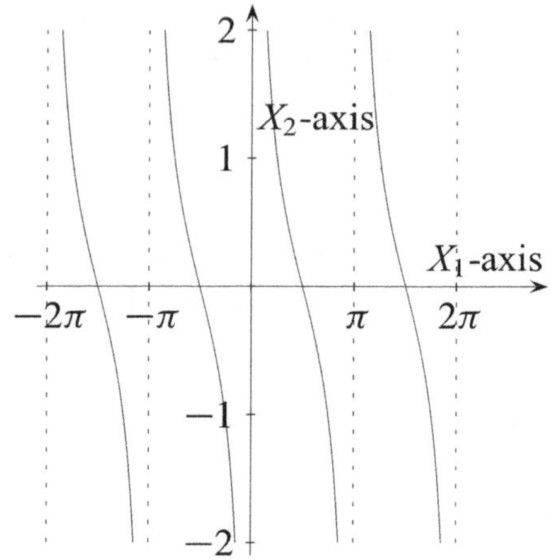

Fig. 3.11 (Part of the) graph of the (vii) cotangent function

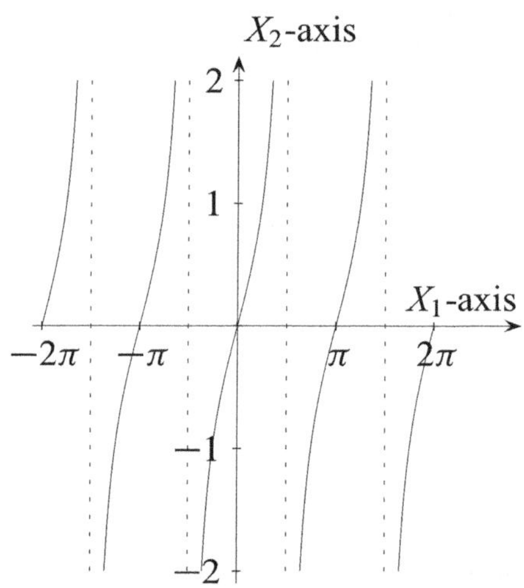

Fig. 3.12 (Part of the) graph of the (viii) tangent function

(vi) $\sin : \mathbb{R} \longrightarrow \mathbb{R}$ ("*sine function*", see Sect. 1.7.2),
(vii) $\cot : \mathbb{R} \setminus \{k\pi \mid k \in \mathbb{Z}\} \longrightarrow \mathbb{R}$ ("*cotangent function*", see Sect. 1.7.4)
(viii) $\tan : \mathbb{R} \setminus \{(2k+1)\frac{\pi}{2} \mid k \in \mathbb{Z}\} \longrightarrow \mathbb{R}$ ("*tangent function*", see Sect. 1.7.4).

We get again the graph of the inverse function or mapping by interchanging the horizontal and vertical axes. The square root is the inverse function of the square, the identity function is its own inverse; the sine, cosine, tangent and cotangent functions have only multivalued functions (mappings) as inverses (but see also Sect. 9.2): These functions and their inverse functions (and mappings) have $\mathbb{R}$ and subsets of $\mathbb{R}$ as domains and ranges. Important subsets of $\mathbb{R}$ are the *intervals* defined as follows

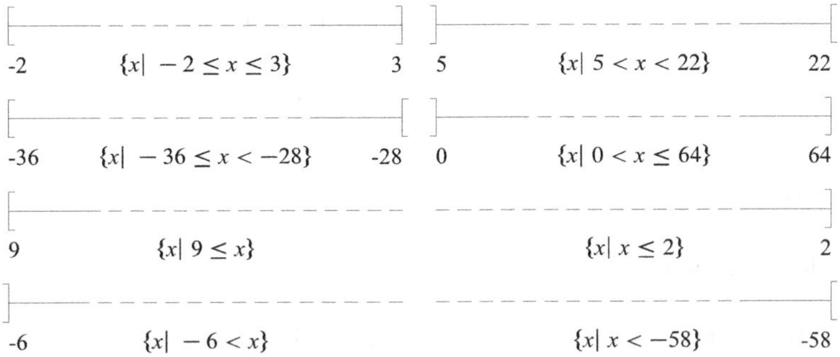

Fig. 3.13 Intervals

(see also Fig. 3.13). Let $a \in \mathbb{R}$, $b \in \mathbb{R}$ be such that $a < b$. Then

$$\begin{array}{lll}
[a, b] & := \{x \in \mathbb{R} \mid a \leq x \leq b\} & \text{(closed interval);} \\
]a, b[& := \{x \in \mathbb{R} \mid a < x < b\} & \text{(finite open interval);} \\
[a, b[& := \{x \in \mathbb{R} \mid a \leq x < b\} & \\
]a, b] & := \{x \in \mathbb{R} \mid a < x \leq b\} & \text{(finite half–closed intervals);} \\
[a, +\infty[& := \{x \in \mathbb{R} \mid x \geq a\} & \\
]-\infty, b] & := \{x \in \mathbb{R} \mid x \leq b\} & \text{(infinite half–closed intervals);} \\
]a, +\infty[& := \{x \in \mathbb{R} \mid x > b\} & \\
]-\infty, b[& := \{x \in \mathbb{R} \mid x < b\} & \text{(infinite open intervals).}
\end{array}$$

(Often $[a, b[$ is denoted by $[a, b)$ and $]a, b[$ by (a, b), and so on. Because of the danger of confusing the latter with the point $(a, b) \in \mathbb{R}^2$, we prefer the above notation).

For instance $\mathbb{R}_+$, $\mathbb{R}_-$ are infinite half-closed intervals, $\mathbb{R}_{++}$, $\mathbb{R}_{--}$ are infinite open intervals. In addition, $\mathbb{R}$ itself is considered an interval (infinite open) and so is a set consisting of a single point. The *singleton* $\{a\}$ may be considered a closed interval $[a, a]$. We see that the image of $\mathbb{R}$ under the *constant function* $s \mapsto c$ ($f(s) = c$ for all $s \in \mathbb{R}$; or on any other set) is the singleton $\{c\}$, the domains of cot and tan are

$$\bigcup_{k=-\infty}^{\infty}]k\pi, (k+1)\pi[\qquad \text{and} \qquad \bigcup_{k=-\infty}^{\infty}](2k-1)\frac{\pi}{2}, (2k+1)\frac{\pi}{2}[,$$

respectively, where

$$\bigcup_{k=-\infty}^{\infty} S_k := \{s \mid s \in S_0 \text{ or } s \in S_1 \text{ or } s \in S_{-1} \text{ or } s \in S_2 \text{ or } s \in S_{-2}, \ldots\}.$$

The images of $\mathbb{R}$ under sin and cos are the closed interval [−1,1]:

$$\sin \mathbb{R} = \cos \mathbb{R} = [-1, 1].$$

3.2.1 Exercises

1. Consider the functions

(a) $f : \mathbb{R} \longrightarrow \mathbb{R},\ x \longmapsto x^3,$ (b) $g : \mathbb{R} \longrightarrow \mathbb{R},\ x \longmapsto x^4,$
(c) $h : \mathbb{R} \longrightarrow \mathbb{R}_+,\ x \longmapsto x^4,$ (d) $\cos : \mathbb{R} \longrightarrow \mathbb{R},$
(e) $\sin : \mathbb{R} \longrightarrow [-1, 1],$ (f) $\cot : \mathbb{R} \setminus \{k\pi \mid k \in \mathbb{Z}\} \longrightarrow \mathbb{R},$
(g) $\phi : \mathbb{R}_{++} \longrightarrow \mathbb{R},\ x \longmapsto 1/x,$

State for each whether it is surjective, injective or bijective, whether its inverse is single-valued or multivalued, and whether its range equals the image of its domain.
2. Determine the inverse functions of the functions

(a) $f : \mathbb{R} \longrightarrow \mathbb{R},\ x \longmapsto -x,$ (b) $g : \mathbb{R} \longrightarrow \mathbb{R},\ x \longmapsto x^3,$
(c) $h : \mathbb{R}_+ \longrightarrow \mathbb{R}_+,\ x \longmapsto x + x^2,$ (d) $p : \mathbb{R}_+ \longrightarrow \mathbb{R}_+,\ x \longmapsto \sqrt{x + x^2},$

3. Draw the graphs of the functions

(a) $f : [-3\pi, 3\pi] \longrightarrow \mathbb{R},\ x \longmapsto x \cos x,$
(b) $g : [0, 3\pi] \longrightarrow \mathbb{R},\ x \longmapsto x - \sin x,$
(c) $h :]-\frac{\pi}{2}, \frac{\pi}{2}[\longrightarrow \mathbb{R},\ x \longmapsto x^2 + \tan x,$
(d) $p : [-2\pi, 2\pi] \longrightarrow \mathbb{R},\ x \longmapsto 4(\cos x)(\sin x).$

4. Determine the images of the functions given in Exercise 3.
5. Draw the graph of the inverse functions determined in Exercise 2.

3.2.2 Answers

1. (a) surjective, injective, bijective, inverse is single valued, range = image,
(b) injective, inverse is multivalued,
(c) surjective, injective, inverse is multivalued, range = image,
(d) injective, inverse is multivalued,
(e) surjective, injective, inverse is multivalued, range = image,
(f) surjective, injective, inverse is multivalued, range = image,
(g) injective, inverse is singlevalued,

2. (a) $f^{-1} : \mathbb{R} \to \mathbb{R},\ x \mapsto -x$,
   (b) $g^{-1} : \mathbb{R} \mapsto \mathbb{R} \to \mathbb{R},\ x \mapsto x^{1/3}$
   (c) $h^{-1} : \mathbb{R}_+ \to \mathbb{R}_+,\ x \mapsto -\frac{1}{2} + \frac{1}{2}(1+4x)^{1/2}$,
   (d) $p^{-1} : \mathbb{R}_+ \to \mathbb{R}_+,\ x \mapsto -\frac{1}{2} + \ (1+4x^2)^{1/2}$.
4. (a) $[-3\pi, 3\pi]$, (b) $[0, 3\pi]$, (c) $\mathbb{R}$, (d) $[-2, 2]$.

3.3 Functions of n Variables, n–Dimensional Intervals, Composition of Functions

As we saw in Sect. 3.1, function values may be, among others, real or complex numbers or vectors. Then we have real-, complex- or vector-valued functions. In the Examples 10, 11, 15 and 16 of Sect. 3.1, a function (or mapping) has assigned to each element of its domain S a *set* as function value. In this case the range is a set of sets (a set whose elements are sets) and we talk about *set-valued functions* or *correspondences*. Such correspondences play important roles in production and utility theory; for examples see Sect. 9.

Also the domains may be different. Of particular interest for us are the cases when they are subsets of $\mathbb{R}$, $\mathbb{R}^n$ or $\mathbb{C}$: in these cases we speak about *functions of one* or *n real variables* (or *of one n-component vector variable*) or *of a complex variable*, respectively. Functions of *n real* variables are particular cases of functions defined on a subset of the Cartesian product $S_1 \times S_2 \times \ldots \times S_n$ (see Sect. 1.4),which are also called *functions of n variables* or *n-place functions*. Vector-valued functions of vector variable(s) are often called *vector–vector functions*.

The role of intervals as domains of functions of a real variable are often played by *n-dimensional intervals* for the functions of n real variables. Let $\mathbf{a} = (a_1, \ldots, a_n) \in \mathbb{R}^n$, $\mathbf{b} = (b_1, \ldots, b_n) \in \mathbb{R}^n$ be such that $\mathbf{a} < \mathbf{b}$. *Closed n-dimensional intervals* are the sets

$$[\mathbf{a}, \mathbf{b}] = [(a_1, \ldots, a_n), (b_1, \ldots, b_n)] \ := \{(x_1, \ldots, x_n) \mid a_k \le x_k \le b_k\};$$

half-open, open, and/or *infinite n-dimensional intervals* are similarly defined (compare Sects. 3.5 and 3.13). Note that an n-dimensional interval may be open in one variable, half open or closed in another, finite in one variable, infinite in another. The two-dimensional closed interval $[(a, c), (b, d)]$ is the rectangle with vertices (a, c), (a, d), (b, c), (b, d).

Real-valued functions of two real variables ($F : S \to T$, $S \subset \mathbb{R}^2$, $T \subset \mathbb{R}$) can be represented in $\mathbb{R}^3$ (their graph is a subset of $\mathbb{R}^3$), where the point (x_1, x_2, y) represents the value y of the function F at $(x_1, x_2) \in \mathbb{R}^2$. While such a representation by a three-dimensional model is quite possible (but cumbersome, even as a hologram), one prefers drawings in the plane by perspective as in Fig. 3.14. The "production surface" is the graph of the maximal output quantity as function of the two-component input vector (in the two-dimensional interval $[(0, 0), (b, d)]$) which produces it (in the production period).

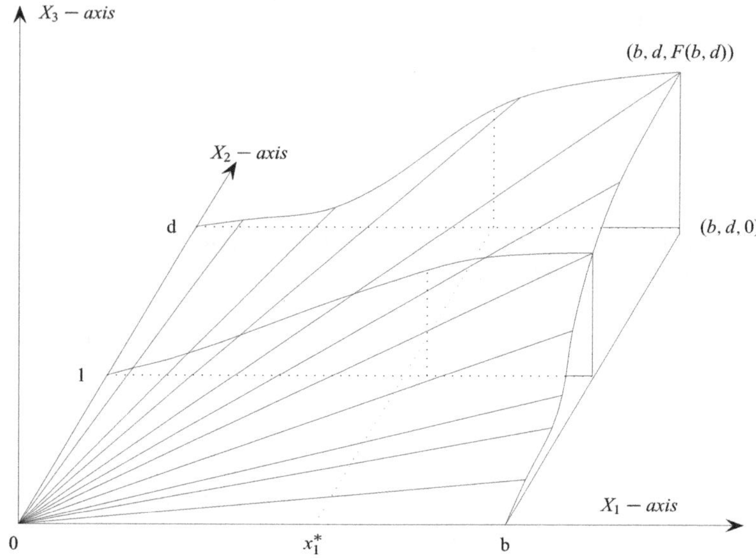

Fig. 3.14 A production surface. Keeping x_1 or x_2 constant, that is, intersecting the surface parallel to the (X_1, Y)–plane or the (X_2, Y)–plane, respectively, gives total product curves similar to those of Fig. 3.18

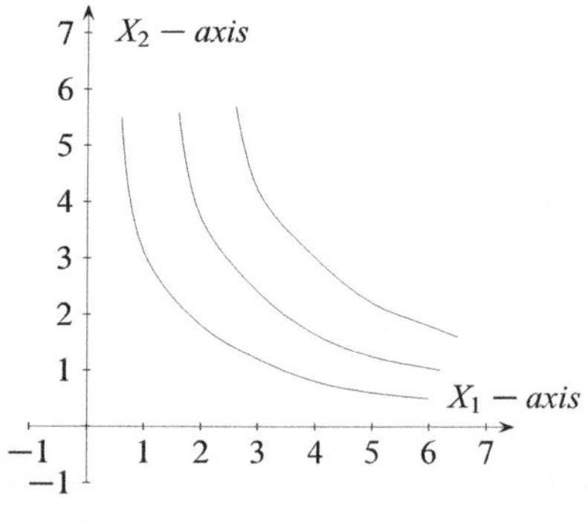

Fig. 3.15 Contour-line representation of a real-valued function ($F : \mathbb{R}^2_+ \to \mathbb{R}_+$) of two variables

Another geometric representation of real-valued functions of two real variables is done by "contour-lines" (Fig. 3.15); these are defined for $F : S \to \mathbb{R}$ $(S \subset \mathbb{R}^2)$ by

$$\{(x_1, x_2) \in S \mid F(x_1, x_2) = c \in F(S)\}.$$

All these "lines" (for all $c \in F(S)$) *together* form the *contour-line representation*. Of course, in Fig. 3.15 we could draw only finitely many of them but, with some practice, one gets from them an impression of the behaviour of F. They play an important role in *nomography*. If F is a utility function (assigning to inputs x_1, x_2

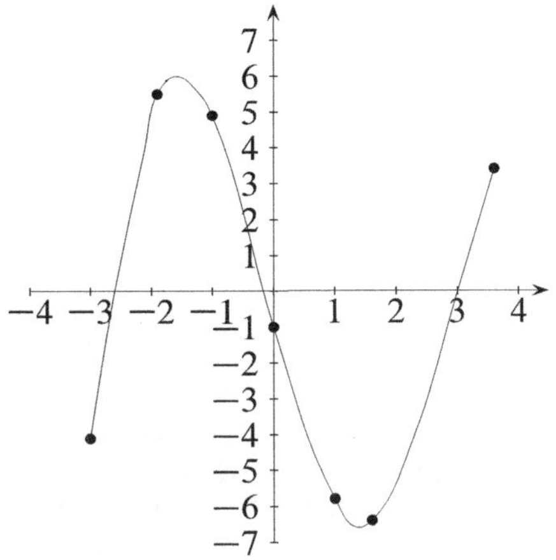

Fig. 3.16 Extension of the graph in Fig. 3.6 (and of the function which it represents)

the utility $F(x_1, x_2)$ then the contour-lines are called "*indifference curves*", if F is a *production function* then they are called "*isoquants*".

The functions whose graphs were drawn in Figs. 3.6 and 3.7 are somewhat different from those in Figs. 3.8, 3.9, 3.10, 3.11, 3.12, 3.13, 3.14 and 3.15 because their domain consists of finitely many points (numbers) rather than (unions of) intervals. While, as mentioned, points ("singletons") may be considered as ("degenerated") intervals, one often connects the points (elements) of the (plane representation of the) graph by a *curve* as, for instance, in Fig. 3.16. We emphasise that there are many possibilities of connecting the *same* points even by pretty "smooth" curves. But when we connected them in some way then we extended therewith also the function, in this case to the whole closed interval. A function $\hat{f}$ defined on a domain $\hat{S}$, is an *extension* of a function f defined on a set $S \subset \hat{S}$, if $\hat{f}(s) = f(s)$ for all $s \in S$. But one has to be careful because the extended function may not make the same sense as the original. For instance, the output of textile produced by 3 factories makes sense, that produced by $\sqrt{2}$ factories does not.

Exactly in economics, the "instructions" describing a function (or mapping) are often themselves *not formulas* but ("empirical" or machine-generated) curves or surfaces representing the graph of this function (or mapping) (see for instance Figs. 3.14, 3.17, 3.18 and 3.19). As those mentioned above, these are also extensions and should be handled with the caution just emphasised.

The *identity function* $s \mapsto s$ can be defined on *any* domain S. The importance of domains and ranges (in particular images or codomains) becomes highly visible in *composition of functions* as in Fig. 3.20. If we have two functions $f : S \to T$ and $g : U \to V$ then they can be composed, that is, the *composite function* $g \circ f : S \to V$ can be defined, by

$$s \longmapsto g \circ f(s) := g[f(s)] \quad (s \in S),$$

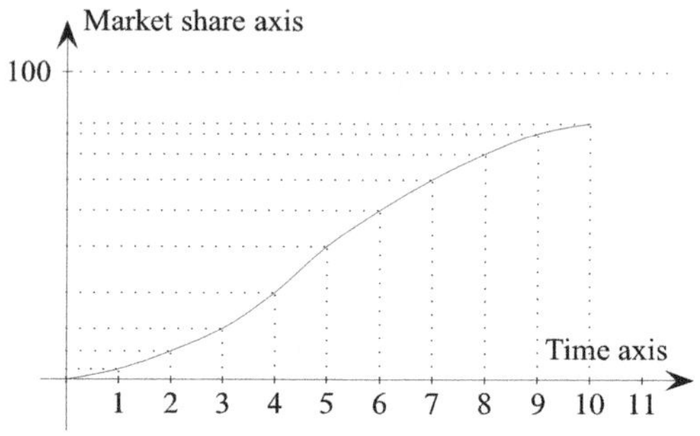

Fig. 3.17 Curve describing the market share of an improved product in percent as function of time. The curve extends the graph of the "market share function" that is only defined at $t = 1, \ldots, 10$ (time periods, for instance months). We assume that the market share was actually measured only for these t

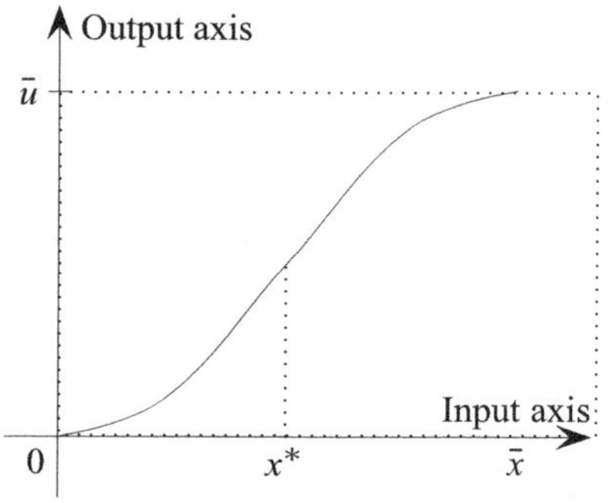

Fig. 3.18 Total product curve showing the maximal output quantity which can be obtained from the quantity x of one input factor when all other inputs are held constant. If the curve is the graph of the function f then the domain of f is $S = [0, \bar{x}]$, its range is $f(S) = [0, \bar{u}] \subset \mathbb{R}_+$

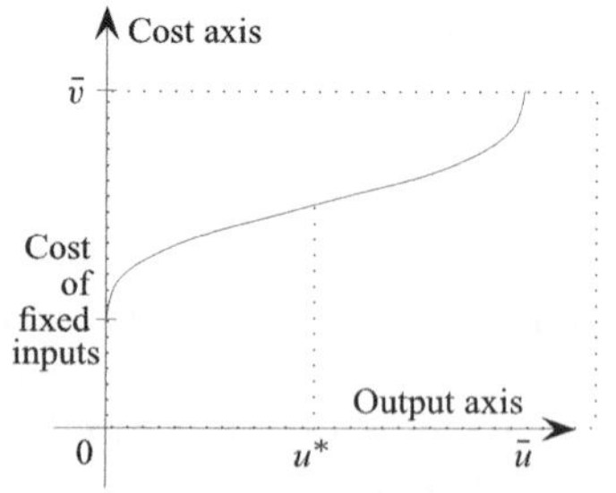

Fig. 3.19 The total cost curve corresponding to the total product curve in Fig. 3.18. In a situation, where every variable, except the input quantity x is fixed, it shows how the minimal cost of production of an output depends on its quantity u. The domain is $S = [0, \bar{u}]$ and the range is $T = [0, \bar{v}]$

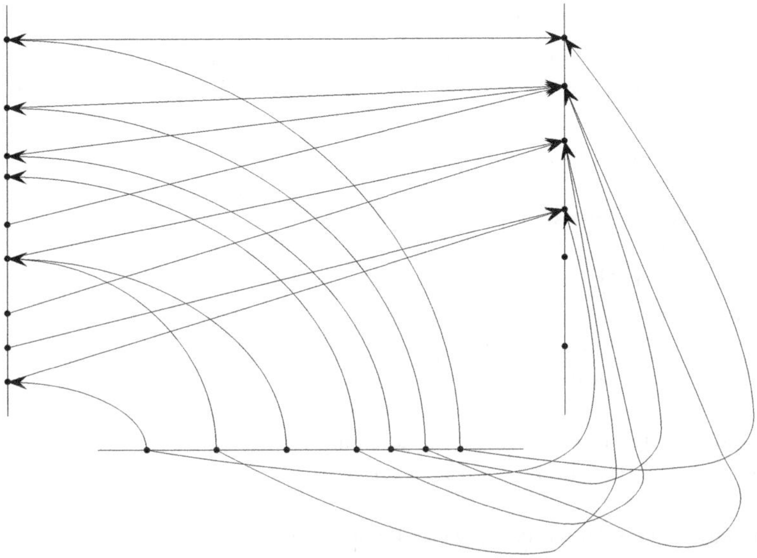

Fig. 3.20 Composition of mappings $f: S \to T$ and $g: U \to V$ The sets S, $T \cup U$ and V consist of the points indicated on the three segments. Obviously, t does not belong to U and therefore $s = f^{-1}(t)$ does not belong to the domain of $g \circ f$

only if $f(S) \subset U$ (if the image of S under f is a subset of U). Actually, also if $f(S)$ and U have one or more common elements, that is, if

$$I := f(S) \cap U \neq \emptyset,$$

then $g \circ f$ can be defined, but only on the *set* $f^{-1}(I)$, where f^{-1} is the inverse mapping, that is, the inverse, possibly multivalued function of f. Even if $g \circ f$ exists, the function $f \circ g$, defined by $f \circ g(u) := f[g(u)] \quad (u \in U)$ *may not exist*, among others because S and U may be completely different sets and, even if both $f \circ g$ and $g \circ f$ exist, *they are usually different*. For example for $f(s) = -\cos s$, $g(u) = u^2$ we have $f \circ g \neq g \circ f$ because for instance

$$f \circ g(0) = -\cos(0^2) = -1 \neq 1 = (-\cos 0)^2 = g \circ f(0).$$

In the rest of this chapter we introduce some particular classes of functions, domains, and ranges, which are important for applications.

3.3.1 Exercises

1. Draw three contour–lines for each of the following functions F, G, H:
   (a) $F : \mathbb{R}^2_+ \longrightarrow \mathbb{R}_+$, $(x, y) \longmapsto F(x, y) = \sqrt{xy/(x + y)}$,

(b) $G : \mathbb{R}_+^2 \longrightarrow \mathbb{R}_+, (x, y) \longmapsto G(x, y) = x\sqrt{y}/(x + y)$,
(c) $H : \mathbb{R}_+^2 \longrightarrow \mathbb{R}_+, (x, y) \longmapsto H(x, y) = xy/(\sqrt{x} + y)$.

2. Represent the functions F, G, H defined in Exercise 1 by drawings in the plane by perspective.
3. Consider the functions $f : \mathbb{R} \longrightarrow \mathbb{R},\ x \longmapsto x^3$,
$g : \mathbb{R} \longrightarrow \mathbb{R},\ x \longmapsto x^4$,
$h : \mathbb{R} \longrightarrow \mathbb{R}_+,\ x \longmapsto x^4$,
$\cos : \mathbb{R} \longrightarrow \mathbb{R}$,
$\sin : \mathbb{R} \longrightarrow [-1, 1]$,
$\phi : \mathbb{R}_{++} \longrightarrow \mathbb{R},\ x \longmapsto 1/x$,
$\psi : \mathbb{R}_{++}^2 \longrightarrow \mathbb{R}_{++},\ (x, y) \longmapsto \frac{1}{x+y}$,
Determine the composite functions of functions

(a) f and g, (b) g and cos, (c) h and sin,
(d) cos and ϕ, (e) ϕ and h, (f) ψ and f,

on the largest sets on which they can be defined.
4. For the function $F : \mathbb{R}^2 \to \mathbb{R},\ (x, y) \mapsto x^2 + y^2 + 3x + 2$, determine a function $g : \mathbb{R}_{++} \to \mathbb{R}$ such that the composite function $g \circ F : \mathbb{R}^2 \to \mathbb{R}$ is $(x, y) \mapsto 1/((x + 1)(y + 2) + y^2)$.
5. For the function $g : \mathbb{R}_+ \to \mathbb{R}_+, x \mapsto \sqrt{|\sin x|}$, determine a function $f : \mathbb{R} \to \mathbb{R}_+$ such that the composite function $g \circ F : \mathbb{R}_+ \to \mathbb{R}_+$ is $x \mapsto (\sin x)^2$.

3.3.2 Answers

3. (a) $g \circ f : \mathbb{R} \to \mathbb{R},\ x \mapsto x^{12}$,
(b) $\cos \circ g : \mathbb{R} \to \mathbb{R},\ x \mapsto \cos x^4$,
(c) $\sin \circ h : \mathbb{R} \to [-1, 1],\ x \mapsto \sin x^4$,
(d) $\phi \circ \cos : \{x | \cos x > 0\} \to \mathbb{R},\ x \mapsto \frac{1}{\cos x}$,
(e) $h \circ \phi : \mathbb{R}_{++} \to \mathbb{R}_+,\ x \mapsto \frac{1}{x^4}$,
(f) $f \circ \psi : \mathbb{R}_{++}^2 \to \mathbb{R},\ (x_1, x_2) \mapsto \frac{1}{(x_1+x_2)^3}$.
4. $g : \mathbb{R}_{++} \to \mathbb{R}, x \mapsto 1/x$.
5. $f : \mathbb{R} \to \mathbb{R}_+, x \mapsto x^4$.

3.4 Monotonic and Linearly Homogeneous Functions. Maxima and Minima

A function $f : S \to \mathbb{R}$, where S is a set of real numbers ($S \subset \mathbb{R}$) *is increasing* on a subset X of S ($X \subset S$) *if, for all* $x \in X,\ x' \in X$ *with* $x < x'$,

$$f(x) \leq f(x'); \tag{3.1}$$

it *is decreasing* on X *if*, again *for all* $x \in X$, $x' \in X$ *with* $x < x'$ we have

$$f(x) \geq f(x'). \tag{3.2}$$

If in (3.1) or (3.2) we have $<$ *resp.* $>$ *for all* $x \in X$, $x' \in X$ *with* $x < x'$ *then* f *is strictly increasing* or *strictly decreasing*, respectively. Figures 3.17, 3.18 and 3.19 are examples of graphs of strictly increasing functions on $[0, 10]$, $[0, \bar{x}]$, $[0, \bar{u}]$, respectively. Often strictly increasing (strictly decreasing) functions are called "increasing" ("decreasing") while, what we called increasing (decreasing) is called "nondecreasing" ("non-increasing"). There exist functions which are neither increasing (or strictly increasing) nor decreasing (or strictly decreasing) as, for instance, the functions the graphs of which are drawn in Figs. 3.6, 3.7, 3.9, 3.10 and 3.16. So "nondecreasing" and "non-increasing" may lead to misunderstanding and we will not use these words. Note that constant functions ($x \mapsto c$, c a constant real number) and *only those are both increasing and decreasing*, but neither strictly increasing nor strictly decreasing.

A function is *monotonic* on $X \subset \mathbb{R}$ if it is either increasing on all of X or decreasing on all of X. It is *strictly monotonic* on X if it is either strictly increasing or strictly decreasing on X (*these* names are universally accepted). In Fig. 3.7 the function f whose graph is represented there is strictly increasing on $\{-3, -2\}$ and $\{2, 3\}$, and strictly decreasing on $\{-2, -1, 0, 1, 2\}$. The extension of (the graph of) this function represented in Fig. 3.16 obviously has its maximum at some real number in the interval $[-2, -1]$ and its minimum at some real number in the interval $[1, 2]$.

The general definition is the following. A function $f : X \to \mathbb{R}$, where $X \subset \mathbb{R}$, *has at* x_M *a maximum on* X *if*

$$f(x_M) \geq f(x) \quad \textit{for all} \quad x \in X. \tag{3.3}$$

It has at x_m *a minimum on* X *if*

$$f(x_m) \leq f(x) \quad \textit{for all} \quad x \in X. \tag{3.4}$$

If we have $>$ in (3.3) or $<$ in (3.4) for all $x \neq x_M$ resp. for all $x \neq x_m$ then the maximum or minimum is *sharp* (or *strict*). Of course, *if the (non-sharp) maximum and minimum of* f *on* X *are equal, then* f *is constant on* X.

If a function on an interval $I \in \mathbb{R}$ first increases till x_M, where it has a maximum, then decreases thereafter, or if it first decreases till x_m, where it has a minimum, and increases thereafter, then the function is *unimodal* on I. Unimodal functions play an important role for instance in statistics. The functions represented in Figs. 3.21 and 3.22 are unimodular, those in Figs. 3.23 and 3.24 are not.

Note however that, for instance on a closed interval I, the maximum or minimum needs not be inside the interval, it can be at the left or right end: in Fig. 3.23 both the maximum and the minimum are at the ends, in Fig. 3.24 the minimum is at the left end, the maximum inside. There is *no complete analogue of monotonicity for*

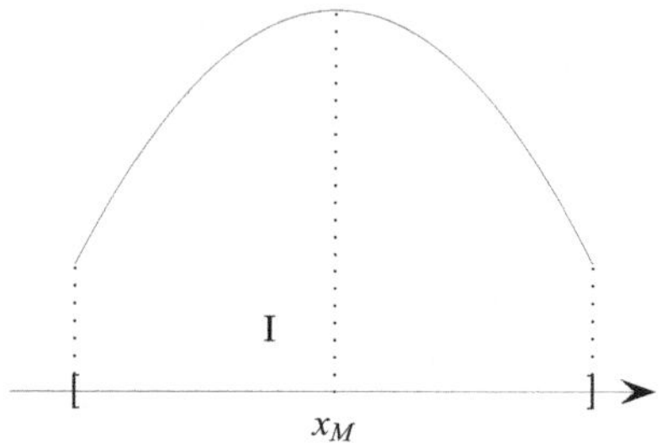

Fig. 3.21 Graph of a unimodal function. Maximum at x_M

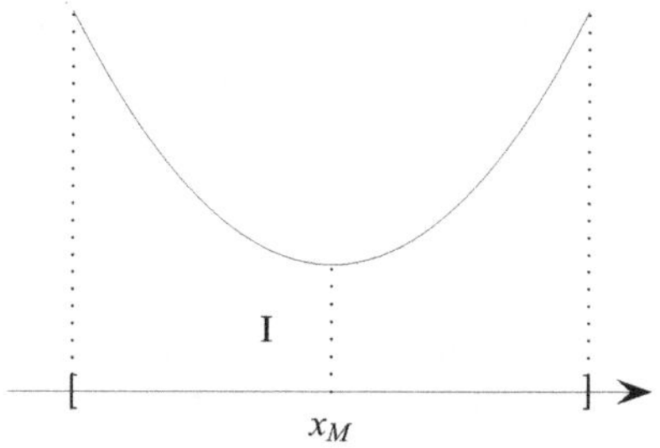

Fig. 3.22 Graph of a unimodal function. Minimum at x_M

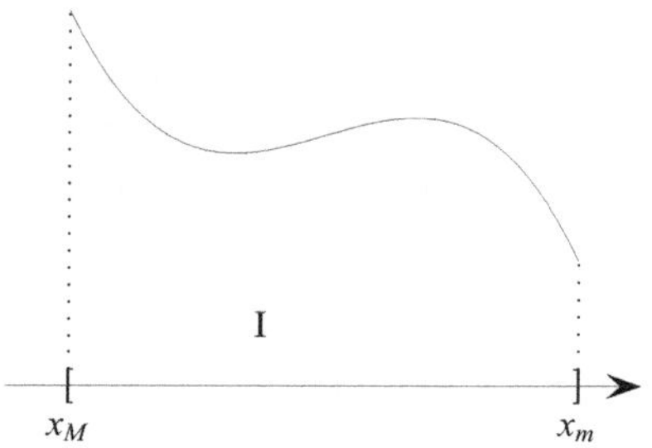

Fig. 3.23 Graph of a function which has maximum at the *left end* of I and minimum at the *right end*

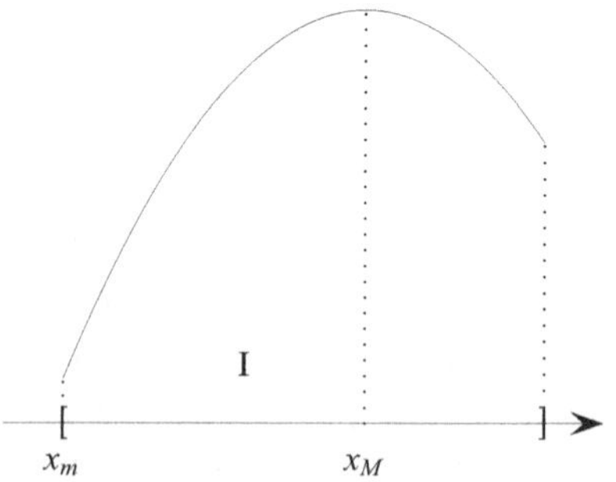

Fig. 3.24 Graph of a function with minimum at the left end of I and maximum in the interior of I, at x_M

multi-place functions, not even for functions of two real variables, because, as we have seen in Sect. 1.3, $\mathbb{R}^n$ *for* $n > 1$ *is not totally ordered* under either of the usual orderings ($>$, $\geq$, $\geqq$). However, there are several partial analogues.

We will deal here with real-valued functions of $n \geq 2$ real variables, in other words, with real-valued multi-place functions. (For m-component vector-valued functions ($m \geq 2$), in particular complex-valued functions, the above mentioned lack of total ordering, this time on the range, would make things so complicated that one usually does not define monotonicity for them.)

Let $S \subset \mathbb{R}^n$. *The function* $F : S \to \mathbb{R}$ *is increasing on* $X \subset S$ *if*

$$F(\mathbf{x}) \leq F(\mathbf{x}') \tag{3.5}$$

whenever $\mathbf{x} \in X$, $\mathbf{x}' \in X$ *and* $\mathbf{x} \leq \mathbf{x}'$ (as defined in Sect. 1.3: $\mathbf{x} = (x_1, \ldots, x_n) \leq \mathbf{x}' = (x'_1, \ldots, x'_n)$ if $x_k \leq x'_k$ for all $k \in \{1, \ldots, n\}$ but, at least for one ℓ, $x_\ell < x'_\ell$, $\ell \in \{1, \ldots, n\}$). *The function* $F : S \to \mathbb{R}$ (often, but not always, one denotes multi-place functions by capital letters) is *decreasing on* $X \subset S$ *if*

$$F(\mathbf{x}) \geq F(\mathbf{x}') \tag{3.6}$$

whenever $\mathbf{x} \in X$, $\mathbf{x}' \in X$ *and* $\mathbf{x} \leq \mathbf{x}'$. *If, in (3.5) or (3.6) < resp. > stands then F is strictly increasing* or *strictly decreasing*, respectively.

Again the name covering both increasing and decreasing is "*monotonic*", that covering both strictly increasing and strictly decreasing is "*strictly monotonic*". An example of a monotonic (here: increasing) scalar-valued function on $\mathbb{R}^2_+$ is partly given by the graph in Fig. 3.25. Notice that this function is strictly increasing on $\mathbb{R}^2_{++}$.

Of course, most functions are not monotonic on most domains. For instance, the function $F : [(0, 0), (b, d)] \to \mathbb{R}_+$ whose graph ("production surface") is drawn in Fig. 3.14 is *not* increasing (or decreasing) on its domain $[(0, 0), (b, d)]$. But notice

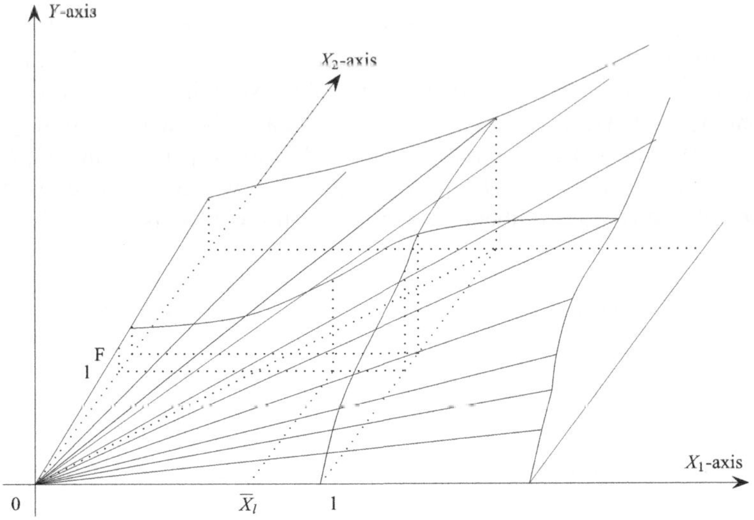

Fig. 3.25 Part of the graph of an increasing function F defined on $\mathbb{R}^2_+$ (strictly increasing on $\mathbb{R}^2_{++}$)

that it is strictly increasing on every ray running from the origin (0,0) within $\mathbb{R}_{++} \times \mathbb{R}_{++}$ to the boundary of the interval $[(0,0),(b,d)]$.

Again we define the maximum and minimum: The function $F : X \to \mathbb{R}$, $X \subset \mathbb{R}^n$, has at $\mathbf{x}_M$ a *maximum*, at $\mathbf{x}_m$ a *minimum* on X if

$$F(\mathbf{x}_M) \geq F(\mathbf{x}), \quad F(\mathbf{x}) \geq F(\mathbf{x}_m) \quad \text{for all} \quad \mathbf{x} \in X,$$

respectively. If here $>$ stands instead of $\geq$ for all $\mathbf{x} \neq \mathbf{x}_M$ or $\mathbf{x} \neq \mathbf{x}_m$, respectively, then the *maximum* or *minimum* is *strict* (or *sharp*).

If F is (strictly) increasing on X, $X \subset \mathbb{R}^n$, then $-F$ $(\mathbf{x} \mapsto -F(\mathbf{x}))$ is (strictly) decreasing there and the same is true the other way round (this follows from the definition because, if $F(\mathbf{x}) > F(\mathbf{x}')$ then $-F(\mathbf{x}) < -F(\mathbf{x}')$).

Another way of formulating monotonicity, say increasing, is the requirement that, *holding all but one variable fixed, the function of this one variable is increasing*. In formula , F is increasing on X if

$$F(x_1,\ldots,x_{k-1},x_k,x_{k+1},\ldots,x_n) \leq F(x_1,\ldots,x_{k-1},x_k',x_{k+1},\ldots,x_n)$$
$$(k = 1,\ldots,n),$$

whenever $(x_1,\ldots,x_k,\ldots,x_n) \in X$, $(x_1,\ldots,x_K',\ldots,x_n) \in X$ and $x_k < x_k'$. (Show that this is equivalent to the above definition!) Often this is expressed by saying that the functions

$$x_k \mapsto F(x_1,\ldots,x_{k-1},x_k,x_{k+1},\ldots,x_n) \quad \text{or} \quad F(x_1,\ldots,x_{k-1},\cdot,x_{k+1},\ldots,x_n)$$

are increasing $(k = 1,\ldots,n)$. Geometrically, in the case $n = 2$, this means that all "cuts" of the graph, "parallel" to the (X_1, Y) and to the (X_2, Y)-planes are graphs of increasing functions of one real variable. Note, however that for F to be monotonic $x_k \mapsto F(x_1,\ldots,x_{k-1},x_k,x_{k+1},\ldots,x_n)$ has to be either increasing for *all* $k(= 1,\ldots,n)$ or decreasing for *all* k. Figure 3.26 shows (part of) the graph of the function $F : \mathbb{R}^2 \to \mathbb{R}$ given by $F(x_1,x_2) = x_1^2 - x_2^2$. If we are only interested in the restriction of F to $\mathbb{R}_+^2$, say $\tilde{F}$, then $x_1 \mapsto x_1^2 - x_2^2$ is strictly increasing for all $x_2 \in \mathbb{R}_+$,

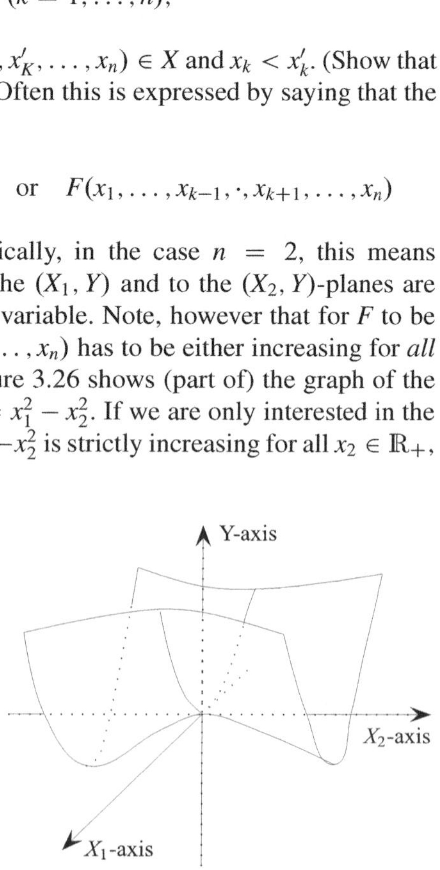

Fig. 3.26 Graph of (part of) $(x_1,x_2) \mapsto x_1^2 - x_2^2$ on $\mathbb{R}^2$. For the restriction of this function and its graph to $\mathbb{R}_+^2$, all cuts parallel to the (X_1, Y)-plane are graphs of strictly increasing functions, all cuts parallel to the (X_2, Y)-plane are graphs of strictly decreasing functions. Notice the "saddle point" at (0,0)

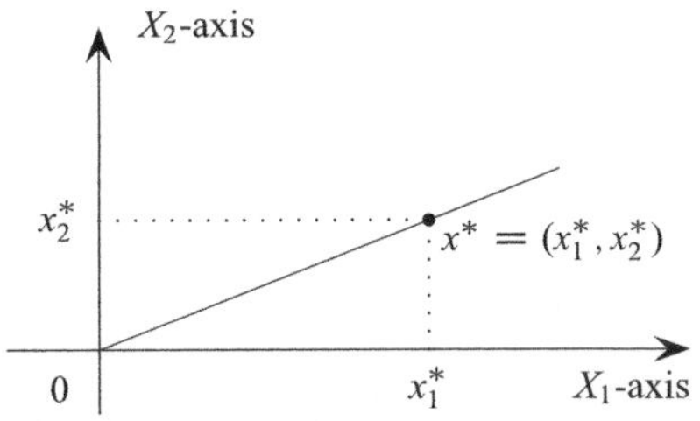

Fig. 3.27 The ray going through $\mathbf{x}^* = (x_1^*, x_2^*)$

while $x_2 \mapsto x_1^2 - x_2^2$ is strictly decreasing for all $x_1 \in \mathbb{R}_+$, but $\tilde{F}$ is *not* monotonic. In connection with Fig. 3.14 we introduced the name "ray" (see also Fig. 3.25). For $\mathbf{x}^* \in \mathbb{R}^n$ the set

$$\{t\mathbf{x}^* \mid t \in \mathbb{R},\ t > 0\}$$

is a *ray* (the name "*ray going through* $\mathbf{x}^*$"; for $n = 2$ see Fig. 3.27). If the function

$$t \mapsto F(t\mathbf{x}^*),\ \mathbf{x}^* \in \mathbb{R}^n,\ t \in \mathbb{R}_{++} \tag{3.7}$$

is *monotonic (strictly monotonic, increasing, strictly increasing, decreasing, strictly decreasing) on every ray in* $\mathbb{R}^n$ *or on the intersection* (see Sect. 1.2) *of every ray with a set* $X \subset \mathbb{R}^n$ *then* F *is ray-monotonic (strictly ray-monotonic, ray-increasing, strictly ray-increasing, ray-decreasing, strictly ray-decreasing) on* $\mathbb{R}^n$ *or on* X*, respectively*. As we have mentioned already, the function F, whose graph is in Fig. 3.14, is *strictly ray–increasing* on the interval $X = [(0,0), (b,d)]$ (but *not* increasing there). In particular, for this F, and also for that in Fig. 3.25, the function (3.7) is linear (compare Sect. 4.1) for all $\mathbf{x}^*$ in the interval. Functions F, for which the mapping (3.7) has this property, are called linearly homogeneous.

The general definition is: *A function* $F : X \to \mathbb{R}$ $(X \subset \mathbb{R}^n)$ *is positively linearly homogeneous, "linearly homogeneous" for short, on* $X \subset \mathbb{R}^n$ *if*

$$F(t\mathbf{x}) = tF(\mathbf{x}) \quad \textit{whenever} \quad t \in \mathbb{R}_{++},\ \mathbf{x} \in X \quad \textit{and} \quad t\mathbf{x} \in X. \tag{3.8}$$

In the case of *production functions* we speak about "*constant returns to scale*" if (3.8) is satisfied. There is a quite imaginative argument that, *if all variables ("productions factors", "input quantities") are taken into consideration then every production function is linearly homogeneous*, at least for $t \in \mathbb{N}$ or even $t \in \mathbb{Q}_{++}$ in (3.8): If the production process were developed in exactly the same way in $t\ (\in \mathbb{N})$ identical factories (say) then *all* input quantities would increase t-fold, and so would the production output, that is, (3.8) would hold for $t \in \mathbb{N}$. But, if (3.8) holds for $t = n$, then it holds also for $t = 1/n$. To see this just put into

$$F(n\mathbf{x}) = nF(\mathbf{x}) \quad (\mathbf{x} \in X,\ n\mathbf{x} \in X) \tag{3.9}$$

$\mathbf{y} = n\mathbf{x}$ to get

$$F(\mathbf{y}) = nF\left(\frac{1}{n}\mathbf{y}\right) \quad \left(\frac{1}{n}\mathbf{y} \in X,\ \mathbf{y} \in X\right)$$

Combining this with (3.9) (for m in place of n) we indeed get

$$F\left(\frac{m}{n}\mathbf{x}\right) = \frac{m}{n}F(\mathbf{x}) \quad \left(m \in \mathbb{N},\ n \in \mathbb{N};\ \mathbf{x} \in X,\ \frac{m}{n}\mathbf{x} \in X\right),$$

that is, (3.8) for $t \in \mathbb{Q}_{++}$. About linearly homogeneous and, more generally, homogeneous functions see also Sects. 4.2, 4.3 and 6.12.

3.4.1 Exercises

1. Consider the functions
   (a) $f : [-1, 1] \longrightarrow \mathbb{R},\ x \longmapsto x^2$,
   (b) $f : [-1, 1] \longrightarrow \mathbb{R},\ x \longmapsto x^3$,
   (c) $f : [1, 3] \longrightarrow \mathbb{R},\ x \longmapsto 1/x$,
   (d) $f : [-3, -1] \longrightarrow \mathbb{R},\ x \longmapsto x^3 - x + 1$,
   (e) $f : [1, 4] \longrightarrow \mathbb{R},\ x \longmapsto x^3 - x + 1$,
   (f) $f : [-3, 0] \longrightarrow \mathbb{R},\ x \longmapsto x^3 - x + 1$.
   State which are unimodal, which have maximum inside domain, which have minimum inside domain, which are strictly increasing, and which are strictly decreasing.
2. Consider the functions
   (a) $F : \mathbb{R}^2 \longrightarrow \mathbb{R},\ (x_1, x_2) \longmapsto 2x_1 + 3x_2$,
   (b) $F : \mathbb{R}^2 \longrightarrow \mathbb{R},\ (x_1, x_2) \longmapsto 2x_1 - 3x_2$,
   (c) $F : \mathbb{R}^2_{++} \longrightarrow \mathbb{R},\ (x_1, x_2) \longmapsto x_1^2/x_2$,
   (d) $F : \mathbb{R}^2_{++} \longrightarrow \mathbb{R},\ (x_1, x_2) \longmapsto x_1x_2$,
   (e) $F : \mathbb{R}^2_{+} \longrightarrow \mathbb{R},\ (x_1, x_2) \longmapsto \sqrt{x_1x_2}$,
   (f) $F : \mathbb{R}^2_{++} \longrightarrow \mathbb{R},\ (x_1, x_2) \longmapsto 1/(x_1 + x_2)$.
   State which are linearly homogeneous, monotonic, strictly monotonic.
3. Draw the graph of a function $F : X \to \mathbb{R}$ where $X \subset \mathbb{R}^2$, such that
   (a) F is strictly increasing,
   (b) F is strictly decreasing,
   (c) F is strictly ray-increasing, but not strictly increasing,
   (d) F is positively linearly homogeneous, but not monotonic,
   (e) $x_1 \longmapsto F(x_1, x_2)$ and $x_2 \longmapsto F(x_1, x_2)$ are unimodal.
4. Draw the graph of a function $F : X \to \mathbb{R}$, where $x \subset \mathbb{R}^2$ is a two–dimensional closed interval, such that
   (a) F has both a sharp maximum and a sharp minimum inside X,
   (b) F has both a maximum and a minimum on the boundary of X.

5. Draw the graph of a function $F : X \to \mathbb{R}$, where $X \subset \mathbb{R}^2$, which has a maximum at exactly two points $\mathbf{x}^1 \in X$, $\mathbf{x}^2 \in X$ and a minimum at exactly three points $\mathbf{x}_1 \in X$, $\mathbf{x}_2 \in X$, $\mathbf{x}_3 \in X$. Are these maxima and minima sharp?

3.4.2 Answers

1. (a) unimodal, minimum inside domain,
   (b) strictly increasing, (c) strictly decreasing,
   (d) strictly increasing, (e) strictly increasing,
   (f) unimodal, maximum inside domain.
2. (a) linearly homogeneous, strictly monotonic (increasing),
   (b) linearly homogeneous, (c) linearly homogeneous,
   (d) strictly monotonic (increasing),
   (e) monotonic (increasing), linearly homogeneous,
   (f) strictly monotonic (decreasing).

3.5 Convex (Concave) Functions. Convex Sets

For many real valued functions of one or several real variables in economics, questions of convexity are of great importance. *A function* $f : S \to \mathbb{R}\,(S \subset \mathbb{R})$ *is convex from below on an interval* $I \subset S$ *if*

$$f(\lambda u + (1 - \lambda)v) \leq \lambda f(u) + (1 - \lambda)f(v)$$

for all $u \neq v$ *in* I *and for all* $\lambda \in]0, 1[$.

If here $\leq$ *is replaced by* $<$, $\geq$ or $>$ *then we have* the definitions of *functions strictly convex from below, convex from above* or *strictly convex from above* on I, respectively. As we see from Fig. 3.28, the graphs of functions strictly convex from below or above are kind of "arched" downwards or upwards, respectively (the arc between any two points is below or above the chord, respectively), while convex but not strictly convex ones may have straight stretches. As we just did, we *call a function convex on I if it is convex either from below on I or from above on I, strictly convex on I if it is either strictly convex from below on I or strictly convex from above on I.* This is one of the advantages of our way of using the word convex. Actually, in mathematics and economics functions convex (strictly convex) from below are often called simply *convex (strictly convex)* while those convex (strictly convex) from above are called *concave (strictly concave)*. We may use these expressions occasionally but there are two kinds of troubles with them: In some fields, for instance in engineering, these names are used exactly in the opposite sense (concave = convex from below, etc.) and in either case there is no common name covering both convex and concave.

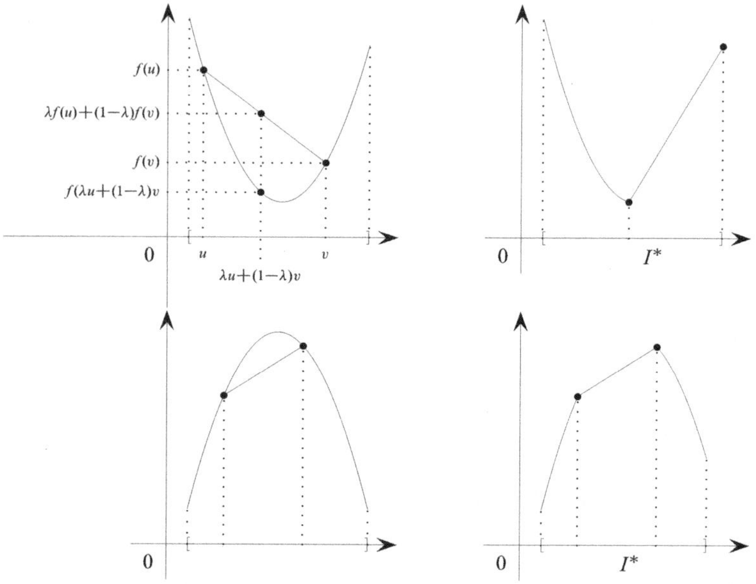

Fig. 3.28 Graphs of functions strictly convex and convex from below (first two graphs) and above (last two graphs) (strictly convex, convex, strictly concave, concave) on an interval in $\mathbb{R}$

Affine functions $x \mapsto ax + b$ (see also Sect. 4.1) whose graphs are straight lines (straight line segments) and only these *are convex both from below and from above* (see also the graphs over I^* in Fig. 3.28) but they are, of course, *not strictly convex.*

Again, *most functions* on most intervals *are not convex either from above or from below.* Those in Figs. 3.17 and 3.18 are strictly convex from below on $[0, t^*]$, $[0, x^*]$, respectively, strictly convex from above on the rest of their domain. The function whose graph is in Fig. 3.16 is strictly convex from above on $[-3, 0]$, strictly convex from below on $[0, 3]$. *The points* 0, t^*, x^*, u^* (see Figs. 3.16, 3.17, 3.18 and 3.19), *where a segment convex from one side meets a segment convex from the other side, are "points of inflection".*

For functions of several real variables, the role played above by intervals is taken over by convex sets. *A set* $X \subset \mathbb{R}^n$ *is convex if, with any two of its points* $\mathbf{u} \in X$, $\mathbf{v} \in X$, *the whole "straight line segment connecting* $\mathbf{u}$ *and* $\mathbf{v}$", that is (Fig. 3.29, see also Fig. 1.7 in Sect. 1.5) the set

$$\{\mathbf{x} = \lambda\mathbf{u} + (1 - \lambda)\mathbf{v} \mid \lambda \in [0, 1]\}$$

"*belongs to* X" (meaning here that it is a subset of X). For $n = 1$ the only convex sets are the (one-dimensional) intervals (including single points ("singletons") and the whole $\mathbb{R}$). The following *examples* show that there are many more kinds of convex sets in $\mathbb{R}^n$ for $n > 1$.

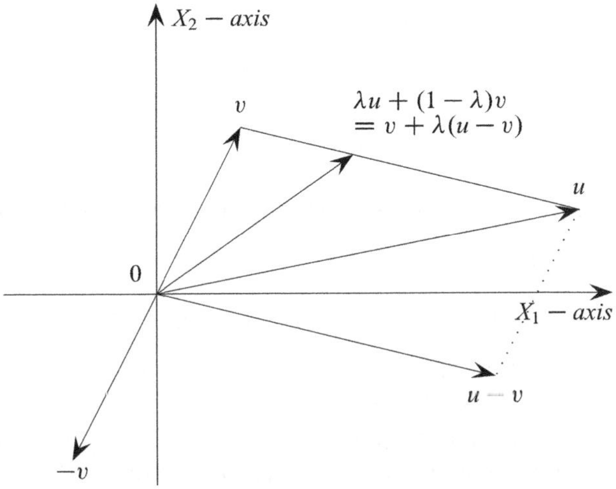

Fig. 3.29 The point $\lambda\mathbf{u} + (1-\lambda)\mathbf{v}$, where $\lambda \in [0, 1]$, lies on the straight line connecting **u** and **v**. As λ goes through [0,1], the whole segment is covered

1. *Open balls* in $\mathbb{R}^n$ (for $n = 2$: *interior of a circle*; compare to Sect. 6.11):

$$\{\mathbf{x} \in \mathbb{R}^n \mid ||\mathbf{x} - \mathbf{a}|| < r\}$$

(for the distance $||\mathbf{x} - \mathbf{a}||$ see Sects. 1.3 and 1.5), where $\mathbf{a} \in \mathbb{R}^n$ (centre of the ball), $r \in \mathbb{R}_{++}$ (radius) are constants.

2. *Closed balls* in $\mathbb{R}^n$ (for $n = 2$: *circle and its interior*):

$$\{\mathbf{x} \in \mathbb{R}^n \mid ||\mathbf{x} - \mathbf{a}|| \leq r\},$$

where $\mathbf{a} \in \mathbb{R}^n$, $r \in \mathbb{R}_{++}$ are again constants.

3. *Interiors of ellipses* in $\mathbb{R}^2$, *of ellipsoids* in $\mathbb{R}^3$ (with centres at **0**):

$$\left\{\mathbf{x} = (x_1, x_2) \in \mathbb{R}^2 \;\middle|\; \frac{x_1^2}{a_1^2} + \frac{x_2^2}{a_2^2} < 1\right\},$$

$$\left\{\mathbf{x} = (x_1, x_2, x_3) \in \mathbb{R}^3 \;\middle|\; \frac{x_1^2}{a_1^2} + \frac{x_2^2}{a_2^2} + \frac{x_3^2}{a_3^2} < 1\right\}$$

with constant a_1, a_2, a_3 in $\mathbb{R}_{++}$.

4. *The empty set* $\emptyset$: Note that the definition of a convex set X says that "for any pair $\mathbf{u} \in X$, $\mathbf{v} \in X$ we should have $\{\lambda\mathbf{u} + (1-\lambda)\mathbf{v} \mid \lambda \in [0, 1]\} \subset X$". But $X = \emptyset$ *has no element* so the statement in parentheses is trivially (vacuously) true.

5. *n-dimensional intervals* (compare Sect. 3.1):
   (i) *closed intervals* as $[\mathbf{a}, \mathbf{b}] := \{\mathbf{x} \in \mathbb{R}^n \mid \mathbf{a} \leq \mathbf{x} \leq \mathbf{b}\}$,
   (ii) *open intervals* as $]\mathbf{a}, \mathbf{b}[:= \{\mathbf{x} \in \mathbb{R}^n \mid \mathbf{a} < \mathbf{x} < \mathbf{b}\}$,
   (iii) *half open intervals* as $[\mathbf{a}, \mathbf{b}[:= \{\mathbf{x} \in \mathbb{R}^n \mid \mathbf{a} \leqq \mathbf{x} < \mathbf{b}\}$
   or $]\mathbf{a}, \mathbf{b}] := \{\mathbf{x} \in \mathbb{R}^n \mid \mathbf{a} < \mathbf{x} \leqq \mathbf{b}\}$,

(iv) *other finite intervals* as, for instance,

$$\{(x_1, x_2, \ldots, x_n) \in \mathbb{R}^n \mid a_1 \le x_1 \le b_1,\ a_j < x_j < b_j\ (j = 2, \ldots, n)\},$$

(v) *infinite intervals* as, for instance,

$$[\mathbf{a}, \infty[:= \{\mathbf{x} \in \mathbb{R}^n \mid \mathbf{x} \geqq \mathbf{a}\},\ ]\infty, \mathbf{b}[:= \{\mathbf{x} \in \mathbb{R}^n \mid \mathbf{x} < \mathbf{b}\},$$

and

$$\{(x_1, x_2, \ldots, x_n) \in \mathbb{R}^n \mid x_1 > a_1,\ x_2 \le b_2,\ a_\ell \le x_\ell < b_\ell\ (\ell = 3, \ldots, n)\},$$

where $\mathbf{a} = (a_1, \ldots, a_n) \in \mathbb{R}^n$, $\mathbf{b} = (b_1, \ldots, b_n) \in \mathbb{R}^n$ are constants, $a_k < b_k\ (k = 1, \ldots, n)$.

6. *The convex hull of p points* (vectors) $\mathbf{x}_1, \ldots, \mathbf{x}_p$ in $\mathbb{R}^n$ (for $n = 2$: *hull of polygons*, see Fig. 3.30; for $n = 1$: *closed intervals*):

$$\{\mathbf{x} = \lambda_1\mathbf{x}_1 + \ldots + \lambda_p\mathbf{x}_p \in \mathbb{R}^n \mid \lambda_j \in [0, 1], (j = 1, \ldots, p), \lambda_1 + \ldots + \lambda_p = 1\}.$$

In Sect. 1.4 we called $\lambda_1\mathbf{x}_1 + \ldots + \lambda_p\mathbf{x}_p$ linear combinations of $\mathbf{x}_1, \ldots, \mathbf{x}_p$; there $\lambda_1, \ldots, \lambda_p$ were arbitrary real numbers. If we restrict $\lambda_1, \ldots, \lambda_p$ so that $\lambda_j \in [0, 1]\,(j = 1, \ldots, p)$ and $\lambda_1 + \ldots + \lambda_p = 1$ we have the *convex linear combinations* of $\mathbf{x}_1, \ldots, \mathbf{x}_p$. So *the convex hull of the p points* (vectors) $\mathbf{x}_1, \mathbf{x}_2, \ldots, \mathbf{x}_p$ is *the set of all convex linear combinations of* $\mathbf{x}_1, \mathbf{x}_2, \ldots, \mathbf{x}_p$. For this set we prove that it is a convex set. Indeed, for any two of its points, $\mathbf{u} = \mu_1\mathbf{x}_1 + \ldots + \mu_p\mathbf{x}_p$ and $\mathbf{v} = \nu_1\mathbf{x}_1 + \ldots + \nu_p\mathbf{x}_p$ $(\mu_1 + \ldots + \mu_p = \nu_1 + \ldots + \nu_p = 1)$ we have

$$\begin{aligned}
&\lambda\mathbf{u} + (1 - \lambda)\mathbf{v} \\
&= \lambda(\mu_1\mathbf{x}_1 + \ldots + \mu_p\mathbf{x}_p) + (1 - \lambda)(\nu_1\mathbf{x}_1 + \ldots + \nu_p\mathbf{x}_p) \\
&= (\lambda\mu_1 + (1 - \lambda)\nu_1)\mathbf{x}_1 + \ldots + (\lambda\mu_p + (1 - \lambda)\nu_p)\mathbf{x}_p.
\end{aligned}$$

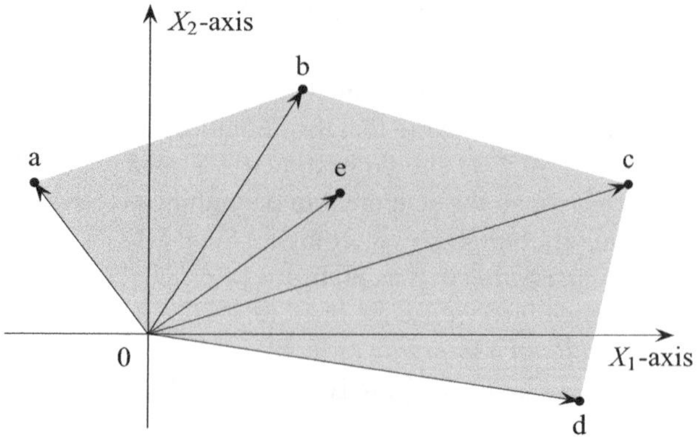

Fig. 3.30 Convex hull of the six points (vectors) **0, a, b, c, d, e** in the plane $\mathbb{R}^2$. The convex hull H is the shaded area. Notice that the polygon limiting H belongs to H, and that H is also the convex hull of **o, a, b, c, d**, that is, the point (vector) **e** which lies in the interior of H has no influence on the shape of H

But here the scalar coefficients $\lambda\mu_j + (1-\lambda)\nu_j$ $(j = 1,\ldots,p)$ are in $[0,1]$ (because $\lambda,\ \mu_1,\ldots,\mu_p,\nu_1,\ldots,\nu_p$ are there) and they add up to 1:

$$\begin{aligned}
&(\lambda\mu_1 + (1-\lambda)\nu_1) + \ldots + (\lambda\mu_p + (1-\lambda)\nu_p) \\
&= \lambda(\mu_1 + \ldots + \mu_p) + (1-\lambda)(\nu_1 + \ldots + \nu_p) \\
&= \lambda + (1-\lambda) = 1,
\end{aligned}$$

so $\lambda\mathbf{u} + (1-\lambda)\mathbf{v}$ belongs to the same set, as asserted. In Sect. 3.2 we defined rays as sets $\{\lambda\mathbf{x}_j \mid \lambda \in \mathbb{R}_{++}\}$. The following example is a generalisation of rays. It is the set of all linear combinations with *nonnegative* coefficients of p vectors $\mathbf{x}_1,\ldots,\mathbf{x}_p$, that is:

7. The *cone generated by the vectors* $\mathbf{x}_1,\ldots,\mathbf{x}_p$:

$$\{\mathbf{x} = \lambda_1\mathbf{x}_1 + \ldots + \lambda_p\mathbf{x}_p \mid \lambda \in \mathbb{R}_+ \ (j = 1,\ldots,p)\}$$

(Notice that here neither $\lambda_j \leq 1$ $(j = 1,\ldots,p)$ nor $\lambda_1 + \ldots + \lambda_p = 1$ is supposed. Prove that, nevertheless, this is a convex set).

Having defined convex sets in $\mathbb{R}^n$, we can now define convex functions of n real variables: *A function $F : S \to \mathbb{R}$ $(S \subset \mathbb{R}^n)$ is convex from below on the convex set $X \subset S$ if*

$$F(\lambda\mathbf{u} + (1-\lambda)\mathbf{v}) \leq \lambda F(\mathbf{u}) + (1-\lambda)F(\mathbf{v}) \tag{3.10}$$

for all $\mathbf{u} \neq \mathbf{v}$ in X and for all $\lambda \in]0,1[$.

(We see from the left hand side why we needed that the set X be convex.) *If here $\leq$ is replaced by $>$, $\geq$ or $>$, we get* the definitions of *functions strictly convex from below, convex from above, strictly convex from above on X, respectively.* Again, functions (strictly) convex from above are sometimes called (strictly) concave. For $n = 1$ these definitions, of course, reduce to those given above for functions of one real variable. Moreover, if we remember (Fig. 3.29) that the geometric meanings of the set $\{\lambda\mathbf{u} + (1-\lambda)\mathbf{v} \mid \lambda \in [0,1]\}$ is the straight line segment connecting the points $\mathbf{u}$ and $\mathbf{v}$, we see that (3.10) reduces the definition of convex functions of n variables to that of convex functions in a single variable λ (from above and similarly, from below, strictly or otherwise). For $n = 2$, for instance, this means that F is strictly convex from below on $X \subset \mathbb{R}^2$ if, and only if, on all "vertical cuts" of its graph, the arc between any two points is under the chord.

It is easy to show (do it!) that, *if on a convex set $X \subset \mathbb{R}^n$ the functions* $F_1 : S_1 \to \mathbb{R}$ and $F_2 : S_2 \to \mathbb{R}$ $(X \subset S_1,\ X \subset S_2)$ *are convex from below* (that is, they satisfy (3.10)) *then so is their linear combination with nonnegative coefficients* (compare Sect. 1.4 for a similar concept) $F = a_1F_1 + a_2F_2 : S_1 \cap S_2 \to \mathbb{R}$ (where a_1, a_2 are arbitrary nonnegative constants), that is, also F satisfies (3.10) on X. Notice that F is defined on $S_1 \cap S_2$ by $F(\mathbf{x}) := a_1F_1(\mathbf{x}) + a_2F_2(\mathbf{x})$. *Similar* *statements hold*, of course, *for functions strictly convex from below or convex or strictly convex from above* (as long as F_1, F_2 belong to the same class).

We can write (3.10) as

$$F\left(\frac{q_1\mathbf{u}_1 + q_2\mathbf{u}_2}{q_1 + q_2}\right) \leq \frac{q_1F(\mathbf{u}_1) + q_2F(\mathbf{u}_2)}{q_1 + q_2} \tag{3.11}$$

for all $\mathbf{u}_1 \in X, \mathbf{u}_2 \in X$ and for all $q_1 \in \mathbb{R}_{++}, q_2 \in \mathbb{R}_{++}$.

Through $\mathbf{u} := \mathbf{u}_1, \mathbf{v} := \mathbf{u}_2$, $\lambda := q_1/(q_1 + q_2)$ this is reduced to (3.10). This can be connected to the convex hull of polyhedra, which we introduced in Example 6 above. Indeed we get from (3.10) or from (3.11), which is called *two–term Jensen inequality, by induction* (see Appendix) *the p-term Jensen inequality*

$$F\left(\frac{q_1\mathbf{u}_1 + \ldots + q_p\mathbf{u}_p}{q_1 + \ldots + q_p}\right) \leq \frac{q_1F(\mathbf{u}_1) + q_2F(\mathbf{u}_2)}{q_1 + \ldots + q_p} \tag{3.12}$$

for all $\mathbf{u}_1 \in X$, $\mathbf{u}_p \in X$ and for all $q_1, \ldots, q_p \in \mathbb{R}_{++}$,

or, equivalently ($\lambda_j = q_j/(q_1 + \ldots + q_p)$, $j = 1, 2, \ldots, p$),

$$F(\lambda_1\mathbf{u}_1 + \ldots + \lambda_p\mathbf{u}_p) \leq \lambda_1F(\mathbf{u}_1) + \ldots + \lambda_pF(\mathbf{u}_p) \tag{3.13}$$

whenever $\mathbf{u}_1 \in X, \ldots, \mathbf{u}_p \in X$, $\lambda_1 \in]0, 1[, \ldots, \lambda_p \in]0, 1]$, $\lambda_1 + \ldots + \lambda_p = 1$.

Indeed, (3.12) holds for $p = 2$ (that was our initial inequality (3.11)) and, if it holds for 2 and for p then it holds also for $p + 1$:

$$\begin{aligned}
&F\left(\frac{q_1\mathbf{u}_1 + \ldots + q_p\mathbf{u}_p + q_{p+1}\mathbf{u}_{p+1}}{q_1 + \ldots + q_p + q_{p+1}}\right) \\
&= F\left(\frac{(q_1 + \ldots + q_p)\dfrac{q_1\mathbf{u}_1 + \ldots + q_p\mathbf{u}_p}{q_1 + \ldots + q_p} + q_{p+1}\mathbf{u}_{p+1}}{(q_1 + \ldots + q_p) + q_{p+1}}\right) \\
&\leq \frac{(q_1 + \ldots + q_p)F\left(\dfrac{q_1\mathbf{u}_1 + \ldots + q_p\mathbf{u}_p}{q_1 + \ldots + q_p}\right) + q_{p+1}F(\mathbf{u}_{p+1})}{(q_1 + \ldots + q_p) + q_{p+1}} \\
&\leq \frac{(q_1 + \ldots + q_p)\dfrac{q_1F(\mathbf{u}_1) + \ldots + q_pF(\mathbf{u}_p)}{q_1 + \ldots + q_p} + q_{p+1}F(\mathbf{u}_{p+1})}{q_1 + \ldots + q_p + q_{p+1}} \\
&= \frac{q_1F(\mathbf{u}_1) + \ldots + q_pF(\mathbf{u}_p) + q_{p+1}F(\mathbf{u}_{p+1})}{q_1 + \ldots + q_p + q_{p+1}},
\end{aligned}$$

as asserted (for the first $\leq$ we used (3.11), the second $\leq$ followed from (3.12)). So (3.11) indeed implies (3.12) (and (3.10) implies (3.13)). Of course, again *similar results hold for functions strictly convex from below and for functions convex from above, strictly or otherwise.*

The only functions of n variables convex both from above and from below are the affine functions

$$(x_1, \ldots, x_n) \mapsto a_1x_1 + \ldots + a_nx_n \quad (a_1, \ldots, a_n \text{ real constants})$$

(compare to Sects. 4.1 and 4.2). They are *not strictly convex*. For $n = 2$ their graphs are planes.

It is clear from the definition that, if F is (strictly) convex from below on X then $-F : \mathbf{x} \to -F(\mathbf{x})$ is (strictly) convex from above on X and the other way round.

Surfaces separating sets where a function is strictly convex from one side from sets where it is strictly convex from the other side are called *surfaces of inflection*, in particular for $n = 2$ *lines of inflection* (see, for instance, Fig. 3.31).

Convex functions play important roles in the social sciences. However, as we have seen just now and before, the functions whose graphs are in Figs. 3.14, 3.17, 3.18, 3.19 and 3.25 and which are important in economics, are not convex (neither from below nor from above) or are convex only on a part of their domains. But they are quasi-convex, as will be defined in the next section.

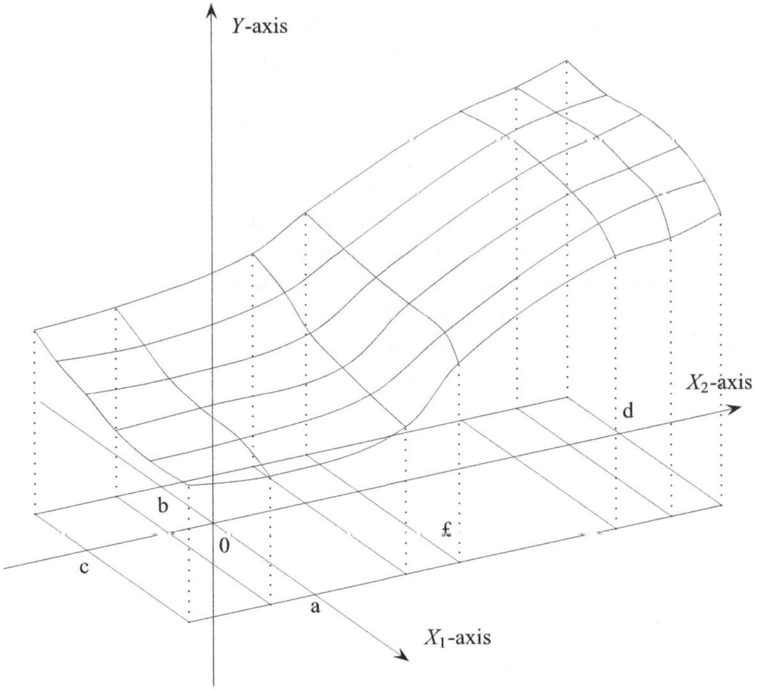

Fig. 3.31 Graph of a function $F : [(a, c), (b, d)] \to \mathbb{R}$. On the *left* from the line $\mathcal{L}$ it is strictly convex from below, on the *right* of $\mathcal{L}$ strictly convex from above, that is, $\mathcal{L}$ is a line of inflection

3.5.1 Exercises

1. Consider the functions

(a) $f : \mathbb{R} \to \mathbb{R},\ x \mapsto 3x - 1,$
(b) $f : \mathbb{R} \to \mathbb{R},\ x \mapsto x^2,$
(c) $f : \mathbb{R} \to \mathbb{R},\ x \mapsto x^3,$
(d) $f : \mathbb{R}_+ \to \mathbb{R},\ x \mapsto \sqrt{x},$
(e) $f : \mathbb{R} \to \mathbb{R},\ x \mapsto \sin x,$
(f) $f : [0, \pi] \to \mathbb{R},\ x \mapsto \sin x,$
(g) $f :]0, \pi[\to \mathbb{R},\ x \mapsto \cot x,$
(h) $f : [-3, 1] \to \mathbb{R},\ x \mapsto x^3 - 8x - 1,$
(i) $f : \mathbb{R}^2 \to \mathbb{R},\ (x_1, x_2) \mapsto x_1^2 + x_2^2,$
(j) $f : \mathbb{R}^2_+ \to \mathbb{R},\ (x_1, x_2) \mapsto \sqrt{x_1 x_2}.$

State which of them are convex from below, convex from above, strictly convex from below, strictly convex from above.

2. Which of the following sets are convex:
   (a) $[-3, 1[\cup [0, 1],$
   (b) $[-3, 1[\cap [2, 5],$
   (c) $\{\mathbf{x} \in \mathbb{R}^n \mid |\mathbf{x} - \mathbf{a}| < r,\ \mathbf{a} \in \mathbb{R}^n,\ r \in \mathbb{R}_{++}\}$
   $\cap\ \{\mathbf{x} \in \mathbb{R}^n \mid |\mathbf{x} - \mathbf{b}| < s,\ \mathbf{b} \in \mathbb{R}^n,\ s \in \mathbb{R}_{++}\},$
   (d) $[(0, 0), (2, 2)]\ \cup\ [(0, 0), (1, 3)],$
   (e) $[(0, 0), (2, 2)]\ \cup\ [(0, 0), (2, 3)].$
3. Show that, if the functions $F : X \to \mathbb{R}$ and $G : X \to \mathbb{R}$ are convex from below on the convex set $X \subset \mathbb{R}^n$ then so is their linear combination $aF + bG =: H$ on X ($a \in \mathbb{R}_+$, $b \in \mathbb{R}_+$ constants).
4. Draw the graph of a function $F : X \to \mathbb{R}$, where $X \subset \mathbb{R}^2$ is convex, such that F is
   (a) strictly convex from below and strictly increasing,
   (b) strictly convex from below and strictly decreasing,
   (c) strictly convex from above and strictly increasing,
   (d) strictly convex from above and strictly decreasing,
   (e) strictly convex from below and not monotonic,
   (f) strictly convex from above and not monotonic.
5. For which values of the real parameter a are the following functions convex from below:
   (a) $f : \mathbb{R} \to \mathbb{R}, x \mapsto 1 + x + ax^2,$
   (b) $F : \mathbb{R}^n \to \mathbb{R}^n, \mathbf{x} \mapsto ax_1^2 + x_2^2 + \ldots + x_n^2.$

3.5.2 Answers

1. (a) convex from below, convex from above,
   (b) convex from below, strictly convex from below,
   (d) convex from above, strictly convex from above,
   (f) convex from above, strictly convex from above,

(i) convex from below, strictly convex from below,
(j) convex from above.

2. The sets (a), (c), (e) are convex.
3. If $F : X \to \mathbb{R}$ and $G : X \to \mathbb{R}$ are convex from below on the convex set $X \subset \mathbb{R}^n$, that is, if for all $\mathbf{u} \neq \mathbf{v}$ in X and all $\lambda \in]0, 1[$ we have

$$F(\lambda\mathbf{u} + (1-\lambda)\mathbf{v}) \le \lambda F(\mathbf{u}) + (1-\lambda)F(\mathbf{v}),$$
$$G(\lambda\mathbf{u} + (1-\lambda)\mathbf{v}) \le \lambda G(\mathbf{u}) + (1-\lambda)G(\mathbf{v}),$$

then the linear combination $aF + bG =: H$ ($a \in \mathbb{R}_+$, $b \in \mathbb{R}_+$ constants) is also convex from below:

$$\begin{aligned} H(\lambda\mathbf{u} + (1-\lambda)\mathbf{v}) &= aF(\lambda\mathbf{u} + (1-\lambda)\mathbf{v}) + bG(\lambda\mathbf{u} + (1-\lambda)\mathbf{v}) \\ &\le a\lambda F(\mathbf{u}) + a(1-\lambda)F(\mathbf{v}) + b\lambda G(\mathbf{u}) + b(1-\lambda)G(\mathbf{v}) \\ &= \lambda(aF(\mathbf{u}) + bG(\mathbf{u})) + (1-\lambda)(aF(\mathbf{v}) + bG(\mathbf{v})) \\ &= \lambda H(\mathbf{u}) + (1-\lambda)H(\mathbf{v}). \end{aligned}$$

5. (a) $a \in \mathbb{R}_+$, (b) $a \in \mathbb{R}_+$.

3.6 Quasi-convex Functions

A consequence of the definition (3.10) of functions convex from below on a convex set $X \subset \mathbb{R}^n$ is

$$\begin{gathered} F(\lambda\mathbf{u} + (1-\lambda)\mathbf{v}) \le \max\{F(\mathbf{u}), F(\mathbf{v})\} \\ \text{for all } \mathbf{u} \neq \mathbf{v} \text{ in } X \text{ and for all } \lambda \in]0, 1[, \end{gathered} \tag{3.14}$$

where, in $\mathbb{R}$ (or in any other totally ordered set, compare Sect. 1.3) $\max\{a, b\}$ is the greater of a and b. (Similarly $\max\{a_1, \dots, a_n\}$ is the greatest among $a_1, a_2, \dots, a_n$; in general, $\max S$ is the greatest, $\min S$ the smallest element of $S \subset \mathbb{R}$—according to the order of magnitude of reals—if there exist a greatest and/or smallest element; in sets with infinitely many elements there may be but need not be a maximal and/or a minimal element, for instance $\mathbb{N}$ or $\{(n-1)/n \mid n \in \mathbb{N}\}$ have no greatest element—1 is *not* an element of the latter set—but they have a minimal element $\min \mathbb{N} = 1$, $\min\{(n-1)/n \mid n \in \mathbb{N}\} = 0$.) The *inequality (3.14) is the definition of functions quasi-convex from below on* X. In (3.14) (just as in (3.10)) it would not be necessary to exclude $\mathbf{u} = v$ and $\lambda = 0$ or $\lambda = 1$, in which cases obviously equality holds, but the definition in the above form is more convenient to make the following definitions short. *If, in (3.14) $\le$ is replaced by $<$ then F is strictly quasi-convex from below. A* *function* $F : S \to \mathbb{R}$ $(S \subset \mathbb{R}^n)$ *is quasi-convex from above on the convex set* $X \subset S$ *if*

$$F(\lambda \mathbf{u} + (1-\lambda)\mathbf{v}) \geq \min\{F(\mathbf{u}), F(\mathbf{v})\} \qquad \text{for all } \mathbf{u} \neq \mathbf{v} \text{ in } X \text{ and for all } \lambda \in]0,1[\,, \tag{3.15}$$

and *strictly quasi-convex from above on* X *if (3.15) holds with* $>$ *in place of* $\geq$. Here too, all functions (strictly) convex from above are also (strictly) quasi-convex from above. But the converse is not true. For instance, the functions represented in Figs. 3.17, 3.18, 3.19 and 3.14 are strictly quasi-convex from above on $[0, 10]$, $[0, \bar{x}]$, $[0, \bar{u}]$ and on $[(0, 0), (b, d)]$, respectively, but not convex from above. Again one can say instead of (strictly) quasi-convex from below or from above just *(strictly) quasi-convex* and *(strictly) quasi-concave*, respectively, but we prefer the above names because then we can use (strictly) quasi-convex as covering name for both. Here too, if F is (strictly) quasi-convex from above then $-F$ is (strictly) quasi-convex from below and the converse is also true.

Not only are, as we have seen, the convex functions particular cases of quasi-convex ones (from above or below, strictly or otherwise) but, as we can see by comparing (3.14) and (3.15) to (3.1) and (3.2), *every (strictly) increasing and every (strictly) decreasing function of one real variable on a real interval* I *is (strictly) quasi-convex both from above and from below.* (The unimodal functions on an interval $I \subset \mathbb{R}$, also defined in Sect. 3.2 as first increasing to a single maximum and decreasing thereafter or first decreasing to a single minimum and increasing thereafter are quasi-convex from above or below, respectively, see Figs. 3.21 and 3.22.) But *real-valued monotonic functions of more than one real variable are not necessarily quasi-convex* (say, from above) *anymore*. The reason for this is that the definition, say, of decreasing functions, (3.6), does not say anything about those pairs $\mathbf{x} \in \mathbb{R}^n$, $\mathbf{x}' \in \mathbb{R}^n$ which cannot be ordered in the sense of Sect. 1.5. For instance, the function $F : \mathbb{R}^2_{++} \to \mathbb{R}$ defined by $F(x_1, x_2) = 1 - x_1x_2$ is decreasing ($x_1 \mapsto 1 - x_1x_2$ and $x_2 \mapsto 1 - x_1x_2$ both decrease for all $x_1 > 0$, $x_2 > 0$) but, for instance for $\mathbf{u} = (\frac{3}{2}, \frac{1}{2})$, $\mathbf{v} = (\frac{1}{2}, \frac{3}{2})$, $\lambda = \frac{1}{2}$, we have

$$\begin{aligned} F(\lambda \mathbf{u} + (1-\lambda)\mathbf{v}) &= 1 - \left(\frac{1}{2}\frac{3}{2} + \frac{1}{2}\frac{1}{2}\right)\left(\frac{1}{2}\frac{1}{2} + \frac{1}{2}\frac{3}{2}\right) \\ &= 0 < \frac{1}{4} = \min\{\frac{1}{4}, \frac{1}{4}\} = \min\{F(\mathbf{u}), F(\mathbf{v})\}, \end{aligned}$$

so, by (3.15), this F is *not quasi-convex from above* (Fig. 3.32).

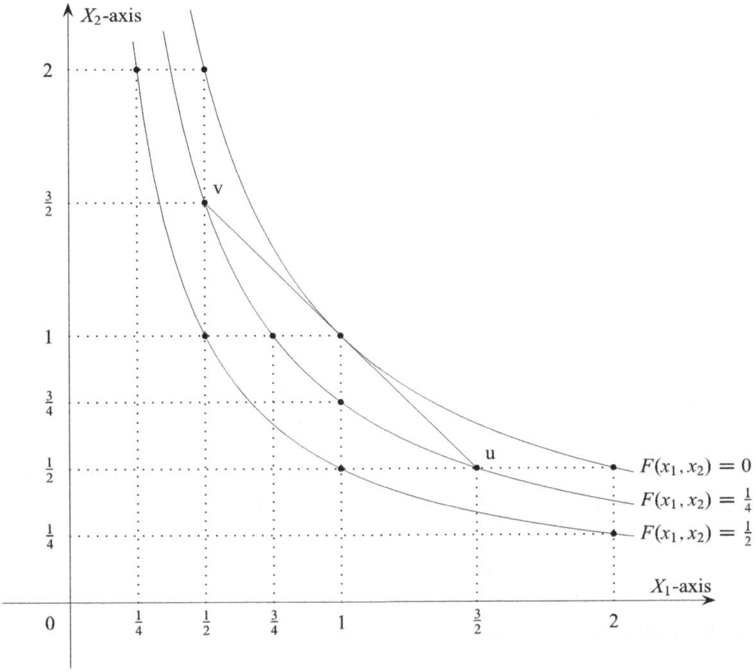

Fig. 3.32 Contour-line representation of the function $F : \mathbb{R}^2_{++} \rightarrow \mathbb{R}$ given by $F(x_1, x_2) = 1 - x_1x_2$ (see also Fig. 3.15). The restriction of F to the segment from $\mathbf{u} = (3/2, 1/2)$ to $\mathbf{v} = (1/2, 3/2)$ has its maximum 1/4 at $\mathbf{u}$ and $\mathbf{v}$ and its minimum 0 at (1,1). Hence, F is not quasi-convex from above

Quasi-convex functions relate to convex sets in more ways than one: *A function $F : S \rightarrow \mathbb{R}$ $(S \subset \mathbb{R}^n)$ is quasi-convex from above on the convex set $X \subset S$ if, and only if, the upper level sets*

$$L(t) = \{\mathbf{x} \in X \mid F(\mathbf{x}) \geq t\} \tag{3.16}$$

are convex sets for all $t \in \mathbb{R}$. (Notice the connection and difference between these upper level sets and the contour lines defined at the end of Sect. 3.3.) These are the sets of points in X for which the function value is not smaller than t (for $n = 2$). The projection to the (X_1, X_2)-plane of the function values above the "horizontal plane" $y = t$; see Fig. 3.33. (Production or utility functions are frequently assumed to have such a form).

We prove first *that the upper level sets (3.16) are convex sets for all functions quasi-convex from above on* the convex set X. Of course, for some $t \in \mathbb{R}$, the set $L(t)$ may be empty or consist of a single point (singleton) but, as we have seen in Sect. 3.5 (Example 4 and the first case of $\mathbf{S}$ with $\mathbf{a} = \mathbf{b}$), the empty set and the singleton are convex sets. Let now $L(t)$ have at least two points (elements), $\mathbf{u}$ and $\mathbf{v}$, that is, $F(\mathbf{u}) \geq t$, $F(\mathbf{v}) \geq t$. Since X is a convex set, $\lambda\mathbf{u} + (1 - \lambda)\mathbf{v} \in X$ for all

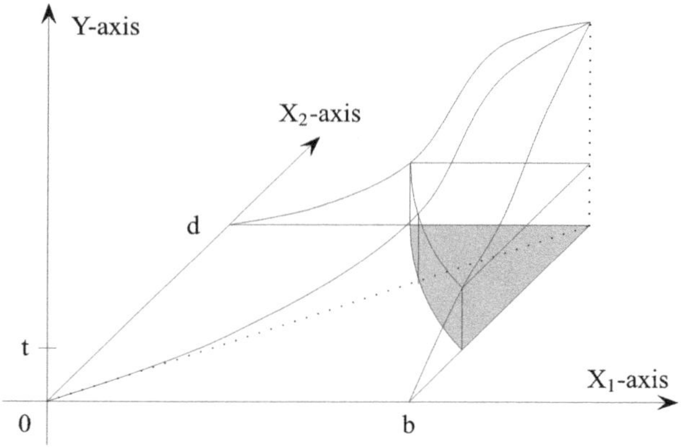

Fig. 3.33 Upper level set $L(t)$ (=shaded convex area) for the graph of a function $F : [(0,0), (b,d)] \to \mathbb{R}_+$ which is quasi-convex from above

$\lambda \in [0,1]$, and, since F is quasi-convex from above, by (3.15) we have (for $\lambda \in]0,1[$, but also for $\lambda = 1$ and $\lambda = 0$ because $F(\mathbf{u}) \geq t,\ F(\mathbf{v}) \geq t$):

$$F(\lambda\mathbf{u} + (1-\lambda)\mathbf{v}) \geq t.$$

So the whole straight line segment

$$\{\lambda\mathbf{u} + (1-\lambda)\mathbf{v} \mid \lambda \in [0,1]\}$$

belongs to $L(t)$ which thus is a convex set for each $t \in \mathbb{R}$, as asserted.

Now we show that the convexity of all upper level sets $L(t)$ implies the quasi-convexity of F from above. Let $\mathbf{u} \neq v$ be any two points of X (if X had just one point, then the statement would be trivially true) and define

$$\tau = \min\{F(\mathbf{u}, F(\mathbf{v}))\}, \quad \text{which implies} \quad F(\mathbf{u}) \geq \tau,\ F(\mathbf{v}) \geq \tau,$$

so that $\mathbf{u}$ and $\mathbf{v}$ are in $L(\tau)$. Since $L(\tau)$ is a convex set, also $\lambda\mathbf{u} + (1-\lambda)\mathbf{v}$ is in $L(\tau)$ for all $\lambda \in]0,1[$, that is,

$$F(\lambda\mathbf{u} + (1-\lambda)\mathbf{v}) \geq \tau = \min\{F(\mathbf{u}, F(\mathbf{v}))\} \quad \text{for all} \quad \lambda \in]0,1[,$$

and F is quasi-convex from above on X, as asserted.

One proves similarly that $F : S \to \mathbb{R}$ ($S \in \mathbb{R}^n$) is *quasi-convex from below on the convex set $X \subset S$ if, and only if, the lower level sets, defined by*

$$\Lambda(t) := \{\mathbf{x} \in X \mid F(\mathbf{x}) \leq t\},$$

are convex sets for all $t \in \mathbb{R}$. Notice that the intersection $\Lambda(t) \cap L(t)$ of a lower and an upper level set belonging to the same t is a contour line as defined in Sect. 3.3.

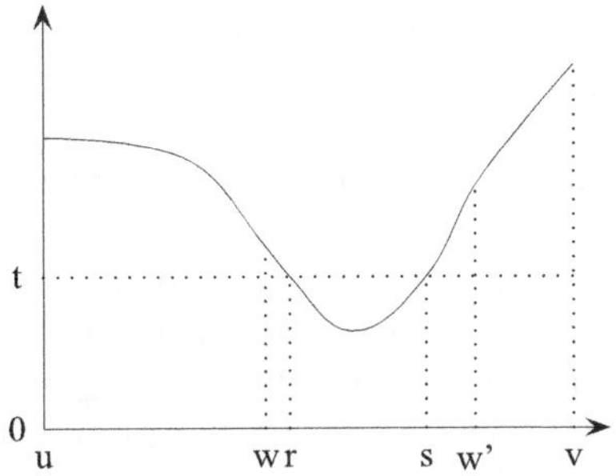

Fig. 3.34 Graph of a function, which is strictly quasi-convex from below on [u,v]; the set $\Lambda(t) = [r, s]$ is convex, the set $L(t) = [u, r] \cup [s, v]$ is not convex. The function is strictly convex from below on $[w, w']$, from above on $[u, w]$ and $[w', v]$

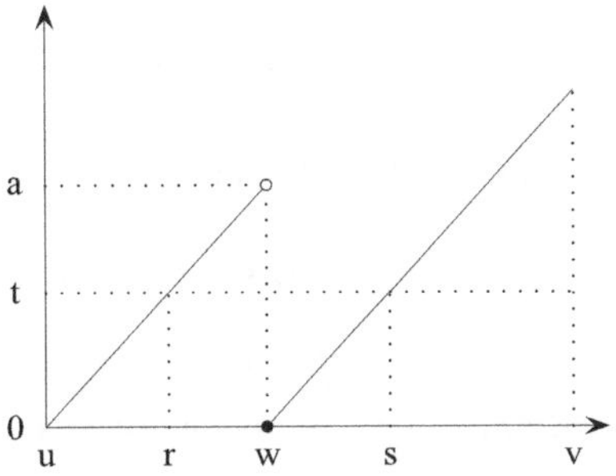

Fig. 3.35 The function with this graph (note: at w its value is 0, not a) is not quasi-convex either from above or below: $\Lambda(t) = [u, r] \cup [w, s]$ and $L(t) = [r, w] \cup [s, v]$ are not convex sets. But on $[u, w[$ and $[w, v]$ it is convex both from above and below

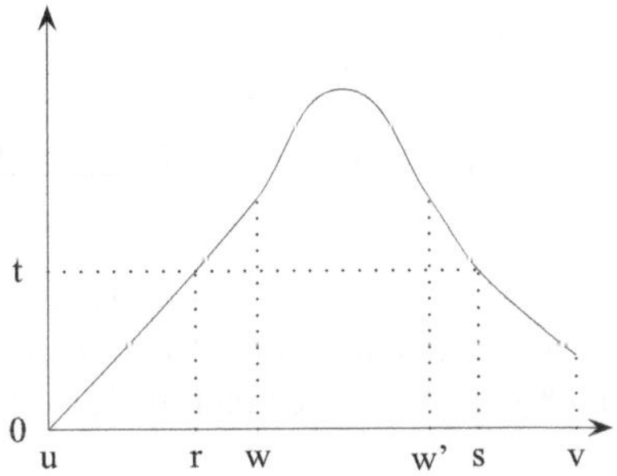

Fig. 3.36 Graph of a function, which is strictly quasi-convex from above on $[u, v]$, strictly convex from above on $[w, w']$, from below on $[u, w]$ and on $[w', v]$; $L(t) = [r, s]$ is convex, $\Lambda(t) = [u, r] \cup [s, v]$ is not convex

It may be worthwhile to draw the graphs of a few more quasi-convex (and not quasi-convex) functions from below and from above and some of their (lower, upper) levels sets, see Figs. 3.34, 3.35, and 3.36.

In production theory there has been a long dispute whether there exist linearly homogeneous functions (see Sect. 3.2) $F : \mathbb{R}_+^n \to \mathbb{R}_+$ such that all "cuts" (compare Sect. 3.2 and Figs. 3.14, 3.25)

$$x_k \mapsto F(x_1, \ldots, x_{k-1}, x_k, x_{k+1}, \ldots, x_n) \quad (k = 1, \ldots, n) \tag{3.17}$$

(called "*partial factor variations*" there) are

(i) strictly convex from below on $[0, \bar{x}_k]$, where $\bar{x}_k$ depends on the variables: $\bar{x}_k = f_k(x_1, \ldots, x_{k-1}, x_{k+1}, \ldots x_n)$, and
(ii) strictly convex from above on $[\bar{x}_k, \infty[$.

(Note that then the functions (3.17) would be strictly quasi-convex from above). It turned out that *such functions do not exist*. This is suggested already by the examples in Figs. 3.14 and 3.25 which represent the graphs of linearly homogeneous functions, but, of course, *examples*, no matter how many, *cannot give a general proof of this statement*.

We prove by contradiction (see Appendix), for the sake of simplicity for $n = 2$, that such functions cannot exist. Indeed, suppose that there would be a linearly homogeneous function $F : \mathbb{R}^2_+ \to \mathbb{R}_+$ such that at least $t \mapsto F(t, 1)$ has the properties (i) and (ii). For illustration see Fig. 3.25. Properties (i) and (ii) say that for all $\lambda \in]0, 1[$

$$F(\lambda s + (1 - \lambda)t, 1) < \lambda F(s, 1) + (1 - \lambda)F(t, 1) \tag{3.18}$$

for all $s \neq t$ in $[0, \bar{x}_1]$, and

$$F(\lambda s + (1 - \lambda)t, 1) > \lambda F(s, 1) + (1 - \lambda)F(t, 1) \tag{3.19}$$

for all $s \neq t$ in $[\bar{x}_1, \infty]$.

By the linear homogeneity we have

$$\begin{aligned} F(1, \mu u + (1 - \mu)v) &= (\mu u + (1 - \mu)v)F\left(\frac{1}{\mu u + (1 - \mu)v}, 1\right) \\ &= (\mu u + (1 - \mu)v)F\left(\frac{\mu u}{\mu u + (1 - \mu)v}\frac{1}{u} + \frac{(1 - \mu)v}{\mu u + (1 - \mu)v}\frac{1}{v}, 1\right) \end{aligned} \tag{3.20}$$

for all $u > 0$, $v > 0$, $\mu \in [0, 1]$. We choose now $u \neq v$ so that $1/u$, $1/v$ are both in $]0, \bar{x}_1]$ (that is, $u \neq v$ are both in $]1/\bar{x}_1, \infty[$). Then, by the right-hand side of (3.20) and by (3.18) (notice that the factors of $1/u$ and $1/v$ in (3.20) add up to 1),

$$\begin{aligned} &F(1, \mu u + (1 - \mu)v) \\ &\quad < (\mu u + (1 - \mu)v)\left(\frac{\mu u}{\mu u + (1 - \mu)v}F\left(\frac{1}{u}, 1\right) + \frac{(1 - \mu)v}{\mu u + (1 - \mu)v}F\left(\frac{1}{v}, 1\right)\right) \\ &\quad = \mu u F\left(\frac{1}{u}, 1\right) + (1 - \mu)vF\left(\frac{1}{v}, 1\right) = \mu F(1, \mu) + (1 - \mu)F(1, v) \end{aligned}$$

(the latter again by the linear homogeneity). This shows that $u \mapsto F(1, u)$ is *strictly convex from below on* $]1/\bar{x}_1, \infty[$ which contradicts (ii) and proves our statement.

One can get similarly a contradiction to (i) from (3.19). Figure 3.25 shows that the linear homogeneity of the graph drawn there

- with the convexity from below of $x_1 \mapsto F(x_1, 1)$ on $[0, \bar{x}_1]$ imply the convexity from below of $x_2 \mapsto F(1, x_2)$ on $[1/\bar{x}_1, \infty[$ (what we proved generally),

- with the convexity from above of $x_2 \mapsto F(1, x_2)$ on $[0, 1/\bar{x}_1]$ imply the convexity from above of $x_1 \mapsto F(x_1, 1)$ on $[\bar{x}_1, \infty[$.

Notice that in Fig. 3.25 the function $x_1 \mapsto F(x_1, 1)$ and $x_2 \mapsto F(1, x_2)$ have the points of inflection $(\bar{x}_1, 1)$ and $(1, 1/\bar{x}_1)$, respectively, and are strictly quasi-convex from above. Compare, in this connection, Fig. 3.14 showing a similar graph of a linearly homogeneous function whose "vertical cuts" $x_1 \mapsto F(x_1, x_2)$ and $x_2 \mapsto F(x_1, x_2)$ are also strictly quasi-convex from above (but not all are strictly increasing as those of the graph in Fig. 3.25). This does not necessarily mean that F itself is quasi-convex from above, for the same reason that monotonic functions were not necessarily quasi-convex: for functions $F : \mathbb{R}^n_+ \to \mathbb{R}_+$ to be quasi-convex from above not only the functions (3.17) but also all functions ("vertical cuts")

$$\lambda \mapsto F(\lambda \mathbf{u} + (1 - \lambda)\mathbf{v}) \quad \text{for all} \quad \mathbf{u} \in \mathbb{R}^n_+, \mathbf{v} \in \mathbb{R}^n_+$$

have to be quasi-convex from above. In Sect. 6.12 we will present linearly homogeneous functions $F : \mathbb{R}^n_+ \to \mathbb{R}_+$ whose "cuts" (3.17) all satisfy (i) and:

(iii) The cuts (3.17) are strictly increasing up to a maximum at a unique x_k^* (which depends on $x_1, \ldots, x_{k-1}, x_{k+1}, \ldots, x_n$), then strictly decreasing for all $x_k \geq x_k^*$ and, after a point of inflection, strictly convex from below.

As an example of a cut having the properties (i) and (ii) simultaneously, see in Fig. 3.14 the vertical cut $x_1 \mapsto F(x_1, 1)$ which assumes its maximum at x_1^*.

3.6.1 Exercises

1. (a) Which of the functions (a)–(h) in Exercise 3.1 are strictly quasi-convex from below or above?
   (b) Is $f : [-3, 3] \to \mathbb{R}\; x \mapsto x^3 - 8x - 1$ quasi-convex from below or from above?
2. Which of the following functions are quasi-convex from below:
   (a) $F : \mathbb{R}^2_{++} \to \mathbb{R}$, $(x_1, x_2) \mapsto 1 - x_1 x_2$,
   (b) $F : \mathbb{R}^2_{++} \to \mathbb{R}$, $(x_1, x_2) \mapsto x_1^2 - x_2$,
   (c) $F : \mathbb{R}^2_{++} \to \mathbb{R}$, $(x_1, x_2) \mapsto x_2^2 - x_1^2$.
3. Draw the graphs of functions
   (a) $f : [0, 10] \to \mathbb{R}$,
   (b) $F : [0, 10] \times [0, 10] \to \mathbb{R}$.
   Which are strictly quasi-convex from below, non-convex, and non-monotonic?
4. Draw the graphs of functions
   (a) $f : [0, 10] \to \mathbb{R}$,
   (b) $F : [0, 10] \times [0, 10] \to \mathbb{R}$.
   Which are strictly quasi-convex from above, non-convex, and non-monotonic?

5. Draw the graph of a function $F : [0, 10] \times [0, 10] \to \mathbb{R}_+$ which is linearly homogeneous, quasi-convex from above, and non-monotonic.

3.6.2 Answers

1. (a) Functions (c) and (g) (Exercises 3.3.1) are quasiconvex from below *and* from above, function (h) is quasiconvex from above.
   (b) No.
2. Functions (a) and (b) are quasiconvex from below, but not from above. Function (c) is quasiconvex from above, but not from below.

3.7 Functions in the "Statistical Theory" of Price Indices

We mentioned in Example 13 of Sect. 3.1 the most frequently used price index which was introduced (or the importance of which was recognised) by E. Laspeyres (1834–1913) in 1871. It compares the cost of the usually consumed quantities $q_1^0, \ldots, q_n^0$ of n "typical" goods and services (the basket of goods) at a base time with their cost at a comparison time (usually the present). If these quantities are united into a vector

$$\mathbf{q}^0 = (q_1^0, \ldots, q_n^0) \in \mathbb{R}^n_{++}$$

and so are their prices $p_1^0, \ldots, p_n^0$ and $p_1, \ldots, p_n$ at the base time and at the comparison time, respectively:

$$\mathbf{p}^0 = (p_1^0, \ldots, p_n^0) \in \mathbb{R}^n_{++}, \qquad \mathbf{p} = (p_1, \ldots, p_n) \in \mathbb{R}^n_{++}$$

then the costs are

$$q_1^0 p_1^0 + \ldots + q_n^0 p_n^0 = \mathbf{q}^0 \cdot \mathbf{p}^0 \in \mathbb{R}_+ \quad \text{and} \quad q_1^0 p_1 + \ldots + q_n^0 p_n = \mathbf{q}^0 \cdot \mathbf{p} \in \mathbb{R}_+$$

respectively, and *Laspeyres's price index value* is defined by

$$\frac{\mathbf{q}^0 \cdot \mathbf{p}}{\mathbf{q}^0 \cdot \mathbf{p}^0} = \frac{q_1^0 p_1 + \ldots q_n^0 p_n}{q_1^0 p_1^0 + \ldots q_n^0 p_n^0} \in \mathbb{R}_{++}.$$

This is the function value $L(\mathbf{q^0}, \mathbf{p^0}, \mathbf{p}) = q^0 \cdot \mathbf{p}/q^{0.0}$ of *Laspeyres's price index* (function) $L : \mathbb{R}^n_{++} \times \mathbb{R}^n_{++} \times \mathbb{R}^n_{++} \to \mathbb{R}_{++}$.

Of course, one may ask whether the quantities of "typical" goods remain unchanged between the base and the comparison time; we know that they do not: circumstances and tastes change (in particular prices influence the quantities consumed): Actually even some "typical" goods, like top-hats and services like horse-shoeing may disappear or become atypical and new ones like compact disc players and computing emerge; we disregard this, supposing for convenience that only a relatively short time passed between the base and the comparison points in time. However, it seems just as justified to take the quantities $q_1, \ldots, q_n$ at comparison time, united again into a vector

$$\mathbf{q} = (q_1, \ldots, q_n) \in \mathbb{R}^n_{++}$$

and take

$$\frac{\mathbf{q} \cdot \mathbf{p}}{\mathbf{q} \cdot \mathbf{p}^0} = \frac{q_1 p_1 + \ldots + q_n p_n}{q_1 p_1^0 + \ldots + q_n p_n^0} \in \mathbb{R}_{++}$$

as price index value. That is what H. Paasche (1851–1925) has done in 1874, so this is called the (value of the) *Paasche price index* (function). Clearly, it also has a blemish, opposite to that of the Laspeyres index: here the quantities at the base time are ignored. As we will see the two can be combined in more than one "reasonable" way.

Before we give, as further examples, such "reasonable" price indices, we say what we mean by "reasonable". We can do this in common sense mathematical terms by stating those requirements (assumptions) which seem natural for a reasonable price index to fulfill.

Here are some of these assumptions ("axioms") for the *price index* function

$$P : \mathbb{R}^n_{++} \times \mathbb{R}^n_{++} \times \mathbb{R}^n_{++} \times \mathbb{R}^n_{++} \to \mathbb{R}_{++}$$

with function value

$$P(\mathbf{q^0}, \mathbf{p^0}, \mathbf{q}, \mathbf{p}) \in \mathbb{R}_{++} \quad (\mathbf{q^0} \in \mathbb{R}^n_{++}, \mathbf{p^0} \in \mathbb{R}^n_{++}, \mathbf{q} \in \mathbb{R}^n_{++}, \mathbf{p} \in \mathbb{R}^n_{++}) :$$

A1. *Monotonicity. The price index function is strictly increasing in* $\mathbf{p}$, *strictly decreasing in* $\mathbf{p}^0$ (compare to Sects. 1.5 and 3.4):

$$P(\mathbf{q}^0, \mathbf{p}^0, \mathbf{q}, \mathbf{p}) > P(\mathbf{q}^0, \mathbf{p}^0, \mathbf{q}, \tilde{\mathbf{p}})$$

for all $\mathbf{q}^0, \mathbf{p}^0, \mathbf{q}, \mathbf{p}, \tilde{\mathbf{p}}$ in $\mathbb{R}^n_{++}$ with $\mathbf{p} \geq \tilde{\mathbf{p}}$, and

$$P(\mathbf{q}^0, \mathbf{p}^0, \mathbf{q}, \mathbf{p}) < P(\mathbf{q}^0, \tilde{\mathbf{p}}^0, \mathbf{q}, \mathbf{p})$$

for all $\mathbf{q}^0, \mathbf{p}^0, \tilde{\mathbf{p}}^0, \mathbf{q}, \mathbf{p}$ in $\mathbb{R}^n_{++}$ with $\mathbf{p}^0 \geq \tilde{\mathbf{p}}^0$.

A2. *Proportionality. If all prices change* (usually increase, unfortunately) λ–fold between the base time and the comparison time, then the (value of the) price index equals just this λ, whatever $\mathbf{q^0}, \mathbf{p^0}, \mathbf{q}$ are:

$$P(\mathbf{q^0}, \mathbf{p^0}, \mathbf{q}, -\mathbf{p^0}) = \lambda \text{ for all } \lambda \in \mathbb{R}_{++}, \ \mathbf{q^0} \in \mathbb{R}_{++}, \mathbf{p^0} \in \mathbb{R}_{++}, \mathbf{q} \in \mathbb{R}_{++},$$

A3. *Price extension invariance* (often called "price dimensionality axiom"). *If the prices both at the base time and at the comparison time change* λ*-fold then the value of the price index (function) remains unchanged*:

$$P(\mathbf{q}^0, -\mathbf{p}^0, \mathbf{q}, -\mathbf{p}) = P(\mathbf{q}^0, \mathbf{p}^0, \mathbf{q}, \mathbf{p})$$
$$\text{for all } \lambda \in \mathbb{R}_{++}, \mathbf{q}^0 \in \mathbb{R}_{++}, \mathbf{p}^0 \in \mathbb{R}_{++}, \mathbf{q} \in \mathbb{R}_{++}.$$

A4. *Price-quantity compensation* (often called "commensurability axiom"). *If the prices* p_k^0, p_k *change* λ_k*-fold, but* also *the quantities* q_k^0, q_k *change* $(1/\lambda_k)$*-fold* $(k = 1, \dots, n)$ *then* (the costs and therefore) *the (value of the) price index (function) remains unchanged*:

$$P(q_1^0/\lambda_1, \dots, q_n^0/\lambda_n, \lambda_1 p_1^0, \dots, \lambda_n p_n^0, q_1/\lambda_1, \dots, q_n/\lambda_n, \lambda_1 p_1, \dots, \lambda_n p_n)$$
$$= P(q_1^0, \dots, q_n^0, p_1^0, \dots, p_n^0, q_1, \dots, q_n, p_1, \dots, p_n)$$

for all positive $\lambda_k, q_k^0, p_k^0, q_k, p_k$ $(k = 1, \dots, n)$.

It is easy to check that both the Laspeyres and the Paasche indices satisfy these requirements, but so do several others, for instance the *Marshall–Edgeworth index* given by

$$P(\mathbf{q}^0, \mathbf{p}^0, \mathbf{q}, \mathbf{p}) = \frac{(\mathbf{q}^0 + \mathbf{q}) \cdot \mathbf{p}}{(\mathbf{q}^0 + \mathbf{q}) \cdot \mathbf{p}^0}$$

and *Fischer's ideal index* given by

$$P(\mathbf{q}^0, \mathbf{p}^0, \mathbf{q}, \mathbf{p}) = \left(\frac{\mathbf{q}^0 \cdot \mathbf{p}}{\mathbf{q}^0 \cdot \mathbf{p}^0} \frac{\mathbf{q} \cdot \mathbf{p}}{\mathbf{q} \cdot \mathbf{p}^0} \right)^{1/2},$$

the "*geometric mean value*" of the values of the Laspeyres and the Paasche index.

The above requirements (assumptions) were, however, chosen so that other reasonable requirements follow, for instance we prove now that **A1** and **A2** together imply

$$\min\{p_1/p_1^0, \dots, p_n/p_n^0\} \le P(\mathbf{q}^0, \mathbf{p}^0, \mathbf{q}, \mathbf{p}) \le \max\{p_1/p_1^0, \dots, p_n/p_n^0\} \tag{3.21}$$

($\min\{x_1, \dots, x_n\}$ is the smallest, $\max\{x_1, \dots, x_n\}$ the greatest among the real numbers $x_1, \dots, x_n$). This means that the *value of the price index is between the smallest and the greatest of the individual price quotients*, a very reasonable requirement indeed.

In order to show that (3.21) follows from **A1** and **A2** we write for short

$$\mu = \min\{p_1/p_1^0, \ldots, p_n/p_n^0\}, \qquad M = \max\{p_1/p_1^0, \ldots, p_n/p_n^0\}$$

and note that p_k^0 multiplied by the smallest (resp. largest) of $p_1/p_1^0, \ldots, p_n/p_n^0$ cannot be larger (resp. smaller) than p_k $(k = 1, \ldots, n)$:

$$\mu\mathbf{p^0} = \mu(p_1^0, \ldots, p_n^0) \leqq \mathbf{p} = (p_1, \ldots, p_n) \leq M(p_1^0, \ldots, p_n^0) = M\mathbf{p^0}. \qquad (3.22)$$

We now apply each of the proportionality assumption **A2** (with μ and M in place of λ) and the monotonicity **A1** twice:

$$\mu = P(\mathbf{q}^0, \mathbf{p}^0, \mathbf{q}, \mu\mathbf{p}^0) \leq P(\mathbf{q}^0, \mathbf{p}^0, \mathbf{q}, \mathbf{p}) \leq P(\mathbf{q}^0, \mathbf{p}^0, \mathbf{q}, M\mathbf{p}^0) = M,$$

($\leq$ not $<$ because in (3.22) we had $\leqq$, not $\leq$; for the definitions see Sect. 1.5), which is exactly (3.21).

In our opinion, every reasonable price index should satisfy (3.21), so any set of assumptions from which (3.21) does not follow, is not complete. Price indices which satisfy **A1**, **A2** (whence (3.21)), **A3** and **A4** are called *statistical price indices* to distinguish them from "*economic price indices*". The latter take into consideration the change in demand caused by the change of prices, that is, it is presumed that people change their demands for goods and services so that the *utility* of those which they can afford be maximal under the new prices.

3.7.1 Exercises

1. Show that, if the values of Laspeyres' and Paasche's index are different, then the value of Fischer's ideal index is smaller than the arithmetic mean of these two values.
2. Show that Laspeyres' index is *not* monotonic as a function of $\mathbf{q}^0$.
3. Show that Fischer's ideal index is linearly homogeneous as a function of $\mathbf{p}$.
4. Show that the indices of Laspeyres, Paasche, Marshall–Edgeworth and Fischer satisfy the requirements **A1**, **A2**, **A3** and **A4**.
5. Show that the so-called Walsh index defined by

$$P(\mathbf{q}^0, \mathbf{p}^0, \mathbf{q}, \mathbf{p}) = \frac{\sqrt{q_1^0 q_1}p_1 + \ldots + \sqrt{q_n^0 q_n}p_n}{\sqrt{q_1^0 q_1}p_1^0 + \ldots + \sqrt{q_n^0 q_n}p_n}$$

satisfies the requirements **A1**, **A2**, **A3** and **A4**.

3.7.2 Answers

1. Let us denote the values of Laspeyres' and Paasche's index by a and b, respectively. Then the value of Fisher's ideal index is $\sqrt{ab}$. We have to show that $\sqrt{ab} < (a+b)/2$ if $a \neq b$. Let $a \neq b$. Then $0 < (a-b)^2$, that is $4ab < (a-b)^2 + 4ab = (a+b)^2$, whence $2\sqrt{ab} < a+b$. □
2. Let $n = 2$, $\mathbf{p} = (2, 3)$, $\mathbf{p}^0 = (1, 4)$. Then the value of Laspeyres' index is $(2q_1^0 + 3q_2^0)/(q_1^0 + 4q_2^0)$, and this is strictly increasing with q_1^0 and strictly decreasing with q_2^0.
3.

$$P(\mathbf{q}^0, \mathbf{p}^0, \mathbf{q}, \lambda\mathbf{p}) = \sqrt{\frac{\mathbf{q}^0 \cdot (\lambda\mathbf{p})}{\mathbf{q}^0 \cdot \mathbf{p}^0} \frac{\mathbf{q} \cdot (\lambda\mathbf{p})}{\mathbf{q} \cdot \mathbf{p}^0}} = \sqrt{\lambda^2 \frac{\mathbf{q}^0 \cdot \mathbf{p}}{\mathbf{q}^0 \cdot \mathbf{p}^0} \frac{\mathbf{q} \cdot \mathbf{p}}{\mathbf{q} \cdot \mathbf{p}^0}}$$

$$= \lambda \sqrt{\frac{\mathbf{q}^0 \cdot \mathbf{p}}{\mathbf{q}^0 \cdot \mathbf{p}^0} \frac{\mathbf{q} \cdot \mathbf{p}}{\mathbf{q} \cdot \mathbf{p}^0}} = \lambda P(\mathbf{q}^0, \mathbf{p}^0, \mathbf{q}, \mathbf{p}).$$

Affine and Linear Functions and Transformations (Matrices), Linear Economic Models, Systems of Linear Equations and Inequalities

4

Happiness is thinking everything is linear.

ED ADAMS
ADAMS STATE COLLEGE PROFESSOR

4.1 Introduction

Consider the following question. A production plant (factory) produces with maximal operating performance 56 units in 7 hours. How much does it produce in 3 hours? What comes first to mind is: 8 units in 1 hour, so 24 unit in 3 hours and, by extension,

$$y = 8t \tag{4.1}$$

units in t hours. Such relations or mappings (of t into y) or functions (y as function of t; compare Chap. 3) are called *linear*. But we did not say that the relation between length of time and units produced during that time has to be linear (it probably is not during very short or very long time intervals). So the above question may have different answers, depending on the circumstances.

But, in absence of other information, we often do suppose that such a relationship is linear or, at least, can be *approximated* by a linear function. In many economic situations, for instance in the following, the assumption of linearity is more justified than in others.

A supermarket chain is willing to buy for an extended time period two kinds of detergent from a factory, say x_1 weight units per week of the first detergent and x_2 of the second, but not more than 100 weight units per week altogether:

$$x_1 + x_2 \leq 100. \tag{4.2}$$

W. Eichhorn, W. Gleißner, *Mathematics and Methodology for Economics*, Springer Texts in Business and Economics, DOI 10.1007/978-3-319-23353-6_4

The factory initially charged \$6 per weight unit on the first, \$9 on the second kind of detergent and this contributes 60 and 90 cents per weight unit, respectively, to its profit. The supermarket chain makes it clear that it does not want to spend more than \$720 a week for detergents, that is,

$$6x_1 + 9x_2 \leq 720 \tag{4.3}$$

which means $60x_1 + 90x_2 \leq 7200$ cents profit contribution. One could aim at maximal quantity and profit for x_1, x_2 which then satisfy:

$$x_1 + x_2 = 100 \quad \text{and} \quad 6x_1 + 9x_2 = 720.$$

Such (and more general) *systems of linear equations* will be solved in Sects. 4.6 and 4.7. *The solution is* $x_1 = 60, x_2 = 40$.

The above involves both linear equation and inequalities. Its inequality aspect leads to linear optimisation, which will be object of Sects. 5.1 and 5.2 in the following Chap. 5. That will further expand the argument above. Section 4.5 in the present Chap. 4 also deals with linear inequalities in the context of a couple of important economic models.

The word *linear* is used often and in several contexts, both in mathematics and in economics (and also in other sciences). We devote Chap. 4 to linear and to the somewhat more general *affine* functions.

Actually, we have encountered *linear* objects in this book before: *linear combinations of vectors*, their *linear dependence and independence* (Sect. 1.5), *linearly homogeneous and linear technologies and production models, linear optimisation* problems (Sect. 2.3) and general *linearly homogeneous functions* (Sect. 3.3).

As we will see in Sect. 4.3 (for $m = n = 1$ in Sect. 4.2), vector-vector functions (mappings) $f : \mathbb{R}^n \to \mathbb{R}^m$, which are linearly homogeneous, that is,

$$\mathbf{f}(\lambda\mathbf{x}) = \lambda f(\mathbf{x}) \tag{4.4}$$

(for all $\mathbf{x} \in \mathbb{R}^n$, $\lambda \in \mathbb{R}$) and which are also additive, that is,

$$\mathbf{f}(\mathbf{x} + \mathbf{y}) = \mathbf{f}(\mathbf{x}) + \mathbf{f}(\mathbf{y}) \tag{4.5}$$

(for all $\mathbf{x}, \mathbf{y} \in \mathbb{R}^n$), are called linear. In this case it turns out (Sect. 4.3) that there exist $m \cdot n$ real constants $a_{11}, \ldots, a_{1n}, \ldots, a_{m1}, \ldots, a_{mn}$ (forming a *matrix*, see Sects. 4.3 and 4.4), such that

$$f_j(x_1, \ldots, x_n) = a_{j1}x_1 + \ldots + a_{jn}x_n \quad (j = 1, \ldots, m) \tag{4.6}$$

for all $x_k \in \mathbb{R}$ $(k = 1, \ldots, n)$, where $f_1, \ldots, f_m$ and $x_1, \ldots, x_n$ are the components of $\mathbf{f}$ and $\mathbf{x}$, respectively:

$$\mathbf{f}(\mathbf{x}) = \mathbf{f}(x_1, \ldots, x_n) = (f_1(x_1, \ldots, x_n), \ldots, f_m(x_1, \ldots, x_n)). \tag{4.7}$$

However, the linear homogeneity (4.4), additivity (4.5) and linearity make sense also in more general spaces (as long as addition and multiplication by scalar are defined), where (4.6) does not necessarily follow.

Somewhat more general are the *affine functions*, which are defined by $\mathbf{g}(\mathbf{x}) = \mathbf{f}(\mathbf{x}) + \mathbf{b}$, *where* $\mathbf{b}$ *is an arbitrary constant and* $\mathbf{f}$ *an (arbitrary) linear function*. So, in the above situation

$$\mathbf{g}(\mathbf{x}) = \mathbf{g}(x_1, \ldots, x_n) =$$
$$= (a_{11}x_1 + \ldots + a_{1n}x_n + b_1, \ldots, a_{m1}x_1 + \ldots + a_{mn}x_n + b_n).$$

Linear and affine functions will be applied in Sect. 4.5, as mentioned, to *linear (Leontief, von Neumann) models* and, in Sect. 4.8, to *aggregation* in economics, respectively. Later (Sects. 6.9 and 6.10) they will serve to explain *differentials*.

4.2 Proportionality, Linear and Affine Functions. Additivity, Linear Homogeneity, Linearity

If there is a *proportionality* between the amount $x \in \mathbb{R}_+$ of the output of a production process and the amount $y \in \mathbb{R}_+$ of the input, that is,

$$y = ax \qquad (x \in \mathbb{R}_+), \tag{4.8}$$

where a is a positive constant, the *production coefficient*, then y is a linear function of x. We speak of a *linear function* (in older terminology "homogeneous linear function") also if the above equation holds between other quantities, on intervals other than $\mathbb{R}_+$ (also multidimensional domains, see the next section) and with the constant a not necessarily in $\mathbb{R}_{++}$. The *graph* of the function given by

$$y = ax \qquad (x \in \mathbb{R}) \tag{4.9}$$

is clearly (Fig. 4.1) a straight line through the origin of the coordinate system. If the domain is a subset of $\mathbb{R}$ then the graph consists of the part(s) of the straight line above (or below) that subset. The coefficient a is the *slope* of the line.

Also the graph of the *affine function* (in older terminology "linear function"), given by

$$y = ax + b$$

for x in $\mathbb{R}$ (or in parts thereof) is a straight line (or part thereof) with slope a (Fig. 4.2) but not through the origin if $b \neq 0$.

As substitution of $f(x) = ax$ immediately shows, *the linear functions are additive*

$$f(x_1 + x_2) = f(x_1) + f(x_2)$$

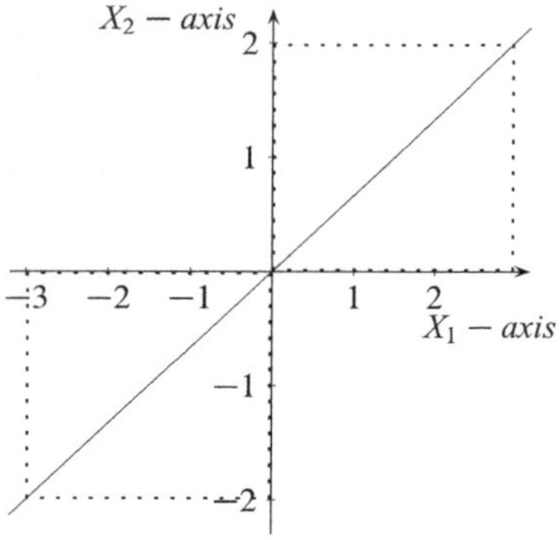

Fig. 4.1 (Part of the) graph of the linear function with slope $a = \frac{2}{3}$

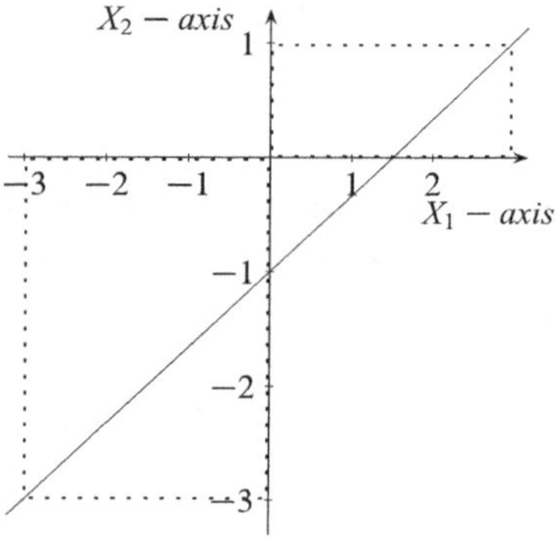

Fig. 4.2 (Part of the) graph of the affine function described by $y = \frac{2}{3}x - 1\,(a = \frac{2}{3},\, b = -1)$

for all $x_1 \in \mathbb{R}_+$, $x_2 \in \mathbb{R}_+$ in case (4.8) and for all $x_1 \in \mathbb{R}$, $x_2 \in \mathbb{R}$ in case (4.9). An economic interpretation of this equation is the following. The input quantity necessary to produce the sum $x_1 + x_2$ of the output quantities x_1 and x_2 is the sum of the input quantities y_1 and y_2 necessary to produce the output quantities x_1 and x_2, respectively. The linear functions are also *linearly homogeneous*

$$f(\lambda x) = \lambda f(x)$$

for all $x \in \mathbb{R}_+$, $\lambda \in \mathbb{R}_+$ in case (4.8) and for all $x \in \mathbb{R}$, $\lambda \in \mathbb{R}$ or all $x \in \mathbb{R}$, $\lambda \in \mathbb{R}_{++}$ (*positive linear homogeneity*) in case (4.9). Additivity and linear homogeneity can be condensed into the *linearity* equation

$$f(\lambda_1 x_1 + \lambda_2 x_2) = \lambda_1 f(x_1) + \lambda_2 f(x_2) \tag{4.10}$$

with $x_1 \in \mathbb{R}$, $x_2 \in \mathbb{R}$ or $x_1 \in \mathbb{R}_+$, $x_2 \in \mathbb{R}_+$ and $\lambda_1 \in \mathbb{R}$, $\lambda_2 \in \mathbb{R}$ or $\lambda_1 \in \mathbb{R}_+$, $\lambda_2 \in \mathbb{R}_+$ or $\lambda_1 \in \mathbb{R}_{++}$, $\lambda_2 \in \mathbb{R}_{++}$.

The question is natural whether the additivity and/or linear homogeneity *characterise* the linear functions or are there other functions with these properties.

Concerning linear homogeneity the answer is pretty easy. Suppose it first in the form

$$f(\lambda x) = \lambda f(x) \quad \text{for all} \quad \lambda > 0, x > 0.$$

Then, putting here $x = 1$ and calling a the constant $f(1)$, we already have

$$f(\lambda) = a\lambda \quad \text{for all} \quad \lambda > 0, \quad \text{that is,} \quad f(x) = ax \quad \text{for all} \quad x \in \mathbb{R}_{++},$$

so the *linear functions are the only linearly homogeneous functions on* $\mathbb{R}_{++}$. Obviously *the result is the same for*

$$f(\lambda x) = \lambda f(x) \quad \text{for all} \quad \lambda \in \mathbb{R}, x \in \mathbb{R} \quad \text{or for all} \quad \lambda \in \mathbb{R}_+, x \in \mathbb{R}_+$$

on $\mathbb{R}$ *or* $\mathbb{R}_+$, respectively. But for *positive linear homogeneity*

$$f(\lambda x) = \lambda f(x) \quad (\lambda \in \mathbb{R}_{++}, x \in \mathbb{R}), \tag{4.11}$$

while we still get $x = 1, f(1) = a$

$$f(\lambda) = a\lambda \quad \text{for} \quad \lambda \in \mathbb{R}_{++}, \quad \text{that is,} \quad f(x) = ax \quad \text{for} \quad x > 0,$$

this does not follow anymore for $x < 0$. For $x = 0$, Eq. (4.11) gives $f(0) = \lambda f(0)$ so, since λ can be any positive number, $f(0) = 0$, which still fits in with $f(x) = ax$. But for $x < 0$ this does not follow at all: Putting $x = -1, f(-1) = a'$ into (4.11), we have

$$f(-\lambda) = a'\lambda \quad (\lambda > 0), \quad \text{that is,} \quad f(x) = -a'x \quad \text{if} \quad x < 0.$$

So *the solution of* (4.11) *is given* (with $\tilde{a} = -a'$) *as*

$$f(x) = \begin{cases} ax \text{ for } x \geq 0, \\ \tilde{a}x \text{ for } x < 0, \end{cases} \tag{4.12}$$

where a and $\tilde{a}$ may be different (they *may* also be equal). Substitution shows that this function indeed satisfies (4.11), whatever a and $\tilde{a}$ are:

$$f(\lambda x) = a\lambda x = \lambda f(x) \quad \text{if} \quad x \geq 0, \quad f(\lambda x) = \tilde{a}\lambda x = \lambda f(x) \quad \text{if} \quad x < 0.$$

(Since $\lambda > 0$, if $x \geq 0$ also $\lambda x \geq 0$, if $x < 0$, also $\lambda x < 0$.) The *graph* of this function (Fig. 4.3) is a (possibly) broken line (broken at 0). *The linear functions* $f(x) = ax$ *are restored* as only solutions *if we suppose* Eq. (4.10) (*linearity*) in the following form:

$$f(\lambda_1 x_1 + \lambda_2 x_2) = \lambda_1 f(x_1) + \lambda_2 f(x_2) \quad (x_1, x_2 \in \mathbb{R}, \lambda_1, \lambda_2 \in \mathbb{R}_{++})$$

Fig. 4.3

$$f(x) = \begin{cases} ax \text{ for } x \geq 0, \\ \tilde{a}x \text{ for } x < 0, \end{cases}$$

with $a \neq \tilde{a}$ satisfies positive linear homogeneity. (In the figure we have $a > \tilde{a} > 0$.)

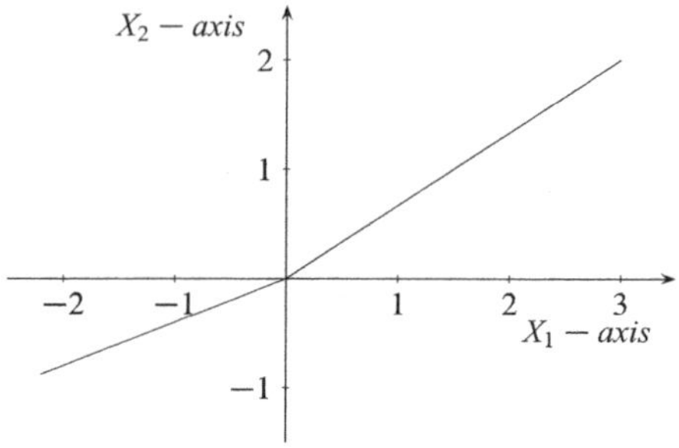

or, what is the same, *positive linear homogeneity*

$$f(\lambda x) = \lambda f(x) \qquad (x \in \mathbb{R},\ \lambda \in \mathbb{R}_{++})$$

and additivity

$$f(x_1 + x_2) = f(x_1) + f(x_2) \qquad \text{for all} \qquad x_1 \in \mathbb{R},\ x_2 \in \mathbb{R} \tag{4.13}$$

(not if (4.13) is supposed only for $x_1 \in \mathbb{R}_+, x_2 \in \mathbb{R}_+$).

Indeed substitute (4.13) into (4.12) with $x_1 = 2,\ x_2 = -1$:

$$a(2-1) = 2a + (-1)\tilde{a}, \text{ that is } a = 2a - \tilde{a},\ \tilde{a} = a,\ f(x) = ax \text{ for all } x \in \mathbb{R}.$$

Now we look at the additivity (4.13) alone. This is also called the *Cauchy functional equation*. (A *functional equation* is an equation where the unknown is a function). Without further supposition it certainly does not characterise the *linear function*, it has much crazier solutions than (4.12): their graphs cannot even be drawn because they are "everywhere dense" in the plane. But very weak conditions (to which we can attribute the purpose to eliminate these "crazy" solutions) already guarantee that the linear functions are the only solutions.

One such condition is that there exist numbers M_1 and M_2 (no matter how large) so that on an interval (no matter how small), say on $[0, 1]$, we have

$$-M_1 \leq f(x) \leq M_2 \qquad \text{for all} \qquad x \in [0, 1] \tag{4.14}$$

(that is, f is *locally bounded*, in this particular case *bounded on* $[0, 1]$, compare to Sect. 6.2). *If* (4.13) *and* (4.14) *are satisfied then there exists a constant a such that*

$$f(x) = ax \qquad \textit{for all} \qquad x \in \mathbb{R}.$$

Of course, *the converse is also true*: this function obviously satisfies (4.13) and also (4.14), say with $M_1 = 0, M_2 = a \geq 0$ or $M_1 = a < 0, M_2 = 0$.

If it exists, what would this a be? From $f(x) = ax$ we would clearly have

$$a = f(1),$$

so we have to *prove that*

$$f(x) = f(1)x,$$

that is, *the function* g, defined by

$$g(x) = f(x) - f(1)x \tag{4.15}$$

is identically 0. Clearly *also* g *satisfies the Cauchy functional equation* (is *additive*):

$$\begin{aligned} g(x_1 + x_2) &= f(x_1 + x_2) - f(1)(x_1 + x_2) \\ &= f(x_1) + f(x_2) - f(1)x_1 - f(1)x_2 = g(x_1) + g(x_2), \end{aligned}$$

(where we applied (4.13)). *The function* g *is also bounded on* $[0, 1]$: with

$$M_1' = M_1 + |f(1)|, \; M_2' = M_2 + |f(1)|$$

we get

$$-M_1 - |f(1)| \leq f(x) - |f(1)| \leq f(x) - f(1)x \leq M_2 + |f(1)| \quad \text{for} \quad x \in [0, 1]$$

(because of (4.14) and since $0 \leq x \leq 1$, $-|f(1)| \leq f(1) \leq |f(1)|$), so

$$-|f(1)| \leq f(1)x \leq |f(1)| \quad \text{for} \quad x \in [0, 1]$$

that is,

$$-M_1' \leq g(x) \leq M_2' \quad \text{for} \quad x \in [0, 1]. \tag{4.16}$$

Now, a consequence of the definition (4.15) is that $g(1) = 0$ and, applying the *additivity* of g

$$g(x + 1) = g(x) + g(1) = g(x) \quad \text{for all} \quad x \in \mathbb{R},$$

that is, g *is a periodic function with period* 1, in other words, the stretch of values of g on $[0, 1]$ keeps repeating on $[1, 2]$, $[2, 3], \ldots$ and also on $[-1, 0]$, $[-2, -1], \ldots$ (just as the values of $\sin x$ on $[0, 2\pi]$ keep repeating on $[2\pi, 4\pi], [4\pi, 6\pi], \ldots, [-2\pi, 0], [-4\pi, -2\pi], \ldots$). Therefore $g(x)$ *cannot have values for any* $x \in \mathbb{R}$ *which it does not have already on* $[0,1]$. So (4.16) has to be on true all $\mathbb{R}$, *g is bounded on all* $\mathbb{R}$:

$$-M_1' \leq g(x) \leq M_2' \quad \text{for all} \quad x \in \mathbb{R}.$$

On the other hand, from the additivity $g(x_1 + x_2) = g(x_1) + g(x_2)$ of g we get

$$g(2x) = 2g(x)\ (x_1 = x_2 = x),\ \ g(3x) = 3g(x)\ (x_1 = 2x,\ x_2 = x),$$

and so on (by induction, see Appendix), *for all positive integer n and for all* $x \in \mathbb{R}$

$$g(nx) = ng(x).$$

We prove now $g(x) = 0$ *by contradiction*: if there were even one $x_0 \in \mathbb{R}$ such that

$$g(x_0) \neq 0, \quad \text{say} \quad g(x_0) > 0$$

then, by what we have just proved, we would have

$$g(nx_0) = ng(x_0) \quad \text{for all} \quad n = 1, 2, 3, \ldots$$

If we chose n large enough (in particular $n > M_2'/g(x_0)$) then this would give

$$g(nx_0) > M_2'$$

which is a *contradiction* to the $g(x) \leq M_2'$ for all $x \in \mathbb{R}$ part of (4.16). (If we had $g(x_0) < 0$ then we would get into contradiction with the $-M_1' \leq g(x)$ for all $x \in \mathbb{R}$ part). So we have to have $g(x) \equiv 0$ and, by (4.15), $f(x) = f(1)x = ax$ $(x \in \mathbb{R})$.

Thus, the only additive functions, bounded on an interval, are the linear functions.

4.2.1 Exercises

1. The graphs of each of six affine functions given by $y = ax + b$ contain the following pairs of points. Determine a and b.

   (a) $(0, 4), (-1, 1)$, (b) $(-1, -3), (-5, 5)$,
   (c) $(2, 3), (1, -3)$, (d) $(1, 7), (9, -1)$,
   (e) $(-1, -3), (1, 5)$, (f) $(0, 0), (2, -6)$.

2. Determine a for the affine function given by $y = ax + 2$ whose graph contains the point

   (a) $(-1, 6)$, (b) $(1, 9)$, (c) $(-2, 8)$, (d) $(2, -4)$, (e) $(0, 2)$.

3. Determine b for the affine function given by $y = 5x + b$ whose graph contains the point

   (a) $(1, 2)$, (b) $(-2, 6)$, (c) $(1, -2)$, (d) $(-3, -6)$, (e) $(0, 3)$.

4. Determine the point which belongs both to the graph of $x \mapsto ax + b$ and of $x \mapsto cx + d$ $(a \neq c)$.
5. For which values of the real parameters a, b, c, d is the function $f : \mathbb{R} \to \hat{\mathbb{R}}$, $x \mapsto a + bx + cx^2 + dx^3$
   (a) additive,
   (b) not linear?

4.2.2 Answers

1. (a) $a = 3, b = 4$, (b) $a = -2, b = -5$,
   (c) $a = 6, b = -9$, (d) $a = -1, b = 8$,
   (e) $a = 4, b = 1$, (f) $a = -3, b = 0$.
2. (a) $a = -4$, (b) $a = 7$, (c) $a = 5$,
   (d) $a = -3$, (e) a any arbitrary number.
3. (a) $b = -3$, (b) $b = 4$, (c) $b = -7$,
   (d) $b = 9$, (e) $b = 3$.
4. $\left(\dfrac{d-b}{a-c}, \dfrac{ad-bc}{a-c}\right)$
5. (a) $b \in \mathbb{R}$ an arbitrary constant, $a = c = d = 0$.
   (b) f is not linear if at least one of the parameters a, b, c is different from zero.

4.3 Additivity, Linear Homogeneity, Linearity of Vector-Vector Functions, Matrices

In the previous section we considered functions where both the variables and the function values were real numbers (*scalars*). For many applications this is not enough. So we will also deal with *vector-vector functions*, that is with functions $\mathbf{f} : \mathbb{R}^n \to \mathbb{R}^m$ defined on the n-dimensional real space with values in the m-dimensional real space, in other words with n-component vectors as variables and m component vectors as function values. We can consider

$$\mathbf{f}(\mathbf{x})$$

also as compressing the values of m *functions* of n real *variables* (n-*place functions*)

$$f_1(x_1, x_2, \ldots, x_n)$$

$$f_2(x_1, x_2, \ldots, x_n)$$

$$\vdots$$

$$f_m(x_1, x_2, \ldots, x_n)$$

into one symbol.

Such a vector-vector function is *additive* if

$$\mathbf{f}(\mathbf{x} + \mathbf{y}) = \mathbf{f}(\mathbf{x}) + \mathbf{f}(\mathbf{y}) \quad \text{for} \quad \mathbf{x} \in \mathbb{R}^n,\ \mathbf{y} \in \mathbb{R}^n \tag{4.17}$$

(sometimes $\mathbf{f}$ is defined or this equation required only for $\mathbf{x}$, $\mathbf{y}$ in a subset of $\mathbb{R}^n$). It is *linearly homogeneous* if

$$\mathbf{f}(\lambda\mathbf{x}) = \lambda\mathbf{f}(\mathbf{x}) \quad \text{for} \quad \mathbf{x} \in \mathbb{R}^n,\ \lambda \in \mathbb{R}$$

(if this equation is required only for $\lambda > 0$ then we speak again of *positive linear homogeneity*). These equations can be condensed into the *linearity* equation

$$\mathbf{f}(\lambda\mathbf{x} + \mu\mathbf{y}) = \lambda\mathbf{f}(\mathbf{x}) + \mu\mathbf{f}(\mathbf{y}) \quad (\lambda \in \mathbb{R},\ \mu \in \mathbb{R};\ \mathbf{x} \in \mathbb{R}^n,\ \mathbf{y} \in \mathbb{R}^n). \tag{4.18}$$

We determine now all vector-vector functions satisfying (4.18). If we introduce the unit vectors of $\mathbb{R}^n$ (the *"basis"* of $\mathbb{R}^n$):

$$\mathbf{e}_1 = \begin{pmatrix} 1 \\ 0 \\ 0 \\ \vdots \\ 0 \\ 0 \end{pmatrix}, \quad \mathbf{e}_2 = \begin{pmatrix} 0 \\ 1 \\ 0 \\ \vdots \\ 0 \\ 0 \end{pmatrix}, \ldots, \quad \mathbf{e}_n = \begin{pmatrix} 0 \\ 0 \\ 0 \\ \vdots \\ 0 \\ 1 \end{pmatrix}$$

(it will be convenient to write column vectors here and in what follows), we can represent (see Sect. 1.4) every vector

$$\mathbf{x} = \begin{pmatrix} x_1 \\ x_2 \\ \vdots \\ x_n \end{pmatrix} \in \mathbb{R}^n$$

as

$$\mathbf{x} = x_1\mathbf{e}_1 + x_2\mathbf{e}_2 + \ldots + x_n\mathbf{e}_n.$$

Since $\mathbf{f}$ maps every vector in $\mathbb{R}^n$ into a vector in $\mathbb{R}^m$, we have in particular $\mathbf{f}(\mathbf{e}_j) \in \mathbb{R}^m$ ($j = 1, 2, \ldots, n$). We write also the unit vectors in $\mathbb{R}^m$ (the basis of $\mathbb{R}^m$) as column vectors:

$$\mathbf{v}_1 = \begin{pmatrix} 1 \\ 0 \\ \vdots \\ 0 \\ 0 \end{pmatrix}, \quad \mathbf{v}_2 = \begin{pmatrix} 0 \\ 1 \\ \vdots \\ 0 \\ 0 \end{pmatrix}, \dots, \quad \mathbf{v}_m = \begin{pmatrix} 0 \\ 0 \\ \vdots \\ 0 \\ 1 \end{pmatrix}.$$

The $\mathbf{f}(\mathbf{e}_j)$ (as all m-component vectors) can be written as linear combinations (with scalar coefficients) of $\mathbf{v}_1, \mathbf{v}_2, \dots, \mathbf{v}_m$:

$$\begin{aligned} \mathbf{f}(\mathbf{e}_1) &= a_{11}\mathbf{v}_1 + a_{21}\mathbf{v}_2 + \dots + a_{m1}\mathbf{v}_m, \\ \mathbf{f}(\mathbf{e}_2) &= a_{12}\mathbf{v}_1 + a_{22}\mathbf{v}_2 + \dots + a_{m2}\mathbf{v}_m, \\ &\vdots \\ \mathbf{f}(\mathbf{e}_n) &= a_{1n}\mathbf{v}_1 + a_{2n}\mathbf{v}_2 + \dots + a_{mn}\mathbf{v}_m \end{aligned}$$

$(a_{ij} \in \mathbb{R};\ i = 1, 2, \dots, m;\ j = 1, 2, \dots, n)$.

From the linearity equation (4.18)

$$\mathbf{f}(\lambda_1\mathbf{z}_1 + \lambda_2\mathbf{z}_2 + \dots + \lambda_p\mathbf{z}_p) = \lambda_1 f(\mathbf{z}_1) + \lambda_2 f(\mathbf{z}_2) + \dots + \lambda_p f(\mathbf{z}_p)$$
$$(\mathbf{z}_j \in \mathbb{R}^n,\ \lambda_j \in \mathbb{R};\ j = 1, 2, \dots, p)$$

follows (by induction, see Appendix), so

$$\begin{aligned} \mathbf{f}(\mathbf{x}) &= \mathbf{f}(x_1\mathbf{e}_1 + x_2\mathbf{e}_2 + \dots + x_n\mathbf{e}_n) \\ &= x_1\mathbf{f}(\mathbf{e}_1) + x_2\mathbf{f}(\mathbf{e}_2) + \dots + x_n\mathbf{f}(\mathbf{e}_n) \\ &= (a_{11}x_1 + a_{12}x_2 + \dots + a_{1n}x_n)\mathbf{v}_1 \\ &+ (a_{21}x_1 + a_{22}x_2 + \dots + a_{2n}x_n)\mathbf{v}_2 \\ &\vdots \\ &+ (a_{m1}x_1 + a_{m2}x_2 + \dots + a_{mn}x_n)\mathbf{v}_m \end{aligned}$$

$$= \begin{pmatrix} a_{11}x_1 + a_{12}x_2 + \dots + a_{1n}x_n \\ a_{21}x_1 + a_{22}x_2 + \dots + a_{2n}x_n \\ \vdots \\ a_{m1}x_1 + a_{m2}x_2 + \dots + a_{mn}x_n \end{pmatrix}.$$

This, with arbitrary constants a_{ij} $(i = 1, 2, \dots, m;\ j = 1, 2, \dots, n)$, *is the general solution of* (4.18). (One can easily verify by substitution that it indeed satisfies (4.18).)

We write the result in the short form

$$\mathbf{f}(\mathbf{x}) = \mathbf{A}\mathbf{x},$$

where

$$\mathbf{A} = \begin{pmatrix} a_{11} & a_{12} & \dots & a_{1n} \\ a_{21} & a_{22} & \dots & a_{2n} \\ \vdots & \vdots & & \vdots \\ a_{m1} & a_{m2} & \dots & a_{mn} \end{pmatrix}$$

is a *matrix* and, by definition,

$$\mathbf{A}\mathbf{x} = \begin{pmatrix} a_{11} & a_{12} & \dots & a_{1n} \\ a_{21} & a_{22} & \dots & a_{2n} \\ \vdots & \vdots & & \vdots \\ a_{m1} & a_{m2} & \dots & a_{mn} \end{pmatrix} \begin{pmatrix} x_1 \\ x_2 \\ \vdots \\ x_n \end{pmatrix}$$

$$= \begin{pmatrix} a_{11}x_1 + a_{12}x_2 + \dots + a_{1n}x_n \\ a_{21}x_1 + a_{22}x_2 + \dots + a_{2n}x_n \\ \vdots \\ a_{m1}x_1 + a_{m2}x_2 + \dots + a_{mn}x_n \end{pmatrix}. \tag{4.19}$$

The function $\mathbf{f} : \mathbb{R}^n \to \mathbb{R}^m$, described by $\mathbf{f}(\mathbf{x}) = \mathbf{A}\mathbf{x}$ is called a *linear function* or *linear transformation* of $\mathbb{R}^n$ into $\mathbb{R}^m$, both because the above form is a generalisation of the linear function $f(x) = ax$ from $\mathbb{R}$ into $\mathbb{R}$ and because, as we have just proved, it is the general solution of the linearity equation (4.18) or, what is the same, *the linear functions are the only additive and linearly homogeneous functions.* (We note that, just as in the previous section for $m = n = 1$, here too *the linear function is* the only locally bounded solution of (4.17)), that is, *the only locally bounded additive function*; see (Sect. 4.8).

So, once the unit vectors (the bases) in $\mathbb{R}^n$ and $\mathbb{R}^m$ are chosen, the matrix $\mathbf{A}$ completely determines the linear transformation and vice versa. The a_{ij} $(i = 1, 2, \dots, m;\ j = 1, 2, \dots, n)$ are the *components of the matrix* $\mathbf{A}$. *Two matrices are equal if their respective components are equal.*

Here the components were real numbers but one can form matrices also from complex numbers or elements of more general sets.

Again the functions given by $\mathbf{f}(\mathbf{x}) = \mathbf{A}\mathbf{x} + \mathbf{b}$ ($\mathbf{b}$ a vector) are called *affine*.

4.3.1 Exercises

1. Determine the vector $\mathbf{f}(\mathbf{x}) = \mathbf{A}\mathbf{x} + \mathbf{b}$ for

(a) $\mathbf{A} = \begin{pmatrix} -2 & 4 & -3 & 5 \\ 7 & -6 & 1 & -8 \\ 9 & -5 & -2 & 0 \end{pmatrix}$, $\mathbf{x} = \begin{pmatrix} 3 \\ 2 \\ -1 \\ 5 \end{pmatrix}$, $\mathbf{b} = \begin{pmatrix} -28 \\ 35 \\ -12 \end{pmatrix}$,

(b) $\mathbf{A} = \begin{pmatrix} u & v & w \\ r & s & t \end{pmatrix}$, $\mathbf{x} = \begin{pmatrix} -2 \\ 6 \\ 3 \end{pmatrix}$, $\mathbf{b} = \begin{pmatrix} 2u - 3w \\ 4r - 6s - 3t \end{pmatrix}$.

2. Determine the vector $\mathbf{x}$ for which $\mathbf{A}\mathbf{x} = \mathbf{b}$ holds if

(a) $\mathbf{A} = \begin{pmatrix} 1 & 2 \\ 3 & 4 \end{pmatrix}$, $\mathbf{b} = \begin{pmatrix} 1 \\ 5 \end{pmatrix}$, $\mathbf{x} = \begin{pmatrix} x_1 \\ x_2 \end{pmatrix}$,

(b) $\mathbf{A} = \begin{pmatrix} -2 & 3 \\ 5 & -6 \end{pmatrix}$, $\mathbf{b} = \begin{pmatrix} 4 \\ -7 \end{pmatrix}$, $\mathbf{x} = \begin{pmatrix} x_1 \\ x_2 \end{pmatrix}$,

(c) $\mathbf{A} = \begin{pmatrix} -3 & -4 \\ -5 & -6 \end{pmatrix}$, $\mathbf{b} = \begin{pmatrix} -1 \\ -3 \end{pmatrix}$, $\mathbf{x} = \begin{pmatrix} x_1 \\ x_2 \end{pmatrix}$.

3. Determine at least two vectors $\mathbf{x}$ for which $\mathbf{A}\mathbf{x} = \mathbf{b}$ holds if

(a) $\mathbf{A} = \begin{pmatrix} 4 & 3 & 2 \\ 1 & 6 & 5 \end{pmatrix}$, $\mathbf{b} = \begin{pmatrix} 9 \\ 12 \end{pmatrix}$, $\mathbf{x} = \begin{pmatrix} x_1 \\ x_2 \\ x_3 \end{pmatrix}$,

(b) $\mathbf{A} = \begin{pmatrix} 3 & 6 \\ 2 & 4 \\ 4 & 8 \end{pmatrix}$, $\mathbf{b} = \begin{pmatrix} 0 \\ 0 \\ 0 \end{pmatrix}$, $\mathbf{x} = \begin{pmatrix} -2 \\ 1 \end{pmatrix}$.

4. Show that $\mathbf{A}\mathbf{x} = \mathbf{b}$ does not have any solution $\mathbf{x}$ if

$$\mathbf{A} = \begin{pmatrix} 1 & 2 \\ 3 & 4 \\ 0 & 5 \end{pmatrix}, \quad \mathbf{b} = \begin{pmatrix} 1 \\ 5 \\ 2 \end{pmatrix}, \quad \mathbf{x} = \begin{pmatrix} x_1 \\ x_2 \end{pmatrix}.$$

5. For which values of x_1, x_2, y_1, y_2 does there exist a solution of the equation

$$\begin{pmatrix} 2 & -3 \\ 4 & 1 \\ 5 & 6 \\ 7 & 8 \end{pmatrix} \begin{pmatrix} x_1 \\ x_2 \end{pmatrix} = \begin{pmatrix} 8 \\ 2 \\ y_1 \\ y_2 \end{pmatrix}?$$

4.3.2 Answers

1. (a) $\mathbf{f}(\mathbf{x}) = \begin{pmatrix} 2 \\ 3 \\ 7 \end{pmatrix}$, (b) $\mathbf{f}(\mathbf{x}) = \begin{pmatrix} 6v \\ 2r \end{pmatrix}$.

2. (a) $\mathbf{x} = \begin{pmatrix} 3 \\ -1 \end{pmatrix}$, (b) $\mathbf{x} = \begin{pmatrix} 1 \\ 2 \end{pmatrix}$, (c) $\mathbf{x} = \begin{pmatrix} -1 \\ -3 \end{pmatrix}$.

3. (a) $\mathbf{x} = \begin{pmatrix} 1 \\ 1 \\ 1 \end{pmatrix}, \mathbf{x} = \begin{pmatrix} 8/7 \\ 1/7 \\ 2 \end{pmatrix}$, (b) $\mathbf{x} = \begin{pmatrix} -2 \\ 1 \end{pmatrix}, \mathbf{x} = \begin{pmatrix} 0 \\ 0 \end{pmatrix}$.

4. $\mathbf{x} = \begin{pmatrix} 3 \\ -2 \end{pmatrix}$ is the unique solution of $\begin{pmatrix} 1 & 2 \\ 3 & 4 \end{pmatrix}\begin{pmatrix} x_1 \\ x_2 \end{pmatrix} = \begin{pmatrix} 1 \\ 5 \end{pmatrix}$ (see Exercise 4.2.1 2. (a)). With this $\mathbf{x}$ one obtains

$$\begin{pmatrix} 1 & 2 \\ 3 & 4 \\ 0 & 5 \end{pmatrix}\begin{pmatrix} 3 \\ -1 \end{pmatrix} = \begin{pmatrix} 1 \\ 5 \\ -5 \end{pmatrix} \neq \begin{pmatrix} 1 \\ 5 \\ 2 \end{pmatrix}.$$

5. $x_1 = 1, x_2 = -2, y_1 = -7, y_2 = -9.$

4.4 Matrix Algebra

By (4.19) we have already defined the *product of a matrix and a vector*. In this section we define other operations for matrices. Most will be derived from (4.19).

1. Product of matrices. Let $\mathbf{f} : \mathbb{R}^n \to \mathbb{R}^m$ and $\mathbf{g} : \mathbb{R}^p \to \mathbb{R}^n$ be *linear* (vector-vector) functions, say

$$\mathbf{f}(\mathbf{y}) = \mathbf{A}\mathbf{y} = \begin{pmatrix} a_{11}y_1 + a_{12}y_2 + \ldots + a_{1n}y_n \\ a_{21}y_1 + a_{22}y_2 + \ldots + a_{2n}y_n \\ \vdots \\ a_{m1}y_1 + a_{m2}y_2 + \ldots + a_{mn}y_n \end{pmatrix},$$

$$\mathbf{g}(\mathbf{x}) = \mathbf{B}\mathbf{x} = \begin{pmatrix} b_{11}x_1 + b_{12}x_2 + \ldots + b_{1p}x_p \\ b_{21}x_1 + b_{22}x_2 + \ldots + b_{2p}x_p \\ \vdots \\ b_{n1}x_1 + b_{n2}x_2 + \ldots + b_{np}x_p \end{pmatrix},$$

that is,

$$\begin{aligned} g_1(\mathbf{x}) &= b_{11}x_1 + b_{12}x_2 + \ldots + b_{1p}x_p \\ g_2(\mathbf{x}) &= b_{21}x_1 + b_{22}x_2 + \ldots + b_{2p}x_p \\ \vdots \quad & \quad \vdots \qquad\quad \vdots \qquad\qquad \vdots \\ g_n(\mathbf{x}) &= b_{n1}x_1 + b_{n2}x_2 + \ldots + b_{np}x_p \end{aligned}$$

for the components of $\mathbf{g}(\mathbf{x})$.

Then their *composition* (see Sect. 3.2) $\mathbf{f} \circ \mathbf{g} : \mathbb{R}^p \to \mathbb{R}^m$ is defined by

$$\mathbf{f} \circ \mathbf{g}(\mathbf{x}) = \mathbf{f}[\mathbf{g}(\mathbf{x})] = \mathbf{A}(\mathbf{B}\mathbf{x}).$$

Since **f** and **g** are additive, $\mathbf{f} \circ \mathbf{g}$ must be additive too:

$$\begin{aligned} \mathbf{f} \circ \mathbf{g}(\mathbf{x} + \mathbf{y}) &= \mathbf{f}[\mathbf{g}(\mathbf{x} + \mathbf{y})] = \mathbf{f}[\mathbf{g}(\mathbf{x}) + \mathbf{g}(\mathbf{y})] \\ &= \mathbf{f}[\mathbf{g}(\mathbf{x})] + \mathbf{f}[\mathbf{g}(\mathbf{y})] = \mathbf{f} \circ \mathbf{g}(\mathbf{x}) + \mathbf{f} \circ \mathbf{g}(\mathbf{y}). \end{aligned}$$

Similarly, since both **f** and **g** are linearly homogeneous, so is $\mathbf{f} \circ \mathbf{g}$:

$$\mathbf{f} \circ \mathbf{g}(-\mathbf{x}) = \mathbf{f}[\mathbf{g}(-\mathbf{x})] = \mathbf{f}[-\mathbf{g}(\mathbf{x})] = -\mathbf{f}[\mathbf{g}(\mathbf{x})] = -\mathbf{f} \circ \mathbf{g}(\mathbf{x}).$$

As we have seen at the end of the previous section, it follows that also $\mathbf{f} \circ \mathbf{g} : \mathbb{R}^p \to \mathbb{R}^m$ is a linear function:

$$\begin{aligned} \mathbf{f} \circ \mathbf{g}(\mathbf{x}) = \mathbf{C}\mathbf{x} &= \begin{pmatrix} c_{11} & c_{12} & \dots & c_{1p} \\ c_{21} & c_{22} & \dots & c_{2p} \\ \vdots & \vdots & & \vdots \\ c_{m1} & c_{m2} & \dots & c_{mp} \end{pmatrix} \begin{pmatrix} x_1 \\ x_2 \\ \vdots \\ x_p \end{pmatrix} \\ &= \begin{pmatrix} c_{11}x_1 + c_{12}x_2 + \dots + c_{1p}x_p \\ c_{12}x_1 + c_{22}x_2 + \dots + c_{2p}x_p \\ \vdots \\ c_{m1}x_1 + c_{m2}x_2 + \dots + c_{mp}x_p \end{pmatrix}. \end{aligned}$$

On the other hand, from the definition of **f**, **g** and $\mathbf{f} \circ \mathbf{g}$:

$$\begin{aligned} \mathbf{f} \circ \mathbf{g}(\mathbf{x}) = \mathbf{f}[\mathbf{g}(\mathbf{x})] &= \mathbf{f}\left[\begin{pmatrix} b_{11}x_1 + b_{12}x_2 + \dots + b_{1p}x_p \\ b_{21}x_1 + b_{22}x_2 + \dots + b_{2p}x_p \\ \vdots \\ b_{n1}x_1 + b_{n2}x_2 + \dots + b_{np}x_p \end{pmatrix}\right] \\ &= \begin{pmatrix} a_{11}g_1(\mathbf{x}) + a_{12}g_2(\mathbf{x}) + \dots + a_{1n}g_n(\mathbf{x}) \\ a_{21}g_1(\mathbf{x}) + a_{22}g_2(\mathbf{x}) + \dots + a_{2n}g_n(\mathbf{x}) \\ \vdots \\ a_{m1}g_1(\mathbf{x}) + a_{m2}g_2(\mathbf{x}) + \dots + a_{mn}g_n(\mathbf{x}) \end{pmatrix} \end{aligned}$$

$$= \begin{pmatrix} a_{11}(b_{11}x_1 + \ldots + b_{1p}x_p) + \ldots + a_{1n}(b_{n1}x_1 + \ldots + b_{np}x_p) \\ a_{21}(b_{11}x_1 + \ldots + b_{1p}x_p) + \ldots + a_{2n}(b_{n1}x_1 + \ldots + b_{np}x_p) \\ \vdots \\ a_{m1}(b_{11}x_1 + \ldots + b_{1p}x_p) + \ldots + a_{mn}(b_{n1}x_1 + \ldots + b_{np}x_p) \end{pmatrix}$$

$$= \begin{pmatrix} (a_{11}b_{11} + \ldots + a_{1n}b_{n1})x_1 + \ldots + (a_{11}b_{1p} + \ldots + a_{1n}b_{np})x_p \\ \vdots \\ (a_{m1}b_{11} + \ldots + a_{mn}b_{n1})x_1 + \ldots + (a_{m1}b_{1p} + \ldots + a_{mn}b_{np})x_p \end{pmatrix}$$

$$= \begin{pmatrix} a_{11}b_{11} + \ldots + a_{1n}b_{n1} & \ldots & a_{11}b_{1p} + \ldots + a_{1n}b_{np} \\ \vdots & & \vdots \\ a_{m1}b_{11} + \ldots + a_{mn}b_{n1} & \ldots & a_{m1}b_{1p} + \ldots + a_{mn}b_{np} \end{pmatrix} \begin{pmatrix} x_1 \\ \vdots \\ x_p \end{pmatrix}.$$

Comparing the two expressions for $\mathbf{f} \circ \mathbf{g}$ we see that

$$c_{ik} = a_{i1}b_{1k} + a_{i2}b_{2k} + \ldots + a_{in}b_{nk} = \mathbf{a}_i \cdot \mathbf{b}_k \text{ for } i = 1, 2, \ldots, m;\ k = 1, 2, \ldots, p,$$

where $\mathbf{a}_i \cdot \mathbf{b}_k$ is the inner product (scalar product) of $\mathbf{a}_i$ and $\mathbf{b}_k$ as defined in Sect. 1.5.

We say that the matrix $\mathbf{C}$ is the *product* of the matrices $\mathbf{A}$ and $\mathbf{B}$, $\mathbf{C} = \mathbf{AB}$, that is,

$$\mathbf{A}(\mathbf{Bx}) = \mathbf{Cx} = (\mathbf{AB})\mathbf{x} \tag{4.20}$$

that is,

$$\begin{pmatrix} c_{11} & \ldots & c_{1p} \\ \vdots & & \vdots \\ c_{m1} & \ldots & c_{mp} \end{pmatrix} = \begin{pmatrix} a_{11} & \ldots & a_{1n} \\ \vdots & & \vdots \\ a_{m1} & \ldots & a_{mn} \end{pmatrix} \begin{pmatrix} b_{11} & \ldots & b_{1p} \\ \vdots & & \vdots \\ b_{n1} & \ldots & b_{np} \end{pmatrix}$$

$$= \begin{pmatrix} a_{11}b_{11} + \ldots + a_{1n}b_{n1} & \ldots & a_{11}b_{1p} + \ldots + a_{1n}b_{np} \\ \vdots & & \vdots \\ a_{m1}b_{11} + \ldots + a_{mn}b_{n1} & \ldots & a_{m1}b_{1p} + \ldots + a_{mn}b_{np} \end{pmatrix}, \tag{4.21}$$

which *defines the product of matrices.*

Matrices with m rows and n columns are also called $m \times n$ matrices. Notice that an $m \times n$ and a $q \times p$ *matrix can be multiplied only* if $n = q$. Actually, if we consider n-component column vectors as $n \times 1$ matrices then the formula (4.20) of the previous section was already an example of matrix multiplication. So is the inner product of two n-component vectors $\mathbf{a}$ and $\mathbf{b}$ but only if $\mathbf{a}$ is written as row vector (as $1 \times n$

matrix) and $\mathbf{b}$ as column vector (as $n \times 1$ matrix):

$$\mathbf{a} = (a_1, \ldots, a_n),\ \mathbf{b} = \begin{pmatrix} b_1 \\ \vdots \\ b_n \end{pmatrix},\ \mathbf{a} \cdot \mathbf{b} = (a_1, \ldots, a_n) \begin{pmatrix} b_1 \\ \vdots \\ b_n \end{pmatrix} = a_1 b_1 + \ldots + a_n b_n.$$

We will now compose functions repeatedly and use the notations (compare Sect. 3.2)

$$(\mathbf{f} \circ \mathbf{g})\mathbf{x} = (\mathbf{f} \circ \mathbf{g})(\mathbf{x}) = \mathbf{f} \circ \mathbf{g}(\mathbf{x}) = \mathbf{f}[\mathbf{g}(\mathbf{x})].$$

From

$$[(\mathbf{f} \circ \mathbf{g}) \circ \mathbf{h}]\mathbf{x} = (\mathbf{f} \circ \mathbf{g})[\mathbf{h}(\mathbf{x})] = \mathbf{f}(\mathbf{g}[\mathbf{h}(\mathbf{x})]) = \mathbf{f}[\mathbf{g} \circ \mathbf{h}(\mathbf{x})] = [\mathbf{f} \circ (\mathbf{g} \circ \mathbf{h})]\mathbf{x}$$

and from the definition

$$(\mathbf{f} \circ \mathbf{g})\mathbf{x} = \mathbf{ABx} = \mathbf{A}(\mathbf{Bx}) = \mathbf{f}[\mathbf{g}(\mathbf{x})]$$

it follows that

$$[(\mathbf{AB})\mathbf{C}]\mathbf{x} = (\mathbf{AB})(\mathbf{Cx}) = [\mathbf{A}(\mathbf{BC})]\mathbf{x}$$

for all vectors $\mathbf{x}$. Developing this according to (4.21) and comparing the coefficients of $x_1, x_2, \ldots, x_n$ we get

$$(\mathbf{AB})\mathbf{C} = \mathbf{A}(\mathbf{BC}),$$

that is, *matrix multiplication is associative*. However, it is in general *not commutative*, that is, $\mathbf{AB} = \mathbf{BA}$ does *not* always hold. Take for instance

$$\mathbf{A} = \begin{pmatrix} 0 & 1 \\ 2 & 3 \end{pmatrix},\ \mathbf{B} = \begin{pmatrix} 3 & 0 \\ 1 & 2 \end{pmatrix}:$$

Then

$$\mathbf{AB} = \begin{pmatrix} 0 \cdot 3 + 1 \cdot 1 & 0 \cdot 0 + 1 \cdot 2 \\ 2 \cdot 3 + 3 \cdot 1 & 2 \cdot 0 + 3 \cdot 2 \end{pmatrix} = \begin{pmatrix} 1 & 2 \\ 9 & 6 \end{pmatrix},$$

$$\mathbf{BA} = \begin{pmatrix} 3 \cdot 0 + 0 \cdot 2 & 3 \cdot 1 + 0 \cdot 3 \\ 1 \cdot 0 + 2 \cdot 2 & 1 \cdot 1 + 2 \cdot 3 \end{pmatrix} = \begin{pmatrix} 0 & 3 \\ 4 & 7 \end{pmatrix} \neq \mathbf{AB}.$$

Actually, if $\mathbf{A}$ is an $m \times n$ and $\mathbf{B}$ a $q \times p$ matrix, *the equality* $\mathbf{AB} = \mathbf{BA}$ *would make even formal sense only if* $m = n = q = p$ that is, $\mathbf{A}$ and $\mathbf{B}$ have the same number of elements and both are *square matrices*, that is, *the number of rows and columns is equal*. That $\mathbf{AB} \neq \mathbf{BA}$ *is possible even under these circumstances, shows also that it is possible that both composite functions* $f \circ g$ *and* $g \circ f$ *exist but they are not equal.*

2. *Product of a scalar and a matrix.* From the linear homogeneity

$$\mathbf{f}(\lambda\mathbf{x}) = \lambda\mathbf{f}(\mathbf{x})$$

of the linear function $\mathbf{f}(\mathbf{x}) = \mathbf{A}\mathbf{x}$ we get

$$\mathbf{A}(\lambda\mathbf{x}) = \lambda(\mathbf{A}\mathbf{x}).$$

The left hand side is, according to (4.19),

$$\begin{pmatrix} a_{11} & \dots & a_{1n} \\ \vdots & & \vdots \\ a_{m1} & \dots & a_{mn} \end{pmatrix} \begin{pmatrix} \lambda x_1 \\ \vdots \\ \lambda x_n \end{pmatrix} = \begin{pmatrix} \lambda a_{11}x_1 + \dots + \lambda a_{1n}x_n \\ \vdots \\ \lambda a_{m1}x_1 + \dots + \lambda a_{mn}x_n \end{pmatrix}$$

If, similarly to (4.20), we *define the product of a scalar* λ *and a matrix* $\mathbf{A}$ by

$$(\lambda\mathbf{A})\mathbf{x} = \lambda(\mathbf{A}\mathbf{x})$$

then, by the previous two equations, we have as *definition*

$$\lambda\mathbf{A} = \lambda \begin{pmatrix} a_{11} & \dots & a_{1n} \\ \vdots & & \vdots \\ a_{m1} & \dots & a_{mn} \end{pmatrix} = \begin{pmatrix} \lambda a_{11} & \dots & \lambda a_{1n} \\ \vdots & & \vdots \\ \lambda a_{m1} & \dots & \lambda a_{mn} \end{pmatrix}. \tag{4.22}$$

Actually, this too is a special case of (4.21) if we consider *scalars* λ *as diagonal matrices* with λ at all places of the main diagonal (a matrix $\mathbf{B}$ is *diagonal* if $b_{ij} = 0$ for $i \neq j$, that is, if all components of $\mathbf{B}$ are 0 except those in the main diagonal, going from the left top to the right bottom):

$$\lambda\mathbf{A} = \begin{pmatrix} \lambda & 0 & \dots & 0 \\ 0 & \lambda & \dots & 0 \\ \vdots & \vdots & \ddots & \vdots \\ 0 & 0 & \dots & \lambda \end{pmatrix} \begin{pmatrix} a_{11} & a_{12} & \dots & a_{1n} \\ a_{21} & a_{22} & \dots & a_{2n} \\ \vdots & \vdots & & \vdots \\ a_{m1} & a_{m2} & \dots & a_{mn} \end{pmatrix}$$

$$= \begin{pmatrix} \lambda a_{11} & \lambda a_{12} & \dots & \lambda a_{1n} \\ \lambda a_{21} & \lambda a_{22} & \dots & \lambda a_{2n} \\ \vdots & \vdots & & \vdots \\ \lambda a_{m1} & \lambda a_{m2} & \dots & \lambda a_{mn} \end{pmatrix}$$

$$= \begin{pmatrix} a_{11} & a_{12} & \dots & a_{1n} \\ a_{21} & a_{22} & \dots & a_{2n} \\ \vdots & \vdots & & \vdots \\ a_{m1} & a_{m2} & \dots & a_{mn} \end{pmatrix} \begin{pmatrix} \lambda & 0 & \dots & 0 \\ 0 & \lambda & \dots & 0 \\ \vdots & \vdots & \ddots & \vdots \\ 0 & 0 & \dots & \lambda \end{pmatrix} = \mathbf{A}\lambda.$$

(Notice that first we identified λ with an $m \times m$ then with an $n \times n$ diagonal matrix. In this sense, this special product $\lambda\mathbf{A}$ is commutative.)

3. *Sums and linear combination of matrices.* In general one *defines the sum of two functions* (which have a domain in common), in particular of two linear functions by

$$(f + g)(x) = f(x) + g(x).$$

So, for *all* vectors $\mathbf{x}$, by the rules for adding vectors, we have

$$\begin{aligned}(\mathbf{A} + \mathbf{B})\mathbf{x} &= \mathbf{A}\mathbf{x} + \mathbf{B}\mathbf{x} \\ &= \begin{pmatrix} a_{11}x_1 + \dots + a_{1n}x_n \\ \vdots \\ a_{m1}x_1 + \dots + a_{mn}x_n \end{pmatrix} + \begin{pmatrix} b_{11}x_1 + \dots + b_{1n}x_n \\ \vdots \\ b_{m1}x_1 + \dots + b_{mn}x_n \end{pmatrix} \\ &= \begin{pmatrix} (a_{11} + b_{11})x_1 + \dots + (a_{1n} + b_{1n})x_n \\ \vdots \\ (a_{m1} + b_{m1})x_1 + \dots + (a_{mn} + b_{mn})x_n \end{pmatrix}\end{aligned}$$

($\mathbf{A}$ and $\mathbf{B}$ have to be both $m \times n$ matrices). Comparing the coefficients of $x_1, \dots, x_n$, we get

$$\begin{aligned}\mathbf{A} + \mathbf{B} &= \begin{pmatrix} a_{11} & \dots & a_{1n} \\ \vdots & & \vdots \\ a_{m1} & \dots & a_{mn} \end{pmatrix} + \begin{pmatrix} b_{11} & \dots & b_{1n} \\ \vdots & & \vdots \\ b_{m1} & \dots & b_{mn} \end{pmatrix} \\ &= \begin{pmatrix} a_{11} + b_{11} & \dots & a_{1n} + b_{1n} \\ \vdots & & \vdots \\ a_{m1} + b_{m1} & \dots & a_{mn} + b_{mn} \end{pmatrix}.\end{aligned}$$

That is, *by definition, matrices are added component wise*. This nicely conforms with (4.22), since from both

$$2\mathbf{A} = 2\begin{pmatrix} a_{11} & \dots & a_{1n} \\ \vdots & & \vdots \\ a_{m1} & \dots & a_{mn} \end{pmatrix} = \begin{pmatrix} 2a_{11} & \dots & 2a_{1n} \\ \vdots & & \vdots \\ 2a_{m1} & \dots & 2a_{mn} \end{pmatrix}.$$

Combined with (4.22) we get for *linear combinations of two $m \times n$ matrices*

$$\begin{aligned} \lambda\mathbf{A} + \mu\mathbf{B} &= \lambda\begin{pmatrix} a_{11} & \dots & a_{1n} \\ \vdots & & \vdots \\ a_{m1} & \dots & a_{mn} \end{pmatrix} + \mu\begin{pmatrix} b_{11} & \dots & b_{1n} \\ \vdots & & \vdots \\ b_{m1} & \dots & b_{mn} \end{pmatrix} \\ &= \begin{pmatrix} \lambda a_{11} + \mu b_{11} & \dots & \lambda a_{1n} + \mu b_{1n} \\ \vdots & & \vdots \\ \lambda a_{m1} + \mu b_{m1} & \dots & \lambda a_{mn} + \mu b_{mn} \end{pmatrix}. \end{aligned}$$

(this is similar but not the same as (4.18)!). In particular, for the *difference of two matrices*:

$$\mathbf{A} - \mathbf{B} = \mathbf{A} + (-1)\mathbf{B} = \begin{pmatrix} a_{11} - b_{11} & \dots & a_{1n} - b_{1n} \\ \vdots & & \vdots \\ a_{m1} - b_{m1} & \dots & a_{mn} - b_{mn} \end{pmatrix}.$$

4.4.1 Exercises

1. For the matrices

$$\mathbf{A} = \begin{pmatrix} 3 & -5 & 4 \\ -2 & 1 & 0 \\ 4 & 2 & -6 \end{pmatrix}, \mathbf{B} = \begin{pmatrix} 4 & 3 & 1 \\ 2 & 4 & 6 \\ 1 & 5 & 8 \end{pmatrix}, \mathbf{C} = \begin{pmatrix} 1 & 0 & 1 \\ 0 & 1 & 0 \\ 1 & 0 & 1 \end{pmatrix}$$

calculate

(a) **AB**, (b) **AC**, (c) **BC**, (d) **BA**,
(e) **CA**, (f) **CB**, (g) **(AB)C**, (h) **A(BC)**,
(i) $9\mathbf{A} - 8\mathbf{B} + 7\mathbf{C}$, (j) $3(\mathbf{AB})\mathbf{C} - 5\mathbf{BA}$.

2. For the matrices

$$\mathbf{A} = \begin{pmatrix} 2 & 3 & -4 \\ -6 & 1 & 0 \end{pmatrix}, \mathbf{B} = \begin{pmatrix} 7 & 3 \\ 1 & 9 \\ 4 & 8 \end{pmatrix}, \mathbf{C} = \begin{pmatrix} 2 & 1 \\ -1 & -1 \end{pmatrix}$$

calculate

(a) **AB**, (b) **BA**, (c) **BC**, (d) **CA**,
(e) **(AB)C**, (f) **A(BC)**.

3. Determine x, y, z such that
(a) $\begin{pmatrix} 2 & x \\ y & 3 \end{pmatrix} \begin{pmatrix} x & y & 0 \\ z & z & z \end{pmatrix} = \begin{pmatrix} 7 & 13 & 5 \\ 19 & 31 & 15 \end{pmatrix}$, [1.0ex]
(b) $\begin{pmatrix} x & 2 \\ 1 & z \\ -2 & y \end{pmatrix} \begin{pmatrix} 5 & z \\ z & 3 \end{pmatrix} = \begin{pmatrix} 6 & 6 \\ 14 & 12 \\ 14 & 18 \end{pmatrix}$, [1.0ex]
(c) $\begin{pmatrix} x & y \\ -y & x \end{pmatrix} \begin{pmatrix} x & -y \\ y & x \end{pmatrix} = \begin{pmatrix} z^2 & 0 \\ 0 & z^2 \end{pmatrix}$. [1.0ex]
(d) Calculate x in (c) if $z = 5$ and $y = 4$.

4. From the definition of $\mathbf{A} - \mathbf{B}$ at the end of Sect. 4.4, prove that $(\mathbf{A} - \mathbf{B}) + \mathbf{B} = \mathbf{A}$.

5. Construct two matrices $\mathbf{A} - \begin{pmatrix} a & b \\ c & d \end{pmatrix}$, $\mathbf{B} = \begin{pmatrix} r & s \\ t & u \end{pmatrix}$ where the components a, b, c, d, r, s, t, u are real numbers none of them 0, no two of them equal, such that $\mathbf{AB} - \mathbf{BA}$.

4.4.2 Answers

1. (a) $\begin{pmatrix} 6 & 9 & 5 \\ -6 & -2 & 4 \\ 14 & -10 & -32 \end{pmatrix}$, (b) $\begin{pmatrix} 7 & -5 & 7 \\ -2 & 1 & -2 \\ -2 & 2 & -2 \end{pmatrix}$, (c) $\begin{pmatrix} 5 & 3 & 5 \\ 8 & 4 & 8 \\ 9 & 5 & 9 \end{pmatrix}$,

(d) $\begin{pmatrix} 10 & -15 & 10 \\ 22 & 6 & -28 \\ 25 & 16 & -44 \end{pmatrix}$, (e) $\begin{pmatrix} 7 & -3 & -2 \\ -2 & 1 & 0 \\ 7 & -3 & -2 \end{pmatrix}$, (f) $\begin{pmatrix} 5 & 8 & 9 \\ 2 & 4 & 6 \\ 5 & 8 & 9 \end{pmatrix}$,

(g) $\begin{pmatrix} 11 & 9 & 11 \\ -2 & -2 & -2 \\ -18 & -10 & -18 \end{pmatrix}$, (h) $\begin{pmatrix} 11 & 9 & 11 \\ -2 & -2 & -2 \\ -18 & -10 & -18 \end{pmatrix}$,

(i) $\begin{pmatrix} 2 & -69 & 35 \\ -34 & -16 & -48 \\ 35 & -22 & -111 \end{pmatrix}$, (j) $\begin{pmatrix} -17 & 102 & -17 \\ -116 & -36 & 134 \\ -179 & -110 & -166 \end{pmatrix}$.

2. (a) $\begin{pmatrix} 1 & 1 \\ -41 & -9 \end{pmatrix}$, (b) $\begin{pmatrix} -4 & 24 & -28 \\ -52 & 12 & -4 \\ -40 & 20 & -16 \end{pmatrix}$, (c) $\begin{pmatrix} 11 & 4 \\ -7 & -8 \\ 0 & -4 \end{pmatrix}$,

(d) $\begin{pmatrix} -2 & 7 & -8 \\ 4 & -4 & 4 \end{pmatrix}$, (e) $\begin{pmatrix} 1 & 0 \\ -73 & -32 \end{pmatrix}$, (f) $\begin{pmatrix} 1 & 0 \\ -73 & -32 \end{pmatrix}$.

3. (a) $x = 1, y = 4, z = 5,$ (b) $x = 0, y = 8, z = 3,$

(c) $z^2 = x^2 + y^2$, whence, e.g., $x = 3, y = 4, z = 5$

(d) $x = 3$.

4. For $\mathbf{A} = \begin{pmatrix} a_{11} & \dots & a_{1n} \\ \vdots & & \vdots \\ a_{n1} & \dots & a_{nn} \end{pmatrix}$, $\mathbf{B} = \begin{pmatrix} b_{11} & \dots & b_{1n} \\ \vdots & & \vdots \\ b_{n1} & \dots & b_{nn} \end{pmatrix}$ we have

$$(\mathbf{A} - \mathbf{B}) + \mathbf{B} = \begin{pmatrix} a_{11} - b_{11} & \dots & a_{1n} - b_{1n} \\ \vdots & & \vdots \\ a_{n1} - b_{n1} & \dots & a_{nn} - b_{nn} \end{pmatrix} + \begin{pmatrix} b_{11} & \dots & b_{1n} \\ \vdots & & \vdots \\ b_{n1} & \dots & b_{nn} \end{pmatrix}$$

$$= \begin{pmatrix} a_{11} - b_{11} + b_{11} & \dots & a_{1n} - b_{1n} + b_{1n} \\ \vdots & & \vdots \\ a_{n1} - b_{n1} + b_{n1} & \dots & a_{nn} - b_{nn} + b_{nn} \end{pmatrix}$$

$$= \begin{pmatrix} a_{11} & \dots & a_{1n} \\ \vdots & & \vdots \\ a_{n1} & \dots & a_{nn} \end{pmatrix} = \mathbf{A}.$$

5. $\mathbf{A} = \begin{pmatrix} 1 & 2 \\ 3 & 4 \end{pmatrix}$, $\mathbf{B} = \begin{pmatrix} 5 & 6 \\ 9 & 14 \end{pmatrix}$. $\mathbf{AB} = \begin{pmatrix} 23 & 34 \\ 51 & 74 \end{pmatrix} = \mathbf{BA}$.

4.5 Linear Economic Models: Leontief, von Neumann

In what follows, we introduce the *input-output model* of W. A. Leontief (1906; Nobel laureate of 1973).

We start with an '*input-output table*' (Table 4.1). The table informs of the values (say in \$) put into and taken out of the individual producing and service industries ('sectors') in an *open economy*, that is an economy with exports (Ex) and imports (Im). The j-th column represents the *input vector* of the j-th sector ($j = 1, 2, \dots, n$) and the i-th row lists how much value the i-th sector ($i = 1, 2, \dots, n$) has supplied to the different user industries and to the final demand of the economy (private and government consumption, delivered and self-produced products, changes in inventories, exports). The sum of numbers in the i-th row, X_i, is the value of the

Table 4.1 Input–output table of an economy

	Demand for inputs from the sectors				Final demand for goods and services from the economy					
Receiving sectors → Supplying sectors ↓	Sector 1	Sector 2	$\cdots$	Sector n	C	C*	P	Ch	Ex	Gross output
Sector 1	A_{11}	A_{12}	$\cdots$	A_{1n}	C_1	C_1^*	P_1	Ch_1	Ex_1	X_1
Sector 2	A_{21}	A_{22}	$\cdots$	A_{2n}	C_2	C_2^*	P_2	Ch_2	Ex_2	X_2
$\vdots$	$\vdots$	$\vdots$	$\ddots$	$\vdots$	$\vdots$	$\vdots$	$\vdots$	$\vdots$	$\vdots$	$\vdots$
Sector n	A_{n1}	A_{n2}	$\cdots$	A_{nn}	C_n	C_n^*	P_n	Ch_n	Ex_n	X_n
Imports	Im_1	Im_2	$\cdots$	Im_n						
Depreciation	D_1	D_2	$\cdots$	D_n						
Indirect taxes minus subsidies	T_1 $-S_1$	T_2 $-S_2$	$\cdots$	T_n $-S_n$						
Wages, salaries	W_1	W_2	$\cdots$	W_n						
Other income	I_1	I_2	$\cdots$	I_n						
Sum of amounts per column = gross output	X_1	X_2	$\cdots$	X_n						

C = private consumption
C^* = government consumption
P = delivered and self-produced products
Ch = changes in inventories
Ex = exports

gross output of the i-th sector; *the sum of numbers in the i-th column has to be the same*. This is achieved by adding, for bookkeeping purposes, the *"other income"* to the j-th *input* values in order to get X_j as the sum of the numbers in the j-th column $(j = 1, \dots, n)$.

Leontief's production model for any economy with n sectors (industries) consists of n production processes, written as column vectors (see Sects. 1.4 and 2.3):

$$\mathbf{v}_1 = \begin{pmatrix} A_{11} \\ A_{21} \\ \vdots \\ A_{n1} \\ X_1 \\ 0 \\ \vdots \\ 0 \end{pmatrix}, \mathbf{v}_2 = \begin{pmatrix} A_{12} \\ A_{22} \\ \vdots \\ A_{n2} \\ 0 \\ X_2 \\ \vdots \\ 0 \end{pmatrix}, \dots, \mathbf{v}_n = \begin{pmatrix} A_{1n} \\ A_{2n} \\ \vdots \\ A_{nn} \\ 0 \\ 0 \\ \vdots \\ X_n \end{pmatrix}. \tag{4.23}$$

The first n components of the j-th vector represent the inputs which flow from all n industries into the j-th ($j = 1, \ldots, n$). The $(n+j)$-th component of the same vector is X_j, the gross output of the j-th industry (see Table 4.1).

An example of distinguishing $n = 14$ sectors could be: (1) agriculture and fishery, (2) power-supply industry and mining, (3) chemical industry, rock, stone and related minerals, (4) iron and steel industry, nonferrous metals, (5) mechanical engineering, vehicles construction, (6) electrical engineering, (7) timber, paper, leather, textile industry, (8) food, beverages, and tobacco industry, (9) construction industry, (10) trade, (11) transport(ation) and communication, (12) other services, (13) government, (14) private households and private no-gain organisations.

Leontief assumes that in that particular economy not only the production processes (4.23) but also their linear combinations (compare Sects. 1.5 and 2.3) with nonnegative coefficients

$$\lambda_1 \mathbf{v}_1 + \ldots + \lambda_n \mathbf{v}_n = \begin{pmatrix} \lambda_1 A_{11} + \ldots + \lambda_n A_{1n} \\ \vdots \\ \lambda_1 A_{n1} + \ldots + \lambda_n A_{nn} \\ \lambda_1 X_1 \\ \vdots \\ \lambda_n X_n \end{pmatrix} \quad (\lambda_j \in \mathbb{R}_+; j = 1, 2, \ldots, n)$$

and only these can be "run" as production processes. So this is a *linear technology* generated by the above n production processes and thus we have a *linear production model* (compare to Sect. 2.3).

We call *Leontief processes* the (proportionally) reduced production processes (each component divided by the row sum in Table 4.1):

$$\boldsymbol{\ell}_1 = \frac{\mathbf{v}_1}{X_1} = \begin{pmatrix} A_{11}/X_1 \\ \vdots \\ A_{n1}/X_1 \\ 1 \\ 0 \\ \vdots \\ 0 \end{pmatrix} = \begin{pmatrix} a_{11} \\ \vdots \\ a_{n1} \\ 1 \\ 0 \\ \vdots \\ 0 \end{pmatrix}, \ldots, \boldsymbol{\ell}_n = \frac{\mathbf{v}_n}{X_n} = \begin{pmatrix} A_{1n}/X_n \\ \vdots \\ A_{nn}/X_n \\ 0 \\ 0 \\ \vdots \\ 1 \end{pmatrix} = \begin{pmatrix} a_{1n} \\ \vdots \\ a_{nn} \\ 0 \\ 0 \\ \vdots \\ 1 \end{pmatrix},$$

where we wrote $a_{ij} = A_{ij}/X_j$ ($i = 1, \ldots, n; j = 1, \ldots, n$), that is, $A_{ij} = a_{ij}X_j$. Just as a was in (4.8), also these a_{ij} are called *production coefficients*. There a showed how much input was needed for a unit value of output, here a_{ij} is the value of input needed from the industry or sector i in order to produce a unit value of (gross) output in industry or sector j.

If we want to produce in sector j a gross output of value x_j (*"intensity"*, which at this stage, is a *variable*) instead of 1, we have to multiply the column vector

(*Leontief process*) $\boldsymbol{\ell}_j$ by x_j $(j = 1, \ldots, n)$:

$$x_j \boldsymbol{\ell}_j = \begin{pmatrix} a_{1j}x_j \\ \vdots \\ a_{nj}x_j \\ 0 \\ \vdots \\ x_j \\ \vdots \\ 0 \end{pmatrix} \qquad (j = 1, \ldots, n).$$

This vector represents the contributions of all sectors $i = 1, \ldots, n$ to the gross output value x_j of sector j $(j = 1, \ldots, n)$. The sum

$$x_1 \boldsymbol{\ell}_1 + \ldots + x_n \boldsymbol{\ell}_n = x_1 \frac{\mathbf{v}_1}{X_1} + \ldots + x_n \frac{\mathbf{v}_n}{X_n} = \begin{pmatrix} a_{11}x_1 + \ldots + a_{1n}x_n \\ \vdots \\ a_{n1}x_1 + \ldots + a_{nn}x_n \\ x_1 \\ \vdots \\ x_n \end{pmatrix} \tag{4.24}$$

of these vectors represents therefore, in its first n components, the *contributions of all sectors $i = 1, \ldots, n$ to the gross output values $x_1, \ldots, x_n$ of all sectors.*

That part of the gross output of industry i, which is needed by all sectors as input from sector i, is given in (4.24) by $a_{i1}x_1 + \ldots + a_{in}x_n$. The "rest" of the gross output of sector i satisfies other demands, namely the *final demand* for sector i's goods and services (see Table 4.1).

Let the final demands $c_1, \ldots, c_n$ (measured, say, in \$) be *given*. Remember that we did not fix the "*intensities*" $x_1, \ldots, x_n$ but considered them as *variables*. In production theory, the question is with what intensities $x_1, \ldots, x_n$ must the production process (4.24) "run" so that the gross output values of the individual sectors minus the deliveries to all sectors, be not smaller than the final demands $c_1, \ldots, c_n$ of the economy.

The corresponding mathematical problem is determining $x_1, \ldots, x_n \in \mathbb{R}_+$, so that

$$\begin{aligned} x_1 - a_{11}x_1 - \ldots - a_{1n}x_n &\geq c_1, \\ &\vdots \\ x_n - a_{n1}x_1 - \ldots - a_{nn}x_n &\geq c_n. \end{aligned}$$

If there should be no surplus, then this *system of equalities* is replaced by the *system of equations*

$$\begin{aligned} x_1 - a_{11}x_1 - \ldots - a_{1n}x_n &= c_1, \\ &\vdots \\ x_n - a_{n1}x_1 - \ldots - a_{nn}x_n &= c_n. \end{aligned} \tag{4.25}$$

The first question to ask is whether for every final demand vector

$$\begin{pmatrix} c_1 \\ \vdots \\ c_n \end{pmatrix} \qquad (c_k \geq 0;\ k = 1, \ldots, n),$$

there exist solution vectors

$$\begin{pmatrix} x_1 \\ \vdots \\ x_n \end{pmatrix} \qquad (x_k \geq 0;\ k = 1, \ldots, n)$$

of the above systems of Eqs. (4.25). We will answer this question for equations in the next two sections.

With the notations

$$\mathbf{A} = \begin{pmatrix} a_{11} & \ldots & a_{1n} \\ \vdots & & \vdots \\ a_{n1} & \ldots & a_{nn} \end{pmatrix}, \ \mathbf{x} = \begin{pmatrix} x_1 \\ \vdots \\ x_n \end{pmatrix}, \ \mathbf{c} = \begin{pmatrix} c_1 \\ \vdots \\ c_n \end{pmatrix}, \ \mathbf{I} = \begin{pmatrix} 1 & 0 & \ldots & 0 \\ 0 & 1 & \ldots & 0 \\ \vdots & \vdots & \ddots & \vdots \\ 0 & 0 & \ldots & 1 \end{pmatrix}$$

(the latter called a *unit matrix*), and with the operations defined in the previous section, we can write the above systems of inequalities and equations as

$$(\mathbf{I} - \mathbf{A})\mathbf{x} \geqq \mathbf{c} \quad \text{or} \quad (\mathbf{I} - \mathbf{A})\mathbf{x} = \mathbf{c}.$$

The matrix $(\mathbf{I} - \mathbf{A})$ is called the *Leontief matrix*. In the terminology of Sect. 4.2, we are looking for those $\mathbf{x} \in \mathbb{R}^n_+$ for which the values of the linear function defined by $\mathbf{f}(\mathbf{x}) = (\mathbf{I} - \mathbf{A})\mathbf{x} \in \mathbb{R}^n$ will be $\geqq \mathbf{c}$ or $= \mathbf{c}$, respectively.

In the following model too, *the input vector* $\mathbf{a}_k$ *and the output vector* $\mathbf{b}_k$ *have the same number of components*. In particular a_{jk} and b_{jk} are amounts (this time *not* values) of the *same* good or service ($j = 1, 2, \ldots, n$). Such processes are called *von Neumann production processes* (J. VON NEUMANN, 1903–1957). They are not necessarily representing sectors (industries) like in Leontief's model. It is assumed that the economy consists of r such production processes:

$$\begin{pmatrix} \mathbf{a}_1 \\ \mathbf{b}_1 \end{pmatrix} = \begin{pmatrix} a_{11} \\ \vdots \\ a_{n1} \\ b_{11} \\ \vdots \\ b_{n1} \end{pmatrix} \in \mathbb{R}^{2n}_{+} \dots \begin{pmatrix} \mathbf{a}_r \\ \mathbf{b}_r \end{pmatrix} = \begin{pmatrix} a_{1r} \\ \vdots \\ a_{nr} \\ b_{1r} \\ \vdots \\ b_{nr} \end{pmatrix} \in \mathbb{R}^{2n}_{+}.$$

So a_{jk} and b_{jk} are the quantities of the j-th good or service entering in or emerging from the k-th process ($j = 1, \dots, n$; $k = 1, \dots, r$).

The linear combination (see Sects. 1.5 and 2.3) of these processes, that is, the linear technology (see Sect. 2.3) generated by these vectors with arbitrary "intensities" $x_1, \dots, x_r \in \mathbb{R}_+$, namely

$$x_1 \begin{pmatrix} \mathbf{a}_1 \\ \mathbf{b}_1 \end{pmatrix} + \dots + x_r \begin{pmatrix} \mathbf{a}_r \\ \mathbf{b}_r \end{pmatrix} = \begin{pmatrix} \mathbf{A} \\ \mathbf{B} \end{pmatrix} \mathbf{x}, \tag{4.26}$$

where

$$\mathbf{A} = \begin{pmatrix} a_{11} & \dots & a_{1r} \\ \vdots & & \vdots \\ a_{n1} & \dots & a_{nr} \end{pmatrix}, \ \mathbf{B} = \begin{pmatrix} b_{11} & \dots & b_{1r} \\ \vdots & & \vdots \\ b_{n1} & \dots & b_{nr} \end{pmatrix}, \ \mathbf{x} = \begin{pmatrix} x_1 \\ \vdots \\ x_r \end{pmatrix}$$

is a *von Neumann technology*. The production model thus represented is, under some further assumptions, a *von Neumann model of an expanding economy*.

If the row vector $\mathbf{p} = (p_1, \dots, p_n)$ is the *price vector* ("price system") for the n goods, then the row vectors $\mathbf{pA}$ and $\mathbf{pB}$ give the values of the inputs and outputs of the individual processes, respectively (with "intensity 1") while the scalars $\mathbf{pAx}$ and $\mathbf{pBx}$ are the values of the *combined inputs* and *combined outputs* of the process (4.26).

Now we take also the "period" (denoted by a positive integer, $t \in \mathbb{N}$), that is the time interval of always equal length into consideration in which the process runs, and assume that the intensity vector $\mathbf{x}$ depends upon $t : \mathbf{x}(t)$ and so do the input and output quantities $a_j(t)$ and $b_j(t)$ of the j-th good or service. Then we have

$$a_j(t) = \mathbf{A}_j\mathbf{x}(t) \quad \text{and} \quad b_j(t) = \mathbf{B}_j\mathbf{x}(t) \quad (j = 1, 2, \dots, n), \tag{4.27}$$

where $\mathbf{A}_j$ and $\mathbf{B}_j$ are the j-th row vectors (the row vectors containing the components in the j-th rows of the above matrices $\mathbf{A}$ and $\mathbf{B}$, respectively; remember: a row vector times a column vector is a scalar).

In the next period, $t + 1$, the input $a_j(t + 1)$ is less than or equal to the previous output $b_j(t)$. So, by (4.27),

$$a_j(t+1) \leq \mathbf{B}_j\mathbf{x}(t) \quad (j = 1, 2, \ldots, n). \tag{4.28}$$

The *growth rate* (= growth rate in percent divided by 100) of the j-th good or service is defined by

$$\hat{a}_j(t) := \frac{a_j(t+1) - a_j(t)}{a_j(t)},$$

so that, by (4.28) we get for the "*growth factor*" $1 + \hat{a}_j(t)$ the inequality

$$1 + \hat{a}_j(t) = \frac{a_j(t+1)}{a_j(t)} \leq \frac{\mathbf{B}_j\mathbf{x}(t)}{\mathbf{A}_j\mathbf{x}(t)}.$$

One sees that, if $\mathbf{x}(t)$ is replaced by $\gamma\mathbf{x}(t)$ with arbitrary $\gamma \in \mathbb{R}_+$, then the right hand side of this inequality does not change, so that the *direction rather than the length of the vector* $\mathbf{x}(t)$ *determines how large the grow rate* $\hat{a}_j(t)$ *can get.*

The following are two problems originating from the von Neumann model:

Problem 1 Determine the intensity vector $\mathbf{x}(t) \in \mathbb{R}^n_+$ so that we have a *balance of growth*, that is *the input of all goods and services grows with the same constant growth rate*:

$$1 + \hat{a}_j(t) = \frac{a_j(t+1)}{a_j(t)} = \alpha \text{ (constant) for all } t \in \mathbb{N} \text{ and for } j = 1, \ldots, n. \tag{4.29}$$

J. von Neumann was also interested in finding the greatest possible growth factor α. By (4.27) and (4.28) relation (4.29) implies for all $t \in \mathbb{N}$

$$\alpha a_j(t) = \alpha\mathbf{A}_j\mathbf{x}(t) \leq \mathbf{B}_j\mathbf{x}(t) \quad (j = 1, \ldots, n),$$

that is,

$$\alpha\mathbf{A}\mathbf{x}(t) \leqq \mathbf{B}\mathbf{x}(t) \tag{4.30}$$

or

$$(\mathbf{B} - \alpha\mathbf{A})\mathbf{x}(t) \geqq \mathbf{0}. \tag{4.31}$$

Obviously, *one has to look for that set* $S_\alpha \subseteq \mathbb{R}^r_+$ *which is mapped by the linear function*

$$\mathbf{x} \mapsto \mathbf{f}_\alpha(\mathbf{x}) = (\mathbf{B} - \alpha\mathbf{A})\mathbf{x}$$

into $\mathbb{R}^n_+$. Notice that condition (4.30) or (4.31) makes the model closed, that is, α times the inputs for a given period are not larger than the outputs of the previous period.

Problem 2 *Which price vector* $\mathbf{p} = (p_1, \ldots, p_n) \in \mathbb{R}^n_+$ *is compatible with the existence of a constant* $\beta \in \mathbb{R}_+$ *such that*

$$\beta \mathbf{pA} \geqq \mathbf{pB}, \qquad \text{that is,} \qquad \mathbf{p}(\mathbf{B}-\beta\mathbf{A}) \leqq \mathbf{0}. \tag{4.32}$$

(a row vector, that is, a $1 \times n$ matrix multiplied by an $n \times r$ matrix is a $1 \times r$ row vector). Here it is also of interest to find the smallest β for which there exists a $\mathbf{p} \geqq 0$ satisfying the inequality (4.32).

The constant β may be considered an interest factor (= 1 + interest rate = 1 + (interest rate in percent)/100). If β is given the prices $(p_1, \ldots, p_n) = \mathbf{p}$ have to be established so that the vector $\beta \mathbf{pA}$ of the input values increased by interest should not be smaller than the vector $\mathbf{pB}$ of output values. Then there does not exist an extra profit in the economy. One can see that, whenever $\mathbf{p}$ satisfies (4.32), so does $\gamma\mathbf{p}$ with arbitrary $\gamma \in \mathbb{R}_+$, so that the length of the vector $\mathbf{p}$ is again irrelevant for the validity of (4.32).

We can give Problem 2 a meaning similar to what we said about Problem 1: *We are looking for the set* $S_\beta \subseteq \mathbb{R}^n_+$ *which is mapped by the linear function* $\mathbf{p} \mapsto \mathbf{f}_\beta(\mathbf{p}) - \mathbf{p}(\mathbf{B} - \beta\mathbf{A})$ *into* $\mathbb{R}^r_-$.

(Systems of)*linear inequalities*, like those in Problems 1 and 2 will be considered again in Chap. 5.

4.5.1 Exercises

1. Take (4.23) with $n = 2$ (two sectors) and with

$$\mathbf{v}_1 = \begin{pmatrix} 50 \\ 75 \\ 250 \\ 0 \end{pmatrix}, \quad \mathbf{v}_2 = \begin{pmatrix} 15 \\ 60 \\ 0 \\ 150 \end{pmatrix}.$$

   (a) Determine the Leontief processes $\boldsymbol{\ell}_1$ and $\boldsymbol{\ell}_2$.
   (b) Determine all linear combinations (with nonnegative intensities x_1, x_2) of these Leontief processes.
   (c) Determine the Leontief matrix.
   (d) Determine the intensities so that the final demand vector $\mathbf{c} = \begin{pmatrix} 215 \\ 60 \end{pmatrix}$ will be reached exactly.

2. For the matrices $\mathbf{A} = \begin{pmatrix} 3 & 4 \\ 2 & 5 \end{pmatrix}$, $\mathbf{B} = \begin{pmatrix} 3.6 & 4.0 \\ 5.0 & 5.4 \end{pmatrix}$ find vectors $\begin{pmatrix} x_1 \\ x_2 \end{pmatrix} \in \mathbb{R}^2_+$ and $(p_1, p_2) \in \mathbb{R}^2_+$ such that $(\mathbf{B} - \alpha\mathbf{A})\mathbf{x} \geqq \mathbf{0}$ and $\mathbf{p}(\mathbf{B} - \beta\mathbf{A}) \leqq \mathbf{0}$ for

   (a) $\alpha = 1.05,\ \beta = 1.05$, (b) $\alpha = 1.1,\ \beta = 1.15$,
   (c) $\alpha = 1.15,\ \beta = 1.05$, (d) $\alpha = 1.05,\ \beta = 1.1$.

3. Determine the maximal α and the minimal β such that the inequalities in Exercise 2 have solutions

$$\begin{pmatrix} x_1 \\ x_2 \end{pmatrix} \in \mathbb{R}^2_+ \quad \text{and} \quad (p_1, p_2) \in \mathbb{R}^2_+$$

   that are different from $\begin{pmatrix} 0 \\ 0 \end{pmatrix}$.
4. Do Exercise 3 (with the inequalities in Exercise 2) for the following matrices
   (a) $\mathbf{A} = \begin{pmatrix} 1 & 2 \\ 3 & 4 \end{pmatrix}$, $\mathbf{B} = \begin{pmatrix} 1.1 & 2.2 \\ 3.3 & 4.4 \end{pmatrix}$,
   (b) $\mathbf{A} = \begin{pmatrix} 1 & 2 \\ 3 & 4 \end{pmatrix}$, $\mathbf{B} = \begin{pmatrix} 1.1 & 2.2 \\ 3.6 & 4.8 \end{pmatrix}$.
5. Do Exercise 3 (with the inequalities in Exercise 2) for the following matrices
   (a) $\mathbf{A} = \begin{pmatrix} 2 & 3 \\ 4 & 1 \end{pmatrix}$, $\mathbf{B} = \begin{pmatrix} 2.2 & 3.6 \\ 4.2 & 1.2 \end{pmatrix}$,
   (b) $\mathbf{A} = \begin{pmatrix} 2 & 3 \\ 4 & 1 \end{pmatrix}$, $\mathbf{B} = \begin{pmatrix} 2.2 & 3.6 \\ 4.2 & 1.1 \end{pmatrix}$.

4.5.2 Answers

1. (a) $\boldsymbol{\ell}_1 = \begin{pmatrix} 0.2 \\ 0.3 \\ 1 \\ 0 \end{pmatrix}$, $\boldsymbol{\ell}_2 = \begin{pmatrix} 0.1 \\ 0.4 \\ 0 \\ 1 \end{pmatrix}$, (b) $x_1\boldsymbol{\ell}_1 + x_2\boldsymbol{\ell}_2 = \begin{pmatrix} 0.2x_1 + 0.1x_2 \\ 0.3x_1 + 0.4x_2 \\ x_1 \\ x_2 \end{pmatrix}$,

   (c) $(\mathbf{I} - \mathbf{A}) = \begin{pmatrix} 1 - 0.2 & -0.1 \\ -0.3 & 1 - 0.4 \end{pmatrix} = \begin{pmatrix} 0.8 & -0.1 \\ -0.3 & 0.6 \end{pmatrix}$,

   (d) $\begin{pmatrix} x_1 \\ x_2 \end{pmatrix} = \begin{pmatrix} 300 \\ 250 \end{pmatrix}$.
3. $\alpha = \beta = 1.2$.
4. (a) $\alpha = \beta = 1.1$, (b) $\alpha = \beta = 1.1$.
5. (a) $\alpha = \beta = 1.2$, (b) $\alpha = \beta = 1.1$.

4.6 Systems of Linear Equations. Solution by Elimination. Rank. Necessary and Sufficient Conditions

We saw in the previous Sect. 4.5 that satisfying "without surplus" the "final demand" of the economy leads in the Leontief model to the problem of finding that vector (or those vectors) $\mathbf{x} \in \mathbb{R}^n_+$ for which the value of the linear function $f : \mathbb{R}^n_+ \to \mathbb{R}^n_+$ defined by $\mathbf{f}(\mathbf{x}) = (\mathbf{I} - \mathbf{A})\mathbf{x}$ is the given final demand vector $\mathbf{c} \in \mathbb{R}^n_+$. This was found to be equivalent to solving the system of linear equations (4.25), that is, finding those nonnegative numbers $x_1, \ldots, x_n$ for which all equations in (4.25) are satisfied.

There we had the same number of equations (n) as the number of the unknown numbers $x_1, \ldots, x_n$ ("unknowns" for short). In general in a (real) *system of linear equations* there are m equations for n unknowns:

$$\begin{aligned} a_{11}x_1 + a_{12}x_2 + \cdots + a_{1n}x_n &= b_1 \\ a_{21}x_1 + a_{22}x_2 + \cdots + a_{2n}x_n &= b_2 \\ \vdots \qquad\qquad & \quad \vdots \\ a_{m1}x_1 + a_{m2}x_2 + \cdots + a_{mn}x_n &= b_m. \end{aligned}$$

Here the $a_{jk} \in \mathbb{R}$ ($j = 1, 2, \ldots, m$; $k = 1, 2, \ldots, n$) are the *coefficients*, $b_1, \ldots, b_m$ the "*constant terms*" or "*coefficients of -1*" and, in general, we look also for the "*solutions*" $x_1, \ldots, x_n$ among the real numbers. (The $x_1, \ldots, x_n$ are "unknowns" before we determined their value, "solutions" when we know which numbers they are but the distinction is not very important and the two words are often used interchangeably.)

In vector-matrix notation, with

$$\mathbf{x} = \begin{pmatrix} x_1 \\ \vdots \\ x_n \end{pmatrix}, \mathbf{A} = \begin{pmatrix} a_{11} & \ldots & a_{1n} \\ \vdots & & \vdots \\ a_{m1} & \ldots & a_{mn} \end{pmatrix}, \mathbf{b} = \begin{pmatrix} b_1 \\ \vdots \\ b_m \end{pmatrix},$$

the above system of linear equations can be written as

$$\mathbf{A}\mathbf{x} = \mathbf{b},$$

where $\mathbf{A}$ is the "*matrix of coefficients*" or "*coefficient matrix*", $\mathbf{x}$ the "*unknown vector*" and $\mathbf{b}$ the "constant vector". If we introduce the "*column vectors*"

$$\mathbf{a}_k = \begin{pmatrix} a_{1k} \\ \vdots \\ a_{mk} \end{pmatrix} \qquad (k = 1, \ldots, n)$$

of $\mathbf{A}$, then we obtain another equivalent form of these equations:

$$\mathbf{a}_1 x_1 + \ldots + \mathbf{a}_n x_n = \mathbf{b}.$$

(using, from Sect. 1.5, the rules for forming linear combinations of vectors). If in these equations $\mathbf{b} = \mathbf{0}$, that is $b_1 = \ldots = b_m = 0$, then the system of *equations* is "*homogeneous*", otherwise "*inhomogeneous*".

We are looking for *necessary and sufficient conditions* for the *existence* (and *uniqueness*) of solutions $\mathbf{x}$ (or $x_1, \ldots, x_n$) and for ways to *determine* these solutions.

We begin with several examples and analyse them in order to get general conditions and methods.

Example 1 is a system of three equations with three unknowns:

$$\begin{array}{rcl|r|r|r} x_1 + 2x_2 \phantom{{}+ x_3} &=& 2 & 2 & \frac{-1}{2} & \frac{-5}{4} \\ x_1 + 4x_2 + x_3 &=& 1 & -1 & \frac{1}{2} & \frac{1}{2} \\ 3x_1 + 2x_2 + 2x_3 &=& 12 & & & \frac{1}{4} \end{array} \tag{4.33}$$

(Notice that not every unknown has to figure in every equation, for instance, there is no x_3 in the first equation; there its coefficient is 0.) A time honoured method is the "elimination of unknowns". This consists of trying, by multiplying equations by constants and adding them (that is, taking their linear combinations), to transform this system into another, consisting of equations, each of which contains just one unknown (later: as few unknowns as possible). Then we can solve them individually as single linear equations of one unknown each (we will try even to get the solutions explicitly in the last step).

First we want to eliminate x_2 from the first two equations. For this we multiply the first equation by 2 and the second by (-1) and add the resulting equations. We can also eliminate x_1 from these equations by multiplying the first by $(-1/2)$ and the second by $(1/2)$ and again add the equations thus obtained. (We could also multiply by (-1) and by 1; we chose $(-1/2)$ and $(1/2)$ in order to get x_2 with coefficient 1. We listed the multipliers on the right of the equations above for better understanding and easier checking.) So we get

$$\begin{array}{rcr} 2x_1 + 4x_2 \phantom{{}- x_3} &=& 4 \\ -x_1 - 4x_2 - x_3 &=& -1 \\ \hline x_1 + 0x_2 - x_3 &=& 3 \end{array} \qquad \begin{array}{rcr} -\frac{1}{2}x_1 - x_2 \phantom{{}+ \frac{1}{2}x_3} &=& -1 \\ \frac{1}{2}x_1 + 2x_2 + \frac{1}{2}x_3 &=& \frac{1}{2} \\ \hline 0x_1 + x_2 + \frac{1}{2}x_3 &=& \frac{1}{2} \end{array}.$$

Getting x_3 with coefficient 1 and as few unknowns as possible is somewhat more complicated but multiplying the first equation by $(-5/4)$, the second by $(1/2)$, the third by $(1/4)$ and adding up will do the trick (we will show later why we chose the multipliers as we did):

$$\begin{array}{rcl}
-\frac{5}{4}x_1 - \frac{5}{2}x_2 & = & -\frac{5}{2} \\
\frac{1}{2}x_1 + 2x_2 + \frac{1}{2}x_3 & = & \frac{1}{2} \\
\frac{3}{4}x_1 + \frac{1}{2}x_2 + \frac{1}{2}x_3 & = & 3 \\
\hline
0x_1 + 0x_2 + x_3 & = & 1\,.
\end{array}$$

(The trick was to eliminate both x_1 and x_2, that is why we needed all three equations in (4.33). The multipliers could have been (-5), 2, 1; then we would have obtained $4x_3 = 4$, we divided the multipliers by 4 just for aesthetic reasons, to get x_3 with coefficient 1. But it is important that *all* equations in (4.33) were used in an essential way.)

So we have already that $x_3 = 1$ but we need this also to determine x_1 and x_2. We have three new equations (those under the lines above):

$$\begin{array}{rcl|c|c|c}
x_1 \quad - x_3 & = & 3 & 1 & & \\
x_2 + \frac{1}{2}x_3 & = & -\frac{1}{2} & & 1 & \\
x_3 & = & 1 & 1 & -\frac{1}{2} & 1\,.
\end{array} \tag{4.34}$$

This time, in order to eliminate x_3 from the first and third equation it is enough to add them; to do the same from the second and third equation we add $\left(-\frac{1}{2}\right)$ times the third equation to the second:

$$\begin{array}{rcl}
x_1 + 0x_2 + 0x_3 & = & 4 \\
x_2 + 0x_3 & = & -1 \\
x_3 & = & 1\,.
\end{array}$$

(The third equation we just copied from (4.34). Again we indicated in (4.34) the multipliers on the right of the equations.) But *now we have the solutions* "served on a plate":

$$x_1 = 4,\ x_2 = -1,\ x_3 = 1.$$

One checks readily that *they satisfy the equations* (4.33): $4 + 2 \cdot (-1) = 2$, $4 + 4 \cdot (-1) + 1 = 1$, $3 \cdot 4 + 2 \cdot (-1) + 2 \cdot 1 = 12$; of course they satisfy also (4.34) and the process shows that there are no other solutions. In order to get just the right solutions, it was important that all the time *the "old" k'th equation should not be omitted from the transformation yielding the "new"* k'th equation (that is, *in the linear combination of old equations yielding the new k'th equation, the multiplier of the old k'th equation should not be* 0 $(k = 1, 2, \ldots, n)$).

Let us register what happened during these transformations to the matrix of coefficients in (4.33):

$$\begin{pmatrix} 1 & 2 & 0 \\ 1 & 4 & 1 \\ 3 & 2 & 2 \end{pmatrix} \mapsto \begin{pmatrix} 1 & 0 & -1 \\ 0 & 1 & 1/2 \\ 0 & 0 & 1 \end{pmatrix} \mapsto \begin{pmatrix} 1 & 0 & 0 \\ 0 & 1 & 0 \\ 0 & 0 & 1 \end{pmatrix}.$$

So we *transformed it into the matrix* **I**. The same transformations on the "constant vector" on the right hand side of (4.33) had the effect of yielding the "solution vector":

$$\begin{pmatrix} 2 \\ 1 \\ 12 \end{pmatrix} \mapsto \begin{pmatrix} 3 \\ -1/2 \\ 1 \end{pmatrix} \mapsto \begin{pmatrix} 4 \\ -1 \\ 1 \end{pmatrix} =: \mathbf{b}^*.$$

That is, $\mathbf{Ax} = \mathbf{b}$ has been transformed into $\mathbf{x} = \mathbf{Ix} = \mathbf{b}^*$. This shows again that by "solution vector" we mean the "unknown vector" after the value of its components have been calculated. The first step in both chains of transformations was to multiply the *first* row *by* 2 and add to it the second row multiplied by -1 (and the third row multiplied by 0) in order to get the new *first* row, multiply the *second* row *by* $(1/2)$ and add to it the first row multiplied by $(-1/2)$ (and the third row multiplied by 0) in order to get the new *second* row, and multiply the *third* row *by* $(1/4)$ and add to it $(1/2)$ times the second row and $(-5/4)$ times the first row to get the new *third* row.

We can register these operations as

(i) *replacing a row by a linear combination of rows as long as the coefficient (multiplier) of the original row is not* 0,
but it is more customary to break this in two:
(i′) *multiplying any row by a nonzero number* and
(i″) *adding to a row a linear combination* (also 0 coefficients permitted) *of other rows*.

We did the same things in the second step: Added the first and third row to get the new first row and added the second row to $(-1/2)$ times the third row to get the new second row; the third row was left unchanged. We did not need any more steps (transformations) since we already got the unit matrix and the solution vector. In more complicated cases we may need more steps but the results would be similar with the following variations.

Example 2 First we consider a slight variation of (4.33)

$$\begin{array}{rcl|r|r|r} x_1 \phantom{{}+x_2} + 2x_3 &=& 2 & 2 & \frac{-1}{2} & \frac{-5}{4} \\ x_1 + \ x_2 + 4x_3 &=& 1 & -1 & \frac{1}{2} & \frac{1}{2} \\ 3x_1 + 2x_2 + 2x_3 &=& 12 & 0 & 0 & \frac{1}{4}\,. \end{array} \tag{4.35}$$

By the same transformations as for (4.33) (we wrote the multipliers again to the right of the equations) we get in succession

$$\begin{array}{rcr|r|r|l} x_1 - \ x_2 \phantom{{}+x_3} &=& 3 & 1 & 0 & 0 \\ \frac{1}{2}x_2 + x_3 &=& -\frac{1}{2} & 0 & 1 & 0 \\ x_2 \phantom{{}+x_3} &=& 1 & 1 & -\frac{1}{2} & 1\,, \end{array}$$

$$\begin{aligned} x_1 &= \ x_1 + 0x_2 + 0x_3 = \ 4 \\ x_3 &= 0x_1 + 0x_2 + \ x_3 = -1 \\ x_2 &= 0x_1 + \ x_2 + 0x_3 = \ 1\,, \end{aligned}$$

and the transformations on the "constant vector" are the same, while the coefficient matrix undergoes these:

$$\begin{pmatrix} 1 & 0 & 2 \\ 1 & 1 & 4 \\ 3 & 2 & 2 \end{pmatrix} \mapsto \begin{pmatrix} 1 & -1 & 0 \\ 0 & \frac{1}{2} & 1 \\ 0 & 1 & 0 \end{pmatrix} \mapsto \begin{pmatrix} 1 & 0 & 0 \\ 0 & 0 & 1 \\ 0 & 1 & 0 \end{pmatrix}$$

and the result is not quite the unit matrix. If we multiply *this* matrix by the "unknown vector" we see that it changes the order of components

$$\begin{pmatrix} 1 & 0 & 0 \\ 0 & 0 & 1 \\ 0 & 1 & 0 \end{pmatrix} \begin{pmatrix} x_1 \\ x_2 \\ x_3 \end{pmatrix} = \begin{pmatrix} x_1 \\ x_3 \\ x_2 \end{pmatrix},$$

so $\mathbf{Ax} = \mathbf{b}$ is changed into

$$\begin{pmatrix} x_1 \\ x_3 \\ x_2 \end{pmatrix} = \begin{pmatrix} 1 & 0 & 0 \\ 0 & 0 & 1 \\ 0 & 1 & 0 \end{pmatrix} \begin{pmatrix} x_1 \\ x_2 \\ x_3 \end{pmatrix} = \mathbf{b}^* = \begin{pmatrix} 4 \\ -1 \\ 1 \end{pmatrix}.$$

No big deal, we still got the (unique) solutions of (4.35). But we can also perform a "cosmetic surgery" in order to get, at the end of the chain of transformations of the coefficient matrix, the "nice" unit matrix. We "rename" the unknown x_3 to x_2 and vice versa, which leads us back to (4.33) and on the matrix of coefficients of (4.35) results in the new operation (which, *connected with interchanging also the corresponding two components of the unknown vector* will give us the "right" x_1, x_2, x_3):

(ii) *interchanging two columns.*

So the transformations of the coefficient matrix and of the constant vector in (4.35) will now be:

$$\begin{pmatrix} 1 & 0 & 2 \\ 1 & 1 & 4 \\ 3 & 2 & 2 \end{pmatrix} \mapsto \begin{pmatrix} 1 & 2 & 0 \\ 1 & 4 & 1 \\ 3 & 2 & 2 \end{pmatrix} \mapsto \begin{pmatrix} 1 & 0 & -1 \\ 0 & 1 & \frac{1}{2} \\ 0 & 0 & 1 \end{pmatrix} \mapsto \begin{pmatrix} 1 & 0 & 0 \\ 0 & 1 & 0 \\ 0 & 0 & 1 \end{pmatrix},$$

$$\begin{pmatrix} 2 \\ 1 \\ 12 \end{pmatrix} \mapsto \begin{pmatrix} 3 \\ -\frac{1}{2} \\ 1 \end{pmatrix} \mapsto \begin{pmatrix} 4 \\ -1 \\ 1 \end{pmatrix},$$

so that $\mathbf{Ax} = \mathbf{b}$ is transformed into

$$\begin{pmatrix} x_1 \\ x_3 \\ x_2 \end{pmatrix} = \mathbf{I} \begin{pmatrix} x_1 \\ x_3 \\ x_2 \end{pmatrix} = \begin{pmatrix} 4 \\ -1 \\ 1 \end{pmatrix},$$

that is, $x_1 = 4, x_2 = 1, x_3 = -1$, as it should be. We check (4.35): $4 + 2 \cdot (-1) = 2$, $4 + 1 + 4 \cdot (-1) = 1$, $3 \cdot 4 + 2 \cdot 1 + 2 \cdot (-1) = 12$; one should *always check* whether the obtained "solutions" satisfy the equations.

Example 3 Now we examine another variation on the theme of (4.33), the system of equations

$$\begin{array}{rcl|r|r|r|r} x_1 + 2x_2 & = & 2 & 2 & -\frac{1}{2} & -\frac{5}{4} & \\ x_1 + 4x_2 + x_3 & = & 1 & -1 & \frac{1}{2} & \frac{1}{2} & 1 \\ 3x_1 + 2x_2 + 2x_3 & = & 12 & & & \frac{1}{4} & -1 \\ 2x_1 - 2x_2 + x_3 & = & 11 & & & & 1\,. \end{array} \tag{4.36}$$

We proceed as indicated by the multipliers at the right and get (the first three rows are the same as for (4.33), we have only to check the fourth):

$$\begin{array}{rcr|r|r|r|l} x_1 + 0x_2 - x_3 & = & 3 & 1 & & & \\ 0x_1 + x_2 + \frac{1}{2}x_3 & = & -\frac{1}{2} & & 1 & & \\ 0x_1 + 0x_2 + x_3 & = & 1 & 1 & -\frac{1}{2} & 1 & \\ 0x_1 + 0x_2 + 0x_3 & = & 0 & & & & 1, \end{array}$$

(continued)

$$\begin{aligned} x_1 + 0x_2 + x_3 &= 4 \\ 0x_1 + x_2 + 0x_3 &= -1 \\ 0x_1 + 0x_2 + x_3 &= 1 \\ 0x_1 + 0x_2 + 0x_3 &= 0\,, \end{aligned}$$

that is $x_1 = 4$, $x_2 = -1$, $x_3 = 1$. After the first step, the last equation is $0 = 0$, even though we have used the last equation of (4.36) essentially (with multiplier 1 and just once). Nevertheless $x_1 = 4$, $x_2 = -1$, $x_3 = 1$ which we got previously from (4.33) here from the first three equations, satisfies also this fourth in (4.36): $2 \cdot 4 - 2 \cdot (-1) + 1 = 11$. This means that the fourth equation is *redundant*: it follows from the first three equations of (4.36): it is obtained by subtracting the second equation from the third.

Example 4 We change now just one thing in (4.36): the right hand side of the fourth equation:

$$\begin{array}{rcl|r|r|r|l} x_1 + 2x_2 \phantom{{}+ 2x_3} &=& 2 & 2 & -\frac{1}{2} & -\frac{5}{4} & \\ x_1 + 4x_2 + x_3 &=& 1 & -1 & \frac{1}{2} & \frac{1}{2} & 1 \\ 3x_1 + 2x_2 + 2x_3 &=& 12 & & & \frac{1}{4} & -1 \\ 2x_1 - 2x_2 + x_3 &=& 14 & & & & 1\,. \end{array} \tag{4.37}$$

We could get $x_1 = 4$, $x_2 = -1$, $x_3 = 1$ (from the first three equations), but they would not satisfy the fourth equation: this shows again how important checking is. But, applying the same transformations as before, we get $x_1 - x_3 = 3$, $x_2 + (1/2)x_3 = -1/2$, $x_3 = 1$ and $0 = 3$ (!).

The last equation is *nonsense*. That is because the system (4.37) is *contradictory*. Indeed, if we subtract the second equation of (4.37) from the third, we get $2x_1 - 2x_2 + x_3 = 11$ which contradicts the fourth equation. The transformations on the matrices of coefficients (they are the same in (4.36) and (4.37)) and on the constant vectors (which are different) are now

$$\begin{pmatrix} 1 & 2 & 0 \\ 1 & 4 & 1 \\ 3 & 2 & 2 \\ 2 & -2 & 1 \end{pmatrix} \mapsto \begin{pmatrix} 1 & 0 & -1 \\ 0 & 1 & \frac{1}{2} \\ 0 & 0 & 1 \\ 0 & 0 & 0 \end{pmatrix} \mapsto \begin{pmatrix} 1 & 0 & 0 \\ 0 & 1 & 0 \\ 0 & 0 & 1 \\ 0 & 0 & 0 \end{pmatrix}$$

(continued)

(the unit matrix extended by a (0, 0, 0) row),

$$\begin{pmatrix} 2 \\ 1 \\ 12 \\ 11 \end{pmatrix} \mapsto \begin{pmatrix} 3 \\ -\frac{1}{2} \\ 1 \\ 0 \end{pmatrix} \mapsto \begin{pmatrix} 4 \\ -1 \\ 1 \\ 0 \end{pmatrix}$$

and

$$\begin{pmatrix} 2 \\ 1 \\ 12 \\ 14 \end{pmatrix} \mapsto \begin{pmatrix} 3 \\ -\frac{1}{2} \\ 1 \\ 3 \end{pmatrix} \mapsto \begin{pmatrix} 4 \\ -1 \\ 1 \\ 3 \end{pmatrix},$$

respectively. So $\mathbf{A}\mathbf{x} = \mathbf{A}\begin{pmatrix} x_1 \\ x_2 \\ x_3 \end{pmatrix} = \mathbf{b}$ was transformed into

$$\begin{pmatrix} x_1 \\ x_2 \\ x_3 \\ 0 \end{pmatrix} = \begin{pmatrix} 1 & 0 & 0 \\ 0 & 1 & 0 \\ 0 & 0 & 1 \\ 0 & 0 & 0 \end{pmatrix} \mathbf{x} = \mathbf{b}',$$

possible if $\mathbf{b}' = \begin{pmatrix} 4 \\ -1 \\ 1 \\ 0 \end{pmatrix}$ but not if $\mathbf{b}' = \begin{pmatrix} 4 \\ -1 \\ 1 \\ 3 \end{pmatrix}$.

Contrary to popular belief even if we have as many equations as variables, the solution may not be unique.

Example 5 Take the system

$$\begin{array}{rcl|r|r|r|r} x_1 + 2x_2 \phantom{{}+ x_3} + 4x_4 &=& 2 & 2 & -\frac{1}{2} & -\frac{5}{4} & \\ x_1 + 4x_2 + x_3 + 3x_4 &=& 1 & -1 & \frac{1}{2} & \frac{1}{2} & 1 \\ 3x_1 + 2x_2 + 2x_3 + 2x_4 &=& 12 & & & \frac{1}{4} & -1 \\ 2x_1 - 2x_2 + x_3 - x_4 &=& 11 & & & & 1 \end{array} \tag{4.38}$$

(continued)

which also bears some similarity with the previous ones. Again with the multipliers indicated on the right, we get

$$\begin{array}{rcl|c|c|c|c} x_1 + 0x_2 - \phantom{\tfrac{1}{2}}x_3 + 5x_4 &=& 3 & 1 & & & \\ 0x_1 + x_2 + \tfrac{1}{2}x_3 - \tfrac{1}{2}x_4 &=& -\tfrac{1}{2} & & 1 & & \\ 0x_1 + 0x_2 + \phantom{\tfrac{1}{2}}x_3 - 3x_4 &=& 1 & 1 & -\tfrac{1}{2} & 1 & \\ 0x_1 + 0x_2 + 0x_3 + 0x_4 &=& 0 & & & & 1\,, \end{array}$$

$$\begin{array}{rcl} x_1 \qquad\qquad + 2x_4 &=& 4 \\ x_2 \qquad + x_4 &=& -1 \\ x_3 - 3x_4 &=& 1 \\ 0 &=& 0\,. \end{array}$$

This shows that the fourth equation of (4.38) is again redundant and *there is an arbitrary "parameter" (variable)* $x_4 =: \lambda$ *in the solution*. Indeed

$$x_1 = 4 - 2\lambda, \qquad x_2 = -1 - \lambda, \qquad x_3 = 1 + 3\lambda, \qquad x_4 = \lambda$$

satisfies (4.38) for all $\lambda \in \mathbb{R}$:

$$\begin{aligned} (4-2\lambda) + 2(-1-\lambda) + 4\lambda &= 2\,, \\ (4-2\lambda) + 4(-1-\lambda) + (1+3\lambda) + 3\lambda &= 1\,, \\ 3(4-2\lambda) + 2(-1-\lambda) + 2(1+3\lambda) + 2\lambda &= 12\,, \\ 2(4-2\lambda) - 2(-1-\lambda) + (1+3\lambda) - \lambda &= 11\,. \end{aligned}$$

We look again at the transformations of the matrices of coefficients and of the constant vector:

$$\begin{pmatrix} 1 & 2 & 0 & 4 \\ 1 & 4 & 1 & 3 \\ 3 & 2 & 2 & 2 \\ 2 & -2 & 1 & -1 \end{pmatrix} \mapsto \begin{pmatrix} 1 & 0 & -1 & 5 \\ 0 & 1 & \tfrac{1}{2} & -\tfrac{1}{2} \\ 0 & 0 & 1 & -3 \\ 0 & 0 & 0 & 0 \end{pmatrix} \mapsto \begin{pmatrix} 1 & 0 & 0 & 2 \\ 0 & 1 & 0 & 1 \\ 0 & 0 & 1 & -3 \\ 0 & 0 & 0 & 0 \end{pmatrix},$$

$$\begin{pmatrix} 2 \\ 1 \\ 12 \\ 11 \end{pmatrix} \mapsto \begin{pmatrix} 3 \\ -\tfrac{1}{2} \\ 1 \\ 0 \end{pmatrix} \mapsto \begin{pmatrix} 4 \\ -1 \\ 1 \\ 0 \end{pmatrix} \quad \text{(the same as for (4.36)).}$$

(continued)

So $\mathbf{Ax} = \mathbf{A}\begin{pmatrix} x_1 \\ x_2 \\ x_3 \\ x_4 \end{pmatrix} = \mathbf{b}$ was transformed into

$$\begin{pmatrix} x_1 + 2x_4 \\ x_2 + \ x_4 \\ x_3 - 3x_4 \\ 0 \end{pmatrix} = \begin{pmatrix} 1\,0\,0 & 2 \\ 0\,1\,0 & 1 \\ 0\,0\,1 & -3 \\ 0\,0\,0 & 0 \end{pmatrix}\begin{pmatrix} x_1 \\ x_2 \\ x_3 \\ x_4 \end{pmatrix} = \begin{pmatrix} 4 \\ -1 \\ 1 \\ 0 \end{pmatrix}$$

which again gives

$$x_1 = 4 - 2x_4 = 4 - 2\lambda, \qquad x_2 = -1 - x_4 = -1 - \lambda,$$
$$x_3 = 1 + 3x_4 = 1 + 3\lambda, \qquad x_4 = \lambda.$$

In most of these transformations we used the operations (i′) and (i″) (or, equivalently, (i)). In our last example we will apply, in addition to (i) also (ii). It will be a system of four equations for five unknowns. *If the number of unknowns is larger than the number of equations* (and the system is not *contradictory*) then there are always parameters in the solution (though, as we saw, this can happen also when there are as many unknowns as equations).

Example 6 Take the system

$$\begin{array}{rcl|r|r|r|r} x_1 + 2x_2 - \ x_3 \qquad + 4x_5 &=& 2 & 2 & -\frac{1}{2} & -\frac{5}{4} & \\ x_1 + 4x_2 - \ 5x_3 + \ x_4 + 3x_5 &=& 1 & -1 & \frac{1}{2} & \frac{1}{2} & 1 \\ 3x_1 + 2x_2 + \ 5x_3 + 2x_4 + 2x_5 &=& 12 & & & \frac{1}{4} & -1 \\ 2x_1 - 2x_2 + 10x_3 + \ x_4 - \ x_5 &=& 11 & & & & 1\,. \end{array} \tag{4.39}$$

With the indicated multipliers we get in succession

$$\begin{array}{rcl|r|r|r|r} x_1 + 0x_2 + 3x_3 - \ x_4 + 5x_5 &=& 3 & 1 & & & \\ 0x_1 + \ x_2 - 2x_3 + \frac{1}{2}x_4 - \frac{1}{2}x_5 &=& -\frac{1}{2} & & 1 & & \\ 0x_1 + 0x_2 + 0x_3 + \ x_4 - 3x_5 &=& 1 & 1 & -\frac{1}{2} & 1 & \\ 0x_1 + 0x_2 + 0x_3 + 0x_4 + 0x_5 &=& 0 & & & & 1\,, \end{array}$$

(continued)

$$\begin{aligned} x_1 \quad + 3x_3 \quad + 2x_5 &= 4 \\ x_2 - 2x_3 \quad + \ x_5 &= -1 \\ x_4 - 3x_5 &= 1 \\ 0 &= 0\,. \end{aligned}$$

So the fourth *equation* is *redundant* also in (4.39) and in the *solution* of (4.39) there are this time *two arbitrary parameters* $x_3 = \lambda_1$ and $x_5 = \lambda_2$. Indeed

$$\begin{aligned} &x_1 = 4 - 3\lambda_1 - 2\lambda_2, \quad x_2 = -1 + 2\lambda_1 - \lambda_2, \\ &x_3 = \lambda_1, \quad x_4 = 1 + 3\lambda_2, \quad x_5 = \lambda_2 \end{aligned} \tag{4.40}$$

satisfy (4.39), whatever $\lambda_1 \in \mathbb{R}$, $\lambda_2 \in \mathbb{R}$ are (check!). With transformation of the coefficient matrix we run into a situation as we had for (4.33):

$$\begin{pmatrix} 1 & 2 & -1 & 0 & 4 \\ 1 & 4 & -5 & 1 & 3 \\ 3 & 2 & 5 & 2 & 2 \\ 2 & -2 & 10 & 1 & -1 \end{pmatrix} \mapsto \begin{pmatrix} 1 & 0 & 3 & -1 & 5 \\ 0 & 1 & -2 & \frac{1}{2} & -\frac{1}{2} \\ 0 & 0 & 0 & 1 & -3 \\ 0 & 0 & 0 & 0 & 0 \end{pmatrix}.$$

We interrupt the chain here because we see already that we cannot get this way our accustomed

$$\begin{pmatrix} 1 & 0 & 0 \\ 0 & 1 & 0 \\ 0 & 0 & 1 \end{pmatrix}$$

in the upper left corner since the third row starts with three 0's. But we can remedy this situation, as we saw for (4.33): We interchange the third and fourth column (and simultaneously also the corresponding two components of the unknown vector):

$$\begin{pmatrix} 1 & 0 & -1 & 3 & 5 \\ 0 & 1 & \frac{1}{2} & -2 & -\frac{1}{2} \\ 0 & 0 & 1 & 0 & -3 \\ 0 & 0 & 0 & 0 & 0 \end{pmatrix} \mapsto \begin{pmatrix} 1 & 0 & 0 & 3 & 2 \\ 0 & 1 & 0 & -2 & 1 \\ 0 & 0 & 1 & 0 & -3 \\ 0 & 0 & 0 & 0 & 0 \end{pmatrix},$$

$$\begin{pmatrix} 2 \\ 1 \\ 12 \\ 11 \end{pmatrix} \mapsto \begin{pmatrix} 3 \\ -\frac{1}{2} \\ 1 \\ 0 \end{pmatrix} \mapsto \begin{pmatrix} 4 \\ -1 \\ 1 \\ 0 \end{pmatrix}.$$

So $\mathbf{Ax} = \mathbf{b}$ is transformed into

$$\begin{pmatrix} x_1 + 3x_3 + 2x_5 \\ x_2 - 2x_3 + x_5 \\ x_4 - 3x_5 \\ 0 \end{pmatrix} = \begin{pmatrix} 1 & 0 & 0 & 3 & 2 \\ 0 & 1 & 0 & -2 & 1 \\ 0 & 0 & 1 & 0 & -3 \\ 0 & 0 & 0 & 0 & 0 \end{pmatrix} \begin{pmatrix} x_1 \\ x_2 \\ x_4 \\ x_3 \\ x_5 \end{pmatrix} = \begin{pmatrix} 4 \\ -1 \\ 1 \\ 0 \end{pmatrix},$$

which, again with $x_3 =: \lambda_1, x_5 =: \lambda_2$, gives (4.40).

All the above suggests the concept of *rank of a matrix*. This is *the number r of linearly independent row vectors* of the $m \times n$ matrix

$$\mathbf{A} = \begin{pmatrix} a_{11} & \dots & a_{1n} \\ \vdots & & \vdots \\ a_{m1} & \dots & a_{mn} \end{pmatrix}.$$

The *row vectors* of $\mathbf{A}$ are, of course, the vectors consisting of the rows of $\mathbf{A}$ which played such an important role above:

$$\mathbf{a}'_1 = (a_{11}, \dots, a_{1n}), \dots, \mathbf{a}'_m = (a_{m1}, \dots, a_{mn}).$$

One can show that *the rank r equals* also *the number of linearly independent column vectors* in the matrix $\mathbf{A}$. Clearly, $r \leq m$ *and* $r \leq n$. The important thing is that *the rank of a matrix* $\mathbf{A}$ *does not change if the matrix undergoes the operations* (i) (or equivalently (i′) and (i″)) and (ii) or, more generally,

(I) *replacing a row (or column) by a linear combination of the rows (columns) in which the coefficient of the original row (column) is not different from* 0,

which can be split into the following two:

(I′) *multiplying any row or column by a nonzero number,*
(II″) *adding to a row (or column) a linear combination of other rows (columns)*,

and

(II) *interchanging two rows or two columns.*

Of course, (I), (I′); (I″), (II) differ from (i), (i′), (i″), (ii) only insofar that the same operations are permitted either for rows or for columns. That the rank of the matrix does not change (is "*invariant*") under these operations, follows from the fact, important in itself, that *the rank* of $\mathbf{A}$ *is* just *the dimension of the space spanned* (see Sect. 1.5 **2**) *by the row vectors* of $\mathbf{A}$ (and also *by its column vectors*) and from the fact, equally easy to see, that (I′), (I″) and (II) do not change this space.

We have seen in our examples and it is true also in general, that *every matrix can be brought by* (several applications of) *the operations* (I) *and* (II) *to the form*

$$\mathbf{C} =: \begin{pmatrix} 1 & 0 & \dots & 0 & c_{1,r+1} & \dots & c_{1,n} \\ 0 & 1 & \dots & 0 & c_{2,r+1} & \dots & c_{2,n} \\ \vdots & \vdots & \ddots & \vdots & \vdots & & \vdots \\ 0 & 0 & \dots & 1 & c_{r,r+1} & \dots & c_{r,n} \\ 0 & 0 & \dots & 0 & 0 & \dots & 0 \\ \vdots & \vdots & & \vdots & \vdots & & \vdots \\ 0 & 0 & \dots & 0 & 0 & \dots & 0 \end{pmatrix} \tag{4.41}$$

(we put commas between the subscripts of c in order to avoid ambiguity).

Now we can give the promised *necessary and sufficient condition for the existence of a solution* $\mathbf{x}$ *of* $\mathbf{Ax} = \mathbf{b}$:

The system of linear equations

$$\begin{aligned} a_{11}x_1 + \dots + a_{1n}x_n &= b_1 \\ \vdots \qquad\qquad & \quad \vdots \\ a_{m1}x_1 + \dots + a_{mn}x_n &= b_m \end{aligned} \tag{4.42}$$

has a solution, if and only if, the matrices

$$\mathbf{A} = \begin{pmatrix} a_{11} & \dots & a_{1n} \\ \vdots & & \vdots \\ a_{m1} & \dots & a_{mn} \end{pmatrix} \quad \text{and} \quad \mathbf{B} := \begin{pmatrix} a_{11} & \dots & a_{1n} & b_1 \\ \vdots & & \vdots & \vdots \\ a_{m1} & \dots & a_{mn} & b_m \end{pmatrix}$$

have the same rank.

As mentioned at the beginning of this section, with help of the column vectors $\mathbf{a}_1, \dots, \mathbf{a}_n$ of $\mathbf{A}$ we can always write (4.42) as

$$\mathbf{a}_1 x_1 + \dots + \mathbf{a}_n x_n = \mathbf{b} \tag{4.43}$$

Note also that *the space* U *spanned by the* n *vectors* $\mathbf{a}_1, \dots, \mathbf{a}_n$ *is always a subset of the space* V *spanned by* $\mathbf{a}_1, \dots, \mathbf{a}_n, \mathbf{b}$ (because of the additional vector $\mathbf{b}$): $U \subset V$.

A condition is *sufficient* for a statement to be true if the statement follows from the condition, it is *necessary* if the condition follows from the statement. The condition which follows "if" is sufficient, that which follows "only if" is necessary (see also Appendix).

So, *we prove first that, if the ranks of* $\mathbf{A}$ *and* $\mathbf{B}$ *are equal, then* (4.42) *has a solution* $(x_1, \dots, x_n)$. The equality of the ranks of $\mathbf{A}$ and $\mathbf{B}$ means that *the dimensions of* U *and of* V *are equal*. But, as we have seen, U is a subset of V. It is easy to see that then $U = V$ (remember that a set is its own subset, so $U \subset V$ allows $U = V$ as special case). But then the vector $\mathbf{b}$, which is in V, is also in U, the space spanned by $\mathbf{a}_1, \ldots, \mathbf{a}_n$, so $\mathbf{b}$ is a linear combination of $\mathbf{a}_1, \ldots, \mathbf{a}_n$, that is, there exist $x_1, \ldots, x_n$ such that $\mathbf{b} = x_1\mathbf{a}_1 + \ldots + x_n\mathbf{a}_n$ which is the (4.43) we wanted.

Now we prove that, *if* (4.42) *has a solution* $(x_1, \ldots, x_n)$, *then the ranks of* $\mathbf{A}$ *and* $\mathbf{B}$ *are equal.* Indeed "(4.42) has a solution" means that there exist $x_1, \ldots, x_n$ satisfying (4.43) which means that $\mathbf{b}$ is a linear combination of $\mathbf{a}_1, \ldots, \mathbf{a}_n$, that is, $\mathbf{b}$ is in U. Since also $\mathbf{a}_1, \ldots, \mathbf{a}_n$ are in U, so V, the space spanned by $\mathbf{a}_1, \ldots, \mathbf{a}_n, \mathbf{b}$ is a subset of U. But we know already that U is a subset of V so $U = V$, therefore the rank of $\mathbf{A}$ which is the dimension of U equals the rank of $\mathbf{B}$, which is the dimension of V, as asserted.

What we proved can be written briefly as: $\mathbf{Ax} = \mathbf{b}$ *has a solution* $\mathbf{x}$ *if, and only if,* $\operatorname{rank}\mathbf{A} = \operatorname{rank}\mathbf{B}$.

After having established this necessary and sufficient condition for the existence of a solution, we proceed to *determine* these solutions. As we have seen in the examples and is true also in general, *the equation* $\mathbf{Ax} = \mathbf{b}$ (an equation between vectors or, equivalently, a system of equations between scalars) *is equivalent to* $\mathbf{Cx}^* = \mathbf{d}$, where $\mathbf{C}$ is the matrix (4.41) obtained from $\mathbf{A}$ by applying (several times) operations (I) and (II), $\mathbf{x}^*$ is a "rearrangement" of $\mathbf{x}$: has the same components as $\mathbf{x}$ but in a different order made necessary by the operations (II), and $\mathbf{d}$ is the vector we get by applying the same operations to $\mathbf{b}$ which we applied to $\mathbf{A}$ in order to get $\mathbf{C}$.

We write the first r components of $\mathbf{Cx}^* = \mathbf{d}$:

$$\begin{array}{cccccccc} x_1^* & & & + c_{1,r+1}x_{r+1}^* & + \ldots + & c_{1,n}x_n^* & = & d_1 \\ & x_2^* & & + c_{2,r+1}x_{r+1}^* & + \ldots + & c_{2,n}x_n^* & = & d_2 \\ & & \ddots & \vdots & \vdots & \vdots & & \vdots \\ & & x_r^* & + c_{r,r+1}x_{r+1}^* & + \ldots + & c_{r,n}x_n^* & = & d_r. \end{array} \tag{4.44}$$

The remaining $n - r$ equations are either the trivial $0 = 0$ or *if at least one of the* $d_{r+1}, \ldots, d_n$ *is not* 0 *then the system was contradictory*: $\operatorname{rank}\mathbf{A} \neq \operatorname{rank}\mathbf{B}$ (since the ranks of the transformed matrices are different). Setting $x_{r+1}^* =: \lambda_1, \ldots, x_n^* =: \lambda_{n-r}$ arbitrary, we have the *general solution of* $\mathbf{Cx}^* = \mathbf{d}$, if it exists, as

$$\begin{aligned} x_1^* &= d_1 - \lambda_1 c_{1,r+1} - \ldots - \lambda_{n-r}c_{1,n} \\ x_2^* &= d_2 - \lambda_1 c_{2,r+1} - \ldots - \lambda_{n-r}c_{2,n} \\ &\vdots \qquad\qquad\qquad \vdots \\ x_r^* &= d_r - \lambda_1 c_{r,r+1} - \ldots - \lambda_{n-r}c_{1,n} \\ x_{r+1}^* &= \lambda_1 \\ &\vdots \\ x_n^* &= \lambda_{n-r} \end{aligned} \tag{4.45}$$

where $\lambda_1, \ldots, \lambda_{n-r}$ *are arbitrary parameters* and *the general solution of* $\mathbf{Ax} = \mathbf{b}$ *is the* $\mathbf{x}$ *obtained by rearranging the components of* $\mathbf{x}^*$ *to their original order.*

We said that **d** is obtained from **b** by the same operations as **C** from **A**. The two transformations can be made into one by *transforming the "extended matrix"*

$$\begin{pmatrix} a_{11} & \dots & a_{1n} & b_1 \\ \vdots & & \vdots & \vdots \\ a_{m1} & \dots & a_{mn} & b_m \end{pmatrix} \tag{4.46}$$

into

$$\begin{pmatrix} 1 & 0 & \dots & 0 & c_{1,r+1} & \dots & c_{1,n} & d_1 \\ 0 & 1 & \dots & 0 & c_{2,r+1} & \dots & c_{2,n} & d_2 \\ \vdots & \vdots & \ddots & \vdots & \vdots & & \vdots & \vdots \\ 0 & 0 & \dots & 1 & c_{r,r+1} & \dots & c_{r,n} & d_r \\ 0 & 0 & \dots & 0 & 0 & \dots & 0 & d_{r+1} \\ \vdots & \vdots & \ddots & \vdots & \vdots & & \vdots & \vdots \\ 0 & 0 & \dots & 0 & 0 & \dots & 0 & d_n \end{pmatrix}. \tag{4.47}$$

We will show this transformation and how it gives the general solution (4.40) in the Example 6, equations (4.39) in a moment, but first we state some important *special cases*:

If $r = n$ then the solution (which, as we have seen *exists exactly if* $r = \operatorname{rank} \mathbf{B}$) *is unique* ($x_1^* = d_1, \dots, x_n^* = d_n$, no parameters).

If $\mathbf{b} = \mathbf{0}$ (homogeneous system of equations) *and* $r = n$ *then the "trivial solution"* $x_1 = \dots = x_n = 0$ *is the only solution* (since $\mathbf{x} = \mathbf{0}$ satisfies $\mathbf{Ax} = \mathbf{0}$ and as we have just seen, in the case $r = n$ there is only one solution vector).

In the homogeneous case $\mathbf{b} = \mathbf{0}$, *if* $r < n$ *then* $\mathbf{Ax} = \mathbf{0}$ *has also other solutions than* 0 ("nontrivial solutions").

One sees immediately, both from the above and directly, that *the general solution of the inhomogeneous equation* $\mathbf{Ax} = \mathbf{b}$ *is the sum of one ("particular") solution of the inhomogeneous equation and of the general solution of the corresponding homogeneous equation* $\mathbf{Ax} = \mathbf{0}$. In other words, *given one solution* $\mathbf{x}_0$ *of* $\mathbf{Ax} = \mathbf{b}$, *every solution* $\mathbf{x}$ *of this equation is the sum of* $\mathbf{x}_0$ *and a solution of* $\mathbf{Ax} = \mathbf{0}$. Indeed, $\mathbf{x}_0$ being a solution of $\mathbf{Ax} = \mathbf{b}$, we have

$$\mathbf{Ax}_0 = \mathbf{b}.$$

If $\mathbf{x}$ is *any* solution of

$$\mathbf{Ax} = \mathbf{b}$$

then, subtracting the first equation from the second, we get (because we can multiply a difference of vectors by a matrix term by term, since $\mathbf{f(x)} = \mathbf{Ax}$ satisfies (4.18) in Sect. 4.3) $\mathbf{A(x - x_0)} = \mathbf{0}$, that is, $\mathbf{x}^\dagger := \mathbf{x} - \mathbf{x}_0$ satisfies $\mathbf{Ax}^\dagger = \mathbf{0}$ and so, for every solution $\mathbf{x}$ of $\mathbf{Ax} = \mathbf{b}$, there exists a solution $\mathbf{x}^{\dagger}$ of $\mathbf{Ax} = \mathbf{0}$ such that $\mathbf{x} = \mathbf{x}_0 + \mathbf{x}^{\dagger}$, as asserted.

Had we chosen to solve homogeneous equations first, this would give us the general solution of inhomogeneous equations. But we solved both in one svelte (or not so svelte) argument.

Notice that the general solutions (4.45) of (4.44) (which is a "rearrangement" of the general solution $\mathbf{x}$ of $\mathbf{Ax} = \mathbf{b}$) can be written as

$$\mathbf{x}^* = \begin{pmatrix} d_1 \\ \vdots \\ d_r \\ 0 \\ 0 \\ \vdots \\ 0 \end{pmatrix} + \lambda_1 \begin{pmatrix} -c_{1,r+1} \\ \vdots \\ -c_{r,r+1} \\ 1 \\ 0 \\ \vdots \\ 0 \end{pmatrix} + \lambda_2 \begin{pmatrix} -c_{1,r+2} \\ \vdots \\ -c_{r,r+2} \\ 0 \\ 1 \\ \vdots \\ 0 \end{pmatrix} + \cdots + \lambda_{n-r} \begin{pmatrix} -c_{1,n} \\ \vdots \\ -c_{r,n} \\ 0 \\ 0 \\ \vdots \\ 1 \end{pmatrix}.$$

Here (check!) the first vector on the right is a ("particular") solution of (4.44) (and after rearrangement, a particular solution of $\mathbf{Ax} = \mathbf{b}$), while the rest is the general solution of the corresponding system of homogeneous equations (which we get from (4.44) by writing 0's on the right hand sides) and; after rearrangement, the general solution of $\mathbf{Ax} = \mathbf{0}$. Notice further that this *general solution of the homogeneous equation is a linear combination of the* $(n-r)$ *column vectors of the* $n \times (n-r)$ *matrix formed from the upper right* $r \times (n-r)$ *matrix in* $\mathbf{C}$ (see (4.41)), *multiplied by* (-1) *and from the* $(n-r) \times (n-r)$ *unit matrix put under it.* Since the $\lambda_1 \in \mathbb{R}, \ldots, \lambda_r \in \mathbb{R}$ are arbitrary, we can also say that *the solutions of the homogeneous equation form an r-dimensional space spanned by these r column vectors.*

In the case of (4.39), this representation of the general solution (4.40) is

$$\begin{pmatrix} x_1^* \\ x_2^* \\ x_4^* \\ x_3^* \\ x_5^* \end{pmatrix} = \begin{pmatrix} 4 \\ -1 \\ 1 \\ 0 \\ 0 \end{pmatrix} + \lambda_1 \begin{pmatrix} -3 \\ 2 \\ 0 \\ 1 \\ 0 \end{pmatrix} + \lambda_2 \begin{pmatrix} -2 \\ -1 \\ 3 \\ 0 \\ 1 \end{pmatrix} \tag{4.48}$$

(we have already appropriately rearranged the left hand side).

We had obtained above the general solution by manipulating the *equations* but we have also been "mirroring" these manipulations by transformations of the "coefficient matrix" and the "constant vector". We mentioned also that the latter two steps can be reduced to one by *manipulating the "extended matrix"* (4.46) instead, *with the aim of bringing it to the form* (4.47). Actually this alone gives the general solution of the original system of linear equations, without having to manipulate the equations themselves.

We show now on the equations (4.39) our most complicated example, Example 6, how this is done and explain also how the "multipliers", which we used to write to the right of the equations, can be found: For the system of linear equations (4.39) the extended matrix is

$$\begin{pmatrix} 1 & 2 & -1 & 0 & 4 & 2 \\ 1 & 4 & -5 & 1 & 3 & 1 \\ 3 & 2 & 5 & 2 & 2 & 12 \\ 2 & -2 & 10 & 1 & -1 & 11 \end{pmatrix}. \tag{4.49}$$

In order to bring it to the form (4.47) the important thing is to obtain a unit matrix in the upper left corner (the rest will take care of itself) through transformation of the types (I) (or (I′), (I″)) and (II). We will use actually only (i) and (II), which we repeat in the form needed here:

(i) *Replace a row by a linear combination of the rows in which the coefficient of the original row is not 0,*
(II) *interchange two columns (or two rows).*

While the first row of (4.49) starts already with a 1, we want a 0 in its second place. After some trial and error we see that multiplying it by 2 and the second row by (-1) and adding will do the trick (transformation (i)):

$$\begin{pmatrix} 1 & 0 & 3 & -1 & 5 & 3 \\ 1 & 4 & -5 & 1 & 3 & 1 \\ 3 & 2 & 5 & 2 & 2 & 12 \\ 2 & -2 & 10 & 1 & -1 & 11 \end{pmatrix}.$$

We want the second row to start with 0 and 1. Now we have the advantage that the first row starts with 1 0. Subtracting it from the second gives a 0 4 start, so a multiplication by $\frac{1}{4}$ seems to be appropriate. Therefore we do (i) by taking (1/4) times the second row plus $(-1/4)$ times the first row. Everything else remains unchanged, so we have now

$$\begin{pmatrix} 1 & 0 & 3 & -1 & 5 & 3 \\ 0 & 1 & -2 & \frac{1}{2} & -\frac{1}{2} & -\frac{1}{2} \\ 3 & 2 & 5 & 2 & 2 & 12 \\ 2 & -2 & 10 & 1 & -1 & 11 \end{pmatrix}.$$

(Consider how we obtained the new first row from (4.49): we really have added $(-1/2)$ of the first row to $(1/2)$ of the second row in (4.49).) So long, so good. We want now the third to start with 0 0 followed, we hope, by 1. This is a bit more complicated since we need in (i) the linear combination of three rows. But we have again the advantage that the first two rows already start with 1 0 and 0 1, respectively.

So we should subtract 3 times the first row and 2 times the second row from the third row (again an (i) transformation):

$$\begin{pmatrix} 1 & 0 & 3 & -1 & 5 & 3 \\ 0 & 1 & -2 & \frac{1}{2} & -\frac{1}{2} & -\frac{1}{2} \\ 0 & 0 & 0 & 4 & -12 & 4 \\ 2 & -2 & 10 & 1 & -1 & 11 \end{pmatrix}.$$

(This third row is the linear combination (-5) times the first row plus 2 times the second plus the third in (4.49).) Oh - oh: the third row starts with 0 0 0: there is no multiplier which would make 1 of the third 0. But here we can use the operation (II): interchange the third and fourth column (remembering also to exchange, at the end the third and fourth component of $\mathbf{x}$). We do this because the fourth column has a nonzero in the third row. (If everything were 0 in the third row, we would exchange it with the fourth using the second part of (II) this time; this means only exchanging the third and fourth equation; if both the third and the fourth row consisted of 0's only then we had already a matrix with all 0's in the last two lines and leave them alone). We think the reader will agree, to save time and space by multiplying the new third row at the same time by $(1/4)$ in order to get 1 in its place (this results in $(-5/4)$, $(1/2)$, $(1/4)$ as multipliers in (4.39)):

$$\begin{pmatrix} 1 & 0 & -1 & 3 & 5 & 3 \\ 0 & 1 & \frac{1}{2} & -2 & -\frac{1}{2} & -\frac{1}{2} \\ 0 & 0 & 1 & 0 & -3 & 1 \\ 2 & -2 & 1 & 10 & -1 & 11 \end{pmatrix}.$$

Now we try to start the fourth row with 0 0 0 and maybe have 1 in the fourth place. Adding 2 times the second row and (-2) times the first to the fourth row would give 0 0 at the first two places but 4 at the third. So we also add (-4) times the third row (which has 0 at the first two places); this still is a transformation (i) (in (4.49) it would mean adding the second and fourth row and subtracting the third):

$$\begin{pmatrix} 1 & 0 & -1 & 3 & 5 & 3 \\ 0 & 1 & \frac{1}{2} & -2 & -\frac{1}{2} & -\frac{1}{2} \\ 0 & 0 & 1 & 0 & -3 & 1 \\ 0 & 0 & 0 & 0 & 0 & 0 \end{pmatrix}.$$

Now we have a situation mentioned above, where the *last row consists of* 0's *only*. So at *no place* of the last row could we get 1 by applying transformations (I) and/or (II) (really (ii), excluding exchanges of rows). This last row is already as in (4.47) (we have even $d_n = 0$), so we leave it alone and just try to get 0's at the third places of the first and second row (in place of -1 and $1/2$, respectively). Of course we do this with aid of the third row, adding it to the first *row* and subtracting $(1/2)$ times the third row from the second. We dare to do these two transformations of type (i) at once and get

$$\begin{pmatrix} 1 & 0 & 0 & 3 & 2 & 4 \\ 0 & 1 & 0 & -2 & 1 & -1 \\ 0 & 0 & 1 & 0 & -3 & 1 \\ 0 & 0 & 0 & 0 & 0 & 0 \end{pmatrix}. \tag{4.50}$$

Incredible as this may sound, we are at the end of our journey: this is already of the form (4.47).

Now our general solution (4.45) of (4.44) (see p. 150) gives instantly the general solution of (4.39):

$$\begin{pmatrix} x_1^* \\ x_2^* \\ x_4^* \\ x_3^* \\ x_5^* \end{pmatrix} = \mathbf{x}^* = \begin{pmatrix} d_1 \\ d_2 \\ d_3 \\ 0 \\ 0 \end{pmatrix} + \lambda_1 \begin{pmatrix} -c_{14} \\ -c_{24} \\ -c_{34} \\ 1 \\ 0 \end{pmatrix} + \lambda_2 \begin{pmatrix} -c_{15} \\ -c_{25} \\ -c_{35} \\ 0 \\ 1 \end{pmatrix}$$

$$= \begin{pmatrix} 4 \\ -1 \\ 1 \\ 0 \\ 0 \end{pmatrix} + \lambda_1 \begin{pmatrix} -3 \\ 2 \\ 0 \\ 1 \\ 0 \end{pmatrix} + \lambda_2 \begin{pmatrix} -2 \\ -1 \\ 3 \\ 0 \\ 1 \end{pmatrix}$$

which is exactly (4.48) (or (4.40) for that matter). We have $n - r = 2$ arbitrary λ's in our solution since $n = 5$ (number of variables) and $r = 3$ (rank).

We repeat that, *had we got $d_4 = d_n = 0$, then the system of linear equations would have been contradictory.*

Example 7 If the right hand side of the last equation in (4.39) were 14 instead of 11, everything else being unchanged; this would give $d_4 = 3$ and the system

$$\begin{aligned} x_1 + 2x_2 - x_3 \phantom{{}+2x_4} + 4x_5 &= 2 \\ x_1 + 4x_2 - 5x_3 + x_4 + 3x_5 &= 1 \\ 3x_1 + 2x_2 + 5x_3 + 2x_4 + 2x_5 &= 12 \\ 2x_1 - 2x_2 + 10x_3 + x_4 - x_5 &= 14 \end{aligned}$$

is indeed contradictory, because subtracting the second equation from the third would give

(continued)

$$2x_1 - 2x_2 + 10x_3 + x_4 - x_5 = 11$$

in contradiction to the fourth equation. (Note that, as it is, the fourth equation in (4.39) is *redundant*, because the difference of the third and second gives the same. That is why we had all 0's in the last row of (4.50).)

The method which we presented here in considerable detail is, as we have seen, both practical and leads to theoretical results. It is also both old and new: It used to be called "elimination of unknowns" and that was how systems of equations were solved for centuries. Then "determinants" were discovered which gave a simple explicit formula ("Cramer's rule") for the solutions (we will present it briefly in the next section). But *calculating* solutions with this formula turned out to be much lengthier than the above method. So nowadays people and computers merrily use this method again for solving systems of linear equations.

4.6.1 Exercises

1. Determine the rank of the matrices

$$\text{(a)} \begin{pmatrix} 3 & -4 & 5 \\ 1 & 7 & -6 \\ 8 & -2 & -3 \end{pmatrix}, \quad \text{(b)} \begin{pmatrix} 3 & -4 & 5 & 1 \\ 1 & 7 & -6 & 2 \\ 8 & -2 & -3 & 3 \end{pmatrix}, \quad \text{(c)} \begin{pmatrix} 5 & 6 & -2 \\ 7 & -3 & 1 \\ 8 & -4 & -5 \end{pmatrix},$$

$$\text{(d)} \begin{pmatrix} 5 & 5 & -2 & -24 \\ 7 & -3 & 1 & 31 \\ 8 & -4 & -5 & 7 \end{pmatrix}, \quad \text{(e)} \begin{pmatrix} 1 & 2 & 3 & -9 & -15 & 18 & 28 \\ -1 & 1 & -1 & -5 & 21 & -18 & -16 \\ 0 & 7 & 1 & -29 & 41 & -33 & 1 \\ 4 & 0 & -2 & 10 & -4 & -6 & 8 \end{pmatrix},$$

$$\text{(f)} \begin{pmatrix} 5 & 6 & -2 \\ 7 & -3 & 1 \\ 8 & -4 & -5 \\ 4 & 7 & 4 \end{pmatrix}, \quad \text{(g)} \begin{pmatrix} 5 & 6 & -2 & -24 \\ 7 & -3 & 1 & 31 \\ 8 & -4 & -5 & 7 \\ 4 & 7 & 4 & 0 \end{pmatrix},$$

$$\text{(h)} \begin{pmatrix} 5 & 6 & -2 & -24 \\ 7 & -3 & 1 & 31 \\ 8 & -4 & -5 & 7 \\ 4 & 7 & 4 & 1 \end{pmatrix}, \quad \text{(i)} \begin{pmatrix} 3 & -4 & 5 \\ 1 & 7 & -6 \\ 8 & -2 & -3 \\ 1 & 1 & 1 \end{pmatrix},$$

$$\text{(j)} \begin{pmatrix} 3 & -4 & 5 & 1 \\ 1 & 7 & -6 & 2 \\ 8 & -2 & -3 & 3 \\ 1 & 1 & 1 & 4 \end{pmatrix}, \quad \text{(k)} \begin{pmatrix} 1 & 2 & 3 & -9 & -15 & 18 \\ -1 & 1 & -1 & -5 & 21 & -18 \\ 0 & 7 & 1 & -29 & 41 & -33 \\ 4 & 0 & -2 & 10 & -4 & -6 \end{pmatrix}.$$

2. Determine the rank of the matrix

$$\begin{pmatrix} 1 & 1 & 1 \\ 1 & 2 & 3 \\ 2 & 3 & a \end{pmatrix} \quad \text{for} \quad a \in \mathbb{R}.$$

3. Solve

(a) $$\begin{aligned} 3x_1 - 4x_2 + 2x_3 &= 1 \\ x_1 + 7x_2 - 6x_3 &= 2 \\ 8x_1 - 2x_2 - 3x_3 &= 3\,, \end{aligned}$$

(b) $$\begin{aligned} 5x_1 + 6x_2 - 2x_3 &= -24 \\ 7x_1 - 3x_2 + x_3 &= 31 \\ 8x_1 - 4x_2 - 5x_3 &= 7\,, \end{aligned}$$

(c) $$\begin{aligned} 5x_1 + 6x_2 - 2x_3 &= -24 \\ 7x_1 - 3x_2 + x_3 &= 31 \\ 8x_1 - 4x_2 - 5x_3 &= 7 \\ 4x_1 + 7x_2 + 4x_3 &= 0\,. \end{aligned}$$

4. Solve

(a) $$\begin{aligned} x_1 + x_2 + x_3 &= 0 \\ x_1 + 2x_2 + 3x_3 &= 0 \\ 2x_1 + 3x_2 + ax_3 &= 0 \quad \text{for } a \in \mathbb{R}, \end{aligned}$$

(b) $$\begin{aligned} 3x_1 - 4x_2 + 5x_3 &= 1 \\ x_1 + 7x_2 - 6x_3 &= 2 \\ 8x_1 - 2x_2 - 3x_3 &= 3 \\ x_1 + x_2 + x_3 &= 4\,, \end{aligned}$$

(c) $$\begin{aligned} 5x_1 + 6x_2 - 2x_3 &= -24 \\ 7x_1 - 3x_2 + x_3 &= 31 \\ 8x_1 - 4x_2 - 5x_3 &= 7 \\ 4x_1 + 7x_2 + 4x_3 &= 1\,. \end{aligned}$$

5. Solve

$$\begin{aligned} x_1 + 2x_2 + 3x_3 - 9x_4 - 15x_5 + 18x_6 &= 28 \\ -x_1 + x_2 - x_3 - 5x_4 + 21x_5 - 18x_6 &= -16 \\ 7x_2 + x_3 - 29x_4 + 41x_5 - 33x_6 &= 1 \\ 4x_1 - 2x_3 + 10x_4 - 4x_5 - 6x_6 &= 8\,. \end{aligned}$$

4.6.2 Answers

1. (a) $r = 3$, (b) $r = 3$, (c) $r = 3$, (d) $r = 3$,
   (e) $r = 3$, (f) $r = 3$, (g) $r = 3$, (h) $r = 4$,
   (i) $r = 3$, (j) $r = 4$, (k) $r = 3$.
2. $r = 3$ for $a \neq 4$, $r = 2$ for $a = 4$.

3. (a) $\begin{pmatrix} x_1 \\ x_2 \\ x_3 \end{pmatrix} = \begin{pmatrix} 1 \\ 1 \\ 1 \end{pmatrix}$, (b) $\begin{pmatrix} x_1 \\ x_2 \\ x_3 \end{pmatrix} = \begin{pmatrix} 2 \\ -4 \\ 5 \end{pmatrix}$, (c) $\begin{pmatrix} x_1 \\ x_2 \\ x_3 \end{pmatrix} = \begin{pmatrix} 2 \\ -4 \\ 5 \end{pmatrix}$.

4. (a) $\begin{pmatrix} x_1 \\ x_2 \\ x_3 \end{pmatrix} = \begin{pmatrix} 0 \\ 0 \\ 0 \end{pmatrix}$ for $a \neq 0$, $\begin{pmatrix} x_1 \\ x_2 \\ x_3 \end{pmatrix} = \begin{pmatrix} 1 \\ -2 \\ 1 \end{pmatrix}$ for $a = 4$,

(b) und (c) have no solutions.

5.

$$\begin{pmatrix} x_1 \\ x_2 \\ x_3 \\ x_4 \\ x_5 \\ x_6 \end{pmatrix} = \begin{pmatrix} 6 \\ -1 \\ 8 \\ 0 \\ 0 \\ 0 \end{pmatrix} + \lambda_1 \begin{pmatrix} -2 \\ 4 \\ 1 \\ 1 \\ 0 \\ 0 \end{pmatrix} + \lambda_2 \begin{pmatrix} 5 \\ -7 \\ 8 \\ 0 \\ 1 \\ 0 \end{pmatrix} + \lambda_3 \begin{pmatrix} -3 \\ 6 \\ -9 \\ 0 \\ 0 \\ 1 \end{pmatrix}.$$

4.7 Determinant, Cramer's Rule, Inverse Matrix

Let us try to do in general, what we derived only from examples (though from several examples) in the previous section: solve systems of linear equations explicitly; at least when $m = n$ (same number of equations as unknowns), first for $m = n = 2$ then for $m = n = 3$ and try to generalise to arbitrary $m = n$. We solve the two equations with two unknowns

$$\begin{array}{rcl|r|r} a_{11}x_1 + a_{12}x_2 &=& b_1 & a_{22} & -a_{21} \\ a_{21}x_1 + a_{22}x_2 &=& b_2 & -a_{12} & a_{11}, \end{array}$$

that is,

$$\mathbf{Ax} = \mathbf{b} \quad \text{with} \quad \mathbf{A} = \begin{pmatrix} a_{11} & a_{12} \\ a_{21} & a_{22} \end{pmatrix}, \mathbf{b} = \begin{pmatrix} b_1 \\ b_2 \end{pmatrix}, \mathbf{x} = \begin{pmatrix} x_1 \\ x_2 \end{pmatrix}$$

again by eliminating unknowns. Multiplying by the first or second column of multipliers and adding we get

$$\begin{aligned} (a_{11}a_{22} - a_{12}a_{21})x_1 &= b_1a_{22} - b_2a_{12}, \\ (a_{11}a_{22} - a_{12}a_{21})x_2 &= a_{11}b_2 - a_{21}b_1, \end{aligned} \tag{4.51}$$

respectively. We call $a_{11}a_{22} - a_{12}a_{21}$ the *determinant* of the 2×2 matrix $\mathbf{A} = \begin{pmatrix} a_{11} & a_{12} \\ a_{21} & a_{22} \end{pmatrix}$, in symbols

$$\det \mathbf{A} = \det \begin{pmatrix} a_{11} & a_{12} \\ a_{21} & a_{22} \end{pmatrix} = a_{11}a_{22} - a_{12}a_{21}. \tag{4.52}$$

Then, similarly,

$$b_1 a_{22} - b_2 a_{12} = \det \begin{pmatrix} b_1 & a_{12} \\ b_2 & a_{22} \end{pmatrix} = \det \mathbf{A_1},$$

$$a_{11} b_2 - a_{21} b_1 = \det \begin{pmatrix} a_{11} & b_1 \\ a_{21} & b_2 \end{pmatrix} = \det \mathbf{A_2},$$

where

$$\mathbf{A_1} := \begin{pmatrix} b_1 & a_{12} \\ b_2 & a_{22} \end{pmatrix}, \ \mathbf{A_2} := \begin{pmatrix} a_{11} & b_1 \\ a_{12} & b_2 \end{pmatrix}.$$

If $\det \mathbf{A} = a_{11}a_{22} - a_{12}a_{21} \neq 0$ then we get x_1 and x_2 by dividing the two equations (4.51) by $\det \mathbf{A}$.

So *for* $n = 2$ *we have proved* the following:

Cramer's rule *If* $\mathbf{A}$ *is an* $n \times n$ *square matrix and* $\det \mathbf{A} \neq 0$ then, *for any column vector* $\mathbf{b}$, $\mathbf{Ax} = \mathbf{b}$ *has a unique solution*

$$\mathbf{x} = \frac{1}{\det \mathbf{A}} \begin{pmatrix} \det \mathbf{A_1} \\ \vdots \\ \det \mathbf{A}_n \end{pmatrix}, \tag{4.53}$$

where the $n \times n$ *matrix* $\mathbf{A}_k$ *is formed by replacing the k'th column vector of* $\mathbf{A}$ *by* $\mathbf{b}$ $(k = 1, 2, \ldots, n)$.

Let us see also the case $n = 3$ in detail. Now $\mathbf{Ax} = \mathbf{b}$ reads as follows:

$$\begin{array}{l|l|l|l} a_{11}x_1 + a_{12}x_2 + a_{13}x_3 = b_1 & a_{22}a_{33} - a_{23}a_{32} & a_{23}a_{31} - a_{21}a_{33} & a_{21}a_{32} - a_{22}a_{31} \\ a_{21}x_1 + a_{22}x_2 + a_{23}x_3 = b_2 & a_{13}a_{32} - a_{12}a_{33} & a_{11}a_{33} - a_{13}a_{31} & a_{12}a_{31} - a_{11}a_{32} \\ a_{31}x_1 + a_{32}x_2 + a_{33}x_3 = b_3 & a_{12}a_{23} - a_{13}a_{22} & a_{13}a_{21} - a_{11}a_{23} & a_{11}a_{22} - a_{12}a_{21} \ . \end{array}$$

With the multipliers on the right and summing up, we get three equations each containing just one of the unknowns x_1, x_2, x_3 and, defining

$$\det \mathbf{A} = \det \begin{pmatrix} a_{11} & a_{12} & a_{13} \\ a_{21} & a_{22} & a_{23} \\ a_{31} & a_{32} & a_{33} \end{pmatrix} := \begin{matrix} a_{11}a_{22}a_{33} + a_{12}a_{23}a_{31} + a_{13}a_{21}a_{32} \\ -a_{11}a_{23}a_{32} - a_{12}a_{21}a_{33} - a_{13}a_{22}a_{31} \end{matrix}, \tag{4.54}$$

$$\mathbf{A}_1 := \begin{pmatrix} b_1 & a_{12} & a_{13} \\ b_2 & a_{22} & a_{23} \\ b_3 & a_{32} & a_{33} \end{pmatrix}, \mathbf{A}_2 := \begin{pmatrix} a_{11} & b_1 & a_{13} \\ a_{21} & b_2 & a_{23} \\ a_{31} & b_3 & a_{33} \end{pmatrix}, \mathbf{A}_3 := \begin{pmatrix} a_{11} & a_{12} & b_1 \\ a_{21} & a_{22} & b_2 \\ a_{31} & a_{32} & b_3 \end{pmatrix},$$

we obtain

$$x_1 = \frac{1}{\det \mathbf{A}} \det \mathbf{A}_1, \qquad x_2 = \frac{1}{\det \mathbf{A}} \det \mathbf{A}_2, \qquad x_3 = \frac{1}{\det \mathbf{A}} \det \mathbf{A}_3.$$

This *verifies Cramer's rule* (4.53) also for $n = 3$. We could proceed to $n = 4$ and so on but it just may get tedious.

As we have seen here (and in the previous section), eliminating unknowns is relatively easy, the problem is to find for the determinant $\det \mathbf{A}$ of the $n \times n$ matrix $\mathbf{A}$ a unified and reasonably understandable expression for all n. We give here such an expression, not with general proof but by analysing the $n = 2$ and $n = 3$ cases (4.52) and (4.54): In each term of these formulas the numbers $1, 2$ or $1, 2, 3$ (in general this would be $1, 2, \ldots, n$) figure as subscripts once in natural (increasing) order as first subscript and once in the second subscript not always in increasing order anymore, but still *each number just figuring once*. What we just described is a "*permutation*" (or "*rearrangement*") of the numbers $1, 2, \ldots, n$ and is usually denoted by Π, which can be considered as a bijection (see Sect. 3.2) with the domain $\{1, \ldots, n\}$, whose range (or codomain) is the same set. If at the k'th place of the rearrangement the number ℓ stands then $\Pi(k) := \ell$ (of course, permutations on the other sets can be similarly defined). In (4.52) and (4.54) we have *sums* and *differences* of terms with the numbers $1, 2, \ldots$ in their natural order as first subscripts and a permutation of these numbers as second subscript. We notice also that *all* possible permutations of $\{1, 2, \ldots, n\}$ figure as second subscripts, at least for $n = 2$ and $n = 3$ in (4.52) and (4.54), respectively. We notice further that some terms in these expressions are added, others subtracted: our expressions are so far of the form

$$\sum_{\Pi} (\pm) a_{1\Pi(1)} a_{2\Pi(2)} \ldots a_{n\Pi(n)}$$

(the Π under the summation sign means that we have to take the terms with all possible permutations as sets of subscripts and sum up). The remaining question is the $\pm$ sign: which terms should be added and which subtracted.

We seek the answer by looking at the permutations in the second subscripts of both the positive and the negative terms of, say, (4.54):

(+) second subscripts in + terms: 123, 231, 312,
(−) second subscripts in − terms: 132, 213, 321.

Maybe nothing is apparent at first sight. In order to see more, we introduce the notion of *inversion* in a permutation Π: if $\Pi(j) > \Pi(k)$ for $j < k$ then we have an inversion. The number of inversions in the three (+)-terms above is 0 ($\Pi(1) < \Pi(2) < \Pi(3)$), 2 ($2 = \Pi(1) > \Pi(3) = 1$, $3 = \Pi(2) > \Pi(3) = 1$), and 2 ($\Pi(1) > \Pi(2)$, $\Pi(1) > \Pi(3)$) and in the (−)-terms 1 ($3 = \Pi(2) > \Pi(3) = 2$), 1 ($\Pi(1) > \Pi(2)$), and 3 ($\Pi(1) > \Pi(2)$, $\Pi(1) > \Pi(3)$, $\Pi(2) > \Pi(3)$), respectively. After some inspection we may notice that, *if the number of inversions of the permutation in the second subscripts in a term is an even number then we have a "+" sign in front of the term* and *if it is an odd number then there is a "−" in front*. The same holds for (4.52): in $(+a_{11}a_{22})$ there is 0 inversion, in $(-a_{12}a_{21})$ 1 inversion in the second subscripts. since $(-1)^1 = (-1)^3 = (-1)^5 = \ldots = -1$ and $(-1)^0 = (-1)^2 = (-1)^4 = \ldots = 1$, we will *define the determinant by*

$$\det \mathbf{A} = \det \begin{pmatrix} a_{11} & \ldots & a_{1n} \\ \vdots & & \vdots \\ a_{n1} & \ldots & a_{nn} \end{pmatrix} := \sum (-1)^{N(\Pi)} a_{1\Pi(1)} a_{2\Pi(2)} \ldots a_{n\Pi(n)},$$

where the summation is taken over all permutations Π of $(1, 2, \ldots, n)$ and $N(\Pi)$ is the number of inversions in Π.

One can prove that, *with this definition of the determinant, the Cramer rule holds for all n.*

We do not give examples for the use of Cramer's rule to solve systems of linear equations, because the method described in some detail in Sect. 4.6 is shorter (fewer equations). Indeed the number of multiplications and additions needed to evaluate the determinant of an $n \times n$ matrix becomes so large when n is large that, rather than using determinants to solve systems of equations, one often uses the transformations (I) and (II), applied to equations and matrices in the previous section, to evaluate determinants. But the fact that the solution of systems of equations (when the number of equations equals the number of unknowns) and a condition for the existence of solutions can be written explicitly with help of determinants, is often of importance.

In particular, comparing results of this and of the previous section, we see that a system of *n linear equations with n unknowns has exactly then a unique solution if the coefficient matrix has the rank n* ($r = n$) *or equivalently, if the determinant of the coefficient matrix is not* 0.

A third way of determining the (unique) solutions of $\mathbf{Ax} = \mathbf{b}$ for $n \times n$ matrices $\mathbf{A}$ with rank n, that is, with $\det \mathbf{A} \neq 0$, is by the use of the *inverse matrix* $\mathbf{A}^{-1}$ of $\mathbf{A}$. This is defined as the (unique) $n \times n$ matrix which satisfies ($\mathbf{I}$ being the $n \times n$ unit matrix)

$$\mathbf{A}^{-1}\mathbf{A} = \mathbf{I} \tag{4.55}$$

We verify a little bit later that such a $\mathbf{A}^{-1}$ indeed exists but we first show how it helps to solve $\mathbf{Ax} = \mathbf{b}$: Multiply $\mathbf{Ax} = \mathbf{b}$ from the left by $\mathbf{A}^{-1}$, we get $\mathbf{A}^{-1}(\mathbf{Ax}) = \mathbf{A}^{-1}\mathbf{b}$ or, since by (4.20) "the bracket can be moved", $(\mathbf{A}^{-1}\mathbf{A})\mathbf{x} = \mathbf{A}^{-1}\mathbf{b}$. But, by (4.55), $(\mathbf{A}^{-1}\mathbf{A})\mathbf{x} = \mathbf{Ix} = \mathbf{x}$, therefore

$$\mathbf{x} = \mathbf{A}^{-1}\mathbf{b}.$$

So, *if* $\mathbf{A}$ *is an* $n \times n$ *matrix of rank* n *then the only solution of* $\mathbf{Ax} = \mathbf{b}$ *is* $\mathbf{x} = \mathbf{A}^{-1}\mathbf{b}$.

We prove first that

$$\mathbf{AX} = \mathbf{I} \tag{4.56}$$

has a solution $\mathbf{X}$ *and that it is unique if* $\det \mathbf{A} \neq 0$. This equation clearly breaks up into n equations of the type $\mathbf{Ax}_k = \mathbf{b}_k$ ($k = 1, \ldots, n$):

$$\mathbf{Ax}_1 = \begin{pmatrix} 1 \\ 0 \\ \vdots \\ 0 \end{pmatrix}, \quad \mathbf{Ax}_2 = \begin{pmatrix} 0 \\ 1 \\ \vdots \\ 0 \end{pmatrix}, \ldots, \mathbf{Ax}_n = \begin{pmatrix} 0 \\ 0 \\ \vdots \\ 1 \end{pmatrix}. \tag{4.57}$$

Since $\det \mathbf{A} \neq 0$, by Cramer's rule, each of these equations has a unique solution. These unique $\mathbf{x}_1, \mathbf{x}_2, \ldots, \mathbf{x}_n$ are column vectors of the "unknown matrix" $\mathbf{X}$ which is now known, so we have proved the existence and uniqueness of the solution $\mathbf{X}$ of (4.56).

Now we show that this $\mathbf{X}$ satisfies also $\mathbf{XA} = \mathbf{I}$, so it can serve as $\mathbf{A}^{-1}$ in (4.55). Indeed, multiplying (4.56) from the right by $\mathbf{A}$ and using the associativity of matrix multiplication (see Sect. 4.4 **1**), we get

$$\mathbf{A}(\mathbf{XA}) = (\mathbf{AX})\mathbf{A} = \mathbf{IA} = \mathbf{A}.$$

Of course, also $\mathbf{AI} = \mathbf{A}$ holds. But, since $\det \mathbf{A} \neq \mathbf{0}$, the solution $\mathbf{Y}$ of $\mathbf{AY} = \mathbf{A}$ is *unique*, that is,

$$\mathbf{XA} = \mathbf{I}$$

as asserted. Again there can be no two such $\mathbf{X}$ belonging to $\mathbf{A}$. If there were another, say $\mathbf{Z}$, such that $\mathbf{ZA} = \mathbf{I}$ then, since $(\mathbf{X} - \mathbf{Z})\mathbf{A} = \mathbf{XA} - \mathbf{ZA}$ can easily be checked (distributivity of matrix multiplication upon addition and subtraction), we would have

$$(\mathbf{X} - \mathbf{Z})\mathbf{A} = \mathbf{0}.$$

Multiplied by $\mathbf{X}$ from the right we get, using the associativity of matrix multiplication (see Sect. 4.4 **1**)

$$\mathbf{0X} = ((\mathbf{X} - \mathbf{Z})\mathbf{A})\mathbf{X} = (\mathbf{X} - \mathbf{Z})(\mathbf{AX})$$

But $\mathbf{0X} = \mathbf{0}$ and by (4.56), $\mathbf{AX} = \mathbf{I}$. Furthermore, of course $(\mathbf{X} - \mathbf{Z})\mathbf{I} = \mathbf{X} - \mathbf{Z}$, so we have $\mathbf{0} = \mathbf{X} - \mathbf{Z}$, that is, $\mathbf{Z} = \mathbf{X}$. Thus there *exists* a *unique* $\mathbf{A}^{-1}$ (the $\mathbf{X}$ which we have just determined), satisfying (4.55) and we see that this inverse $\mathbf{A}^{-1}$ satisfies also

$$\mathbf{AA}^{-1} = \mathbf{I}$$

(from (4.56)—not from (4.55) because, as seen in Sect. 4.4 **1**, matrix multiplication is not always commutative; what we have just proved shows that $\mathbf{A}$ and $\mathbf{A}^{-1}$ *commute*: $\mathbf{AA}^{-1} = \mathbf{A}^{-1}\mathbf{A}$—and that $\mathbf{A}$ is the inverse of $\mathbf{A}^{-1}$: *inverse of the inverse is the original matrix*: $(\mathbf{A}^{-1})^{-1} = \mathbf{A}$).

In order to *calculate the inverse matrix*, we have to remember that its column vectors are the solution of (4.57). As we have seen in Sect. 4.6, we calculate them by transforming the matrices

$$\begin{pmatrix} a_{11} & \dots & a_{1n} & 1 \\ a_{21} & \dots & a_{2n} & 0 \\ \vdots & & \vdots & \vdots \\ a_{n1} & \dots & a_{nn} & 0 \end{pmatrix}, \begin{pmatrix} a_{11} & \dots & a_{1n} & 0 \\ a_{21} & \dots & a_{2n} & 1 \\ \vdots & & \vdots & \vdots \\ a_{n1} & \dots & a_{nn} & 0 \end{pmatrix}, \dots, \begin{pmatrix} a_{11} & \dots & a_{1n} & 0 \\ a_{21} & \dots & a_{2n} & 0 \\ \vdots & & \vdots & \vdots \\ a_{n1} & \dots & a_{nn} & 1 \end{pmatrix}$$

to the forms (no 0-rows, because, as we know, the rank of th $n \times n$ matrix $\mathbf{A}$ has to be n in order for $\mathbf{A}^{-1}$ to exist):

$$\begin{pmatrix} 1 & 0 & \dots & 0 & a'_{11} \\ 0 & 1 & \dots & 0 & a'_{21} \\ \vdots & \vdots & \ddots & \vdots & \vdots \\ 0 & 0 & \dots & 1 & a'_{n1} \end{pmatrix}, \begin{pmatrix} 1 & 0 & \dots & 0 & a'_{12} \\ 0 & 1 & \dots & 0 & a'_{22} \\ \vdots & \vdots & \ddots & \vdots & \vdots \\ 0 & 0 & \dots & 1 & a'_{n2} \end{pmatrix}, \dots, \begin{pmatrix} 1 & 0 & \dots & 0 & a'_{1n} \\ 0 & 1 & \dots & 0 & a'_{2n} \\ \vdots & \vdots & \ddots & \vdots & \vdots \\ 0 & 0 & \dots & 1 & a'_{nn} \end{pmatrix},$$

respectively. Note that the d_j's from Sect. 4.6 are now denoted by a'_{jk} in the case of the k'th matrix ($k = 1, 2, \ldots, n$). Transforming the whole matrix

$$\begin{pmatrix} a_{11} & \ldots & a_{1n} & 1 & 0 & \ldots & 0 \\ a_{21} & \ldots & a_{2n} & 0 & 1 & \ldots & 0 \\ \vdots & & \vdots & \vdots & \vdots & \ddots & \vdots \\ a_{n1} & \ldots & a_{nn} & 0 & 0 & \ldots & 1 \end{pmatrix} \quad \text{to the form} \quad \begin{pmatrix} 1 & 0 & \ldots & 0 & a'_{11} & \ldots & a'_{1n} \\ 0 & 1 & \ldots & 0 & a'_{21} & \ldots & a'_{2n} \\ \vdots & & \ddots & \vdots & \vdots & & \vdots \\ 0 & 0 & \ldots & 1 & a'_{n1} & \ldots & a'_{nn} \end{pmatrix}$$

clearly gives the same a'_{jk}'s ($j = 1, 2, \ldots, n$; $k = 1, 2, \ldots, n$) and is much simpler. The transformation goes through the same chain of operations which we saw in Sect. 4.6.

Example We determine the inverse of

$$\mathbf{A} = \begin{pmatrix} 1 & 2 & 0 \\ 1 & 4 & 1 \\ 3 & 2 & 2 \end{pmatrix}$$

(the coefficient-matrix of (4.33)) by transforming, as indicated, the following extended matrix:

$$\begin{pmatrix} 1 & 2 & 0 & 1 & 0 & 0 \\ 1 & 4 & 1 & 0 & 1 & 0 \\ 3 & 2 & 2 & 0 & 0 & 1 \end{pmatrix} \mapsto \begin{pmatrix} 1 & 0 & -1 & 2 & -1 & 0 \\ 1 & 4 & 1 & 0 & 1 & 0 \\ 3 & 2 & 2 & 0 & 0 & 1 \end{pmatrix} \mapsto \begin{pmatrix} 1 & 0 & -1 & 2 & -1 & 0 \\ 0 & 4 & 2 & -2 & 2 & 0 \\ 3 & 2 & 2 & 0 & 0 & 1 \end{pmatrix}$$

$$\mapsto \begin{pmatrix} 1 & 0 & -1 & 2 & -1 & 0 \\ 0 & 1 & 1/2 & -1/2 & 1/2 & 0 \\ 3 & 2 & 2 & 0 & 0 & 1 \end{pmatrix} \mapsto \begin{pmatrix} 1 & 0 & -1 & 2 & -1 & 0 \\ 0 & 1 & 1/2 & -1/2 & 1/2 & 0 \\ 0 & 0 & 4 & -5 & 2 & 1 \end{pmatrix}$$

$$\mapsto \begin{pmatrix} 1 & 0 & -1 & 2 & -1 & 0 \\ 0 & 1 & 1/2 & -1/2 & 1/2 & 0 \\ 0 & 0 & 1 & -5/4 & 1/2 & 1/4 \end{pmatrix} \mapsto \begin{pmatrix} 1 & 0 & 0 & 3/4 & -1/2 & 1/4 \\ 0 & 1 & 0 & 1/8 & 1/4 & -1/8 \\ 0 & 0 & 1 & -5/4 & 1/2 & 1/4 \end{pmatrix}.$$

So

$$\mathbf{A}^{-1} = \begin{pmatrix} 3/4 & -1/2 & 1/4 \\ 1/8 & 1/4 & -1/8 \\ -5/4 & 1/2 & 1/4 \end{pmatrix}$$

(continued)

and indeed, calculating the product of matrices as in Sect. 4.4 **1**,

$$\mathbf{A}\mathbf{A}^{-1} = \begin{pmatrix} 1 & 2 & 0 \\ 1 & 4 & 1 \\ 3 & 2 & 2 \end{pmatrix} \begin{pmatrix} 3/4 & -1/2 & 1/4 \\ 1/8 & 1/4 & -1/8 \\ -5/4 & 1/2 & 1/4 \end{pmatrix} = \begin{pmatrix} 1 & 0 & 0 \\ 0 & 1 & 0 \\ 0 & 0 & 1 \end{pmatrix} = \mathbf{I},$$

$$\mathbf{A}\mathbf{A}^{-1} = \begin{pmatrix} 3/4 & -1/2 & 1/4 \\ 1/8 & 1/4 & -1/8 \\ -5/4 & 1/2 & 1/4 \end{pmatrix} \begin{pmatrix} 1 & 2 & 0 \\ 1 & 4 & 1 \\ 3 & 2 & 2 \end{pmatrix} = \begin{pmatrix} 1 & 0 & 0 \\ 0 & 1 & 0 \\ 0 & 0 & 1 \end{pmatrix} = \mathbf{I}.$$

We note that the solution $\mathbf{X}$ of a *matrix equation* of the form ($\mathbf{A}$ an $m \times n$, $\mathbf{B}$ an $m \times p$ matrix)

$$\mathbf{A}\mathbf{X} = \mathbf{B}$$

can be calculated (and conditions for existence and uniqueness determined) in the way we calculated here the inverse matrix as the solution of $\mathbf{A}\mathbf{X} = \mathbf{I}$.

Inverse matrices help us to determine, in the Leontief production model (see Sect. 4.5 and the beginning of Sect. 4.6), the intensity vector $\mathbf{x}$ which satisfies the given final demand $\mathbf{c}$ "without surplus". This means (compare (4.25)) that we want to solve

$$(\mathbf{I} - \mathbf{A})\mathbf{x} = \mathbf{c}$$

for $\mathbf{x}$. Since $\mathbf{A}$ and thus $\mathbf{I} - \mathbf{A}$ are $n \times n$ matrices, *a unique solution*

$$\mathbf{x} = (\mathbf{I} - \mathbf{A})^{-1}\mathbf{c} \tag{4.58}$$

exists if the rank of $(\mathbf{I} - \mathbf{A})$ *is* n. In applications to economics this is usually the case because the "production coefficients" a_{jk} are *small* compared to 1, which is in the diagonal of the unit matrix

$$\mathbf{I} = \begin{pmatrix} 1 & 0 & \dots & 0 \\ 0 & 1 & \dots & 0 \\ \vdots & \vdots & \ddots & \vdots \\ 0 & 0 & \dots & 1 \end{pmatrix},$$

and, it can be proved that this is enough to make rank $(\mathbf{I} - \mathbf{A}) = n$. Moreover, it can be shown that all components of $(\mathbf{I} - \mathbf{A})^{-1}$ are nonnegative; they cannot be all zero, since $\mathbf{0}$ is not the inverse matrix of any matrix ($\mathbf{0B} = \mathbf{0}$, not $\mathbf{I}$, for *all* $\mathbf{B}$). Since the components of the final demand $\mathbf{c}$ are, of course, positive, from (4.58) the intensities are nonnegative and not all 0, as it should be.

4.7.1 Exercises

1. Write the sum (of the 24 terms) on the right-hand side of

$$\det\begin{pmatrix} a_{11} & a_{12} & a_{13} & a_{14} \\ a_{21} & a_{22} & a_{23} & a_{24} \\ a_{31} & a_{32} & a_{33} & a_{34} \\ a_{41} & a_{42} & a_{43} & a_{44} \end{pmatrix} = \sum_{\Pi}(-1)^{N(\Pi)} a_{1\Pi(1)}a_{2\Pi(2)}a_{3\Pi(3)}a_{4\Pi(4)}.$$

2. Apply Cramer's rule to solve $\mathbf{Ax} = \mathbf{b}$ for $\mathbf{x} = \begin{pmatrix} x_1 \\ x_2 \end{pmatrix}$ and

(a) $\mathbf{A} = \begin{pmatrix} 3 & 4 \\ 2 & -5 \end{pmatrix}, \quad \mathbf{b} = \begin{pmatrix} 10 \\ -1 \end{pmatrix},$

(b) $\mathbf{A} = \begin{pmatrix} -1 & -6 \\ 7 & 9 \end{pmatrix}, \quad \mathbf{b} = \begin{pmatrix} 14 \\ 1 \end{pmatrix}.$

3. Apply Cramer's rule to solve $\mathbf{Ax} = \mathbf{b}$ for $\mathbf{x} = \begin{pmatrix} x_1 \\ x_2 \\ x_3 \end{pmatrix}$ and

(a) $\mathbf{A} = \begin{pmatrix} 5 & -2 & 4 \\ -6 & 3 & 2 \\ 7 & -5 & 9 \end{pmatrix}, \quad \mathbf{b} = \begin{pmatrix} 9 \\ 7 \\ 4 \end{pmatrix},$

(b) $\mathbf{A} = \begin{pmatrix} -6 & 7 & 2 \\ 3 & -8 & -5 \\ 4 & -9 & -3 \end{pmatrix}, \quad \mathbf{b} = \begin{pmatrix} 7 \\ 2 \\ 5 \end{pmatrix}.$

4. Determine $\mathbf{A}^{-1}$ for the matrices $\mathbf{A}$ in Exercises 2 and 3.

5. For $\mathbf{I} = \begin{pmatrix} 1 & 0 & 0 \\ 0 & 1 & 0 \\ 0 & 0 & 1 \end{pmatrix}, \mathbf{A} = \begin{pmatrix} 1/3 & 1/6 & 2/7 \\ 1/5 & 1/6 & 3/8 \\ 1/3 & 2/9 & 1/4 \end{pmatrix}, \mathbf{c} = \begin{pmatrix} 15 \\ 12 \\ 10 \end{pmatrix}$ determine $\mathbf{x} = \begin{pmatrix} x_1 \\ x_2 \\ x_3 \end{pmatrix}$ such that $(\mathbf{I} - \mathbf{A})\mathbf{x} = \mathbf{c}$.

4.7.2 Answers

1.

$$\begin{aligned}
&a_{11}a_{22}a_{33}a_{44} + a_{11}a_{23}a_{34}a_{42} + a_{11}a_{24}a_{32}a_{43}\\
&- a_{11}a_{22}a_{34}a_{43} - a_{11}a_{23}a_{32}a_{44} - a_{11}a_{24}a_{33}a_{42}\\
&+ a_{12}a_{21}a_{34}a_{43} + a_{12}a_{23}a_{31}a_{44} + a_{12}a_{24}a_{33}a_{41}\\
&- a_{12}a_{21}a_{33}a_{44} - a_{12}a_{23}a_{34}a_{41} - a_{12}a_{24}a_{31}a_{43}\\
&+ a_{13}a_{21}a_{32}a_{44} + a_{13}a_{22}a_{34}a_{41} + a_{13}a_{24}a_{31}a_{42}\\
&- a_{13}a_{21}a_{34}a_{42} - a_{13}a_{22}a_{31}a_{44} - a_{13}a_{24}a_{32}a_{41}\\
&+ a_{14}a_{21}a_{33}a_{42} + a_{14}a_{22}a_{31}a_{43} + a_{14}a_{23}a_{32}a_{41}\\
&- a_{14}a_{21}a_{32}a_{43} - a_{14}a_{22}a_{33}a_{41} - a_{14}a_{23}a_{31}a_{42}.
\end{aligned}$$

2. (a) $\begin{pmatrix} x_1 \\ x_2 \end{pmatrix} = \begin{pmatrix} 2 \\ 1 \end{pmatrix}$, (b) $\begin{pmatrix} x_1 \\ x_2 \end{pmatrix} = \begin{pmatrix} 4 \\ -3 \end{pmatrix}$.

3. (a) $\begin{pmatrix} x_1 \\ x_2 \\ x_3 \end{pmatrix} = \begin{pmatrix} 3 \\ 7 \\ 2 \end{pmatrix}$, (b) $\begin{pmatrix} x_1 \\ x_2 \\ x_3 \end{pmatrix} = \begin{pmatrix} -4 \\ -3 \\ 2 \end{pmatrix}$.

4. $\frac{1}{23}\begin{pmatrix} 5 & 4 \\ 2 & -3 \end{pmatrix}$, $\frac{1}{33}\begin{pmatrix} 9 & 6 \\ -7 & -1 \end{pmatrix}$, $\frac{1}{85}\begin{pmatrix} 37 & -2 & -16 \\ 68 & 17 & -34 \\ 9 & 11 & 3 \end{pmatrix}$, $\frac{1}{59}\begin{pmatrix} -21 & 3 & -19 \\ -11 & 10 & -24 \\ 5 & -26 & 27 \end{pmatrix}$.

5. $x_1 = 60, x_2 = 54, x_3 = 56$.

4.8 Applications of Functions of Vector Variables: Aggregation in Economics

Before proceeding from systems of linear equations (Sects. 4.6, 4.7) to systems of linear inequalities and to linear optimisation in Chap. 5, we show applications of functions of vector variables which, under certain conditions turn out to be *linear or affine*. They have to do with *aggregation* in economics and in other social sciences. The first is an *aggregation* result for an *allocation problem.* Suppose a certain (fixed) amount s of money or of some resource (for example energy and/or some materials) has to be *allocated* (distributed) among n projects. Each member of a group of *m decision makers* ("advisers") makes recommendations, the j-th *allocating* the amount x_{jk} to the k-th project ($j = 1, \dots, m;\ k = 1, \dots, n$), see Table 4.2. These should be *aggregated* ("synthesised", unified) into a final *allocation.*

For the sake of easier notation, we introduce the column vectors

$$\mathbf{x}_k = \begin{pmatrix} x_{1k} \\ \vdots \\ x_{mk} \end{pmatrix} \qquad (k = 1, \dots, n)$$

Table 4.2 Aggregating recommendations by m decision makers on allocating the amount s among n projects

Decision	Projects						
makers	1	2	...	k	...	n	Sums
1	x_{11}	x_{12}	...	x_{1k}	...	x_{1n}	s
$\vdots$	$\vdots$	$\vdots$		$\vdots$		$\vdots$	$\vdots$
j	x_{j1}	x_{j2}	...	x_{jk}	...	x_{jn}	s
$\vdots$	$\vdots$	$\vdots$		$\vdots$		$\vdots$	$\vdots$
m	x_{m1}	x_{m2}	...	x_{mk}	...	x_{mn}	s
Column vectors	$\mathbf{x}_1$	$\mathbf{x}_2$	...	$\mathbf{x}_k$	...	$\mathbf{x}_n$	$\begin{pmatrix} s \\ \vdots \\ s \end{pmatrix} = s\mathbf{1}$
Aggregated allocations	$g_1(\mathbf{x}_1)$	$g_2(\mathbf{x}_2)$	...	$g_k(\mathbf{x}_k)$	...	$g_n(\mathbf{x}_n)$	

and suppose that the *aggregated allocation* for the k-th project is

$$g_k(\mathbf{x}_k) \qquad (k = 1, \ldots, n).$$

This notation shows the assumption that the *aggregated allocation* for the k-th project depends *only* upon the allocations recommended by the advisers for *that* project. But the *aggregator functions* g_k

$$g_k : [0, s]^m \longrightarrow [0, s] \tag{4.59}$$

may at this stage be different for each project. This notation again shows another assumption, though a pretty natural one: *neither the recommended nor the aggregated allocation can or should be negative*. The only remaining assumption is the "*consensus on rejection*":

$$g_k(\mathbf{0}) = 0 \qquad (k = 1, \ldots, n), \tag{4.60}$$

that is, "if all advisers recommend rejection of a project, then no resource will be allocated to that project". This is again rather plausible, but see later reservations about both.

The essential and also reasonable assumption is that *the entire amount s will be allocated*, so from

$$\mathbf{x}_1 + \ldots + \mathbf{x}_n = s\mathbf{1} \quad \text{it follows that} \quad g_1(\mathbf{x}_1) + \ldots + g_n(\mathbf{x}_n) = s. \tag{4.61}$$

We will first deal with the case of *at least three projects*: $n > 2$. The implication (4.61) can be written as a functional equation which contains n unknown functions $g_1, \ldots, g_n$:

$$g_1(s\mathbf{1} - \mathbf{x}_2 - \ldots - \mathbf{x}_n) = s - g_2(\mathbf{x}_2) - \ldots - g_n(\mathbf{x}_n). \tag{4.62}$$

Putting herein $\mathbf{x}_2 = \ldots = \mathbf{x}_n = \mathbf{0}$ and using (4.60) we get

$$g_1(s\mathbf{1}) = s \quad \text{and similarly} \quad g_k(s\mathbf{1}) = s \quad \text{for all} \quad k = 1, \ldots, n,$$

which could be called "*consensus on overwhelming merit*". We take (4.60) also into consideration when putting $\mathbf{x}_3 = \ldots = \mathbf{x}_n = \mathbf{0}$ and, say, $\mathbf{x}_2 = \mathbf{z}$ into (4.62):

$$g_1(s\mathbf{1} - \mathbf{z}) = s - g_2\mathbf{z}. \tag{4.63}$$

Now we use both this and (4.60) when substituting $\mathbf{x}_4 = \ldots = \mathbf{x}_n = \mathbf{0}$ and, say, $\mathbf{x}_2 = \mathbf{y}$, $\mathbf{x}_3 = \mathbf{z}$ into (4.62):

$$s - g_2(\mathbf{y}) - g_3(\mathbf{z}) = g_1(s\mathbf{1} - \mathbf{y} - \mathbf{z}) = s - g_2(\mathbf{y} + \mathbf{z}),$$

(here is where we made use of $n > 2$), that is,

$$g_2(\mathbf{y} + \mathbf{z}) = g_2(\mathbf{y}) + g_3(\mathbf{z}).$$

Now, this is an interesting equation because, putting here $\mathbf{y} = \mathbf{0}$, we get, in view of $g_2(\mathbf{0}) = 0$,

$$g_2(\mathbf{z}) = g_3(\mathbf{z}),$$

that is, g_2 and g_3 is the same function. But the subscripts 2 and 3 have no privileged role in (4.61) (all subscripts $1, \ldots, n$ in (4.61) are interchangeable), so $g_1, \ldots, g_n$ *are all equal*:

$$g_1 = \ldots = g_n =: g, \tag{4.64}$$

that is, we can omit the subscripts, also in our above "interesting equation":

$$g(\mathbf{y} + \mathbf{z}) = g(\mathbf{y}) + g(\mathbf{z}). \tag{4.65}$$

This means that g is *additive* (compare Sect. 4.3).

But *all additive functions* $g : [0, s]^m \mapsto [0, s]$ (compare (4.59) and (4.64)) *are linear that is, of the form*

$$g(\mathbf{x}) = g(x_1, x_2, \ldots, x_m) = a_1x_1 + a_2x_2 + \ldots + a_mx_m, \tag{4.66}$$

where $a_1, a_2, \ldots, a_m$ are nonnegative constants ((4.66) can also be written as $g(\mathbf{x}) = \mathbf{a} \cdot \mathbf{x}$ using the inner product, see Sect. 1.5 **3**). We have mentioned this in Sect. 4.3 (under the condition of local boundedness) but we can also reduce it here right away to the result in Sect. 4.2:

From (4.65),

$$\begin{aligned} g(x_1, x_2, \ldots, x_m) &= g(x_1 + 0, 0 + x_2, \ldots, 0 + x_m) \\ &= g(x_1, 0, \ldots, 0) + g(0, x_2, \ldots, x_m) = \ldots \\ &= g(x_1, 0, \ldots, 0) + g(0, x_2, 0, \ldots, 0) + \ldots + g(0, \ldots, 0, x_m) \\ &= \sum_{j=1}^{m} g(0, \ldots, 0, x_j, 0, \ldots, 0) \end{aligned} \tag{4.67}$$

and

$$g(0, \ldots, 0, y_j + z_j, 0, \ldots, 0) = g(0, \ldots, 0, y_j, 0, \ldots, 0) + g(0, \ldots, 0, z_j, 0, \ldots, 0).$$

So $x_j \mapsto g(0, \ldots, 0, x_j, 0, \ldots, 0)$ is an additive real-valued function of a real variable, bounded on $[0, s]$ from below by 0 and from above by s because of (4.59). So, by the result in Sect. 4.2, $g(0, \ldots, 0, x_j, 0, \ldots, 0) = a_j x_j$ and we have to have $a_j \geq 0$ in order that $a_j x_j$ be nonnegative ($j = 1, \ldots, m$). Further, by (4.67), we have indeed (4.66):

$$g(\mathbf{x}) = g(x_1, \ldots, x_m) = \sum_{j=1}^{m} a_j x_j \ . \quad (a_j \geq 0).$$

Finally we use the "consensus on overwhelming merit" property which becomes, in view of (4.64),

$$g(s\mathbf{1}) = g(s, s, \ldots, s) = s.$$

So, from (4.66), we have

$$a_1 s + a_2 s + \ldots + a_m s = s$$

and (since $s \neq 0$)

$$a_1 + a_2 + \ldots + a_m = 1.$$

Linear functions (4.66) with $a_j \geq 0$ ($j = 1, \ldots, m$) and $a_1 + a_2 + \ldots + a_m = 1$ are *weighted arithmetic means* ($a_1, a_2, \ldots, a_m$ are the *weights*). Using (4.64) and (4.66) again we get

$$g_1(x_1, \ldots, x_m) = \ldots = g_n(x_1, \ldots, x_m) = a_1 x_1 + \ldots + a_m x_m$$

$$\text{with} \quad a_j \geq 0 \ (j = 1, \ldots, m) \quad \text{and} \quad a_1 + \ldots + a_m = 1.$$

Straightforward checking shows that, conversely, if each g_j $(j = 1, \ldots, m)$ is the *same* weighted arithmetic mean, then all our conditions (4.59), (4.60) and (4.61) are satisfied.

The result in Sect. 4.2, which we used here, is about functions additive for all *reals* (there (4.13) had to hold for all $x_1 \in \mathbb{R}$, $x_2 \in \mathbb{R}$). Here all variables (and also the function values) have to stay in the interval $[0, s]$. However, that result is true also in this case. (This is not very difficult to show but we will not do it here, the reader may wish to prove it).

So we obtained here the result that *under the* rather plausible *assumptions* (4.59), (4.60) *and* (4.61), *for* $n > 2$ *projects, the aggregator function for each project has to be the same weighted arithmetic mean—and every weighted arithmetic mean will do*. (The $a_1, \ldots, a_m$ can be considered to be *the "weights of influence" of the individual adviser, which may be different but*, as a consequence of our result, they *cannot change from project to project*).

The "hidden assumption" that the aggregated allocation for the k-th project depends only upon the recommended all cations for that project, which, as we mentioned is implicit in the notation $g_k(\mathbf{x}_k)$ is rather restrictive. It is not that we cannot solve the problem if each aggregated allocation may depend upon the whole matrix

$$\mathbf{X} = \begin{pmatrix} x_{11} & \ldots & x_{1n} \\ \vdots & & \vdots \\ x_{m1} & \ldots & x_{mn} \end{pmatrix}$$

of recommended allocations, on the contrary, we get too many solutions: In this situation we have in place of (4.61) and (4.59)

$$g_1(\mathbf{X}) + g_2(\mathbf{X}) + \ldots + g_n(\mathbf{X}) = s \tag{4.68}$$

for all $m \times n$ matrices $\mathbf{X}$ with *components* in $[0, s]$, for which the sum of each row is s. We have also $g_k(\mathbf{X}) \in [0, s]$ $(k = 1, 2, \ldots, n)$.

With the latter restriction (and further two to follow), we can choose $g_2, \ldots, g_n$ arbitrarily and just define

$$g_1(\mathbf{X}) = s - g_2(\mathbf{X}) - \ldots - g_n(\mathbf{X})$$

to get a solution. Actually, as we see, we have to have also $g_2(\mathbf{X}) + \ldots + g_n(\mathbf{X}) \leq s$ but this does not restrict the choice of $g_2, \ldots, g_n$ too much. Neither does the assumption corresponding to (4.60): if the k-th column of $\mathbf{X}$ is $\mathbf{0}$ then $g_k(\mathbf{X}) = 0$ $(k = 1, 2, \ldots, n)$. Denoting the column vectors of $\mathbf{X}$ by $\mathbf{x}_1, \mathbf{x}_2, \ldots, \mathbf{x}_n$ this means

$$g_l(\mathbf{x}_1, \ldots, \mathbf{x}_{l-1}, \mathbf{0}, \mathbf{x}_{l+1}, \ldots, \mathbf{x}_n) = 0 \qquad (l = 2, \ldots, n)$$

and, in view of (4.68),

$$g_2(\mathbf{0}, \mathbf{x}_2, \ldots, \mathbf{x}_n) + \ldots + g_n(\mathbf{0}, \mathbf{x}_2, \ldots, \mathbf{x}_n) = s,$$

which still leaves plenty of freedom in the choice of $g_2, \ldots, g_n$. Even in our original formulation, it is possible that the assumption (4.59) has to be modified and (4.60) even omitted. The former is the case when all allocations are constrained to be between a prescribed minimum and a maximum. This changes the domain and range in (4.59) but not the boundedness which was essential in solving (4.65). On the other hand, the "consensus on rejection" condition (4.60) is not so self-evident anymore if the final decision is made by an external person (or persons) who may ignore even such a categorical recommendation of rejection. Then also the "consensus on overwhelming merit" equation does not follow and therefore $a_1 + \ldots + a_n$ need not be equal 1 anymore. But, by a somewhat different method one still gets that $g_1, \ldots, g_n$ will be, if not linear, at least affine functions $g_k(x_1, \ldots, x_n) = a_1x_1 + \ldots + a_nx_n + b_k$ $(k = 1, \ldots, n)$ (as we see $g_1 = g_2 = \ldots = g_n$ may get lost too; see also below).

All this, except (4.68), was for $n > 2$. The case $n = 1$ (just one project) is completely trivial. We show that in the case $n = 2$, even under the original assumptions, about as much freedom of choice is left as, under different circumstances, in the solution of (4.68). Then (4.61) reduces to the statement that

$$\text{from} \quad \mathbf{x}_1 + \mathbf{x}_2 = s\mathbf{1} \quad \text{it follows that} \quad g_1(\mathbf{x}_1) + g_2(\mathbf{x}_2) = s$$

and the solution is again simple: choose $g_2 : [0, s]^m \to [0, s]$ arbitrarily and $g_1 : [0, s]^m \to [0, s]$, as determined by (4.62):

$$g_1(\mathbf{x}_1) = s - g_2(s\mathbf{1} - \mathbf{x}_1).$$

So, also the values of the function g_1 will automatically be in $[0, s]$ and, clearly, (4.59) and (4.60) are satisfied. In order to satisfy also (4.60), the only restrictions on the choice of g_2 will be $g_2(\mathbf{0}) = 0$, and $g_2(s\mathbf{1}) = s$ ("consensus" both "on rejection" and "on overwhelming merit") which leaves it still pretty arbitrary but establishes also $g_1(\mathbf{0}) = 0$. And that is all there is to it.

We now sketch a second *aggregation problem*. Suppose that m "agents" for instance producers, use (at least) n kinds of goods and services. Let x_{jk} be the input quantity used by the j-th producer from the k-th good or service $(k = 1, \ldots, n)$ and let y_j be the maximal output value (in market prices) which this j-th producer $(j = 1, \ldots, m)$ can establish from these goods and services (other inputs fixed). Or let m households buy n kinds of goods and services, the j-th household the quantity x_{jk} of the k-th good or service $(k = 1, \ldots, n)$ and let y_j be the utility for the j-th household $(j = 1, \ldots, m)$ of all these quantities $x_{j1}, \ldots, x_{jn}$ of goods and services bought. Many more situations in economics and other social sciences follow this scheme, described by Table 4.3.

Table 4.3 Aggregation of input or purchase quantities which establish output value or utility

Agents (producers or households)	Goods and services 1	…	k	…	n	Row vectors	Maximal output values or utilities (microeconomic production or utility functions)
1	x_{11}	…	x_{1k}	…	x_{1n}	$\mathbf{x}_1$	$y_1 = f_1(\mathbf{x}_1)$
⋮	⋮		⋮		⋮	⋮	⋮
j	x_{j1}	…	x_{jk}	…	x_{jn}	$\mathbf{x}_j$	$y_j = f_j(\mathbf{x}_j)$
⋮	⋮		⋮		⋮	⋮	⋮
m	x_{m1}	…	x_{mk}	…	x_{mn}	$\mathbf{x}_m$	$y_m = f_m(\mathbf{x}_m)$
Column vectors	$\mathbf{x}'_1$	…	$\mathbf{x}'_k$	…	$\mathbf{x}'_n$		$\mathbf{y}'$
Aggregates (aggregator functions)	$z_1 = g_1(\mathbf{x}'_1)$	…	$z_k = g_k(\mathbf{x}'_k)$	…	$z_n = g_n(\mathbf{x}'_n)$	$\mathbf{z}$	$F(\mathbf{z}) = G(\mathbf{y}')$ "Aggregation equation"

It will again be of advantage to introduce the column vectors but this time we denote them by

$$\mathbf{x}'_k = \begin{pmatrix} x_{1k} \\ \vdots \\ x_{mk} \end{pmatrix} \qquad (k = 1, \ldots, n),$$

because we will denote the row vectors by $\mathbf{x}_j$:

$$\mathbf{x}_j = (x_{j1}, \ldots, x_{jn}) \qquad (j = 1, \ldots, m).$$

In both the above examples the supposition that y_j depends only upon the quantities $x_{j1}, \ldots, x_{jn}$ of goods and services, that is, upon $\mathbf{x}_j$:

$$y_j = f_j(\mathbf{x}_j) \qquad (j = 1, \ldots, m),$$

is more plausible than in the allocation problem. The problem is, whether *aggregates* of the quantities in the columns, that is

$$z_k = g_k(\mathbf{x}'_k) = g_k(x_{1k}, \ldots, x_{mk}) \qquad (k = 1, \ldots, n)$$

and

$$Y = G(\mathbf{y}') = G(y_1, \dots, y_m)$$

can be determined (again $g_1, \dots, g_n$ and $G : \mathbb{R}_+^m \to \mathbb{R}_+$ are the aggregator functions) so that $z_1, \dots, z_n$ act as "*aggregate quantities*" producing the "*aggregate maximal output*" or having the "*aggregate utility*" Y. If so, then clearly one more function $F : \mathbb{R}_+^n \to \mathbb{R}_+$ has to exist so that the functional equation ("*aggregation equation*") $F(\mathbf{z}) = G(\mathbf{y}')$, that is,

$$F(g_1(\mathbf{x}'_1), \dots, g_n(\mathbf{x}'_n)) = G(f_1(\mathbf{x}_1), \dots, f_m(\mathbf{x}_m)) \tag{4.69}$$

be satisfied.

In this case we say that, for the aggregator functions $g_1, \dots, g_n$ and the "*microeconomic correlations*" *(functions)* $f_1, \dots, f_m$, there exist a "*macroeconomic aggregator function*" G and a "*macroeconomic correlation*" (*function*) F such that F assigns to the aggregates $z_1 = g_1(\mathbf{x}'_1), \dots, z_n = g_n(\mathbf{x}'_n)$ exactly the value $Y = F(\mathbf{z})$, aggregated from the microeconomic function (correlation) values $y_1 = f_1(\mathbf{x}_1), \dots, y_m = f_m(\mathbf{x}_m)$, whatever the original x_{jk} ($j = 1, \dots, m$; $k = 1, \dots, n$) were. In our above example of producers, this common value Y is the *maximal total output value*, in the example about households it is the *total utility* (equal, by (4.69), to $G(\mathbf{y}')$, the *aggregate maximal output* or the *aggregate utility*, respectively). In the case of producers the functions $f_1, \dots, f_m, F$ are "*production functions*", in the case of households they are "*utility functions*".

If all inputs (goods and services) considered in Table 4.3 could be "totally separated" so that there is no overlap, then it would seem reasonable to take

$$g_k(\mathbf{x}'_k) = g_k(x_{1k}, \dots, x_{mk}) = x_{1k} + \dots + x_{mk} \quad (k = 1, \dots, n), \tag{4.70}$$

that is, we would have the case where *all aggregator functions are sums* (where aggregation is done by adding up the quantities). A further assumption could be that

$$G(\mathbf{y}') = G(y_1, \dots, y_m) = y_1 + \dots + y_m, \tag{4.71}$$

that is, the maximal output values or utilities also add up. Then (4.69) becomes

$$F(x_{11} + \dots + x_{m1}, \dots, x_{1n} + \dots + x_{mn})$$

$$= f_1(x_{11}, \dots, x_{1n}) + \dots + f_m(x_{m1}, \dots, x_{mn}),$$

that is,

$$F(\mathbf{x}_1 + \dots + \mathbf{x}_m) = f_1(\mathbf{x}_1) + \dots + f_m(\mathbf{x}_m). \tag{4.72}$$

This equation is similar to (4.62) (put into (4.62) $F(\mathbf{x}) := s - g_1(s\mathbf{1} - \mathbf{x})$) and is solved in a similar way: Remembering that all functions $f_1, \ldots, f_m$, F map $\mathbb{R}_+^n$ into $\mathbb{R}_+$, we put into (4.72) $\mathbf{x}_2 = \ldots = \mathbf{x}_m = \mathbf{0}$ and get

$$F(\mathbf{x}_1) = f_1(\mathbf{x}_1) + f_2(\mathbf{0}) + \ldots + f_m(\mathbf{0}).$$

We write $b_j := f_j(\mathbf{0})$ $(j = 1, \ldots, m)$ (this time we do *not* necessarily have $f_j(\mathbf{0}) = 0$, since, for instance, inputs not considered in Table 4.3 may be used with (fixed) positive quantities). Then (4.72) with $\mathbf{x}_1 = \mathbf{x}_2 = \ldots = \mathbf{x}_m = \mathbf{0}$ gives

$$F(\mathbf{0}) = b_1 + b_2 + \ldots + b_m =: b. \tag{4.73}$$

So we have $f_2(\mathbf{0}) + \ldots + f_m(\mathbf{0}) = b - b_1$ and $f_1(\mathbf{x}) = F(\mathbf{x}) + b_1 - b$. Similarly,

$$f_j(\mathbf{x}) = F(\mathbf{x}) + b_j - b \quad (j = 1, 2, \ldots, m). \tag{4.74}$$

Putting this back into (4.72) we obtain

$$\begin{aligned} &F(\mathbf{x}_1 + \mathbf{x}_2 + \ldots + \mathbf{x}_m) \\ &= F(\mathbf{x}_1) + b_1 - b + F(\mathbf{x}_2) + b_2 - b + \ldots + F(\mathbf{x}_m) + b_m - b \\ &= F(\mathbf{x}_1) + F(\mathbf{x}_2) + \ldots + F(\mathbf{x}_m) + b - mb \end{aligned}$$

which, with

$$h(\mathbf{x}) := F(\mathbf{x}) - b, \tag{4.75}$$

becomes

$$h(\mathbf{x}_1 + \mathbf{x}_2 + \ldots + \mathbf{x}_m) = h(\mathbf{x}_1) + h(\mathbf{x}_2) + \ldots + h(\mathbf{x}_m).$$

From (4.73) and (4.75), $h(\mathbf{0}) = 0$ so that, putting $\mathbf{x}_3 = \ldots = \mathbf{x}_m = \mathbf{0}$ gives

$$h(\mathbf{x}_1 + \mathbf{x}_2) = h(\mathbf{x}_1) + h(\mathbf{x}_2) \quad (\mathbf{x}_1 \in \mathbb{R}_+^n,\ \mathbf{x}_2 \in \mathbb{R}_+^n).$$

This is essentially the same equation as (4.65) (but the domain is different). Moreover, by definition, $F(\mathbf{x}) \geq 0$ $(\mathbf{x} \in \mathbb{R}_+^n)$ so, by (4.75), $h(\mathbf{x}) \geq -b$, that is, h is bounded from below on $\mathbb{R}_+^n$. But then we know that

$$h(\mathbf{x}) = a_1x_1 + \ldots + a_nx_n = \mathbf{a} \cdot \mathbf{x}$$

and, by (4.75), (4.74) and (4.73),

$$\begin{aligned} F(\mathbf{x}) &= \mathbf{a} \cdot \mathbf{x} + b = \mathbf{a} \cdot \mathbf{x} + b_1 + \ldots + b_m, \\ f_j(\mathbf{x}) &= \mathbf{a} \cdot \mathbf{x} + b_j \quad (j = 1, \ldots, m), \end{aligned} \tag{4.76}$$

where $\mathbf{a} \cdot \mathbf{x} = a_1x_1 + \ldots + a_nx_n$ is again the inner product. Now $f_j(\mathbf{x}) \geq 0$ for $\mathbf{x} \in \mathbb{R}^n_+$, in particular $f_j(\mathbf{0}) = b_j \geq 0$ $(j = 1, \ldots, m)$. If any of the $a_1, \ldots, a_n$ were negative, say $a_\ell = -\alpha < 0$ $(\alpha > 0)$, then we would have

$$fj(0, \ldots, 0, x_\ell, 0, \ldots, 0) = -\alpha x_\ell + b_j < 0 \quad \text{for} \quad x_\ell > \frac{b_j}{\alpha} \in \mathbb{R}_+$$

while f_j should be nonnegative on $\mathbb{R}^n_+$. So $a_k \geq 0$ for all $k = 1, \ldots, n$. On the other hand, (4.76) satisfies (4.72). Furthermore, if $\mathbf{a} = (a_1, \ldots, a_n)$ with $a_k \geq 0$ $(k = 1, \ldots, n)$ and if also $b_j \geq 0$ $(j = 1, \ldots, m)$, then the $f_1, \ldots, f_m$, F in (4.76) are nonnegative, as required.

So we proved that, *if all aggregator functions are sums (including the macroeconomic G) then the microeconomic functions $f_1, \ldots, f_m$ and the macroeconomic F are affine functions*:

$$f_j(\mathbf{x}) = a_1x_1 + \ldots + a_nx_n + b_j \quad (j = 1, \ldots, m),$$
$$F(\mathbf{x}) = a_1x_1 + \ldots + a_nx_n + b_1 + \ldots + b_m$$

with $a_k \geq 0$ $(k = 1, \ldots, n)$, $b_j \geq 0$ $(j = 1, \ldots, m)$. Notice that here the $f_1, \ldots, f_m$ are not equal anymore but they are "almost equal": they differ only in constants. In many cases in practice the empirically determined functions (for instance the production functions in our example of maximal output of producers) are *not* affine, so the above assumptions, in particular (4.70) and (4.71), may be too restrictive.

The aggregation equation (4.69) is much more general and has many non-affine solutions. Even so, there is no guarantee that to given (empirical) production functions there exist aggregation functions which satisfy (4.69). If there exist such aggregation functions they are not always what we would have expected (for instance not sums). In these cases one can say that there is a "deficit": a nonzero difference between the two sides of (4.69).

4.8.1 Exercises

1. Four decision makers A, B, C, D allocate the amount 100 K (= \$100, 000) among ten projects as follows:

	Projects									
	1	2	3	4	5	6	7	8	9	10
A	10	10	10	10	10	10	10	10	10	10
B	8	8	9	9	10	10	11	11	12	12
C	2	2	6	6	10	10	14	14	18	18
D	1	3	5	7	9	11	13	15	17	19

The aggregation process follows the assumption made in Sect. 4.8.The aggregated allocations are
(a) 5.25 5.75 7.5 8 9.75 10.25 12 12.5 14.25 14.75,
(b) 4.4 5.2 7 7.8 9.6 10.4 12.2 13 14.8 15.6,
(c) 3.6 4.4 6.6 7.4 9.6 10.4 12.6 13.4 15.6 16.4.
Determine the aggregator functions.

2. For the aggregator function $g(x_1, x_2, x_3, x_4) = a_1x_1 + a_2x_2 + a_3x_3 + a_4x_4$, where $a_1 = a_2 = 1/4$, determine a_3, a_4 so that in the situation described in Exercise 1
(a) projects 1 and 10 get 5.1 and 14.9, respectively,
(b) projects 3 and 8 get 7.25 and 12.75, respectively.
3. In Table **4.3** let

$$y_j = f_j(\mathbf{x}_j) = c_j \prod_{k=1}^{n} x_{jk} \qquad (c_j \in \mathbb{R}_{++}, j = 1, \ldots, m)$$

and

$$z_k = g_k(\mathbf{x'}_k) = d_k \prod_{j=1}^{m} x_{jk} \qquad (d_k \in \mathbb{R}_{++}, k = 1, \ldots, n).$$

Determine a pair F, G of aggregator functions such that $F(z_1, \ldots, z_n) = G(y_1, \ldots, y_m)$ for all $x_{jk} \in \mathbb{R}_+$.
4. In Table **4.3** let $m = 2$, $n = 3$ and

$$y_1 = x_{11} + 2x_{12} + 3x_{13}, \quad y_2 = x_{21} + 4x_{22} + 9x_{23},$$

$$z_1 = x_{11} + \frac{1}{4}x_{21}, \quad z_2 = x_{12} + \frac{1}{2}x_{22}, \quad z_3 = x_{13} + \frac{3}{4}x_{23}.$$

(a) Do there exist aggregator functions F, G of the form

$$F(z_1, z_2, z_3) = z_1 + az_2 + bz_3 \qquad (a \in \mathbb{R}_{++}, b \in \mathbb{R}_{++}),$$

$$G(y_1, y_2) = cy_1 + dy_2 \qquad (c \in \mathbb{R}_{++}, d \in \mathbb{R}_{++})$$

such that

$$F(z_1, z_2, z_3) = G(y_1, y_2) \quad \text{for all} \quad x_{jk} \in \mathbb{R}_+ \qquad (j = 1, 2;\ k = 1, 2, 3)?$$

(b) Same question, if the coefficient of x_{12} in the first equation is different from 2.
5. Compare Tables **4.2** and **4.3**. Specify the rows and columns of Table **4.3** so that it gets transformed into Table **4.2**.

4.8.2 Answers

1. (a) $g(x_1, x_2, x_3, x_4) = \frac{1}{4}(x_1 + x_2 + x_3 + x_4)$,
   (b) $g(x_1, x_2, x_3, x_4) = \frac{1}{5}(x_1 + x_2 + x_3 + 2x_4)$,
   (c) $g(x_1, x_2, x_3, x_4) = \frac{1}{10}(x_1 + 2x_2 + 3x_3 + 4x_4)$.
2. (a) $a_3 = 1/10$, $a_4 = 4/10$, (b) $a_3 = 0$, $a_4 = 1/2$.
3. $F(z_1, \ldots, z_n) = (cz_1z_2 \cdots z_n)$, $G(y_1, \ldots, y_m) = (dy_1y_2 \cdots y_m)$, where $c = c_1c_2 \cdots c_md/d_1d_2 \cdots d_n$.
4. (a) Yes, if and only if $a = 2$, $b = 3$, $c = 1$, $d = 1/4$,
   (b) No, there exist no such aggregator functions.
5. The sum of the elements of each of the m rows should equal s, the $f_j(\mathbf{x}_j)$ in Table **4.3** should be $f_j(\mathbf{x}_j) = x_{j_1} + x_{j_2} + \cdots + x_{j_n} = s$ $(j = 1, 2, \ldots, m)$, $\mathbf{Y}' = s\mathbf{1}$ should hold, where $\mathbf{1} = (1, 1, \ldots, 1)$.

Linear Optimisation, Duality: Zero-Sum Games 5

For when the One Great Scorer comes
To write against your name,
He marks—not that you won or lost—
But how you played the game.

GRANTLAND RICE (1880–1954)
AMERICAN SPORTS JOURNALIST AND POET

5.1 Introduction

In Sect. 4.1 we discussed (and solved) the following simple linear optimisation problem. A supermarket chain intends to keep buying some, say x_1, weight units of one kind and x_2 units of a second kind of detergent but not more than 100 weight units for not more then \$720 a week. The factory originally charged \$6 and \$9 per weight unit of the first respectively the second detergent which would have contributed 60 or 90 cents, respectively, all together $60x_1 + 90x_2 \leq 7200$ cents to its weekly profit. This leads to the inequalities

$$x_1 + x_2 \leq 100, \tag{5.1}$$

$$6x_1 + 9x_2 \leq 720. \tag{5.2}$$

If these are the only constraints then maximal quantity would give maximal profit and we had to solve the system of linear equations

$$x_1 + x_2 = 100,$$

$$6x_1 + 9x_2 = 720,$$

which we did. We got $x_1 = 60$, $x_2 = 40$ (7200 cents = \$72 per week contribution to the profit of the factory).

W. Eichhorn, W. Gleißner, *Mathematics and Methodology for Economics*, Springer Texts in Business and Economics, DOI 10.1007/978-3-319-23353-6_5

But now the factory owners make a new offer. They give a discount of 20 and 10 cents per weight unit of the first or second detergent (that is, the profit contribution from them to the profit of the factory is just 40 and 80 cents, respectively) if they (the factory owners) can determine the quantities x_1 and x_2 of the two kinds of detergents within the confines of condition (5.1). (That is, if they deliver x_1 units of the first kind then they cannot deliver more then $100-x_1$ units of the second kind.) Since the factory would clearly gain most by delivering 100 units of the second and none of the first detergent, the supermarket chain specifies that it accepts at most 60 weight units of the second detergent:

$$x_2 \leq 60. \tag{5.3}$$

Moreover, to be one the safe side, the supermarket owners want also the condition (5.2) upheld (notice that $x_1 = 40$, $x_2 = 60$ would contradict this condition). The question is, what quantities x_1 and x_2 should the factory deliver of the two kinds of detergent in order to maximise its profit (in cents)

$$H(x_1, x_2) = 40x_1 + 80x_2 \tag{5.4}$$

under conditions (5.1), (5.2) and (5.3).

This is clearly a linear optimisation problem of the kind we discussed in Sect. 2.4 and will discuss in more detail in Sects. 5.2 and 5.3. In Sect. 5.2 we will solve a problem equivalent to maximising (5.4) under conditions (5.2), (5.3), (5.4) (and $x_1 \in \mathbb{R}_+, x_2 \in \mathbb{R}_+$). This will yield *a solution different from the one above*: $x_1 = 30$, $x_2 = 60$ as optimal quantities and thus $40 \cdot 30 + 80 \cdot 60 = 6000$ (cent = \$60) as maximal profit.

We discussed in Sect. 2.2 the "*economic efficiency rule*", according to which one strives to achieve a goal with lowest cost or to maximise the output with given amounts of inputs. As we saw there this leads to linear optimisation problems. There, as here, this is the case in such simple production systems as linear technologies. In Sect. 2.4 we have shown by means of an example how linear optimisation problems in two variables can be solved geometrically. The same would be difficult for such problems in three variables and, for lack of more than three dimensional geometric intuition, for practical purposes all but impossible in the case of four or more variables.

Therefore we introduced already there a process of numerical approximation, the "*method of steepest ascent*" for solving linear optimisation problems in any number of variables (the name still has a geometric connotation). This method leads to "close to optimal" values—if there exist optimal values at all.

By now we have acquired mathematical tools which make the application and understanding of further methods for solving linear approximation problems possible. The classical and still fundamental method is called "*simplex algorithm*". We will see in Sect. 5.2 on an example how and why it works.

This will be followed by the notion of *duality* in linear optimisation (Sect. 5.3) which is useful, among others, in the theory of *two-person zero-sum games* and will give us occasion to have an insight into that theory (Sect. 5.4).

5.2 Linear Optimisation Problems

Linear optimisation problems 1, 2 and 3 in Sect. 2.2 are of the following form.

A problem of *linear optimisation* is to determine where a *linear function* $F : \mathbb{R}^s \to \mathbb{R}$, given by

$$F(x_1, \ldots, x_s) = c_1x_1 + \ldots + c_sx_s, \tag{5.5}$$

is maximal (or minimal), if also the conditions

$$a_{j1}x_1 + \ldots + a_{js}x_s \leq b_j \quad (j = 1, \ldots, m_1), \tag{5.6}$$

$$a_{j1}x_1 + \ldots + a_{js}x_s \geq b_j \quad (j = m_1 + 1, \ldots, m_2), \tag{5.7}$$

$$a_{j1}x_1 + \ldots + a_{js}x_s = b_j \quad (j = m_2 + 1, \ldots, m_3), \tag{5.8}$$

are satisfied (c_k, b_j, a_{jk} $(k = 1, \ldots, s)$ are real constants). The function F is the *objective function.*

Sometimes some variables are supposed to be nonnegative, say

$$x_k \geq 0 \quad (k = 1, \ldots, t;\ t \leq s). \tag{5.9}$$

These are clearly of the form (5.7) too (but often listed separately). Also, affine functions, given by

$$\tilde{F}(x_1, \ldots, x_s) = c_1x_1 + \ldots + c_sx_s + d$$

(d a real constant), have their maxima and minima at the same place where (5.5), so admitting them would not be an essential generalisation. Moreover, one can simplify the above formulation by asking only where (5.5) is *maximal under the conditions* or *restrictions*

$$a_{j1}x_1 + \ldots + a_{js}x_s \leq b_j \quad (j = 1, \ldots, 2m_3 - m_2). \tag{5.10}$$

Indeed, if the problem were to *minimise* (5.5) we can instead maximise

$$(-c_1)x_1 + \ldots + (-c_s)x_s$$

which is of the same form. Similarly, (5.7) can be replaced by

$$(-a_{j1})x_1 + \ldots + (-a_{js})x_s \leq -b_j \quad (j = m_1 + 1, \ldots, m_2),$$

which are of the form (5.10) and, finally, Eqs. (5.8) are equivalent to twice as many inequalities

$$a_{j1}x_1 + \ldots + a_{js}x_s \leq b_j \quad \textit{and} \quad -a_{j1}x_1 - \ldots - a_{js}x_s \leq b_j,$$

also of the form (5.10). (We see now why we let j in (5.10) go from 1 to $m_2 + 2(m_3 - m_2) = 2m_3 - m_2$.) Often those $(x_1, \ldots, x_s)$ which satisfy (5.10) are called the *feasible solutions* and those among them which maximise (5.5) the *optimal solutions* of the linear optimisation problem.

We stated above the obvious way how to write, in form of inequalities (5.10), the restrictions which had been given as Eqs. (5.8). We will use them in this form later. Now we note, however, that conversely, *one can replace all inequalities* (5.10) *by equations and by special inequalities of the form* (5.9), *if one increases the number of variables*. Indeed, by introducing $q := 2m_3 - m_2$ "*slack variables*"

$$x_{s+j} := b_j - a_{j1}x_1 - \ldots - a_{js}x_s \quad (j = 1, \ldots, q)$$

one can reduce (5.10) to

$$x_{s+j} \geq 0 \quad \text{and} \quad a_{j1}x_1 + \ldots + a_{js}x_s + x_{s+j} = b_j \quad (j = 1, \ldots, q),$$

that is, to inequalities of the form (5.9) and to equations of the form (5.8). One may even require (5.9) to hold for *all* variables by increasing the number of equations of the form (5.8) and also again the number of variables. Indeed if, for instance, x_k is permitted to take any real value, we can substitute $x_k = x'_k - x''_k$ and add the two inequalities $x'_k \geq 0$, $x''_k \geq 0$ which are of the form (5.9). This gives the following "*canonical form*" of linear optimisation problems.

Maximise

$$G(x_1, \ldots, x_n) = c_1x_1 + \ldots + c_nx_n \tag{5.11}$$

under the conditions (restrictions)

$$a_{j1}x_1 + \ldots + a_{jn}x_n = b_j \quad (j = 1, \ldots, m), \tag{5.12}$$

$$x_k \geq 0 \quad (k = 1, \ldots, n). \tag{5.13}$$

With the vector and matrix notations of Sects. 1.3, 1.4, 4.2 and 4.3 (including the inner product in 1.3 **3**), this can be written as follows. *Find*

$$\mathbf{x} = \begin{pmatrix} x_1 \\ \vdots \\ x_n \end{pmatrix} \in \mathbb{R}^n$$

which maximises

$$G(\mathbf{x}) = \mathbf{c} \cdot \mathbf{x}$$

under the conditions

$$\mathbf{Ax} = \mathbf{b}, \qquad \mathbf{x} \geqq \mathbf{0}, \tag{5.14}$$

where

$$\mathbf{c} = \begin{pmatrix} c_1 \\ \vdots \\ c_n \end{pmatrix} \in \mathbb{R}^n, \quad \mathbf{b} = \begin{pmatrix} b_1 \\ \vdots \\ b_m \end{pmatrix} \in \mathbb{R}^m, \quad \mathbf{A} = \begin{pmatrix} a_{11} & \dots & a_{1n} \\ \vdots & & \vdots \\ a_{m1} & \dots & a_{mn} \end{pmatrix} \in \mathbb{R}^{mn}$$

are constant vectors and matrices.

The set of feasible solutions of this problem is a polyhedron (since all conditions are affine) and *is convex* (see Sect. 3.3), since $\mathbf{x}^1$ and $\mathbf{x}^2$ satisfying

$$\mathbf{Ax}^1 = \mathbf{b},\ \mathbf{x}^1 \geqq \mathbf{0},\ \mathbf{Ax}^2 = \mathbf{b},\ \mathbf{x}^2 \geqq \mathbf{0}$$

implies, for all $\lambda \in]0, 1[$,

$$\begin{aligned} \mathbf{A}(\lambda\mathbf{x}^1 + (1-\lambda)\mathbf{x}^2) &= \mathbf{A}\lambda\mathbf{x}^1 + \mathbf{A}(1-\lambda)\mathbf{x}^2 = \\ &= \lambda\mathbf{Ax}^1 + (1-\lambda)\mathbf{Ax}^2 = \lambda\mathbf{b} + (1-\lambda)\mathbf{b} = \mathbf{b} \end{aligned}$$

and of course,

$$\lambda\mathbf{x}^1 + (1-\lambda)\mathbf{x}^2 \geqq \mathbf{0}.$$

Example 1 The example on which we introduce the simplex algorithm will again, as in Sect. 2.3, contain only two (genuine) variables, in order that we can check the results on a figure, but we will be careful to describe the algorithm so that its applicability to any number of variables become eventually clear. The problem (equivalent to that in Sect. 4.1) is: *Maximise*

$$F(x_1, x_2) = x_1 + 2x_2 \tag{5.15}$$

under the conditions

$$x_1 + x_2 \leq 100, \tag{5.16}$$

(continued)

$$6x_1 + 9x_2 \leq 720, \tag{5.17}$$

$$x_2 \leq 60, \tag{5.18}$$

$$x_1 \geq 0, \tag{5.19}$$

$$x_2 \geq 0. \tag{5.20}$$

As expected, we see in Fig. 5.1 that the set of *feasible* solutions is a convex polygon. We see also that if this polygon is finite (bounded) then *at least one of the vertices* of this polygon *will be an optimal solution*: Just move the contour lines

$$\{(x_1, x_2) \mid F(x_1, x_2) = x_1 + 2x_2 = c\}$$

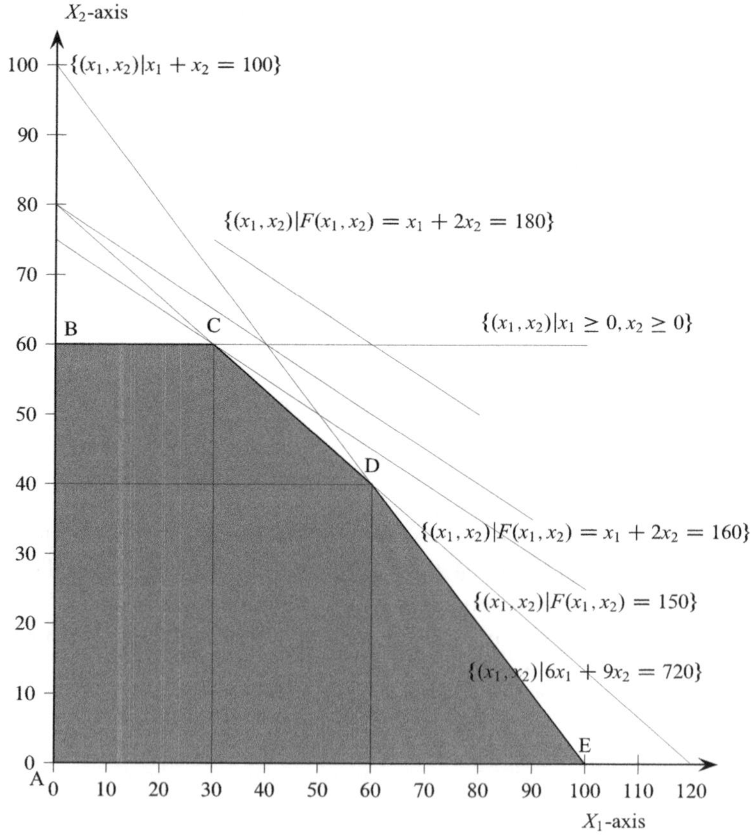

Fig. 5.1 Set of feasible solutions (*shaded area*) and contour lines (the three parallel lines) of the linear optimisation problem(5.15), (5.16), (5.17), (5.18), (5.19), and (5.20)

parallely by increasing c. (Note that, if one of the sides of the polygon were parallel to the contour lines, then more than one vertex and not only vertices would be optimal solutions.)

The canonical form (5.11), (5.12), (5.13) of this problem is as follows. *Maximise*

$$\begin{aligned} G(x_1, x_2, x_3, x_4, x_5) &= c_1x_1 + c_2x_2 + c_3x_3 + c_4x_4 + c_5x_5 \\ &= x_1 + 2x_2 + 0x_3 + 0x_4 + 0x_5 \qquad (5.21) \\ &(= F(x_1, x_2)) \end{aligned}$$

under the restrictions

$$x_1 + x_2 + x_3 = 100, \qquad (5.22)$$

$$6x_1 + 9x_2 + x_4 = 720, \qquad (5.23)$$

$$x_2 + x_5 = 60, \qquad (5.24)$$

$$x_1 \geq 0, x_2 \geq 0, x_3 \geq 0, x_4 \geq 0, x_5 \geq 0. \qquad (5.25)$$

In order to decide which vertex or vertices (in case of more than one also the sides between them) are *optimal*, let us tabulate their impact (Table 5.1).

This shows that C is an optimal solution *among the vertices*. But F is *strictly increasing* in both variables x_1 and x_2, so *no other* (non-vertex) *points of the feasible set can be optimal*. Actually, *in this case the "method of steepest ascent"*, presented in Sect. 2.3, happens *to yield the optimal solution* of the linear optimisation problem. Indeed, the direction of steepest ascent for $F(x_1, x_2) = x_1 + 2x_2$ is (compare Sect. 2.3) that of the vector $(1, 2)$, so the *lines of steepest ascent* through $(0, 0)$ consists of the points

$$(x_1, x_2) = (\lambda, 2\lambda).$$

Here $\lambda \geq 0$ by (5.19) and $24\lambda \leq 720$, that is, $\lambda \leq 30$ by (5.17). Since $F(x_1, x_2) = x_1 + 2x_2$ strictly increases both with x_1 and x_2, thus with λ, the maximal value of F on this line of steepest ascent is reached for $\lambda = 30$, that is at

$$C = (30, 60)$$

Table 5.1 Slack variables and function values at the vertices in Fig. 5.1

Vertex	Values of variables different from 0	Variables equal to 0	Feasible solution point $(x_1, x_2, x_3, x_4, x_5)$	Value of G there
$A = (0, 0)$	$x_3 = 100, x_4 = 720, x_5 = 60$	x_1, x_2	$(0, 0, 100, 720, 60)$	0
$B = (0, 60)$	$x_2 = 60, x_3 = 40, x_4 = 180$	x_1, x_5	$(0, 60, 40, 180, 0)$	120
$C = (30, 60)$	$x_1 = 30, x_2 = 60, x_3 = 10$	x_4, x_5	$(30, 60, 10, 0, 0)$	150
$D = (60, 40)$	$x_1 = 60, x_2 = 40, x_5 = 20$	x_3, x_4	$(60, 40, 0, 0, 20)$	140
$E = (100, 0)$	$x_1 = 100, x_4 = 120, x_5 = 60$	x_2, x_3	$(100, 0, 0, 120, 60)$	100

which clearly satisfies also (5.16), (5.18) and (5.20). As pointed out in Sect. 2.3, the line of steepest ascent from a given point does not always reach the (or an) optimal point. Actually it is the single line ascent from the origin, not the advance orthogonal to the contour lines (planes, hyperplanes) what is restrictive. In order to find a more broadly efficient method, let us look more thoroughly at the conditions (5.16), (5.17), (5.18), (5.19), and (5.20) in the form (5.22), (5.23), (5.24), and (5.25), as illustrated by Fig. 5.1 and Table 5.1.

On the sides *BC*, *CD* and *DE there has to be equality in one of the inequalities* (5.16), (5.17), and (5.18)—in exactly one, because *none of Eqs.* (5.22), (5.23), and (5.24) *is redundant or contradictory*, since the matrix of coefficients in their system of linear equations,

$$\mathbf{A} = \begin{pmatrix} 1\ 1\ 1\ 0\ 0 \\ 6\ 9\ 0\ 1\ 0 \\ 0\ 1\ 0\ 0\ 1 \end{pmatrix},$$

has rank 3 (see Sect. 4.5), for instance the last three column vectors are linearly independent. But equality in (5.16), (5.17) or (5.18) means, by the definition of slack variables, that

$$x_3 := 100 - x_1 - x_2 \quad \text{or} \quad x_4 := 720 - 6x_1 - 9x_2 \quad \text{or} \quad x_5 := 60 - x_2,$$

respectively, is zero. On the sides *AB* and *AE*, of course, $x_1 = 0$ or $x_2 = 0$, respectively. Accordingly, *at each of the vectors A, B, C, D, E* exactly *two among* x_1, x_2, x_3, x_4 *and* x_5 *are zero.*

In general, for (5.12) and (5.13), $n - m$ variables will be 0 at the "vertices". It can be shown that *linear optimisation problems have their solutions at* these *vertices*. More exactly: *If a linear function G has a maximum* at all *on the convex set described by* (5.12) *and* (5.13) *then it assumes it on one of its vertices*. (We did not say *only* there. Actually, *the set of optimal solutions is convex*, since $G(\mathbf{x}^1) = G(\mathbf{x}^2)$ implies

$$\begin{aligned} G(\lambda\mathbf{x}^1 + (1-\lambda)\mathbf{x}^2) &= c_1(\lambda x_1^1 + (1-\lambda)x_1^2) + \ldots + c_n(\lambda x_n^1 + (1-\lambda)x_n^2) \\ &= \lambda G(\mathbf{x}^1) + (1-\lambda)G(\mathbf{x}^2) = G(\mathbf{x}^1) \end{aligned}$$

for all $\lambda \in [0, 1]$). So we may restrict ourselves to vertices. This is another reason why we considered only vertices in Table 5.1. In general, however, it would also be prohibitively too much work to calculate, as we have done in Table 5.1, the values of the objective function even at all vertices and find the largest among them.

The *simplex algorithm permits to move from the* (*trivial*) *vertex* $A = (0, 0)$ *with less calculation* (in the case (5.21), (5.22), (5.23), (5.24), and (5.25) in "three easy steps") *right away to the* (*or one*) *vertex yielding the optimal solution* (in this case to $C = (30, 60)$ and so to 150 maximum).

On (5.21), (5.22), (5.23), (5.24), and (5.25) it works as follows. Since the slack variables have been defined there by

$$x_3 = 100 - x_1 - x_2, \qquad x_4 = 720 - 6x_1 - 9x_2, \qquad x_5 = 60 - x_2 \tag{5.26}$$

(see (5.22), (5.23) and (5.24), respectively), in the point A, that is for $x_1 = x_2 = 0$, we get

$$x_3 = 100, \quad x_4 = 720, \quad x_5 = 60 \quad \text{and} \quad G(0, 0, 100, 720, 60) = 0. \tag{5.27}$$

Of course, we want to do better (we hardly could have done worse) in getting a larger value of G in (5.21). That is not difficult. For instance, keeping $x_1 = 0$, we may increase x_2 (the "*entering variable*") and get larger values of G. How far can we increase x_2? Because of (5.25), x_1, x_3, x_4, x_5 are nonnegative, so we have by (5.24) $x_2 \leq 60$, by (5.22) $x_2 \leq 100$, and, by (5.23) $9x_2 \leq 720$, that is, $x_2 \leq 80$. Clearly, $x_2 \leq 60$ is the most stringent of these restrictions and thus $x_2 = 60$ is the best (gives the greatest G-value) among the feasible solutions with $x_1 = 0$. This yields the next solution (use also (5.26) and (5.21)):

$$\begin{aligned} &x_1 = 0, x_2 = 60, x_3 = 40, x_4 = 180, x_5 = 0 \\ &\text{and} \quad G(0, 60, 40, 180, 0) = 120, \end{aligned} \tag{5.28}$$

certainly much better. (Actually we happened to arrive at vertex B in Fig. 5.1, see also Table 5.1.) The variable x_5, which became 0, now "*leaves*".

What is the simplest way to improve this further? In the first step it was helpful that in (5.27) two of the variables (x_1 and x_2) were 0. We kept one zero and increased the second so far as we could (keeping the solution feasible) thus increasing the value of G. But there are two 0 variables in (5.28) too: x_1 and x_5 (there *have to be* since we decreased at least one of the quantities in (5.26) to 0). The essential thing is to *treat all variables in the canonical form* (5.21), (5.22), (5.23), (5.24), and (5.25) *in the same way*. Only now we must replace (5.26) by equations containing x_1 and x_5 rather than x_1 and x_2 on the right. This is easy, the following equations are clearly equivalent to (5.26):

$$x_2 = 60 - x_5, \qquad x_3 = 40 - x_1 + x_5, \qquad x_4 = 180 - 6x_1 + 9x_5. \tag{5.29}$$

With these we get a new form of the objective function, namely,

$$G(x_1, x_2, x_3, x_4, x_5) = 120 + x_1 - 2x_5. \tag{5.30}$$

Note that the increase in x_5 would decrease the value of G (that is why x_5 had to "leave"), so we have no choice (we had in the first step: we could have increased x_1 rather than x_2; we chose x_2 because (5.21) grows faster in x_2 than in x_1): the right-hand side variable to increase is x_1 (this is the *entering* variable). So, keeping $x_5 = 0$, we increase G by increasing x_1. How much can we increase x_1? Since now

$x_5 = 0$ and still $x_3 \geq 0$, $x_4 \geq 0$, we get from (5.29) $x_1 \leq 40$ and $x_1 \leq 180/60 = 30$. The latter being the stronger restriction, the greatest feasible value of x_1 is 30. This gives $x_4 = 0$, so x_4 is the next (and last) *leaving* variable. With (5.29) and (5.30) we get

$$\begin{aligned} x_1 = 30, x_2 = 60, x_3 = 10, x_4 = 0, x_5 = 0 \\ \text{and} \quad G(30, 60, 10, 0, 0) = 150. \end{aligned} \tag{5.31}$$

Actually, we arrived at the vertex C in Fig. 5.1 (compare Table 5.1), so we know that we got the optimal solution: We needed just the three vertices A,B and C. They span a triangle, which is a *simplex* in $\mathbb{R}^2$. A *simplex* in $\mathbb{R}^n$ is a convex polyhedron (compare Sect. 3.3 **6**) spanned by $(n+1)$ points. That is where the name *simplex algorithm* comes from. But we are not finished yet, because we know only from inspection of Fig. 5.1 or from the longer calculation above that we reached the (or an) optimal solution. However, we can recognise this also by the same procedure which we used already twice to improve the solution: Also in (5.31) there stand 0's for two variables, x_4 and x_5. Expressing first x_1 then x_2, x_3 and finally, the function value in terms of x_4 and x_5, we get from (5.29) and (5.30)

$$x_1 = 30 - \frac{1}{6}x_4 + \frac{3}{2}x_5, \qquad x_2 = 60 - x_5, \qquad x_3 = 10 + \frac{1}{6}x_4 - \frac{1}{2}x_5, \tag{5.32}$$

$$G(x_1, x_2, x_3, x_4, x_5) = 150 - \frac{1}{6}x_4 - \frac{1}{2}x_5. \tag{5.33}$$

It is clear from the last equation that, among nonnegative x_4, x_5, the values $x_4 = 0$, $x_5 = 0$ (and so $x_1 = 30$, $x_2 = 60$, $x_3 = 10$) give the greatest value of G and that is 150. But this was (5.31), so no improvement is possible, *the linear approximation problem is solved*. This procedure, which is considerably shorter than putting Table 5.1 together and at least as simple to describe, is the *simplex algorithm*.

What we will do now is to describe, using appropriate terminology, this simplex algorithm for the general linear optimisation problem written, with the aid of "slack variables", in the form (5.11), (5.12), and (5.13). If r is the rank of the matrix

$$\mathbf{A} = \begin{pmatrix} a_{11} & \dots & a_{1n} \\ \vdots & & \vdots \\ a_{m1} & \dots & a_{mn} \end{pmatrix}$$

of coefficients in the system of linear equations

$$a_{j1}x_1 + \ldots + a_{jn}x_n = b_j \quad (j = 1, \ldots, m) \tag{5.34}$$

then (see Sect. 4.5) $r \leq m$, $r \leq n$. A particular solution of this system, in which $n - r$ of the variables $x_1, \ldots, x_n$ are 0 (and the rest is nonnegative) is called a *basic feasible solution*. The r variables, which we did not choose to be 0 are the *basic variables* in this set-up or "dictionary". If we let the other variables vary again then the basic variables can be expressed with them, if we solve the system of linear equations (5.34) with respect to these r (basic) variables (note that rank $\mathbf{A} = r$, compare Sect. 4.5). In the systems of equations thus obtained (such as (5.26), (5.29) and (5.32)), the r basic variables are on the left hand side and linear combinations (with constant coefficients) of the non-basic variables and of 1 are on the right (compare (4.45) here the non-basic variables are the parameters). Systems of equations set up like this are called *dictionaries*. Clearly *the* different *dictionaries belonging to the same problem* (differing only in the choice of the basic variables) *are equivalent*, because they are all equivalent to (5.34).

So, after writing the general linear optimisation problem in the form (5.11), (5.12), and (5.13), we choose, say, the last $n - r$ variables as basic variables, express them with aid of the first r, non-basic, variables, thus get our first dictionary and substitute this into the objective function. If we choose 0's as the values of the non-basic variables, we get the *first basic feasible solution*. Looking at the objective function in terms of varying non-basic variables again, we choose one of the (non-basic) variables, that with the largest positive coefficient as *entering variable* and keep it varying, while putting 0's for the remaining non-basic variables. From the dictionary we get the largest feasible value of our non-basic variable, making (at least) one other, the *leaving variable* 0. We obtain also the values of all basic variables and of the objective function. This gives our *second basic feasible solution*. Necessarily (at least) $n - r$ variables will be 0 in it. We choose these as the new non-basic variables of our problem and keep repeating this procedure till all (non-basic) variables in the objective function have nonpositive coefficients (as in (5.33)). Then the linear optimisation problem is *solved*.

We do not dwell here upon the details of when and why this happens, the example may be instructive enough. However we present a simplified writing of (5.21), (5.22), (5.23), (5.24), (5.25), (5.26), (5.27), (5.28), (5.29), (5.30), (5.31), (5.32), and (5.33). We repeat these equations in a slightly changed but clearly equivalent form while indicating in the left column the basic variable or function value which the equation serves to determine (Table 5.2): (Because of the way we wrote the operation(s) determining G, the algorithm ends when all coefficients in it are *nonnegative*). In Table 5.3 we convert the above into a skeleton array by omitting the variables (keeping just the coefficients) and the second occurrences of G. We list also the basic solutions.

The maximum is 150 *and so, in the problem in Sect. 4.1, the maximum profit contribution is* $40 \cdot 150 = 6000$ (cents).

Table 5.3 consists of the (three) *simplex tableaus*; it forms the *tableau format* belonging to the linear optimisation problem (5.21), (5.22), (5.23), (5.24), and (5.25).

Table 5.2 Simplex tableau for a zero-sum game

x_3			x_1	$+$	x_2	$+$	x_3					$=$	100
x_4			$6x_1$	$+$	$9x_2$			$+$	x_4			$=$	720
x_5					x_2					$+$	x_5	$=$	60
G	G	$-$	x_1	$-$	$2x_2$							$=$	0
x_3			x_1			$+$	x_3			$-$	x_5	$=$	40
x_4			$6x_1$					$+$	x_4	$+$	$9x_5$	$=$	180
x_2					x_2					$+$	x_5	$=$	60
G	G	$-$	x_1							$+$	$2x_5$	$=$	120
x_3							x_3	$-$	$\frac{1}{6}x_4$	$+$	$\frac{1}{2}x_5$	$=$	10
x_1			x_1					$+$	$\frac{1}{6}x_4$	$-$	$\frac{3}{2}x_5$	$=$	30
x_2					x_2					$+$	x_5	$=$	60
G	G							$+$	$\frac{1}{6}x_4$	$+$	$\frac{1}{2}x_5$	$=$	150

Table 5.3 Simplex tableaus: the tableau format and its use for solving the linear optimisation problem (5.21), (5.22), (5.23), (5.24), and (5.25)

Basic variables and function values determined	Coefficients of x_1	x_2	x_3	x_4	x_5	Constants on the right hand side	Basic feasible solutions
x_3	1	1	1			100	First: $x_3 = 100$,
x_4	6	9		1		720	$x_4 = 720, x_5 = 60$;
x_5		1			1	60	$x_1 = x_2 = 0$;
G	-1	-2				0	$G(0, 0, 100, 720, 60) = 0$
x_3	1		1		-1	40	Second: $x_2 = 60$,
x_4	6			1	-9	180	$x_3 = 40, x_4 = 180$,
x_2		1			1	60	$x_1 = x_5 = 0$;
G	-1				2	120	$G(0, 60, 40, 180, 0) = 120$
x_3			1	$-\frac{1}{6}$	$\frac{1}{2}$	10	Third: $x_1 = 30$,
x_1				$\frac{1}{6}$	$-\frac{3}{2}$	30	$x_2 = 60, x_3 = 10$,
x_2					1	60	$x_4 = x_5 = 0$;
G				$\frac{1}{6}$	$\frac{1}{2}$	150	$G(30, 60, 10, 0, 0) = 150$

Solving the problem (5.21), (5.22), (5.23), (5.24), and (5.25) can be completely mechanised (or computerised) by transforming each, in our case four-line tableau—of which the first immediately corresponds to the original problem with slack variables—into the next, as follows.

Step 1. Ignore the last two columns ("Basic feasible solutions" and "Constants on the right hand side"). *If* (as in the third tableau in Table 5.3) *all numbers in the last* (fourth) *row are nonnegative then stop*: the tableau describes an optimal solution. Otherwise find (one of) its minimal numbers (-2 for the first, -1 for the second tableau in Table 5.3). The column in which it appears is that of the *entering variable* (x_2, resp. x_1 in Table 5.3) called *pivot column*. We frame it. For the first and second tableau of our example we get (see Table 5.3)

$$\begin{array}{r|r|ccc} 1 & 1 & 1 & 0 & 0 \\ 6 & 9 & 0 & 1 & 0 \\ 0 & 1 & 0 & 0 & 1 \\ -1 & -2 & 0 & 0 & 0 \end{array} \qquad \begin{array}{|r|cccc} 1 & 0 & 1 & 0 & -1 \\ 6 & 0 & 0 & 1 & -9 \\ 0 & 1 & 0 & 0 & 1 \\ -1 & 0 & 0 & 0 & 2 \end{array} .$$

Step 2. For each row (except the last, G), whose entry, say α, in the pivot column is positive, look up the entry, say β, in the (previously omitted) "constants in the right hand side" column. The row with the *smallest ratio* β/α is that of the *leaving* variable (x_5, resp. x_4) called the *pivot row*. (If all entries of the pivot column are nonpositive then the problem is *unbounded*, see Example 3). We frame this row too. The number at the intersection of the *pivot row* and pivot column is the *pivot number*. Divide every entry in the pivot row by the pivot number.

The pivot numbers of our first and second tableau in Table 5.3 are 1 and 6, respectively:

$$\begin{array}{r|r|cccr} 1 & 1 & 1 & 0 & 0 & 100 \\ 6 & 9 & 0 & 1 & 0 & 720 \\ 0 & [1] & 0 & 0 & 1 & 60 \\ -1 & -2 & 0 & 0 & 0 & 0 \end{array} \qquad \begin{array}{|r|ccccr} 1 & 0 & 1 & 0 & -1 & 40 \\ [6] & 0 & 0 & 1 & -9 & 180 \\ 0 & 1 & 0 & 0 & 1 & 60 \\ -1 & 0 & 0 & 0 & 2 & 120 \end{array} .$$

Step 3. Add (compare Sect. 4.5 (**I**)) a (positive or negative) multiple of the pivot row to each other row so that 0's should stand as entries in the pivot column (except for the pivot number). Now apply Step 1 to the new tableau.
In our example we obtain so from the first tableau written down with Step 2 the second one and from that the third and the last simplex tableau of Table 5.3:

$$\begin{array}{cccccc} 0 & 0 & 1 & -1/6 & 1/2 & 10 \\ 1 & 0 & 0 & 1/6 & -3/2 & 30 \\ 0 & 1 & 0 & 0 & 1 & 60 \\ 0 & 0 & 0 & 1/6 & 1/2 & 150. \end{array}$$

It is the last tableau since the entries in the last row are all nonnegative.It describes the optimal solution of our example; see Step 1. The reader can easily check how the operations on the tableaus correspond to those on the equations.

The application of this algorithm, that is the *simplex algorithm* or *simplex method*, to the cases of more equations and of more variables should be equally clear. It is not always free of problems, however. Three kinds of problems may occur:

(i) The "*origin*" (zero vector); in the above example $x_1 = 0, x_2 = 0$ *may not be a feasible solution*. In this case (and if there exists a feasible solution at all) one starts with a feasible solution point, preferably close to the suspected optimal solution point. (We can do this even if the origin *is* a feasible solution point). Often introducing further dummy variables helps. We do not go into details.

(ii) *The simplex algorithm may "go in circles"* (compare Sects. 6.7 and 11.1), the first dictionary reappears, an optimal solution is not reached even if an optimal solution exists. This happens rarely but it can happen. For instance *all* variables of a basic (feasible) solution may have the value 0. Changing ("perturbing") the right hand sides of the conditions by different (independent) quantities $\epsilon_1, \ldots, \epsilon_m$, solving the new linear optimisation problem and then ignoring $\epsilon_1, \ldots, \epsilon_m$ in the solution (replacing them by 0) is a possible way to get out of this dilemma. We do not go into these details either.

(iii) *There may not exist an optimal solution or even a feasible solution* at all.

Example 2 The linear optimisation problem: maximise

$$2x_1 + 3x_2 \tag{5.35}$$

under the conditions

$$x_1 + 2x_2 \leq \quad 4, \tag{5.36}$$

$$-x_1 - 4x_2 \leq -12, \tag{5.37}$$

$$x_1 \geq 0, \quad x_2 \geq 0 \tag{5.38}$$

has no feasible solution, so also no optimal solution: Indeed (see Fig. 5.2) there exist no nonnegative x_1, x_2 which satisfy both (5.36) and (5.37).

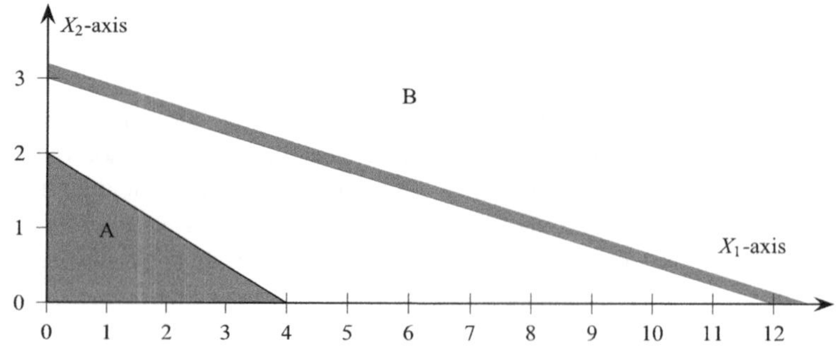

Fig. 5.2 The two sets $A = \{(x_1, x_2) \in \mathbb{R}^2_+ \mid x_1 + 2x_2 \leq 4\}$ and $B = \{(x_1, x_2) \in \mathbb{R}^2_+ \mid x_1 + 4x_2 \geq 12\}$ have no point in common, so the pair of inequalities $x_1 + 2x_2 \leq 4, -x_1 - 4x_2 \leq -12$ has no nonnegative solutions

Example 3 The linear optimisation problem: maximise

$$F(x_1, x_2) = 5x_1 - 2x_2$$

under the conditions

$$-x_1 + x_2 \leq -2,$$
$$-x_1 - 3x_2 \leq -3,$$
$$x_1 \geq 0, x_2 \geq 0$$

has feasible solutions (for instance $x_1 = 4$, $x_2 = 1$) *but no optimal solution.* Indeed the pairs (x_1, x_2) satisfying all three conditions are represented in Fig. 5.3 by the points (elements) of the sets $L \cap M$. The contour lines (see Sect. 3.1) of F are given by $5x_1 - 2x_2 = c$. In Fig. 5.3 these are drawn for $c = 5, 20$ and 35. Clearly, arbitrary large c-values greater than or equal to 10.75 ($= F(9/4, 1/4)$), compare Fig. 5.3 can be reached within $L \cap M$.

Of course nonexistence or existence of an optimal solution and its value can be determined without recourse to figures (we showed this explicitly for Example 1 by the simplex method) and similar statements can be proved for more than two variables, but the proofs are often more difficult.

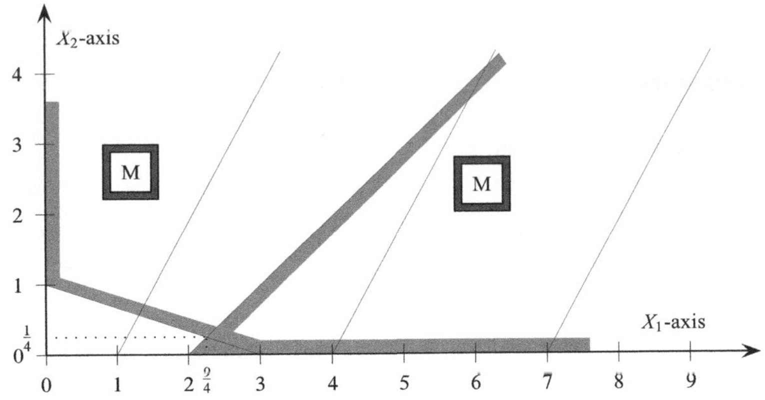

Fig. 5.3 The set $L \cap M$, where $L = \{(x_1, x_2) \in \mathbb{R}^2_+ \mid -x_1 + x_2 \leq -2\}$ and $M = \{(x_1, x_2) \in \mathbb{R}^2_+ \mid x_1 + 3x_2 \geq 3\}$, represents all feasible solutions of the linear optimisation problem in Example 3. The three parallel straight lines through $(1, 0)$, $(4, 0)$ and $(7, 0)$ are the contour lines $\{(x_1, x_2) \in \mathbb{R}^2_+ \mid 5x_1 - 2x_2 = c\}$ of F for $c = 5$, $c = 20$, and $c = 35$, respectively

In all such problems consideration of the *dual problem* may help and lead further to the solution if it exists.

5.2.1 Exercises

Solve, by establishing the simplex tableaus, the following linear optimisation problems.

1. Maximise $G(x_1, x_2, x_3, x_4) = 6x_1 + 4x_2$ under the restrictions

$$x_1 + 2x_2 \leq 8, 3x_1 + x_2 \leq 9, x_1 \geq 0, x_2 \geq 0.$$

2. Minimise $G(x_1, x_2, \ldots, x_6) = -2x_1 + x_2$ under the restrictions

$$-3x_1 + 2x_2 \leq 6, x_1 + 5x_2 \leq 32, x_1 + x_2 \leq 12, 3x_1 + x_2 \leq 30,$$
$$x_1 \geq 0, x_2 \geq 0.$$

3. Maximise $G(x_1, x_2, \ldots, x_6) = 2x_1 + x_2$ under the restrictions in Exercise 2.
4. Minimise $G(x_1, x_2, \ldots, x_5) = x_1 + 2x_2$ under the restrictions

$$x_1 + x_2 \geq 4, -2x_1 + x_2 \geq 1, -x_1 + 2x_2 \leq 8, x_1 \geq 0, x_2 \geq 0.$$

5. Maximise $G(x_1, x_2, \ldots, x_5) = 20x_1 + 10x_2$ under the restrictions

$$x_1 + x_2 \leq 100, 9x_1 + 6x_2 \leq 720, x_1 \leq 60, x_1 \geq 0, x_2 \geq 0.$$

5.2.2 Answers

1. It follows the simplex tableau:

	x_1	x_2	x_3	x_4		Basic feasible solutions
x_3	1	2	1		8	First: $x_3 = 8, x_4 = 9;$
x_4	[3]	1		1	9	$x_1 = x_2 = 0;$
G	-6	-4			0	$G(0, 0, 8, 9) = 0$
x_3		$[\frac{5}{3}]$	1	$-\frac{1}{3}$	5	Second: $x_1 = 3, x_3 = 5;$
x_1	1	$\frac{1}{3}$		$\frac{1}{3}$	3	$x_2 = x_4 = 0;$
G		-2		2	18	$G(3, 0, 5, 0) = 18$
x_2		1	$\frac{3}{5}$	$-\frac{1}{5}$	3	Third (= optimal solution):
x_1	1		$-\frac{1}{5}$	$\frac{2}{5}$	2	$x_1 = 2, x_2 = 3, x_3 = x_4 = 0;$
G			$\frac{6}{5}$	$\frac{8}{5}$	24	$G(2, 3, 0, 0) = 24$

2. First basic feasible solution:

$$x_3 = 6, \quad x_4 = 32, \quad x_5 = 12, \quad x_6 = 30; \quad x_1 = x_2 = 0;$$
$$G(0, 0, 6, 32, 12, 30) = 0.$$

   Second basic feasible solution (= optimal solution):

$$x_1 = 10, \quad x_3 = 32, \quad x_4 = 22, \quad x_5 = 2; \quad x_2 = x_6 = 0;$$
$$G(10, 0, 36, 22, 2, 0) = -20.$$

3. First basic feasible solution: same as in Exercise 2.
   Second basic feasible solution: same as in Exercise 2, but this time we do not get an optimal solution.
   Third basic feasible solution (= optimal solution):

$$x_1 = 9, \quad x_2 = 3, \quad x_3 = 27, \quad x_4 = 8; \quad x_5 = x_6 = 0;$$
$$G(9, 3, 27, 8, 0, 0) = 21.$$

4. First basic feasible solution:

$$x_3 = -4, \quad x_4 = -1, \quad x_5 = 8; \quad x_1 = x_2 = 0;$$
$$G(0, 0, -4, -1, 8) = 0.$$

   Second basic feasible solution:

$$x_1 = 4, \quad x_4 = -9, \quad x_5 = 12; \quad x_2 = x_3 = 0;$$
$$G(4, 0, 0, 9, 12) = 4.$$

   Third basic feasible solution (= optimal solution):

$$x_1 = 1, \quad x_2 = 3, \quad x_5 = 3; \quad x_3 = x_4 = 0;$$
$$G(1, 3, 0, 0, 3) = 7.$$

5. First basic feasible solution:

$$x_3 = 100, \quad x_4 = 720, \quad x_5 = 60; \quad x_1 = x_2 = 0;$$
$$G(0, 0, 100, 720, 60) = 0.$$

   Second basic feasible solution:

$$x_1 = 60, \quad x_3 = 40, \quad x_4 = 180; \quad x_2 = x_5 = 0;$$
$$G(60, 0, 40, 180, 0) = 1200.$$

Third basic feasible solution (= optimal solution):

$$x_1 = 60, \quad x_2 = 30, \quad x_3 = 10; \quad x_4 = x_5 = 0;$$
$$G(60, 30, 10, 0, 0) = 1500.$$

5.3 Duality

We introduce *duality* using the example of the linear optimisation problem (5.15), (5.16), (5.17), (5.18), (5.19), and (5.20). Going back to the interpretation we gave in Sect. 4.1 to this problem (there (4.1), (4.2), (4.3), and (4.4) for $(x_1, x_2) \in \mathbb{R}^2_+$), suppose that, as a first approach, the factory owners want only a realistic estimation of the maximal profit they can attain under the conditions set by the supermarket chain. Of course, *any feasible solution gives a lower estimate* of the maximum. For instance, as we have seen, $x_1 = 60$, $x_2 = 40$ satisfy (5.16), (5.17), (5.18), (5.19), and (5.20). They give

$$H(60, 40) = 40F(60, 40) = 40(60 + 80) = 5600$$

as a lower estimate of the maximum. (We happen to know that the maximum is 6000). A better (because larger) lower estimate of the maximum is furnished by $x_1 = 45$, $x_2 = 50$, which also satisfy (5.16), (5.17), (5.18), (5.19), and (5.20) and give

$$H(45, 50) = 40(45 + 100) = 5800.$$

As to the estimation of the maximum from above, (5.16) gives $2x_1 + 2x_2 \leq 200$, so

$$H(x_1, x_2) = 40(x_1 + 2x_2) \leq 40(2x_1 + 2x_2) \leq 8000$$

(we used also $x_1 \geq 0$, that is, (5.19)). We get a better (because smaller) upper estimate of the maximum, if we multiply (5.17), that is $6x_1 + 9x_2 \leq 720$, by 2/9: $(4/3)x_1 + 2x_2 \leq 160$. This and $x_1 \geq 0$ give

$$H(x_1, x_2) = 40(x_1 + 2x_2) \leq 40\left(\frac{4}{3}x_1 + 2x_2\right) \leq 6400.$$

Clearly we can use *a linear combination with coefficients* of the inequalities (5.16), (5.17) and (5.18) to get upper estimates of the maximum. The question is, *which coefficients are best. These will give* not just an upper estimate, but *exactly the maximum*.

For easier reference, we restate the problem. Maximise

$$F(x_1, x_2) = x_1 + 2x_2 \tag{5.39}$$

(or $H(x_1, x_2) = 40F(x_1, x_2)$) under the conditions

$$x_1 + x_2 \leq 100, \tag{5.40}$$

$$6x_1 + 9x_2 \leq 720, \tag{5.41}$$

$$x_2 \leq 60, \tag{5.42}$$

$$x_1 \geq 0, \tag{5.43}$$

$$x_2 \geq 0. \tag{5.44}$$

We now multiply (5.40), (5.41) and (5.42) by $y_1 \geq 0$, $y_2 \geq 0$, and $y_3 \geq 0$, respectively, and add the inequalities so obtained (which we can do), that is, we take a linear combination of (5.40), (5.41) and (5.42) with nonnegative coefficients:

$$y_1(x_1 + x_2) + y_2(6x_1 + 9x_2) + y_3 x_2 \leq 100y_1 + 720y_2 + 60y_3. \tag{5.45}$$

Rearrangement gives

$$(y_1 + 6y_2)x_1 + (y_1 + 9y_2 + y_3)x_2 \leq 100y_1 + 720y_2 + 60y_3. \tag{5.46}$$

We want to choose y_1, y_2, y_3 so that $F(x_1, x_2) = x_1 + 2x_2$ be smaller than or equal to the left hand side of (5.46). When y_1, y_2, y_3 will be chosen accordingly, we will use the right hand side as an upper estimate (really the value) of the maximum of $x_1 + 2x_2$. So we look for y_1, y_2, y_3 such that

$$\begin{aligned} F(x_1, x_2) = x_1 + 2x_2 \leq (y_1 + 6y_2)x_1 + (y_1 + 9y_2 + y_3)x_2 \\ (\leq 100y_1 + 720y_2 + 60y_3). \end{aligned} \tag{5.47}$$

This can hold for all (or sufficiently many) x_1 and x_2 only if

$$\begin{aligned} y_1 + 6y_2 \quad\quad &\geq 1, \\ y_1 + 9y_2 + y_3 &\geq 2. \end{aligned}$$

So, if the nonnegative y_1, y_2, y_3 satisfy these inequalities then $100y_1 + 720y_2 + 60y_3$ is an upper bound of $F(x_1, x_2)$. We try of course to make this as small as possible which then gives the maximum of $F(x_1, x_2)$ (the equality of the least upper bound to the maximum is not completely obvious, compare Sect. 6.2). So our problem is now: *minimise*

$$f(y_1, y_2, y_3) = 100y_1 + 720y_2 + 60y_3 \tag{5.48}$$

under the conditions

$$y_1 + 6y_2 \geq 1, \tag{5.49}$$

$$y_1 + 9y_2 + y_3 \geq 2, \tag{5.50}$$

$$y_1 \geq 0, y_2 \geq 0, y_3 \geq 0. \tag{5.51}$$

As we see, this is a linear optimisation problem too, the *dual problem* to the *primal problem* (5.39), (5.40), (5.41), (5.42), (5.43), and (5.44). Notice that *the coefficients in* (5.48) *are the right hand sides in* (5.40), (5.41), and (5.42) *while the right hand sides in* (5.49), (5.50) *are the coefficients in* (5.39). *The coefficient matrix of* (5.49) and (5.50) *is the transposed of that in* (5.40), (5.41), and (5.42) (rows and columns interchanged).

We present an intuitive way to the situation of the dual problem (5.48), (5.49), (5.50), and (5.51) which is faster than that of the primal problem. Since we want to minimise (5.48), the smaller y_1, y_2, y_3 are the better. So let us replace $\geq$ in (5.49) and (5.50) by $=$ and see whether there are nonnegative solutions. The system of linear equations

$$y_1 + 6y_2 = 1, \tag{5.52}$$

$$y_1 + 9y_2 + y_3 = 2 \tag{5.53}$$

can be solved by the method in Sects. 4.5 and 4.6, for instance by elimination. Subtract (5.52) from (5.53):

$$3y_2 + y_3 = 1.$$

Now multiply (5.52) by $3/2$ and subtract (5.53):

$$\frac{1}{2}y_1 - y_3 = -\frac{1}{2}.$$

These give

$$y_1 = 2\lambda - 1, \quad y_2 = \frac{1-\lambda}{3} \quad \text{with} \quad y_3 = \lambda, \tag{5.54}$$

which indeed satisfy (5.52) and (5.53) for every λ.

But, by (5.51), $y_1 = 2\lambda - 1 \geq 0$ and $y_2 = (1-\lambda)/3 \geq 0$, that is,

$$\frac{1}{2} \leq \lambda \leq 1$$

(then also $y_3 = \lambda > 0$). Substitute (5.54) into (5.48):

$$\begin{aligned} f(y_1, y_2, y_3) &= f\left(2\lambda - 1, \frac{1-\lambda}{3}, \lambda\right) \\ &= 100(2\lambda - 1) + 720\frac{1-\lambda}{3} + 60\lambda = 140 + 20\lambda . \end{aligned}$$

Since $\lambda \geq 1/2$, the right hand side will be smallest for $\lambda = 1/2$, that is (see (5.54)), for

$$y_1 = 0, \quad y_2 = \frac{1}{6}, \quad y_3 = \frac{1}{2} \tag{5.55}$$

and the minimum of $f(y_1, y_2, y_3)$, thus the maximum of $F(x_1, x_2)$ will be $140 + 10 = 150$, in accordance with what we have found before. (However, the x_1, x_2 which give $F(x_1, x_2) = 150$ and satisfy (5.16), (5.17), (5.18), (5.19), and (5.20) have still to be determined. We saw before that they are $x_1 = 30, x_2 = 60$.)

The ($y_1 = 0$ and) $y_2 = 1/6$, $y_3 = 1/2$ of (5.55) in the above solution are called "*shadow prices*" or "*opportunity costs*" for the following reason. We go back to the original formulation of our linear optimisation problem in Sect. 4.1 and ask what would happen if the supermarket chain were willing to increase, say by t, the \$720 bound on what it intended to spend per week for the two kinds of detergents (see (5.17) and (5.2)).

Then system (5.16), (5.17), (5.18), (5.19), and (5.20) of inequalities in our original problem changes to

$$x_1 + x_2 \leq 100, \tag{5.56}$$

$$6x_1 + 9x_2 \leq 720 + t \tag{5.57}$$

$$x_2 \leq 60 \tag{5.58}$$

$$x_1 \geq 0, \tag{5.59}$$

$$x_2 \geq 0, \tag{5.60}$$

while we still want to maximise the factory's profit (really $(1/40)$ times the profit)

$$F(x_1, x_2) = x_1 + 2x_2. \tag{5.61}$$

Every feasible solution of this problem satisfies

$$\begin{aligned} F(x_1, x_2) &= x_1 + 2x_2 \\ &= 0(x_1 + x_2) + \tfrac{1}{6}(6x_1 + 9x_2) + \tfrac{1}{2}x_2 \\ &\leq \tfrac{1}{6}(720 + t) + \tfrac{1}{2}60 \\ &= 150 + \tfrac{1}{6}t. \end{aligned} \tag{5.62}$$

(Compare this to (5.45) and (5.47) with (5.55)). The connection is not accidental, as we will see below in rule (**v**) of duality theory. So the extra profit will never exceed (1/6). In fact, the factory has the *opportunity* to increase its profit by $(1/6)t$ with $0 \leq t \leq 60$ (in order to satisfy (5.56)) by choosing

$$x_1 = 30 + \frac{1}{6}t, \qquad x_2 = 60.$$

(Remember, $x_1 = 30$, $x_2 = 60$ was the solution of our original problem). Indeed, then $x_2 > 0$, $x_1 > 0$, $x_2 \leq 60$ and

$$\begin{aligned} x_1 + \ x_2 &= 30 + \tfrac{1}{6}t + 60 = 90 + \tfrac{1}{6}t \leq 100 \qquad (\text{because } t \leq 60), \\ 6x_1 + 9x_2 &= 6\left(30 + \tfrac{1}{6}t\right) + 540 = 720 + t, \end{aligned}$$

so (5.60), (5.59), (5.58), (5.56) and (5.57) are satisfied, while $F(x_1, x_2) = x_1 + 2x_2 = 150 + (1/6)t$ for (5.61) (compare (5.62)). A similar result applies if we want to change 100 in (5.56) or 60 in (5.58) (or two or all three).

In (5.45) and (5.46), as in our original problem, x_1 and x_2 were weight units, so their coefficients y_1, y_2, y_3 can be considered *prices* (as 6 and 9 in (5.56), (5.41) and (5.2)). If, as in (5.62) and (5.55), they are chosen as solutions of the dual problem then they are called "*shadow prices*". Of course, linear optimisation and duality applies to many other practical matters, not just to prices.

In general, *duality* can be formulated as follows. *The primal linear optimisation problem: maximise*

$$F(x_1, \ldots, x_n) = c_1x_1 + \ldots + c_nx_n \tag{5.63}$$

under the conditions

$$a_{j1} + \ldots + a_{jn}x_n \leq b_j \quad (j = 1, \ldots, m), \tag{5.64}$$

$$x_k \geq 0 \quad (k = 1, \ldots, n) \tag{5.65}$$

has the dual problem: minimise

$$f(y_1, \ldots, y_m) = b_1y_1 + \ldots + b_my_m \tag{5.66}$$

under the conditions

$$a_{1k}y_1 + \ldots + a_{mk}y_m \geq c_k \quad (k = 1, \ldots, n) \tag{5.67}$$

$$y_j \geq 0 \quad (j = 1, \ldots, m). \tag{5.68}$$

So, also in this general situation, *in the dual problem we have to minimise a linear function* (5.66) *whose coefficients are the right hand sides (upper bounds) in the conditions* (5.64) *of the primal problem, while the right hand sides (lower bounds) in the conditions* (5.67) *of the dual problem are the coefficients of the linear function* (5.63) which was *to be maximised in the primal problem.* Finally, *the matrix of coefficients in the conditions* (5.67) *of the dual problem is the transposed matrix*

$$\mathbf{A}^T = \begin{pmatrix} a_{11} & \dots & a_{m1} \\ \vdots & & \vdots \\ a_{1n} & \dots & a_{mn} \end{pmatrix} \quad \textit{of the matrix} \quad \mathbf{A} = \begin{pmatrix} a_{11} & \dots & a_{1n} \\ \vdots & & \vdots \\ a_{m1} & \dots & a_{mn} \end{pmatrix}$$

in the conditions (5.64) *of the primal problem.* In vector-matrix form (Sects. 1.4, 4.2, and 4.3) with the usual notations

$$\mathbf{x} = \begin{pmatrix} x_1 \\ \vdots \\ x_n \end{pmatrix}, \quad \mathbf{y} = \begin{pmatrix} y_1 \\ \vdots \\ y_n \end{pmatrix}, \quad \mathbf{b} = \begin{pmatrix} b_1 \\ \vdots \\ b_n \end{pmatrix}, \quad \mathbf{c} = \begin{pmatrix} c_1 \\ \vdots \\ c_n \end{pmatrix},$$

the primal problem is

$$\textit{maximise} \quad F(\mathbf{x}) = \mathbf{c} \cdot \mathbf{x} \quad \textit{under the conditions} \quad \mathbf{Ax} \leqq \mathbf{b}, \mathbf{x} \geqq \mathbf{0}, \tag{5.69}$$

while *the dual problem is*

$$\textit{minimise} \quad f(\mathbf{y}) = \mathbf{b} \cdot \mathbf{y} \quad \textit{under the conditions} \quad \mathbf{A}^T\mathbf{y} \geqq \mathbf{c}, \mathbf{y} \leqq \mathbf{0}. \tag{5.70}$$

Of course, *one may interchange "maximise" with "minimise" and "upper bound" with "lower bound".*

The *duality theory* consists of results like the following.

(i) *The dual problem of the dual problem is the primal problem.*
(ii) *If the dual problem has a feasible solution* $\hat{\mathbf{y}}$ *then the primal problem has a feasible solution* $\hat{\mathbf{x}}$ *and they satisfy* (see (5.69) and (5.70))

$$F(\hat{\mathbf{x}}) = \mathbf{c} \cdot \hat{\mathbf{x}} \leq \mathbf{b} \cdot \hat{\mathbf{y}} = f(\hat{\mathbf{y}}). \tag{5.71}$$

(iii) *If the dual problem has an optimal solution* $\hat{\mathbf{y}}$ *then the primal problem has an optimal solution* $\hat{\mathbf{x}}$ *and they satisfy*

$$F(\hat{\mathbf{x}}) = \mathbf{c} \cdot \hat{\mathbf{x}} = \mathbf{b} \cdot \hat{\mathbf{y}} = f(\hat{\mathbf{y}}). \tag{5.72}$$

(iv) *If feasible solutions* $\hat{\mathbf{x}}$, $\hat{\mathbf{y}}$ *of the primal and dual problems satisfy* (5.72) *then they are optimal.*
(v) *If the primal problem* (5.69) *has a (nondegenerate) optimal solution* $\mathbf{x} = \hat{\mathbf{x}}$ *then there is a positive* ϵ *with the following property:* If $|t_k| \leq \epsilon$ for $k = 1, \dots, n$ then the problem maximise $F(\mathbf{x}) = \mathbf{c} \cdot \mathbf{x}$ under the conditions $\mathbf{Ax} \leqq \mathbf{b} + \mathbf{t}$, $\mathbf{x} \geqq \mathbf{0}$,

(where $\mathbf{t} = (t_1, \ldots, t_n)$) has an optimal solution and its value (maximum) is

$$F(\hat{\mathbf{x}}) + \hat{\mathbf{y}} \cdot \mathbf{t},$$

where $\mathbf{y} = \hat{\mathbf{y}}$ *is the optimal solution of the dual problem* (5.70) *of* (5.69).

The result (iii) is often called the *duality theorem.* It is proved in general essentially the same way as we did for the duality between (5.39), (5.40), (5.41), (5.42), (5.43), and (5.44) and (5.48), (5.49), (5.50), and (5.51). Of course, (ii) and (iii) imply that, *if the primal problem has no feasible or no optimal solution then the dual problem has no such solution either.*

The result (i) is obvious ($(\mathbf{A}^T)^T = \mathbf{A}$). We prove here only a weaker form of (ii) (and (iv) and (v) not at all), namely, *we suppose that* (5.69) *and* (5.70) *have feasible solutions* $\hat{\mathbf{x}}$ *and* $\hat{\mathbf{y}}$ *and show that they satisfy* (5.72). The scalar product $\hat{\mathbf{x}} \cdot \mathbf{z} = \hat{x}_1 z_1 + \ldots + \hat{x}_n z_n$ of two nonnegative vectors $\hat{\mathbf{x}} \geqq \mathbf{0}$ (see (5.69)) and $\mathbf{z} = \mathbf{A}^T\mathbf{y} - \mathbf{c} \geqq \mathbf{0}$ (see (5.70)) is nonnegative: $0 \leq \hat{\mathbf{x}} \cdot \mathbf{z} = \hat{\mathbf{x}} \cdot (\mathbf{A}^T\hat{\mathbf{y}} - \mathbf{c}) = \hat{\mathbf{x}} \cdot \mathbf{A}^T\hat{\mathbf{y}} - \hat{\mathbf{x}} \cdot \mathbf{c}$, so

$$\hat{\mathbf{x}} \cdot \mathbf{A}^T\hat{\mathbf{y}} \geqq \hat{\mathbf{x}} \cdot \mathbf{c} = \mathbf{c} \cdot \hat{\mathbf{x}} = F(\hat{\mathbf{x}}). \tag{5.73}$$

On the other hand, by (5.69), $\mathbf{A}\hat{\mathbf{x}} \leqq \mathbf{b}$, that is

$$a_{j1}\hat{x}_1 + \ldots + a_{jn}\hat{x}_n \leq b_j \quad (j = 1, \ldots, m).$$

But then (we use also $\hat{\mathbf{y}} \geqq \mathbf{0}$, see (5.70))

$$\begin{aligned}
\hat{\mathbf{x}} \cdot \mathbf{A}^T\hat{\mathbf{y}} &= + \begin{pmatrix} \hat{x}_1 \\ \vdots \\ \hat{x}_n \end{pmatrix} \cdot \begin{pmatrix} a_{11}\hat{y}_1 + \ldots + a_{m1}\hat{y}_m \\ \vdots \\ a_{1n}\hat{y}_1 + \ldots + a_{mn}\hat{y}_m \end{pmatrix} \\
&= \hat{x}_1(a_{11}\hat{y}_1 + \ldots + a_{m1}\hat{y}_m) + \ldots + \hat{x}_n(a_{1n}\hat{y}_1 + \ldots + a_{mn}\hat{y}_m) \\
&= (a_{11}\hat{x}_1 + \ldots + a_{1n}\hat{x}_n)\hat{y}_1 + \ldots + (a_{m1}\hat{x}_1 + \ldots + a_{mn}\hat{x}_n)\hat{y}_m \\
&\leq b_1\hat{y}_1 + \ldots + b_m\hat{y}_m = \mathbf{b} \cdot \hat{\mathbf{y}} = f(\hat{\mathbf{y}}).
\end{aligned}$$

In view of (5.73), we proved (5.72):

$$F(\hat{\mathbf{x}}) = \mathbf{c} \cdot \hat{\mathbf{x}} \leq \hat{\mathbf{x}} \cdot \mathbf{A}\hat{\mathbf{y}} \leq \mathbf{b} \cdot \hat{\mathbf{y}} = f(\hat{\mathbf{y}}).$$

5.3.1 Exercises

Formulate the dual problem of the linear optimisation problem presented in

1. Exercise 1 (Sect. 5.1),

2. Exercise 2 (Sect. 5.1),
3. Exercise 3 (Sect. 5.1),
4. Exercise 4 (Sect. 5.1),
5. Exercise 5 (Sect. 5.1).
6. Show that the minima or maxima determined as solutions of Exercises 1–5 are the maxima or minima of Exercises 1–5 in Sect. 5.1, respectively.

5.3.2 Answers

1. Minimize $g(y_1, y_2) = 8y_1 + 9y_2$ under the restrictions

$$y_1 + 3y_2 \geq 6, \quad 2y_1 + y_2 \geq 4, \quad y_1 \geq 0, \quad y_2 \geq 0.$$

2. Maximize $g(y_1, y_2, y_3, y_4) = -6y_1 - 32y_2 - 12y_3 - 30y_4$ under the restrictions

$$3y_1 - y_2 - y_3 - 3y_4 \leq -2, \quad -2y_1 - 5y_2 - y_3 - y_4 \leq 1,$$
$$y_1 \geq 0, \quad y_2 \geq 0, \quad y_3 \geq 0, \quad y_4 \geq 0.$$

3. Minimize $g(y_1, y_2, y_3, y_4) = 6y_1 + 32y_2 + 12y_3 + 30y_4$ under the restrictions

$$-3y_1 + y_2 + y_3 + 3y_4 \geq 2, \quad 2y_1 + 5y_2 + y_3 + y_4 \geq 1,$$
$$y_1 \geq 0, \quad y_2 \geq 0, \quad y_3 \geq 0, \quad y_4 \geq 0.$$

4. Maximize $g(y_1, y_2, y_3) = 4y_1 + y_2 - 8y_3$ under the restrictions

$$y_1 - 2y_2 + y_3 \leq 1, \quad y_1 + y_2 - 2y_3 \leq 2, \quad y_1 \geq 0, \quad y_2 \geq 0, \quad y_3 \geq 0.$$

5. Minimize $g(y_1, y_2, y_3) = 100y_1 + 720y_2 + 60y_3$ under the restrictions

$$y_1 + 9y_2 + y_3 \geq 20, \quad y_1 + 6y_2 \geq 10, \quad y_1 \geq 0, \quad y_2 \geq 0, \quad y_3 \geq 0.$$

5.4 Two-Person Zero-Sum Games

The methods of linear optimisation and, in particular, duality theory have important applications in *game theory*. The following describes a *simple "game"*. There are two "players", P and Q, who have the sets of *"strategies"* $S = \{s_1, \ldots, s_m\}$ and $T = \{t_1, \ldots, t_n\}$, respectively. Each *strategy* is a sequence (ordered set, see Sects. 1.3 and 3.1) of "moves" which take into consideration the prior moves of both players. If P applied the strategy $s_j \in S$ and Q the strategy $t_k \in T$ then, at the end of the game, P receives the *payoff* $a_{jk} \in \mathbb{R}$ (it may be positive, 0, or negative) and Q receives the payoff $-a_{jk}$ ($j = 1, \ldots, m; k = 1, \ldots, n$). Since there are just two players and the sum of the payoff is 0, this is called a *two-person zero-sum game*. It can be

Table 5.4 Matrix of payoffs a_{jk} for the player P. The payoffs for the player Q are $-a_{jk}$

Strategies of P	Strategies of Q t_1	t_2	$\dots$	t_n
s_1	a_{11}	a_{12}	$\dots$	a_{1n}
s_2	a_{21}	a_{22}	$\dots$	a_{2n}
$\vdots$	$\vdots$	$\vdots$		$\vdots$
s_m	a_{m1}	a_{m2}	$\dots$	a_{mn}

Table 5.5 Example of a payoff matrix of a deterministic game

	t_1	t_2	t_3	α_j
s_1	-2	1	-3	-3 $min\{a_{jk}\|k = 1, 2, 3\}$
s_2	5	4	6	4 α_j
A_k	5	4	6	
	$A_k = max\{a_{jk}\|j = 1, 2\}$			

represented by the *matrix* $\mathbf{A} = (a_{jk})$ *of payoffs* for P (see Table 5.4). The matrix of payoffs for Q is $-\mathbf{A}$.

If the player P chooses the strategy s_j, he gets at least the payoff

$$a_j := min\{a_{jk}|k = 1, \dots, n\}.$$

He can obtain the maximum of these minima by a strategy s_{j*} for which

$$\alpha_{j*} = max\{\alpha_j|j = 1, \dots, m\}.$$

Such an s_{j*} is called a *maximin-strategy* for P and the payoff a_{j*} which this strategy offers to P is the *lower game value*. In the example in Table 5.5

$$\alpha_1 = -3, \qquad \alpha_2 = 4, \qquad \text{and} \qquad \alpha_{j*} = \alpha_2 = 4.$$

Similarly, choosing the strategy t_k, the player Q *looses* at most

$$A_k := max\{a_{jk}|j = 1, \dots, m\}.$$

She will have to pay the least if she chooses the strategy t_{k*} for which

$$A_{k*} = min\{A_k|k = 1, \dots, n\}.$$

Such a strategy, which minimises the maximal loss is called a *minimax-strategy* for Q and A_{k*} is the *upper game value*. In Table 5.5

$$A_1 = 5, \; A_2 = 4, \; A_3 = 6, \qquad \text{and} \qquad A_{k*} = A_2 = 4.$$

If P uses the strategy s_{j^*} *and* Q the strategy t_{k^*} in the same play then, by the definition of the $\alpha_j's$ and $A_k's$ as minima and maxima, respectively,

$$\alpha_{j^*} = a_{j^*k^*} \leq A_{j^*}. \tag{5.74}$$

If, as in Table 5.5,

$$\alpha_{j^*} = a_{j^*k^*} \leq A_{k^*} \tag{5.75}$$

then we say that the game is *deterministic*, $a_{j^*k^*}$ is the *game value in pure strategies* and the pair (s_{k^*}, t_{k^*}) of strategies a *saddle point of the game*. The name "saddle point" is used because

$$a_{j^*k^*} \leq a_{j^*k} \quad \textit{for} \quad k = 1, \ldots, n \quad \textit{and} \quad a_{j^*k^*} \geq a_{jk^*} \quad \textit{for} \quad j = 1, \ldots, m,$$

so that, as with (horizontal) saddle points of functions of two variables (see Fig. 3.26 in Sect. 3.2 and, later, Fig. 8.6 in Sect. 8.3) in one direction (here the row) of the matrix **A** there are no smaller values than $a_{j^*k^*}$, in another (here the columns) there are no greater values than $a_{j^*k^*}$. The strategies (s_{j^*}, t_{k^*}) (in Table 5.5 (s_2, t_2)) themselves are called *equilibrium strategies* because it is of no advantage to move away from them (as long as the other sticks with it).

We now consider the game represented by Table 5.6.

It is *not deterministic* because the lower game value $\alpha_1 = -2$ is not equal to the upper game value $A_1 = 4$. Clearly there exist no equilibrium strategies. In each of the possible confrontations (s_1, t_1), (s_2, t_2), (s_2, t_1), (s_2, t_2) there exists a more favourable strategy for at least one player ((s_1, t_2) better for Q than (s_1, t_1), (s_2, t_2) better for P than (s_1, t_2), (s_1, t_1) better for P than (s_2, t_1), (s_2, t_1) better for Q than (s_2, t_2)).

The pairs of strategies (s_1, t_1),(s_1, t_2), (s_2, t_1), (s_2, t_2), can be described also by $(0, 0)$, $(0, 1)$, $(1, 0)$, $(1, 1)$. These are *pure strategies*. If the game is played several times then it is reasonable to assume that strategies are used with certain *probabilities*. If Q plays always her strategy t_1 while P plays his strategy s_1 with probability $p \in [0, 1]$ and so his strategy s_2 with probability $1-p$, then the expected value of his payoff will be $4p + (-7)(1-p)$. This expected value as function of p is represented by the straight line segment connecting $(0, -7)$ with $(1, 4)$ (Fig. 5.4).

Similarly, if Q always plays strategy t_2 and P plays as before, then the expected value of P's payoff is $(-2)p + 8(1 - p)$, represented by the segment connecting $(0, 8)$ and $(1, -2)$.

Table 5.6 The payoff matrix of a non-deterministic game

	t_1	t_2	
s_1	4	−2	−2
s_2	−7	8	−7
A_k	4	8	

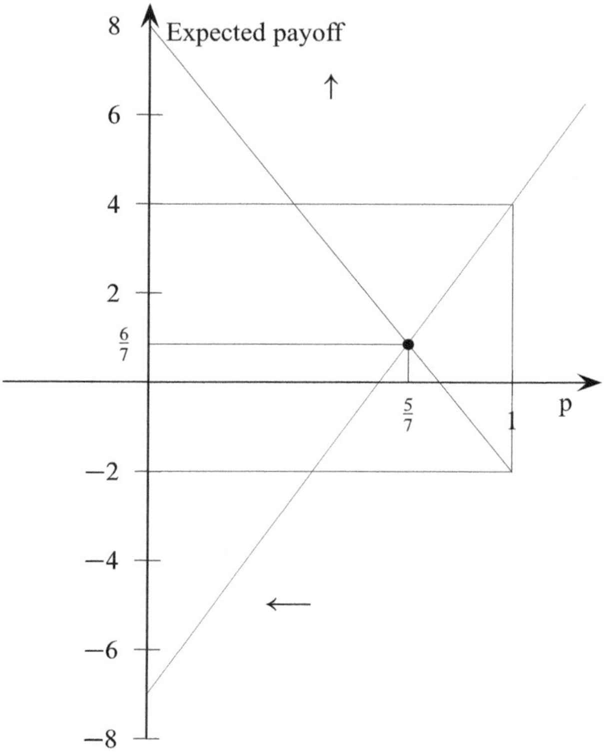

Fig. 5.4 Expected value of payoff as function of p for the first player in Table 5.6, when first player plays s_1 with probability p, s_2 with probability $1-p$ while the second player sticks to t_1 or to t_2

The two straight lines intersect at $p = 5/7$ since $4(5/7) + (-7)(2/7) = (-2)(5/7) + 8(2/7) = 6/7$. So, if P plays his strategies s_1 and s_2 with probabilities $5/7$ and $2/7$, respectively, while Q sticks to t_1 or to t_2, then the expected value of his minimal payoff will be maximal (equal to $6/7$).

With aid of duality theory we will prove that the greatest (expected) minimal payoff for the second player, Q, who plays her strategies t_1 and t_2 with probabilities q and $1 - q$, respectively (while P sticks to s_1 or to s_2) will be $-6/7$, so her smallest maximal loss will be $6/7$, attained for $q = 10/21$. Of course, we could show this by the same direct way, but the duality theorem will show that Q's "maximal expected minimal payoff" in the second situation necessarily equals (-1) times that of P in the first. In such situations we say that P and Q play *mixed strategies* (for pure strategies the probabilities are 0 or 1) and $6/7$ is *the value of the game played by mixed strategies*, while the pairs of probabilities $(5/7, 2/7)$ or $(10/21, 11/21)$, which yield them, are the *equilibrium probabilities (strategies) under mixed strategy*.

We set up the two linear optimisation problems as follows. The first player, P, wants to *maximise his minimal expected payoff* $x_1 - x_2$ (written so because the payoff may be negative but the variables have to be nonnegative) by playing his strategies s_1 and s_2 with probabilities p_1 and $p_2 = 1 - p$, respectively, (we wrote above $p_1 = p$).

So, in the first linear optimisation problem the variables are p_1, p_2 (we use them as separate variables with $p_1 + p_2 = 1$ as condition; in the case of n probabilities this makes the calculation easier) and x_1, x_2. The "objective function" F depends only upon the third and fourth variables

$$F(p_1, p_2, x_1, x_2) = x_1 - x_2. \tag{5.76}$$

This has to be maximised under the conditions (see Table 5.6)

$$4p_1 - 7p_2 \geq x_1 - x_2 \tag{5.77}$$

(for the second player, Q, playing t_1),

$$-2p_1 + 8p_2 \geq x_1 - x_2 \tag{5.78}$$

(for Q playing t_2) and

$$p_1 + p_2 = 1, \tag{5.79}$$

$$p_1 \geq 0, p_2 \geq 0, x_1 \geq 0, x_2 \geq 0. \tag{5.80}$$

We have to bring the conditions (5.77), (5.78) to the form (5.64):

$$-4p_1 + 7p_2 + x_1 - x_2 \leq 0,$$
$$2p_1 - 8p_2 + x_1 - x_2 \leq 0.$$

We replace the equality (5.79), as we did with (5.8), by two inequalities:

$$\begin{aligned} p_1 + p_2 &\leq \ \ 1, \\ -p_1 - p_2 &\geq -1. \end{aligned}$$

The condition (5.80) is already in the form (5.65):

$$p_1 \geq 0, p_2 \geq 0, x_1 \geq 0, x_2 \geq 0.$$

We could solve this, for instance, by the simplex method, but we have solved it above and got

$$\hat{p}_1 = \frac{5}{7}, \quad \hat{p}_2 = \frac{2}{7}, \quad F(\hat{p}_1, \hat{p}_2, \hat{x}_1, \hat{x}_2) = \hat{x}_1 - \hat{x}_2 = \frac{6}{7}$$

as optimal solutions.

On the other hand *the second player* also *wants to maximise her minimum expected payoff, say $y_2 - y_1$, that is, minimise her maximum expected loss $y_1 - y_2$*, by playing her strategies t_1 and t_2 with probabilities q_1 and $q_2 = 1 - q_1$, respectively. So the variables are q_1, q_2, y_1, y_2 and we have to *minimise*

$$f(q_1, q_2, y_1, y_2) = y_1 - y_2$$

under the conditions (we write the losses, that is, the negatives of the payoffs)

$$4q_1 - 2q_2 \leq y_1 - y_2, -7q_1 + 8q_2 \leq y_1 - y_2, q_1 + q_2 = 1, \tag{5.81}$$

that is,

$$\begin{aligned} -4q_1 + 2q_2 + y_1 - y_2 &\geq 0, \\ 7q_1 - 8q_2 + y_1 - y_2 &\geq 0, \\ q_1 + q_2 &\geq 1, \\ -q_1 - q_2 &\geq -1, \\ q_1 \geq 0, q_2 \geq 0, y_1 \geq 0, y_2 &\geq 0. \end{aligned}$$

We see (compare (5.63), (5.64), (5.65), (5.66), (5.67), and (5.68)) that this *second linear optimisation problem is dual to the first*, so, by the "duality theorem" (iii), for their optimal solution we have, as asserted,

$$\hat{y}_1 - \hat{y}_2 = f(\hat{q}_1, \hat{q}_2, \hat{y}_1, \hat{y}_2) = F(\hat{p}_1, \hat{p}_2, \hat{x}_1, \hat{x}_2) = \hat{x}_1 - \hat{x}_2 = \frac{6}{7}. \tag{5.82}$$

(The minimal loss of the second player equals the maximal gain of the first). The values of $\hat{q}_1, \hat{q}_2$ can be determined from (5.81) and (5.82):

$$4\hat{q}_1 - 2\hat{q}_2 \leq \frac{6}{7}, \quad -7\hat{q}_1 + 8\hat{q}_2 \leq \frac{6}{7}, \quad \hat{q}_1 + \hat{q}_2 = 1.$$

If we write for simplicity $\hat{q}_1 = q$ then $\hat{q}_2 = 1 - q$ and we have

$$4q - 2(1-q) \leq \frac{6}{7}, \quad \text{that is,} \quad q \leq \frac{10}{21}$$

and

$$-7q + 8(1-q) \leq \frac{6}{7}, \quad \text{that is,} \quad q \geq \frac{10}{21}.$$

So indeed

$$\hat{q}_1 = q = \frac{10}{21}, \quad \hat{q}_2 = (1-q) = \frac{11}{21},$$

as we stated above.

This method works also with arbitrary (but finite) numbers m, n of strategies (compare Table 5.4) for two-person zero-sum games and leads to similar results, whether they are *deterministic or not deterministic* (as in Table 5.6). One finds that the following result holds:

Every two-person zero-sum game with m strategies for the first and n for the second player (these are "pure" strategies) *has a value* $\hat{z}$ *in mixed strategies, so that there exists at least one mixed strategy with probabilities* $(\hat{p}_1, \ldots, \hat{p}_m)$ *for the first and with probabilities* $(\hat{q}_1, \ldots, \hat{q}_n)$ *for the second player which guarantees a minimal expected payoff ("gain")* $\hat{z}$ *for the first and* $-\hat{z}$ *for the second player. This means "maximal expected loss"* $\hat{z}$ *for the second player* but notice that $\hat{z}$ may be negative, in which case the second expects to gain at least $-\hat{z}$ and the first expects to loose at most $\hat{z}$.

We have used different methods above even for problems which could be solved in the same way in order to acquaint the reader with several methods.

While two-person zero-sum games are important basic notions in game theory and have applications, for instance to parlour games and to very simple situations in economics, competitive situations in economics are better described by more general, *not* (or not necessarily) *zero-sum* "games" with two or more "players". Their discussion is done by methods beyond linear optimisation theory. We will deal with some of them in Sect. 8.6.

5.4.1 Exercises

Consider the matrix $\mathbf{A} = \begin{pmatrix} 3 & 2 & 4 \\ 1 & 4 & 0 \end{pmatrix}$ as the payoff matrix of player P in a two-person zero-sum game.

1. Determine the lower and the upper game value.
2. Change one of the elements (payoffs) of $\mathbf{A}$ so that the changed game is deterministic with game value $= 2$.
3. With the above $\mathbf{A}$ determine equilibrium probabilities (strategies) under mixed strategies for player P.
4. Do the same for the second player Q.
5. Determine the value of the game.

5.4.2 Answers

1. Lower game value $= 2$, upper game value $= 3$.
2. Insert, for instance, 2 for the number 3 in $\mathbf{A}$.
3. $p_1 = p = 3/4, p_2 = (1 - p) = 1/4$.
4. $q_1 = 1/2, q_2 = 1/2, q_3 = 0$.
5. Game value $= 5/2$.

Functions, Their Limits and Their Derivatives

6

Hold infinity in the palm of your hand.
WILLIAM BLAKE (1757–1827)

6.1 Introduction

Many concepts in the social sciences, in particular in economics, such as marginal productivity, marginal costs, marginal profit, marginal rate of substitution, elasticity (for instance price–elasticity of demand, elasticity of substitution) cannot be well defined without the notion of *derivative* (or, what is the same, *differential quotient*). The following example may show what we mean.

Example A nursery produces strawberries in its greenhouses. It calculated that, other conditions being equal, the dependence of the volume y of strawberry production upon the amount x of a fertiliser is described by the curve from A to D in Fig. 6.1. (Of course, in practice only a finite number of points on the curve are obtained from the data, the curve is then drawn to connect, "interpolate" them smoothly, see also Sect. 3.2). The increases by four or by one metric ton of the output quantity at $x = 10$ or $x = 11$, respectively, could be called the *marginal returns for increase by one hundred gallons* each of the fertiliser amount from ten to eleven hundred or from eleven to twelve hundred gallons, respectively. This notion is inadequate since the amount of marginal return *would depend upon the volume unit* of fertiliser (here hundred gallons). In order to avoid this unpleasantness, we consider, for instance at the point of the curve belonging to $x = 10$, instead of the

(continued)

W. Eichhorn, W. Gleißner, *Mathematics and Methodology for Economics*,
Springer Texts in Business and Economics, DOI 10.1007/978-3-319-23353-6_6

slope of the chord BC $(4/1 = 4)$, the *slope of the tangent* to the curve at the point $B = (10, 28)$ as the *marginal yield at* $x = 10$. If this tangent exists (it does not always; see Sect. 6.4) then its slope is the "*limit*" of the slopes of the chords BC, BC_1, $BC_2, \dots$ as the points C, C_1, $C_2, \dots$ of the curve "*converge*" to B. In Fig. 6.1 *the slope of the tangent* at B *is the "limit" of the "difference quotients"*

$$\frac{f(10+1)-f(10)}{(10+1)-10} = f(11)-f(10),$$

$$\frac{f(10+h_1)-f(10)}{(10+1)-1} = \frac{f(10+h_1)-f(10)}{h_1},$$

$$\frac{f(10+h_2)-f(10)}{(10+h_2)-10} = \frac{f(10+h_2)-f(10)}{h_2}, \dots$$

as the "*sequence*" of quantities $h_0 = 1 = 100$ gallons, h_1, h_2 "*tends to* 0". Economists call this limit the *marginal product* (for the production function f) on applying 1,000 gallons of the fertiliser. Mathematicians call it the *derivative* of the function f at $x = 10$. Economists call the j-th term of the above defined sequence the *marginal productivity* of f at $x = 10$ *with step* h_j $(j = 0, 1, 2, \dots)$.

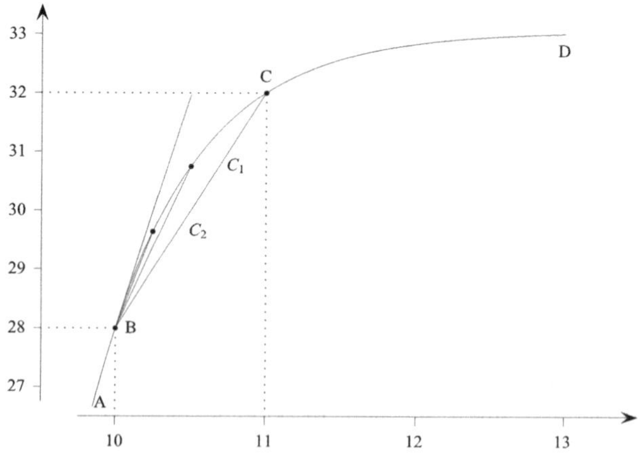

Fig. 6.1 Graph describing the production of strawberries as function f of the quantity x of the fertiliser

6.2 Limits, Infinity as Limit, Limit at Infinity, Sequences: Trigonometric Functions, Polynomials, Rational Functions

As we have just seen, in order to calculate derivatives, we first have to know how to handle limits.

On the other hand, in Chap. 3 we have seen that functions may be defined on several different kinds of sets, their domains. If the domain of a function is a (possibly infinite) real interval or the union of real intervals, then we will call open intervals (intervals which do not contain their endpoints, if any) *neighbourhoods*. Notice that we did not exclude infinite intervals. If the open interval is not bounded from above (it goes to infinity on the right), in symbol $]q, \infty[$ (short for $]q, +\infty[$), it is a *neighbourhood of* $+\infty$ (Fig. 6.2a), if not bounded from below (infinite to the left; in symbol $]-\infty, q)$ then it is a *neighbourhood of* $-\infty$ (Fig. 6.2b) if the open interval is finite, we usually consider it a neighbourhood of its midpoint p, in particular, if its length is 2ε (that is, we deal with the open interval $]p-\varepsilon, p+\varepsilon[$, Fig. 6.2c) then we call it the *ε-neighbourhood* of p.

We often permit a slight anomaly: we remove the point p itself from its neighbourhood (which thus splits into two open intervals) and speak about a *punctured neighbourhood* (to be exact, the *punctured ε-neighbourhood* of p, Fig. 6.2d). As defined, every neighbourhood of $+\infty$ or of $-\infty$ is punctured.

Functions defined on the positive (or nonnegative) *integers* (sometimes also those defined on all integers) are called *sequences*. For the set $\mathbb{N}$ of positive integers we also consider *the set of all integers greater than a given M a neighbourhood of* $+\infty$.

As far as the graph representing a function is concerned, we are still on the "X–axis". Neighbourhoods can, of course, be defined also on the "Y–axis". Let p be a point on the X–axis and ℓ one on the Y–axis. Take a function f, which may *not* be defined at p but *is defined on some punctured neighbourhood* of p. We say that ℓ is the *limit of f at p*, and write

$$\ell = \lim_{x \to p} f(x)$$

Fig. 6.2 Neighbourhoods

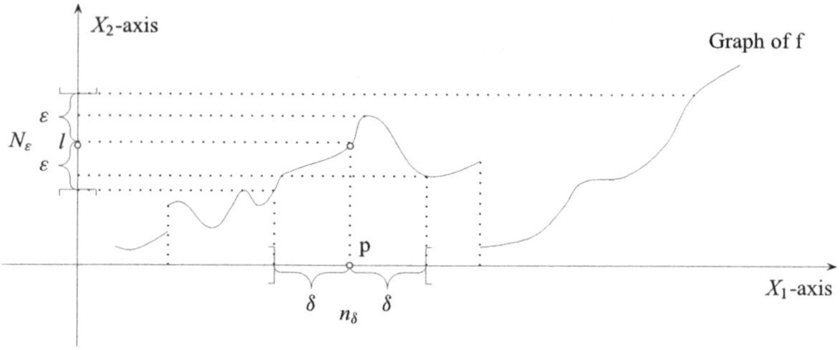

Fig. 6.3 Continuity

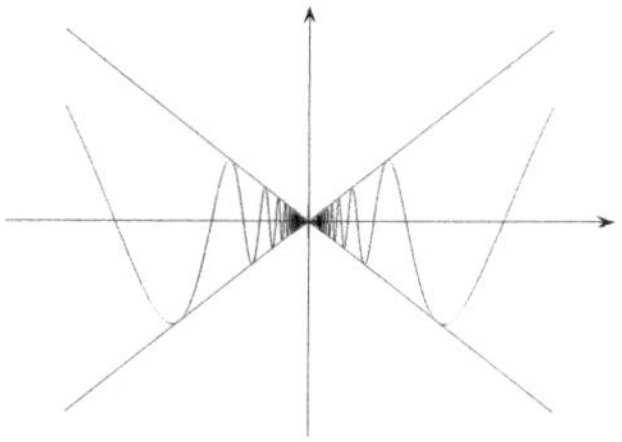

Fig. 6.4 $\mathrm{f(x)} = 2x \sin(\frac{1}{x})$ $(x \neq 0)$

if, for every ε*–neighbourhood of* ℓ*, say* N_ε*, there exists a punctured* δ*–neighbourhood* n_δ *of* p *where* f *is defined and such that* $f(x)$ *is in* N_ε *whenever* x *is in* n_δ (Fig. 6.3).

In formulas:

$$\begin{aligned} &\forall \varepsilon\, \exists \delta : \qquad x \in n_\delta \quad \Longrightarrow \quad f(x) \in N_\varepsilon \qquad \text{or} \\ &\forall \varepsilon\, \exists \delta : 0 < |x - p| < \delta \Longrightarrow |f(x) - \ell| < \varepsilon. \end{aligned} \tag{6.1}$$

Of course, δ may be (and usually is) different for different ε. We always suppose $\varepsilon > 0$, $\delta > 0$. We can verbalise (6.1) as "$f(x)$ *tends to* ℓ *as* x *approaches* p".

Example 1 We show our point on the nontrivial example of the function defined by

$$f(x) = 2x \sin \frac{1}{x} \qquad \text{for all real } x \text{ except } x = 0$$

(see Sect. 1.7 **2** and Fig. 6.4). The above formula shows that $f(0)$ is not defined (while we could define $f(0)$ to be 0, or anything else, it is even less clear how

(continued)

to define $g(x) = \sin(1/x)$ for $x = 0$: its graph undergoes infinitely many fluctuations in every (punctured) neighbourhood of 0, see Fig. 6.5).
But this function has the limit $\ell = 0$ at 0. In formula

$$\lim_{x\to 0} 2x \sin\frac{1}{x} = 0.$$

Indeed, if

$$0 < |x - 0| < \delta$$

then

$$|f(x) - 0| = \left|2x \sin\frac{1}{x}\right| = 2|x| \left|\sin\frac{1}{x}\right| \leq 2|x| < 2\delta,$$

so that (6.1) is satisfied for $\delta = \frac{\varepsilon}{2}$ (here $\ell = 0, p = 0$), that is, there indeed exists a $\delta > 0$ to every $\varepsilon > 0$ so that $0 < |x - 0| < \delta \Rightarrow \left|2x \sin\frac{1}{x} - 0\right| < \varepsilon$, namely $\delta = \frac{\varepsilon}{2}$ (also every smaller positive δ). The function given by $g(x) = \sin(1/x)$ has, as mentioned above, clearly *no limit* at 0 (Fig. 6.5).

Using neighbourhoods in the above generality also has the advantage that we can define *limits at infinity* and $+\infty$ *or* $-\infty$ *as limit* in a quite similar manner: If some function f is defined on a punctured neighbourhood of p then *f has the limit $+\infty$ at p*, that is,

$$\lim_{x\to p} f(x) = +\infty$$

if, for every neighbourhood $N_q =]q, \infty[$ of $+\infty$, there exists a punctured neighbourhood n_δ of p so that $f(x)$ is in N_q whenever x is in n_δ. In formulas this can be written again as

$$\forall q\, \exists \delta : x \in n_\delta \Rightarrow f(x) \in N_q,$$

which now means

$$\forall q\, \exists \delta : 0 < |x - p| < \delta \Rightarrow f(x) > q.$$

The limit $-\infty$ can be defined similarly.

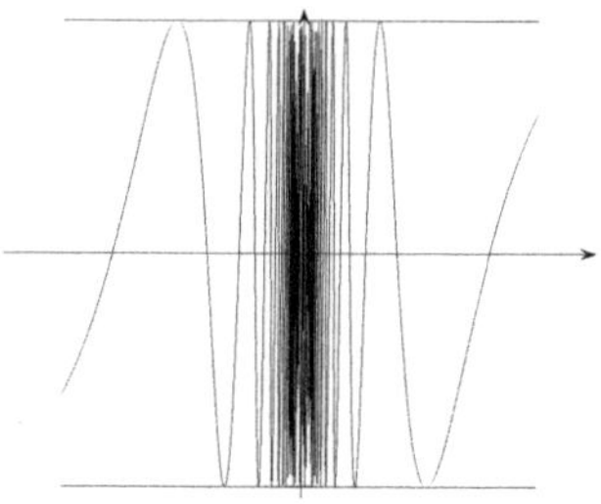

Fig. 6.5 $g(x) = \sin(1/x)$ $(x \neq 0)$

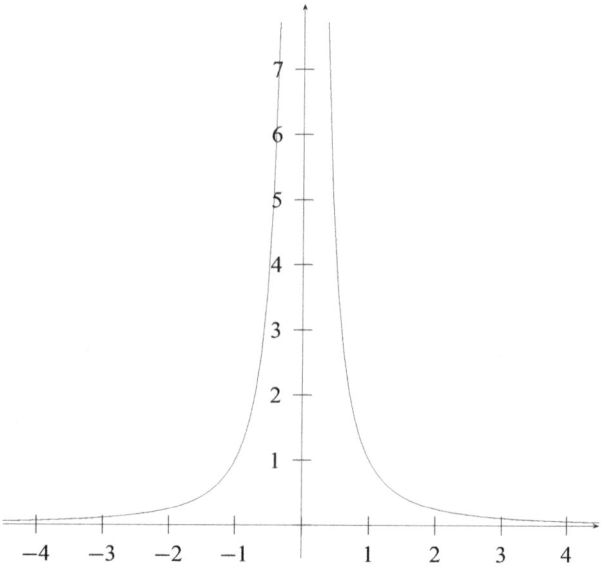

Fig. 6.6 $f(x) = x^{-2}$

Example 2 $f(x) = 1/x$ (see Fig. 6.6); $\lim_{x \to 0} 1/x^2 = +\infty$, because

$$\text{if} \quad 0 < |x - 0| < \delta, \qquad \text{then} \quad \frac{1}{x^2} > \frac{1}{\delta^2}$$

(here $p = 0$) so that $f(x) = 1/x^2 > q$ is satisfied if $q \geq 1/\delta^2$, that is, there indeed exists a δ ($\leq 1/\sqrt{q}$) for every q such that

$$0 < |x - 0| < \delta \Longrightarrow \frac{1}{x^2} > q.$$

Similarly, if $f(x)$ is defined at least for large enough x, then it has *the limit* ℓ at $+\infty$ *if, for every neighbourhood* N_ε *of* ℓ *there exists a neighbourhood* n_q

(continued)

of $+\infty$ *so that* $f(x)$ *is in* N_ε *if* x *is in* n_q. In formula:

$$\forall \varepsilon\, \exists q : x \in n_q \Rightarrow f(x) \in N_\varepsilon.$$

This can clearly be elaborated in the following form:

$$\forall \varepsilon\, \exists q : x > q \Rightarrow |f(x) - \ell| < \varepsilon. \tag{6.2}$$

Then we write

$$\ell = \lim_{x\to\infty} f(x),$$

where we wrote for short $x \to \infty$ instead of $x \to +\infty$. The limit at $-\infty$, $\lim_{x\to-\infty} f(x)$ can be defined similarly.

Example 3 Let $f(x) = \frac{2x}{x-3}$ be defined for $x > 3$ ($f(x)$ can be defined also for $x < 3$ but we are not obliged to do so):

$$\lim_{x\to\infty} \frac{2x}{x-3} = \lim_{x\to\infty} \left(2 + \frac{6}{x-3}\right) = 2,$$

because, if $x > q > 3$, then

$$|f(x) - 2| = \frac{6}{x-3} < \frac{6}{q-3}$$

(here $\ell = 2$; for $|x - 3|$ we wrote $x - 3$ because $x > 3$) so that $|f(x) - 2| < \varepsilon$ is satisfied if $\frac{6}{q-3} < \varepsilon$ or, what is the same, $q - 3 > \frac{\varepsilon}{6}$, that is, $q > \frac{\varepsilon}{6} + 3$. Thus there indeed exists a q for every $\varepsilon > 0$ such that

$$x > q \Longrightarrow \left|\frac{2x}{x-3} - 2\right| < \varepsilon.$$

The functions $f_2(x) = \frac{1}{x^2}, f_3(x) = \frac{2x}{x-3}$ in Examples **2**, **3** (the subscripts serve to distinguish the two functions) are examples of *rational functions*. The general form of a rational functions is

$$x \mapsto \frac{a_p x^p + a_{p-1}x^{p-1} + \cdots + a_1 x + a_0}{b_q x^q + b_{q-1}x^{q-1} + \cdots + b_1 x + b_0} = R_{p,q}(x).$$

Here the *coefficients* $a_0, a_1, \ldots, a_p$ and $b_0, b_1, \ldots, b_q$ are real or complex numbers. We assume $a_p \neq 0$, $b_q \neq 0$. The rational functions are *defined everywhere, where the denominator is not zero*. Both the numerator and the denominator are *polynomials*. As polynomial of degree q, $b_q x^q + b_{q-1}x^{q-1} + \cdots + b_1 x + b_0$ has at most q different real *zeros* (x-values for which the value of the polynomial is 0). The *fundamental theorem of algebra* states that *every polynomial has at least one (real or complex) zero*. The zero may be complex, even if the coefficients, $b_q, b_{q-1}, \ldots, b_1$, b_0 of the above polynomial are real. (For instance the only zeros of the polynomial $x^2 + 1$ are i and $-$i.) It is a consequence of the fundamental theorem of algebra that every polynomial or *rational function* with $p < q$ can be written in the following forms (products or partial fractions):

$$b_q x^q + \cdots + b_1 x + b_0 = (x - x_1)^{q_1}(x - x_2)^{q_2} \cdots (x - x_s)^{q_s}$$

and

$$\begin{aligned} R_{p,q}(x) = {} & \frac{\alpha_1}{(x - x_1)} + \frac{\alpha_2}{(x - x_1)^2} + \cdots + \frac{\alpha_{q_1}}{(x - x_1)^{q_1}} \\ & + \frac{\beta_1}{(x - x_2)} + \cdots + \frac{\beta_{q_2}}{(x - x_2)^{q_2}} + \cdots \\ & + \frac{\rho_1}{(x - x_s)} + \cdots + \frac{\rho_{q_s}}{(x - x_s)^{q_s}}, \end{aligned}$$

respectively, where $q_1 + \cdots + q_s = q$, and $x_1, \ldots, x_s$ are the (possible complex, possibly multiple) zeros of "multiplicities" $q_1, \ldots, q_s$ of the polynomial $b_q x^q + \cdots + b_1 x + b_0$ (or of the denominator of the rational function $R_{p,q}$ satisfying $p < q$). Here we prove neither the fundamental theorem of algebra—which would be quite difficult—nor this consequence.

In Sects. 6.3, 6.5, 6.7, and 6.9 we will return to polynomials and rational functions, giving in Sect. 6.9 also a method to approximate the real zeros of polynomials (with real coefficients), and of the other functions.

The above definitions of neighbourhoods of infinity, adapted to sequences, permits us also to define the *limit of a sequence*. As mentioned above, sequences are functions defined on the positive (or nonnegative) integers. They are denoted by $\{f(n)\}$ or $\{a_n\}$ or $\{a_1, a_2, \ldots\}$ (or $\{a_0, a_1, a_2, \ldots\}$). That the sequence $\{a_n\}$ has ℓ as limit, in symbols

$$\lim_{n \to \infty} a_n = \ell,$$

can be written in the same way as (6.2):

$$\forall \varepsilon\ \exists q : n > q \Rightarrow |a_n - \ell| < \varepsilon.$$

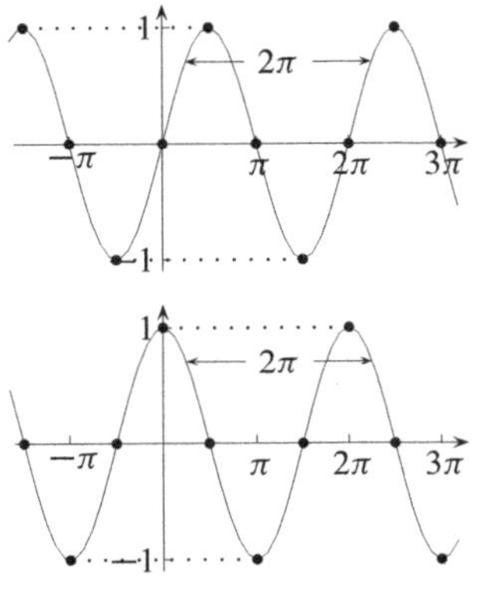

Fig. 6.7 Graphs of $\sin x$, $\cos x$

For example,

$$\lim_{n\to\infty} \frac{2n}{n+3} = 2,$$

The proof is the same as in Example **3** above.

The $a_1, a_2, \ldots, a_n, \ldots$ (or $a_0, a_1, a_2, \ldots$) are the *terms* of the sequence $\{a_n\}$. If a sequence has a finite limit, we say it is *convergent*. We see from all these definitions and examples that, intuitively, *the limit is a value to which the function* (*or sequence*) *gets as close as we want it to get.*

In Example **1**, the *sine function* ($\sin x$) figured prominently. We may remember (Sect. 1.7 **2**) that both it and the *cosine function* ($\cos x$) can be defined *for every real* x (Fig. 6.7; see also Fig. 1.12, where the function of the sine and the cosine function values are illustrated for $x = 0, \psi, \pi/2, \pi, \pi - \psi, -\psi, (\pi/2) - \psi$).

As in Fig. 1.12, x can be regarded as an angle. If we add to an angle x another full period of 2π (and also if we subtract one), both the sine and the cosine function (values) remain unchanged.

As mentioned in Sect. 1.7 **2** and as we will see, it is very convenient to measure the angles in *radians*, that is, by the *length of the arc belonging to them on the unit circle*. For instance, the right angle will be $\pi/2$, the full angle (full turn) is 2π. (The last sentence of the previous paragraph can then be stated as: sin and cos are *periodic* with period 2π.) If we want to measure the angle in degrees (right angle $= 90°$, full angle $= 360°$ or "decimal degrees" (right angle $= 100°$), we have to use other symbols in place of sin and cos (otherwise e.g. cos 7 would yield completely different values in radians, degrees and decimal degrees), for instance sin and cos. We also define

$$\tan x = \sin x / \cos x, \qquad \cot x = \cos x / \sin x = 1/ \tan x$$

(and similarly $\tan x = \sin x/\cos x$, $\cot x = \cos x/\sin x$ if x is measured in degrees). Notice that tan *and* cot *are not defined where the denominators* cos *and* sin *are* 0.

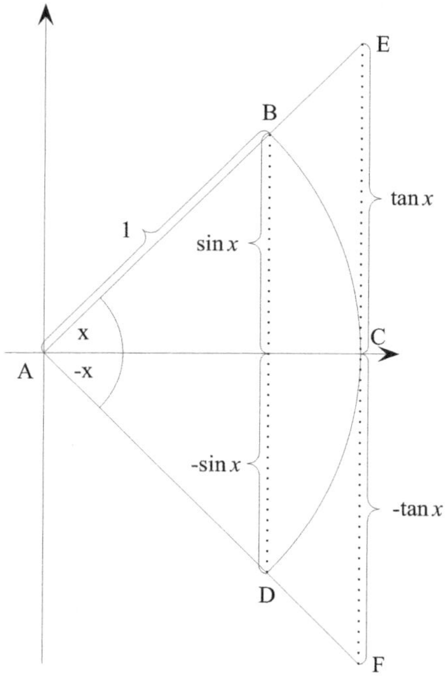

Fig. 6.8 $\sin x \leq x \leq \tan x$ $(x \geq 0)$

An important limit is

$$\lim_{x \to 0} \frac{\sin x}{x} = 1. \tag{6.3}$$

(This again would not be true if the angle x were measured in anything but radians.) The usual proof of this relies on Fig. 6.8.

Since the area of the union of the two triangles ABC and ACD is $\frac{2\sin x \cdot 1}{2} = \sin x$ if x is positive (we do not have to worry about x's greater than $\pi/2$ since the limit at 0 is influenced only by small neighbourhoods of 0), which is not greater (actually, smaller, except if $x = 0$) than the area of sector $ABCD$ of the circle, that is $\frac{1 \cdot 2x}{2} = x$, and this is not greater (is smaller) than the area $\frac{2\tan x \cdot 1}{2} = \tan x$ of the triangle AEF, we have $\sin x \leq x \leq \tan x = \frac{\sin x}{\cos x}$, that is, $\frac{\sin x}{x} \leq 1$ and $\frac{\sin x}{x} \geq \cos x$ or, in a single chain of inequalities,

$$\cos x \leq \frac{\sin x}{x} \leq 1 \tag{6.4}$$

if x is positive. If x is *negative* then, as we saw in Sect. 1.7, Fig. 1.12, the following holds $\cos(-x) = \cos x$, $\sin(-x) = -\sin x$, so (6.4) *remains valid also for negative* x. Now, it is obvious that *the limit of a constant function* c (here $c = 1$) *is* c and one can prove that $\lim_{x \to 0} \cos x = 1$ (while we emphasised above that the function *need not* be defined at the place where we take its limit, it *may* be defined; in this case $\cos 0 = 1$ is defined). Furthermore, if

$$f(x) \leq g(x) \leq h(x)$$

in a neighbourhood of p and both $\lim_{x \to p} f(x)$ and $\lim_{x \to p} h(x)$ exist and are equal:

$$\lim_{x \to p} f(x) = \lim_{x \to p} h(x) = L$$

then also $\lim_{x \to p} g(x)$ exists and is the same:

$$\lim_{x \to p} g(x) = L$$

("*squeeze rule*"), so we get from the chain (6.4) of inequalities

$$\lim_{x \to 0} \frac{\sin x}{x} = 1.$$

(of course also $\lim_{x \to 0} 1 = 1$; the limit of the constant function c is c *everywhere*), which proves (6.3). Actually, also negative x's in punctured neighbourhoods of 0 should be considered but these give the same since $\frac{\sin(-x)}{-x} = \frac{\sin x}{x}$.

As hinted above, if the angle x were measured in any other unit than radians, say in degrees, $\lim_{x \to 0} \frac{\sin x}{x}$ would still exist but not be 1.

There is a certain circularity in the above reasoning: Equation (6.3) states roughly that "*the arc approximately equals the chord, the smaller the arc, the more so*". We "proved" it by use of the area of circular sectors ($x/2$) of arc length x in unit circle. Most readers, if at all, have seen this (just as the area of the whole circle) "proved" at high school exactly by use of this "small arcs are approximately equal to chords" principle (approximating areas of circles and sectors by those of polygons with many small sides). There are ways to get around this difficulty but we will not go into their details here.

We will need the following facts, which are easy to prove:

> If the limits of two functions exist at a point then so does, at the same point, the limit of their sum, difference, product and quotient (if the limit of the denominator is not 0) and they equal the sum, difference, product or quotient, respectively, of the limits of those two functions.

For functions of several real or one (or several) complex variables, limits can be defined similarly. Only the neighbourhood will be the interiors of rectangles, parallelepipeds, of circles, spheres, etc. For instance, *a function f of two variables,* defined on a punctured (C, D)-neighbourhood of (a, b), that is, for all $x \in]a-C, a+C[$, $y \in]b-D, b+D[$ except possibly at (a, b), *has the limit* ℓ *at* (a, b), in symbol

$$\ell = \lim_{\substack{x \to a \\ x \to b}} f(x, y),$$

if, for every ε-neighbourhood of ℓ (Fig. 6.2c), say N_ε, there exists a punctured (γ, δ)-neighbourhood $n_{\gamma,\delta}$ of (a, b) $(\gamma \le C, \delta \le D)$ such that $f(x, y)$ is in N_ε whenever (x, y) is in $n_{\gamma,\delta}$. In formulas:

$$\forall \varepsilon \, \exists \gamma, \delta : 0 < |x - a| < \gamma,\ 0 < |y - b| < \delta \Rightarrow |f(x) - \ell| < \varepsilon.$$

For the use of circles and spheres as neighbourhoods to define limits, see Sect. 6.10.

6.2.1 Exercises

1. Determine
   (a) $\lim_{x \to 1} \dfrac{x^2 - 1}{x - 1}$, (b) $\lim_{x \to -2} \dfrac{x^2 - x - 6}{x + 2}$, (c) $\lim_{x \to 0} \dfrac{\sin 2x}{\sin x}$,
   (d) $\lim_{x \to 0} \dfrac{\tan x}{\sin x}$, (e) $\lim_{x \to 1} \dfrac{x^3 - 7x + 6}{x^2 + 2x - 3}$, (f) $\lim_{x \to -3} \dfrac{x^3 - 7x + 6}{x^2 + 2x - 3}$,
   (g) $\lim_{x \to 0} \dfrac{(2 + x)^2 - 4}{x}$, (h) $\lim_{x \to \pi/4} \sin x$, (i) $\lim_{x \to \infty} \dfrac{3x^2 + 2}{10x^2 - 3x}$,
   (j) $\lim_{x \to -\infty} \dfrac{x^2 - 7}{3 + 4x^2}$, (k) $\lim_{x \to \infty} \dfrac{\cos x}{x}$, (l) $\lim_{x \to -\infty} \dfrac{x^4 - 3 \sin x}{3x + 5x^5}$.
2. Write the first four terms of the sequence $\{f(n)\}$, $n = 1, 2, 3, \ldots$, when
   (a) $f(n) = \dfrac{3n^2 - 4}{n^2 + 2n + 5}$, (b) $f(n) = -3 + \dfrac{n - 1}{n^2 - 1}$,
   (c) $f(n) = \dfrac{1}{1 + n + n^2 + n^3}$, (d) $f(n) = \dfrac{5n^3 + 7}{n^4 - n + 3}$.
3. Determine the limits of the sequences given in Exercise 2.
4. Let the first three terms of a sequences be 3, 5, 7. Obviously, the sequence $\{f(n)\}$, where $f(n) = 2n + 1$ $(n = 1, 2, 3, \ldots)$, starts with these terms. Find two other sequences $\{g(n)\}$, $\{h(n)\}$ whose first three terms are also 3, 5, 7.
5. Write the following polynomials in their product form:
   (a) $D(x) = x^4 + 4x^3 - 16x - 16$,
   (b) $N(x) = x^3 + x^2 - x - 1$.
   Hint: Obviously, $D(2) = D(-2) = 0$, $N(1) = 0$.
6. Write the rational function $R_{3,4}(x) = \dfrac{N(x)}{D(x)}$ (see Exercise 5) in its partial fraction form.

6.2.2 Answers

1. (a) 2, (b) -5, (c) 2, (d) 1, (e) -1, (f) -5,
   (g) 4, (h) $\sqrt{2}/2$, (i) $3/10$, (j) $1/4$, (k) 0, (l) 0.
2. (a) $-\frac{1}{8}, \frac{8}{13}, \frac{23}{20}, \frac{44}{29}$, (b) $-\frac{5}{2}, -\frac{8}{3}, -\frac{11}{4}, -\frac{14}{5}$,
   (c) $\frac{1}{4}, \frac{1}{15}, \frac{1}{40}, \frac{1}{85}$, (d) $4, \frac{47}{17}, \frac{142}{81}, \frac{109}{85}$.
3. (a) 3, (b) -3, (c) 0, (d) 0.
4. For example, $g(n) = n^3 - 6n^2 + 13n - 5$, $h(n)$ = the n-th prime number following 2.
5. (a) $D(x) = (x+2)^3(x-2)$,
   (b) $N(x) = (x+1)^2(x-1)$.
6. $R_{3,4}(x) = \dfrac{55/64}{(x+2)} - \dfrac{25/16}{(x+2)^2} + \dfrac{3/4}{(x+2)^3} + \dfrac{9/64}{(x-2)}$.

6.3 Continuity, Sectional Continuity, Left and Right Limits

If f has a limit at a point, is defined at that point, and its value is equal to that limit, then f is continuous at that point. This definition works equally well for functions of one or several real or complex or vectors variables.

When a function is continuous at all points of a set (for one real variable usually one or several intervals, for complex, vector variables, or several real variables usually one or several connected domains) *then we say that the function is continuous on that set.* One can prove that *polynomials, the sine and cosine functions are continuous on the whole real line, rational functions are continuous at every point where the denominator is not* 0 and, similarly, *the tangent and cotangent functions are continuous everywhere, where the cosine or the sine is not* 0*, respectively. If two functions are continuous at a point, so are their sums, differences, products and, if the denominator is not* 0 *there, also their quotient. If f is continuous at p and g at $f(p)$ then the composite functions $g \circ f$ ($x \mapsto g[f(x)]$) is continuous at p.*

From the above definition and the preceding ones in Sect. 6.2 we should get the intuitive meaning that *a function is continuous (on a set) if we can keep the changes of its values arbitrarily small as long as we confine the variable(s) to sufficiently small changes*. However, maybe the word "*continuous*" suggests a "continuous flow", for functions of one real variable a curve "which can be drawn without lifting the pen". This is not so bad but one has to be careful: the function similar to one mentioned before (in Sect. 6.2, Example **1**), defined by

$$f_1(x) = \begin{cases} 2x \sin \dfrac{1}{x} & \text{for all} \quad x \neq 0, \\ 0 & \text{for} \quad x = 0, \end{cases}$$

makes (Fig. 6.4) infinitely many waves, so "cannot be drawn without lifting the pen" but it is *everywhere continuous* on the real line: at points $x \neq 0$ because of the above rules on products, quotients and substitution of continuous functions into continuous functions and at $x = 0$ because we have shown in Example **1** of Sect. 6.2 that the limit

$$\lim_{x \to 0} 2x \sin \frac{1}{x}$$

exists and is 0 and now we defined $f_1(x)$ so that its value should be 0 at $x = 0$ which by the above definition means exactly f_1 is *continuous also at* 0.

Functions which are not everywhere continuous, can also have important roles in economics (some may play even more important roles that continuous ones), for example the following.

Example 1 (Cost function) In a factory 1,000 pieces of a commodity are produced during a shift. The "fixed cost" for a shift is \$500. The "variable cost" is \$1.5 per piece. Then the *cost function* will be given by

$$C(x) = \begin{cases} 5P{,}000 + 15x & \text{for} \quad 0 \leq x \leq 1{,}000, \\ 10{,}000 + 15x & \text{for} \quad 1{,}000 < x \leq 2{,}000, \\ 15{,}000 + 15x & \text{for} \quad 2{,}000 < x \leq 3{,}000, \\ \ldots\ldots\ldots\ldots\ldots\ldots & \end{cases}$$

(see Fig. 6.9). This function is *discontinuous* at the "jump points" 1,000, 2,000, 3,000, ...

This function and similar ones in applications are "not very discontinuous" (a function is *discontinuous* at a point if it not continuous there) it is *sectionally* (or piecewise) *continuous*: every finite interval for the variable can be divided into finitely many parts in the interior of which the function is continuous and even at the dividing and endpoints left and right limits exist. Finite open intervals with a on their left end or b on their right end are *right neighbourhoods* of a or *left neighbourhoods* of b, respectively. A function has the *right limit* ℓ at the point a (*left limit* ℓ at the point b), in symbols

$$\ell = \lim_{x \to a+} f(x) = \lim_{\substack{x \to a \\ x > a}} f(x) \qquad (\text{or } \ell = \lim_{x \to b-} f(x) = \lim_{\substack{x \to b \\ x < b}} f(x)),$$

if, for every ε-neighbourhood of ℓ, say N_ε, there exists a right-neighbourhood of a (left neighbourhood of b) n_δ so that

$$f(x) \quad \text{is in } N_\varepsilon \text{ whenever } x \text{ is in } n_\delta.$$

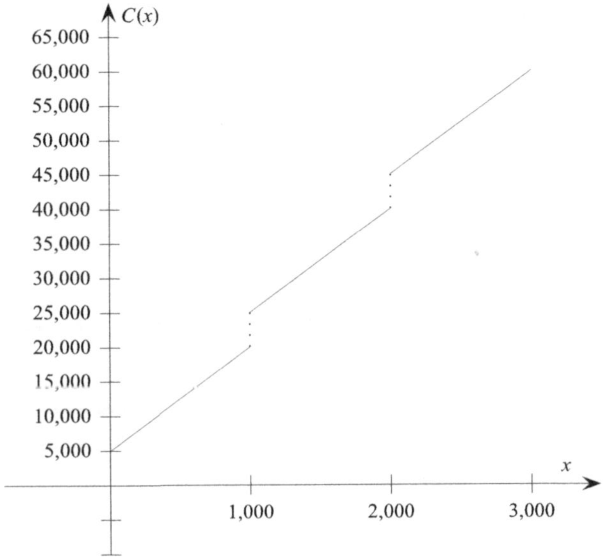

Fig. 6.9 A discontinuous cost function

In formulas, $\ell = \lim_{x \to a+} f(x)$ if

$$\forall \varepsilon \, \exists \delta : 0 < x - a < \delta \Rightarrow |f(x) - \ell| < \varepsilon$$

and $\ell = \lim_{x \to b-} f(x)$ if

$$\forall \varepsilon \, \exists \delta : 0 < b - x < \delta \Rightarrow |f(x) - \ell| < \varepsilon$$

Notice that these differ from the definition (6.1) in Sect. 6.2 of the (two-sided) limit, since there the condition with δ also contained an absolute value sign ($|x - p| < \delta$). Notice also that f *needs not be defined at* a *or* b, respectively.

A function is *left continuous* at a or *right continuous* at b if $f(a), f(b)$ exist and

$$f(a) = \lim_{x \to a+} f(x) \quad \text{or} \quad f(b) = \lim_{x \to b-} f(x), \tag{6.5}$$

respectively. Functions defined (considered) on an interval starting with (and including) a point a or ending with (and including) a point b (that is, not stretching to $-\infty$ or $+\infty$, respectively), are called continuous on that interval if they are continuous in the interior of the interval and right continuous at a and/or left continuous at b, respectively. (This is just a clarification of the definition of continuity on an interval finite and closed on at least one side: since the function may not be defined outside the interval, two-sided continuity at the boundary may make no sense.)

Fig. 6.10 $[x]$ for $1 \leq x \leq 4$

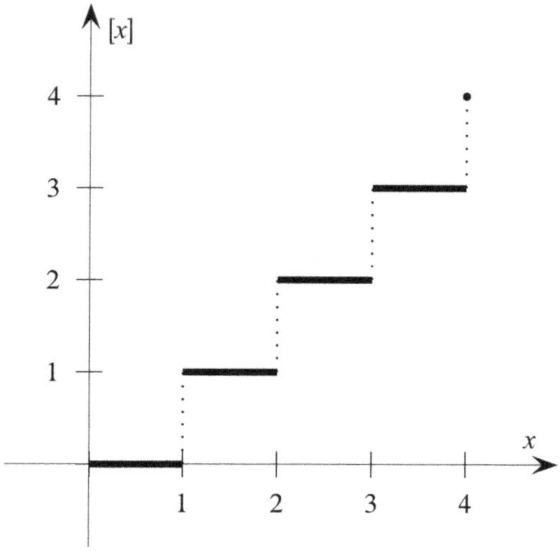

Example 2 The "*integer part function*" ("entire"), ordering to each real number x *the largest integer not greater than* x (Fig. 6.10), denoted by $[x]$, is another example of a sectionally continuous function. At 2 the left and right limits are

$$\lim_{x\to 2-} f(x) = 1, \qquad \lim_{x\to 2+} f(x) = 2 \quad \text{while} \quad f(2) = 2$$

(the function is clearly discontinuous at 2 and at all other integers, it has not even a limit there but, by (6.5), it is left continuous at all integers).

Functions *continuous on closed intervals* have particularly attractive properties. We state them here without proof but point out why they are not so obvious as they may sound (for 1 and 2 see Fig. 6.11).

1. *Every function continuous on a closed interval is bounded both from above and from below on that interval* (that is, there exist numbers m and M such $m \leq f(x) \leq M$ for all x in the interval). This is *not true for open or half-open intervals*: For instance $f(x) = 1/x$ is continuous on $]0, 1[$ (or on $]0, 1]$), but not bounded from above (Fig. 6.12).
2. *Every function continuous on a closed interval assumes its greatest and smallest value on that interval*, that is, there exist x_1 and x_2 such that $f(x_1) \leq f(x) \leq f(x_2)$ for all x in that interval. This again is *not true for open or half-open intervals*: Even such a simple function as that given by $f(x) = x$ does assume neither

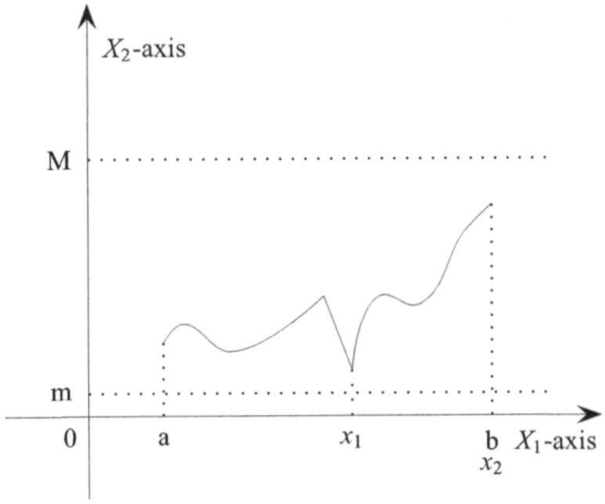

Fig. 6.11 For the continuous function $f : [a, b] \to \mathbb{R}$ given by this graph, m and M are lower and upper bounds, respectively, and f assumes its smallest value at x_1 and its greatest value at $x_2 = b$

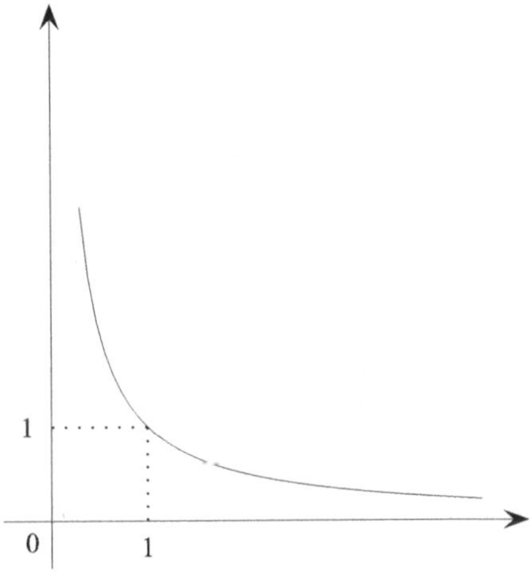

Fig. 6.12 $f(x) = 1/x$ is continuous on $]0, 1]$, but there exist no M such that $M > 1/x$ for all $x \in]0, 1]$

its greatest nor its smallest value on any open interval, for instance on $]0, 1[$ (Fig. 6.13) because neither 0 nor 1 belongs to the intervals.
Even the following simple fact (Fig. 6.14) would need proof.

3. *Every function continuous on* $[a, b]$ *assumes every value between* $f(x)$ *and* $f(b)$ (that is, if $f(a) \leq C \leq f(b)$, then there exists a $c \in [a, b]$ for which $f(c) = C$). Continuous functions can also be defined just on the rational numbers, say in $[0, 1]$, but for those this is *not true*.

There are similar results for functions of several variables.

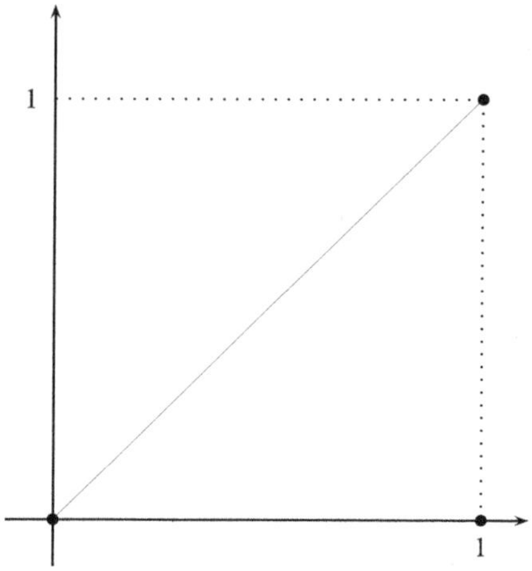

Fig. 6.13 $f(x) = x$ is continuous on $]0, 1[$, but assumes neither its greatest nor its smallest value

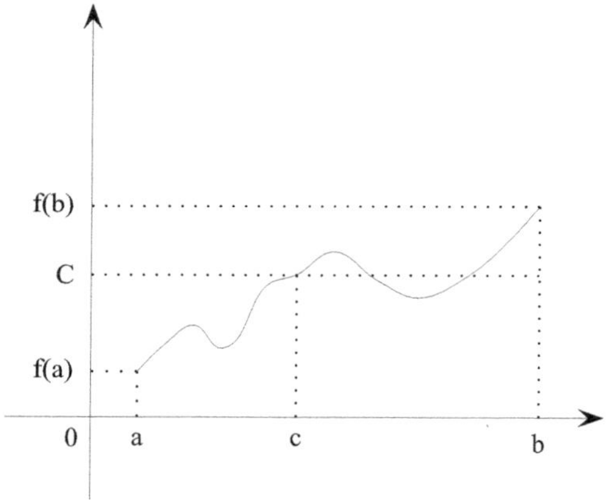

Fig. 6.14 Property 3

6.3.1 Exercises

1. In which points of the real line are the functions (a)–(d) not continuous?

(a) $x \mapsto \begin{cases} \dfrac{1+x}{2x-3x^2+x^3} & \text{if} \quad x \notin \{0,1,2,3\} \\ 2/3 & \text{if} \quad x = 0 \text{ or } x = 1 \text{ or } x = 2, \end{cases}$

(b) $x \mapsto \begin{cases} \dfrac{1}{\sin x} & \text{if} \quad x \neq \frac{k}{2}\pi \\ 1 & \text{if} \quad x = \frac{k}{2}\pi, \end{cases} \qquad (k \in \mathbb{Z})$

(c) $x \mapsto \begin{cases} \tan x & \text{if} \quad x \neq \frac{k}{2}\pi \\ 0 & \text{if} \quad x = \frac{k}{2}\pi, \end{cases} \qquad (k \in \mathbb{Z})$

(d) $x \mapsto \begin{cases} \cot x & \text{if} \quad x \neq \frac{k}{2}\pi \\ 0 & \text{if} \quad x = \frac{k}{2}\pi, \end{cases} \qquad (k \in \mathbb{Z})$

2. Which of the following functions are continuous at $x = 1$?
   (a) $f : \mathbb{R} \to \mathbb{R}, x \mapsto \dfrac{x^2 + x + 1}{x - 1}, 1 \mapsto 3,$
   (b) $g : \mathbb{R} \to \mathbb{R}, x \mapsto \dfrac{x^3 + 4x^2 + x - 6}{(x - 1)(x + 2)}, 1 \mapsto 4,$
   (c) $h : \mathbb{R} \to \mathbb{R}, x \mapsto x^3 - x \cos x,$
   (d) $x \mapsto \dfrac{x^2 + x - 3}{x - 1}, 1 \mapsto 2.$
3. (a) Draw a function $f : [0, 4] \to \mathbb{R}$ which is discontinuous at $x = 1$, $x = 2$ and $x = 3$.
   (b) Draw a function $F : [(0, 0), (3, 4)] \to \mathbb{R}$ which is discontinuous at $(x_1, x_2) = (1, 1)$ and $(x_1, x_2) = (2, 2)$.
4. Determine, for $a \in \mathbb{R}_{++}$, $\lim\limits_{x \to a+} f(x)$ for
   (a) $f :]a, \infty[\to \mathbb{R}, x \mapsto (x - a)/\sqrt{x - a},$
   (b) $f :]a, 3a[\to \mathbb{R}, x \mapsto \dfrac{x^3 + ax^2 - 5a^2x + 3a^3}{(x - a)^2}.$
5. Determine $\lim\limits_{x \to \frac{\pi}{2}-} \dfrac{x - x(\sin x)^2}{1 - \sin x}.$

6.3.2 Answers

1. (a) $x = 0, \quad x = 1, \quad x = 2,$
   (b) $x = k\pi$ and $x = \dfrac{2k + 3}{2}\pi \quad (k \in \mathbb{Z}),$
   (c) $x = k\dfrac{\pi}{2} \quad (k \in \mathbb{Z}),$
   (d) $x = k\pi \quad (k \in \mathbb{Z}).$
2. The functions f, g and h defined in (a), (b) and (c), respectively, are continuous, the function given in (d) is not.
4. (a) 0, (b) $4a$.
5. π.

6.4 Derivative, Derivation

Having got acquainted with limits, we can now *better understand* and make exact the notion of *derivatives* introduced in Sect. 6.1. Moreover, we have now tools to calculate derivatives.

Let the real-valued function f of a real variable be defined on a neighbourhood of x_0. Then

$$f'(x_0) = \lim_{x \to x_0} \frac{f(x) - f(x_0)}{x - x_0}$$

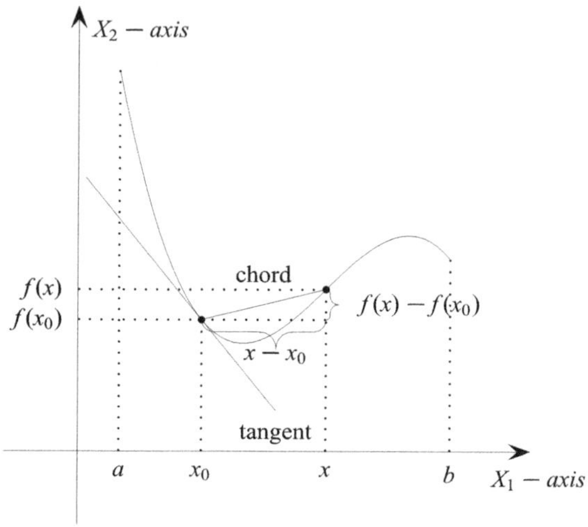

Fig. 6.15 Graph of a function $f : [a, b] \to \mathbb{R}$. Difference quotient $\frac{f(x)-f(x_0)}{x-x_0}$ and derivative at x_0: the slope of the tangent at x_0

is the *derivative of* f *at* x_0, *if the (finite) limit on the right hand side exists.* The fraction $\frac{f(x)-f(x_0)}{x-x_0}$ is the *difference quotient.*

As we saw in Sect. 6.1, the difference quotient is the slope of the chord connecting $(x_0, f(x_0))$ with $(x, f(x))$ (Fig. 6.15) and, as x approaches x_0, it tends to $f'(x_0)$, the slope of the tangent at x_0, if it exists.

Example 1 $f(x) = x^2$. The difference quotient is

$$\frac{f(x) - f(x_0)}{x - x_0} = \frac{x^2 - x_0^2}{x - x_0} = \frac{(x - x_0)(x + x_0)}{x - x_0} = x + x_0.$$

(The last step is valid only for $x \neq x_0$—we must not divide by 0—but, as we have seen in Sect. 6.2, the value of a function at x_0 or whether it is defined there at all, does not interfere with the existence and value of the limit. Therefore *in calculating the derivative at* x_0 *we may always suppose* $x \neq x_0$ *in the difference quotient.*)

The limit of the right hand side as x approaches x_0 exists and is $2x_0$. So

$$f'(x_0) = 2x_0.$$

The above function f is defined for all real x and, as we have just seen, its derivative can be determined at every $x_0 \in \mathbb{R}$. Of course, its value depends on x_0, we can consider it a function of x_0. This function is the *derivative function* (*derivative*, for short); writing x for x_0 we have $f'(x) = 2x$ as derivative function in this case. In general the *derivative function assigns the values*

(continued)

Fig. 6.16 $f(x) = |x|$ is not differentiable at 0

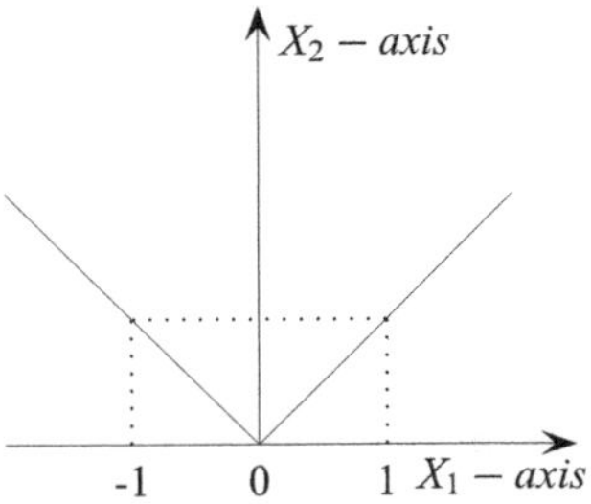

Fig. 6.17 $|x|/x = 1$ for $x > 0, = -1$ for $x < 0$, not defined and no limit at 0

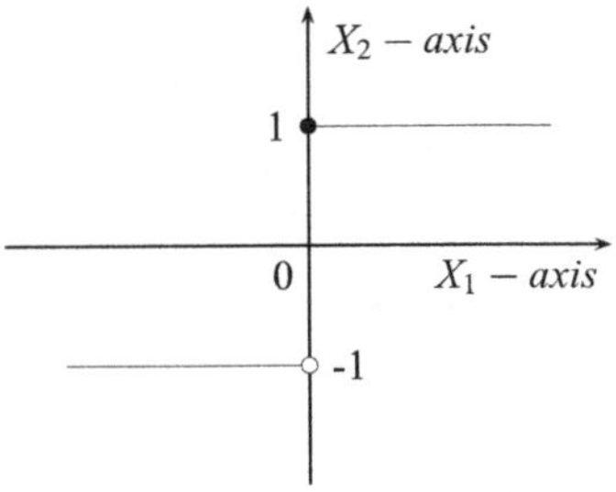

of the derivative to those points where the derivative exists, if any. There may be points where a function has no derivative, is *not differentiable*.

Example 2 $f(x) = |x|$ (Fig. 6.16). Let us try to calculate the derivative at 0. The difference quotient

$$\frac{f(x) - f(0)}{x - 0} = \frac{|x|}{x} = \begin{cases} 1 & \text{if} \quad x > 0, \\ -1 & \text{if} \quad x < 0. \end{cases}$$

This clearly has no limit (the left limit is -1, the right limit is 1; see Fig. 6.17). So this $f(x)$ is not differentiable at 0. (It is differentiable at every other point: $f'(x) = -1$ if $x < 0, f'(x) = 1$ if $x > 0$, see also Example **4**.) But it is easy to see that it is *continuous*.

Another example of a continuous function which is not everywhere differentiable comes from *taxation* (Fig. 6.18).

There even exist function which are nowhere differentiable, indeed also functions which are (everywhere defined but) nowhere continuous:

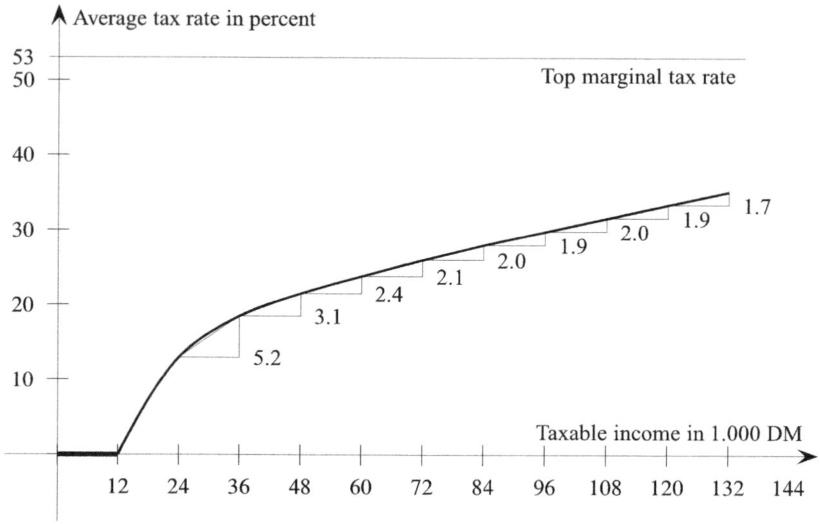

Fig. 6.18 Germany's 1998 average tax rate approximatively represented as a continuous function of taxable income. This function is not differentiable at 12,096

Example 3

$$f(x) = \begin{cases} 1 & \text{if } x \text{ is } \textit{rational}, \\ -1 & \text{if } x \text{ is } \textit{irrational}. \end{cases}$$

is a such function. Of course, its graph cannot be drawn but a little contemplation shows that it indeed can not have even a limit *anywhere* (in every neighbourhood, no matter how small, of every point, there are both rational and irrational numbers, so $f(x)$ could not stay in an ε-neighbourhood with $\varepsilon < 1$ of either 1 or -1 or of any other number). As consequence, this function is *nowhere differentiable*. Indeed, we have the following result:

Theorem *If a function is differentiable at a point then it is also continuous there.*

(So, a function which is not continuous at one point or at many, is not differentiable there either.)

Proof If f is differentiable at x_0 then it is defined on a neighbourhood of x_0 and

$$\lim_{x \to x_0} \frac{f(x) - f(x_0)}{x - x_0} = f'(x_0).$$

Define $\eta(x) = \frac{f(x)-f(x_0)}{x-x_0} - f'(x_0)$ for $x \neq x_0$. As a consequence (since the limit of a constant is itself),

$$\lim_{x \to x_0} \eta(x) = 0 \quad \text{and} \quad f(x) = f(x_0) + (f'(x_0) + \eta(x))(x - x_0).$$

So, by the rules on limits,

$$\lim_{x \to x_0} f(x) = f(x_0) + f'(x_0) \cdot 0 = f(x_0)$$

which means exactly that f is continuous at x_0, as asserted.

The above Example **2** shows that *the converse is not true*: *a function can be continuous but not differentiable*.

We give now two trivial examples, where the derivative function is constant.

Example 4 $f(x) = x$. The limit in the definition of the derivative clearly exists everywhere:

$$f'(x_0) = \lim_{x \to x_0} \frac{f(x) - f(x_0)}{x - x_0} = \lim_{x \to x_0} \frac{x - x_0}{x - x_0} = \lim_{x \to x_0} 1 = 1$$

(remember, the value of $\frac{x-x_0}{x-x_0}$ at $x = x_0$ its existence or no existence does not influence the limits). So *for* $f(x) = x$ *the derivative function is* $f'(x) = 1$ (at every point x; the derivative function is constant).

Example 5 $f(x) = c$ (any constant). The difference quotient

$$\frac{f(x) - f(x_0)}{x - x_0} = \frac{c - c}{x - x_0} = 0 \quad \text{for all} \quad x \neq x_0,$$

so its limit is 0,

$$f'(x) = 0 \quad \text{for } all \quad x.$$

The operation assigning to a function its derivative function is called derivation. We often write the results of Examples **1**, **4**, and **5** as

$$(x^2)' = 2x, \ (x)' = 1, \ (c)' = 0,$$

and similarly for other derivatives.

We move now to somewhat more sophisticated examples.

Example 6 $f(x) = 1/x$. This function is not defined (even less continuous) at 0 so it cannot be differentiable there. For all other x_0 the difference quotient is

$$\frac{\frac{1}{x} - \frac{1}{x_0}}{x - x_0} = \frac{\frac{x_0 - x}{x x_0}}{x - x_0} = -\frac{1}{x x_0} \qquad (x \neq x_0, x_0 \neq 0, x \neq 0).$$

So the derivative at x_0 is $-\frac{1}{x_0{}^2}$ and the derivative function is

$$\left(\frac{1}{x}\right)' = -\frac{1}{x^2} \qquad (x \neq 0).$$

Since $x^{-1} = \frac{1}{x}$, $x^0 = x$ and $x^1 = x$, Examples **6**, **5**, **4** and **1** suggest

$$(x^n)' = nx^{n-1}.$$

This is indeed true for all (positive, 0, negative) integers, also for rational n (see Sect. 6.5) and even for irrational ones (see Sect. 7.2, where x^n will be defined for irrational n in the first place).

For our next example we need from Sect. 6.2 the result (6.3) (we use t in place of x; that makes no difference):

$$\lim_{t \to 0} \frac{\sin t}{t} = 1. \tag{6.6}$$

We will need also some consequences of $\sin(\alpha + \beta) = \sin\alpha\cos\beta + \cos\alpha\sin\beta$. The first is

$$\sin(\alpha - \beta) = \sin\alpha\cos\beta - \cos\alpha\sin\beta$$

(because $\sin(-\beta) = -\sin(\beta)$, $\cos(-\beta) = \cos\beta$). The second results from subtraction of these two equations:

$$\sin(\alpha + \beta) - \sin(\alpha - \beta) = 2\cos\alpha\sin\beta.$$

If we write here $\alpha = \frac{x+y}{2}$, $\beta = \frac{x-y}{2}$, then we get

$$\sin x - \sin y = 2\cos\frac{x+y}{2}\sin\frac{x-y}{2} \qquad \text{for all real} \quad x, y.$$

Now we are ready for the derivative of the sine function:

Example 7 $f(x) = \sin x$. The difference quotient is, by what we have just shown,

$$\frac{\sin x - \sin x_0}{x - x_0} = \frac{2\cos\frac{x+x_0}{2}\sin\frac{x-x_0}{2}}{x-x_0} = \cos\frac{x+x_0}{2}\,\frac{\sin\frac{x-x_0}{2}}{\frac{x-x_0}{2}}.$$

By (6.6), the limit, as x approaches x_0, of the second factor on the right is 1 (put $t = \frac{x-x_0}{2}$); by the cosine function, the limit of the first factor is $\cos x_0$ as x approaches x_0 and so $\frac{x+x_0}{2}$ approaches x_0. As stated near the end of Sect. 6.2, the limit of a product of two functions is the product of their limits, so

$$f'(x_0) = \lim_{x\to x_0} \frac{\sin x - \sin x_0}{x - x_0} = \cos x_0$$

and

$$(\sin x)' = \cos x.$$

Similarly one could prove

$$(\cos x)' = -\sin x$$

but we will derive this in the next section from a general rule.

In general, nice as these proofs are, it would be tiresome to calculate the derivative of each function as it comes up. The general rules in the next section make the determination of derivatives (the process of derivation) easier and almost mechanical.

We note here also that a derivative function may be differentiable too, giving the *second derivative* $f''(x)$ and further

$$f'''(x),\ f^{(4)}(x),\ \ldots,\ f^{(n)}(x).$$

6.4.1 Exercises

1. In which points of their domains are the following functions *not* differentiable.
   (a) $f : \mathbb{R} \to \mathbb{R}, x \mapsto |\sin x|$,
   (b) $g : \mathbb{R} \to \mathbb{R}, x \mapsto \left|x^2 - x - 6\right|$,

(c) $h : \mathbb{R} \to \mathbb{R}, x \mapsto |x| + x^2$,
(d) $\varphi : \mathbb{R} \to \mathbb{R}, x \mapsto |x^3| + |x - 1|$.

2. Define four functions which are continuous in their domains but not differentiable at all points of the domains.
3. Determine the derivative of the following functions at $x = 2$.
(a) $x \mapsto x^3$, (b) $x \mapsto x^{-2}$.
4. Determine the derivative f' of the function $f : \mathbb{R} \to \mathbb{R}$, $x \mapsto a_0 + a_1x + a_2x^2 + a_3x^3$ (a_0, a_1, a_2, a_3 real constants).
5. Determine the derivative g' of the function $g : \mathbb{R}_{++} \to \mathbb{R}$, $x \mapsto b_0 + b_1x^{-1} + b_2x^{-2}$ (b_0, b_1, b_2 real constants).

6.4.2 Answers

1. (a) $x = k\pi \quad (k \in \mathbb{Z})$, (b) $x = -2,\ x = 3$,
(c) $x = 0$, (d) $x = 1$.
3. (a) 12, (b) $-\frac{1}{4}$.
4. $f'(x) = a_1 + 2a_2x + 3a_3x^2$.
5. $g'(x) = -b_1x^{-2} - 2b_2x^{-3}$.

6.5 Rules Which Make Derivation Easier

1. *Derivation of linear combinations*. We leave the easy proof of our first rule to the reader as an exercise:

$$(c_1f_1(x) + c_2f_2(x) + \cdots + c_nf_n(x))' = c_1f_1'(x) + c_2f_2'(x) + \cdots + c_nf_n'(x)$$

($c_1, c_2, \ldots, c_n$ and c below are constants). In particular

$$\begin{aligned}(f(x) + g(x))' &= f'(x) + g'(x),\\ (f(x) - g(x))' &= f'(x) - g'(x),\\ (cf(x))' &= cf'(x).\end{aligned}$$

We emphasise that these and all following rules *hold where all derivatives on the right hand sides of the equations are defined*. If we know already that

$$(x^n)' = nx^{n-1} \qquad (n = 0, 1, 2, \ldots) \tag{6.7}$$

then our first rule gives the derivatives of all *polynomials*:

$$(a_n x^n + a_{n-1}x^{n-1} + \cdots + a_1 x_1 + a_0)' = na_n x^{n-1} + (n-1)a_{n-1}x^{n-2} + \cdots + a_1.$$

2. We can obtain (6.7) from the *derivation rule for products*:

$$[f(x)g(x)]' = f'(x)g(x) + f(x)g'(x).$$

We prove this by forming, as usual, the difference quotient, and then transforming it a bit:

$$\begin{aligned}\frac{f(x)g(x) - f(x_0)g(x_0)}{x - x_0}\\ &= \frac{f(x)g(x) - f(x_0)g(x) + f(x_0)g(x) - f(x_0)g(x_0)}{x - x_0}\\ &= \frac{f(x) - f(x_0)}{x - x_0}g(x) + f(x_0)\frac{g(x) - g(x_0)}{x - x_0}.\end{aligned}$$

The rules near the end of Sect. 6.2 guarantee that this has a limit:

$$\begin{aligned}\lim_{x\to x_0}\frac{f(x)g(x) - f(x_0)g(x_0)}{x - x_0}\\ &= \lim_{x\to x_0}\frac{f(x) - f(x_0)}{x - x_0}\lim_{x\to x_0} g(x) + f(x_0)\lim_{x\to x_0}\frac{g(x) - g(x_0)}{x - x_0}\\ &= f'(x_0)g(x_0) + f(x_0)g'(x_0),\end{aligned}$$

as asserted (g being differentiable, it is also continuous at x_0, so indeed $\lim_{x\to x_0} g(x) = g(x_0)$).

We apply this to derive further cases of (6.7) from Examples **1** and **4** of Sect. 6.4:

$$\begin{aligned}(x^3)' &= (x^2 \cdot x)' = 2x \cdot x + x^2 \cdot 1 = 3x^2,\\ (x^4)' &= (x^3 \cdot x)' = 3x^2 \cdot x + x^3 \cdot 1 = 4x^2,\end{aligned}$$

and so on; it should be clear by now that (6.7) indeed holds for all positive n (an exact proof would use induction). For $n = 0$, (6.7) is the $c = 1$ case of the trivial rule (Example **5** in Sect. 6.4) that *the derivative of the constant function* $f(x) = c$ *is everywhere zero*. Moreover, as mentioned before, a rule similar to (6.7) holds also for negative exponents. Indeed, applying Example **6** from the previous section and our present rule, we get

$$(x^{-2})' = \left(\frac{1}{x}\cdot\frac{1}{x}\right)' = \left(-\frac{1}{x^2}\right)\cdot\frac{1}{x} + \frac{1}{x}\cdot\left(-\frac{1}{x^2}\right) = \frac{2}{x^3} = -2x^{-3},$$
$$(x^{-3})' = \left(\frac{1}{x^2}\cdot\frac{1}{x}\right)' = \left(-\frac{2}{x^3}\right)\cdot\frac{1}{x} + \frac{1}{x^2}\cdot\left(-\frac{1}{x^2}\right) = -\frac{3}{x^4} = -3x^{-4},$$

and so on, $(x^{-m})' = -mx^{-m-1}$ $(m = 1, 2, \ldots)$.

3. *Derivation of fractions.* A little thinking can reduce the amount of calculation: Let

$$f(x) = \frac{h(x)}{g(x)}, \qquad \text{then} \qquad h(x) = f(x)g(x),$$

(of course, at places where $g(x) \neq 0$).
By the above rule **2** (derivation of products),

$$h'(x) = f'(x)g(x) + f(x)g'(x), \qquad \text{that is,} \qquad f'(x) = \frac{h'(x) - f(x)g'(x)}{g(x)}.$$

Recalling $f(x) = \frac{h(x)}{g(x)}$, we get

$$\left(\frac{f(x)}{g(x)}\right)' = \frac{h'(x)g(x) - h(x)g'(x)}{g(x)^2},$$

and this is the derivation rule for fractions (quotients).
This rule permits the *derivation of all rational functions* (at the points where their denominators are not 0). A further application is the following *derivation of the tangent*

$$(\tan x)' = \left(\frac{\sin x}{\cos x}\right)' = \frac{(\sin x)'\cos x - \sin x(\cos x)'}{(\cos x)^2}$$
$$= \frac{\cos x\cos x - \sin x(-\sin x)}{\cos^2 x} = \frac{\cos^2 x + \sin^2 x}{\cos^2 x} = \frac{1}{\cos^2 x},$$

(The cotangent can be similarly derived but we will do it in another way below.)

4. *Chain rule (derivation of composite functions).* Let f be differentiable at x_0 and g be (defined and) differentiable at $f(x_0)$. Then the difference quotient of the composite function $x \mapsto g[f(x)]$ (often denoted by $g \circ f$) at x_0 is

$$\frac{g[f(x)] - g[f(x_0)]}{x - x_0} = \frac{g[f(x)] - g[f(x_0)]}{f(x) - f(x_0)}\frac{f(x) - f(x_0)}{x - x_0}.$$

Since f is differentiable at x_0, it is also continuous, so $t = f(x)$ tends to $f(x_0)$ as x approaches x_0. So the first factor will tend to $g'[f(x_0)]$ and

$$\lim_{x \to x_0} \frac{g[f(x)] - g[f(x_0)]}{x - x_0} = g'[f(x_0)]f'(x_0).$$

Thus we get the *chain rule*

$$(g[f(x)])' = g'[f(x)]f'(x).$$

Often the derivative $f'(x)$ is denoted by $\frac{df(x)}{dx}$, or by $\frac{dt}{dx}$ if $t = f(x)$, and called the "*differential quotient*". Then the chain rule can be written in the following form, if $t = f(x)$, $y = g(t) = g[f(x)]$:

$$\frac{dy}{dx} = \frac{dy}{dt}\frac{dt}{dx}.$$

While the "*differentials*" dx, dy, df, dg can be defined exactly (compare Sect. 6.8), here they serve rather as a memory aid and for actual calculations we recommend returning to the complete form.

Example 1 The derivative of the cosine. Since

$$\cos x = \sin\left(\frac{\pi}{2} - x\right) \quad \text{and} \quad (\sin t)' = \cos t, \quad \left(\frac{\pi}{2} - x\right)' = -1,$$

therefore

$$(\cos x)' = -\cos\left(\frac{\pi}{2} - x\right) = -\sin x.$$

Example 2 Derivative of the cotangent:

$$(\cot x)' = \left(\frac{1}{\tan x}\right)' = \left(-\frac{1}{(\tan x)^2}\right)\frac{1}{(\cos x)^2}$$
$$= -\frac{(\cos x)^2}{(\sin x)^2}\frac{1}{(\cos x)^2} = -\frac{1}{(\sin x)^2}.$$

5. *Derivatives of inverse functions.* Let f be differentiable on a interval I (not necessarily closed). Then, by the Theorem of Sect. 6.4, it is also continuous there and, by the result **3** of Sect. 6.3, *the function assumes all values of an*

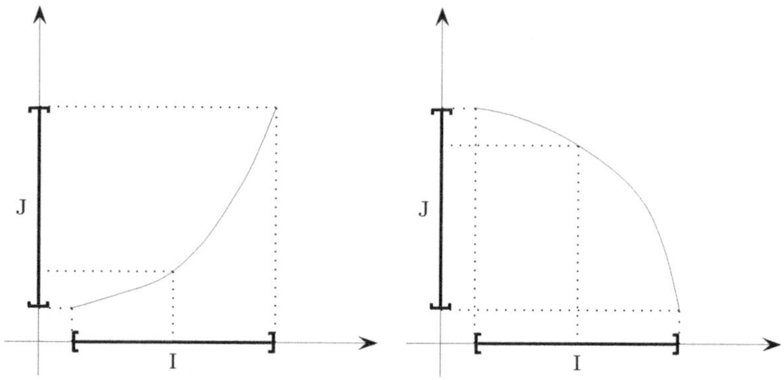

Fig. 6.19 Strictly increasing and strictly decreasing continuous functions assume every value just once

interval, say J. This is so because, by property **3**, with every pair of values also every value in between belongs to J. However, as the Fig. 6.14 accompanying property **3** shows, numbers in J may be values of f at several points in I. If f is *strictly monotonic* on I (*strictly increasing*: $x_1 < x_2 \Rightarrow f(x_1) < f(x_2)$ or *strictly decreasing*: $x_1 < x_2 \Rightarrow f(x_1) > f(x_2)$ for x_1, x_2 in I) then f assumes every value in J exactly once (Fig. 6.19). So, if $y = f(x)$ is given in J, the value x can be uniquely determined. This assigns to every y in J a unique x (in I), so every x in I can be considered a function value for a y in J. This new function g, described by $x = g(y)$, is the *inverse function* of f and is denoted by f^{-1} : $g(x) = f^{-1}(x)$ is equivalent to $f[g(x)] = x$ (cf. Sect. 3.2).

We differentiate both sides of the last equation using the chain rule:

$$f'[g(x)]g'(x) = 1, \qquad \text{that is,} \qquad g'(x) = \frac{1}{f'[g(x)]}.$$

So the derivative of the inverse function is

$$[f^{-1}(x)]' = \frac{1}{f'[f^{-1}(x)]}$$

or, with, the "differential quotient" notation:

$$\frac{dx}{dy} = \frac{1}{\frac{dy}{dx}}.$$

This is sometimes stated as "the derivative of the inverse function is the reciprocal of the derivate of the original function" but one has to be careful: $(x^3)' = 3x^2$ and

$\sqrt[3]{x}$ is the inverse of x^3 but $(\sqrt[3]{x})'$ is not $\frac{1}{3x^2}$ but

$$(\sqrt[3]{x})' = \frac{1}{3(\sqrt[3]{x})^2} = \frac{1}{3}x^{-2/3}$$

(see the following Example **3**).

Example 3 The root $\sqrt[n]{x}$ for $x \in \mathbb{R}_{++}$. This is the inverse of x^n for $x \in \mathbb{R}_{++}$ (if $x = y^n$ then $y = \sqrt[n]{x}$). So

$$(\sqrt[n]{x})' = \frac{1}{n(\sqrt[n]{x})^{n-1}} = \frac{1}{nx^{(n-1)/n}} = \frac{1}{n}x^{\frac{1}{n}-1},$$

where we used also (6.7). Combined with the chain rule **4** we get the following.

Example 4 Powers with rational exponents:

$$(x^{m/n})' = ((\sqrt[n]{x})^m)' = m(\sqrt[n]{x})^{m-1}\frac{1}{n}x^{\frac{1}{n}-1} = \frac{m}{n}x^{\frac{m-1}{n}}x^{\frac{1}{n}-1} = \frac{m}{n}x^{\frac{m}{n}-1}$$

($x \in \mathbb{R}_{++}$) and this is true also if m is a negative integer. So a rule similar to (6.7) indeed holds for all rational exponents.

Example 5 Inverse sine (Arc sine) *function*. The sine function is strictly monotonic (increasing) on $[-\frac{\pi}{2}, \frac{\pi}{2}]$ (but not, for instance, on $[0, \pi]$; see Fig. 6.20). Its inverse function, defined on $[-1, 1]$ is denoted by arc sin, that is, $y = \arcsin x$ implies $x = \sin y$ for $y \in [-\frac{\pi}{2}, \frac{\pi}{2}]$ so $\sin(\arcsin x) = x$. Therefore

$$(\arcsin x)' = \frac{1}{\cos(\arcsin x)} = \frac{1}{\sqrt{1 - (\sin(\arcsin x))^2}} = \frac{1}{\sqrt{1 - x^2}}.$$

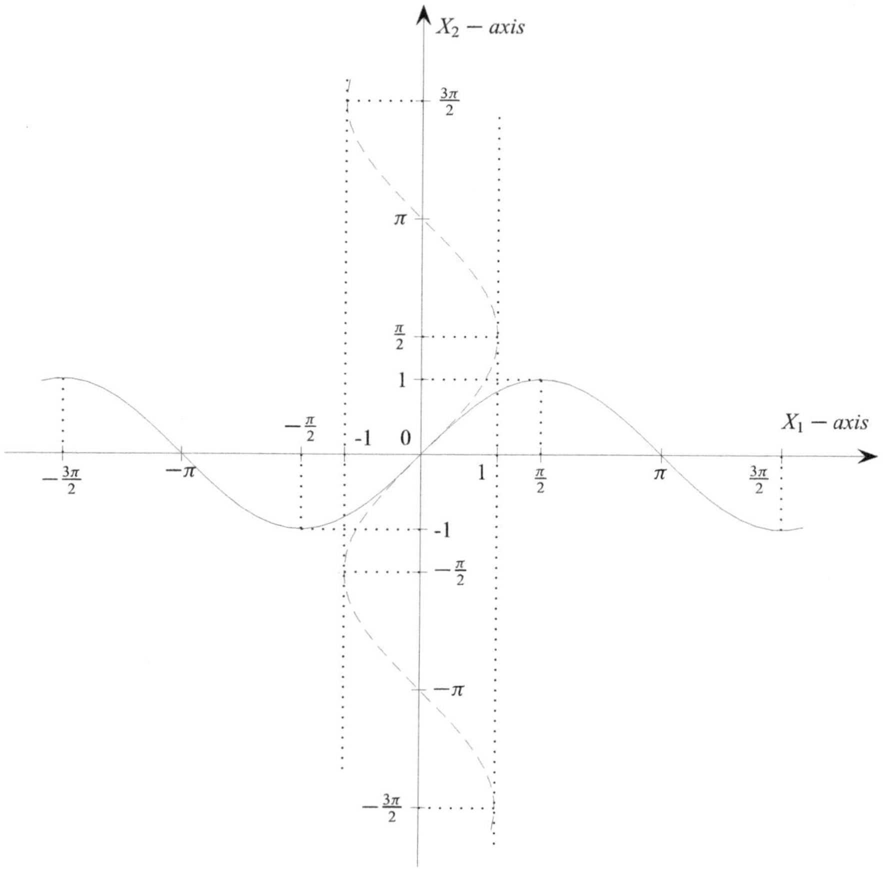

Fig. 6.20 The sine, the Arc sine and other inverse sine (arc sine) functions. The (graphs of the) arc sine functions are parts of the dotted curve, the Arc sine function is the bold-faced part

arc sin is the "*main branch of the inverse sine function*". The sine (Fig. 6.20) is strictly monotonic (decreasing) also on $[\frac{\pi}{2}, \frac{3\pi}{2}]$, on $[-\frac{3\pi}{2}, \frac{\pi}{2}]$ and so on. Different inverse functions (all called arc sin) belong to these, but each has either $-1/\sqrt{1-x^2}$ or $1/\sqrt{1-x^2}$ as derivative.

Since *we get the inverse function* f^{-1} *by exchanging* x *and* y *in* $y = f(x)$ $(x = f(y) \Leftrightarrow y = f^{-1}(x))$, f^{-1} *is represented by a graph which is a reflexion, on the* $45°$ $(\frac{\pi}{4})$*-line, of the graph of* f.

Fig. 6.21 The cosine (strictly decreasing on $[0, \pi]$) and the (*dotted* graph of the Arc cosine)

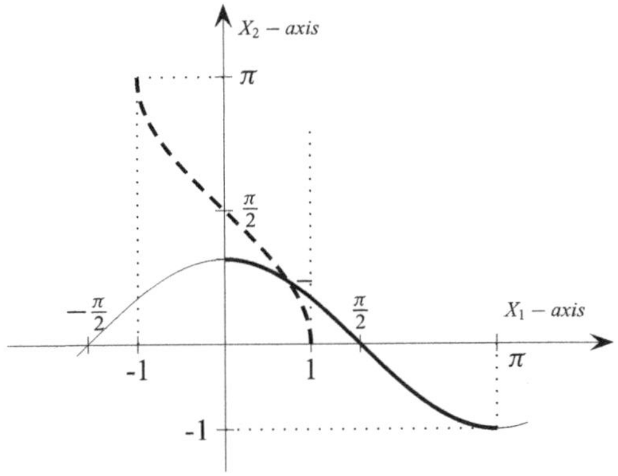

Example 6 *Inverse cosine* (Arc cosine) *function*. The cosine is strictly monotonic decreasing on $[0, \pi]$. Its inverse function, defined on $[-1, 1]$, is denoted by arc cos (Fig. 6.21); $y = \arccos x \Rightarrow x = \cos y$ for $y \in [0, \pi]$, so $\cos(\arccos x) = x$ and

$$(\arcsin x)' = \frac{1}{-\sin(\arccos x)} = -\frac{1}{\sqrt{1-(\cos(\arccos x))^2}} = -\frac{1}{\sqrt{1-x^2}}.$$

Again, there are other inverse cosine functions corresponding to the cosine on $[\pi, 3\pi], [3\pi, 5\pi], \ldots$. arc cos is the *main branch*.

Example 7 *Inverse tangent* (Arc tan) *function*. The tangent is *strictly increasing* on $]-\frac{\pi}{2}, \frac{\pi}{2}[$ and assumes all real values (Fig. 6.22). So the inverse function, arc tan, is defined on all $\mathbb{R}$. In this case $y = \arctan x$ implies $x = \tan y$ for $y \in]-\frac{\pi}{2}, \frac{\pi}{2}[$, so $\tan(\arctan x) = x$ and

$$(\arctan x)' = \frac{1}{1/(\cos(\arctan x))^2} = (\cos(\arctan x))^2 = \frac{1}{1+x^2}$$

(because $\frac{1}{1+(\tan y)^2} = \frac{(\cos y)^2}{(\cos y)^2+(\sin y)^2} = (\cos y)^2$).

Again there are other inverse functions of the tangent function on $]\frac{\pi}{2}, \frac{3\pi}{2}[$, $]-\frac{3\pi}{2}, -\frac{\pi}{2}[$, etc. They differ only in constants $k\pi$ and all have as derivative $1/(1+x^2)$.

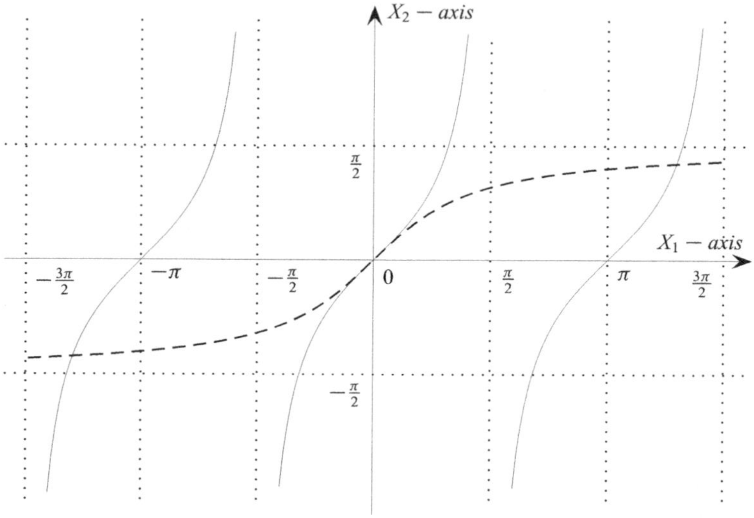

Fig. 6.22 The tangent (strictly increasing on $]-\frac{\pi}{2}, \frac{\pi}{2}[$) and (*dotted graph*) the Arc tan function

6.5.1 Exercises

1. Let a, b, c, c_1, c_2 be real constants, where $a < b$. Let $f_1 :]a,b[\to \mathbb{R}$ and $f_2 :]a,b[\to \mathbb{R}$ be differentiable functions. Prove that
   (a) $(cf_1(x))' = cf_1'(x)$,
   (b) $(c_1 f_1(x) + c_2 f_2(x))' = c_1 f_1'(x) + c_2 f_2'(x)$.
2. Calculate the first and second derivatives of the functions f given by
   (a) $f(x) = x^2 + x\cos x$, (b) $f(x) = (x^4 - x)\sin x$,
   (c) $f(x) = x^3 \sin x \cos x$, (d) $f(x) = (\sin x)^2 - (\cos x)^2$.
3. Calculate the first derivatives of the functions given by $h(x)/g(x)$, $g(x) \neq 0$, where $h : \mathbb{R} \to \mathbb{R}$ and $g : \mathbb{R} \to \mathbb{R}$ are given by
   (a) $h(x) = x^3 - x^2 - 4x + 4$, $g(x) = x^2 + x - 2$,
   (b) $h(x) = 1 - x^2 \cos x$, $g(x) = x \sin x$,
   (c) $h(x) = \sin x \cos x$, $g(x) = x^3$.
4. Determine the first derivative of the composite function
   (a) $g \circ h$, (b) $h \circ g$,
   where g and h are the functions given in Exercise 3 (c).
5. Find the derivative of the inverse function f^{-1} of the function f given by
   (a) $f : \mathbb{R}_+ \to \mathbb{R}, x \mapsto 1 + x^3$,
   (b) $f : \mathbb{R}_+ \to \mathbb{R}, x \mapsto x + x^2$.

6.5.2 Answers

1. (a) $(cf_1(x))' = \lim_{w\to x} \frac{cf_1(w)-cf_1(x)}{w-x} = \lim_{w\to x} c\frac{f_1(w)-f_1(x)}{w-x}$
$= c\lim_{w\to x} \frac{f_1(w)-f_1(x)}{w-x} = cf_1'(x) \quad (w \in]a,b[, \ x \in]a,b[)$,
(b) $(c_1 f_1(x) + c_2 f_2(x))'$
$= \lim_{w\to x} \frac{(c_1 f_1(w)+c_2 f_2(w))-(c_1 f_1(x)+c_2 f_2(x))}{w-x}$
$= \lim_{w\to x} \frac{(c_1 f_1(w)-c_1 f_1(x))+(c_2 f_2(w)-c_2 f_2(x))}{w-x}$
$= \lim_{w\to x} c_1\frac{f_1(w)-f_1(x)}{w-x} + \lim_{w\to x} c_2\frac{f_2(w)-f_2(x)}{w-x}$
$= c_1 \lim_{w\to x} \frac{f_1(w)-f_1(x)}{w-x} + c_2 \lim_{w\to x} \frac{f_2(w)-f_2(x)}{w-x}$
$= c_1 f_1'(x) + c_2 f_2'(x) \quad (w \in]a,b[, \ x \in]a,b[)$.
2. (a) $2x - x\sin x + \cos x$, (b) $(x^4 - x)\cos x + (4x^3 - 1)\sin x$,
(c) $x^3((\cos x)^2 - (\sin x)^2) + 3x^2 \sin x \cos x$, (d) $4 \sin x \cos x$.
3. (a) 1, (b) $x - 2\cot x - x^2(\cot x)^2 + \frac{\cot x}{\sin x} - \frac{x}{\sin x} + \frac{1}{x \sin x}$,
(c) $\frac{(\cos x)^2 - (\sin x)^2}{x^3} - \frac{3 \sin x \cos x}{x^4}$.
4. (a) $3(\sin x)^2(\cos x)^2[(\cos x)^2 - (\sin x)^2]$,
(b) $3x^2[(\cos x^3)^2 - (\sin x^3)^2]$.
5. (a) $1/3(x-1)^{2/3}$ ($x > 1$, because $f^{-1} :]1,\infty[\to \mathbb{R}_+$),
(b) $1/2\sqrt{x + 1/4}$ ($x > 0$, because $f^{-1} : \mathbb{R}_+ \to \mathbb{R}_+$).

6.6 An Application: Price-Elasticity of Demand

We look at the dependence of the amount q of sale of a product upon its price p during a fixed time period (a day, a week, a month, a season, a year, etc.). Let p be in an interval $]a,b[$ of nonnegative numbers and $f :]a,b[\to \mathbb{R}_{++}$ the function describing this dependence, the so-called price-demand function. We will need to change p by a (small) positive or negative number h so that $p + h$ is still in $]a,b[$, that is why we took $]a,b[$ to be an *open* interval. The *price elasticity* $\eta(p,h)$ *of the demand at price p under change by h* is the relative change of quantity of sold products (goods) caused by the change of their price from p to $p + h$ divided by the relative change of price. In formula:

$$\eta(p,h) = \frac{f(p+h) - f(p)}{f(p)} \Big/ \frac{h}{p}.$$

Note the dependence on the price increase (or decrease) $(p + h) - p = h$. In practice $h = 0.01p$, a price increase of one percent, is of particular interest. Then, of course, $\eta(p, p/100)$ is the ratio by which the sale quantity changes when the price is increased by 1 %. The choice of h can nevertheless be arbitrary.

The instantaneous change of the quantity of sold goods under small changes of price clearly describes the tendency of this dependence, the smaller h is the better. So the following *price elasticity at p* is of importance:

$$\varepsilon(p) = \lim_{h\to 0} \frac{f(p+h) - f(p)}{f(p)} \frac{p}{h} = \lim_{h\to 0} \frac{f(p+h) - f(p)}{f(p)} \frac{p}{f(p)} = f'(p)\frac{p}{f(p)}$$

(under the supposition that the limit, that is the derivative, exists and, of course, $f(p) \neq 0$). One often writes

$$\varepsilon = \frac{df}{dp}\frac{p}{f} = \frac{df}{f}\Big/\frac{dp}{p} = \frac{dq}{dp}\Big/\frac{q}{p} = \frac{dq}{q}\Big/\frac{dp}{p}$$

(since $q = f(p)$) but with the mental reservation about "differentials" mentioned in Sect. 6.8 **4** on occasion of the chain rule.

Example If the price-demand function $f : \mathbb{R}_{++} \to \mathbb{R}_{++}$ is given by

$$f(p) = \frac{\alpha}{\beta + p},$$

where α, β are positive constants, then (by calculating the derivative of a fraction or applying the chain rule)

$$\eta(p,h) = \frac{\alpha/(\beta+p+h) - \alpha/(\beta+p)}{\alpha/(\beta+p)}\frac{h}{p} = -\frac{p}{\beta+p+h},$$

$$\eta(p,p/100) = -\frac{p}{\beta + 1.01p},$$

$$\varepsilon(p) = f'(p)\frac{p}{f(p)} = -\frac{\alpha}{(\beta+p)^2}\frac{p}{\alpha/(\beta+p)} = -\frac{p}{\beta+p}.$$

(It is not surprising that the price elasticity is negative since, with increasing price, the quantity of sold goods usually diminishes.)

Remark In economics often *the price p is considered a function of the quantity q*: $p = g(q)$ rather than the other way round. Then, by the derivation rule of inverse functions (**5** in Sect. 6.5),

$$\varepsilon(g(q)) = \varepsilon(p) = \frac{dq}{dp}\Big/\frac{q}{p} = \frac{1}{dp/dq}\frac{p}{q} = \frac{p}{q}\frac{1}{g'(q)}.$$

6.6.1 Exercises

1. Calculate the price elasticity $\varepsilon(p)$ for the price-demand function $f : \mathbb{R}_{++} \to \mathbb{R}_{++}, p \mapsto \alpha p^{-1/2}$, where α is a positive constant.
2. Same problem for $p \mapsto \alpha/(\beta + p^{1/2})$, α and β positive constants.
3. Same problem for $p \mapsto \alpha/(\beta + p^2)$, α and β positive constants.
4. Same problem for $p \mapsto (1 + p)/(2 + p^2)$.
5. Let the price-demand function $g : \mathbb{R}_{++} \to \mathbb{R}_{++}$ be given in the following "inverse form": $p = \alpha/(\beta q + q^2)$, α and β positive constants. Determine $\varepsilon(p) = \varepsilon(g(q))$.

6.6.2 Answers

1. $-1/2.$
2. $\dfrac{-p^{1/2}}{2(\beta + p^{1/2})}.$
3. $-\dfrac{2p^2}{\beta + p^2}.$
4. $-\dfrac{2p^2}{2 + p^2} + \dfrac{p}{1 + p}.$
5. $-\dfrac{\beta + q}{\beta + 2q}.$

6.7 Laws of the Mean, Taylor Series, Bernoulli–L'Hospital Rule

The *law of the mean* states that, *if f is differentiable on the finite open interval $]a, b[$ and continuous at a and at b (from the right or from the left, respectively), then there exists at least one $\xi \in]a, b[$ such that*

$$f'(\xi) = \frac{f(b) - f(a)}{b - a}.$$

This sounds pretty obvious (if f is differentiable on an interval then to every chord of the graph over that interval there exists at least one parallel tangent, see Fig. 6.23), so we do not prove it here.

However, as in Sect. 6.3, we give examples that *even a slight relaxing of the conditions may render the law of the mean invalid.* The function (Fig. 6.24) $f_1(x) = |x|$ is continuous at -1 and at $\frac{1}{2}$ and *differentiable* on $] - 1, \frac{1}{2}[$ *everywhere but at* 0, still the graph has *no tangent parallel to the chord between* $(-1, 1)$ *and* $\left(\frac{1}{2}, \frac{1}{2}\right)$. On the other hand, the function $f_2(x) = x - [x]$, considered on the interval $[0, 1]$ (Fig. 6.25) is differentiable on $]0, 1[$, continuous at 0 (from the right; if $x - [x]$ would be considered also left from 0, it would not be continuous from the left at 0, but that does not matter here), it is however, *not continuous at* 1 *from the left* ($x - [x]$, considered for $x \geq 1$ would be continuous from the right at 1, but this does not help)

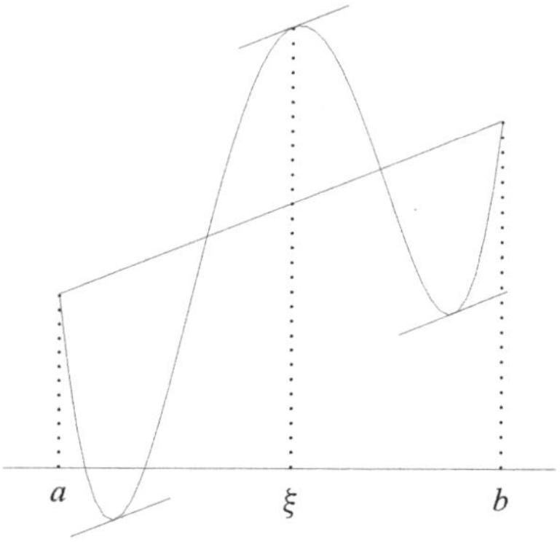

Fig. 6.23 Law of the mean

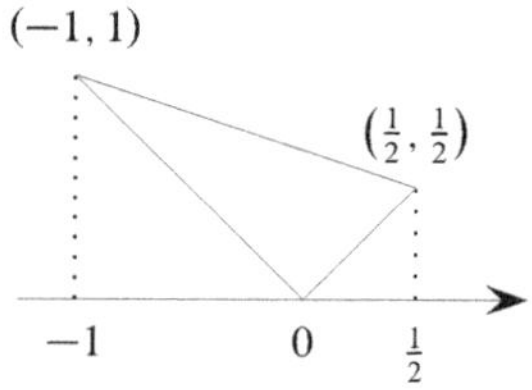

Fig. 6.24 $f_1(x) = |x|$, not differentiable at 0, no tangent parallel to chord between $(-1, 1)$ and $(\frac{1}{2}, \frac{1}{2})$

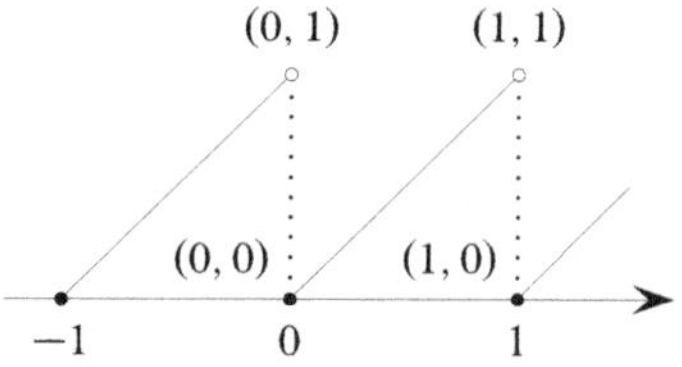

Fig. 6.25 $f_2(x) = x - \lfloor x \rfloor$, not continuous from the left at 1, no tangent parallel to the chord between $(0, 0)$ and $(1, 0)$

and there is *no* (horizontal) tangent *parallel to the chord between the points* $(0, 0)$ and $(1, 0)$ of the graph.

Applying the law of the mean to

$$\phi(x) = f(x) - \frac{[g(x) - g(a)][f(b) - f(a)]}{g(b) - g(a)}$$

we get *Cauchy's law of the mean*:

If f and g are differentiable on $]a, b[$, *continuous at a* from the right *and at b from the left*, $g'(t) \neq 0$ *for all* $t \in]a, b[$, *then there exists at least one* $\xi \in]a, b[$ *such that*

$$\frac{f'(\xi)}{g'(\xi)} = \frac{f(b) - f(a)}{g(b) - g(a)}.$$

The laws of the mean, simple as they sound, have far reaching consequences, of which we give here two. One yields the Taylor formula and the Taylor series (Brook Taylor (1685–1731)).

We first give an argument indicating what form this formula and series ought to take. Take a polynomial

$$f(x) = b_0 + b_1x + b_2x^2 + \cdots + b_nx^n$$

and rearrange it along powers of $(x - a)$:

$$f(x) = c_0 + c_1(x-a) + c_2(x-a)^2 + c_3(x-a)^3 + \cdots + c_n(x-a)^n. \tag{6.8}$$

What will the new coefficients $c_0, c_1, \ldots, c_n$ be? First substitute $x = a$ to get

$$c_0 = f(a).$$

Now differentiate both sides of (6.8) (polynomials are differentiable):

$$f'(x) = c_1 + 2c_2(x-a) + 3c_3(x-a)^2 + \cdots + nc_n(x-a)^{n-1}$$

and substitute $x = a$ again:

$$c_1 = f'(a).$$

Repeating this procedure we get in succession:

$$\begin{aligned} f''(x) &= 2c_2 + 2\cdot 3(x-a) + \cdots + n(n-1)c_n(x-a)^{n-2}, \\ f'''(x) &= 3!c_3 + \cdots + n(n-1)(n-2)c_n(x-a)^{n-3}, \\ &\vdots \\ f^{(n)}(x) &= n(n-1)(n-2)\cdot 3\cdot 2\cdot 1\cdot c_n, \end{aligned}$$

with

$$c_2 = \frac{f''(a)}{2} = \frac{f''(a)}{2!}, \quad c_3 = \frac{f'''(a)}{3!}, \quad \ldots, \quad c_n = \frac{f^{(n)}(x)}{n!} = \frac{f^{(n)}(a)}{n!},$$

($n! = 1\cdot 2\cdot 3\ldots\cdot n$; $0! = 1$ by definition), so that (6.8) becomes

$$\begin{aligned} f(x) &= \frac{f(a)}{0!} + \frac{f'(a)}{1!}(x-a) + \frac{f''(a)}{2!}(x-a)^2 \\ &\quad + \frac{f'''(a)}{3!}(x-a)^3 + \cdots + \frac{f^{(n)}(a)}{n!}(x-a)^n \\ &= \sum_{k=0}^{n} \frac{f^{(k)}(a)}{k!}(x-a)^k. \end{aligned}$$

Of course, this holds only for polynomials of n-th degree. But let us see *how well*

$$f(a) + \frac{f'(a)}{1!}(x-a) + \frac{f''(a)}{2!}(x-a)^2 + \cdots + \frac{f^{(n)}(a)}{n!}(x-a)^n$$

approximates $f(x)$ for any n times differentiable function f. We denote the difference between $f(x)$ and this polynomial by

$$R_n(x) = f(x) - f(a) - \frac{f'(a)}{1!}(x-a) - \cdots - \frac{f^{(n)}(a)}{n!}(x-a)^n$$

and call it the *remainder*. Of course $R_n(a) = 0$. There are several ways to calculate the remainder at other places. Here is one: In analogy to the law of the mean, we want to prove here that, for $a \neq b$,

$$\begin{aligned} R_n(b) - R_n(a) &= R_n(b) \\ &= f(b) - f(a) - \frac{f'(a)}{1!}(b-a) - \cdots - \frac{f^{(n)}(a)}{n!}(b-a)^n \\ &= \frac{f^{(n+1)}(\xi)}{(n+1)!}(b-a)^{n+1} \end{aligned} \tag{6.9}$$

meaning that there exists a ξ between a and b so that (6.9) holds. There certainly is no difficulty in finding a K such that

$$\begin{aligned} f(b) - f(a) - \frac{f'(a)}{1!}(b-a) - \cdots - \frac{f^{(n)}(a)}{n!}(b-a)^n \\ = \frac{K}{(n+1)!}(b-a)^{n+1}, \end{aligned} \tag{6.10}$$

namely

$$K = (n+1)!\left[\frac{f(b) - f(a)}{(b-a)^{n+1}} - \frac{f'(a)}{1!}(b-a)^{-n} - \cdots - \frac{f^{(n)}(a)}{n!}(b-a)^{-1}\right].$$

We define now

$$\begin{aligned} F(t) = f(b) - f(t) - \frac{f'(t)}{1!}(b-t) - \frac{f''(t)}{2!}(b-t)^2 - \cdots \\ - \frac{f^{(n)}(t)}{n!}(b-t)^n - \frac{K}{(n+1)!}(b-t)^{n+1}. \end{aligned}$$

Of course, $F(b) = 0$, from (6.10) we have also $F(a) = 0$ and F is continuous at a and b, so, since F is differentiable (everywhere), we can apply the law of the mean: There exists a ξ between a and b such that

$$0 = \frac{F(b) - F(a)}{b - a} = F'(\xi).$$

However, differentiating F (with respect to t), we get

$$F'(t) = -f'(t) + f'(t) - f''(t)(b - t) + f''(t)(b - t) - \cdots$$
$$- \frac{f^{(n+1)}(t)}{n!}(b - t)^n - \frac{K}{n!}(b - t)^n.$$

Therefore, $F'(\xi) = 0$ means

$$K = f^{(n+1)}(\xi)$$

and (6.10) becomes (6.9), so that we have

$$f(b) = f(a) + \frac{f'(a)}{1!}(b - a) + \frac{f''(a)}{2!}(b - a)^2 + \cdots$$
$$+ \frac{f^{(n)}(a)}{n!}(b - a)^n + \frac{f^{(n+1)}(\xi)}{(n + 1)!}(b - a)^{n+1}$$

with *some* ξ between a and b. Thus, according to (6.9), with x in place of b, the remainder is

$$R_n(x) = \frac{f^{(n+1)}(\xi)}{(n + 1)!}(x - a)^{n+1}$$

with *some* ξ between a and x, furthermore we get the "*Taylor formula with remainder in the Lagrange form*"

$$\begin{aligned} f(x) = f(a) &+ \frac{f'(a)}{1!}(x - a) + \frac{f''(a)}{2!}(x - a)^2 + \cdots \\ &+ \frac{f^{(n)}(a)}{n!}(x - a)^n + \frac{f^{(n+1)}(\xi)}{(n + 1)!}(x - a)^{n+1}. \end{aligned} \tag{6.11}$$

From this, the *polynomial part* $P_n(x) := f(x) - R_n(x)$ is

$$P_n(x) = f(a) + \frac{f'(a)}{1!}(x - a) + \frac{f''(a)}{2!}(x - a)^2 + \cdots + \frac{f^{(n)}(a)}{n!}(x - a)^n.$$

For polynomials of n-th degree, as we saw at the beginning of this argument, this was all of $f(x)$ and our present question was how well P_n approximates f.

We answer this question in two ways. First we notice that (by differentiating the last equality 0, 1, 2, ..., n-times and substituting $x = a$).

$$P_n(a) = f(a), \;\; P_n'(a) = f'(a), \;\; P_n''(a) = f''(a), \;\; \ldots, \;\; P_n^{(n)}(a) = f^{(n)}(a).$$

Therefore, P_n has a graph that is similar to that of f for points near a. In particular, the graph of P_n has at a the same height, the same slope, the same curvature (which depends upon the first and second derivative at a), ... This gives the impression that the values of f and P_n are pretty close if we stay near a.

Our second answer estimates $|R_n(x)| = |f(x) - P_n(x)|$ in a neighbourhood of a for the purpose to see *in what neighbourhood of a will this difference be small*— and even tend to 0 as $n \to \infty$: in these neighbourhoods, if any, we talk about the *Taylor expansion* of f around a. Clearly this has to be estimated for the individual functions.

Example 1 (Sine and cosine functions) For the functions $f_1(a) = \sin x$ and $f_2(x) = \cos x$ we have

$$\begin{aligned} f_1'(x) &= \cos x, \; f_1''(x) = -\sin x, \; f_1'''(x) = -\cos x, \; f_1^{(4)} = \sin x, \ldots \\ f_2'(x) &= -\sin x, \; f_2''(x) = -\cos x, \; f_2'''(x) = \sin x, \; f_2^{(4)} = \cos x, \ldots \end{aligned}$$

respectively. In any case, from (6.11), both for $f(x) = f_1(x)$ and for $f(x) = f_2(x)$,

$$|f(x) - P_n(x)| = \frac{|f^{(n+1)}(\xi)|}{(n+1)!}|x-a|^{n+1} \le \frac{1}{(n+1)!}|x-a|^{n+1}$$

since we have in the numerator either $|\sin \xi|$ or $|\cos \xi|$ which are always ≤ 1, no matter what x (and thus ξ) is. We show that, for fixed x, we can make the right hand side as small as we want to. By the definition of limit this means

$$\lim_{n\to\infty} P_n(x) = f(x).$$

In order to show that we can make $|x-a|^{n+1}/(n+1)!$ as small as we want to, we write $A = |x-a| > 0$ (remember, x is now fixed). Let n be larger than $[2A]$ (the integer part of $2A$). Then

$$\frac{|x-a|^{n+1}}{(n+1)!} = \frac{A}{1}\frac{A}{2}\cdots\frac{A}{[2A]}\frac{A}{[2A]+1}\frac{A}{[2A]+2}\cdots\frac{A}{n+1}.$$

(continued)

While $B = (A/1)(A/2)\cdots(A/[2A])$ is a fixed number,

$$\frac{A}{[2A]+k} < \frac{1}{2} \quad (k = 1, 2, \ldots), \frac{A}{[2A]+1}\frac{A}{[2A]+2}\cdots\frac{A}{n+1} < \left(\frac{1}{2}\right)^{n-[2A]+1}$$

and

$$\frac{|x-a|^{n+1}}{(n+1)!} = B\left(\frac{1}{2}\right)^{n-[2A]+1},$$

which can indeed be made as small as we want it to ($\lim_{n\to\infty}(1/2)^n = 0$). So $\lim_{n\to\infty} P_n(x) = f(x)$, that is,

$$\lim_{n\to\infty}\left(f(a) + \frac{f'(a)}{1!}(x-a) + \frac{f''(a)}{2!}(x-a)^2 + \cdots + \frac{f^{(n)}(a)}{n!}(x-a)^n\right) = f(x).$$

The left hand side is the *Taylor series* around a (an infinite series), written as

$$\sum_{n=0}^{\infty}\frac{f^{(n)}(a)}{n!}(x-a)^n$$

(we write $f^{(0)}(a) = f(a)$ and $0! = 1$, as before) and what we got is, that *in the cases* $f(x) = \cos x$ *and* $f(x) = \sin x$ *the Taylor series converges to* $f(x)$. (This is not always so: *the Taylor series may converge, but not to* $f(x)$ *or it may not converge at all.*) If we choose $a = 0$, we get a Taylor series around 0 called also a *MacLaurin series*. In the cases of $\sin x$ and $\cos x$, from the derivatives calculated at the start of this Example **1**, we have, *for all real* x (x is a variable again),

$$\sin x = x - \frac{x^3}{3!} + \frac{x^5}{5!} - \frac{x^7}{7!} + \ldots \quad \text{and} \quad \cos x = 1 - \frac{x^2}{2!} + \frac{x^4}{4!} - \frac{x^6}{6!} + \ldots.$$

These Taylor and, in particular, MacLaurin series are very useful. We observe that, while we were able to make $|f(x) - P_n(x)|$ as small as we wanted to for large enough $n > N$, this $N = [2A] = [2(x-a)]$ depended on x (not only upon the ε below which we wanted to bring $|f(x) - P_n(x)|$). For many uses of these power series (every $\sum_{n=0}^{\infty} c_n(x-a)^n$ is a *power series*), for instance for derivation and integration, it helps if they are *uniformly convergent*, that is, *the same* N *is good for all* x, say on an interval. In the present example, both series are *uniformly convergent to* $f(x)$ (to $\sin x$ or to $\cos x$) *on any closed interval* $[a-r, a+r]$: just choose in the above argument $A = 2r$ ($|x-a| \le 2r = A$ if x is in $[a-r, a+r]$). This again is not always the case.

Example 2 (Taylor series of $1/x$ around 1) For $f(x) = x^{-1}$,

$$f'(x) = -x^{-2}, \quad f''(x) = 2x^{-3},$$
$$f'''(x) = -3!x^{-4}, \ldots, f^{(n)}(x) = (-1)^n n! x^{-(n+1)}.$$

So, in this case,

$$f(1) = 1, \; f'(1) = -1, \; f''(1) = 2, \; f'''(1) = -3!, \; \ldots, \; f^{(n)}(1) = (-1)^n n!$$

and

$$P_n(x) = 1 - (x-1) + (x-1)^2 + \cdots + (-1)^{n+1}(x-1)^n.$$

Instead of using the form (6.11) of the Taylor formula, we calculate the remainder

$$R_n(x) = x^{-1} - P_n(x)$$

directly: The simple trick is (as in establishing the sum of the geometric series) to multiply

$$R_n(x) = \frac{1}{x} - 1 + (x-1) - (x-1)^2 + \cdots + (-1)^{n+1}(x-1)^n$$

by $(x-1)$:

$$(x-1)R_n(x) = \frac{x-1}{x} - (x-1) + (x-1)^2 + \cdots + (-1)^n(x-1)^n + (-1)^{n+1}(x-1)^{n+1}$$

and add up the last two equations:

$$xR_n(x) = (-1)^{n+1}(x-1)^{n+1}. \text{ So } |R_n(x)| = \frac{|x-1|^{n+1}}{|x|} \quad \text{if } x \neq 0.$$

It is easy to see (and will be shown formally in Sect. 7.2) that n-th powers of real numbers greater than 1 tend to ∞, while n-th powers of positive numbers smaller than 1 converge to 0 as $n \to \infty$. Therefore

$$\lim_{n\to\infty} |R_n(x)| = \infty \quad \text{if } |x-1| > 1, \text{ that is, } x > 2 \text{ or } x < 0,$$

$$\lim_{n\to\infty} |R_n(x)| = 0 \quad \text{if } |x-1| < 1, \text{ that is, } 0 < x < 2.$$

(continued)

Thus we have the Taylor series of $1/x$ around 1:

$$\frac{1}{x} = 1 - (x-1) + (x-1)^2 - (x-1)^3 + \cdots \quad \textit{if} \quad 0 < x < 2,$$

while this formula is not true for $x > 2$ and for $x < 0$ (then the series on the right *does not converge* to any finite number at all). Of the two remaining cases, $x = 0$ makes no sense in our case ($1/x$ is not defined for $x = 0$) but, anyway, $P_n(0) = n + 1$ goes to ∞, while, for $x = 2$, $|R_n(2)| = \frac{1}{2}1^{n+1} = \frac{1}{2}$, independently from n (constant sequence), so $R_n(2)$ does *not* tend to 0 and therefore $P_n(2)$ does not converge to $f(2) = \frac{1}{2}$. (Actually, $P_n(2)$ is 1 if n is odd and $P_n(2)$ is 0 if n is even, so $\{P_n(2)\} = \{1, 0, 1, 0, 1, 0, \ldots\}$ is divergent.) So *the above Taylor expansion of* $1/x$ *holds exactly on* $]0, 2[$. We get another form, a *MacLaurin series* of $1/(1 + t)$ around 0 with $x = 1 + t$:

$$(1+t)^{-1} = 1 - t + t^2 - t^3 + \cdots \quad \textit{if and only if} \quad -1 < t < 1 \tag{6.12}$$

(Here again, the convergence is uniform on $[-r, r]$ with any $r \in]0, 1[$.)

Our second application of the laws of the mean is the Bernoulli-L'Hospital rule. First about the name. Most English language textbooks call it the L'Hôpital rule. While today hôpital is the French spelling of the word hospital, the French mathematician Marquis de L'Hospital (1661–1704) spelled his name this way or even as "L'Hospital". More importantly, the famous Swiss mathematician Johann Bernoulli (1667–1748) claimed in 1704 that he discovered first the rule, which L'Hospital published it in his Analysis monograph in 1696. The late claim (after L'Hospital's death) made it suspicious and disputed. But the publication in 1955(!) of a letter of L'Hospital to Johann Bernoulli made the claim more credible and also explained why Bernoulli staked it only after L'Hospital's death: it is because L'Hospital *bought* several of Bernoulli's results. Indeed, L'Hospital *wrote* to Bernoulli on March 17, 1694, among others:

> I shall give you with pleasure an annuity of three hundred livres... I shall send two hundred livres for the first half of the year because of the notebooks that you have sent, and it will be one hundred and fifty livres for the other half of the year and so on in the future. I promise to increase this annuity soon... I am not so unreasonable as to ask for this all your time, but I shall ask you to give me occasionally some hours of your time to work on what I shall ask you - and also to *communicate to me your discoveries, with the request not to mention them to others...* Monsieur, tout á vous
>
> le M. de L'Hospital.
>
> (translation from French; emphasis added).

So, what is this Bernoulli–L'Hospital rule? In its simplest form it deals with *the limit of fractions* at points where *both the numerator and the denominator tend to* 0 (with "$\frac{0}{0}$ indeterminate forms"). We encountered such limits before, for instance

$$\lim_{x\to 0}\frac{\sin x}{x}$$

in Sect. 6.2,

$$\lim_{x\to x_0}\frac{x^2-x_0^2}{x-x_0},\qquad \lim_{x\to x_0}\frac{(1/x)-(1/x_0)}{x-x_0},\qquad \lim_{x\to x_0}\frac{\sin x-\sin x_0}{x-x_0}$$

in Sect. 6.3 and, also in Sect. 6.4,

$$\lim_{x\to 0}\frac{|x|}{x}$$

which turned out *not to exist* (though the left limit -1 and the right limit $+1$ *do* exist). All those limits *served* indirectly or directly *to determine derivatives or to show that the derivative does not exist.* We now turn this around and use derivatives to determine such limits. Of course we could not have used derivatives to determine those among the above limits, which do exist, because that would be circular reasoning (using the derivative, which we have not established yet, to determine the limit which gives the derivative).

We will formulate and prove the *Bernoulli–L'Hospital rule* in a somewhat stronger form: *Suppose that there exist a right neighbourhood of a on which f and g are differentiable and $g'(x)\neq 0$ and that the following limits exist*:

$$\lim_{x\to a+} f(x)=\lim_{x\to a+} g(x)=0 \quad\text{and}\quad \lim_{x\to a+}\frac{f'(x)}{g'(x)}=A \quad \text{(finite or infinite).}$$

Then also $\lim_{x\to a+}[f(x)/g(x)]$ *exists and equals* A:

$$\lim_{x\to a+}\frac{f(x)}{g(x)}=\lim_{x\to a+}\frac{f'(x)}{g'(x)}=A.$$

A similar rule (with similar proof) *holds also for left limits* $\lim_{x\to a-}$ *and so also for limits* $\lim_{x\to a}$ (throughout, in place of $\lim_{x\to a+}$). We did not exclude infinity as limit: $A=\infty$ is permissible and so is $A=-\infty$.

All conditions have to be carefully checked, in particular the existence of $\lim_{x\to a+}[f'(x)/g'(x)]$. It is even possible that $\lim_{x\to a+}[f'(x)/g'(x)]$ *does not exist* (neither finite nor infinite) *but* $\lim_{x\to a+}[f(x)/g(x)]$ *exists* (this does *not* contradict the above rule; why?):

Example 3 By Sects. 6.5 **1** and 6.2 Example **1**, the following limit *exists*:

$$\lim_{x\to 0}\frac{x+2x^2\sin(1/x)}{x}=\lim_{x\to 0}\left(1+2x\sin\frac{1}{x}\right)=1+\lim_{x\to 0}\left(2x\sin\frac{1}{x}\right)=1.$$

But

$$\lim_{x\to 0}\frac{[x+2x^2\sin(1/x)]'}{(x)'}=\lim_{x\to 0}\frac{1+4x\sin(1/x)-2\cos(1/x)}{1}=1-2\lim_{x\to 0}\cos\frac{1}{x}$$

(again by Sects. 6.5 **1** and 6.2 Example **1**) *does not exist*, because $\lim_{x\to 0}\cos\frac{1}{x}$ does not exist for the same reason as $\lim_{x\to 0}\sin\frac{1}{x}$ does not exist (compare Sect. 6.2 Example **1**).

Proof of the Bernoulli–L'Hospital rule As we have seen in the definition of (right) limits (Sect. 6.7) neither f nor g needs to be defined at a. But we can *define* them (or redefine if $f(a)$ or $g(a)$ were already defined otherwise) as follows:

$$f(a)=\lim_{x\to a+}f(x)=0,\qquad g(a)=\lim_{x\to a+}g(x)=0.$$

Then (compare Sect. 6.3) f and g will be *right continuous* at a. So the conditions of Cauchy's law of the mean hold in the right neighbourhood of a on which f and g are differentiable and $g\neq 0$, that is, there exists a ξ between a and x (x being in that right neighbourhood) such that

$$\frac{f(x)}{g(x)}=\frac{f(x)-f(a)}{g(x)-g(a)}=\frac{f'(\xi)}{g'(\xi)}.$$

As x tends to a, so does ξ which is between x and a. Therefore

$$\lim_{x\to a+}\frac{f(x)}{g(x)}=\lim_{\xi\to a+}\frac{f'(\xi)}{g'(\xi)}=\lim_{x\to a+}\frac{f'(x)}{g'(x)}$$

if the right hand limit exists (of course the limit in the middle is the same as the limit on the right hand side). This concludes the proof of the Bernoulli–L'Hospital rule.

Example 4

$$\lim_{x\to 0}\frac{1-\cos x}{x}=\lim_{x\to 0}\frac{\sin x}{1}=\sin 0=0$$

(continued)

since the sine is continuous at 0. Here we applied the Bernoulli–L'Hospital rule to calculate a limit rather than a right limit.

The rule can also be applied repeatedly; if you do not succeed first, try again:

Example 5

$$\lim_{x\to 0}\frac{1-\cos x}{x^2}=\lim_{x\to 0}\frac{\sin x}{2x}.$$

On the right hand side, numerator and denominator still tend to 0 (we still have a "$\frac{0}{0}$–form") if $x = 0$. So we do "it" again:

$$\lim_{x\to 0}\frac{1-\cos x}{x^2}=\lim_{x\to 0}\frac{\sin x}{2x}=\lim_{x\to 0}\frac{\cos x}{2}=\frac{1}{2},$$

since also the cosine is continuous at 0.

But we have to be careful not to overdo it; it can be continued only while each limit (except the last) is "$\frac{0}{0}$–form":

Example 6

$$\lim_{x\to 2}\frac{x^2-x-2}{x^2-2x}=\lim_{x\to 2}\frac{2x-1}{2x-2}\neq\lim_{x\to 2}\frac{2}{2}=1.$$

because $\lim_{x\to 2}\frac{2x-1}{2x-2}$ is no "$\frac{0}{0}$–form" anymore. Correctly:

$$\lim_{x\to 2}\frac{x^2-x-2}{x^2-2x}=\lim_{x\to 2}\frac{2x-1}{2x-2}=\frac{3}{2},$$

because $2x-1$ and $2x-2$ are continuous.

The Bernoulli–L'Hospital rule (or a consequence) can be applied also "$\frac{0}{0}$–forms" generated by limits at infinity: *if* $\lim_{x\to\infty} f(x) = \lim_{x\to\infty} g(x) = 0$ *and, for some* M, $g'(x)\neq 0$ *if* $x > M$, *furthermore* $\lim_{x\to\infty}\frac{f'(x)}{g'(x)} = A$ *exist, then*

$$\lim_{x\to\infty}\frac{f(x)}{g(x)}=\lim_{x\to\infty}\frac{f'(x)}{g'(x)}=A.$$

Indeed, write $t = 1/x$. Then in the following calculation (which uses also the chain rule Sect. 6.5 **4**) the second limit is of the previous form:

$$\lim_{x\to\infty} \frac{f(x)}{g(x)} = \lim_{t\to 0+} \frac{f(1/t)}{g(1/t)} = \lim_{t\to 0+} \frac{-f'(1/t)/t^2}{-g'(1/t)/t^2}$$

$$= \lim_{t\to 0+} \frac{f'(1/t)}{g'(1/t)} = \lim_{x\to\infty} \frac{f'(x)}{g'(x)}$$

if the limit on the right exists. Also other "indeterminate forms" like $\frac{\infty}{\infty}$ and $\infty - \infty$ can be reduced to the $\frac{0}{0}$–form and the Bernoulli–L'Hospital rule applies. We show only an example of the latter:

Example 7 (The second limit is already of the $\frac{0}{0}$–form. We apply the Bernoulli–L'Hospital rule twice.)

$$\lim_{x\to 0+} \left(\frac{1}{\sin x} - \frac{1}{x}\right) = \lim_{x\to 0+} \frac{x - \sin x}{x \sin x} = \lim_{x\to 0+} \frac{1 - \cos x}{\sin x + x\cos x}$$

$$= \lim_{x\to 0+} \frac{\sin x}{\cos x + \cos x - x \sin x} = \frac{0}{2} = 0.$$

6.7.1 Exercises

1. Determine the ξ in $f'(\xi) = \dfrac{f(b) - f(a)}{b - a}$ for the functions
   (a) $f :]a, b[\to \mathbb{R}, x \mapsto x^2$, where $a = -1, b = 3$,
   (b) $f :]a, b[\to \mathbb{R}, x \mapsto 1 - x^2 + x^4$, where $a = -3, b = 3$.
2. Determine the ξ in $\dfrac{f'(\xi)}{g'(\xi)} = \dfrac{f(b) - f(a)}{g(b) - g(a)}$ for the functions
   $f :]a, b[\to \mathbb{R}, x \mapsto 1 + x - x^2 + x^3/3$,
   $f :]a, b[\to \mathbb{R}, x \mapsto 1 + x^2/2$,
   where $a = 1, b = 3$.
3. Determine the Taylor series of $f : (\mathbb{R} - \{0\}) \to \mathbb{R}, x \mapsto \dfrac{1}{x}$, around $x = -1$. For which values of x does this series converge?
4. Calculate the following limits by applying the Bernoulli–L'Hospital rule.
   (a) $\lim_{x\to 0} \dfrac{1 - \cos\frac{x}{2}}{1 - \cos x}$, (b) $\lim_{x\to 0} \dfrac{2\tan x}{\tan(2x)}$,
   (c) $\lim_{x\to 0} \dfrac{\sin x - x\cos x}{x \sin x}$, (d) $\lim_{x\to\pi/2} (\pi - 2x)\tan x$.

5. (a) Why is the following application of the Bernoulli–L'Hospital rule wrong:

$$\lim_{x\to 1}\frac{x^3+x^2-x-1}{x^2-1}=\lim_{x\to 1}\frac{3x^2+2x-1}{2x}=\lim_{x\to 1}\frac{6x+2}{2}=4?$$

(b) Determine the true value of the limit.

6.7.2 Answers

1. (a) $\xi = 1$, (b) $\xi_1 = 0$, $\xi_2 = \sqrt{2}/2$, $\xi_3 = -\sqrt{2}/2$.
2. $\xi_1 = (4 + \sqrt{7})/3$, $\xi_2 = (4 - \sqrt{7})/3$.
3. $f(x) = \dfrac{1}{x} = -1 - (x+1) - (x+1)^2 - (x+1)^3 - \cdots$ converges if $-2 < x < 0$.
4. (a) $\dfrac{1}{4}$, (b) 1, (c) 0, (d) 2.
5. (a) $\lim_{x\to 1}\dfrac{3x^2+2x-1}{2x}$ is not of the form $\dfrac{0}{0}$ $(\lim_{x\to 1}(3x^2+2x-1) = 4 \neq 0)$,
   (b) $\lim_{x\to 1}\dfrac{3x^2+2x-1}{2x} = 2$.

6.8 Monotonicity, Local Maxima, Minima and Convexity of Differentiable Functions

We can use (6.11), the "Taylor formula with remainder in the Lagrange form" (6.11), to find conditions, sometimes necessary, sometimes sufficient, sometimes both, for a function to be monotonic, strictly monotonic (see Sect. 3.3), which we will do here, or to be convex or strictly convex (see Sect. 3.5), which we will do in Sect. 7.2. Since the derivative at a point (Sect. 6.4) is the slope of the tangent of the graph, it seems intuitive, that a function strictly increases on an interval, if the derivative is positive there. This is actually true, it follows from the $n = 0$ case of the Taylor formula, which is really just the *law of the mean*

$$f(x) = f(a) + f'(\xi)(x - a).$$

Indeed, *if* $x > a$ *and* $f'(\xi)$ *is positive for* $\xi \in]a, x[$ (meaning also that f' exists on that interval and f is continuous at a and at x; we will suppose here that f is differentiable on $[a, b]$; $x \leq b$) *then* $f(x) > f(a)$, that is, f *strictly increases* and does so as long as f' remains positive. By the same argument, *if* $f'(\xi) \geq 0$ *on an interval then* f *increases there in the wider sense*. Similar rules hold for strictly decreasing ($f'(\xi) < 0$) and decreasing in the wider sense ($f'(\xi) \leq 0$).

How about the *converse*? Does *the increasing of a differentiable function on an interval imply that the derivative is nonnegative* there? The answer is *yes*: If $f(x) \geq$

$f(x_0)$ for all $x > x_0$ on an interval then, by the definition of the derivative in Sect. 6.4,

$$f'(x_0) = \lim_{x \to x_0} \frac{f(x) - f(x_0)}{x - x_0} \geq 0.$$

(We took $x > x_0$ but the proof goes the same way if we approach x_0 by $x < x_0$.) Similarly, *if f is decreasing (in the broader sense) on an interval then $f' \leq 0$* there.

But, if f is *strictly decreasing* on an interval, does it follow that $f'(x) < 0$ everywhere on the interval, except may be at its ends (where it still should be continuous)? The answer is *no*.

Example $f(x) = -x^3$ strictly decreases on $[-1, 1]$, but

$$f'(0) = -3x^2|_{x=0} = 0.$$

On the other hand, $f'(x) = 0$ on a whole subinterval implies, as noted right after (6.11), that f is constant on that subinterval, so f cannot be strictly decreasing or strictly increasing on an interval where $f'(x) = 0$. Conversely, if f is monotonic but *not* strictly monotonic then there exist $x_1, x_2, x_1 < x_2$ such that $f(x_1) = f(x_2) =: c$ (say). Since f is monotonic (in the wider sense) this is possible only if $f(x) = c$ for all $x \in [x_1, x_2]$, that is on an interval of positive length (on a "proper interval").

Thus *a function f differentiable on the interior of an interval $I \subset \mathbb{R}$ and continuous on its ends is strictly increasing (respectively. strictly decreasing) if, and only if, $f'(x) \geq 0$ (respectively. $f'(x) \leq 0$) on the interior of I and there is no proper subinterval on which $f'(x) = 0$.*

As we know from Sect. 3.3, if the function f increases before x_0, say on $[x_0 - \delta_1, x_0]$, and decreases after x_0, on $[x_0, x_0 + \delta_2]$, then f has a local maximum at x_0. If f is differentiable on $[x_0 - \delta_1, x_0 + \delta_2]$ then $f'(x) \geq 0$ for $x \leq x_0$ and $f'(x) \leq 0$ for $x \geq x_0$, so $f'(x_0) = 0$ *at local maxima and*, by the same argument, *also at local minima*. However, the converse is *not* true: $f'(x_0) = 0$ is possible also if x_0 is neither a local maximum, nor a local minimum, as $f(x) = -x^3$ shows at $x = 0$ (Fig. 6.26). If also $f'(x)f'(\tilde{x}) > 0$ for $x < x_0 < \tilde{x}$ in a neighbourhood of x_0 then x_0 is called a *horizontal point of inflection*. (The word "horizontal" is often omitted, but we saw in Sect. 3.4 and will see in Sect. 7.4 also other kinds of "points of inflection":) So how can we decide whether at x_0 with $f'(x_0) = 0$ the function f has a local maximum, minimum or horizontal point of inflection? As we saw, right *before a local maximum x_0 we have $f'(x) \geq 0$*, right after it $f'(x) \leq 0$. So f' *decreases* (in the wider sense) *on a neighbourhood of a local maximum*. If f' is differentiable, that is, f'' *exists on* such *a neighbourhood of a local maximum, then $f''(x) \leq 0$ there*. The converse is clearly also true: *if $f'(x_0) = 0$ and $f''(x) \leq 0$ on a neighbourhood of x_0, then f has a local maximum at x_0*. But it is not enough

(continued)

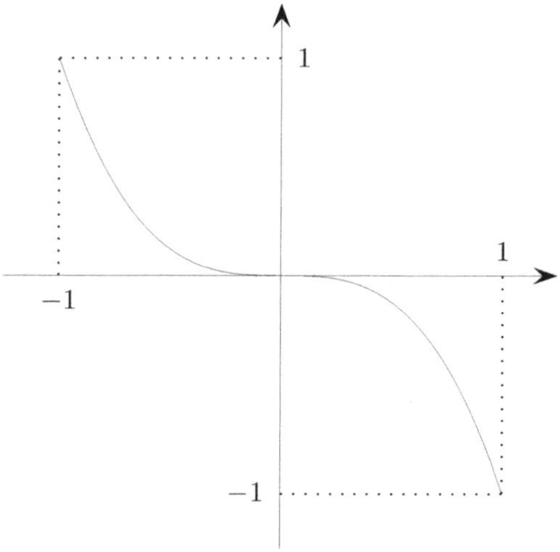

Fig. 6.26 $x \mapsto -x^3$ is strictly decreasing on $[-1, 1]$, but the derivative is 0 at $x = 0$

to have, in addition to $f'(x_0) = 0$ just $f''(x_0) \leq 0$ as again the example $f(x) = -x^3$ (Fig. 6.26) shows. *If, however, f' and f'' exist, f'' is continuous at x_0, $f'(x_0) = 0$ and $f''(x_0) < 0$* (not just ≤ 0) then *f has indeed a local maximum at x_0* (why?). But again this sufficient condition is not necessary: for $f(x) = 1 - x^4$, f' and f'' everywhere exist, are continuous, $f'(0) = f''(0) = 0$ but f has a maximum at 0 (show by calculation or drawing).

Similar statements hold for local minima (local maxima and minima are called collectively "local extrema"). We remind the reader, however (see Sect. 3.4), that a (global) maximum or minimum (global "extremum" for short) can be also on a closed end of an interval (Figs. 3.23 and 3.24). This gives the following test for extrema on an interval: *Find the points where $f'(x) = 0$ and check there $f''(x)$. If it is positive then there is a local minimum, if negative then a local maximum.* (This test is not decisive where $f'(x_0) = f''(x_0) = 0$; one can prove that when the first nonzero derivative is of even order then we have a local extremum and if it is of odd order, then a horizontal point of inflection.) In order to determine the global maximum (minimum) on the interval, *calculate the local maximum (minimum) values* of the function on that interval *and the function values at the closed ends of the interval* (if any). Now *the largest (smallest) among* all *these will be the* global maximum (minimum) value *on that interval and the point or points where it is assumed the global maximum (minimum) point or points.* (We have seen in Sect. 6.3, property 2 that on a *closed bounded* interval every continuous function assumes both its maximum and its minimum.)

Example Determine the global and local maxima, minima and the horizontal points of inflection of the function given by

$$f(x) = 0.15x^5 - 0.25x^3 + 0.1$$

on $[-2, 2]$. We differentiate:

$$f'(x) = 0.75x^4 - 0.75x^2 = 0.75x^2(x^2 - 1) = 0.75x^2(x - 1)(x + 1)$$

so $f'(x) = 0$ at -1, 0 and 1. We differentiate again:

$$f''(x) = 3x^3 - 1.5x, \quad f''(-1) = -1.5 < 0, \quad f''(0) = 0, \quad f''(1) = 1.5$$

so -1 is a local maximum point, 1 a local minimum point. This test is indecisive concerning $x = 0$ but $f'''(0) = -1.5 \neq 0$ so 0 is a horizontal point of inflection. The local maximum and minimum values are

$$f(-1) = 0.2 \quad \text{and} \quad f(1) = 0,$$

respectively. But the function values at the end of the closed interval $[-2, 2]$ are

$$f(-2) = -2.7 \quad \text{and} \quad f(2) = 2.9$$

so the global maximum value on $[-2, 2]$ is 2.9 (and not 0.2) and the global minimum value is -2.7 (not 0), attained at 2 and -2, respectively (Fig. 6.27).

(continued)

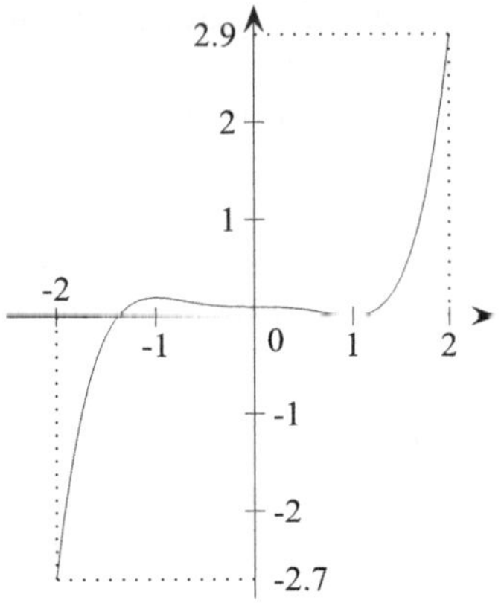

Fig. 6.27 Global and local extrema and horizontal point of inflection of $x \mapsto 0.15x^5 - 0.25x^3 + 0.1$ on $[-2, 2]$

Note that it is also possible that $f'(a) = 0$ but there is neither a local extremum nor a point of inflection at a. For instance:

$$f(x) = \begin{cases} x^4 \cos \frac{1}{x} & (x \neq 0) \\ 0 & (x = 0) \end{cases},$$

$$f'(x) = \begin{cases} 4x^3 \cos \frac{1}{x} + x^2 \sin \frac{1}{x} & (x \neq 0) \\ 0 & (x = 0) \end{cases},$$

$$f''(x) = \begin{cases} (12x^2 - 1) \cos \frac{1}{x} + 6x \sin \frac{1}{x} & (x \neq 0) \\ 0 & (x = 0) \end{cases}$$

(check!) but $f'''(0)$ is *not defined*:

$$\frac{f''(x) - f''(0)}{x} = (12x - \frac{1}{x}) \cos \frac{1}{x} + 6 \sin \frac{1}{x}$$

has *no limit* as $x \to 0$.

In Sect. 8.3 we will obtain from the $n = 2$ case of the Taylor formula (6.11) similar conditions for convexity (from above or from below) as we did for monotonicity. Actually, compare Figs. 6.11, 6.14, and 6.27, in the neighbourhood of a local minimum (maximum) the function is convex from below (from above).

6.8.1 Exercises

1. (a) Draw the graph of the function $f : [-1, 2] \to \mathbb{R}, x \mapsto x^3 - x^2$. Determine
   (b) the local extrema (minima, maxima) of f in the interior of $[-1, 2]$,
   (c) the global extrema of f,
   (d) its points of inflection.
2. (a) Draw the graph of the function $f : \mathbb{R} \to \mathbb{R}, x \mapsto -2x^3 + 9x^2 - 12x - 6$ for $-2 \leq x \leq 3$.
   (b) Where are the local minima of f?
   (c) Where are the local maxima of f?
   (d) Where are the points of inflection of f?
3. Present a function $f : [-3, 3] \to \mathbb{R}$ which is strictly increasing, four times differentiable and satisfies $f'(1) = f''(1) = f'''(1) = f^{(IV)}(1) = 0$.
4. Present a continuous function $f :]-3, 3[\to \mathbb{R}$ which is decreasing and differentiable everywhere on $]-3, 3[$ up to the points $-2, -1, 0, 1, 2$, but not strictly decreasing in $[0, 1]$.

5. Is the function $f :]0,3[\to \mathbb{R}$ given by

$$f(x) = \begin{cases} \sqrt{x} & \text{for } x \in]0,1] \\ x^2 & \text{for } x \in [1,2] \\ -4+4x & \text{for } x \in [2,3[\end{cases}$$

continuous, strictly increasing and differentiable?

6.8.2 Answers

1. (b) Local minimum at $x = 2/3, f(2/3) = -4/27$, local maximum at $x = 0$, $f(0) = 0$,
   (c) global minimum at $x = -1, f(-1) = -2$, global maximum at $x = 2, f(2) = 4$,
   (d) point of inflection $x = 1/3, f(1/3) = -2/27$.
2. (b) Local minimum at $x = 1, f(1) = -11$,
   (c) local maximum at $x = 2, f(2) = -10$,
   (d) point of inflection $x = 3/2, f(3/2) = 21/2$.
3. Take, for instance, $f(x) = (x-1)^5$.
4. Take, for instance,

$$f(x) = \begin{cases} -2x & \text{for} & x \in]-3,-2] \\ 2-x & \text{for} & x \in [-2,-1] \\ -3x & \text{for} & x \in [-1,\ 0] \\ 0 & \text{for} & x \in [0,1] \\ 1-x & \text{for} & x \in [1,2] \\ 7-4x & \text{for} & x \in [2,3[. \end{cases}$$

5. The function is continuous and strictly increasing. It is differentiable at any point different from $x = 1$. The left limit of the difference quotient at $x = 1$ is $\lim\limits_{\substack{x\to 1\\ x<1}} \dfrac{\sqrt{x}-\sqrt{1}}{x-1} = \dfrac{1}{2}$, but the right $\lim\limits_{\substack{x\to 1\\ x>1}} \dfrac{x^2-1^2}{x-1} = 2$.

6.9 "Cobweb" Situations in Economics: Points of Intersection of Graphs and Zeros of Functions

We apply now limit, continuity, differentiability and the "law of the mean" to important situations in economics, laying at the same time the groundwork to important algorithms for determining zeros of functions.

If the price of a product at a certain time is relatively *high*, that is an incentive for the producers to *increase the production*. If that can be done, it may raise new problems. By the time the production will have increased, the demand may already be lower but, even if this is not the case, the increased quantity of supply may still push down the price (the original height of which was the incentive for increasing the production) according to the "*law of supply and demand*". The lower price may then lead to lower production which may eventually lead to higher prices again, and so on. Popular examples come from agriculture; one speaks, in particular, about "pork cycles".

In the mathematical *model* we make the following *assumptions*.

(A1) The price p of the product generates a certain quantity y of demand, that is, there exists a *demand function* $f : \mathbb{R}_+ \to \mathbb{R}_+$ such that $y = f(p)$.

(A2) On the other hand the price p determines also the quantity $\bar{y}$ of supply, that is, there exists also a *supply function* $g : \mathbb{R}_+ \to \mathbb{R}_+$ such that $\bar{y} = g(p)$.

(A3) The demand function f is decreasing, the supply function g is increasing; the graphs of these two functions, that is, the demand curve D and the supply curve S *intersect in exactly one point* (p^*, y^*).

Note: It is clear from the above that it is pretty natural to suppose that f decreases and that g increases. *Strict* decreasing and increasing would make the uniqueness of the point of intersection more automatic but are somewhat less natural and rarely occur in practice. Nevertheless, as Fig. 6.28 shows, not-strictly monotonic functions S and D can also produce a single point of intersection. This point (p^*, y^*), which according to (A3) exists and is unique, is called *equilibrium point*, y^* the *equilibrium quantity*, p^* the *equilibrium price* or *market price*. We point out here that in many economic textbooks our price axis (see Figs. 6.28, 6.29, 6.30, and 6.31) is the quantity axis and, accordingly, our quantity axis in the price axis.

At the *market price* p^* the quantities of supply and demand on the market are equal. It is, of course, not fixed in advance but determined (more or less) by the "market forces". Here we are modelling this "market process" according

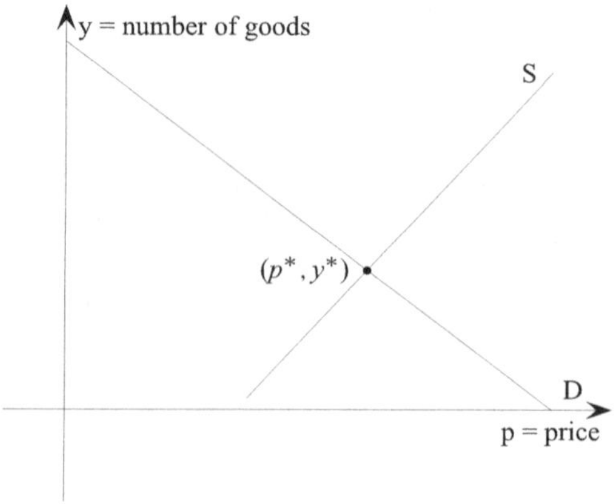

Fig. 6.28 Supply curve S demand curve D, and equilibrium point (p^*, y^*)

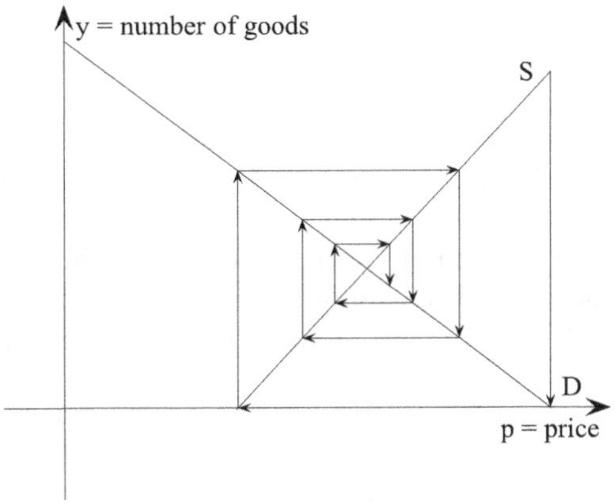

Fig. 6.29 Successive price–quantity points may approach the equilibrium point in a shape reminding of a "cobweb"

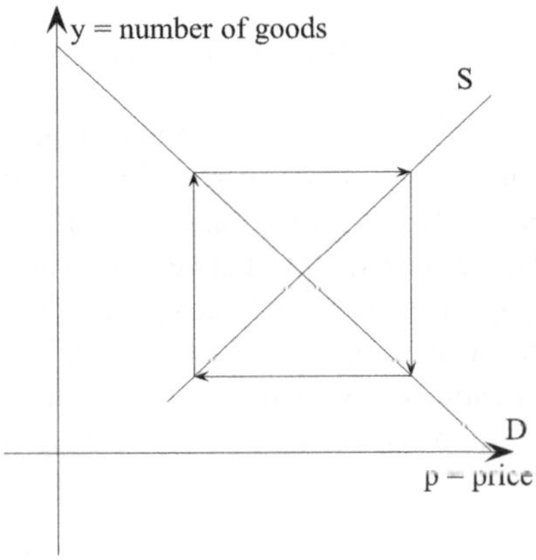

Fig. 6.30 Both $\{p_n\}$ and $\{y_n\}$ oscillate between two fixed values

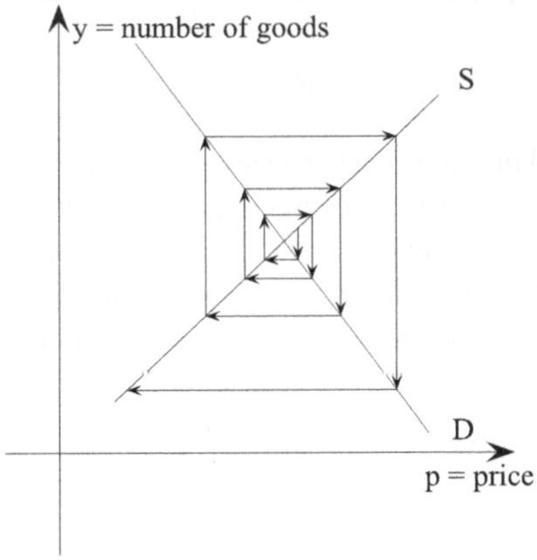

Fig. 6.31 Both $\{p_n\}$ and $\{y_n\}$ "explode"

to our assumptions (A1), (A2), (A3). While one sees that time has a role in the process described above, we will not explicitly refer to it and we will also ignore, the influence of warehousing on the price development. We start with a price p_0. According to (A1) this creates demand in quantity $y_0 = f(p_0)$ while, according to (A2), it generates a (presumably larger) quantity $y_1 = g(p_0)$ of supply. This quantity can be sold only at the (lower) price p_1 (Fig. 6.29), so $y_1 = f(p_1)$. At this price the quantity of supply changes (diminishes) to $y_2 = g(p_1)$ causing a change (increase) of price from p_1 to p_2: $y_2 = f(p_2)$ and so it goes.

As we have just seen, the two sequences $\{p_n\}$ and $\{y_n\}$ are defined by

$$y_n = f(p_n),\ y_{n+1} = g(p_n) \qquad (n = 0, 1, 2, \ldots). \tag{6.13}$$

Actually, if f is strictly monotonic and continuous, we can apply inverse functions and define $\{y_n\}$ and $\{p_n\}$ by

$$p_n = f^{-1}(y_n),\ y_{n+1} = g(f^{-1}(y_n)) \qquad (n = 0, 1, 2, \ldots). \tag{6.14}$$

If we are lucky, these sequences *converge* to the equilibrium price p^* and the equilibrium quantity y^*, respectively (Fig. 6.29). However, as Figs. 6.30 and 6.31 show, for certain f and g and initial points (p_0, y_0), the sequences $\{p_n\}$ and $\{y_n\}$ may oscillate between two points each or even $|p_n - p_{n-1}|$ and $|y_n - y_{n-1}|$ may, instead of decreasing, increase beyond any bound ("exploding" sequences). Anyway, Figs. 6.29, 6.30, and 6.31 remind economists of cobwebs therefore these phenomena are called "*cobweb situations*" or even "*cobweb theorems*".

In the case where $\{p_n\}$ and $\{y_n\}$ converge and (Fig. 6.29)

$$p^* = \lim_{n\to\infty} p_n, \qquad y^* = \lim_{n\to\infty} y_n,$$

if f and g are continuous at p^* then, from (6.13),

$$f(p^*) = y^* = g(p^*), \qquad \text{that is,} \qquad f(p^*) - g(p^*) = 0.$$

Thus the market price p^* is a point where the value of the function $f - g$ (that is, $p \mapsto f(p) - g(p)$) is zero or, for short, p^* is a *zero* of the function $f - g$. Here we determined the zero of this function by the *iteration process* (6.14). We now regard this in general, not only for the above application.

Not only for monotonic functions but in general, *determining points where two functions f and g are equal (their graphs intersect) is* clearly *equivalent to determining the zeros of $f - g$*:

$$f(p^*) = g(p^*) \iff f(p^*) - g(p^*) = 0.$$

The function g may even be a constant c and then *the function f assumes the value c exactly where $p \mapsto f(p) - c$ is zero*. So, determining zeros of functions is important. As above, it is often done by iteration processes.

In search of the zeros of a function $h : I \to \mathbb{R}$, where I is an *open interval*, the following *iteration process* is often useful: *Choose $x_0 \in I$ arbitrarily, then define*

$$\begin{aligned} x_1 &= x_0 - h(x_0), \\ x_2 &= x_1 - h(x_1), \\ &\vdots \\ x_{n+1} &= x_n - h(x_n) \qquad (n = 0, 1, 2, \ldots). \end{aligned} \tag{6.15}$$

If this sequence converges to x^* and h is continuous at x^* then clearly (taking $n \to \infty$ in the last equation)

$$x^* = x^* - h(x^*), \qquad \text{that is} \qquad h(x^*) = 0$$

so that x^* *is a zero of h.*

We give now *conditions which are sufficient for the sequence $\{x_n\}$*, defined by (6.15), *to converge* whenever we start from an x_0 in an open subinterval I^* of I ($I^* \subset I$). We define a new function $F : I \to \mathbb{R}$ by

$$F(x) = x - h(x).$$

Then (6.15) can be written as

$$x_{n+1} = F(x_n) \qquad (n = 0, 1, 2, \ldots)$$

(compare to the second equation in (6.14)). If *(a) with every $x \in I^*$, also $F(x) \in I^*$ and (b) there exists a (nonnegative) constant $c < 1$ such that*

$$|F(x) - F(y)| \leq c\,|x - y| \qquad \textit{for all} \qquad x, y \in I^*$$

then the sequence $\{x_n\}$, defined by $x_{n+1} = F(x_n)$ starting with an $x_0 \in I^*$, converges (the assumption *(b)* is called a *Lipschitz condition*; Rudolf Lipschitz (1832–1903)). Indeed, first of all, by condition *(a)*, with $x_0 \in I^*$ also $x_1 = F(x_0) \in I^*$ then $x_2 = F(x_1) \in I^*$, and so on, $x_n \in I^*$ for all $n \in \mathbb{N}$. Further, by the Lipschitz condition *(b)*,

$$|x_2 - x_1| = |F(x_1) - F(x_0)| < c\,|x_1 - x_0|\,,$$

$$|x_3 - x_2| < c\,|x_2 - x_1| < c^2\,|x_1 - x_0|$$

and, in general,

$$|x_{n+1} - x_n| < c\,|x_n - x_{n-1}| < \ldots < c^n\,|x_1 - x_0|\,.$$

So, the distance between x_{n+1} and x_n decreases as n increases and

$$\lim_{n\to\infty} |x_{n+1} - x_n| = 0$$

because $\lim_{n\to\infty} c^n = 0$ if $0 \leq c < 1$ (as will be shown in Sect. 7.2 but was used already in Sect. 6.7, Example **2**), that is, the distance between x_{n+1} and x_n *decreases to* 0 as n goes to ∞. So $\{x_n\}$ indeed converges ("squeeze rule", compare Sect. 6.2, proof of (3)).

Let the limit of $\{x_n\}$ be x^*:

$$x^* = \lim_{n\to\infty} x_n.$$

From the Lipschitz condition *(b)*, F *is continuous at every point* $y \in I^*$. Indeed, either $c = 0$, in which case F is constant and thus continuous on I^*, or, by *(b)*,

$$\text{if} \quad |x - y| < \delta \quad \text{then} \quad |F(x) - F(y)| < \varepsilon = c\delta,$$

so for every $\varepsilon > 0$ there exists a $\delta = \varepsilon/c$ such that

$$|F(x) - F(y)| < \varepsilon \quad \text{whenever} \quad |x - y| < \delta$$

which, as we have seen in Sect. 6.3, exactly means that F is continuous at y. This fact with

$$x_{n+1} = F(x_n) \quad \text{and} \quad x^* = \lim_{n\to\infty} x_n$$

means again that

$$x^* = F(x^*).$$

This equation is described by saying that x^* is a *fixed point* of F.

By our definition, $F(x) = x - h(x)$, so

$$x^* = x^* - h(x^*), \quad \text{that is,} \quad h(x^*) = 0$$

and x^* is indeed a zero of h. For h the condition *(b)* translates into

$$|x - y - (h(x) - h(y))| \leq c\,|x - y| \quad (0 \leq c < 1) \quad \text{for} \quad x, y \in I^*.$$

If F is differentiable on I^* and the absolute value of the derivative is not greater than a constant $c < 1$ on I^* then the Lipschitz condition (b) is satisfied (so the Lipschitz condition is something between continuity and differentiability). Indeed, by the law of the mean (Sect. 6.7) for a differentiable F there exists a ξ between x and y such that $(F(x) - F(y))/(x - y) = F'(\xi)$. If $|F'(x)| \leq c$ on I^*, then

$$\left|\frac{F(x) - F(y)}{x - y}\right| = \left|F'(\xi)\right| \leq c \quad \text{so} \quad |F(x) - F(y)| \leq c\,|x - y|,$$

as asserted. With $F(x) = x - h(x)$, the condition $|F'(x)| \leq c < 1$ translates into

$$\left|1 - h'(x)\right| \leq c < 1, \text{ that is } -c \leq h'(x) \leq 1 + c \quad (0 \leq c < 1) \quad \text{for} \quad x \in I^*$$

or, what is the same, $1 - c \leq h'(x) \leq 1 + c$ $(0 \leq c < 1)$ *on* I^* *and* (by *(a)*), $x \in I^* \Rightarrow x - h(x) \in I^*$. *These conditions are sufficient for the sequence* $\{x_n\}$, *defined by*

$$x_0 \in I^*, \qquad x_{n+1} = x_n - h(x_n) \quad (n = 0, 1, 2, \ldots),$$

to converge to a zero of h.

6.9.1 Exercises

1. Let p and y be the price and the quantity of a good, respectively. On a market let $f : \mathbb{R}_+ \to \mathbb{R}_+, p \mapsto \alpha/p$ be the demand function and $g : \mathbb{R}_+ \to \mathbb{R}_+, p \mapsto bp^2$ the supply function. Determine the parameters α and b so that equilibrium point (p^*, y^*) in the market is
(a) $(1, 100)$, (b) $(2, 20)$, (c) $(3, 10)$.
2. Determine the zeros of the functions
(a) $h_1 : \mathbb{R} \to \mathbb{R}, x \mapsto x^2 - \frac{3}{2}x + \frac{1}{2}$,
(b) $h_2 : \mathbb{R}_+ \to \mathbb{R}, x \mapsto \sqrt{x} - 2x + 1$,
(c) $h_3 : \mathbb{R} \to \mathbb{R}, x \mapsto x^3 - 2x^2 - x + 2$.
3. With h_1 from Exercise 2 start the integration process $x_{n+1} = x_n - h_1(x_n)$ $(n = 0, 1, 2, \ldots)$
(a) with $x_0 = 11/10$ and determine x_1, x_2 and x_3,
(b) with $x_0 = 1/4$ and determine x_1, x_2 and x_3.
4. Why does the iteration process in Exercise 3 converge in case *(a)*?
5. Why does the iteration process in Exercise 3 not converge in case *(b)*?

6.9.2 Answers

1. (a) $\alpha = 100, b = 100,$
   (b) $\alpha = 40, b = 5,$
   (c) $\alpha = 30, b = 10/9.$
2. (a) $\frac{1}{2}, 1,$
   (b) $\frac{5}{2} + \frac{\sqrt{29}}{2}, \frac{5}{2} - \frac{\sqrt{29}}{2},$
   (c) $-1, 1, 2.$
3. (a) 1.04, 1.0184, 1.0088,
   (b) 0.0625, −0.347656, −1.490005.
4. Because $F'(x) = (x - h_1(x))' = \left(x - x^3 + \frac{3}{2}x - \frac{1}{2}\right)' = -2x + \frac{5}{2} < 1$ for $x = x_0 = 11/10$.
5. Because $F'(x) = (x - h_1(x))' = -2x + \dfrac{5}{2} > 1$ for $x = x_0 = 1/4$.

6.10 Newton's Algorithm: Differentials (Linear Approximation)

A particularly popular algorithm for determination of zeros of functions is the*Newton algorithm* (Isaac Newton (1643–1727)), because it often *converges fast*.

Here we take, in (6.15), for a function f, differentiable with continuous nonzero derivative on I^*,

$$h(x) := \frac{f(x)}{f'(x)}.$$

So the algorithm defines the sequence $\{x_n\}$ by

$$x_{n+1} = x_n - \frac{f(x_n)}{f'(x_n)} \qquad (n = 0, 1, 2, \ldots). \tag{6.16}$$

If $\{x_n\}$ converges to x^* then accordingly

$$x^* = x^* - \frac{f(x^*)}{f'(x^*)}, \qquad \text{that is,} \qquad f(x^*) = 0,$$

so that in this case x^* is a zero of f.

The geometric meaning of (6.16) is shown in Fig. 6.32: x_{n+1} is the point of intersection of the tangent of the graph of f at x_n and of the X_1–axis. Indeed, since $f'(x_n)$ is the slope $\tan\alpha$, where α is the angle between the tangent and the X_1-axis,

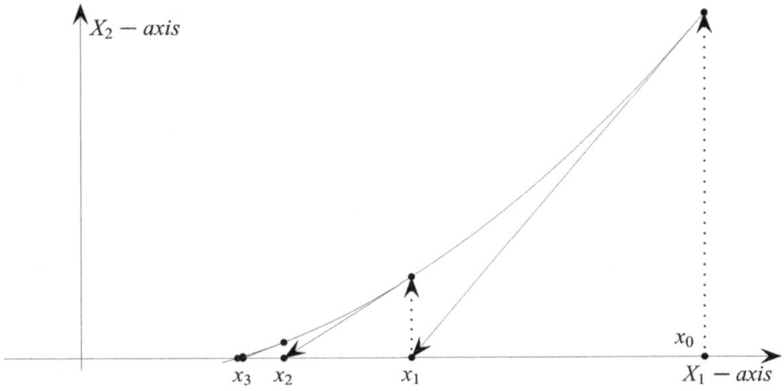

Fig. 6.32 The Newton algorithm

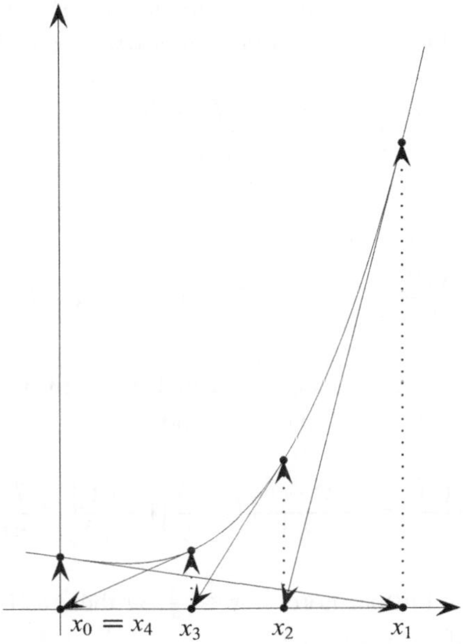

Fig. 6.33 Newton algorithm oscillates between two points

therefore

$$f'(x_n) = \tan\alpha = \frac{f(x_n)}{x_n - x_{n+1}}$$

which is equivalent to (6.16).

But, just as in the "cobweb situation" in the previous section, it is possible that the sequence $\{x_n\}$ of the Newton algorithm "oscillates" between two points (Fig. 6.33) or "explodes" (Fig. 6.34).

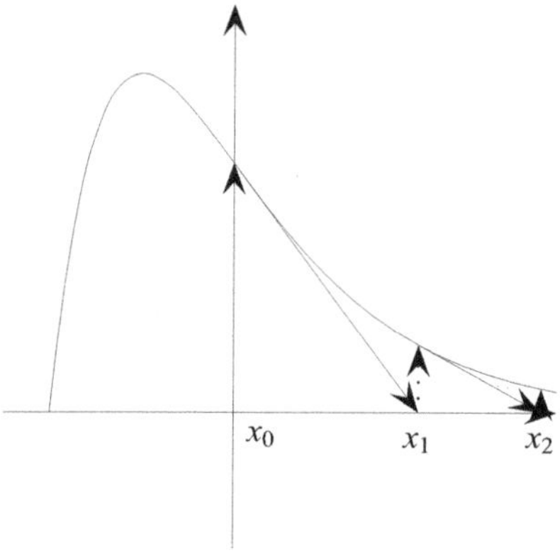

Fig. 6.34 Newton algorithm "ex- plodes"

We get from the last statement of Sect. 6.9 *sufficient conditions for the convergence* of $\{x_n\}$: $x \in I^* \Rightarrow x - f(x)/f'(x) \in I^*$ *and there exists a* $c \in [0, 1[$ *such that*

$$1 - c \leq h'(x) = \left(\frac{f(x)}{f'(x)}\right)' = \frac{f'(x)^2 - f(x)f''(x)}{f'(x)^2} = 1 - \frac{f(x)f''(x)}{f'(x)^2} \leq 1 + c$$

that is,

$$\left|\frac{f(x)f''(x)}{f'(x)^2}\right| \leq c < 1 \quad \text{for all} \quad x \in I^*.$$

An example, where this condition is satisfied, is given by $f(x) = x^3 - 1$, $I^* =]\frac{3}{4}, 3[$. Indeed then $f'(x) = 3x^2, f''(x) = 6x$, and

$$\left|\frac{f(x)f''(x)}{f'(x)^2}\right| = \left|\frac{6x^4 - 6x}{9x^4}\right| = \frac{6}{9}\left|1 - \frac{1}{x^3}\right| \leq \frac{74}{81} < 1$$

so $c = \frac{74}{81} < 1$ will do. Furthermore, if $x \in]\frac{3}{4}, 3[$ then, as one checks easily (for instance on the graph of $\frac{2}{3}x + \frac{1}{3}x^{-2}$),

$$x - \frac{f(x)}{f'(x)} = x - \frac{x^3 - 1}{3x^2} = \frac{2x^3 + 1}{3x^2} \in \left[1, \frac{55}{27}\right[\subset \left]\frac{1}{2}, 3\right[,$$

as required. This, of course, is a test case since it is obvious that $x^* = 1$ is the only zero of $f(x) = x^3 - 1$. Let us see how well and fast the Newton algorithm approaches it, starting, say, with $x_0 = 2$:

$$x_1 = x_0 - \frac{x_0^3 - 1}{3x_0^2} = 2 - \frac{8 - 1}{12} = 1.41666666\ldots,$$

$$x_2 = x_1 - \frac{x_1^3 - 1}{3x_1^2} = 1.11053374\ldots,$$

$$x_3 = 1.01063664\ldots,$$

$$x_4 = 1.00011155\ldots,$$

$$x_5 = 1.0000001\ldots$$

Pretty good!

With the Newton algorithm we have already, in a sense, approximated the function f by a "linear function" (really by a sequence of affine functions): we approximated the zero x^* of f by x_{n+1}, the zero of the affine function whose graph is the tangent of the graph of f at x_n. The idea of the differential rests on a somewhat similar approximation of f by an affine function, represented by the tangent of the graph of f.

As we saw in Sect. 6.4, the derivative at x_0 of a real valued function defined on a neighbourhood of x_0 is given by

$$L = f'(x_0) = \lim_{x \to x_0} \frac{f(x) - f(x_0)}{x - x_0}, \tag{6.17}$$

if this limit exists. Since the limit of the constant L is L, and the limit of a difference is the difference of limits, we can write this also as

$$\lim_{x \to x_0} \left(\frac{f(x) - f(x_0)}{x - x_0} - L \right) = 0 \tag{6.18}$$

(compare also the proof of the Theorem in Sect. 6.4).

Now we introduce a *linear function* ℓ by

$$\ell(t) = Lt$$

and an *affine function* ℓ^* by

$$\ell^*(x) = f(x_0) + \ell(x - x_0).$$

Equation (6.17) shows that *the slope* L *of* (the graph of) ℓ^* *equals that of* (the graph of) f *at* x_0, while (6.18) can be written as

$$\lim_{x \to x_0} \frac{f(x) - f(x_0) - \ell(x - x_0)}{x - x_0} = 0.$$

This equation expresses that (Fig. 6.35) f *can be approximated at* x_0 *by the affine function* ℓ^*. One often gives meaning to the "differential", mentioned in Sect. 6.5 **4**,

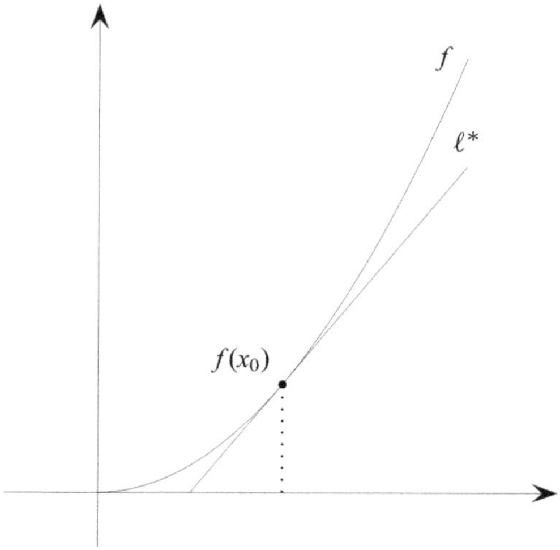

Fig. 6.35 Approximation of f at $(x_0, f(x_0))$ by the affine function ℓ^*

by defining

$$df(x_0) = \ell(x - x_0) = \ell(x) - \ell(x_0)$$

(the last equality being true because ℓ is linear). Usually one writes, in particular if x_0 is fixed,

$$dy = df$$

for the affine function $df(x_0)$ given by

$$df(x_0) = \ell(x) - \ell(x_0),$$

but a consequent definition of the differential $dy = df$ is that it is the following *set* of affine functions:

$$df = \{x \mapsto \ell(x) - \ell(x_0) \mid x_0 \in \text{domain of} f\}.$$

Of course, for the "identity" function $f(x) = x$ we have

$$\ell(x) = x, \qquad \text{because} \qquad \lim_{x \to x_0} \frac{(x - x_0) - (x - x_0)}{x - x_0} = 0,$$

so $dx = x - x_0$ and, since $dy = \ell(x) - \ell(x_0) = Lx - Lx_0$,

$$\frac{dy}{dx} = L = f'(x_0),$$

in accordance with our notation in 6.5 **4**. Another interpretation of differentials is given in *nonstandard analysis*. We do not go into that here.

6.10.1 Exercises

1. Apply the Newton algorithm to determine the zeros of the function $x \mapsto x^2 - \frac{3}{2}x + \frac{1}{2}$ mapping $\mathbb{R}$ into $\mathbb{R}$. Start the iteration

$$x_{n+1} = x_n - f(x_n)/f'(x_n) \qquad (n = 0, 1, 2, \ldots)$$

   (a) with $x_0 = 2$ and determine x_1, x_2, x_3, x_4 and x_5,
   (b) with $x_0 = 1/10$ and determine x_1, x_2, x_3 and x_4.
   (c) Compare your result with the exact solutions of the equation $x^2 - \frac{3}{2}x + \frac{1}{2} = 0$.
2. Why does the iteration process in Exercise 1, case (a), converge to one of the zeros determined in Exercise 1 (c) but not to the other?
3. Why does the iteration process in Exercise 1, case (b), converge to one of the zeros determined in Exercise 1 (c) but not to the other?
4. (a) Draw the graph of the function

$$g : \mathbb{R} \longrightarrow \mathbb{R}, x \longmapsto \begin{cases} \sqrt{x} & \text{for } x \geq 0 \\ -\sqrt{|x|} & \text{for } x \leq 0 \end{cases}.$$

   (b) Show that the Newton algorithm oscillates between x_0 and $-x_0$ for each initial point $x_0 \neq 0$.
5. (a) Draw the graph of the function

$$h : \mathbb{R} \longrightarrow \mathbb{R}, x \longmapsto \begin{cases} x^{1/3} & \text{for } x \geq 0 \\ -|x|^{1/3} & \text{for } x \leq 0 \end{cases}.$$

   (b) Show that the Newton algorithm "explodes" for each initial point $x_0 \neq 0$.

6.10.2 Answers

1. (a) 1.4, 1.12307692…, 1.02030135…,
   1.00076238…, 1.00000115…,
   (b) 0.376923076…, 0.479698652…,
   0.499237603…, 0.499998841….
2. Convergence to $x = 1$, since

$$\left|\frac{f(x)f''(x)}{f'(x)^2}\right| = \left|\frac{2x^2 - 3x + 1}{4x^2 - 6x + \frac{9}{4}}\right| \leq \frac{1}{2} < 1 \quad \text{for all} \quad x \in]0.95, \infty[,$$

but not to $x = \frac{1}{2}$, since in the iteration process

$$x_{n+1} = x_n - \frac{f(x_n)}{f'(x_n)} = x_n - \frac{x_n^2 - \frac{3}{2}x_n + \frac{1}{2}}{2x_n - \frac{3}{2}}$$

$$= \frac{x_n^2 - \frac{1}{2}}{2x_n - \frac{3}{2}} \geq 1 \quad \text{for all} \quad x_n \geq 1 \ (n = 0, 1, 2, \ldots).$$

3. Convergence to $x = \frac{1}{2}$, since

$$\left|\frac{f(x)f''(x)}{f'(x)^2}\right| = \left|\frac{2x^2 - 3x + 1}{4x^2 - 6x + \frac{9}{4}}\right| \leq \frac{1}{2} < 1 \quad \text{for all} \quad x \in]-\infty, 0.55[,$$

but not to $x = 1$, since in the iteration process

$$x_{n+1} = x_n - \frac{f(x_n)}{f'(x_n)} = x_n - \frac{x_n^2 - \frac{3}{2}x_n + \frac{1}{2}}{2x_n - \frac{3}{2}}$$

$$= \frac{x_n^2 - \frac{1}{2}}{2x_n - \frac{3}{2}} \leq \frac{1}{2} \quad \text{for all} \quad x_n \leq \frac{1}{2} \ (n = 0, 1, 2, \ldots).$$

4. (b) Newton algorithm for g:

$$x_{n+1} = x_n - \frac{g(x_n)}{g'(x_n)} = \begin{cases} x_n - \dfrac{\sqrt{x_n}}{\frac{1}{2}x_n^{-\frac{1}{2}}} = x_n - 2x_n & \text{for } x_n \geq 0 \\ x_n - \dfrac{-\sqrt{|x_n|}}{\frac{1}{2}|x_n|^{-\frac{1}{2}}} = x_n + 2\,|x_n| & \text{for } x_n \leq 0, \end{cases}$$

with $x_1 = -x_0$ for $x_0 \geq 0$ or $x_1 = |x_0|$ for $x_0 \leq 0$.

5. (b) Newton algorithm for h:

$$x_{n+1} = x_n - \frac{h(x_n)}{h'(x_n)} = \begin{cases} x_n - \dfrac{x_n^{\frac{1}{3}}}{\frac{1}{3}x_n^{-\frac{2}{3}}} = x_n - 3x_n & \text{for } x_n \geq 0 \\ x_n - \dfrac{-|x_n|^{\frac{1}{3}}}{\frac{1}{3}|x_n|^{-\frac{2}{3}}} = x_n + 3\,|x_n| & \text{for } x_n \leq 0, \end{cases}$$

with $x_1 = -2x_0$ for $x_0 \geq 0$ or $x_1 = 2\,|x_0|$ for $x_0 \leq 0$.

6.11 Linear Approximation: Differentials and Derivatives of Vector-Vector Functions—Partial Derivatives of Higher Orders

We introduced in the previous section linear approximation and differentials of real-valued functions of a real variable in a way that can be generalised right away to vector valued functions of a vector variable (vector-vector functions, compare Sect. 4.8). While the variables are in $\mathbb{R}^n$ and the function value in $\mathbb{R}^m$ in our examples and figures, we will often take $n = 2$, $m = 1$ because, as mentioned in Sect. 3.2, these can still be presented in our three-dimensional space (see Fig. 3.25). Here we will need neighbourhoods in n-dimensional spaces. We touched on them at the end of Sect. 6.2 but choose another definition and relate it to the particular case of one-dimensional neighbourhoods as defined in Sect. 6.2. In our Fig. 6.36 we choose again $n = 2$.

As we saw in Sect. 6.2, an ε-neighbourhood ($\varepsilon > 0$) of a finite point $\mathbf{p}$ (in the one-dimensional space, that is, $\mathbf{p}$ is a real number) is the set of points

$$N_\varepsilon(p) = \{\mathbf{x} \mid |\mathbf{x} - \mathbf{p}| < \varepsilon\}.$$

So we first have to define the analogues of absolute values (or distances, since $|\mathbf{x} - \mathbf{p}|$ is the distance between the points $\mathbf{x}$ and $\mathbf{p}$) in n-dimensional (in particular 2-dimensional) spaces. As we saw in Sects. 1.4, 1.6, and 3.2 this can be done by the (Euclidcan) norm

$$|\mathbf{x}| := (x_1^2 + x_2^2 + \ldots + x_n^2)^{\frac{1}{2}}.$$

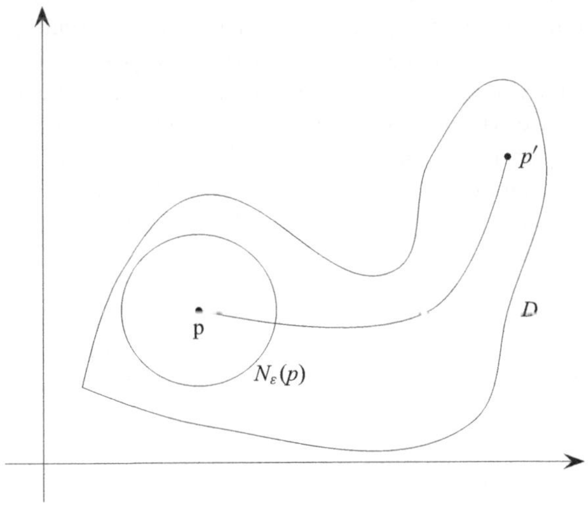

Fig. 6.36 ε-neighbourhood of the point $\mathbf{p}$ in a 2-dimensional space. Open set. Region

Accordingly, the *distance* between the points (vectors) $\mathbf{x}$ and $\mathbf{p}$ of the n-dimensional space is

$$|\mathbf{x} - \mathbf{p}| := ((x_1 - p_1)^2 + (x_2 - p_2)^2 + \ldots + (x_n - p_n)^2)^{\frac{1}{2}}$$

and the n-dimensional *ε-neighbourhood* ($\varepsilon > 0$) is defined by

$$N_\varepsilon(\mathbf{p}) := \{\mathbf{x} \mid |\mathbf{x} - \mathbf{p}| < \varepsilon\}$$

which, for $n = 2$, that is in the plane, is (see Fig. 6.36) the interior of a circle of radius ε around $\mathbf{p}$. Similarly, for $n = 3$ we get the interior of the (3-dimensional) sphere (the "3-ball") of radius ε around $\mathbf{p}$ and, in general, in an n-dimensional space the "n-ball" of radius ε around $\mathbf{p}$ defined by

$$N_\varepsilon(\mathbf{p}) = \{\mathbf{x} \mid |\mathbf{x} - \mathbf{p}| < \varepsilon\}$$
$$= \left\{(x_1, \ldots, x_n) \;\middle|\; ((x_1 - p_1)^2 + (x_2 - p_2)^2 + \ldots + (x_n - p_n)^2)^{\frac{1}{2}} < \varepsilon\right\}.$$

We again define also *punctured ε-neighbourhoods* of $\mathbf{p}$ by

$$N'_\varepsilon(\mathbf{p}) := \{\mathbf{x} \mid |\mathbf{x} - \mathbf{p}| < \varepsilon,\ \mathbf{x} \neq \mathbf{p}\}.$$

Of course, a vector-vector-function need not be defined on all of $\mathbb{R}^n$, only on a *domain* (subset) D of $\mathbb{R}^n$. But it will be of advantage if, at least initially, D is an *open set*, that is, if it contains with every point $\mathbf{p} \in D$ at least one $\varepsilon_\mathbf{p}$-neighbourhood of $\mathbf{p}$ ($\varepsilon_\mathbf{p} > 0$). Then (see Fig. 6.36) it will contain also every ε-neighbourhood of $\mathbf{p}$ with $\varepsilon < \varepsilon_\mathbf{p}$.

In order to make our domain D even more similar to open intervals (whether finite or not) on the real line (that is, in one-dimensional space), we will also suppose that D is *connected*, that is, for any two points $\mathbf{p}$ and $\mathbf{p}'$ in D there is a *path inside D* which connects them (see Fig. 6.36). (Actually, this is the definition of *path-connected sets* but we will not speak about other kinds of connectedness. Strictly speaking, even the concept "path" has to be defined, what we will not do here, since the notion is quite intuitive but the definition would be less so. (See also Sect. 9.2.)

Connected open sets are called *regions*. (Of course, every neighbourhood is a region.) So, let $\mathbf{f}$ be defined on an n-dimensional region $D \subseteq \mathbb{R}^n$ and have its values in $\mathbb{R}^m$, that is, $\mathbf{f} : D \to \mathbb{R}^m$.

A vector $\mathbf{a} \in \mathbb{R}^m$ *is the limit of a function* $\mathbf{f} : D \to \mathbb{R}^m$ *at a point* $\mathbf{p} \in D$ ($D \subseteq \mathbb{R}^n$ is a region or, at least, it contains a neighbourhood of $\mathbf{p}$) *if, for every neighbourhood* $N_\varepsilon(\mathbf{a})$ *of* $\mathbf{a}$, *there exists a punctured neighbourhood* $N'_\delta(\mathbf{p})$ *of* $\mathbf{p}$ *such that, for* $\mathbf{x} \in N'_\delta(\mathbf{p})$, *we have* $\boldsymbol{f}(x) \in N_\varepsilon(\mathbf{a})$. In symbols:

$$\mathbf{a} = \lim_{\mathbf{x} \to \mathbf{p}} \mathbf{f}(\mathbf{x}) \quad \text{if} \quad \forall \varepsilon > 0\ \exists \delta > \mathbf{f}(\mathbf{x}) \in N_\varepsilon(\mathbf{a}) \text{ for all } \mathbf{x} \in N'_\delta(\mathbf{p})$$

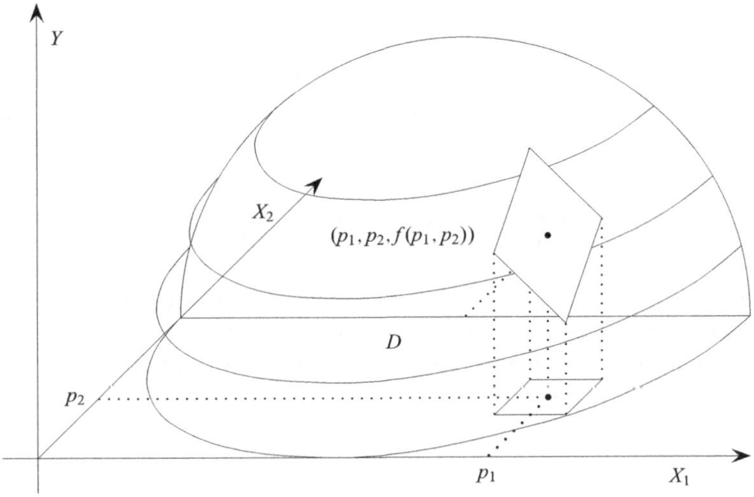

Fig. 6.37 Linear approximation (differentials) of a vector-vector function

or, what is the same, if

$$\forall \varepsilon > 0\ \exists \delta > 0 : |\mathbf{f}(\mathbf{x}) - \mathbf{a}| < \varepsilon \text{ whenever } |\mathbf{x} - \mathbf{p}| < \delta.$$

A function $\mathbf{f} : D \to \mathbb{R}^m$ is *continuous on a set* $S \subseteq D$, if it is continuous at every $\mathbf{p} \in S$. We could (but will not) again define uniform continuity and continuity on the boundary of a region (we even leave the intuitive notion of boundary undefined). We only note that, here too, $\mathbf{a} = \lim_{\mathbf{x}\to\mathbf{p}} \boldsymbol{f}(\mathbf{x})$ *or* $\boldsymbol{f}$ *continuous at* $\mathbf{p}$ *means intuitively that, if* $\mathbf{x}$ *is close enough to* $\mathbf{p}$ *then* $\boldsymbol{f}(\mathbf{x})$ *can get as close to* $\mathbf{a}$ *or to* $\boldsymbol{f}(\mathbf{p})$*, respectively, as we wanted it to get.*

A function $\mathbf{f}$ defined (at least) on a neighbourhood of a point $\mathbf{p} \in \mathbb{R}^n$ that is, $\mathbf{f} : N(\mathbf{p}) \to \mathbb{R}^m$ is *differentiable at* $\mathbf{p}$ *if there exists a linear function* (see Sect. 4.3) $\ell : \mathbb{R}^n \to \mathbb{R}^m$ such that (see Fig. 6.37)

$$\begin{aligned}
&\lim_{\mathbf{x}\to\mathbf{p}} \left(\frac{1}{|\mathbf{x} - \mathbf{p}|} (\mathbf{f}(\mathbf{x}) - \mathbf{f}(\mathbf{p}) - \ell(\mathbf{x} - \mathbf{p}) \right) \\
&= \lim_{\mathbf{x}\to\mathbf{p}} \left(\frac{1}{|\mathbf{x} - \mathbf{p}|} (\mathbf{f}(\mathbf{x}) - \mathbf{f}(\mathbf{p}) - \ell(\mathbf{x}) + \ell(\mathbf{p}) \right) = \mathbf{0},
\end{aligned} \tag{6.19}$$

($\mathbf{0}$ is the 0-vector in $\mathbb{R}^m$). The expression $\ell(\mathbf{x}-\mathbf{p})$ as function of $\mathbf{x}$ (an affine function, in symbols: $\mathbf{x} \mapsto \ell(\mathbf{x}) - \ell(\mathbf{p})$) is often called the ("total" or "exact") *differential of* f *at* $\mathbf{p}$ and denoted by $d\mathbf{f}(\mathbf{p})$. Equation (6.19) states that $\mathbf{x} \mapsto \mathbf{f}(\mathbf{p}) + \ell(\mathbf{x}) - \ell(\mathbf{p})$ *can be considered to be a linear approximation of* f *at* $\mathbf{p}$ (called "linear" for historic reasons, it is really affine, see Sects. 4.2, 4.3, and 6.10) .

A function $\mathbf{f} : D \to \mathbb{R}^m$ is *differentiable* on a region (or, at least, on an open set) $D \subseteq \mathbb{R}^n$ if it is differentiable at every point $\mathbf{p} \in D$. Then we often write $d\mathbf{f}$ for the set of affine functions $\{\mathbf{x} \mapsto \ell(\mathbf{x}) - \ell(\mathbf{p}) \mid \mathbf{p} \in D\}$. We call $d\mathbf{f}$ and $d\mathbf{f}(\mathbf{p})$ *differentials*. Of course, for the identity function $\mathbf{f}(\mathbf{x}) = \mathbf{x}$ on D (in which case $m = n$), $d\mathbf{x} = \{\mathbf{x} \mapsto \ell(\mathbf{x} - \mathbf{p}) = \mathbf{x} - \mathbf{p} \mid \mathbf{p} \in D\}$, since for each fixed $\mathbf{p} \in D$

$$\lim_{\mathbf{x} \to \mathbf{p}} \left(\frac{1}{|\mathbf{x} - \mathbf{p}|} (\mathbf{x} - \mathbf{p} - (\mathbf{x} - \mathbf{p})) \right) = 0,$$

so for the identity function $(\mathbf{x}) = \mathbf{x}$ we have $d= \mathbf{dx}$. If $\mathbf{p} \in D$ is fixed, writing $d(\mathbf{p})\mathbf{dx}(\mathbf{p})$ would be consistent but is not always applied.

In general, as we have seen in Sect. 4.3, with the help of a basis of $\mathbb{R}^n$ the linear function $\ell : \mathbb{R}^n \to \mathbb{R}^n$ can be written as

$$\ell(\mathbf{x}) = \mathbf{L}\mathbf{x} \tag{6.20}$$

where $\mathbf{L}$ is an $m \times n$ matrix.

This matrix $\mathbf{L}$ may be the *derivative of the vector-vector function at* $\mathbf{p}$, denoted by $\mathbf{f}'(\mathbf{p})$ (compare Sect. 6.10). It is also called a *Jacobian matrix*. With the above notations, and with (6.20), we can write

$$d(\mathbf{p}) = \boldsymbol{\ell}(\mathbf{x} - \mathbf{p}) = \mathbf{L}(\mathbf{x} - \mathbf{p}) = \mathbf{f}'(\mathbf{p})\mathbf{dx} \tag{6.21}$$

again as for functions of one (scalar) variable in Sect. 6.9. (Note that $\boldsymbol{\ell}(\mathbf{x} - \mathbf{p})$ is the value of $\boldsymbol{\ell}$ at $\mathbf{x} - \mathbf{p}$ but $\mathbf{L}(\mathbf{x} - \mathbf{p})$ is the product of the matrix $\mathbf{L}$ and the vector $\mathbf{x} - \mathbf{p}$).

We mention that (also in the case $n = 1$), with the notations

$$\Delta\mathbf{f}(\mathbf{p}) = \mathbf{f}(\mathbf{x}) - \mathbf{f}(\mathbf{p}), \qquad \Delta\mathbf{p} = \mathbf{x} - \mathbf{p}, \tag{6.22}$$

(6.19) is often written as

$$\lim_{\Delta\mathbf{p} \to \mathbf{0}} \frac{\Delta\mathbf{f}(\mathbf{p}) - \boldsymbol{\ell}(\Delta\mathbf{p})}{|\Delta\mathbf{p}|} = 0.$$

Also sometimes absolute value is written for the euclidean norm: $|\Delta\mathbf{p}|$ for $|\Delta\mathbf{p}|$ etc. Keeping this distinction may, however, be useful in what follows.

We can express the Jacobian matrix $\mathbf{L}$ *explicitly in terms of the function* $f : D \to \mathbb{R}^m$. Indeed, by the above definition of limits of vector-vector functions (and of connected stes), if (6.19) holds, then the $\mathbf{x}$ in it can *be anywhere* in a certain neighborhood of $\mathbf{p}$ and then approach $\mathbf{p}$ on *any* path. .

Let this path (Fig. 6.38) be a straight line going through $\mathbf{p}$, parallel to the basic unit vector $\mathbf{e}_j$ (a column vector with 1 in the j-th row, 0 in all other rows, compare to Sects. 1.5 and 4.3) and let $\mathbf{x}$ be on this straight line (sufficiently close to $\mathbf{p}$ so that

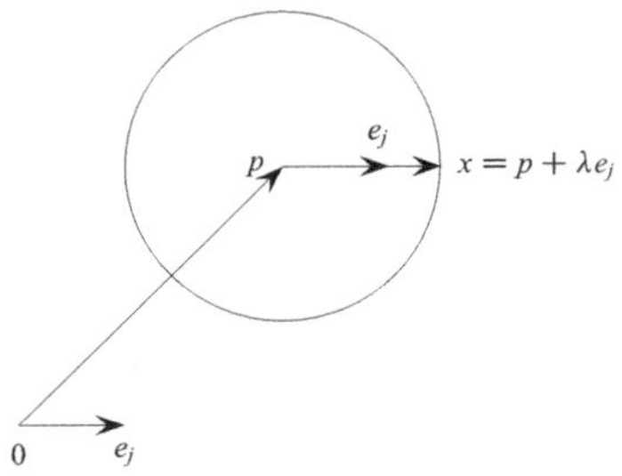

Fig. 6.38 $\mathbf{x}$ is in a neighborhood of $\mathbf{p}$ on a straight line through $\mathbf{p}$, parallel to $\mathbf{e}_j$

$\mathbf{x}$ is in the given neighborhood of $\mathbf{p}$). Then, by the rules of addition of vectors and their multiplication by scalars (Sect. 1.5),

$$\mathbf{x} = \mathbf{p} + \lambda \mathbf{e}_j \quad (\lambda \neq 0 \text{ a real (scalar) variable})$$

and

$$\lim_{\lambda \to 0} \left(\frac{1}{|\lambda \mathbf{e}_j|} [\mathbf{f}(\mathbf{p} + \lambda \mathbf{e}_j) - \mathbf{f}(\mathbf{p}) - \mathbf{L}(\lambda \mathbf{e}_j)] \right) = 0.$$

But, by the definition of euclidean norms (note the norms on the left hand side, absolute value on the right):

$$|\lambda \mathbf{e}_j| = \sqrt{0 + \ldots + 0 + \lambda^2 + 0 + \ldots + 0} = \sqrt{\lambda^2} = |\lambda|$$

and, see Sect. 4.4 **2**,

$$\mathbf{L}(\lambda \mathbf{e}_j) = \lambda \mathbf{L}\, \mathbf{e}_j,$$

so we can write (with $\frac{\mathbf{y}}{\lambda} := \frac{1}{\lambda}\mathbf{y}$ for simplicity)

$$\lim_{\lambda \to 0} \left(\frac{\lambda}{|\lambda|} \left[\frac{\mathbf{f}(\mathbf{p} + \lambda \mathbf{e}_j) - \mathbf{f}(\mathbf{p})}{\lambda} - \mathbf{L}\, \mathbf{e}_j \right] \right) = \mathbf{0}.$$

But $\frac{\lambda}{|\lambda|}$ is either 1 or -1. In either case, the limit can be $\mathbf{0}$ only if the limit of the expression in square brackets is $\mathbf{0}$. Furthermore $\mathbf{L}\mathbf{e}_j$ does not depend on λ and, here too, the limit of the difference is the difference of the limits. So

$$\lim_{\lambda \to 0} \frac{\mathbf{f}(\mathbf{p} + \lambda \mathbf{e}_j) - \mathbf{f}(\mathbf{p})}{\lambda} = \mathbf{L}\, \mathbf{e}_j = \begin{pmatrix} \ell_{1j} \\ \ell_{2j} \\ \vdots \\ \ell_{mj} \end{pmatrix} \tag{6.23}$$

since (see Sect. 4.3)

$$\mathbf{L}\,\mathbf{e}_j = \begin{pmatrix} \ell_{11} \cdots \ell_{1j} \cdots \ell_{1n} \\ \vdots \\ \ell_{m1} \cdots \ell_{mj} \cdots \ell_{mn} \end{pmatrix} \begin{pmatrix} 0 \\ \vdots \\ 1 \\ \vdots \\ 0 \end{pmatrix} = \begin{pmatrix} \ell_{1j} \\ \vdots \\ \ell_{mj} \end{pmatrix}.$$

Consider now the left hand side of (6.23). Since vectors are subtracted and (as is easy to see) their limits are taken componentwise, the i-th component ℓ_{ij} of (6.23) is, written in detail,

$$\lim_{j\to 0} \frac{f_i(p_1,\ldots,p_{j-1},p_j+\lambda,p_{j+1},\ldots,p_n) - f_i(p_1,\ldots,p_{j-1},p_j,p_{j+1},\ldots,p_n)}{\lambda}.$$

If we keep $p_1,\ldots,p_{j-1},p_{j+1},\ldots,p_n$ fixed, this limit is clearly the derivative of the (scalar) function g_j defined by $g_j(x) := f_i(p_1,\ldots,p_{j-1},x,p_{j+1},\ldots,p_n)$ at $x = p_j$. We call this the j-th *partial derivative* of f_i at $\mathbf{p}$ and write it as

$$\frac{\partial f_i}{\partial x_j}(\mathbf{p}) \quad \text{or} \quad \frac{\partial f_i}{\partial x_j} \quad \text{for short.}$$

So $\ell_{ij} = \frac{\partial f_i}{\partial xj}$ and

$$\mathbf{f}'(\mathbf{p}) = \mathbf{L} = \begin{pmatrix} \frac{\partial f_1}{\partial x_1}(\mathbf{p}) \cdots \frac{\partial f_1}{\partial x_n}(\mathbf{p}) \\ \vdots \qquad\qquad \vdots \\ \frac{\partial f_m}{\partial x_1}(\mathbf{p}) \cdots \frac{\partial f_m}{\partial x_n}(\mathbf{p}) \end{pmatrix} = \begin{pmatrix} \frac{\partial f_1}{\partial x_1} \cdots \frac{\partial f_1}{\partial x_n} \\ \vdots \qquad \vdots \\ \frac{\partial f_m}{\partial x_1} \cdots \frac{\partial f_m}{\partial x_n} \end{pmatrix}$$

This is the *explicit form of the Jacobian matrix*, often written as

$$\frac{\partial(f_1,\ldots,f_m)}{\partial(x_1,\ldots,x_m)}.$$

If, in particular, $m = 1$, then f and df are scalars, $d\mathbf{x}$ is as always a column vector and

$$\mathbf{L} = \left(\frac{\partial f}{\partial x_1},\ldots,\frac{\partial f}{\partial x_n}\right)$$

a row vector. So (6.21) becomes (as in Sect. 4.4 **1** we get a scalar product):

$$df(\mathbf{p}) = \left(\frac{\partial f}{\partial x_1}(\mathbf{p}), \ldots, \frac{\partial f}{\partial x_n}(\mathbf{p})\right)\begin{pmatrix} dx_1 \\ \vdots \\ dx_n \end{pmatrix} = \frac{\partial f}{\partial x_1}(\mathbf{p})dx_1 + \ldots + \frac{\partial f}{\partial x_n}(\mathbf{p})dx_n$$

or

$$df = \left(\frac{\partial f}{\partial x_1}, \ldots, \frac{\partial f}{\partial x_n}\right)\begin{pmatrix} dx_1 \\ \vdots \\ dx_n \end{pmatrix} = \frac{\partial f}{\partial x_1}dx_1 + \ldots + \frac{\partial f}{\partial x_n}dx_n$$

for short. This differential is often called the *total differential* of f (whether at one point or on all of D) and the row vector $\mathbf{L}$ the *gradient* (abbreviated grad or ∇) of f (again at one point or on all of D). In notation

$$\operatorname{grad} f = \nabla f = \left(\frac{\partial f}{\partial x_1}, \ldots, \frac{\partial f}{\partial x_n}\right).$$

So, written again for the point $\mathbf{p}$,

$$df(\mathbf{p}) = \operatorname{grad} f(\mathbf{p}) \cdot d\mathbf{x} = \nabla f(\mathbf{p}) \cdot d\mathbf{x}.$$

With the notation (6.22), we have for $m = 1$ that $\Delta f(\mathbf{p}) = f(\mathbf{x}) - f(\mathbf{p})$ is approximated by

$$\begin{aligned} df(\mathbf{p}) &= \frac{\partial f}{\partial x_1}(\mathbf{p})(x_1 - p_1) + \ldots + \frac{\partial f}{\partial x_n}(\mathbf{p})(x_n - p_n) \\ &= \frac{\partial f}{\partial x_1}(\mathbf{p})\Delta x_1 + \ldots + \frac{\partial f}{\partial x_n}(\mathbf{p})\Delta x_n. \end{aligned}$$

If f is constant on a region D then every

$$g_j(x_j) = f(p_1, \ldots, p_{j-1}, x_j, p_{j+1}, \ldots, p_n)$$

is constant, so (see Sect. 6.4, Example **5**)

$$\frac{\partial f}{\partial x_i} = f_j'(x_j) = 0 \qquad (j = 1, \ldots, n)$$

and so

$$df = \frac{\partial f}{\partial x_1}dx_1 + \ldots + \frac{\partial f}{\partial x_n} = 0.$$

Conversely, *if* $df = 0$, that is $\frac{\partial f}{\partial x_1} = \ldots = \frac{\partial f}{\partial x_n} = 0$ *then* f *is constant* (because f is constant in all n of its variables). If $df(\mathbf{p}) = 0$ only *at a certain point* $\mathbf{p}$ *then* f *may have a maximum or minimum there* (see Sect. 8.3).

Just as, at the end of Sect. 6.4 and in Sect. 6.7, derivatives of functions of one real variable could have (but do not always have) derivatives themselves, so partial derivatives may have partial derivatives themselves. We use the notation

$$\frac{\partial^2 f}{\partial x_j \partial x_k}(\mathbf{p}) := \frac{\partial}{\partial x_j}\left(\frac{\partial f}{\partial x_k}(\mathbf{p})\right), \text{ in particular, } \frac{\partial^2 f}{\partial x_k^2}(\mathbf{p}) := \frac{\partial}{\partial x_k}\left(\frac{\partial f}{\partial x_k}(\mathbf{p})\right)$$

for $j, k = 1, \ldots, n$ (sometimes omitting $\mathbf{p}$). These are the *second order partial derivatives*. Similarly one may be able to form third and *higher order partial derivatives*. The above notation and concept means, of course, that $\frac{\partial f}{\partial x_k}(\mathbf{p})$ is a function of $\mathbf{p} = (p_1, \ldots, p_n)$ and *if* the difference quotient of this function with respect to x_j has a limit then *at that point* $\frac{\partial^2 f}{\partial x_j \partial x_k}$ *exists*. This again is a function of $\mathbf{p}$ and may be continuous or differentiable or partially differentiable. If and where it is partially differentiable (at least with respect to one variable), there the third order partial derivative (with respect to that variable) exists, and so on.

Example 1 $f : \mathbb{R}^3 \to \mathbb{R}$, $f(x_1, x_2, x_3) := x_1^2 x_2 + x_2^3 \sin x_3^2$,
(we get $\frac{\partial f}{\partial x_1}(p_1, p_2, p_3) = 2p_1 p_2$, or $\frac{\partial f}{\partial x_1} = 2x_1 x_2$ for short),

$$\frac{\partial f}{\partial x_1} = 2x_1 x_2, \quad \frac{\partial f}{\partial x_2} = x_1^2 + 3x_2^2 \sin x_3^2, \quad \frac{\partial f}{\partial x_3} = 2x_2^3 x_3 \cos x_3^2,$$
$$\frac{\partial^2 f}{\partial x_1^2} = 2x_2, \quad \frac{\partial^2 f}{\partial x_1 \partial x_2} = 2x_1, \quad \frac{\partial^2 f}{\partial x_1 \partial x_3} = 0,$$
$$\frac{\partial^2 f}{\partial x_2 \partial x_1} = 2x_1, \quad \frac{\partial^2 f}{\partial x_2^2} = 6x_2 \sin x_3^2, \quad \frac{\partial^2 f}{\partial x_2 \partial x_3} = 6x_2^2 x_3 \cos x_3^2,$$
$$\frac{\partial^2 f}{\partial x_3 \partial x_1} = 0, \quad \frac{\partial^2 f}{\partial x_3 \partial x_2} = 6x_2^3 x_3 \cos x_3^2, \quad \frac{\partial^2 f}{\partial x_3^2} = 2x_2^3 \cos x_3^2 - 4x_2^3 x_3^2 \sin x_3^2.$$

(We used the differentiation rules from Sects. 6.4 and 6.5).

The attentive reader may have noticed that here

$$\frac{\partial^2 f}{\partial x_1 \partial x_2} = \frac{\partial^2 f}{\partial x_2 \partial x_1}, \quad \frac{\partial^2 f}{\partial x_1 \partial x_3} = \frac{\partial^2 f}{\partial x_3 \partial x_1} \quad \text{and} \quad \frac{\partial^2 f}{\partial x_2 \partial x_3} = \frac{\partial^2 f}{\partial x_3 \partial x_2}$$

at any point $\mathbf{p} \in \mathbb{R}^3$. This "*equality of mixed partial derivatives*" (at point $\mathbf{p}$) *holds* always *if at least one of the two mixed derivatives is continuous* (at that point).

(continued)

We will not prove this result. Instead, as in Sects. 6.3 and 6.7, we show, that the statement *need not be true if (at least) one of the two derivatives is not continuous* at that point. This will also show that *a derivative may exist at a point but may be discontinuous there.*

Example 2 The function $f : \mathbb{R}^2 \to \mathbb{R}$, $f(x_1, x_2)$ is given by

$$f(x_1, x_2) = \begin{cases} \dfrac{x_1 x_2 (x_1^2 - x_2^2)}{x_1^2 + x_2^2} & \text{if} \quad x_1^2 + x_2^2 \neq 0 \\ 0 & \text{if} \quad x_1^2 + x_2^2 = 0, \qquad \text{that is at } (0,0). \end{cases}$$

We calculate the partial derivatives again by the rules in Sects. 6.4 and 6.5 when $x_1^2 + x_2^2 \neq 0$:

$$\begin{aligned} \frac{\partial f}{\partial x_1}(x_1, x_2) &= x_2 \frac{x_1^2 - x_2^2}{x_1^2 + x_2^2} + x_1 x_2 \frac{2x_1(x_1^2 + x_2^2) - (x_1^2 - x_2^2)2x_1}{(x_1^2 + x_2^2)^2} \\ &= \frac{x_1^4 x_2 + x_1^2 x_2^3 - x_2^5}{(x_1^2 + x_2^2)2}, \\ \frac{\partial f}{\partial x_2}(x_1, x_2) &= x_1 \frac{x_1^2 - x_2^2}{x_1^2 + x_2^2} + x_1 x_2 \frac{-2x_2(x_1^2 + x_2^2) - (x_1^2 - x_2^2)2x_2}{(x_1^2 + x_2^2)^2} \\ &= \frac{x_1^5 - 4x_1^3 x_2^2 - x_1 x_2^4}{(x_1^2 + x_2^2)2} \end{aligned}$$

However, at $(0, 0)$, we have to calculate the partial derivatives directly as limits of difference quotients:

$$\frac{\partial f}{\partial x_1}(0,0) = \lim_{x_1 \to 0} \frac{f(x_1, 0) - f(0,0)}{x_1} = \lim_{x_1 \to 0} \frac{0 \cdot x_1^3 / x_1^2 - 0}{x_1} = \lim_{x_1 \to 0} 0 = 0,$$

$$\frac{\partial f}{\partial x_2}(0,0) = \lim_{x_2 \to 0} \frac{f(0, x_2) - f(0,0)}{x_2} = \lim_{x_2 \to 0} \frac{-0 \cdot x_2^3 / x_2^2 - 0}{x_2} = \lim_{x_2 \to 0} 0 = 0.$$

Now we determine, also as limit of difference quotients, the mixed partial derivatives at $(0, 0)$

$$\frac{\partial^2 f}{\partial x_1 \partial x_2}(0,0) = \lim_{x_1 \to 0} \frac{\frac{\partial f}{\partial x_2}(x_1, 0) - \frac{\partial f}{\partial x_2}(0,0)}{x_1} = \lim_{x_1 \to 0} \frac{\frac{x_1^5}{x_1^4}}{x_1} = 1,$$

(continued)

$$\frac{\partial^2 f}{\partial x_2 \partial x_1}(0,0) = \lim_{x_2 \to 0} \frac{\frac{\partial f}{\partial x_1}(0,x_2) - \frac{\partial f}{\partial x_1}(0,0)}{x_2} = \lim_{x_2 \to 0} \frac{\frac{-x_2^5}{x_2^4}}{x_2} = -1.$$

The two are clearly not equal. And indeed, for instance $\frac{\partial^2 f}{\partial x_2 \partial x_1}$, while it exists at $(0,0)$ (as we have just seen, it is -1) it is not continuous there. Indeed, from the above,

$$\frac{\partial f}{\partial x_1}(x_1,0) = \frac{x_1^4 \cdot 0 - 4x_1^2 \cdot 0^3 - 0^5}{(x_1^2 + 0^2)^2} = 0 \quad \text{for} \quad x_1 \neq 0,$$

therefore

$$\frac{\partial^2 f}{\partial x_2 \partial x_1}(x_1,0) = 0 \text{ for } x_1 \neq 0,$$

and

$$\lim_{x_1 \to 0} \frac{\partial^2 f}{\partial x_2 \partial x_1}(x_1,0) = 0 \neq -1 = \frac{\partial^2 f}{\partial x_2 \partial x_1}(0,0).$$

6.11.1 Excercises

1. Determine the Jacobian matrix $\mathbf{f}'(\mathbf{x})$ of the function

$$\mathbf{f} : \mathbb{R}^3 \to \mathbb{R}^2 \quad \text{given by} \quad \mathbf{f}(\mathbf{x}) = (f_1(x_1,x_2,x_3), f_2(x_1,x_2,x_3))$$

with

$$f_1(x_1,x_2,x_3) = 1 + x_1^2 - x_1x_2x_3 + x_2 \cos x_3 \quad \text{and}$$
$$f_2(x_1,x_2,x_3) = x_1x_2(\sin x_3)^2$$

at (a) $\mathbf{x} = \mathbf{p} = (p_1,p_2,p_3)$, (b) $\mathbf{x} = (1,1,0)$, (c) $\mathbf{x} = (1,-1,\frac{\pi}{2})$.

2. Determine the qradient of the functions[-4ex]
(a) $g : \mathbb{R}^n \longrightarrow \mathbb{R}, \quad g(\mathbf{x}) = (1 + x_1^2 + x_2^2 + \ldots + x_n^2)^{-1}$,
(b) $h :]-\frac{\pi}{2},\frac{\pi}{2}[\longrightarrow \mathbb{R}, \quad h(\mathbf{x}) = \tan(x_1 + 2x_2 + \ldots + nx_n)$,
for $x_1 + 2x_2 + \ldots + nx_n \in]-\frac{\pi}{2},\frac{\pi}{2}[$,[-4ex]
at $\mathbf{x} = \mathbf{p} = (p_1,p_2,\ldots,p_n)$ and
$\mathbf{x} = (1,1,\ldots,1)$ in case (a), $\mathbf{x} = (0,0,\ldots,0)$ in case (b).

3. Determine the mixed partial derivatives of second order of the functions
   (a) f_1 given in Exercise 1,
   (b) f_2 given in Exercise 1.
4. For the functions g and h given in Exercise 2 determine, for $j = 1, \ldots, n$, the partial derivatives of the second order

$$\text{(a)}\ \frac{\partial^2 g}{\partial x_j^2}, \qquad \text{(b)}\ \frac{\partial^2 h}{\partial x_j^2}.$$

5. Determine the partial derivatives of the third order of the function $F : \mathbb{R}^2 \to \mathbb{R}$ given by $F(x, y) = xy(\cos x)(\sin y)$.

6.11.2 Answers

1. (a) $\mathbf{f}'(p_1, p_2, p_3) = \begin{pmatrix} 2p_1 - p_2p_3 & -p_1p_3 + \cos p_3 & -p_1p_2 - p_2 \sin p_3 \\ p_2(\sin p_3)^2 & p_1(\sin p_3)^2 & 2p_1p_2 \sin p_3 \cos p_3 \end{pmatrix}$,
   (b) $\mathbf{f}'(1, 1, 0) = \begin{pmatrix} 2 & 1 & -1 \\ 0 & 0 & 0 \end{pmatrix}$,
   (c) $\mathbf{f}'(1, -1, \frac{\pi}{2}) = \begin{pmatrix} 2 + \frac{\pi}{2} & -\frac{\pi}{2} & 2 \\ -1 & 1 & 0 \end{pmatrix}$,
2. (a) $\operatorname{grad} g(\mathbf{p}) = -2(p_1, p_2, \ldots, p_n)(1 + p_1^2 + p_2^2 + \ldots p_n^2)^{-2}$,
   $\operatorname{grad} g(1, 1, \ldots, 1) = -2(n+1)^{-2}(1, 1, \ldots, 1)$,
   (b) $\operatorname{grad} h(\mathbf{p}) = -2(1, 2, \ldots, n)[\cos(x_1 + 2x_2 + \ldots nx_n)]^{-2}$,
   $\operatorname{grad} h(0, 0, \ldots, 0) = (1, 2, \ldots, n)$.
3. (a) $\frac{\partial^2 f_1}{\partial x_1 \partial x_2} = -x_3 = \frac{\partial^2 f_1}{\partial x_2 \partial x_1}$,
   $\frac{\partial^2 f_1}{\partial x_1 \partial x_3} = -x_2 = \frac{\partial^2 f_1}{\partial x_3 \partial x_1}$,
   $\frac{\partial^2 f_1}{\partial x_2 \partial x_3} = -x_1 - \sin x_3 = \frac{\partial^2 f_1}{\partial x_3 \partial x_2}$.
   (b) $\frac{\partial^2 f_2}{\partial x_1 \partial x_2} = (\sin x_3)^2 = \frac{\partial^2 f_1}{\partial x_2 \partial x_1}$,
   $\frac{\partial^2 f_2}{\partial x_1 \partial x_3} = 2x_2(\sin x_3)(\cos x_3) = \frac{\partial^2 f_1}{\partial x_3 \partial x_1}$,
   $\frac{\partial^2 f_2}{\partial x_2 \partial x_3} = 2x_1(\sin x_3)(\cos x_3) = \frac{\partial^2 f_1}{\partial x_3 \partial x_2}$.
4. (a) $\frac{\partial^2 g}{\partial x_j^2} = 8x_j^2(1 + x_1^2 + \ldots + x_n^2)^{-3} - 2(1 + x_1^2 + \ldots + x_n^2)^{-2}$,
   (b) $\frac{\partial^2 h}{\partial x_j^2} = 2j^2[\cos(x_1 + 2x_2 + \ldots + nx_n)]^{-3} \sin(x_1 + 2x_2 + \ldots + nx_n)$.
5. $\frac{\partial^3 F}{\partial x^3} = y \sin y(x \sin x - 3 \cos x)$,
   $\frac{\partial^3 F}{\partial x^2 \partial y} = -(y \cos y + \sin y)(x \cos x + 2 \sin x) = \frac{\partial^3 F}{\partial y \partial x^2}$,
   $\frac{\partial^3 F}{\partial x \partial y^2} = (x \sin x - \cos x)(y \sin y - 2 \cos y) = \frac{\partial^3 F}{\partial y^2 \partial x}$,
   $\frac{\partial^3 F}{\partial y^3} = -x \cos x(y \cos y + 3 \sin y)$,

6.12 Chain Rule: Euler's Partial Differential Equation for Homogeneous Functions

A further way to write the Jacobian matrix, belonging to **f** at $\mathbf{p} \in S \subset \mathbb{R}^n$, is

$$\mathbf{L} = \mathbf{f}'(\mathbf{p}) = \frac{d\mathbf{f}}{d\mathbf{x}}(\mathbf{p}).$$

We also combine (6.19) and (6.20) into

$$\begin{aligned} \mathbf{f}(\mathbf{x}) - \mathbf{f}(\mathbf{p}) &= \mathbf{L}(\mathbf{x}-\mathbf{p}) + \boldsymbol{\phi}(\mathbf{x})\,|\mathbf{x}-\mathbf{p}| \\ &= \frac{d\mathbf{f}}{d\mathbf{x}}(\mathbf{p})(\mathbf{x}-\mathbf{p}) + \boldsymbol{\phi}(\mathbf{x})\,|\mathbf{x}-\mathbf{p}| \\ (\mathbf{x} \in S, \mathbf{p} \in S),&\ \text{ with } \lim\nolimits_{\mathbf{x}\to\mathbf{p}} \boldsymbol{\phi}(\mathbf{x}) = 0. \end{aligned} \tag{6.24}$$

To see this, we just have to define

$$\boldsymbol{\phi}(\mathbf{x}) = \begin{cases} \dfrac{1}{|\mathbf{x}-\mathbf{p}|}[\mathbf{f}(\mathbf{x}) - \mathbf{f}(\mathbf{p}) - \mathbf{L}\mathbf{x} + \mathbf{L}\mathbf{p}] & \text{if } \mathbf{x} \neq \mathbf{p}, \\ 0 \quad \text{(e.g.)} & \text{if } \mathbf{x} = \mathbf{p}. \end{cases}$$

Equation (6.24) makes it clear that, if **f** is differentiable at **p**, then it is also continuous there, just as we saw for functions of one variable in Sect. 6.5. We prove now for vector-vector functions an analogue of the chain rule.

Let $\mathbf{f} : D \to \mathbb{R}^m$ $(D \subset \mathbb{R}^n)$ be differentiable at $\mathbf{p} \in D$ and $\mathbf{g} : S \to \mathbb{R}^k$ $(S \subset \mathbb{R}^m)$ be differentiable at $\mathbf{y} = \mathbf{f}(\mathbf{p})$. For this *we suppose that there exists a neighbourhood* $N(\mathbf{p})$ *of* **p** *such that the set* $\{\mathbf{f}(\mathbf{x}) \mid \mathbf{x} \in N(\mathbf{p})\} \subset S$ *contains a neighbourhood of* **y**. If we denote the variable in **g** by **y** and the Jacobian matrix belonging to **g** at **q** by

$$\mathbf{M} = \mathbf{g}'(\mathbf{q}) = \frac{d\mathbf{g}}{d\mathbf{y}} = \frac{d\mathbf{g}}{d\mathbf{y}}(\mathbf{q}),$$

then, just as in (6.24), we have

$$\mathbf{g}(\mathbf{y}) - \mathbf{g}(\mathbf{q}) = \mathbf{M}(\mathbf{y}-\mathbf{q}) + \boldsymbol{\psi}(\mathbf{y})\,|\mathbf{y}-\mathbf{q}| \qquad \text{with} \qquad \lim_{\mathbf{y}\to\mathbf{q}} \boldsymbol{\psi}(\mathbf{y}) = \mathbf{0}.$$

In particular, for $\mathbf{y} = \mathbf{f}(\mathbf{x})$ (we have supposed that there exists such $\mathbf{x} \in N(\mathbf{p})$) and with $\mathbf{q} = \mathbf{f}(\mathbf{p})$ taking also (6.24) into consideration, we have

$$\begin{aligned} \mathbf{g}(\mathbf{f}(\mathbf{x})) - \mathbf{g}(\mathbf{f}(\mathbf{p})) &= \mathbf{M}\boldsymbol{\psi}(\mathbf{x})\,|\mathbf{x}-\mathbf{p}| + \boldsymbol{\psi}(\mathbf{f}(\mathbf{x}))\,|\mathbf{f}(\mathbf{x}) - \mathbf{f}(\mathbf{p})| \\ = \mathbf{M}\mathbf{L}(\mathbf{x}-\mathbf{p}) &+ \mathbf{M}\boldsymbol{\phi}(\mathbf{x})\,|\mathbf{x}-\mathbf{p}| + \boldsymbol{\psi}(\mathbf{f}(\mathbf{x}))\,|\mathbf{L}(\mathbf{x}-\mathbf{p}) + \boldsymbol{\phi}(\mathbf{x})\,|\mathbf{x}-\mathbf{p}|| \\ &= \frac{d\mathbf{g}}{d\mathbf{y}}(\mathbf{q})\frac{d\mathbf{f}}{d\mathbf{x}}(\mathbf{p})(\mathbf{x}-\mathbf{p}) + \boldsymbol{\chi}(\mathbf{x})\,|\mathbf{x}-\mathbf{p}|\,. \end{aligned}$$

Here we have used the fact that

$$|\mathbf{z}\lambda| = \sqrt{z_1^2\lambda^2 + \ldots + z_n^2\lambda^2} = |\mathbf{z}|\,|\lambda| = |\mathbf{z}|\,\lambda \quad \text{if} \quad \lambda \geq 0$$

(in our case $\lambda = |\mathbf{x} - \mathbf{p}| \geq 0$) and wrote

$$\chi(\mathbf{x}) := \mathbf{M}\phi(\mathbf{x}) + \psi(f(\mathbf{x}))\left|\mathbf{L}(\mathbf{x}-\mathbf{p})\frac{1}{|\mathbf{x}-\mathbf{p}|} + \phi(\mathbf{x})\right|.$$

Since $\frac{\mathbf{x}-\mathbf{p}}{|\mathbf{x}-\mathbf{p}|}$ is a *vector* divided by its *length*, so a unit vector, and limits are taken on punctured neighbourhoods with $\mathbf{x} \neq \mathbf{p}$, therefore the vector-vector function whose norm we are taking here is bounded. Furthermore, since (using also the continuity of $\mathbf{f}$ at $\mathbf{p}$, a consequence of its differentiability)

$$\lim_{\mathbf{x}\to\mathbf{p}} \phi(\mathbf{x}) = 0, \qquad \lim_{\mathbf{x}\to\mathbf{p}} \psi(\mathbf{f}(\mathbf{x})) = \lim_{\mathbf{y}\to\mathbf{q}} \psi(\mathbf{y}) = 0$$

(the first equation we used already in the boundedness argument), therefore we see that, in

$$\mathbf{g}(\mathbf{f}(\mathbf{x})) - \mathbf{g}(\mathbf{f}(\mathbf{p})) = \frac{d\mathbf{g}}{d\mathbf{y}}(\mathbf{q})\frac{d\mathbf{f}}{d\mathbf{x}}(\mathbf{p})(\mathbf{x}-\mathbf{p}) + \chi(\mathbf{x})\,|\mathbf{x}-\mathbf{p}|\,,$$

we have

$$\lim_{\mathbf{x}\to 0} \chi(\mathbf{x}) = 0.$$

But this is an equation of the form (6.24) for $\mathbf{h}(\mathbf{x}) = \mathbf{g}(\mathbf{f}(\mathbf{x})) = (\mathbf{g} \circ \mathbf{f})(\mathbf{x})$, so we have the following:

Chain rule *If f and $\mathbf{g}$ are defined on neighbourhoods of $\mathbf{p} \in \mathbb{R}^m$ and of $\mathbf{q} = f(\mathbf{p}) \in \mathbb{R}^n$, respectively, and f maps a neighbourhood of $\mathbf{p}$ onto a set which contains a neighborhood of $\mathbf{q}$, if further f and $\mathbf{g}$ are differentiable at $\mathbf{p}$ and at $\mathbf{q}$, respectively, then $\mathbf{g} \circ f : N(\mathbf{p}) \to \mathbb{R}^k$ is differentiable at $\mathbf{p}$ and*

$$\frac{d(\mathbf{g} \circ f)}{d\mathbf{x}}(\mathbf{p}) = \frac{d\mathbf{g}}{d\mathbf{y}}(\mathbf{q})\frac{df}{d\mathbf{x}}(\mathbf{p}).$$

em If the conditons hold on regions, so does the chain rule and we write simply

$$\frac{d(\mathbf{g} \circ f)}{d\mathbf{x}} = \frac{d\mathbf{g}}{d\mathbf{y}}\frac{df}{d\mathbf{x}}$$

or, with $\mathbf{y} = \boldsymbol{f}(\mathbf{x})$, $\mathbf{z} = \mathbf{g}(\mathbf{y}) = \mathbf{g}(\boldsymbol{f}(\mathbf{x})) = (\mathbf{g} \circ \boldsymbol{f})(\mathbf{x})$ (compare to Sect. 6.5 **4**):

$$\frac{d\mathbf{z}}{d\mathbf{x}} = \frac{d\mathbf{z}}{d\mathbf{y}}\frac{d\mathbf{y}}{d\mathbf{x}}.$$

If especially $m = k = 1$, then

$$\mathbf{y} = \begin{pmatrix} y_1 \\ \vdots \\ y_n \end{pmatrix} = \mathbf{f}(x) = \begin{pmatrix} f_1(x) \\ \vdots \\ f_n(x) \end{pmatrix},$$

$$z = g(\mathbf{y}) = g(y_1, \ldots, y_n) = g(f_1(x), \ldots, f_n(x)) = g(\mathbf{f}(x)) = (g \circ \mathbf{f})(x)$$

$$\frac{d(g \circ \mathbf{f})}{dx}(p) = \frac{dg}{d\mathbf{y}}(\mathbf{q})\frac{d\mathbf{f}}{dx}(p) \quad \text{or} \quad \frac{dz}{dx} = \frac{dz}{d\mathbf{y}}\frac{d\mathbf{y}}{dx},$$

that is (again a scalar product):

$$\frac{dz}{dx} = \left(\frac{\partial z}{\partial y_1}, \ldots, \frac{\partial z}{\partial y_n}\right)\begin{pmatrix} \dfrac{dy_1}{dx} \\ \vdots \\ \dfrac{dy_n}{dx} \end{pmatrix} = \frac{\partial z}{\partial y_1}\frac{dy_1}{dx} + \ldots + \frac{\partial z}{\partial y_n}\frac{dy_n}{dx} \tag{6.25}$$

(x is a scalar variable so we do not need partial derivatives when we differentiate with respect to x), the most popular and frequently used form of the chain rule.

A function $g : \mathbb{R}^n_+ \to \mathbb{R}$ is *(positively) homogenous of degree* r if (compare to Sect. 4.3 for the case $r = 1$)

$$g(\lambda\mathbf{s}) = \lambda^r g(\mathbf{s}) \qquad (\mathbf{s} = (s_1, \ldots, s_n) \in \mathbb{R}^n_+,\ \lambda \in \mathbb{R}_{++}) \tag{6.26}$$

(we could have taken $\mathbb{R}^n$ or a region $D \subset \mathbb{R}^n$ instead of $\mathbb{R}^n_+$). We know the definition and derivative of $\lambda \mapsto \lambda^r$ for rational r; in Sect. 8.2, $\lambda \mapsto \lambda^r$ will be defined fo any r and we will see that the formula $\frac{d\lambda^r}{d\lambda} = r\lambda^{r-1}$ still holds. We can apply the chain rule to get *conditions necessary and sufficient for a differentiable function to be homogeneous of degree r.*

We differentiate both sides of (6.26) with respect to λ, interchange the two sides and apply the chain rule in the form (6.25):

$$\begin{aligned} r\lambda^{r-1}g(\mathbf{s}) &= \frac{dg(\lambda\mathbf{s})}{d\lambda} \\ &= \frac{\partial g(\lambda\mathbf{s})}{\partial(\lambda s_1)}\frac{(\lambda s_1)}{d\lambda} + \ldots + \frac{\partial g(\lambda\mathbf{s})}{\partial(\lambda s_n)}\frac{(\lambda s_n)}{d\lambda} \\ &= \frac{\partial g(\lambda\mathbf{s})}{\partial(\lambda s_1)}s_1 + \ldots + \frac{\partial g(\lambda\mathbf{s})}{\partial(\lambda s_n)}s_n \end{aligned}$$

(here $x = \lambda$, $y = \lambda \mathbf{s}$). Put now $\lambda = 1$ in order to get

$$rg(\mathbf{s}) = \frac{\partial g(\mathbf{s})}{\partial s_1} s_1 + \ldots + \frac{\partial g(\mathbf{s})}{\partial s_n} s_n \tag{6.27}$$

which is LEONHARD EULER'S (*1707 – †1783) *(partial) differential equation for (positively) homogeneous functions of degree r.*

We have just proved that *all positively homogeneous differentiable functions of degree r satisfy Euler's equation* (6.27), that is, (6.27) is *necessary* for (6.26). *Conversely*, let (6.27) be satisfed (and g, of course, differentiable). We prove that (6.26) follows, so (6.27) is also *sufficient* for (6.26). For this purpose, *we suppose that* (6.27) *holds* and define

$$h(\mathbf{s}, \lambda) = \lambda^{-r} g(\lambda \mathbf{s}) - g(\mathbf{s}).$$

We want to show that $h(\mathbf{s}, \lambda) \equiv 0$ (then we have indeed $g(\lambda \mathbf{s}) = \lambda^r g(\mathbf{s})$, that is (6.26)). We get from the chain rule

$$\begin{aligned} \frac{\partial h(\mathbf{s}, \lambda)}{\partial \lambda} &= -r\lambda^{-r-1} g(\lambda \mathbf{s}) + \lambda^{-r} \frac{\partial g(\lambda \mathbf{s})}{\partial \lambda} - 0 \\ &= \lambda^{-r-1} \left[-rg(\lambda s_1, \ldots, \lambda s_n) + \lambda \frac{\partial g(\lambda \mathbf{s})}{\partial (\lambda s_1)} s_1 + \ldots + \lambda \frac{\partial g(\lambda \mathbf{s})}{\partial (\lambda s_n)} s_n \right] \end{aligned}$$

($g(\mathbf{s})$ is constant in λ, so its partial derivative with respect to λ is 0). The expression in brackets on the right hand side is exactly the difference of the two sides of (6.27) with λs_j in place of s_j ($j = 1, \ldots, n$). Since (6.27) is now supposed to hold for all $\mathbf{s}$, thus this difference has to be identically 0. So $\frac{\partial h(\mathbf{s},\lambda)}{\partial \lambda} \equiv 0$ and therefore $h(\mathbf{s}, \lambda)$ is constant in λ (independent of λ), thus it is some function c of $\mathbf{s}$ alone:

$$\lambda^{-r} g(\lambda \mathbf{s}) - g(\mathbf{s}) = h(\mathbf{s}, \lambda) = c(\mathbf{s}).$$

Putting here $\lambda = 1$ we get $c(\mathbf{s}) = 0$ and so indeed

$$h(\mathbf{s}, \lambda) = \lambda^{-r} g(\lambda \mathbf{s}) - g(\mathbf{s}) = 0 \quad \Rightarrow \quad g(\lambda \mathbf{s}) = \lambda^r g(\mathbf{s}),$$

that is, (6.26) follows from (6.27), *the Euler equation* (6.27) *is necessary and sufficient for the differentiable function g to be positively homogeneous of degree r.*

While above we had $g : \mathbb{R}^n_+ \to \mathbb{R}$, we may have also $g : \mathbb{R}^n_+ \to \mathbb{R}^m$ (*output vectors*), this just means m equations of the form (6.27).

We now apply these results and their proof to the class of production functions which has homogeneity as an essential property. That class plays an important role in several parts of the economic literature. Homogeneity expresses that multiplying each input variable ("production factor") by λ results in multiplying the output (or, if several products are produced, their *value*) by λ^r.

If $g : \mathbb{R}^n_+ \to \mathbb{R}_+$ is a "*microeconomic*" production function, then the function value $g(\mathbf{s})$ is the *maximal* output (or output value) which can be produced (or established, respectively) in an *enterprise* during a given time period by the production factor quantities $s_1, \ldots, s_n$. If g is *homogeneous* and $r = 1$ (that is, g is *linearly homogeneous*) then we speak of *constant returns to scale* (compare Sect. 3.3) while, if $0 < r < 1$ or $r > 1$ then *the returns to scale are decreasing* or *increasing*, respectively. Clearly, $r \leq 0$ would make no economic sense.

The assumption that g is *differentiable* contains an assumption which clearly can hold only approximately, namely that the inputs and outputs *can be divided into arbitrarily small quantities*. The partial derivatives

$$\frac{\partial g(\mathbf{s})}{\partial s_1}, \ldots, \frac{\partial g(\mathbf{s})}{\partial s_n}$$

are the *marginal products* (compare Sect. 6.1) of the production factor quantities $s_1, \ldots, s_n$.

In the *marginal theory of distribution* it is assumed that the production factors, including labour, are rewarded according to their marginal product. (For instance, somewhat simplistically, new workers are employed if it is expected that they will produce more (value of) additional goods than would pay their wages, and workers are laid off after a while if they would have to be paid more than their additional contributions to production). We can consider

$$\frac{\partial g(\mathbf{s})}{\partial s_1} s_1, \ldots, \frac{\partial g(\mathbf{s})}{\partial s_n} s_n$$

as the compensations (measured in quantities of the output or output value) given to the individual production factors. But these are the terms on the right hand side of Euler's differential equation (6.27). So if $r = 1$ then, by (6.27), the sum of compensations uses up the whole production. If $r < 1$ then after all compensations a surplus is left, namely $(1 - r)g(s_1, \ldots, s_n)$. Finally, if $r > 1$ then the output (or output value) $g(s_1, \ldots, s_n)$ is insufficient to pay for the compensations.

These three cases are interesting from the point of view of *production and distribution theory*. In the first case "the distribution problem is solved": the Euler equation describes how to distribute the output (or output value) to compensate the production factors. In the other two cases the employer has surplus or deficit (in this model taxation of enterprises is ignored). In the *dynamical theory of competition* these two situations can persist, at least theoretically, only for a short time, if at all, because, in the first case, additional production is worth while, whereas in the second the enterprise goes bankrupt. There exists also a stronger opinion: absence of linear homogeneity is possible for the duration only if not all production factors have been taken into consideration. If all are considered, then the production function would have to be *linearly* homogeneous (see Sect. 3.3).

On the other hand, g is a "*macroeconomic*" production function if it yields the (gross or net) *national product* (or at least the whole product of a sector of industry) produced during a given time period, say a year, as function of the input values or quantities. For simplicity, one often aggregates the inputs into labour L, capital K, and possibly also energy E (measured, say, in hours of work, money and energy units). Again the case where g is *linearly* homogeneous is of practical importance. Some special linearly homogeneous functions, such as the Cobb-Douglas production functions (see Sect. 8.4), are particularly useful. An example of such a production function and its Euler equation is

$$g(L,K) = cL^{0.7}K^{0.3} \qquad (c \in \mathbb{R}_{++},\ \text{constant}),$$
$$\begin{aligned}\frac{\partial g}{\partial L}L + \frac{\partial g}{\partial K}K &= 0.7cL^{0.7}K^{0.3} + 0.3cL^{0.7}K^{0.3}\\ &= 0.7g(L,K) + 0.3g(L,K)\\ &= g(L,K),\end{aligned}$$

that is, 70 % of the (this time net) national product goes to labour, 30 % to capital. The converse question, whether such distribution implies $g(L,K) = cL^{0.7}K^{0.3}$ at least approximately, is also of interest. The answer is positive. We show that in Sect. 8.4.

6.12.1 Excercises

1. Determine at $(x_1,x_2) = (p_1,p_2)$ the Jacobian matrix $\frac{d\mathbf{f}(\mathbf{x})}{d\mathbf{x}}$ of the function $\mathbf{f}: \mathbb{R}^2 \to \mathbb{R}^2$ given by $\mathbf{f}(\mathbf{x}) = (2x_1 - 3x_2 + 4x_1x_2,\ x_1^3 - x_2^3)$.
2. Determine at $(y_1,y_2) = (q_1,q_2)$ the Jacobian matrix $\frac{d\mathbf{g}(\mathbf{y})}{d\mathbf{y}}$ of the function $\mathbf{g}: \mathbb{R}^2 \to \mathbb{R}^2$ given by $\mathbf{g}(\mathbf{y}) = (y_1 - y_2,\ y_1^2 + 2y_2)$.
3. For the composition $\mathbf{g} \circ \mathbf{f}$ of the functions $\mathbf{f}$ and $\mathbf{g}$ defined in Excercises 1 and 2, respectively,
   (a) determine the Jacobian matrix at $\mathbf{x} = \mathbf{p}$.
   (b) Show that $\frac{d(\mathbf{g}\circ\mathbf{f})}{d\mathbf{x}}(\mathbf{p}) = \frac{d\mathbf{g}}{d\mathbf{y}}(\mathbf{q})\frac{d\mathbf{f}}{d\mathbf{x}}(\mathbf{p})$.
   (Notice that $q_1 = 2p_1 - 3p_2 + 4p_1p_2$, $q_2 = p_1^3 - p_2^3$).
4. Show that the function $h: \mathbb{R}^2_{++} \to \mathbb{R}$ given by $h(x_1,x_2) = (4x_1^{\frac{1}{2}} + 5x_2^{\frac{1}{2}})^3$.
   (a) is homogeneous of degree 1.5 and
   (b) check that it satisfies the equation

$$h(x_1,x_2) = \frac{2}{3}\left(\frac{\partial h(x_1,x_2)}{\partial x_1}x_1 + \frac{\partial h(x_1,x_2)}{\partial x_2}x_2\right).$$

5. Show that the function $f : \mathbb{R}^2_{++} \to \mathbb{R}$ given by $f(x_1, x_2) = \frac{3x_1^4 x_2^5}{6x_1^8 + 7x_2^8}$
   (a) is linearly homogeneous and
   (b) check that it satisfies the equation

$$f(x_1, x_2) = \frac{\partial f(x_1, x_2)}{\partial x_1} x_1 + \frac{\partial f(x_1, x_2)}{\partial x_2} x_2.$$

6.12.2 Answers

1. $\frac{d\mathbf{f}}{d\mathbf{x}}(\mathbf{p}) = \begin{pmatrix} 2 + 4p_2 & -3 + 4p_1 \\ 3p_1^2 & -3p_2^2 \end{pmatrix}$.
2. $\frac{d\mathbf{g}}{d\mathbf{y}}(\mathbf{q}) = \begin{pmatrix} 1 & -1 \\ 2q_1 & 2 \end{pmatrix}$.
3. (a) $\frac{d(\mathbf{g} \circ \mathbf{f})}{d\mathbf{x}}(\mathbf{p}) = \begin{pmatrix} 2 + 4p_2 - 3p_1^2 & -3 + 4p_1 + 3p_2^2 \\ 8p_1 - 24p_2 + 32p_1p_2 & -12p_1 + 36p_2 - 72p_1p_2 \\ +6p_1^2 - 48p_2^2 + 32p_1p_2^2 & +16p_1^2 - 6p_2^2 + 32p_1^2p_2 \end{pmatrix}$.
   (b) Write $2p_1 - 3p_2 + 4p_1p_2$ for q_1 in $\begin{pmatrix} 1 & -1 \\ 2q_1 & 2 \end{pmatrix}$ and multiply by the first matrix from the left. You get the third matrix.
4. (a) $h(\lambda x_1, \lambda x_2) = (4(\lambda x_1)^{1/2} + 5(\lambda x_2)^{1/2})^3 = (\lambda^{1/2} 4x_1^{1/2} + \lambda^{1/2} 5x_2^{1/2})^3$
   $= (\lambda^{1/2}(4x_1^{1/2} + 5x_2^{1/2})^3 = \lambda^{3/2} h(x_1, x_2)$.
   (b) $\frac{2}{3}\left(\frac{\partial h(x_1,x_2)}{\partial x_1} x_1 + \frac{\partial h(x_1,x_2)}{\partial x_2} x_2\right)$
   $= \frac{2}{3}(3(4x_1^{1/2} + 5x_2^{1/2})^2 4 \cdot \frac{1}{2} x_1^{-1/2}) x_1 + 3(4x_1^{1/2} + 5x_2^{1/2})^2 5 \cdot \frac{1}{2} x_2^{-1/2}) x_2)$
   $= (4x_1^{1/2} + 5x_2^{1/2})^2 (4x_1^{1/2} + 5x_2^{1/2}) = h(x_1, x_2)$
5. (a) $f(\lambda x_1, \lambda x_2) = 3(\lambda x_1)^4 (\lambda x_2)^5 / [6(\lambda x_1)^8 + 7(\lambda x_2)^8]$
   $= 3\lambda^9 x_1^4 x_2^5 / \lambda^8 [6x_1^8 + 7x_2^8]$
   $= \lambda f(x_1, x_2)$.
   (b) $\frac{\partial f(x_1,x_2)}{\partial x_1} x_1 + \frac{\partial f(x_1,x_2)}{\partial x_2} x_2$
   $= \frac{3x_1^4 x_2^5 [4(6x_1^8 + 7x_2^8) - 48x_1^8 + 5(6x_1^8 + 7x_2^8) - 56x_2^8]}{(6x_1^8 + 7x_2^8)^2} = f(x_1, x_2)$.

6.13 Implicit Functions

Often a function of one variable is given "implicitly" in terms of a function of two variables (or a function of several variables in terms of functions of more variables). The contour lines (isoquants) in Sect. 3.3 give important examples. They were described by

$$F(x, y) = c. \tag{6.28}$$

If we are lucky, such an equation (with c fixed) determines y as function of x.

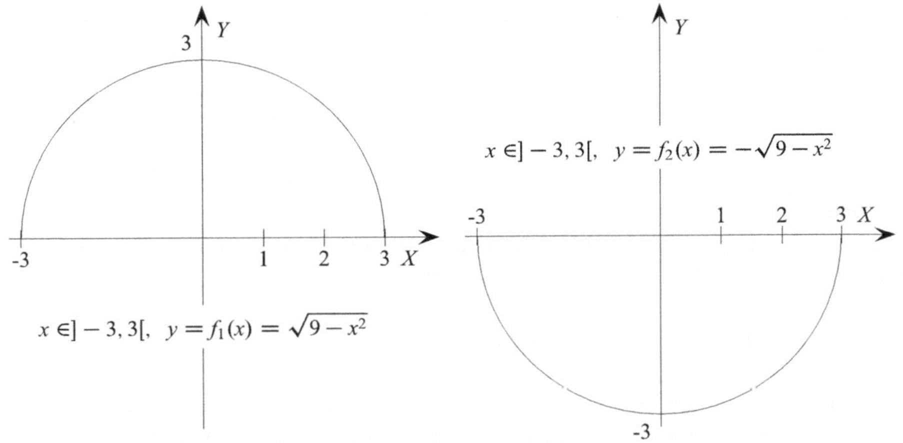

Fig. 6.39 Graphs of the two functions f_1, f_2 satisfying $x^2 + f_k(x) = 9$ ($k = 1, 2$). Both are differentiable on $]-3, 3[$ but not at -3 and 3

The simplest case is illustrated by

$$xy = c \quad \text{which determines} \quad y = \frac{c}{x}$$

for $x, y \in \mathbb{R}_{++}$ if $c \in \mathbb{R}_{++}$. If 0 is permitted as x, y or c, complications start to arise. Somewhat more complicated is

$$x^2 + y^2 = c. \tag{6.29}$$

If $c < 0$, there are *no* real x, y which satisfy such an equation. If $c = 0$ then it is satisfied *only* by $x = 0$ and $y = 0$, which gives a very primitive function indeed (it maps $\{\mathbf{0}\}$ onto $\{\mathbf{0}\}$). Things improve when $c > 0$; actually we get more than we bargained for: *several* functions defined on $]-\sqrt{c}, \sqrt{c}[$ (see Fig. 6.39). For instance, for every $x_0 \in]-\sqrt{c}, \sqrt{c}[$ there exists a function f given by

$$f(x) = \begin{cases} \sqrt{c - x^2} & \text{for } x \in]-\sqrt{c}, x_0] \\ -\sqrt{c - x^2} & \text{for } x \in]x_0, \sqrt{c}[\end{cases}$$

such that (6.29) is fulfilled with $y = f(x)$. This gives infinitely many functions on $]-\sqrt{c}, \sqrt{c}[$ since the choices of x_0 can be made arbitrarily in $]-\sqrt{c}, \sqrt{c}[$. Notice that f is not continuous at x_0. By dividing $]-\sqrt{c}, \sqrt{c}[$ into a finite or infinite number of intervals and by choosing on them alternatively $\sqrt{c - x^2}$ and $-\sqrt{c - x^2}$ as function values, one can get as many discontinuities as one wants (even infinitely many. We also get continuous functions on $]-\sqrt{c},\sqrt{c}[$, but only two, given by

$$y=\sqrt{c-x^2} \quad \text{and} \quad y=-\sqrt{c-x^2}.$$

For $x<-\sqrt{c}$ or $x>\sqrt{c}$ there exists *no* y satisfying (6.29), for $x=\sqrt{c}$ or $x=-\sqrt{c}$ just one: $y=0$.

It is often impossible or inconvenient to express y as function of x from an equation of the form (6.28). Take for instance

$$1+4x-2x^2+15xy-5xy^3-x^2y^3+3xy^5=11. \tag{6.30}$$

Calculating y as a function of x would mean the solution of an equation of fifth degree. If we wanted to determine the derivative of this function, say at $x=5$, from this solution, it would have to be in a form which permits derivation, no mean feat (for solutions of equations of fifth and higher degrees no explicit algebraic formula can be found in general). However, if we guess that for $x=5$ we have, as solution of (6.30), $y=1$ (which is easy) and no others (which is not so easy, but see below), we can use the following result of which we prove here only the second part.

If F is a function of two variables x and y with continuous partial derivatives $\frac{\partial F}{\partial x}$ and $\frac{\partial F}{\partial y}$ at a point (x_0,y_0) for which $F(x_0,y_0)=0$ then there is a neighborhood N of x_0 on which

$$F(x,y)=0 \tag{6.31}$$

determines y uniquely as a differentiable (and thus continuous) function of x:

$$y=f(x) \quad \textit{satisfying} \quad f(x_0)=y_0$$

and we have

$$f'(x)=-\frac{\partial F/\partial x}{\partial F/\partial y}(x,f(x)) \tag{6.32}$$

on N.

(The right hand side means that we calculate

$$-\frac{\partial F}{\partial x}(x,y) \quad \text{and} \quad \frac{\partial F}{\partial y}(x,y),$$

divide them and then substitute $f(x)$ for y).

Accepting without proof the existence and differentiability of f which satisfies (6.31), that is,

$$F(x,f(x))=0, \tag{6.33}$$

on N, we prove (6.32) by differentiating (6.33) with respect to x, using the chain rule:

$$0 = \frac{d}{dx}F(x,f(x)) = \frac{\partial F}{\partial x}(x,f(x)) + \frac{\partial F}{\partial y}(x,f(x))\frac{df(x)}{dx}$$

and dividing by $\frac{\partial F}{\partial y} \neq 0$.

Notice that in (6.30) $F(x,y) = -10 + 4x - 2x^2 + 15xy - 5xy^3 - x^2y^3 + 3xy^5$. So we get

$$f'(x) = -\left.\frac{\partial F/\partial x}{\partial F/\partial y}\right|_{y=f(x)} = -\left.\frac{4 - 4x + 15y - 5y^3 - 2xy^3 + 3y^5}{15x - 15xy^2 - 3x^2y^2 + 15xy^4}\right|_{y=f(x)}.$$

But we are interested in $f'(5)$ and know that $f(5) = 1$, so we get, without having to solve (6.30),

$$f'(5) = \frac{13}{150}.$$

By the way, the fact mentioned above, that Eq. (6.30) for $x = 5$, that is

$$-10 + 20 - 50 + 75y - 25y^3 - 25y^3 + 15y^5 = 0 \tag{6.34}$$

has just one solution ($y = 1$) is true because the derivative of the left hand side of (6.34) with respect to y,

$$75 - 150y^2 + 75y^4 = 75(1 - y^2)^2 > 0 \quad (= 0 \text{ only for } y = 1),$$

so the left hand side of (6.34) strictly increases with y (see Eq. 6.33). Therefore it cannot be 0 for more than one y-value. However, even if (6.31) has (say) $N > 1$ solutions $y_1, \ldots, y_N$ for an $x = x_0$, as is the case for (6.29) (with $N = 2$) in the case $c > 0$ and $x_0 \in]-\sqrt{c}, \sqrt{c}[$, for each solution y_k there is a unique continuous, differentiable f_k with $F(x, f_k(x)) = 0$ ($k = 1, \ldots, N$)—as long as the above conditions are satisfied in particular $\frac{\partial F}{\partial y} \neq 0$. Also if (6.32) holds then for each f_k. However, if $\frac{\partial F}{\partial y} = 0$, as for (6.29) at $(-\sqrt{c}, 0)$ and $(\sqrt{c}, 0)$, then there may be no (finite) derivative (see Fig. 6.39) or no nontrivial function as for $x^2 + y^2 = 0$ at $x = 0$.

Without proof we formulate here a generalization of the result concerning (6.31) since this generalization is often needed for applications in economics (see Sects. 8.7 and 8.7). Instead of (6.31) we consider now the equation

$$\mathbf{F}(\mathbf{x}, \mathbf{y}) = \mathbf{0}, \tag{6.35}$$

where $\mathbf{F}$ is a vector-valued function of the (real) vectors $\mathbf{x} = (x_1, \ldots, x_p)$ and $\mathbf{y} = (y_1, \ldots, y_q)$. In what follows, $\mathbf{F}$ will be an $\mathbb{R}^q$-valued function, that is,

$$\mathbf{F} = \begin{pmatrix} F_1 \\ \vdots \\ F_q \end{pmatrix} \qquad \text{with real-valued functions } F_1, \ldots, F_q.$$

If the partial derivatives in the Jacobian matrices $\frac{\partial \mathbf{F}}{\partial \mathbf{x}}$ *and* $\frac{\partial \mathbf{F}}{\partial \mathbf{y}}$ (see Sect. 6.11) *are continuous at a point* $(\mathbf{x}_0, \mathbf{y}_0)$ *for which* $\mathbf{F}(\mathbf{x}_0, \mathbf{y}_0) = \mathbf{0}$ *and* $\frac{\partial \mathbf{F}}{\partial \mathbf{y}}$ (as a (q, q)-matrix) *is invertible, then there is a neighborhood* N *of* $\mathbf{x}_0$ *on which* $\mathbf{F}(\mathbf{x}, \mathbf{y}) = \mathbf{0}$ *determines* $\mathbf{y}$ *uniquely as a differentiable (and thus continuous) function of* $\mathbf{x}$, $\mathbf{y} = \boldsymbol{f}(\mathbf{x})$, *satisfying* $\boldsymbol{f}(\mathbf{x}_0) = \mathbf{y}_0$ *and we have*

$$\frac{d\boldsymbol{f}}{d\mathbf{x}}(\mathbf{x}) = -\left(\frac{\partial \mathbf{F}}{\partial \mathbf{y}}\right)^{-1} \frac{\partial \mathbf{F}}{\partial \mathbf{x}}(\mathbf{x}, \boldsymbol{f}(\mathbf{x})) \tag{6.36}$$

on N.

Compare this to (6.32). Since in (6.32)

$$\frac{1}{\partial F / \partial y} \qquad \text{can be written} \qquad \left(\frac{\partial F}{\partial y}\right)^{-1}$$

we see that Eqs. (6.36) and (6.32) are of the same form. The right hand side of (6.36) means the same as that in (6.32), but notice that in (6.36) we have to multiply *from the left* by the inverse of the matrix $\frac{\partial \mathbf{F}}{\partial \mathbf{y}}$, while in (6.32)

$$\left(\frac{\partial F}{\partial y}\right)^{-1} \frac{\partial F}{\partial x} = \frac{\partial F}{\partial x} \left(\frac{\partial F}{\partial y}\right)^{-1}.$$

6.13.1 Excercises

1. For each of the equations below determine, whether there exists an implicit function expressing y in terms of x around the point $(x, y) = (1, 2)$.
   (a) $16x^4 + y^4 - 32 = 0$,
   (b) $x^3 + 2x^2y - xy^2 - 1 = 0$,
   (c) $3x^2 + 4xy - y^5 + 21 = 0$.
   If your answer is affirmative, find $\frac{dy}{dx}$ and evaluate it at the said point.
2. Given the equation $x_1^4 + 2x_1 \cos x_2 + \sin y = 0$, is there an implicit function $y = f(x_1, x_2)$ defined around the point $(x_1, x_2, y) = (0, 0, 0)$?
   If your answer is affirmative, find $\frac{\partial y}{\partial x_1}$, $\frac{\partial y}{\partial x_2}$ and evaluate it at the said point.

3. (a) Show that the equation

$$x_1^3 + 2x_1^2 x_2 - 3x_2^3 x_3 + x_3^4 - 2x_1 y + y^2 - 22 = 0$$

implicitly defines a positiv-valued function $y = f(x_1, x_2, x_3)$ around the point $(x_1, x_2, x_3, y) = (1, 2, 3, 4)$.

(b) Find $\frac{\partial y}{\partial x_1}, \frac{\partial y}{\partial x_2}, \frac{\partial y}{\partial x_3}$ and evaluate it at that point.

4. (a) Show that the systems of equations

$$\begin{aligned} x^2 + 2y_1^2 + \ y_2^2 - 4 &= 0 \\ x^2 + \ y_1^2 - 2y_2^2 \quad &= 0 \end{aligned}$$

implicitly defines two positive-valued functions $y_1 = f_1(x)$, $y_2 = f_2(x)$ around the point $(x, y_1, y_2) = (1, 1, 1)$.

(b) Determine $\frac{\partial y_1}{\partial x}$ as an expression involving x, y_1 and evaluate it at $(x, y_1) = (1, 1)$.

(c) Determine $\frac{\partial y_2}{\partial x}$ as an expression involving x, y_2 and evaluate it at $(x, y_2) = (1, 1)$.

5. (a) Show that the systems of equations

$$\begin{aligned} x_1^2 + 2x_2^2 + 3y_1^2 + 4y_2^2 - 10 &= 0 \\ x_1^2 + \ x_2^2 - \ y_1^2 - \ y_2^2 \quad &= 0 \end{aligned}$$

implicitly defines two positive-valued functions $y_1 = f_1(x_1, x_2)$, $y_2 = f_2(x_1, x_2)$ around the point $(x_1, x_2, y_1, y_2) = (1, 1, 1, 1)$.

(b) Determine $\frac{\partial y_1}{\partial x_1}$ and $\frac{\partial y_1}{\partial x_2}$ as expressions involving x_1, y_1 and x_2, y_1, respectively, and evaluate it at $(x_1, x_2, y_1) = (1, 1, 1)$.

(c) Determine $\frac{\partial y_2}{\partial x_1}$ and $\frac{\partial y_2}{\partial x_2}$ as expressions involving x_1, y_2 and x_2, y_2, respectively, and evaluate it at $(x_1, x_2, y_2) = (1, 1, 1)$.

6.13.2 Answers

1. (a) Yes; $\frac{dy}{dx} = -\frac{64x^3}{4y^3} = -\frac{16x^3}{y^3} = -\frac{16}{8} = -2$,

(b) Yes; $\frac{dy}{dx} = -\frac{3x^2+4xy-y^2}{2x^2-2xy} = -\frac{7}{-2} = 3.5$,

(c) Yes; $\frac{dy}{dx} = -\frac{6x+4}{4x-5y^4} = -\frac{10}{-76} = \frac{5}{38}$,

2. Yes; $\frac{dy}{dx_1} = -\frac{4x_1^3+2\cos x_2}{\cos y} = -\frac{2}{1} = -2$, $\frac{dy}{dx_2} = -\frac{-2x_1 \sin x_2}{\cos y} = \frac{0}{1} = 0$.

3. (a) $y = x_1 + (x_1^2 - x_1^3 - 2x_1^2x_2 + 3x_2^3x_3 - x_3^4 + 22)^{1/2}$ (defined for (x_1, x_2, x_3) sufficiently close to $(1, 2, 3)$). Take the *positive* square root and notice that $y = 4$ for $(x_1, x_2, x_3) = (1, 2, 3)$.
   (b) $\frac{dy}{dx_1} = -\frac{3x_1^2+4x_1x_2-2y}{2(y-x_1)} = -\frac{1}{2}$,
   $\frac{dy}{dx_2} = -\frac{2x_1^2-9x_2^2x_3}{2(y-x_1)} = \frac{53}{3}$,
   $\frac{dy}{dx_3} = -\frac{-3x_2^3+4x_3^3}{2(y-x_1)} = -14$,
4. (a) $y_1 = \left(\frac{8-3x^2}{5}\right)^{1/2}$ (defined for x sufficiently close to 1), $y_2 = \left(\frac{4+x2}{5}\right)^{1/2}$. Take the *positive* square roots in both cases.
   (b) $\frac{dy_1}{dx} = -\frac{3x_1}{5y_1} = -\frac{3}{5}$,
   (c) $\frac{dy_2}{dx} = \frac{x}{5y_2} = \frac{1}{5}$.
5. (a) $y_1 = (5x_1^2 + 6x_2^2 - 10)^{1/2}$, $y_2 = (10 - 4x_1^2 - 5x_2^2)^{1/2}$, both defined for (x_1, x_2) sufficiently close to $(1, 1)$. Take the *positive* square roots in both cases and notice that $y_1 = y_2 = 1$ for $(x_1, x_2) = (1, 1)$.
   (b) $\frac{\partial y_1}{\partial x_1} = \frac{5x_1}{y_1} = 5$, $\frac{\partial y_1}{\partial x_2} = \frac{6x_2}{y_1} = 6$.
   (c) $\frac{\partial y_2}{\partial x_1} = -\frac{4x_1}{y_2} = -4$, $\frac{\partial y_2}{\partial x_2} = \frac{-5x_2}{y_2} = -5$.

Nonlinear Functions of Interest to Economics. Systems of Nonlinear Equations 7

Unhappiness is realising almost everything is nonlinear.

Based loosely on ED ADAMS
motto at beginning of Chapter 4

7.1 Introduction

The processes which were objects of the first six chapters were mostly linear or, like differentials served for linear approximation of nonlinear functions of one or several variables.

In Chap. 6 we dealt with many functions; we mentioned also applications. The functions and classes of functions on which we concentrate in this chapter are of particular interest for applications: homogeneous functions and their generalisations (Sect. 7.4), in particular CD (Cobb–Douglas) and other CES (Constant Elasticity of Substitution) functions (Sect. 7.5) as production functions. For this we will need further elementary functions, the logarithms and the exponential functions (Sect. 7.2). The latter will be used also to extend the rule for differentiating power functions to the case of irrational exponents (Sect. 7.2) and will be applied to compounding and discounting (Sect. 7.3). It also provides occasion to the further discussion (Compare Sects. 3.5, 3.6 and 6.7) of convex functions (Sect. 7.2).

In Sect. 7.3 we will see that the question important in everyday life, how much time it takes for a deposit to double, leads to solving the nonlinear equation

$$e^{rt} = 2$$

for t, the doubling time for a deposit under daily compounding or under the smoother "continuous" compounding with compounding rate r.

W. Eichhorn, W. Gleißner, *Mathematics and Methodology for Economics*, Springer Texts in Business and Economics, DOI 10.1007/978-3-319-23353-6_7

Similar nonlinear equations like

$$e^t = t + 2, \qquad e^t = t + 1, \qquad \text{or} \qquad e^t = t$$

have exactly two solutions, one solution, or no solution at all, respectively. Already these examples suggest that the theory of nonlinear equations and their solutions may be difficult. Section 7.6 is devoted to questions of solving not only single nonlinear equations but also systems of them. In this connection, nonlinear vector–valued functions of several variables and Banach's fixed point theorem play an important role.

7.2 Exponential and Logarithm Functions. Powers with Arbitrary Real Exponents. Conditions for Convexity and Applications

In what follows let a (the *base*) be any positive real number. As we know, for *integer exponents* m the power a^m is defined recursively by $a^0 = 1$, $a^1 = a$, $a^{m+1} = a^m a$, $a^{-m} = 1/a^m$ ($m = 1, 2, \ldots$). It is easy to see (by induction, Appendix) that $a^{m+n} = a^m a^n$ and $a^{mn} = (a^m)^n = (a^n)^m$ for $m = 1, 2, \ldots$ and $n = 1, 2, \ldots$.

For *rational exponents* $r = m/n$ (with integer m and positive integer n) the power is

$$a^{m/n} = (\sqrt[n]{a})^m$$

where the nth root is the inverse function of the nth power, applied to $a > 0$. It is easy to see that this definition is unambiguous, that is, $a^{m/n} = a^{m'/n'}$ when $m/n = m'/n'$.

For example,

$$a^{6/4} = \sqrt[4]{a^6} = \sqrt{\sqrt{(a^3)^2}} = \sqrt{a^3} = a^{3/2}.$$

From the $a^m = a^{m-1}a$ part of the above definition

$$\begin{aligned} a^{(m/n)+(m'/n)} &= a^{(m+m')/n} = (\sqrt[n]{a})^{m+m'} = (\sqrt[n]{a})^{m+m'-1}\sqrt[n]{a} \\ &= (\sqrt[n]{a})^{m+m'-2}(\sqrt[n]{a})^2 = \cdots = (\sqrt[n]{a})^m \cdot (\sqrt[n]{a})^{m'} \\ &= a^{m/n} \cdot a^{m'/n} \end{aligned}$$

if m and m' are positive, and the $a^{-m} = 1/a^m$ and $a^0 = 1$ parts of the definition take care of the proof of a similar equality when m and/or m' are negative or 0. So

$$a^{x+x'} = a^x a^{x'} \qquad \text{for all rational } x, x'$$

(we can always choose $x = m/n$ and $x' = m'/n$ with common denominator,—here the positive integer n). Similarly,

$$(a^x)^{x'} = a^{xx'} \qquad \text{for all rational } x, x'.$$

Note that, if $m/n > 0$ then $a^{m/n} > 1$ for $a > 1$ and $a^{m/n} < 1$ for $a < 1$. (*Proof:* $a^n > 1$ if $a > 1$, $n > 0$; $a^n < 1$ if $a < 1$, $n > 0$, so $a^{1/n} = \sqrt[n]{a} > 1$ for $n > 0$, $a > 1$ because $\sqrt[n]{a} \leq 1$ would imply $a = (\sqrt[n]{a})^n \leq 1$, a contradiction; furthermore, $a^{m/n} = (\sqrt[n]{a})^m > 1$ for $a > 1$, $m/n > 0$; similarly $a^{m/n} < 1$ for $a < 1$, $m/n > 0$.) Consequently, if $a > 1$, then a^x *strictly increases with increasing rational* x (if $x'' > x$, say $x'' = x + x'$, $x' > 0$, then $a^{x''} = a^x a^{x'}$) or, what is the same, a^x *strictly decreases with decreasing rational* x.

For *powers with irrational (real) exponents* the definition $a^x = \lim_{n\to\infty} a^{r_n}$ seems appropriate, where $\lim_{n\to\infty} r_n = x$, the r_n $(n = 1, 2, \ldots)$ are rational, and $\{r_n\}$ increases but $\{R_n\} = \{r_n + 1/n\}$ decreases, (the introduction of R_n makes arguments easier).

We have to *prove* that $\lim_{n\to\infty} a^{r_n}$ exists and is unique (the same for different sequences of rational numbers tending to x). For this we need only the limit

$$\lim_{n\to\infty} a^{1/n} = \lim_{n\to\infty} \sqrt[n]{a} = 1 \qquad (a > 0).$$

Proof Let, say $a > 1$. As we have seen, $\{a^{1/n}\}$ decreases with n but each term of this sequence is greater than 1. But then (compare Fig. 7.1) $b = \lim_{n\to\infty} a^{1/n} \geq 1$ exists. Taking the limit of $a^{1/n} = a^{1/(2n)} a^{1/(2n)}$ as $n \to \infty$, we get $b = b \cdot b$, thus $b = 1$ ($b = 0$ is excluded by $b \geq 1$), as asserted.

Furthermore, since $\{r_n\}$ is increasing and $\{R_n\} = \{r_n + 1/n\}$ decreasing, $\{a^{R_n}\}$ decreasing, moreover,

$$a^{r_n} \leq a^{R_n} = a^{r_n + 1/n} = a^{r_n} a^{1/n}.$$

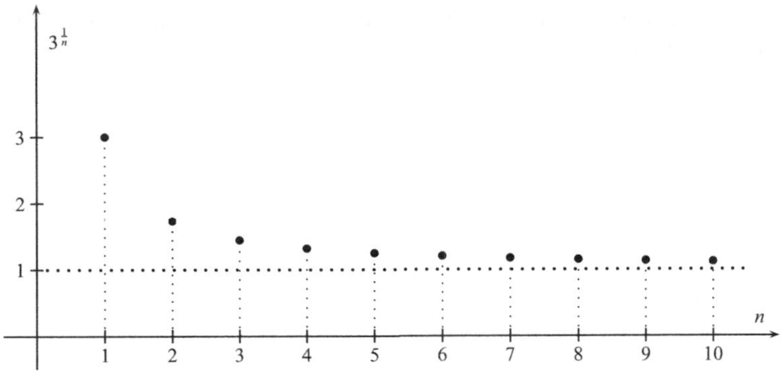

Fig. 7.1 Decreasing sequences bounded from below by c are convergent and their limit is $\geq c$

In particular, $\{a^{R_n}\}$ is bounded from below (by a^{r_1}) and decreasing, so has a limit A and the increasing sequence $\{a^{r_n}\}$ is bounded from above by a^{R_1} and has a limit B. From $\lim_{n\to\infty} a^{1/n} = 1$ and from $a^{R_n} = a^{r_n}a^{1/n}$ we prove $A = B$. Indeed, going to the limit as $n \to \infty$ (limit of product is product of limits), we have $A = B \cdot 1 = B$. By definition, this common limit will be a^x. Since a^x increases with increasing rational x, it is easy to see that *every* sequence $\{r_n\}$ of the above nature ($\{r_n\}$ increasing, $\{r_n + 1/n\}$ decreasing, x in between) leads to the same A. (That such sequences $\{r_n\}$ exist, is easy: Choose r_1 so that $x - 1 < r_1 \leq x$ and choose the further $r_2, \ldots, r_n, r_{n+1}, \ldots$ so that

$$x - \frac{1}{n} < r_n \leq x, \qquad r_n \leq r_{n+1} < r_n + \frac{1}{n} - \frac{1}{n+1}.$$

Then $r_n \leq x$, $r_n \leq r_{n+1}$, $R_{n+1} = r_{n+1} + \frac{1}{n+1} < r_n + \frac{1}{n} = R_n$ and $R_n > x$.)

This definition conforms with the previous one, if r in a^r is rational. We just choose $r_n = r$ ($n = 1, 2, \ldots$), a constant sequence. Then a^x we have just defined for $a > 1$ is still *strictly increasing* in the case of real x: If $x_1 < x_2$, take a lower approximating rational r of x_2 and an upper one, R, of x_1 so that $R < r$ (can be done: R gets as close to x_1 and r to x_2 as we want them to) and get, by the definition of a^{x_1}, a^{x_2} and since a^x is strictly increasing for rational x,

$$a^{x_1} < a^R < a^r < a^{x_2}$$

as asserted. Also, a^x is a *continuous function of* x. Take, for instance, $x > x_0$; then we want

$$0 \leq |a^x - a^{x_0}| = a^x - a^{x_0} < \varepsilon \quad \text{if} \quad 0 < x - x_0 < \delta.$$

Let R be an upper approximation fraction of x_0 for which $a^R - a^{x_0} < \varepsilon$. Now choose $\delta < R - x_0$ and choose x so that $0 < x - x_0 < \delta < R - x_0$. Then $x_0 < x < R$ and $|a^x - a^{x_0}| = a^x - a^{x_0} < a^R - a^{x_0} < \varepsilon$ as required. (The proof for $x < x_0$ is similar). If $a < 1$ then a^x is a strictly decreasing but still a continuous function of x. If $a = 1$ then, of course, $a^x = 1^x = 1$ (Fig. 7.2).

$$\begin{gathered} f((1-q)x_1 + qx_2) < (1-q)f(x_1) + qf(x_2) \\ (0 < q < 1; x_1 \neq x_2, x_1 \in I, x_2 \in I, I = [a,b]). \end{gathered} \tag{7.1}$$

If this is true for all $q \in]0,1[$, $x_1 \neq x_2$, $x_1 \in I$, $x_2 \in I$ with $\leq$ instead of $<$ then f is *convex from below* on I. If we have $\geq$ or $>$, we get the definition of functions *convex* or *strictly convex*, respectively, *from above*.

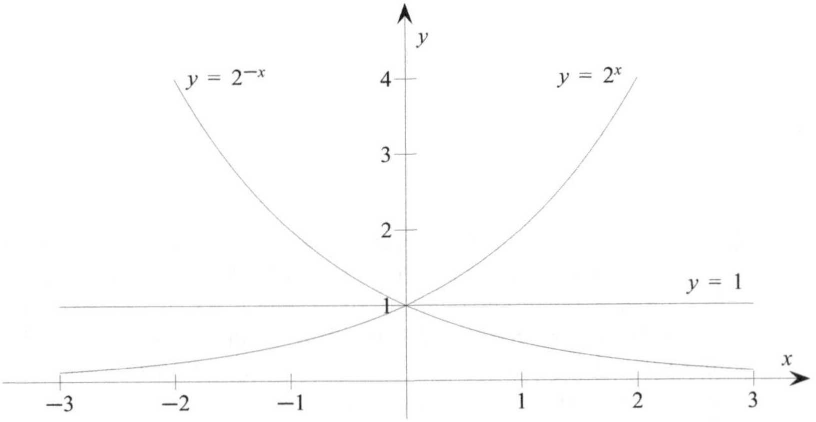

Fig. 7.2 Exponential functions

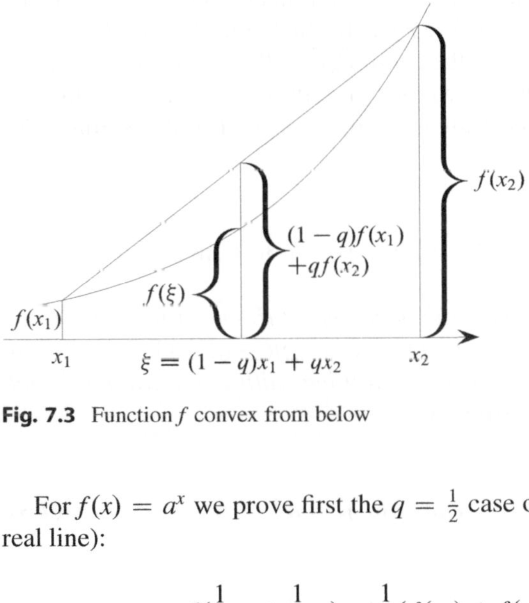

Fig. 7.3 Function f convex from below

The function $x \mapsto a^x$ is the *exponential function* with base a. A further important property of a^x is that it is *convex* (*strictly convex*, if $a \neq 1$) *from below*. As defined in Sect. 3.4, a *function is strictly convex from below* (on an interval I)*if, between its two end points, every chord of its graph is above the graph* (as long as we stay in the interval I). This can be expressed (see Fig. 7.3) by

For $f(x) = a^x$ we prove first the $q = \frac{1}{2}$ case of (7.1) (the interval I is the whole real line):

$$f(\frac{1}{2}x_1 + \frac{1}{2}x_2) < \frac{1}{2}(f(x_1) + f(x_2)) \quad (x1 \neq x_2). \tag{7.2}$$

Indeed $a^{x_1/2+x_2/2} = a^{x_1/2}a^{x_2/2} < \frac{1}{2}(a^{x_1} + a^{x_2})$ because

$$0 < (a^{x_1/2} - a^{x_2/2})^2 = a^{x_1} + a^{x_2} - 2a^{x_1/2}a^{x_2/2} \quad \text{if} \quad x_1 \neq x_2.$$

(The squares of nonzero numbers are positive). This inequality $\sqrt{uv} < \frac{(u+v)}{2}$ ($u \in \mathbb{R}_{++}, v \in \mathbb{R}_{++}, u \neq v$) is the simplest *arithmetic-geometric-inequality*, (we wrote $u = a^{x_1}$, $v = a^{x_2}$). From this, (7.1) follows for all $q \in]0, 1[$ in case of the *continuous* function $f(x) = a^x$.

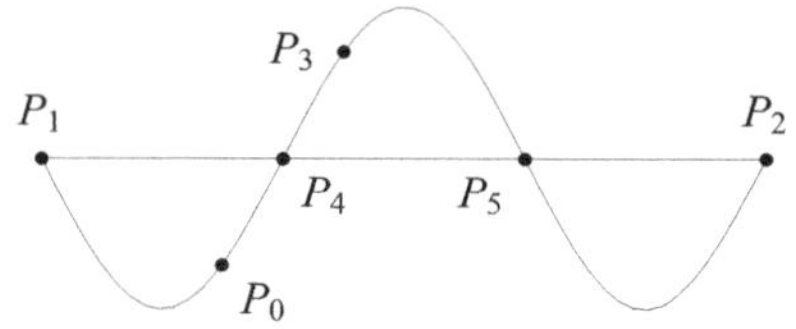

Fig. 7.4 If one point of each chord is above the graph of a continuous function, then all are

Indeed, *if f is continuous and for each pair* $x_1 \neq x_2$, $x_1 \in I$, $x_2 \in I$ *there exists at least one* $q \in]0,1[$ *so that (7.1) holds, then (7.1) holds for all* $q \in]0,1[$, $x_1 = x_2$, $x_1 \in I$, $x_2 \in I$. In other words *if one point of each chord is strictly above the graph of a continuous function, then the function is strictly convex from below.* (Similar statements hold for functions convex from below or above and for functions strictly convex from above). We sketch a proof (Fig. 7.4).

Let P_1, P_2 be the endpoints of a chord and P_0 a point of the graph strictly below that chord. If there existed a P_3 on the graph above the chord $\overline{P_1P_2}$ (say, between P_0 and P_2), then (because of continuity; related to the Property 3 in Sect. 5.3) there would be a last point of the graph, P_4, before P_3, and a first point P_5 after P_3, which would be on the chord. But now look at the chord $\overline{P_4P_5}$: all its points lie under the graph, contrary to supposition. With a little more effort the case where P_3 is both on the graph and on the chord can also be dealt with. So indeed, (7.1) follows from (7.2) for continuous f.

For a^x, inequality (7.1) states that

$$a^{(1-q)x_1+qx_2} < (1-q)a^{x_1} + qa^{x_2} \qquad (q \in]0,1[, u \neq v, \text{ both positive})$$

which is *a more general form of the arithmetic-geometric-mean-inequality*: the inequality between the *weighted arithmetic mean* $(1-q)u + qv$ and the *weighted geometric mean* $u^{1-q}v^q$ $(q \in]0,1[)$. *There are similar inequalities for arithmetic and geometric means of more than two variables.*

The continuity of a^x implies that

$$a^{x+x'} = a^x a^{x'} \quad \text{and} \quad (a^x)^t = a^{xt} \tag{7.3}$$

remain valid for all real x, x', and t.

Since the exponential function (a^x for $a \neq 1$) is strictly monotonic, continuous, and maps the set $\mathbb{R}$ of real numbers onto the set $\mathbb{R}_{++}$ of positive numbers, therefore an inverse function $\log_a$, called the *logarithm* with base a, exists for each $a \neq 1$. The properties (7.3) of a^x imply (with $x_1 = a^x$, $x_2 = a^{x'}$)

$$\log_a(x_1x_2) = \log_a x_1 + \log_a x_2 \quad \text{and} \quad \log_a x_1^t = t\log_a x_1$$

$$(x_1 \in \mathbb{R}_{++},\ x_2 \in \mathbb{R}_{++},\ t \in \mathbb{R}).$$

Also, $\log_a 1 = 0$ (because $a^0 = 1$) and $\log_a$ *is continuous* (for all a), *strictly convex from above and strictly increasing if* $a > 1$, while *strictly decreasing and strictly convex from below if* $a < 1$. The monotonicity statements are obvious consequences of the monotonicity of a^x. We prove the convexity statements. We take the arithmetic-geometric-mean-inequality

$$\frac{x_1 + x_2}{2} > (x_1 x_2)^{1/2} \quad \text{if} \quad x_1 \neq x_2,\ x_1 \in \mathbb{R}_{++},\ x_2 \in \mathbb{R}_{++}. \tag{7.4}$$

Since, for $a < 1$, $\log_a$ is *strictly decreasing*, this is equivalent to

$$\log_a(\tfrac{1}{2}x_1 + \tfrac{1}{2}x_2) < \log_a(x_1 x_2)^{1/2} = \tfrac{1}{2}\log_a x_1 + \tfrac{1}{2}\log_a x_2$$
$$\text{for all} \quad x_1 \in \mathbb{R}_{++},\ x_2 \in \mathbb{R}{++}\ (x_1 \neq x_2;\ a < 1)$$

that is, $\log_a x$ satisfies the last equation, if $a < 1$. As $\log_a x$ is continuous, (7.1) follows and $\log_a$ is strictly convex from below if $a < 1$. But if $a > 1$, the $\log_a$ is strictly increasing, so that taking $\log_a$ on both sides of the inequality (7.4) gives

$$\log_a(\frac{1}{2}x_1 + \frac{1}{2}x_2) > \frac{1}{2}\log_a x_1 + \frac{1}{2}\log_a x_2,$$

and so $\log_a$ is strictly convex from above if $a > 1$.

Now we can *determine* that for the limit $\lim_{x\to\infty} a^x$ $(a > 0)$. First we show that $\lim_{x\to\infty} a^x = 0$ if $0 < a < 1$, that is (see Sect. 6.2), for all $\varepsilon > 0$ there exists an M such that $a^x = |a^x - 0| < \varepsilon$ if $x > M$. For this we choose $M = \log_a \varepsilon$. As we have seen, a^x is *decreasing* with x if $0 < a < 1$, so

$$a^x < a^M - a^{\log_a \varepsilon} = \varepsilon$$

as asserted. If $a > 1$, then $\lim_{x\to\infty} a^x = \infty$, because, for every prescribed (large) M', with $x > \log_a M'$ we have

$$a^x > a^{\log_a M'} = M'$$

since a^x *increases* for $a > 1$. Finally, if $a = 1$, then $\lim_{x\to\infty} 1^x = \lim_{x\to\infty} 1 = 1$ (limit of a constant function). As a consequence (noting also $0^n = 0$),

$$\lim_{n\to\infty} a^n = 0 \text{ if } 0 \le a < 1, \qquad \lim_{n\to\infty} a^n = \infty \text{ if } a > 1, \qquad \lim_{n\to\infty} 1^n = 1,$$

which we needed in Sect. 6.7, Examples 1 and 2. Another consequence is (with $t = -x$)

$$\lim_{x\to-\infty} a^x = \lim_{t\to\infty} (\frac{1}{a})^t = \begin{cases} 0 & \text{if} \quad a > 1 \\ 1 & \text{if} \quad a = 1 \\ \infty & \text{if} \quad 0 \le a < 1. \end{cases}$$

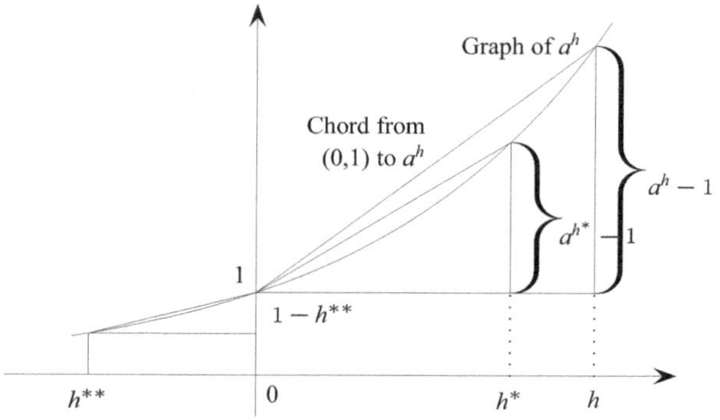

Fig. 7.5 The slopes of the chords, starting from the same point on the graph of a function, convex from below, are increasing

We want now to determine the *derivative* of a^x if it exists, that is, if the following limit exists:

$$\lim_{x \to x_0} \frac{a^x - a^{x_0}}{x - x_0} = \lim_{x \to x_0} a^{x_0} \frac{a^{x-x_0} - 1}{x - x_0} = \lim_{h \to 0} a^{x_0} \frac{a^h - 1}{h}$$

(we have used the rule that, if $\lim f(x)$ exists, then so does $\lim cf(x) = c \lim f(x)$; then wrote $h = x - x_0$). The last limit, if it exists, is *the derivative* of a^x at 0. So *the derivative of a^x exists at every point x_0 if it exists at* 0. But it does exist at 0 since a^x is convex from below, as we have seen, so (Fig. 7.5) the slopes of the chords starting at the point $(0, 1)$ and ending at (h, a^h) are decreasing with decreasing positive h and increasing with increasing negative h (h is increasing, not its absolute value). If $a > 1$ then both kinds of slopes are positive ($a^h > 1 > a^{-h}$ for $h > 0$). So $(a^h - 1)/h$ is decreasing and bounded from below as h decreases to 0 through positive h's; therefore its limit as $h \to 0$ through the positive numbers (the right limit) $\lim_{h \to 0+} \frac{a^h - 1}{h}$ exists (compare Fig. 7.5). If h is negative, $h < 0$, we introduce $h' = -h$.

Then

$$\frac{a^h - 1}{h} = \frac{a^{-h'} - 1}{-h'} = \frac{1}{a^{h'}} \frac{a^{h'} - 1}{h'}.$$

Since $\lim_{h' \to 0} a^{h'} = 1$ (proved the same way as $\lim_{n \to \infty} a^{1/n} = 1$ above) and we have already proved that

$$\lim_{h' \to 0+} \frac{a^{h'} - 1}{h'}$$

exists, we see that also the following left limit exists:

$$\lim_{h \to 0-} \frac{a^h - 1}{h} = \lim_{h' \to 0+} \frac{a^{-h'} - 1}{-h'} = \lim_{h \to 0+} \frac{1}{h'} \lim_{h \to 0+} \frac{a^{h'} - 1}{h'} = \lim_{h \to 0+} \frac{a^{h'} - 1}{h'},$$

and is equal to the right limit. Therefore

$$\lim_{h\to 0} \frac{a^h - 1}{h} = \ell(a) \tag{7.5}$$

exists (it clearly may depend on a, which is why we denote it by $\ell(a)$) and so does

$$(a^x)'_{x=x_0} = \lim_{x\to x_0} \frac{a^x - a^{x_0}}{x - x_0} = a^{x_0}\ell(a), \tag{7.6}$$

that is, a^x is *everywhere differentiable*. The proof is similar if $a \le 1$. It is easy to see that $\ell(a) = 0$ *if and only if* $a = 1$.

We now want to find out more about the function ℓ. By its definition (7.5), $\ell(a)$ *is the slope of the tangent to the graph of* a^x at 0 (look at Fig. 7.5). So $\ell(a^t)$ will be the slope of $(a^t)^x = a^{tx}$ at 0. The graph of a^{tx} is a t-fold horizontal contraction of that of a^x, since we replaced, in a^x, x by tx (Fig. 7.6). So the slope $\ell(a^t)$ of the tangent of a^{tx} at 0 will be t times the slope $\ell(a)$ of that of a^t:

$$\ell(a^t) = t\ell(a). \tag{7.7}$$

If we write $a^t = s$ then $t = \log_a s$, because $\log_a$ is the inverse function of the exponential function with base a. So our equation becomes

$$\ell(s) = \log_a s\ell(a), \tag{7.8}$$

that is, $\ell(s)$ is a constant multiple of a logarithm function. Take any c for which $\ell(c) \ne 0$ (that is, any $c \ne 1$ since $\ell(a) = 0$ only for $a = 1$) and define

$$e = c^{1/\ell(c)}.$$

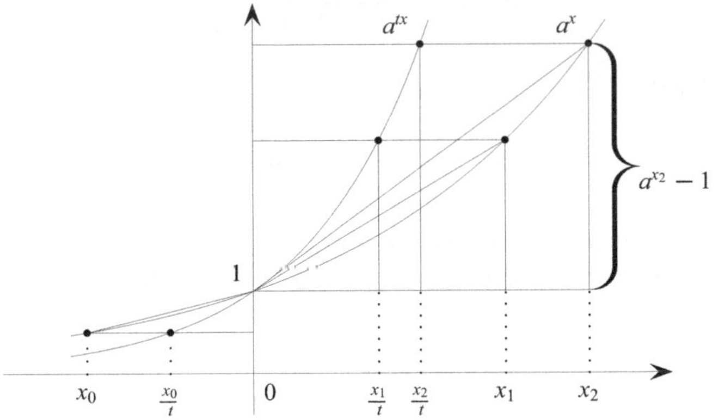

Fig. 7.6 The graph of a^{tx} is a t-fold horizontal contraction of that of a^x

Then, by $\ell(a^t)$ and $\ell(s)$ formulas (7.7) and (7.8) which we have just proved,

$$\ell(e) = \ell(c^{1/\ell(c)}) = \frac{1}{\ell(c)}\ell(c) = 1 \quad \text{and} \quad \ell(s) = \log_e s,$$

that is, ℓ *itself is a logarithm, with base* e. It is easy to see that e does not depend on c, it is simply the value for which $\ell(e) = 1$ (there is only one, because it is easy to see from (7.8) that ℓ is strictly increasing). While $e = c^{1/\ell(c)}$ gives good approximations of e (for instance, if $c = 3$ and $h = 10^{-4} = 0.0001$ then $\ell(3)$ is approximated by $(3^{0.0001} - 1)/0.0001 \approx 1.0987$ and $3^{1/1.0987} \approx 2.718$) there are "nicer" expressions of e, as limits (to be shown later):

$$e = \lim_{n\to\infty}(1 + \frac{1}{n})^n = \lim_{n\to\infty}(1 + \frac{1}{n})^{n+1},$$

It is an irrational number (even *transcendental*, that is, there is no algebraic equation (see Sect. 6.12) of no matter how large degree with rational coefficients of which it is a solution). Its value (to twelve decimals) is

$$e = 2.718281828459\ldots$$

The function $\ell(x) = \log_e x$ is called the *natural logarithm* and is denoted often by $\ln x$. By (7.6),

$$(a^x)' = a^x \ln a,$$

so ln is not so much natural as *unavoidable*: we need it to differentiate a^x—and also to differentiate $\log_a x$—for any $a > 0$, $a \neq 1$ (and $x > 0$): As derivative of an inverse function (Rule 6.5 in Sect. 6.5)

$$(\log_a x)' = \frac{1}{a^{\log_a x}\ln a} = \frac{1}{\ln a}\frac{1}{x}.$$

If $a = e$, we get

$$(e^x)' = e^x \quad \text{and} \quad (\ln x)' = \frac{1}{x}$$

which shows that e^x and $\ln x$ are the "simplest" exponential and logarithm functions, respectively. Moreover, every exponential and logarithmic function can be expressed with their aid:

$$a^x = (e^{\log_e a})^x = e^{x \ln a}$$

and (see also (7.8))

$$\log_a s = \frac{\ln s}{\ln a}.$$

We remind that $\log_a s$ and, in particular, $\ln s$ *are defined only for positive* s.

In addition to exponential and logarithmic functions we can now, extending the rule (6.7) in Sect. 6.5, also differentiate powers with arbitrary real exponents (we have defined them before, at the beginning of this section, only now we denote the base by x and the exponent by p): Since $x = e^{\ln x}$, therefore, by (7.3),

$$x^p = e^{p \ln x} \tag{7.9}$$

and, by the chain rule 6.4 of Sect. 6.5,

$$(x^p)' = e^{p \ln x}(p \ln x)' = x^p p \frac{1}{x} = px^{p-1}.$$

We had above several arguments about convex functions in general. We will now use an argument similar to that which proved the differentiability of a^x, to find conditions for the convexity of differentiable functions in general. As we have seen in Sect. 3.4, *a function is strictly convex from below or from above on an interval if on that interval the arc between two points of its graph is below or above the chord, respectively,* while *for convex functions in the wider sense* we also *permit some points of the chord, other than the endpoints, to be on the arc.* As mentioned there, *one often calls functions convex from below just plainly "convex" while those convex from above are called "concave".*

On the other hand, we saw (Fig. 7.5) that the slopes of the chords, starting from the same point on the graph of a function, convex from below, are increasing. So $h \mapsto (f(x+h)-f(x))/h$ is increasing with increasing positive h or, what is the same, decreasing with h decreasing to 0. But these are the difference quotients of f between x and $x+h$, the limit of which, by definition, is the derivative of x if it exists. Actually, we took $h > 0$ here. The same argument shows that $h \mapsto (f(x+h) - f(x))/h$ increases with negative h's increasing to 0 and its limit is again $f'(x)$ if it exists. So *for differentiable functions f convex from below (in the wider sense),* taking first $x = x_1, h = x_2 - x_1$, then $x = x_2, h = x_1 - x_2 (< 0)$, we get

$$f'(x_1) \leq \frac{f(x_2) - f(x_1)}{x_2 - x_1} \leq f'(x_2),$$

that is, *f' increases.* Similarly, *for differentiable functions f, convex from above ("concave"), f'* decreases.

As in Sect. 6.7 for monotonicity, we want to see whether also the converse is true. Here too we use Taylor's formula with remainder in Lagrange form ((6.11) in Sect. 6.7). So, let us *suppose that*

$$f''(x) \geq 0$$

on a—for simplicity open—*interval* I. By the Taylor formula, just mentioned,

$$f(u) = f(x) + (u - x)f'(x) + \frac{1}{2}(u - x)^2 f''(\xi) \geq f(x) + (u - x)f'(x)$$

and

$$f(v) \geq f(x) + (v - x)f'(x)$$

for all x, u, v (and thus also ξ) in I. We may choose

$$x = (1 - \lambda)u + \lambda v \qquad \text{with any} \qquad \lambda \in]0, 1[$$

since, if u and v are in I, so is this x. Of course, we chose this x because it figures in the definition of convexity in Sect. 3.4 (convexity both from below and from above, let us take here convexity from below). Accordingly, we form, making use of the above,

$$\begin{aligned} &\lambda f(u) + (1 - \lambda)f(v) \\ &\geq (\lambda + (1 - \lambda))f(x) + (\lambda u + (1 - \lambda)v - x)f(x) \\ &\geq f(\lambda u + (1 - \lambda)v). \end{aligned}$$

So f *is* indeed *convex from below on* I. Similarly,

$$f''(x) \leq 0$$

on I *implies that* f *is convex* from above ("concave") *on* I. In both cases convexity was meant in the wider sense. *If* $f''(x) > 0$ *on* I *or* $f''(x) < 0$ *on* I *then* f *is strictly convex there from below or above respectively.* But again f can be strictly convex even if at some points $f''(x) = 0$ (for instance $x \mapsto x^4 - x$ at $x = 0$, see Fig. 7.7) as long as f'' is not 0 on an interval.

As mentioned in Sect. 3.4, the points where a segment convex from one side meets one convex from the other are called *points of inflection*. (The "horizontal points of inflection" in Sect. 6.7 were special cases.) If f is twice differentiable then x_0 is a point of inflection if and only if $f'(x_0) = 0$ and $f''(x)$ changes from positive to negative or vice versa at x_0. (It is not enough that $f''(x_0) = 0$ as again the example $f(x) = x^4 - x$ at $x = 0$ shows, Fig. 7.7.)

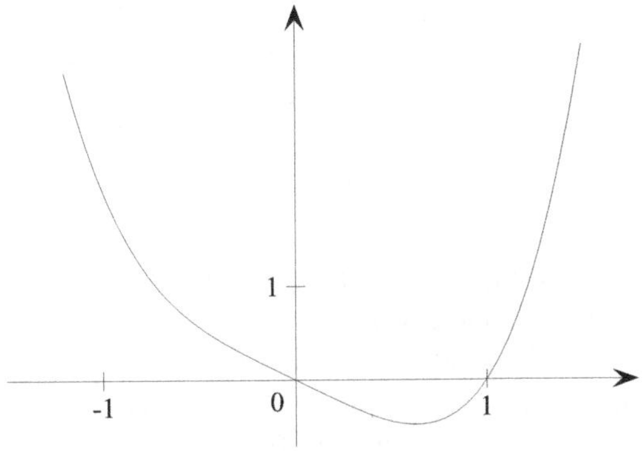

Fig. 7.7 $(x^4 - x)'' \mid_{x=0} = 12x^2 \mid_{x=0} = 0$, but $x \mapsto x^4 - x$ is strictly convex from below everywhere

Another motivation of the exponential function comes from *compound interest.* Banks pay interest annually, semiannually, quarterly, monthly or daily, but in most cases the bank rate stated is annual rate. In what follows the rate r is 0.01 times the bank rate. If the interest is paid every n-th part of the year ($n = 1, 2, 4, 12, 365$ for years, half years, quarter years, months, days; theoretically even smaller units could be considered), then A dollars would grow to $A(1 + r/n)$ during one n-th part of the year, to $A(1 + r/n)^2$ during two n-th parts of a year, ..., and to $A(1 + r/n)^n$ during the whole year, to $A(1 + r/n)^{nt}$ during t years (in practice the formula is slightly altered or fractional years). Both in economics and in mathematics, the limit of this sequence,

$$\lim_{n\to\infty} A(1 + \frac{r}{n})^{nt} \tag{7.10}$$

is of interest.

While this is not of the "form $\frac{0}{0}$" and, anyway, n goes through the positive integers, not through the continuous real (half-)line, the Bernoulli–L'Hospital rule can still be applied after some conversions. First, we take the (natural) logarithm of $(1 + r/n)^{nt}$ and try to calculate its limit

$$\lim_{n\to\infty} \ln(1 + \frac{r}{n})^{nt} = \lim_{n\to\infty} \frac{\ln(1 + \frac{r}{n})}{\frac{1}{n}} t. \tag{7.11}$$

This is now of the form $\frac{0}{0}$ since $\lim_{n\to\infty} \ln(1 + \frac{r}{n}) = 0$ (ln is continuous and $\ln 1 = 0$) and $\lim_{n\to\infty} \frac{1}{n} = 0$. In order to have a continuous variable, we consider (leaving the factor t, which does not depend upon x, aside for now)

$$\lim_{x\to 0} \frac{\ln(1 + rx)}{x}.$$

If the limit exists as x goes to 0 through all reals then it exists (and is the same) when x goes to 0 just through elements of the sequence $\{1/n\}$. But this limit exists by the Bernoulli–L'Hospital rule:

$$\lim_{x\to 0}\frac{\ln(1+rx)}{x}=\lim_{x\to 0}\frac{(1/(1+rx))r}{1}=r,$$

(that is, the limit of the quotient of derivatives indeed exists and therefore also the left hand side exists).

So the limit (7.11) exists and equals rt; and $(1+\frac{r}{n})^{nt}$ (of which $\frac{\ln(1+r/n)}{1/n}t$ is the natural logarithm) *converges* to e^{rt} and the limit (7.10) exists and equals Ae^{rt}, which is thus the *limit amount with accrued continuous interest*. In particular (for $r=1$),

$$\lim_{n\to\infty}(1+\frac{1}{n})^{nt}=e^t$$

(this is our new, second formula for the exponential function) and

$$\lim_{n\to\infty}(1+\frac{1}{n})^n=e,$$

$$\lim_{n\to\infty}(1+\frac{1}{n})^{n+1}=\lim_{n\to\infty}(1+\frac{1}{n})^n\cdot\lim_{n\to\infty}(1+\frac{1}{n})=e\cdot 1=e,$$

as mentioned before.

A third representation of e^x is by its *Taylor series*. Since

$$(e^x)'=e^x,\quad (e^x)''=(e^x)'''=(e^x)^{(4)}=\ldots=e^x,$$

the Taylor formula (6.11) in Sect. (6.7), that is

$$\begin{aligned}f(x)&=f(a)+\tfrac{f'(a)}{1!}(x-a)+\tfrac{f''(a)}{2!}(x-a)^2\\&+\ldots+\tfrac{f^{(n)}(a)}{n!}(x-a)^n+\tfrac{f^{(n+1)}(\xi)}{(n+1)!}(x-a)^{n+1},\end{aligned}\tag{7.12}$$

gives, for $a=0$:

$$e^x=1+\frac{1}{1!}x+\frac{1}{2!}x^2+\ldots+\frac{1}{n!}x^n+\frac{e^\xi}{(n+1)!}x^{n+1}$$

for *some* ξ between 0 and x. If x is in $]-r,r[$, so is ξ and $e^\xi\le e^r$. As we have seen in Sect. 6.7, Example 1, we have

$$\lim_{n\to\infty}\frac{|x|^{n+1}}{(n+1)!}=0,\quad\text{so}\quad\lim_{nto\infty}(1+\frac{1}{1!}x+\frac{1}{2!}x^2+\ldots+\frac{1}{n!}x^n)=e^x$$

(uniform convergence on $]-r, r[$ and this is true for every $r > 0$), which gives the Taylor (really MacLaurin) series expansion

$$e^x = 1 + \frac{x}{1!} + \frac{x^2}{2!} + \ldots + \frac{x^n}{n!} + \ldots \qquad \text{for all real } x.$$

We have also

$$a^x f = e^{x \ln a} = 1 + \frac{\ln a}{1!} x + \frac{(\ln a)^2}{2!} x^2 + \ldots \qquad \text{for all real } x.$$

How about a *Taylor series for* $\ln x$? Since $\ln x$ is not defined at 0, we will look for the *Taylor expansion around* 1 (we could have chosen any other positive number, but the series is nicer around 1 and can be easily transformed into a Taylor series around any other point). In order to get the *Taylor formula* for $\ln x$, note:

$$f(x) = \ln x, \quad f(1) = 0$$

$$f'(1) = \Big(\frac{1}{x}\Big)_{x=1} = 1, \quad f''(1) = \Big(-\frac{1}{x^2}\Big)_{x=1} = -1$$

$$f'''(1) = \Big(2\frac{1}{x^3}\Big)_{x=1} = 2, \quad f^{(4)}(1) = \Big(-2\cdot 3\frac{1}{x^4}\Big)_{x=1} = -3!,$$

$$\ldots$$

$$f^{(n)}(1) = \Big((-1)^{n-1}\frac{(n-1)!}{x^n}\Big)_{x=1} = (-1)^{n-1}(n-1)!$$

$$f^{(n+1)}(\xi) = \Big((-1)^n\frac{n!}{x^{n+1}}\Big)_{x=\xi} = (-1)^n\frac{n!}{\xi^{n+1}}.$$

So, since $(k-1)!/k! = 1/k$, the Taylor formula with $a = 1$ yields

$$\ln x = (x-1) - \frac{1}{2}(x-1)^2 + \frac{1}{3}(x-1)^3 - \frac{1}{4}(x-1)^4$$
$$+ \ldots + (-1)^{n-1}\frac{1}{n}(x-1)^n + (-1)^{n+1}\frac{1}{n+1}\Big(\frac{x-1}{\xi}\Big)^{n+1}.$$

We consider this first for $1 \le x \le 2$. If $1 < x \le 2$, then $1 < \xi \le$ and $0 < |x-1| = x-1 \le 1$. So $0 < c = |x-1| \,/\, |\xi| < 1$ and we know that then $\lim_{n\to\infty} c^{n+1} = \lim_{n\to\infty} c^n c = 0$. Therefore

$$\lim_{n\to\infty} |R_n(x)| = \lim_{n\to\infty} \left|(-1)^{n+1}\frac{1}{n+1}\Big(\frac{x-1}{\xi}\Big)^{n+1}\right|$$
$$= \lim_{n\to\infty}\frac{1}{n+1}\lim_{n\to\infty}\left|\frac{x-1}{\xi}\right|^{n+1} = 0\cdot 0 = 0 \quad \text{for} \quad 1 < x \le 2.$$

Since $R_n(1) = 0$, we have

$$\ln x = (x-1) - \frac{1}{2}(x-1)^2 + \frac{1}{3}(x-1)^3 - \frac{1}{4}(x-1)^4 + \dots \quad \text{for} \quad 1 \le x \le 2$$

(uniform convergence). In particular we have the sum of the following nice, though slowly converging series:

$$\ln 2 = 1 - \frac{1}{2} + \frac{1}{3} - \frac{1}{4} + \dots$$

One can show that the Taylor series of $\ln x$ can be extended to $]0, 2]$, so that

$$\ln x = (x-1) - \frac{1}{2}(x-1)^2 + \frac{1}{3}(x-1)^3 - \frac{1}{4}(x-1)^4 + \dots \quad \text{for} \quad 0 \le x \le 2. \tag{7.13}$$

(Uniform convergence on $[\alpha, 2]$ for every $\alpha > 0$ $(\alpha < 2)$) but that *this equation does not hold for* $x > 2$ (and of course not for $x \le 0$, since $\ln x$ is not defined there). *The series on the right is* even *divergent*, that is,

$$\lim_{n\to\infty} [(x-1) - \frac{1}{2}(x-1)^2 + \frac{1}{3}(x-1)^3 + \dots + (-1)^{n-1}\frac{1}{n}(x-1)^n]$$

does not exist or is not finite for $x > 2$ *and for* $x \le 0$.

We can write (7.13) with $u = ax$ $a \neq 0$ as

$$\ln u = \ln a + \frac{1}{a}(u-a) - \frac{1}{2a^2}(u-a)^2 + \frac{1}{3a^3}(u-a)^3 + \dots \quad \text{for} \quad 0 < u \le 2a,$$

a Taylor series of $\ln u$ around a. Or, with $t = x - 1$,

$$\ln(1+t) = t - \frac{t^2}{2} + \frac{t^3}{3} - \frac{t^4}{4} + \dots \quad \text{for} \quad -1 < t \le 1$$

a Taylor (really MacLaurin) series of $\ln(1 + t)$ around 0. (It does not matter which letter we use for the variable, x, u or t.)

We mention here, without proof, another important MacLaurin series, the *binomial series*:

$$(1+x)^p = 1 + \binom{p}{1}x + \binom{p}{2}x^2 + \binom{p}{3}x^3 + \dots$$

for $-1 < x < 1$ and for all real p (remember that by now $(+x)^p$ is defined for *all real* p and all $x > -1$). Here

$$\binom{p}{n} = \frac{p(p-1)\cdot \ldots \cdot (p-n+2)(p-n+1)}{1 \cdot 2 \cdot 3 \cdot \ldots \cdot n}$$

are the *binomial coefficients*. We have seen the particular case $p = -1$ of the binomial series before:

$$(1+x)^{-1} = 1 - x + x^2 - x^3 + \dots \qquad (-1 < x < 1)$$

(see (6.12) in Sect. 6.7). Another is the binomial formula: *If p is a positive integer then*

$$\binom{p}{p} = \frac{p(p-1)\cdot\ldots\cdot 2\cdot 1}{1\cdot 2\cdot 3\cdot\ldots\cdot p}, \quad \binom{p}{p+1} = \frac{p(p-1)\cdot\ldots\cdot 1\cdot 0}{1\cdot 2\cdot 3\cdot\ldots\cdot p\cdot(p+1)},$$

$$\binom{p}{p+k} = \frac{p(p-1)\cdot\ldots\cdot 0\cdot(-1)\cdot\ldots\cdot(-k+1)}{1\cdot 2\cdot 3\cdot\ldots\cdot p\cdot(p+1)\cdot\ldots\cdot(p+k)} = 0 \quad \text{for} \quad k = 1,\ 2,\ \dots$$

Therefore in this case the binomial series is *finite*, that is, from somewhere (*from the $(p+1)$-st term) on all coefficients are* 0:

$$(1+x)^p = 1 + \binom{p}{1}x + \binom{p}{2}x^2 + \dots + \binom{p}{p-1}x^{p-1} + x^p \quad \text{for positive integer } p.$$

This happens to be true for all real x. Multiplied by u^p after substitution of $x = v/u$ we get the *binomial formula*:

$$(u+v)^p = u^p + \binom{p}{1}u^{p-1}v + \binom{p}{2}u^{p-2}v^2 + \dots + \binom{p}{p-1}uv^{p-1} + v^p$$

for all positive integer p and all real u, v (actually, also for complex u, v).

Comparison of the two series

$$\ln(1+x) = x - \frac{x^2}{2} + \frac{x^3}{3} - \frac{x^4}{4} + \dots \qquad (-1 < x \leq 1)$$

(uniform convergence on $[\gamma, 1]$, $\gamma > -1$) and

$$\frac{1}{1+x} = 1 - x + x^2 - x^3 + \dots \qquad (-1 < x < 1)$$

(uniform convergence on $[-r, r]$, $0 < r < 1$) is an example for the rule that *a series can be differentiated term by term if (where) the series for the derivative is uniformly convergent*. Here the term by term derivation works for every $x \in [-r, r]$, thus, since $r < 1$ is arbitrary, for every $x \in]-1, 1[$, but not for $x = 1$ $(1 - 1 + 1 - 1 + \dots)$ is divergent, as we saw in Sect. 6.7, (Example 2).

7.2.1 Exercises

1. Draw graphs for the functions
   (a) $f_1 : [-4, 4] \to \mathbb{R}_+,\ x \mapsto (3/4)^x$,
   (b) $f_2 : [-4, 4] \to \mathbb{R}_+,\ x \mapsto (3/2)^x$,
   (c) $f_3 : [-3, 2] \to \mathbb{R}_+,\ x \mapsto (5/2)^x$,
   (d) $f_4 : [-2, 3] \to \mathbb{R}_+,\ x \mapsto (2/5)^x$.
2. The functions f_1, f_2, f_3, f_4 from Exercise **1** are convex. Verify that they satisfy $f((1-q)x_1 + qx_2) < (1-q)f(x_1) + qf(x_2)$ for $q = 1/3, x_1 = -3/2, x_2 = 17/10$.
3. Determine the first derivatives of the functions given by
   (a) $g_1(x) = a^{7x} \log_9 x, \quad (x > 0, a > 0, a \neq 1)$,
   (b) $g_2(x) = (\sin a^x) \ln 4x, \quad (x > 0, a > 0, a \neq 1)$,
   (c) $g_3(x) = e^{-1.5x} \cos(2x + 3), \quad (x \in \mathbb{R})$,
   (d) $g_4(x) = \log_5(2 + x^2)/e^{6x}, \quad (x \in \mathbb{R})$,
   (e) $g_5(x) = 1/g_4(x), \quad (x \in \mathbb{R})$.
4. (a) Determine the second derivatives of the functions given by $G_1(x) = x \ln x$ $(x > 0)$ and $G_2(x) = xe^x$ $(x \in \mathbb{R})$.
   (b) Is G_1 strictly convex from below on $\mathbb{R}_{++}$?
   (c) Is G_2 strictly convex from below on $\mathbb{R}_{++}$?
5. Verify that the (convex) functions f_1 and f_3 from Exercise **1** satisfy $f'(x_1) < (f(x_2) - f(x_1))/(x_2 - x_1) < f'(x2)$ for $x_1 = -3/2$ and $x_2 = 4/3$.
6. Verify that the concave (i.e., convex from above) functions given by $F_1(x) = \ln x$ $(x > 0)$, $F_2(x) = \log_1 0x$ $(x > 0)$ satisfy $F'(x_1) > (F(x_2) - F(x_1))/(x_2 - x_1) > F'(x_2)$ for $x_1 = 2$ and $x_2 = 3$.
7. Denote $1 + \frac{1}{1!}\frac{1}{2} + \frac{1}{2!}(\frac{1}{2})^2 + \ldots + \frac{1}{n!}(\frac{1}{2})^n + \ldots = e^{1/2}$ by r. At least how many terms from the beginning of the series on the left has one to add in order to be closer to r than $10^{-6} = 0.000001$?

7.2.2 Answers

2. $(1-q)x_1 + qx_2 = \frac{2}{3}(-3/2) + \frac{1}{3}(17/10) = -0.4333\ldots$
   (a) $(3/4)^{-0.4333\ldots}1.132\ldots 1.230\ldots (2/3)(3/4)^{-3/2}(1/3)(3/4)^{17/10}$,
   (b) $(3/2)^{-0.4333\ldots}0.838\ldots 1.026\ldots (2/3)(3/2)^{-3/2}(1/3)(3/2)^{17/10}$,
   (c) $(5/2)^{-0.4333\ldots}0.672\ldots 1.751\ldots (2/3)(5/2)^{-3/2}(1/3)(5/2)^{17/10}$,
   (d) $(2/5)^{-0.4333\ldots}1.487\ldots 2.705\ldots (2/3)(2/5)^{-3/2}(1/3)(2/5)^{17/10}$.
3. (a) $g_1'(x) = a^{7x}(\ln 9)^{-1}[7(\ln a)(\ln x) + 1/x]$,
   (b) $g_2'(x) = a\cos(ax)\ln(4x) + x^{-1}\sin(ax)$,
   (c) $g_3'(x) = -e^{-1.5x}[1.5\cos(2x + 3) + 2\sin(2x + 3)]$,
   (d) $g_4'(x) = 2e^{-6x}(\ln 5)^{-1}[x(2 + x^2)^{-1} - 3\ln(2 + x^2)^{-1}]$,
   (e) $g_5'(x) = e^{6x}(\ln 5)[\ln(2 + x^2)]^{-2}[6\ln(2 + x^2) - 2x(2 + x^2)^{-1}]$.
4. $G_1''(x) = 1/x, \quad G_2''(x) = (2 + x)e^x$, (a) yes, (b) no.

5. (a) $f_1'(x) = (3/4)^x \ln(3/4)$, $(3/4)^{-3/2} \ln(3/4) = -0.4429\ldots < -0.1960\ldots = (3/4)^{4/3} \ln(3/4)$,
   $[(3/4)^{4/3} - (3/4)^{-3/2}]/[4/3 - (-3/2)] = -0.3028\ldots$
   (c) $f_3'(x) = (5/2)^x \ln(5/2)$,
   $(5/2)^{-3/2} \ln(5/2) = -0.2318\ldots < 3.1089\ldots = (5/2)^{4/3} \ln(5/2)$,
   $[(5/2)^{4/3} - (5/2)^{-3/2}]/[4/3 - (-3/2)] = 1.1082\ldots$
6. $F_1'(x) = (\ln x)' = x^{-1}$, $\quad F_2'(x) = (\log_1 0x)' = (x \ln 10)^{-1}$,
   $F_1'(2) = 1/2 > 1/3 = F_1'(3)$,
   $(F_1(3) - F_1(2))/(3-2) = \ln 3 - \ln 2 = 0.4054\ldots$,
   $F_2'(2) = 1/2 \cdot 2.302585\ldots = 0.2171\ldots > 0.1447\ldots = F_2'(3)$,
   $(F_2(3) - F_2(2))/(3-2) = \log_1 03 - \log_1 02 = 0.1760\ldots$
7. 8 terms.

7.3 Applications: "Discrete" and "Continuous" Compounding, "Effective Interest Rate", Doubling Time, Discounting

As mentioned in the previous section, with "stated yearly interest rate" $r \cdot 100\%$ if paid (or calculated) yearly, semiannually, quarterly, monthly, or, in general every 1th of the year, with interest compounded, a deposit (or loan) amount A grows, by the end of the first year, to

$$A(1+r), A\left(1+\frac{r}{2}\right)^2, \quad A\left(1+\frac{r}{4}\right)^4, \quad A\left(1+\frac{r}{12}\right)^{12},$$
$$A\left(1+\frac{r}{365}\right)^{365}, \quad A\left(1+\frac{r}{n}\right)^n,$$

respectively, (and to the t-th power of these amounts by the end of the t-th year), while in limit (see (7.10)) with "continuous compounding", it will

$$\text{grow to } \lim_{n\to\infty} A\left(1+\frac{r}{n}\right)^n = Ae^r \text{ in a year, to } Ae^{rt} \text{ in t years.} \tag{7.14}$$

An important question is what annual (yearly) interest rate would have the same effect. This is the "effective yearly interest rate", usually denoted by $i \cdot 100\%$. (Of course, this is not to be confused with $i = \sqrt{-1}$ in Sect. 1.7 and elsewhere; we were careful to print *that* i in a different—Roman rather than italic—type.) Of course with this rate i the amount $A > 0$ grows in the year to $A(1+i)$, in t years to $A(1+i)^t$ and we have two simple tasks. The first is to find i such that

$$A(1+i) = A\left(1+\frac{r}{n}\right)^n; \text{ that is } i = \left(1+\frac{r}{n}\right)^n - 1 \tag{7.15}$$

(which satisfies also $A(1+i)^t = A\left(1+\frac{r}{n}\right)^{nt}$.) In many countries this *"effective yearly interest rate" $i \cdot 100\%$ is posted* for saving deposits and *has to be posted* for loans. Notice the subtle difference in the previous sentence. The reason is that $i > r$ (for $r > 0$, $n > 1$, of course), so banks gladly post the higher effective rate for savings, but not-so-gladly for loans. Why is $i > r$? By the *binomial formula* at the end of Sect. 7.2,

$$i = \left(1+\tfrac{r}{n}\right)^n - 1 = \binom{n}{1}\tfrac{r}{n} + \binom{n}{2}\left(\tfrac{r}{n}\right)^2 + \dots$$
$$+ \binom{n}{n-1}\left(\tfrac{r}{n}\right)^n - 1 + \left(\tfrac{r}{n}\right)^n > n$$

Our second task is equally easy: find $i = e^r - 1$ such that

$$A(1+i) = Ae^r, \text{ that is } i = e^r - 1$$

(which of course, satisfies also $(1+i)^t = Ae^{rt}$). This i is clearly not the same as (7.15) which could be denoted by i_n and then, by (7.14),

$$i = e^r - 1 = \lim_{n\to\infty}\left[\left(1+\frac{r}{n}\right)^n - 1\right] = \lim_{n\to\infty} i_n.$$

In the present case we deal with continuous compounding. However, *also this i is greater than r.* (This is not guaranteed by $i_n > r$; for instance $2 + \frac{1}{n} > 2$ but $\lim_{n\to\infty}(2+\frac{1}{n}) = 2$.) This can be seen from the Taylor/MacLaurin series expansion of e^x (ref Sect.7.1):

$$e^x = 1 + x + \frac{x^2}{2!} + \frac{x^3}{3!} + \dots, \qquad \text{so } e^x > 1 + x \text{ if } x > 0,$$

that is, $i = e^r - 1 > r \quad (r > 0)$.

More exactly we can consider the remainder $R_1(x)$ in the Taylor/MacLaurin series if e^x,

$$e^x = 1 + x + R_1(x), \text{ where } \qquad R_1(x) = \frac{e^\xi}{2}x^2 \text{ for some } \xi \text{ between 0 and } x,$$

and this R_1 is, of course, positive if $x > 0$, so again $e^x > 1 + x$, $i = e^r - 1 > r$ for $r > 0$.

Table 7.1 shows the "effective interest rates" in % (percent), that is $100i_n$ corresponding to the "stated interest rate" (also in %, that is, to $100r$) paid every n-th part of a year (n times yearly) for $n = 1, 2, 4, 12, 365$ and also the $100i\%$ corresponding to that $100r$ % stated interest rate in "continuous" *compounding*. Notice that, for small r, i and r hardly differ at all, but for $r = 1$, that is $100r\ \% = 100\ \%$, if $\frac{100}{365}$ % is paid daily and compounded, we have already

Table 7.1 Effective interest corresponding to different stated rates of interest (first line). The first column is the number of payments per year. The last row shows the continuous compounded interest

	1	2	3	4	5	6	7	8	9	10	15	25	50	100
1	1	2	3	4	5	6	7	8	9	10	15	25	50	100
2	1.003	2.010	3.023	4.040	5.063	6.090	7.123	8.160	9.203	10.250	15.563	25.563	56.250	125.000
4	1.004	2.015	3.034	4.060	5.095	6.136	7.186	8.243	9.308	10.381	15.865	27.443	60.181	144.141
12	1.005	2.018	3.042	4.074	5.116	6.168	7.229	8.300	9.381	10.471	16.075	28.073	63.209	161.304
365	1.005	2.020	3.045	4.081	5.127	6.183	7.250	8.328	9.416	10.516	16.180	28.392	64.816	171.457
	1.005	2.020	3.045	4.081	5.127	6.184	7.251	8.329	9.417	10.517	16.183	28.403	64.872	171.828

$100i_{365} = 171.4567\%$ as "effective" (equivalent yearly) interest rate and if, with this $100r\,\% = 100\%$, the compounding is "continuous" then the "effective" interest rate is $100i\,\% = 100(e^r - 1)\,\% = 100(e - 1)\,\% = 171.8281828459\ldots\%$. Surprisingly, this is not much greater then $\frac{100}{365}\,\%$, while from $n = 1$ to $n = 2$ the equivalents $100i\,\%$ and $100i_2\,\%$ of yearly and of semiannual payments with $100i\,\% = 100\,\%$ "stated interest rate" grows from $100i\,\% = 100$ to $100i_2\,\% = 125$ and then, from $n = 2$ to $n = 4$ (from semiannual to quarterly) the equivalent interest grows to $100i_4\,\% = 144.1406$ a smaller, but still respectable growth. the same can be observed (though in somewhat smaller degree) for smaller r in the Table 7.1.

On the other hand, for any yearly (interest) rate $i = i_1$, there is an r such that in any (positive) integer number t of years the continuous compounding with the rate r leads to the same result as they yearly compounding with rate i:

$$Ae^{rt} = A(1 + i)^t \quad (t = 1, 2, \ldots).$$

This r is, of course $r = \ln(1 + i)$. (Clearly, i and r are different, if they are not both 0, because $i = e^r - 1 = r + \frac{r^2}{2!} + \frac{r^3}{3!} + \ldots > r$.)

Now, Ae^{rt} is defined for all real t, in particular for all positive t, so continuous compounding makes sense for any positive (not only integer) length t of time. Similarly, $A(1 + i)^t$ can also be *extended* to (positive) noninteger t at least formally.

It is important and also easy to determine the *doubling time*, that is, *the time during which a deposit doubles under "discrete" or "continuous compounding". For continuous compounding* this means

$$Ae^{rt} = 2A, \quad \text{that is,} \quad e^{rt} = 3, \quad rt = \ln 2$$

or, what is the same, doubling time is

$$t = \frac{\ln 2}{r} = \frac{69.314718\ldots}{100r} = \frac{0.69314718\ldots}{r}.$$

Here this time interval t which is in general of noninteger length, makes sense (but then continuous compounding is an abstraction). On the other hand *for (discrete) yearly compounding*, we have to solve the equation

$$A(1 + i)^t = 2A, \quad \text{that is,} \quad t\ln(1 + i) = \ln 2$$

(where i is the annual effective interest rate), so the doubling time is

$$t = \frac{\ln 2}{\ln(1 + i)} = \frac{0.69314718}{i - \frac{i^2}{2} + \frac{i^3}{3} \ldots},$$

(which is not necessarily integer). We made use of the Taylor/MacLaurin series of $\ln(1+i)$ (from Sect. 7.2, there $\ln(1+t)$), valid for $0 < i \leq 1$. *If i is small, then* the latter formula gives "*approximately*"

$$t \approx \frac{69.314718\ldots}{100i} \tag{7.16}$$

where "$\approx$" reads "*approximately equal*" or "*asymptotically equal*". This is true also in the exact sense that

$$\lim_{t\to 0} \frac{\ln(1+i)}{i} = 1 \tag{7.17}$$

(see Sect. 7.2, right after (10)). So in both cases roughly the "rule of 70" holds for the doubling time:

$$t \sim \frac{70}{100r}, \qquad t \sim \frac{70}{100i}.$$

We wrote $\sim$, not $\approx$, because $\approx$ as in (7.16) makes exactly a *limit statement* like (7.17), what is not the case here. Actually, $t \sim 70/(100i)$ is good approximation up to $i = 0.04 = 4\%$; then, till 10% one uses $t \sim 72/(100i)$; finally $t \sim 74/(100i)$ is appropriate up to $i = 0.16 = 16\%$.

Another important question about compounding is discounting, that is, the *present value* A of an amount Z to become due in t years. Clearly this means solving

$$A(1+i)^t = Z \quad \text{or} \quad Ae^{rt} = Z$$

with respect to A in case of yearly discrete compounding or continuous compounding, respectively. The solutions, that is, the present values of Z, are clearly

$$A = Z(1+i)^{-t} \quad \text{and} \quad A = Ze^{-rt},$$

respectively. Here $(1+i)^{-1}$ is the "*discount factor*", r the "*rate of decay*". (The latter name comes from the natural sciences, where Ze^{-rt} is the formula for instance for chemical or radioactive decay.) Figure 7.8 connects the graphs of the functions $t \mapsto Ae^{rt}$ and $t \mapsto Ze^{-rt}$.

Doubling times and discount factors can, of course, be calculated also for semiannual, quarterly, monthly ... compounding.

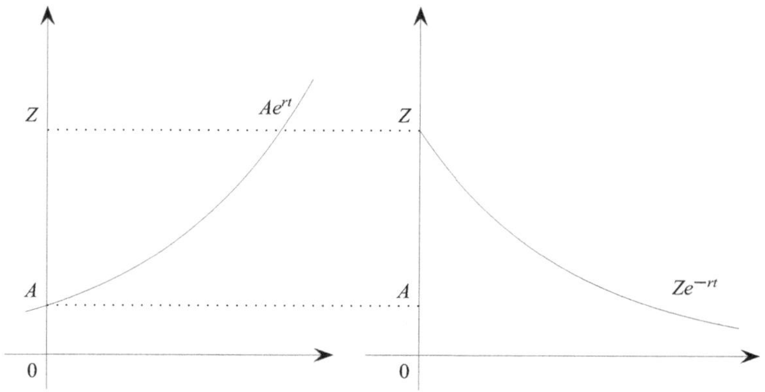

Fig. 7.8 The graphs of growth ($t \mapsto Ae^{rt}$) and decay ($t \mapsto Ze^{-rt}$), $r > 0$

7.3.1 Exercises

1. To which amounts will \$ 1000 grow in 20 years if the stated (yearly) interest rate $r = 0.03$ is paid.
   (a) annually, (b) semiannually, (c) quarterly,
   (d) monthly, (e) daily, (f) continuously?
2. Let the annual effective interest rate i be 0.056. Calculate the number t of years in which a deposit amount A grows to
   (a) $3A$, (b) $5A$, (c) $10A$, (d) $20A$.
3. Calculate the annual effective interest rate $i \cdot 100\%$ with which a deposit amount A grows to
   (a) $5A$, (b) $6A$, (c) $7A$, (d) $8A$
   in 30 years.
4. Let the stated yearly interest rate r be 0.06. Calculate the doubling time t of a deposit amount A if the interest is paid and compounded
   (a) semiannually, (b) quarterly, (c) monthly,
   (d) daily, (e) continuously.
5. Calculate the present value A of 1000 dollars that become due in 10 years if the stated yearly interest rate $r = 0.05$ is assumed to be paid and compounded
   (a) annually, (b) semiannually, (c) quarterly,
   (d) monthly, (e) daily, (f) continuously.

7.3.2 Answers

1. (a) 1806.11, (b) 1814.02. (c) 1818.04,
   (d) 1820.75, (e) 1822.07, (f) 1822.12.
2. (a) 20.162, (b) 29.537. (c) 42.258,
   (d) 54.979.

3. (a) 5.51, (b) 6.15. (c) 6.70,
   (d) 7.18.
4. (a) 11.725, (b) 11.639. (c) 11.581,
   (d) 11.553, (e) 11.552.
5. (a) 613.91, (b) 610.27. (c) 608.41,
   (d) 607.16, (e) 606.55, (f) 606.53.

7.4 Some Interesting Scalar Valued Nonlinear Functions in Several Variables. Homothetic Functions

We have already encountered several particular functions. The linear and affine functions were discussed in Chap. 4 with both variables and function values either vectors or scalars (one-component vectors, real numbers). But we saw also nonlinear functions, such as

- the sine, cosine, tangent, cotangent in Sects. 1.7, 6.2, 6.4, 6.5 and 6.7,
- the polynomials and rational functions in Sects. 6.2, 6.3, 6.5, 6.7 and 6.9,
- the exponential and logarithm in Sects. 7.2 and 7.3.

We got to know them as real valued (scalar valued) functions of *one* real variable.

In what follows, we introduce some classes of scalar valued functions of *several* variables, which play an important role in economics and which are, in general, *nonlinear* (though some special cases may be linear).

1. Polynomials in n real variables. These are functions $\mathbf{P} : \mathbf{R}^n \to \mathbf{R}$ of the form

$$\mathbf{P}(\mathbf{x}) = P_m(x_1, x_2, \ldots, x_n) = \sum_{k_1+k_2+\ldots+k_n=0}^{m} a_{k_1k_2\ldots k_n} x_1^{k_1} x_2^{k_2} \ldots x_n^{k_n}. \tag{7.18}$$

The large sum sign needs explanation. The numbers $k_1, k_2, \ldots, k_n$ are integers, not smaller than 0 and not larger than m. So really n summations take place. But only those terms appear, where $k_1+k_2+\ldots+k_n$ is not larger than m. The (real) constants $a_{k_1k_2\ldots k_n}$ are the *coefficients* while m is the *degree* of the polynomial (if at least one of the coefficients whose subscripts add up to m that is, at least one of the coefficients of products of powers, whose exponents add up to m, is nonzero, then m is the *exact degree*).

The special case $m = 0$ gives simply a *constant* while, in the case of $m = 1$, (7.18) gives the values

$$P_1(x_1, x_2, \ldots, x_n) = a_0 + a_1x_1 + a_2x_2 + \ldots + a_nx_n \tag{7.19}$$

of an *affine function*; in particular, if $a_0 = 0$, then the values of a *linear function*.

For $m = 2$ we get the so-called *quadratic functions* with the values

$$
\begin{aligned}
&P_2(x_1, x_2, \ldots, x_n) \\
&\quad = \quad a_0 + a_1x_1 + a_2x_2 + \ldots + a_nx_n + a_{11}x_1^2 + a_{12}x_1x_2 \\
&\qquad + \ldots + a_{1n}x_1x_n + a_{21}x_2x_1 + a_{22}x_2^2 + \ldots + a_{2n}x_2x_n \\
&\qquad + \ldots + a_{n1}x_nx_1 + a_{n2}x_nx_2 + \ldots + a_{nn}x_n^2.
\end{aligned}
\tag{7.20}
$$

Here we wrote, for the sake of brevity and symmetry, $a_0 := a_{00\ldots0}$, $a_1 := a_{10\ldots0}$, $a_2 := a_{010\ldots0}$, $\ldots$, $a_n := a_{0\ldots0n}$, $a_{11} := a_{20\ldots0}$, $a_{12} = a_{21} := \frac{1}{2}a_{110\ldots0}$, $\ldots$, $a_{1n} = a_{n1} := \frac{1}{2}a_{10\ldots01}$, $\ldots$, $a_{2n} = a_{n2} := \frac{1}{2}a_{010\ldots0n}$, $\ldots$, $a_{nn} := a_{00\ldots02}$. (Of course, $x_1x_2 = x_2x_1$ etc., that is why we wrote $a_{12} = a_{21}$, and comparison to (7.18) gives $a_{110\ldots0} = 2a_{12} = 2a_{21}$, etc.) If, in particular, $a_0 = a_1 = a_2 = \ldots = a_n = 0$ in (7.20) then it is a *quadratic form*.

Polynomials are often used in economics to *approximate* empirically the dependence of a quantity (for instance the cost of output or the utility of some goods) from other quantities (those of outputs or of goods, respectively), as we did already with polynomials of degree 1 (affine functions) in Sect. 6.10. As there, here too, *one gets good approximations by choosing the coefficients appropriately* in (7.18).

For determining the (minimal) cost of producing the desired output $\mathbf{x} = (x_1, \ldots, x_n)$ (the n outputs of different kinds united into an output vector) one often supposes that the cost is a *quadratic function* (7.20) of these output quantities. In this case the *marginal costs* of the production of the j-th output good (compare the marginal product rates in Sect. 6.11) are given by

$$\frac{\partial P_2}{\partial x_j}(x_1, x_2, \ldots, x_n) = a_j + 2(a_{j1}x_1 + \ldots + a_{jj}x_j + \ldots + a_{jn}x_n).$$

With vector and matrix notations (see Sect. 6.4), we can write (7.20) as

$$P_2(x_1, x_2, \ldots, x_n) = P_2(\mathbf{x}) = a_0 + \mathbf{a} \cdot \mathbf{y} + \mathbf{x}\mathbf{A}\mathbf{x}^T, \tag{7.21}$$

where $\mathbf{a} = (a_1, a_2, \ldots, a_n)$ and the matrix

$$\mathbf{A} = \begin{pmatrix} a_{11} & a_{12} & \cdots & a_{1n} \\ a_{21} & a_{22} & \cdots & a_{2n} \\ \vdots & \vdots & \ddots & \vdots \\ a_{n1} & a_{n2} & \cdots & a_{nn} \end{pmatrix}$$

is *symmetric*, that is, $a_{ij} = a_{ji}$ for all $i, j = 1, 2, \ldots, n$ (as we just saw). he vector $\mathbf{x} = (x_1, x_2, \ldots, x_n)$ is a *row vector*, but we need it also in column vector form; this is denoted by

$$\mathbf{x}^T = \begin{pmatrix} x_1 \\ x_2 \\ \vdots \\ x_n \end{pmatrix},$$

the *transpose* of $\mathbf{x}$. Note the inner product in the second term of (7.21) and that, in the third term, which is a *quadratic form*, a $1 \times n$ matrix is multiplied by an $n \times n$ matrix and that by an $n \times 1$ matrix, which can be done and the result is a 1×1 matrix, that is a scalar, as are the other terms in (7.21).

2. Rational functions in n variables. These are *quotients of two polynomials P, Q* of any degree (just as rational functions of one variable are, see Sect. 6.2):

$$R(x_1, x_2, \ldots, x_n) = \frac{P(x_1, x_2, \ldots, x_n)}{Q(x_1, x_2, \ldots, x_n)}. \tag{7.22}$$

They are defined at those points $(x_1, x_2, \ldots, x_n) \in \mathbf{R}^n$, where $Q(x_1, x_2, \ldots, x_n) \neq 0$. If however, the polynomial Q has *some zeros in common with P* then one can cancel in their product representation (see Sect. 6.2) the respective factors and in this way one *may* be able to *extend the definition* of R to these zeros of Q. The following example shows how this is done:

$$R(x_1, x_2) = \frac{P(x_1, x_2)}{Q(x_1, x_2)} = \frac{4 - x_1^2 - 4x_2^3 + x_1^2 x_2^3}{4 - 4x_1 - 4x_2 + x_1^2 + 4x_1x_2 - x_1^2 x_2}.$$

This is *not defined at $x_1 = 2$* (whatever x_2 is) *and at $x_2 = 1$* (whatever x_1 is) because

$$Q(2, x_2) = 4 - 8 - 4x_2 + 4 + 8x_2 - 4x_2 = 0$$

and

$$Q(x_1, 1) = 4 - 4x_1 - 4 + x_1^2 + 4x_1 - x_1^2 = 0.$$

But

$$\begin{aligned} &\frac{4 - x_1^2 - 4x_2^3 + x_1^2 x_2^3}{4 - 4x_1 - 4x_2 + x_1^2 + 4x_1x_2 - x_1^2 x_2} \\ &= \frac{(x_1 - 2)(x_1 + 2)(x_2 - 1)(x_2^2 + x_2 + 1)}{-(x_1 - 2)^2(x_2 - 1)} \\ &= \frac{(x_1 + 2)(x_2^2 + x_2 + 1)}{2 - x_1} \end{aligned}$$

except if $x_2 = 1$ *or* $x_1 = 2$. This equality shows, however, that we can *define* $R(x_1, 1)$ as

$$R(x_1, 1) := \frac{(x_1 + 2)3}{2 - x_1}; \quad \text{we also have} \quad \lim_{x_1 \to 1} R(x_1, x_2) = \frac{(x_1 + 2)3}{2 - x_1},$$

thus the so *extended* R will even be continuous at $x_2 = 1$ for all x_1 except $x_1 = 2$. However, R *cannot be defined at* $x_1 = 2$ *for any* x_2 *so that the extended* R *be continuous at* $x_1 = 2$. The continuity requirement is essential; otherwise we could just assign to R value at $x_1 = 2$ (and also at $x_2 = 1$).

Of course, if the denominator $Q(x_1, x_2, \ldots, x_n)$ is a nonzero constant then the rational function (7.22) equals a polynomial (everywhere).

3. Homogeneous functions. We had encountered (positively) linearly homogeneous functions in Sects. 3.3, 4.2, and 4.3 without any regularity (continuity or differentiability) suppositions. Then, in Sect. 6.11, we dealt with their generalisation, the *(positively) homogeneous functions of degree r*, that is, those functions $F : D \to \mathbf{R}$ ($D \subset \mathbf{R}$) which satisfy

$$F(\lambda \mathbf{x}) = \lambda^r F(\mathbf{x}) \quad (\mathbf{x} \in D,\ \lambda \mathbf{x} \in D;\ \lambda \in \mathbf{R}_{++}), \tag{7.23}$$

under the supposition that F is differentiable. We will look here at (positively) homogeneous functions of degree r and, subsequently, at some of their applications in economics without supposing any regularity. In the case where r is irrational, it would be quit difficult to define λ^r for negative λ (certainly one would have to move from $\mathbf{R}$ to $\mathbb{C}$) and, for $r < 0$, λ^r is not defined for $\lambda = 0$. This, in addition to considerations in economics, explains why λ is supposed to be *positive* in (7.23). We will write in what follows, as usual in economics, *homogeneous* for short in place of "positively homogeneous". s we know from Sect. 7.2 (7.9), λ^r is defined, for all real r and positive λ, by

$$\lambda^r = e^{r \ln \lambda}.$$

Again, $\ln \lambda$ would not be defined for $\lambda \leq 0$.

However, for $r \in \mathbb{N}$, in particular $r = 1$ (linear homogeneity) an $r = 2$, homogeneity, that is (7.23), is often supposed for all $\lambda \in \mathbb{R}$ and, in the case of nonpositive integer r, in particular $r = -1$, for all $\lambda \neq 0$. For instance, the *linear functions* ((7.19) with $a_0 = 0$) are, as we saw in Sect. 4.3, linearly homogeneous, that is, they satisfy (7.23) with $r = 1$ for all $\lambda \in rz$ (and all $D \subset \mathbb{R}^n$) are homogeneous of degree 2, that is, (7.23) is satisfied with $r = 2$, $\lambda \in \mathbb{R}$:

$$P(\lambda \mathbf{x}) = (\lambda \mathbf{x})\mathbf{A}(\lambda \mathbf{x})^T = \lambda \mathbf{x}\mathbf{A}(\lambda \mathbf{x}^T) = \lambda \mathbf{x}\lambda(\mathbf{A}\mathbf{x}^T) = \lambda^2 \mathbf{x}\mathbf{A}\mathbf{x}^T = \lambda^2 P(\mathbf{x}).$$

Here we used the rules of matrix algebra from Sect. 4.4 ((4.20) and (4.21)) and the fact that $(\lambda \mathbf{x})^T = \lambda \mathbf{x}^T$, which is obvious from the definition of the transpose. Finally, special *rational functions* (7.22) where P is linear and Q is a quadratic form are homogeneous of degree -1 in the sense that they satisfy (7.23) with $r = -1$ for all $\lambda \neq 0$ and all D not containing the zeros of Q, as long as $\mathbf{x} \in D$, $\lambda\mathbf{x} \in D$:

$$R(\lambda\mathbf{x}) = \frac{P(\lambda\mathbf{x})}{Q(\lambda\mathbf{x})} = \frac{\lambda P(\mathbf{x})}{\lambda^2 Q(\mathbf{x})} = \lambda^{-1}R(\mathbf{x}).$$

But *from now on we will suppose* λ to be *positive*. *If* also *the domain of* F $\mathbb{R}^n_{++}$ (or a subset of it), then it is easy to give the (or one) *general representation* of the function $F : \mathbb{R}^n_{++} \to \mathbb{R}$, homogeneous of degree r. Indeed, then

$$\begin{aligned} F(x_1, x_2, \ldots, x_n) &= F(x_1 \cdot 1, x_1(x_2/x_1), \ldots, x_1(x_n/x_1)) \\ &= x_1^r F(1, x_1/x_1, \ldots x_n/x_1), \end{aligned}$$

that is, F is of the form

$$F(x_1, x_2, \ldots, x_n) = x_1^r \Phi(x_2/x_1, \ldots, x_n/x_1), \tag{7.24}$$

for some $\Phi(t_2, \ldots, t_n) := F(1, t_2, \ldots, t_n)$. Conversely, for *any* function $\Phi : \mathbb{R}^{n-1}_{++} \to \mathbb{R}$, the function F, given by (7.24) is homogeneous of degree r, that is, satisfies (7.23) with $D = \mathbb{R}^n_{++}$, $\lambda \in \mathbb{R}_{++}$:

$$F(\lambda x_1, \lambda x_2, \ldots, \lambda x_n) = \lambda^r x_1^r \Phi(x_2/x_1, \ldots, x_n/x_1) = \lambda^r F(x_1, x_2, \ldots, x_n).$$

So, *the general homogeneous functions of degree* r, $F : \mathbb{R}^{n-1}_{++} \to \mathbb{R}$ *are given by (7.24) with arbitrary* $\Phi : \mathbb{R}^{n-1}_{++} \to \mathbb{R}$.

Notice, that Φ and thus *the homogeneous function* F *need not be differentiable or even continuous.*

Instead of x_1 *we could have chosen any* x_k ($k = 2, \ldots, n$), giving similar results. So there are several *equivalent* general representations for homogeneous functions of degree r. We give one more, in which none of the variables $x_1, x_2, \ldots, x_n$ has a distinguished role and which works also on $D = \mathbb{R}^n \setminus \{0\}$ (the set of all n-dimensional, nonzero vectors), or on certain subsets thereof. Clearly,

$$F(\mathbf{x}) = F(|\mathbf{x}| \frac{1}{|\mathbf{x}|}\mathbf{x}) = |\mathbf{x}|^r F(\frac{1}{|\mathbf{x}|}\mathbf{x}) \tag{7.25}$$

for homogeneous functions of degree r. Notice that $(1/|\mathbf{x}|)\mathbf{x}$ is a unit vector. Let

$$S := \{\mathbf{z} \mid |\mathbf{z}| = 1\}$$

be the *n-dimensional unit sphere*. Take *any* function $\Psi : S \to \mathbb{R}$. Then

$$F(\mathbf{x}) = |\mathbf{x}|^r \Psi(\frac{1}{|\mathbf{x}|}\mathbf{x}) \tag{7.26}$$

is always a homogeneous function of degree r, since $|\lambda \mathbf{x}| = \lambda\, |\mathbf{x}|$ if $\lambda > 0$:

$$F(\lambda \mathbf{x}) = |\lambda \mathbf{x}|^r\, \Psi(\frac{1}{|\lambda \mathbf{x}|}\lambda \mathbf{x}) = \lambda^r\, |\mathbf{x}|^r\, \Psi(\frac{1}{|\mathbf{x}|}\mathbf{x}) = \lambda^r F(\mathbf{x}).$$

Of course, (7.25) is also of the form (7.26). So *the general homogeneous function of degree r on $\mathbb{R}^n \setminus \{0\}$ is given by (7.26), where $\Psi : S \to \mathbb{R}$ is an arbitrary function.* This is again another (on $\mathbb{R}_{++^n}$ equivalent) representation of homogeneous functions of degree r. If we wish to include also $\mathbf{0}$, then (7.23) becomes

$$F(\lambda \mathbf{x}) = \lambda^r F(\mathbf{x}) \qquad (\mathbf{x} \in \mathbb{R}^n,\ \lambda\ in\, \mathbb{R}_{++}). \tag{7.27}$$

For $\mathbf{x} = \mathbf{0}$ this gives

$$F(\mathbf{0}) = \lambda^r F(\mathbf{0}).$$

If $r \neq 0$, then this is possible only when $F(\mathbf{0}) = 0$. If, however, $r = 0$, then $F(\mathbf{0})$ can be *any* constant. For $\mathbf{x} \neq \mathbf{0}$, F is still given by (7.26). So *the general homogeneous function of degree r on $\mathbb{R}^n$, is given by*

$$F(\mathbf{x}) = \begin{cases} |\mathbf{x}|^r\, \Psi((1/\,|\mathbf{x}|)\mathbf{x}) & \text{for} \quad \mathbf{x} \neq \mathbf{0} \\ c & \text{for} \quad \mathbf{x} = \mathbf{0} \end{cases}$$

where $\Psi : S \to \mathbb{R}$ is an arbitrary function and c is an arbitrary real constant if $r = 1$ but $c = 1$ if $r \neq 0$ (check that all such functions satisfy (7.27)).

The final representation, important for applications, will be presented here in the case $n = 2$ where it is very intuitive. This representation $D = \mathbb{R}^2_{++} \setminus \{(0,0)\}$ or on a subset thereof, in particular on $[0,a] \times [0,b] \setminus \{(0,0)\}$. Its distinguishing feature is that, rather than the unspecific arbitrary functions Φ or Ψ, it features the almost arbitrary (see below) chosen or prescribed values of F on the horizontal and vertical segments $\{(x,b) \mid x \in [0,a]\}$ and $\{(a,y) \mid y \in [0,b]\}$, respectively. If

$$F(x,b) = f(x) \quad \text{for} \quad x \in [0,a] \quad \text{and}$$
$$F(a,y) = g(y) \quad \text{for} \quad y \in [0,b]$$

then

$$F(x,y) = F(x\frac{b}{y}\frac{y}{b}, b\frac{y}{b}) = \left(\frac{y}{b}\right)^r F(x\frac{b}{y}, b) = \left(\frac{y}{b}\right)^r f(x\frac{b}{y}) \quad \text{if} \quad x\frac{b}{y} \in [0,a]$$

and

$$F(x,y) = F(a\frac{x}{a}, y\frac{a}{x}\frac{x}{a}) = \left(\frac{x}{a}\right)^r F(a, y\frac{a}{x}) = \left(\frac{x}{a}\right)^r g(y\frac{a}{y}) \quad \text{if} \quad y\frac{a}{x} \in [0,b].$$

Notice that the two conditions at the end of these two formulas are almost complementary: the second means

$$x > 0, \quad y \geq 0, \quad y/x \leq b/a$$

while the first is satisfied iff

$$y > 0 \text{ and either } x = 0$$
$$\text{or } y/x \geq b/a \text{ with } x \neq 0.$$

As Fig. 7.9 shows, these two, on the figure differently shaded domains cover $\mathbb{R}^2_{++} \setminus \{(0,0)\}$ *with the line* $y = (b/a)x$ $(x > 0)$ *as only possible overlap*. On this line we get both

$$F(x, \frac{b}{a}x) = (\frac{x}{a})^r f(a) \quad \text{and} \quad F(x, \frac{b}{a}x) = (\frac{x}{a})^r g(b).$$

So we have to have $f(a) = g(b)$, that is why we said above that f and g may be chosen *almost* arbitrarily; everywhere else they *can* be prescribed arbitrarily (they need not be differentiable or continuous either). Therefore, *given two functions* $f : [0, a] \to \mathbb{R}$ *and* $g : [0, b] \to \mathbb{R}$ *such that* $f(a) = g(b)$ *but otherwise arbitrary, there exists exactly one homogeneous function of degree r which extends f and g from* $\{(x, b) \mid x \in [0, a]\}$ *and* $\{(a, y) \mid y \in [0, b]\}$, *respectively, to* $\mathbb{R}^2_+ \setminus \{(0,0)\}$ *and this is given by*

$$F(x, y) = \begin{cases} (y/b)^r f(xb/y) & \text{for} \quad y > 0, \ 0 \leq x \leq ay/b \\ (x/a)^r g(ya/x) & \text{for} \quad x > 0, \ 0 \leq y \leq bx/a. \end{cases} \tag{7.28}$$

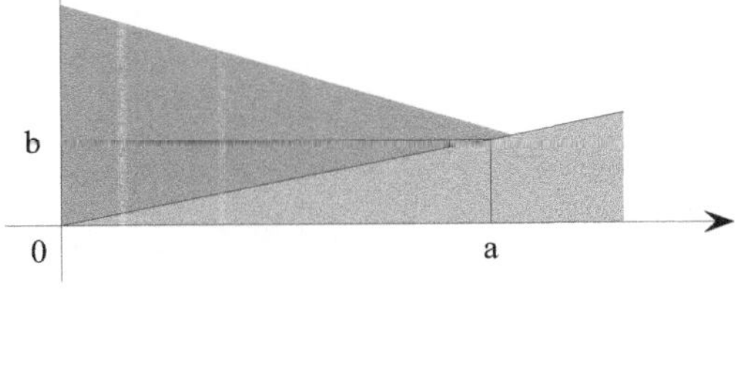

Fig. 7.9 The two parts of formula (7.28) hold on the above two shaded domains. Overlap possible only on the line given by $y := (b/a)x$. No ambiguity if and only if, $f(a) = g(b)$

All that remains to be checked is that this gives a homogeneous function of degree r (do the checking) and that indeed

$$F(x,b) = (b/b)^r f(xb/b) = f(x) \quad \text{for} \quad x \in [0,a] \quad \text{and}$$
$$F(a,y) = (a/a)^r g(ya/a) = g(y) \quad \text{for} \quad y \in [0,b].$$

We remind the reader that in Sect. 3.3 we interpreted the linearly homogeneous functions as *production functions* with constant returns to scale. In the case of linearly homogeneous functions of two variables their graphs, the production surfaces, consist, as we have also seen, of straight lines starting at $\mathbf{0}$. In case of homogeneous functions of degree r these straight lines are replaced by "generalise parabolas", graphs of $\lambda \mapsto \lambda^r F(x_0, y_0)$ (for $r = 2$ we get the usual parabolas). Formula (7.28), which we have just obtained, shows that we can draw these generalised parabolas through two arbitrarily described curves

$$\{(x, b, f(x)) \mid x \in [0,a]\} \quad \text{and} \quad \{(a, y, g(y)) \mid y \in [0,b]\}$$

as long as $f(a) = g(b)$. Note that the intervals $[0,a]$ and $[0,b]$ are *finite*.

In Sect. 3.5 we showed for linearly homogeneous functions the following. If for these functions $x \mapsto F(x,b) = f(x)$ or $y \mapsto F(a,y) = g(y)$ are strictly convex from below "at the beginning" (say on $[0,\bar{x}]$ and $[0,\bar{y}]$, respectively) they may, of course, continue, strictly convex from above ("strictly concave") for a while but "finally" (on $]1/\bar{x}, \infty[$ or $]1/\bar{y}, \infty[$, respectively; these intervals are not finite anymore!), f and g will have to be strictly convex from below again. So the typical "cuts" parallel to the coordinate planes (graphs of the "partial factor variations" in Sect. 3.5 (3.14)) are "bell-shaped curves" (Fig. 7.10).

An *example* of a class of linearly homogeneous functions in n variables for which all cuts of the graph are bell-shaped is given by

$$F(x_1, x_2, \ldots, x_n) = \frac{A x_1^{a_1} x_2^{a_2} \cdot \ldots \cdot x_n^{a_n}}{B_1 x_1^b + B_2 x_2^b + \ldots + B_n x_n^b} \qquad (x_1, x_2, \ldots, x_n \in \mathbb{R}_{++})$$

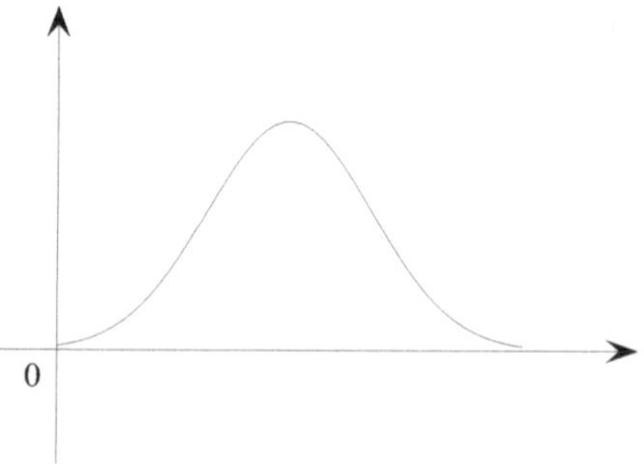

Fig. 7.10 Bell-shaped curve

where $A, b, B_j \in \mathbb{R}_{++}$, $a_j > 1$ $(j = 1, 2, \ldots, n)$ are constants such that $a_1 + a_2 + \ldots + a_n - b = 1$. The last condition guarantees that F is linearly homogeneous (check!) The "partial functions" ("partial factor variations" in Sect. 3.5 (3.14)), for instance $x_1 \mapsto F(x_1, x_{20}, \ldots, x_{n0})$, are given by

$$f(x_1) = Cx_1^{a_1}/(B_1x_1^b + B),$$

where $C = Ax_{20}^{a_2} \cdot \ldots \cdot x_{n0}^{a_n}$, $B = B_2x_{20}^b + \ldots + B_nx_{n0}^b$. From the process described in Sect. 6.7 for finding maxima, it is easy to prove that f has only one maximum, at $x_1 = (Ba_1/B_1(b - a_1))^{1/b}$. Also $f(0) = 0$ (since $B > 0$) and

$$\lim_{x_1 \to \infty} f(x_1) = \lim_{x_1 \to \infty} \frac{C}{B_1x_1^{b-a_1} + Bx_1^{-a_1}} = 0$$

by the rules of limits in Sect. 6.2 and because, from $a_j > 1$ $(j = 2, \ldots, n)$, we have $b - a_1 = a_2 + \ldots + a_{n-1} - 1 > 0$. Now, f'' (calculate it!) is a fraction with $(B_1x_1^b + B)^3$ in the denominator and x_1^{a-2} times a polynomial of second degree of x_1^b in the numerator. So f'' is 0 in at most two points on $\mathbb{R}_{++}$. By what we saw in Sect. 7.2, f (and similarly every $x_j \mapsto F(\ldots, x_j, \ldots)$) thus has at most two points of inflection. But it has to have exactly two because of the maximum and limit established above: f cannot be convex from below at the maximum or convex from above (concave) for large x_1 since it is positive and converges to 0. So there is one point of inflection in between. The proof in Sect. 3.5, quoted above, also shows that, if f is convex from below "at the end", then also "at the beginning". Thus there is another point of inflection between 0 and the maximum and f starts convex from below then turns convex from above (concave), which stretch contains the maximum, then becomes convex from below again and stays so: it is bell-shaped.

Let us note that this F is not necessarily quasi-convex from above ("quasi concave") as defined in Sect. 3.5 (show it for $n = 2$ and, say, $a_1 = 1.4$, $a_2 = 1.6$, $b = 2$).

In the next section we will introduce important production functions with "constant elasticity of substitution" which will turn to be convex from above ("concave") all the way.

4. Generalised homogeneous functions. Homogeneous functions can be generalised in several ways. Some look but really are not much more general, while others are genuine generalisations.

Replacing, in the definition (7.23) of homogeneous functions of degree r, λ^r by $\phi(\lambda)$ is a tempting way to generalise and looks far reaching but really it is not. In order to see this, take this new definition

$$F(\lambda\mathbf{x}) = \phi(\lambda)F(\mathbf{x}) \tag{7.29}$$

($\lambda \in \mathbb{R}_{++}$, $F : \mathbb{R}_{++}^n \to \mathbb{R}$; one may include $\mathbf{0}$ in the domain the same way as right after (7.27), restriction of $\mathbf{x}$ to a subset D, which contains also $\lambda\mathbf{x}$, is not difficult either) and apply it twice (equally three times):

$$F(\lambda\mu) = F(\lambda(\mu\mathbf{x})) = \phi(\lambda)F(\mu\mathbf{x}) = \phi(\lambda)\phi(\mu)F(\mathbf{x}),$$
$$F(\lambda\mu) = F((\lambda\mu)\mathbf{x}) = \phi(\lambda\mu)F(\mathbf{x}).$$

We may have $F(\mathbf{x}) = \mathbf{0}$ for *all* $\mathbf{x}$, which satisfies (7.29) *whatever* ϕ is. This

$$F(\mathbf{x}) \equiv \mathbf{0}, \qquad \phi \text{ arbitrary}$$

is a *trivial solution* which clearly is uninteresting for applications and we will ignore it. So we suppose that there exists an $\mathbf{x}_0$ such that $F(\mathbf{x}_0) \neq 0$. But then comparison of the above two equations gives

$$\phi(\lambda\mu) = \phi(\lambda)\phi(\mu) \qquad (\lambda, \mu \in \mathbb{R}_{++}). \tag{7.30}$$

While this "functional equation" has solutions (functions $\phi : \mathbb{R}_{++} \to \mathbb{R}$ satisfying (7.30)) which are very irregular, the only solution under very weak regularity (continuity, boundedness) conditions is $\phi(\lambda) = \lambda^r$, that is, we get (7.23) again. Here we have ignored $\phi(\lambda) \equiv 0$, as we should, since by (7.29) it would lead to $F(\lambda\mathbf{x}) \equiv 0$ and this F identically 0, which we have already excluded.

We supposed $\phi : \mathbb{R}_{++} \to \mathbb{R}$ to be *real-valued*, however it follows from (7.30) that ϕ can assume only *nonnegative* values. Indeed, choose $\lambda = \mu = \sqrt{\tau}$:

$$\phi(\tau) = \phi(\sqrt{\tau})^2 \geq 0 \qquad \text{for all } \tau \in \mathbb{R}_{++},$$

because the square of any real number is nonnegative. Moreover, since we have just excluded that ϕ be *everywhere* 0, the solutions of (7.30) cannot be 0 *anywhere*. Indeed, if there existed a $\lambda_0 \in \mathbb{R}_{++}$ for which $\phi(\lambda) = 0$ then (7.30) with $\lambda = \lambda_0$, $\mu = \tau 7\lambda_0$ would give for all $\tau \in \mathbb{R}_{++}$

$$\phi(\tau) = \phi(\lambda_0)\phi(\tau/\lambda_0) = 0$$

which was excluded. So ϕ has to be *positive valued*.

Now it is possible to reduce Eq. (7.30) to that of *additivity* (see Sect. 4.2 (4.6)):

$$f(x_1 + x_2) = f(x_1) + f(x_2) \qquad (x_1, x_2 \in \mathbb{R}) \tag{7.31}$$

since we can take now the logarithm of both sides of (7.30) (we have defined ln only for positive arguments) and get

$$\ln \phi(\lambda\mu) = \ln \phi(\lambda) + \ln \phi(\mu) \qquad (\lambda, \mu \in \mathbb{R}_{++}).$$

Putting $\lambda = e^{x_1}$, $\mu = e^{x_2}$ we get every positive λ and μ (and only these) as x_1 and x_2 go trough $\mathbb{R}$ and, defining $f : \mathbb{R} \to \mathbb{R}$ by

$$f(x) = \ln \phi(e^x) \tag{7.32}$$

(ϕ is positive valued but $\ln \phi$ can assume any real value), we get indeed (7.31).

At the end of Sect. 4.2 we noted that the only additive functions, locally bounded from above (bounded from above on an interval of positive length), are linear functions given by, say,

$$f(x) = rx. \tag{7.33}$$

We suppose the same weak regularity condition, *local boundedness from above, about* ϕ. If ϕ is continuous even at one point then this condition is amply satisfied. By the way, it would make no sense to suppose ϕ bounded from below since, as we have see, (7.30) implies that ϕ is always nonnegative, that is, bounded from below by 0. However, we may suppose that ϕ is bounded from below by a positive bound. Anyway, if ϕ is locally bounded from above then, by (7.32), so is f which, as we saw, satisfies (7.31). So (7.33) holds which, by (7.32), gives

$$\phi(e^x) = e^{rx}, \qquad \text{that is,} \qquad \phi(\lambda) = \lambda^r,$$

as asserted.

Thus *all functions* F *satisfying (7.29) with a* ϕ *locally bounded from above are homogeneous of degree* r. (We did not forget about the excluded $F(\mathbf{x}) \equiv 0$ either: it satisfies (7.23) trivially, so it is also homogeneous of any degree r.)

So (7.29) is not really more general than (7.23) except that it unites (7.23) (with $D = \mathbb{R}^n_{++}$) for *all* real r and except for the weak regularity condition on ϕ.

While (7.29) is not a strong enough generalisation of (7.23) to yield new locally bounded functions, the tempting further generalisation

$$F(\lambda\mathbf{x}) = \phi(\lambda, \mathbf{x})F(\mathbf{x}) \quad (\lambda \in \mathbb{R}_{++},\ \mathbf{x} \in D \subset \mathbb{R}^n,\ \lambda\mathbf{x} \in D), \tag{7.34}$$

where ϕ may depend, in addition to λ, also upon $\mathbf{x}$, is so general that it is meaningless. Indeed *to every function* $F : D \to \mathbb{R}$ *there exists a function* $\phi : \mathbb{R}_{++} \times D \to \mathbb{R}$ *such that (7.34) is satisfied. At points where* $F(\mathbf{x}) \neq 0$, *this* ϕ *is simply given by*

$$\phi(\lambda, \mathbf{x}) = F(\lambda\mathbf{x})/F(\mathbf{x}).$$

If $F(\mathbf{x_0}) = 0$ *at a point* $\mathbf{x_0} \in D$ *then*, by (7.34), *also* $F(\lambda\mathbf{x_0}) = 0$ *whenever* $\lambda\mathbf{x_0} \in D$ *and* ϕ *can be anything* (for instance equal to 1) at these points.

Several intermediate generalisations of (7.34) (if $D = \mathbb{R}^n \setminus \{\mathbf{0}\}$) can be formulated. One is

$$F(\lambda\mathbf{x}) = \lambda^{h(\mathbf{x}/|bx|)}F(\mathbf{x}), \qquad (\lambda > 0,\ \mathbf{x} \neq \mathbf{0}) \tag{7.35}$$

which defines *ray-homogeneous functions* ($h : S \to \mathbb{R}_{++}$ where S is again the n-dimensional unit sphere).

The definition

$$F(\lambda\mathbf{x}) = \lambda^{h(\mathbf{x})}F(\mathbf{x}) \qquad (\lambda > 0,\ \mathbf{x} \neq \mathbf{0}) \tag{7.36}$$

seems to lie more naturally between (7.29) [(7.23)] and (7.34), but we show that it is equivalent to (7.35): We substitute into (7.36) $\mathbf{x} = \mathbf{z}/\left|\mathbf{z}\right|$, $\lambda = |\mathbf{z}|$ ($\mathbf{z} \neq \mathbf{0}$) to get

$$F(\mathbf{z}) = \left|\mathbf{z}\right|^{h(\mathbf{z}/|\mathbf{z}|)} F(\mathbf{z}/\left|\mathbf{z}\right|), \tag{7.37}$$

and, as consequence,

$$F(\lambda\mathbf{x}) = \lambda^{h(\mathbf{x}/|\mathbf{x}|)} \left|\mathbf{x}\right|^{h(\mathbf{x}/|\mathbf{x}|)} F(\mathbf{x}/\left|\mathbf{x}\right|).$$

Applying (7.37) to the right hand side we obtain

$$F(\lambda\mathbf{x}) = \lambda^{h(\mathbf{x}/|\mathbf{x}|)}F(\mathbf{x}),$$

that is (7.35). On the other hand, (7.35) is clearly a particular case of (7.36), so *Eqs. (7.35) and (7.36) are equivalent.* At the same time substitution shows that

$$F(\mathbf{z}) = \left|\mathbf{z}\right|^{h(\mathbf{z}/|\mathbf{z}|)} G(\mathbf{z}/\left|bz\right|) \tag{7.38}$$

satisfies (7.35) with *arbitrary* $G : S \to \mathbb{R}$. Since (7.37) is of this form, we have proved that *the general solution of (7.35) and also of (7.36) is given by (7.38) with arbitrary* $G : S \to \mathbb{R}$. The equivalence of (7.35) and (7.36) shows that h *in (7.36) is necessarily homogeneous of degree* 0.

Another but related (look at (7.38)) generalisation is given by

$$F(\mathbf{x}) = \phi(|\mathbf{x}|)F(\mathbf{x}/\left|\mathbf{x}\right|) \qquad (\mathbf{x} \neq \mathbf{0}) \tag{7.39}$$

which defines *quasi homogeneous functions*; here $\phi : \mathbb{R}_{++} \to \mathbb{R}_{++}$. Notice that (7.39) is a generalisation of (7.35) with $\phi(|\mathbf{x}|)$ in place of $|bx|^r$. This may suggest that (7.39) is equivalent to (7.29). However, if we replace $\mathbf{x}$ by $\lambda\mathbf{x}$ in (7.39) we get

$$F(\lambda\mathbf{x}) = \phi(\lambda\left|\mathbf{x}\right|)F(\frac{\mathbf{x}}{|\mathbf{x}|}) = \frac{\phi(\lambda\left|x\right|)}{\phi(|\mathbf{x}|)}F(\mathbf{x}),$$

which is a special case of (7.34), but it is only *then the same as (7.29) if*

$$\phi(\lambda\left|\mathbf{x}\right|) = \phi(\lambda)\phi(|\mathbf{x}|),$$

that is, ϕ *satisfies (7.30)*. Since (7.30) was a consequence of (7.29) but not of (7.39), therefore *quasi homogeneous functions* are genuinely *more general than homogeneous functions of degree r*, to which they clearly reduce iff $\phi(\lambda) = \lambda^r$. It is even more obvious that (7.35) reduces to (7.23) (with $D = \mathbb{R}^n \setminus \{\mathbf{0}\}$), iff $h(t) \equiv r$ (constant).

5. Homothetic functions. Generalisations of homogeneous functions of degree r, of importance for economics, are the *homothetic functions*. These are *compositions of linearly homogeneous functions* $F : D \to \mathbb{R}_+$ $(D \subset \mathbb{R}^n_+)$ *and of strictly increasing functions* $g : F(D) \to \mathbb{R}_+$ (where $F(D)$ is, as we know, the image of D under F), that is, $H = g \circ F$. They reduce to homogeneous functions of degree r exactly when $g(t) = at^r$:

$$H(\lambda\mathbf{x}) = g(F(\lambda\mathbf{x})) = g(\lambda F(\mathbf{x})) = a\lambda^r F(\mathbf{x})^r = \lambda^r g(F(\mathbf{x})) = \lambda^r H(\mathbf{x}).$$

To give an interpretations, we recall the contour lines introduced in Sect. 3.2 and generalise them from two to n variables and to even more general situations. A *level set* of $F : D \to \mathbb{R}_+$ $(D \subset \mathbb{R}^n)$ is defined as

$$\{\mathbf{x} \mid F(\mathbf{x}) = c\} \qquad \text{for a } c \in \mathbb{R}.$$

An example of a homothetic function, which is not homogeneous of any degree, is given by

$$H(\mathbf{x}) = H(x_1, x_2, \ldots, x_n) = (x_1 x_2 \cdot \ldots \cdot x_n)^2 / (1 + (x_1 x_2 \cdot \ldots \cdot x_n)^{2-(1/n^2)})$$

for $\mathbf{x} \in \mathbb{R}^n_+$. Here $F(x_1, x_2, \ldots, x_n) = (x_1 x_2 \cdot \ldots \cdot x_n)^{1/n}$ is the *geometric mean*, which is linearly homogeneous, and

$$g(t) = t^{2n}/(1 + t^{2n-(1/n)}) = (t^{-2n} + t^{-1/n})^{-1}.$$

Since, by the differentiation rules in Sect. 6.5,

$$\begin{aligned} g'(t) &= -(t^{-2n} + t^{-1/n})^{-2}(-2nt^{-2n-1} - (1/n)t^{-(1/n)-1}) \\ &= t^{2n-1}(2n + (1/n)t^{2n-(1/n)})(1 + t^{2n-(1/n)})^{-2} > 0 \end{aligned}$$

for all $t > 0$, therefore g is indeed strictly increasing by what we saw in Sect. 6.5 that the geometric means, including the weighted ones (compare Sect. 7.2) and their powers, there called Cobb–Douglas production functions, play an important role in production theory. The homothetic function in the above example thus has the same contour sets (also called—"*isoquants*" for production functions) as the linearly homogeneous Cobb–Douglas function $F(\mathbf{x}) = (x_1 x_2 \cdot \ldots \cdot x_n)^{1/n}$. However they have a pronounced advantage: while this F is convex from above ("concave") on all of $\mathbb{R}^n_+$, the "*partial functions*"

$$x_j \mapsto H(x_1, x_2, \ldots, x_n) \qquad (j = 1, 2, \ldots, n) \tag{7.40}$$

of this H are first convex from below and then turn into convex from above (concave) at a point of inflection as it is desirable in production theory ("law of eventually diminishing returns"), see for $n = 2$, Fig. 7.4. Such figures are called "*production surfaces*".

Homothetic functions have, among others, the following interesting properties.

The domain of substitution of $H : \mathbb{R}^n_+ \to \mathbb{R}_+$ *is the set* S *of those points in* $\mathbb{R}^n_+$ *at which every partial function (7.40) is strictly increasing. If* H *is partially differentiable with respect to each of its* n *variables then* S *is usually defined by*

$$S := \left\{\mathbf{x} \;\middle|\; \frac{\partial H}{\partial x_j}(\mathbf{x}) > 0;\ j = 1, \ldots, n\right\}.$$

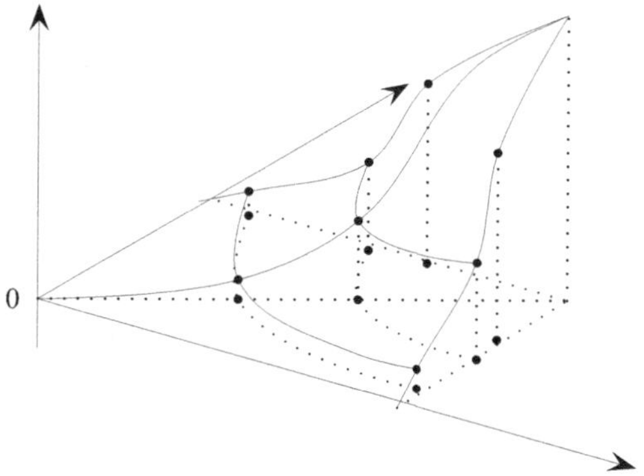

There is a slight difference between these two definitions: we saw in Sect. 6.7 (compare Sect. 6.10) that, while $\partial H(\mathbf{x})/\partial x_j > 0$ indeed guarantees that the functions (7.40) strictly increase at $\mathbf{x}$, they may also strictly increase at certain points where $\partial H(\mathbf{x})/\partial x_j = 0$.

For homothetic functions, if $\mathbf{x}$ *belongs to the domain of substitution* S *then so does* $\lambda\mathbf{x}$ *for all* $\lambda \in \mathbb{R}_{++}$. Indeed, if

$$H(\mathbf{x}') > H(\mathbf{x})$$

then, since $H = g \circ F$, where g is strictly increasing an F linearly homogeneous,

$$\begin{aligned} H(\lambda\mathbf{x}') &= g(F(\lambda\mathbf{x}')) = g(\lambda F(\mathbf{x}')) \\ &> g(\lambda(\mathbf{x})) = g(F(\lambda\mathbf{x})) = H(\lambda\mathbf{x}). \end{aligned} \tag{7.41}$$

Thus *the domain of substitution* S *of a homothetic function is "convex with respect of the origin" ("star shaped") which means* exactly that, *if* $\mathbf{x} \in S$, *then also* $\lambda\mathbf{x} \in S$ *for all* $\lambda \in \mathbb{R}_{++}$. (Similar definitions and statements hold iff H is defined only on a subset of $\mathbb{R}^n_+$ which itself is convex with respect to $\mathbf{0}$.)

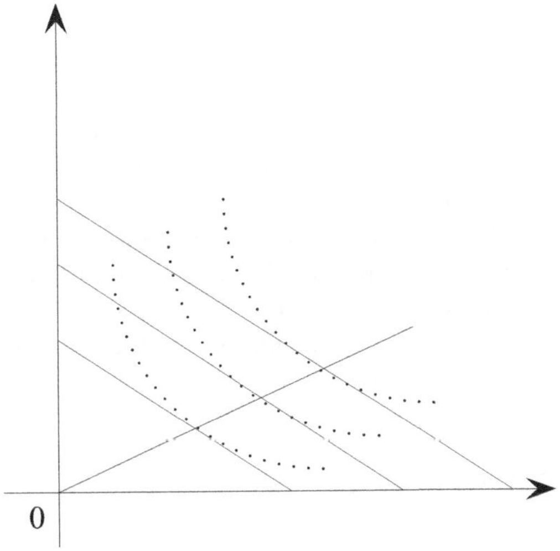

Fig. 7.11 Contour lines of a homothetic production function of two input quantities x_1, x_2. The cost combinations of x_1, x_2 are parallel straight lines. The points where they touch the (*dotted*) isoquants (contour lines of the production function) represent the minimal cost combinations. These are connected by rays starting from the origin

Let now the variables of the homothetic production function be the costs of the n inputs and suppose that the linear combination $\mathbf{a} \cdot \mathbf{x} = a_1x_1 + \ldots + a_nx_n$ with given "weights" $a1, \ldots, a_n$ is the "combination of costs" relevant to us. *If* $\mathbf{x}^* = (x_1^*, \ldots, x_n^*)$ *yields the "minimal cost combination" to produce the output* c *among all* $\mathbf{x}$ *with* $H(\mathbf{x}) = c = H(\mathbf{x}^*)$ (so the $\mathbf{x}$'s are on the isoquant of "hight" $c = H(\mathbf{x}^*)$) *then, by (7.41),* $\lambda\mathbf{x}^* = (\lambda x_1^*, \ldots, \lambda x_n^*)$ *yields the minimal cost combination to produce the output* $H(\lambda\mathbf{x}^*)$. If the situation changes so that $\mathbf{x}^{**}$ yields the minimal cost combination for the output $c = H(\mathbf{x}^{**})$ then every $\lambda\mathbf{x}^{**}$ ($\lambda \in \mathbb{R}_{++}$) will furnish a minimal cost combination. In particular: "enterprises with homothetic production functions expand along rays". See Fig. 7.11 for the case $n = 2$. For linear expansion in the case of vector–valued ("multi product") production functions, see Sect. 7.5.

7.4.1 Exercises

1. Extend the definition (c. Sect. 7.2 **2**) of

$$R(x_1, x_2) = \frac{x_1^2x_2 - x_1^2 - 6x_1x_2 + 6x_1 + 9x_2 - 9}{x_1x_2^2 - 3x_2^2 + x_1 - 3}$$

   to $x_1 = 3$, x_2 arbitrary so that the extended function be continuous.
2. Let $f : [0, a] \to \mathbb{R}_+$, $a \in \mathbb{R}_{++}$, and $g : [0, b) \to \mathbb{R}_+$, $b \in \mathbb{R}_{++}$, be two arbitrary functions satisfying $f(a) = g(b)$. Show that the function F given by (7.28) is
   (a) unambiguous defined on $\mathbb{R}_+^2 \setminus \{(0, 0)\}$,
   (b) homogeneous of degree r on $\mathbb{R}_+^2 \setminus \{(0, 0)\}$.

3. Let $F : \mathbb{R}_{++^2} \to \mathbb{R}_{++}$ be given by $F(x, y) = \frac{x^2y^3}{x^4+y^4}$. The graphs of $x \mapsto F(x, y)$ and $y \mapsto F(x, y)$ are bell–shaped (see Fig. 7.10).
   (a) Determine the abscissae x_1 and x_2 of the turning points of $x \mapsto F(x, 1)$.
   (b) Determine the abscissa x_3 of the maximum of $x \mapsto F(x, 1)$.
   (c) Show that $x \mapsto F(x, 1)$ is convex from below in the intervals $]0, x_1[$, $]x_2, \infty[$ and convex from above in $]x_1, x_2[$.
   (d) Do the corresponding exercises 3(a),(b), (c) for $y \mapsto F(1, y)$.
4. Show that the function F defined in Exercise 3 is not quasi-convex from above ("quasi concave") as defined in Sect. 3.5.
5. Draw three isoquants of the function $H : \mathbb{R}^2_{++} \to \mathbb{R}_{++}$ given by $H(x, y) = x^2y^2/(1 + x^{7/4}y^{7/4})$.

7.4.2 Answers

1.

$$R(x_1, x_2) = \begin{cases} \frac{(x_1-3)^2(x_2-1)}{(x_1-3)(1+x_2^2)} & \text{if} \quad x_1 \neq 3, \\ 0 & \text{if} \quad x_1 = 3. \end{cases}$$

(Continuous because $\lim_{x_1 \to 3} \frac{(x_1-3)^2(x_2-1)}{(x_1-3)(1+x_2^2)} = \lim_{x_1 \to 3} \frac{(x_1-3)(x_2-1)}{1+x_2^2} = 0$.)

2. (a) If the point $(x, y) \in \mathbf{R}_{++}$ lies in one of the two shaded domains in Fig. 7.9 then $F(x, y)$ is defined by one of the two expressions in Sect. 7.4 (7.11). If $(x, y) \in \mathbf{R}_{++}$ lies on the line given by $y = (b/a)x$ then Sect. 7.4 (7.11) yields $F(x, y) = (x/a)^r g(b)$, that is, F is unambiguously defined on this line since $f(a) = g(b)$. For $(x, 0), x \in \mathbf{R}_{++}$, and $(0, y), y \in \mathbf{R}_{++}$, the function F is defined by the second or the first expression in Sect. 7.4 (7.11), respectively.
   (b)

$$F(\lambda x, \lambda y) = \begin{cases} (\lambda y/b)^r f((\lambda xb)/(\lambda y)) & \text{for} \quad \lambda y > 0, 0 \leq \lambda x \leq a\lambda y/b \\ (\lambda x/a)^r g((\lambda ya)/(\lambda x)) & \text{for} \quad \lambda x > 0, 0 \leq \lambda y \leq b\lambda x/a \end{cases}$$

$$= \begin{cases} \lambda^r (y/b)^r f(xb/y) \\ \lambda^r (x/a)^r g(ya/x) \end{cases} = \lambda^r F(x, y).$$

3. (a) $x_1 = 0.5401828\ldots$, $x_2 = 1.4066268\ldots$,
   (b) $x_3 = 1$,
   (c) $F''(x, 1) = \dfrac{2 - 24x^4 + 6x^8}{(1 + x^4)^3} \begin{cases} > 0 & \text{for} \quad 0 < x < x_1,\ x > x_2 \\ < 0 & \text{for} \quad x_1 < x < x_2. \end{cases}$
   (d) $y_1 = 0.5810658\ldots$, $y_2 = 1.4471621\ldots$, $y_3 = 1.3160740\ldots$,

$$F''(1, y) = \frac{y - 54y^5 + 12y^9}{(1 + y^4)^3} \begin{cases} > 0 & \text{for} \quad 0 < y < y_1,\ y > y_2 \\ < 0 & \text{for} \quad y_1 < y < y_2. \end{cases}$$

4. The function $F : \mathbf{R}^2_{++} \to \mathbf{R}_{++}$ given by $x^2y^3/(x^4 + y^4)$ is not quasiconcave since, for example, its function values along the ray given by $y = 1 + x/4$ ($x \in \mathbf{R}_{++}$) strictly increase to a local maximum of $0.6773\ldots$ at $x = 1.52\ldots$, than strictly decrease to a local minimum of $0.413\ldots$ at $x = 7.7\ldots$, and then strictly increases to $+\infty$.

7.5 Fundamental Notions in Production Theory. Production Functions. Elasticity of Substitution

We have encountered *production functions* at the end of the previous Sect. 7.4, in Sect. 6.12 and in Chap. 3. We have been dealing and will continue to deal with production functions of *several variables* (of several "product factors"). In the present section, as in the previous one, we discuss *scalar valued* production functions not because we consider, as some do, single product production, which seems to us too unrealistic, but because, for instance the monetary *value* of the production, in particular the *maximal output value*, may be considered as the function value. This is particularly so for the production function in an enterprise. If the function refers to production of an industry or of a sector of the economy or the entire economy of a nation then the function values are often determined by econometric methods from data of past production (compare Sect. 6.9).

So let $F : \mathbb{R}^n_+ \to \mathbb{R}_+$ be a production function with the input quantities (*input* for short) united into the vector $\mathbf{x} = (x_1, \ldots, x_n)$ as variable and the (maximal) *output value* as function value $F(\mathbf{x})$ (notice the two different senses the word "value" is used in "output value" and "function value"). Some of the fundamental notions in production theory are defined as follows:

- $F(\mathbf{x})/x_j$ is the *average product* at $\mathbf{x}$ *for the j-th production factor* when $x_j > 0$,
- $x_j/F(\mathbf{x})$ its reciprocal with $F(\mathbf{x}) > 0$ is the *product coefficient at* $\mathbf{x}$ *of the j-th product factor*,
- $F'_j(\mathbf{x}) := \frac{\partial F}{\partial x_j}(\mathbf{x})$ is the *marginal product for the j-th production factor* at $\mathbf{x}$,
- $F'_j(\mathbf{x})/F'_k(\mathbf{x})$ is the *marginal rate of (technical) substitution* at $\mathbf{x}$, with $F'_k(\mathbf{x}) > 0$, *of the j-th by the k-th production factor* ($j = 1, \ldots, n$; $k = 1, \ldots, n$).

(Independently of interpretation, $F'_j(\mathbf{x})$ often denotes the partial derivation of F with respect to its j-th variable, at $\mathbf{x}$.)

The average product at $\mathbf{x} = (x_1, \ldots, x_n)$ for x_j shows how many units of output (in value) can be produced in average by one unit of the j-th production factor. Reciprocally, the production coefficient at $\mathbf{x}$ of x_j indicates how many units of the j-th production factor are needed in average to produce one unit of output (in value). In both instances, the total output value is $F(\mathbf{x}) = F(x_1, \ldots, x_n)$.

The above definition of the marginal product clearly *works only if F is partially differentiable with respect to* x_j *at* $\mathbf{x}$. But, in general, the *marginal product* (output) *at* $\mathbf{x} = (x_1, \ldots, x_n)$ *under change by h of the j-th production factor* can be defined by

$$\frac{F(x_1, \ldots, x_{j-1}, x_j + h, x_{j+1}, \ldots, x_n) - F(x_1, \ldots, x_{j-1}, x_j, x_{j+1}, \ldots, x_n)}{h}.$$

If F is partially differentiable with respect to x_j *at* $\mathbf{x}$ *then this has a limit,* when $h \to 0$, the partial derivative

$$F_j'(\mathbf{x}) = \frac{\partial F}{\partial x_j}(\mathbf{x}) \qquad \text{(see Sect. 6.11)},$$

that is, *the marginal product for* x_j *at* $\mathbf{x}$, as defined above. (Compare the definition of price elasticity in Sect. 6.6.)

In order to attach meaning to the marginal rate of substitution, let us calculate what change, say q units of the k-th production factor x_k is needed to have the same effect on the output value $F(\mathbf{x})$ as the change of x_j by h units if all other production factors x_l are unchanged ($l \neq j$, $l \neq k$):

$$F(\ldots, x_j + h, \ldots, x_k, \ldots) = F(\ldots, x_j, \ldots, x_k + q, \ldots)$$

(the values of the variables at the dotted places are the same on both sides). We can write this as

$$\begin{aligned} &F(\ldots, x_j + h, \ldots, x_k, \ldots) - F(\ldots, x_j, \ldots, x_k, \ldots) \\ &\quad = F(\ldots, x_j, \ldots, x_k + q, \ldots) - F(\ldots, x_j, \ldots, x_k, \ldots). \end{aligned}$$

As we saw in Sect. 6.11, *if F is differentiable at* $\mathbf{x}$ then the left and right hand sides are approximately equal to (remember that the x_l's are unchanged for $l \neq j$, $l \neq k$)

$$F_j'(\mathbf{x})h \quad \text{and} \quad F_k'(\mathbf{x})q,$$

respectively, the approximations being the better the smaller h and q are. So

$$F_j'(\mathbf{x})h \approx F_k'(\mathbf{x})q$$

($\approx$ meaning "approximately" or "asymptotically" equal as in Sect. 7.3), that is,

$$\frac{q}{h} \approx \frac{F_j'(\mathbf{x})}{F_k'(\mathbf{x})}.$$

So *the marginal rate of (technical) substitution at* $\mathbf{x}$ *of* x_j *by* x_k *equals approximately the change of the k-th production factor needed to have the same change of the value* $F(\mathbf{x})$ *of the production function as with the change of the j-th production factor by one (small) unit* (see Fig. 7.12).

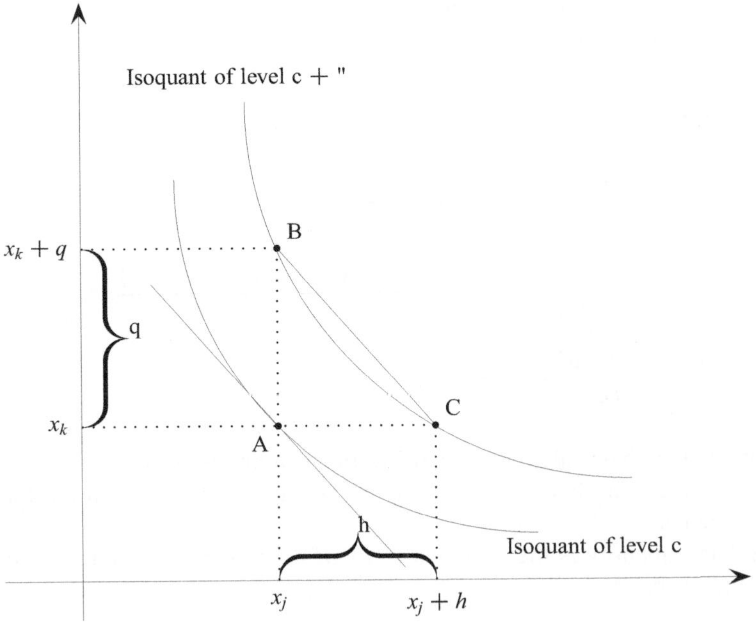

Fig. 7.12 The marginal rate of substitution at $\mathbf{x}$ of x_j by x_k equals approximately q/h, that is q if $h = 1$. Geometrically it equals approximately the slope of the segment from B to C, *exactly* the slope of the tangent at point A

We defined above the marginal rate of substitution of x_j by x_k as the quotient of the marginal products for x_j and x_k. *The marginal product divided by the average product, both for* x_j, both at $\mathbf{x}$, *is the* output elasticity *of the j-th production factor* at $\mathbf{x}$:

$$\varepsilon_j(\mathbf{x}) := \frac{\partial F}{\partial x_j}(\mathbf{x}) \Big/ \frac{F(\mathbf{x})}{x_j} = F_j'(\mathbf{x})x_j/F(\mathbf{x})$$

for $x_j > 0$, $F(\mathbf{x}) > 0$ (notice the similarity to the definition of price elasticity in Sect. 6.6). It is, as we see,

$$\varepsilon_j(\mathbf{x}) = \lim \left(\frac{F(\ldots, x_j + h, \ldots) - F(\ldots, x_j, \ldots)}{F(\ldots, x_j, \ldots)} \Big/ \frac{h}{x_j} \right),$$

the limit, when the change h of the j-th production factor (the only one which changes) *tends to* 0, *of the relative change of the output value divided by the relative change of the j-th production factor.*

The output elasticity of x_j at $\mathbf{x}$ can be written in the form of a so-called "*logarithmic derivative*":

$$\begin{aligned}
&\frac{\partial \ln F}{\partial \ln x_j}(x_1, \dots, x_j, \dots, x_n) \\
&\quad := \left. \frac{\partial \ln F(e^{u_1}, \dots, e^{u_j}, \dots, e^{u_n})}{\partial u_j} \right|_{u_j = \ln x_j} \\
&\quad = \left. \frac{1}{F(e^{u_1}, \dots, e^{u_j}, \dots, e^{u_n})} \frac{\partial F(e^{u_1}, \dots, e^{u_j}, \dots, e^{u_n})}{\partial e^{u_j}} e^{u_j} \right|_{u_j = \ln x_j} \\
&\quad = \frac{1}{F(\mathbf{x})} \frac{\partial F(\mathbf{x})}{\partial x_j} x_j = \varepsilon_j(\mathbf{x}) \qquad (j = 1, 2, \dots, n).
\end{aligned}$$

We applied the chain rule, see Sect. 6.5 **4**; the meaning of $u_j = f_j(\mathbf{x})$ after a vertical bar is that $f_j(\mathbf{x})$—in this case $\ln x_j$—has to be substituted for u_j in the formula in front of the bar.

A related quantity is the *scale elasticity of* F *at* $(\lambda, \mathbf{x})$, defined by (we use the chain rule again)

$$\begin{aligned}
\varepsilon(\lambda, \mathbf{x}) &:= \frac{\partial F(\lambda \mathbf{x})}{\partial \lambda} \frac{\lambda}{F(\lambda \mathbf{x})} = \sum_{j=1}^{n} F_j'(\lambda \mathbf{x}) \frac{\partial (\lambda x_j)}{\partial \lambda} \frac{\lambda}{F(\lambda \mathbf{x})} \\
&= \sum_{j=1}^{n} F_j'(\lambda \mathbf{x}) \frac{\lambda x_j}{F(\lambda \mathbf{x})} = \sum_{j=1}^{n} \varepsilon_j(\lambda \mathbf{x}).
\end{aligned}$$

Now we can define one of the most important notions in production theory, the *elasticity of substitution* $\sigma_{kj}(\mathbf{x})$. It is the *limit of the ratio of relative changes*

- *in the proportion* x_k/x_j *of production factors and*
- *in the marginal rate of substitution* $F_j'(\mathbf{x})/F_k'(\mathbf{x})$

(see Fig. 7.13). We calculate this limit, for each pair (j, k) $(j \neq k)$ at a point $\mathbf{x}^* = (x_1^*, \dots, x_j^*, \dots, x_k^*, \dots, x_n^*)$, and it is the j-th and the k-th variable ("production factor") x_j and x_k which move towards x_j^* and x_k^*, respectively, but staying on the isoquant (contour line, compare Sect. 3.3)

$$\begin{aligned}
&F(x_1^*, \dots, x_{j-1}^*, x_j, x_{j+1}^*, \dots, x_{k-1}^*, x_k, x_{k+1}^*, \dots, x_n^*) = c \\
&\quad = F(x_1^*, \dots, x_j^*, \dots, x_k^*, \dots, x_n^*). \qquad (7.42)
\end{aligned}$$

As we know (Sect. 6.13) under appropriate conditions this equation can be solved yielding x_k as a function of x_j

$$x_k = g_k(x_j) \qquad (7.43)$$

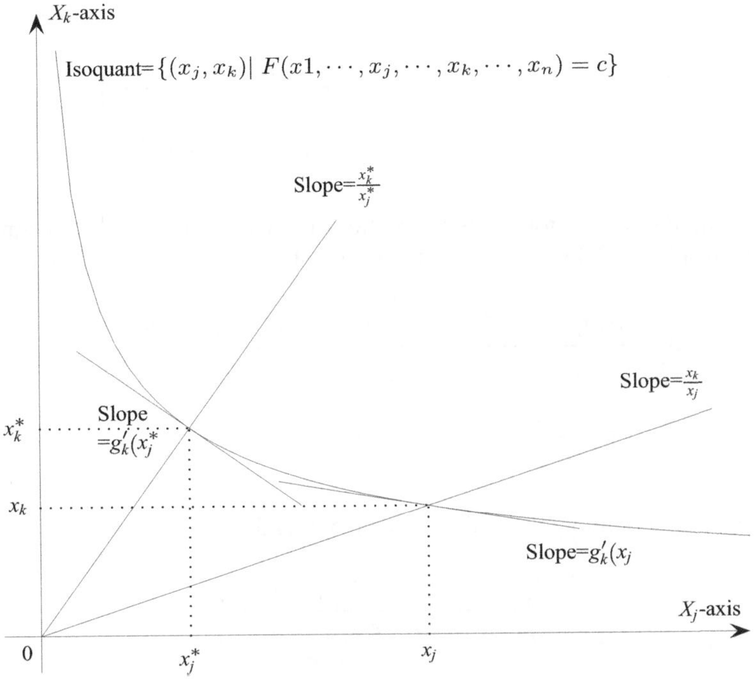

Fig. 7.13 Geometric illustration helping to understand the notion of the elasticity of substitution of a production factor x_j by a production factor x_k. Notice that the slopes of the tangents to the isoquant equal the marginal rates of substitution of x_j by x_k at (x_j, x_k) and (x_j^*, x_k^*)

with the derivative

$$g_k'(x_j) = -\frac{F_j'(\mathbf{x})}{F_k'(\mathbf{x})}.$$

[Notice that g_k' is the derivative of the function g_k of one variable, while F_k' is the derivative of F with respect to its k-th variable.] Here $\mathbf{x} = (x_1^*, \ldots, x_{j-1}^*, x_j, x_{j+1}^*, \ldots, x_{k-1}^*, x_k, x_{k+1}^*, \ldots, x_n^*)$ while, as above, $\mathbf{x}^* = (x_1^*, \ldots, x_j^*, \ldots, x_k^*, \ldots, x_n^*)$ (for the following calculation this is more convenient than the above $x_j + q$, $x_k + h$, x_j, x_k type notation but, of course, equivalent to it.) Since, by (7.42), $\mathbf{x}^*$ is on the same isoquant as $\mathbf{x}$, we have

$$x_k^* = g_k(x_j^*), \qquad g_k'(x_j^*) = -\frac{F_j'(\mathbf{x}^*)}{F_k'(\mathbf{x}^*)}. \tag{7.44}$$

So for $j \neq k$ the *elasticity of substitution of the j-th by the k-th production factor is*

$$\begin{aligned}\sigma_{jk}(\mathbf{x}^*) &= \lim_{x_j \to x_j^*}\left(\frac{x_k/x_j - x_k^*/x_j^*}{x_k^*/x_j^*} \Big/ \frac{F_j'(\mathbf{x})/F_k'(\mathbf{x}) - F_j'(\mathbf{x}^*)/F_k'(\mathbf{x}^*)}{F_j'(\mathbf{x}^*)/F_k'(\mathbf{x}^*)}\right)\\ &= \lim_{x_j \to x_j^*}\left(\frac{g_k(x_j)/x_j - g_k(x_j^*)/x_j^*}{g_k(x_j^*)(x_j^*)} \Big/ \frac{-g_k'(x_j)+g_k'(x_j^*)}{-g_k'(x_j^*)}\right).\end{aligned} \tag{7.45}$$

Dividing the numerators of both fractions (which are the divided again) on the right hand side of (7.45) by $(x_j - x_j^*)$ does not change its value, of course. But

$$\begin{aligned}\lim_{x_j \to x_j^*} \frac{x_k/x_j - x_k^*/x_j^*}{x_j - x_j^*} &= \lim_{x_j \to x_j^*} \frac{g_k(x_j)/x_j - g_k(x_j^*)/x_j^*}{x_j - x_j^*}\\ &= \left(\frac{g_k(x_j)}{x_j}\right)'_{x=x_j^*} = g_k''(x_j^*),\end{aligned} \tag{7.46}$$

by the rule of derivation of fractions Sect. 6.5 **3**, and

$$\lim_{x_j \to x_j^*} \frac{g_k'(x_j) - g_k'(x_j^*)}{x_j - x_j^*} = g_k''(x_j^*),$$

by the definition of the (second) derivative. So (7.45) becomes

$$\sigma_{kj}(\mathbf{x}^*) = \frac{g_k'(x_j^*)(x_j^* g_k'(x_j^*) - x_k^*)}{g_k''(x_j^*)x_j^* x_k^*} \qquad (j \neq k) \tag{7.47}$$

(since $g_k(x_j^*) = x_k^*$) which *expresses the elasticity of substitution* $\sigma_{kj}(\mathbf{x}^*)$ *in terms of the function* g_k *alone.*

It is, of course, more desirable to *express* σ_{kj} *in terms of the production function* F. We can do that too, from the middle term of the chain of equalities (7.45), where we again divide the numerators of both fractions by $(x_j - x_j^*)$. From (7.44) and (7.46) now

$$\begin{aligned}\lim_{x_j \to x_j^*} \frac{x_k/x_j - x_k^*/x_j^*}{x_j - x_j^*} &= \frac{1}{x_j^{*2}}\left(-x_j^* \frac{F_j'(\mathbf{x}^*)}{F_k'(\mathbf{x}^*)} - x_k^*\right)\\ &= -\frac{x_j^* F_j'(\mathbf{x}^*) + x_k^* F_k'(\mathbf{x}^*)}{x_j^{*2} F_k'(\mathbf{x}^*)}.\end{aligned} \tag{7.48}$$

On the other hand, again from the definition of (partial) derivatives and from Sect. 6.5 **3**

$$\lim_{x_j \to x_j^*} \frac{F_j'(\mathbf{x})/F_k'(\mathbf{x}) - F_j'(\mathbf{x}^*)/F_k'(\mathbf{x}^*)}{x_j - x_j^*} = \frac{\partial}{\partial x_j}\left(\frac{F_j'(\mathbf{x})}{F_k'(\mathbf{x})}\right)\Bigg|_{\mathbf{x}=\mathbf{x}^*} = \frac{F_k'(\mathbf{x}^*)\frac{\partial}{\partial x_j}F_j'(\mathbf{x}^*) - F_j'(\mathbf{x}^*)\frac{\partial}{\partial x_j}F_k'(\mathbf{x}^*)}{F_k'(\mathbf{x}^*)^2} \tag{7.49}$$

(in the middle term we wrote the partial derivatives with respect to x_j at $\mathbf{x}^*$).

We now calculate the partial derivatives in the numerator, using (7.43), (7.44) and the chain rule for functions of several variables from Sect. 6.12:

$$\begin{aligned} \frac{\partial}{\partial x_j}F_j'(\mathbf{x}^*) = & \frac{\partial}{\partial x_j}F_j'(x_1^*, \dots, x_{j-1}^*, x_j, x_{j+1}^*, \dots \\ & \dots, x_{k-1}^*, g_k(x_j), x_{k+1}^*, \dots, x_n^*)\Bigg|_{x_j = x_j^*} \\ = & F_{jj}''(\mathbf{x}^*) + F''jk(\mathbf{x}^*)g_k'(x_j^*) \\ = & F_{jj}''(\mathbf{x}^*) - F_{jk}''(\mathbf{x}^*)\frac{F_j'(\mathbf{x}^*)}{F_k'(\mathbf{x}^*)}, \end{aligned} \tag{7.50}$$

and similarly,

$$\frac{\partial}{\partial x_j}F_k'(\mathbf{x}^*) = F_{kj}''(\mathbf{x}^*) - F_{kk}''(\mathbf{x}^*)\frac{F_j'(\mathbf{x}^*)}{F_k'(\mathbf{x}^*)}. \tag{7.51}$$

Here (compare Sect. 6.9) we used the following notation for second partial derivatives, supposing that they exist and are continuous at $\mathbf{x}$:

$$F_{kj}''(\mathbf{x}) = \frac{\partial}{\partial x_j}\left(\frac{\partial}{\partial x_k F(\mathbf{x})}\right) = \frac{\partial}{\partial x_k}\left((\frac{\partial}{\partial x_j F(\mathbf{x})})\right) = F_{jk}''(\mathbf{x}),$$

$$F_{ll}''(\mathbf{x}) = \frac{\partial^2}{\partial x_l^2}F(\mathbf{x}) \quad (l = j; l = k; j = 1, \dots, n; k = 1, \dots, n).$$

So we get from (7.45), (7.48), (7.49), (7.50) and (7.51) for every $\mathbf{x}^* \in \mathbb{R}_{++}^n$ *the expression*

$$\sigma_{kj}(\mathbf{x}^*) = \frac{x_j^*}{x_k^*}\left(-\frac{x_j^* F_j'(\mathbf{x}^*) + x_k^* F_k'(\mathbf{x}^*)}{x_j^{*2} F_k'(\mathbf{x})}\right) \Bigg/ \left(\frac{F_k'(\mathbf{x}^*)}{F_j'(\mathbf{x}^*)} \cdot \frac{F_k'(\mathbf{x}^*)^2 F_{jj}''(\mathbf{x}^*) - 2F_k'(\mathbf{x}^*)F_j'(\mathbf{x}^*)F_{jk(\mathbf{x}^*)}'' + F_j'(\mathbf{x}^*)^2 F_k'(\mathbf{x}^*)}{F_k'(\mathbf{x}^*)^3}\right)$$

or, omitting the asterisks, $(j \neq k)$

$$\sigma_{kj}(\mathbf{x}) = -\frac{F'_j(\mathbf{x})F'_k(\mathbf{x})}{x_j x_k} \cdot \frac{x_j F'_j(\mathbf{x}) + x_k F'_k(\mathbf{x})}{F'_k(\mathbf{x})^2 F''_{jj}(\mathbf{x}) - 2F'_j(\mathbf{x})F'_k(\mathbf{x})''_{jk}(\mathbf{x}) + F'_j(\mathbf{x})^2 F''_{kk}(\mathbf{x})} \tag{7.52}$$

for the elasticity of substitution of the j-th by th k-th production factor. (Not quit simple but we got it; the fact that it is usually obtained by even more complicated calculations may give some comfort to the reader.)

Notice that

$$\sigma_{kj}(\mathbf{x}) = \sigma_{jk}(\mathbf{x}).$$

Therefore one often calls $\sigma_{kj}(\mathbf{x})$ the *elasticity of substitution between the j-th and k-th production factor.*

We calculate the above quantities for an important class of production functions, the Cobb–Douglas functions (Charles W. Cobb (1871–1941), Paul H. Douglas (1892–1976)), for which we saw an example at the end of Sect. 6.12 and which have, as we will see, *constant elasticity of substitution*. They are homogeneous and, if their degree of homogeneity is not greater than 1, they are also convex from above (concave). The *Cobb–Douglas functions* are defined by

$$F(x_1, x_2, \ldots, x_n) = a x_1^{c_1} x_2^{c_2} \cdot \ldots \cdot x_n^{c_n}, \tag{7.53}$$

where a, $c_1, c_2, \ldots, c_n$ are positive constants. Notice that the output (value) is positive if the production factor quantities $x_1, \ldots, x_n$ are positive and that it is then strictly increasing with each x_j $(j = 1, \ldots, n)$. These functions are clearly *homogeneous of degree* $c_1 + c_2 + \ldots + c_n$.

$$F(\lambda x_1, \lambda x_2, \ldots, \lambda x_n) = \lambda^{c_1 + c_2 + \ldots + c_n} F(x_1, x_2, \ldots, x_n).$$

For the Cobb–Douglas functions given by (7.53), the marginal product of the j-th production factor is

$$F'_j(\mathbf{x}) = \frac{\partial F}{\partial x_j}(\mathbf{x}) = c_j a x_1^{c_1} x_2^{c_2} \cdot \ldots \cdot x_{j-1}^{c_{j-1}} x_j^{c_j - 1} x_{j+1}^{c_{j+1}} \cdot \ldots \cdot x_n^{c_n} = c_j \frac{F(\mathbf{x})}{x_j}$$

and so *the marginal rate of substitution of the j-th by the k-th production factor is*

$$\frac{F'_j(\mathbf{x})}{F'_k(\mathbf{x})} = \frac{c_j F(\mathbf{x})}{x_j} \Big/ \frac{c_k F(\mathbf{x})}{x_k} = \frac{c_j}{c_k} \frac{x_k}{x_j}.$$

Notice, that this is *independent of the production factors other than* x_j *and* x_k.

The output elasticity of the j-th production factor at $\mathbf{x}$ *is for Cobb–Douglas function,*

$$\varepsilon(\lambda, \mathbf{x}) = \sum_{j=1}^{n} \varepsilon_j(\lambda \mathbf{x}) = \sum_{j=1}^{n} c_j,$$

is also *constant* and equals its degree of homogeneity.

Finally we calculate the elasticity of substitution $\sigma_{kj}(\mathbf{x})$. First we obtain

$$F''_{kj}(\mathbf{x}) = \frac{\partial^2 F}{\partial x_k \partial x_j}(\mathbf{x}) = c_j \frac{F'_k(\mathbf{x})}{x_j} = c_k c_j \frac{F(\mathbf{x})}{x_k x_j} \quad \text{or} \quad k \neq j$$

and

$$F''_{jj}(\mathbf{x}) = \frac{\partial}{\partial x_j}(c_j a x_1^{c_1} x_2^{c_2} \cdot \ldots \cdot x_{j-1}^{c_{j-1}} x_j^{c_j - 1} x_{j+1}^{c_{j+1}} \cdot \ldots \cdot x_n^{c_n}) = c_j(c_j - 1)\frac{F(\mathbf{x})}{x_j^2}.$$

Plugging these into (7.52) gives

$$\sigma_{kj}(\mathbf{x}) = -\frac{c_j c_k F(\mathbf{x})^2 (c_j F(\mathbf{x}) + c_k F(\mathbf{x}))}{c_j c_k F(\mathbf{x})^3 (c_k(c_j - 1) - 2c_j c_k + c_j(c_k - 1))} = 1, \tag{7.54}$$

that is, *for all Cobb–Douglas functions the elasticity of substitution of the k-th by the j-th production factor is* 1, *independent not only of all production factors but also of the exponents* c_l $(l = 1, \ldots, n)$ *and of the degree of homogeneity* $c_1 + c_2 + \ldots + c_n$.

Our calculations leading to (7.54) show that this result is also true if the constants $a, c_1, c_2, \ldots, c_n$ (see (7.53)) are arbitrary nonzero. If

$$c_1 + \ldots + c_n = 1, \tag{7.55}$$

then the Cobb–Douglas function is *linearly homogeneous*. It is also *concave* (convex from above) on $\mathbb{R}^n_+$ in this case and even if $c_1 + \ldots + c_n \leq 1$. We will show this, at least for $n = 2$, in Sect. 7.5 (Example 3). In the case (7.55) it cannot be *strictly* concave anywhere because

$$\lambda \longmapsto F(\lambda \mathbf{x}_0) = \lambda F(\mathbf{x}_0),$$

so F is *linear* on each *ray* $\{\lambda \mathbf{x}_0 \mid \lambda \in \mathbb{R}_+\}$ (see Sect. 3.4), through every $\mathbf{x}_0 \in \mathbb{R}^n_{++}$. I $r := c_1 + \ldots + c_n > 1$ then F cannot be anywhere concave even in the broader sense: $\lambda \mapsto F(\lambda \mathbf{x}_0) = \lambda^r F(\mathbf{x}_0)$ is strictly convex (from below) for $r > 1$ (see

Sect. 7.2). More generally, if F is homogeneous of degree $r > 1$, then

$$\lambda \longmapsto F(\lambda \mathbf{x}_0) = \lambda^r F(\mathbf{x}_0)$$

is strictly convex (from below) for every $\mathbf{x}_0 \in \mathbb{R}^n_{++}$, that is, F is strictly convex on every ray.

Whatever r is, the *scale elasticity* of the Cobb–Douglas function at $(\lambda, \mathbf{x})$ equals r (constant):

$$\begin{aligned}\varepsilon(\lambda, \mathbf{x}) &= \frac{\partial F(\lambda \mathbf{x})}{\partial \lambda} \frac{\lambda}{F(\lambda \mathbf{x})} = \frac{\partial \lambda^r F(\mathbf{x})}{\partial \lambda} \frac{\lambda}{\lambda^r F(\mathbf{x})} \\ &= r\lambda^{r-1} F(\mathbf{x}) \frac{1}{\lambda^{r-1} F(\mathbf{x})} = r.\end{aligned} \tag{7.56}$$

We say that the *returns to scale on the ray* $\{\lambda \mathbf{x} \mid \lambda \in \mathbb{R}_{++}\}$ *are (strictly) increasing, decreasing or constant if* $\varepsilon(\lambda \mathbf{x}) > 1$, < 1 or $= 1$, respectively. Since (7.56) holds for every *homogeneous function of degree* r, such a function *has increasing, decreasing or constant returns to scale according to whether* $r > 1$, $r < 1$ *or* $r = 1$. *This holds* then in particular also for Cobb–Douglas functions with $r = c_1 + \ldots + c_n$.

We return now to the elasticity of substitution and investigate the reverse question to that settled above. We will determine, at least for $n = 2$, *which linearly homogeneous functions have constant elasticity of substitution* (we do not have to say between which production factors, since now there are only two).

For (twice continuously differentiable) linearly homogeneous functions $F : \mathbb{R}^2_{++} \to \mathbb{R}_{++}$ the formula (7.52) for elasticity of substitution can be simplified by use of Euler's equation (Sect. 6.12 (6.27)):

$$F(\mathbf{x}) = x_1 F_1'(\mathbf{x}) + x_2 F_2'(\mathbf{x}). \tag{7.57}$$

By differentiating it, we get

$$\begin{aligned} F_1'(\mathbf{x}) &= F_1'(\mathbf{x}) + x_1 F_{11}''(\mathbf{x}) + x_2 F_{21}''(\mathbf{x}), \\ F_2'(\mathbf{x}) &= x_1 F_{12}''(\mathbf{x}) + x_2 F_{22}''(\mathbf{x}) + F_2'(\mathbf{x}), \\ F''11(\mathbf{x}) &= -\frac{x_2}{x_1} F_{21}''(\mathbf{x}) = -\frac{x_2}{x_1} F_{12}''(\mathbf{x}), \\ F_{22}''(\mathbf{x}) &= -\frac{x_1}{x_2} F_{12}''(\mathbf{x}) \end{aligned} \tag{7.58}$$

(where we used also the equality of continuous mixed second partial derivatives, $F_{12}'' = F_{21}''$, see the end of Sect. 6.9.) By (7.52), (7.54) and (7.58) the elasticity of substitution is

$$\begin{aligned}\sigma_{21}(\mathbf{x}) &= \sigma_{12}(\mathbf{x}) \\ &= -\frac{F_1'(\mathbf{x})F_2'(\mathbf{x})}{x_1x_2} \\ &\quad\cdot \frac{F(\mathbf{x})}{(F_1'(\mathbf{x})^2(-\frac{x_1}{x_2}) - 2F_1'(\mathbf{x})F_2'(\mathbf{x}) + F_2'(\mathbf{x})^2(-\frac{x_2}{x_1}))F_{12}''(\mathbf{x})} \\ &= \frac{F_1'(\mathbf{x})F_2'(\mathbf{x})F(\mathbf{x})}{(x_1F_1'(\mathbf{x}) + x_2F_2'(\mathbf{x}))^2F_{12}''(\mathbf{x})} = \frac{F_1'(\mathbf{x})F_2'(\mathbf{x})}{F(\mathbf{x})F_{12}''(\mathbf{x})}.\end{aligned} \tag{7.59}$$

But, F being linearly homogeneous, there exists a function $\Phi : \mathbb{R}_{++} \to \mathbb{R}_{++}$ such that

$$F(\mathbf{x}) = F(x_1, x_2) = x_1\Phi(\frac{x_2}{x_1}) = x_1\Phi(u) \quad \text{where} \quad u := \frac{x_2}{x_1} \tag{7.60}$$

(see Sect. 7.5 (7.48)). If F is twice differentiable, so is Φ (since $\Phi(t) = F(1, t)$) and we use (7.60),

$$F_1'(\mathbf{x}) = \frac{\partial}{\partial x_1}(x_1\Phi(\frac{x_2}{x_1})) = \Phi(\frac{x_2}{x_1}) + x_1\Phi'(\frac{x_2}{x_1})(-\frac{x_2}{x_1^2}) = \Phi(u) - u\Phi'(u),$$

$$F_2'(\mathbf{x}) = \frac{\partial}{\partial x_2}(x, \Phi(\frac{x_2}{x_1})) = x_1\Phi'(\frac{x_2}{x_1})\frac{1}{x_1} = \Phi'(u),$$

$$F_{12}''(\mathbf{x}) = \frac{\partial}{\partial x_1}(x, \Phi(\frac{x_2}{x_1})) = \Phi''(\frac{x_2}{x_1})(-\frac{x_2}{x_1^2}) = -\frac{u}{x_1}\Phi''(u)$$

to transform (7.59) in the case $\sigma_{21} = \sigma_{12} = c$ (= constant) into

$$-cu\Phi(u)\Phi''(u) = (\Phi(u) - u\Phi'(u))\Phi'(u). \tag{7.61}$$

This equation, which is supposed to hold for every $u \in \mathbb{R}_{++}$ (since (7.59) was to hold for all $\mathbf{x} \in \mathbb{R}^2_{++}$), is a *second order differential equation*. We will deal with differential equations in Chap. 11 and show methods for their solution. In this case, however, it is quit easy to solve the equation directly. This is especially so if we now assume $c \neq 0$ and $\Phi'(u) > 0$, which suggests itself in broad sections of production theory. (For instance, c = elasticity of substitution = 0 is only possible if the isoquants of F are rays of the kind $\{\lambda(x_1, x_2) \mid (x_1, x_2) \in \mathbb{R}^2_{++},$ fixed, $\lambda \in \mathbb{R}_{++}\}$, see (7.45) and Fig. 7.13; this is not realistic in any production.) We divide both sides in (7.61) by $-cu\Phi(u)\Phi'(u)$ and get

$$\frac{\Phi''(u)}{\Phi'(u)} = \frac{1}{c}\frac{\Phi'(u)}{\Phi(u)} - \frac{1}{c}\frac{1}{u}.$$

If we look carefully, we recognise each term as derivative of easy to guess functions (of u): $1/u$ is the derivative of $\ln u$ (Sect. 7.2), so $\Phi'(u)/\Phi(u)$ is the derivative of $\ln \Phi(u)$ (using also the chain rule Sect. 6.4 **5**) and similarly $\Phi''(u)/\Phi'(u)$ is the derivative of $\ln \Phi'(u)$. Moreover, by the remark after Sect. 6.6 (6.11), $G'(u) = H'(u)$ implies that $G(u) = H(u) + C$ for some constant C. So we have

$$\ln \Phi'(u) = \frac{1}{c} \ln \Phi(u) - \frac{1}{c} \ln u + C.$$

Taking the exponential of both sides (in view of $e^{\ln t} = t$) we get (with $c_2 := e^C$)

$$\Phi'(u) = c_2 \frac{\Phi(u)^{1/c}}{u^{1/c}} \tag{7.62}$$

(compare Sect. 11.2 (11.10)).

We distinguish now *two cases*. *First*, if $c = 1$, then we write (7.62) as

$$\frac{\Phi'(u)}{\Phi(u)} = c_2 \frac{1}{u}.$$

As above, this implies (writing the added constant this time as $\ln a$)

$$\ln \Phi(u) = c_2 \ln u + \ln a,$$

that is,

$$\Phi(u) = a u^{c_2}.$$

Now we substitute this into (7.60) and obtain

$$F(\mathbf{x}) = F(x_1, x_2) = x_1 a (\frac{x_2}{x_1})^{c_2} = a x_1^{1-c_2} x_2^{c_2}, \tag{7.63}$$

that is, writing $c_1 := 1 - c_2$ we get the function (7.53) with $n = 2$, $c_2 > 0$, $c_1 + c_2 = 1$. We have thus proved that *the only linearly homogeneous functions* $F : \mathbb{R}^2_{++} \to \mathbb{R}_{++}$ *which satisfy* $F_1'(x_1, x_2) > 0$, $F_2'(x_1, x_2) > 0$ *and have elasticity of substitution* 1 *are the Cobb–Douglas functions (7.53) with* $n = 2$, $c_2 = 1 - c_1$, $0 < c_1 < 1$, $a > 0$. (We have proved before, that all functions (7.53), with arbitrary nonzero constants $a, c_1, c_2, \ldots, c_n$, have "elasticity of substitution" 1 between any two production factors.)

We now look at the *second case* in (7.62), when $c \neq 1$. Then the notation

$$\gamma := 1 - \frac{1}{c} \neq 0$$

will be of advantage, because it (and multiplication by γ) transforms (7.62) into

$$\gamma\Phi(u)^{\gamma-1}\Phi'(u) = c_2\gamma u^{\gamma-1} \qquad (c_2 = e^C > 0;\ \text{see (7.62)}).$$

In this equation we recognise (see Sects. 6.5 and 7.2) the right and left sides as derivatives (with respect to u) of c_2u^γ and of $\Phi(u)^\gamma$, respectively. As above,

$$\Phi(u)^\gamma = b_2u^\gamma + b_1 \qquad (b_2 := c_2 > 0,\ b_1 \text{ some nonnegative constant})$$

follows (since $\Phi(u)$ is supposed to be > 0, b_1 cannot be negative) and by (7.60),

$$F(x_1, x_2) = x_1\Phi(\frac{x_2}{x_1}) = x_1(b_1(\frac{x_2}{x_1})^\gamma + b_1)^{1/\gamma} = (b_1x_1^\gamma + b_2x_2^\gamma)^{1/\gamma} \qquad (\gamma \neq 0). \tag{7.64}$$

These are called *CES* = *"Constant Elasticity of Substitution" functions* (even though, as we just saw, other functions, like the Cobb–Douglas function, have this property too). And indeed, every step we made can be reversed, so we get that the linearly homogeneous function (7.65) has constant elasticity of substitution:

$$\sigma_{21}(\mathbf{x}) = \sigma_{12}(\mathbf{x}) = c = \frac{1}{1-\gamma} \qquad (\gamma \neq 0,\ \gamma \neq 1). \tag{7.65}$$

Thus we showed that *the only linearly homogeneous production functions* $F : \mathbb{R}^2_{++} \to \mathbb{R}_{++}$ *which satisfy* $F_1'(x_1, x_2) > 0$, $F_2'(x_1, x_2) > 0$ *and have constant elasticity of substitution are given by (7.63) and (7.66)* ($a > 0$, $c_2 \in]0, 1[$, $b_1 > 0$, $b_2 > 0$, $\gamma \neq 0$ *arbitrary constants). If we want the constant elasticity of substitution to be positive, the (see (7.65) in (7.64))* $\gamma < 1$ *or, in the equivalent*

$$F(\mathbf{x}) = F(x_1, x_2) = (b_1x_1^{-\gamma} + b_2x_2^{-\gamma})^{-1/\gamma} \qquad (\gamma \neq 0) \tag{7.66}$$

we have $\gamma > -1$.

Just as (7.53) is a generalisation of (7.40), so (7.66) is generalised to the expressions

$$F(\mathbf{x}) = \left(\sum_{k=1}^{n} b_kx_k^{-\gamma}\right)^{-1/\gamma} \qquad (b_k > 0; k = 1, \ldots, n; \gamma \neq 0) \tag{7.67}$$

which are also called *CES functions* (each $\sigma_{ij}(\mathbf{x})$ elasticity of substitution is indeed constant for them too) and even to

$$F(\mathbf{x}) = \left(\sum_{k=1}^{n} b_kx_k^{-\gamma}\right)^{-r/\gamma}.$$

The latter function is *homogeneous of degree r* (as is (7.53) if $c_1 + \ldots + c_n = 1$).

Since, as we saw, not only the CES functions (7.65) but also the Cobb–Douglas functions (7.53) with $c_1 + \ldots + c_n = 1$ are linearly homogeneous and have constant elasticity of substitution, some effort is made to "include (7.53) into the family (7.67)" as *limit for* $\gamma \to 0$. This *cannot be done with this form of (7.67) when* $b_1 + \ldots + b_n \neq 1$ (why?) but a slightly devious trick seems to make it possible. We introduce

$$c_k := b_k \Big/ \sum_{j=1}^{n} b_j, \qquad a := \left(\sum_{j=1}^{n} b_j\right)^{-1/\gamma} \tag{7.68}$$

in order to transform (7.67) into

$$\begin{aligned} F(\mathbf{x}) &= a\left(\sum_{k=1}^{n} c_k x_k^{-\gamma}\right)^{-1/\gamma} \\ \text{where} \quad &\sum_{k=1}^{n} c_k = 1 \qquad (c_k > 0; k = 1, \ldots, n; a > 0; \gamma \neq 0). \end{aligned} \tag{7.69}$$

This formula, with *constant a*, is equivalent to (7.67) *as long as* γ *remains constant.* But notice that a in (7.68) depends on γ. If a in (7.69) is a constant we prove that *for* $\gamma \to 0$ *the limit of (7.69) is the Cobb–Douglas expression (7.53) with* $c_1 + \ldots + c_n = 1$. Indeed,

$$\begin{aligned} \lim_{\gamma \to 0} \ln \left(a \sum_{k=1}^{n} c_k x_k^{-\gamma}\right)^{-1/\gamma} &= \ln a - \lim_{\gamma \to 0} \frac{\ln(\sum_{k=1})^n c_k x_k^{-\gamma}}{\gamma} \\ &= \ln a - \lim_{\gamma \to 0} \frac{\sum c_k(-x_k^{-\gamma} \ln x_k)}{\sum c_k x_k^{-\gamma}} \Big/ 1 \\ &= \ln a + \sum_{k=1}^{n} c_k \ln x_k. \end{aligned}$$

(We used limits of sums, the Bernoulli–L'Hospital rule, the chain rule and the derivatives of the natural logarithm and of the exponential functions from Sects. 6.1, 6.6, 6.4 and 7.2, respectively, and $\sum c_k = 1$. Be aware that here γ is the variable.) Putting both sides into the exponential function, which is continuous (Sect. 7.2), we get, as asserted,

$$\lim_{\gamma \to 0} \left(a \sum_{k=1}^{n} c_k x_k^{-\gamma}\right)^{-1/\gamma} = a x_1^{c_1} x_2^{c_2} \cdot \ldots \cdot x_n^{c_n}.$$

Finally we note that both (7.53) (with $0 < c_j < 1; j = 1, \ldots, n$) and (7.67) (with $\gamma > -1$) are *strictly concave* functions (convex from above) of each x_j ($j = 1, \ldots, n$). In production theory these properties are called *laws of diminishing marginal returns*. Indeed,

$$\frac{\partial}{\partial x_1}\left(ax_1^{c_1}x_2^{c_2}\cdot\ldots\cdot x_n^{c_n}\right) = ac_1x_1^{c_1-1}x_2^{c_2}\cdot\ldots\cdot x_n^{c_n},$$

$$\frac{\partial^2}{\partial x_1^2}\left(ax_1^{c_1}x_2^{c_2}\cdot\ldots\cdot x_n^{c_n}\right) = ac_1(c_1-1)x_1^{c_1-2}x_2^{c_2}\cdot\ldots\cdot x_n^{c_n} < 0$$

since $0 < c_1 < 1$, $a > 0$. So $x_1 \mapsto ax_1^{c_1}x_2^{c_2}\cdot\ldots\cdot x_n^{c_n}$ and, in the same way (x_1 had no distinguished role) $x_j \mapsto ax_1^{c_1}x_2^{c_2}\cdot\ldots\cdot x_j^{c_j}\cdot\ldots\cdot x_n^{c_n}$ are strictly concave (compare Sect. 7.4 and also Sect. 7.5). Similarly,

$$\frac{\partial}{\partial x_j}\left(\sum_{k=1}^{n} b_k x_k^{-\gamma}\right)^{-1/\gamma} = -\frac{1}{\gamma}(-\gamma)\left(\sum_{k=1}^{n} b_k x_k^{-\gamma}\right)^{(-1/\gamma)-1} b_j x_j^{-\gamma-1}$$

$$\begin{aligned}
\frac{\partial^2}{\partial x_j^2}&\left(\sum_{k=1}^{n} b_k x_k^{-\gamma}\right)^{-1/\gamma} \\
&= \left(-\frac{1}{\gamma}-1\right)(-\gamma)\left(\sum_{k=1}^{n} b_k x_k^{-\gamma}\right)^{(-1/\gamma)-2} b_j x_j^{-\gamma-1} b_j x_j^{-\gamma-1} \\
&\quad + \left(\sum_{k=1}^{n} (b_k x_k)^{(-1/\gamma)-1} b_j(-\gamma-1)\right) x_j^{-\gamma-2} \\
&= (\gamma+1)\left(\sum_{k=1}^{n} b_k x_k^{-\gamma}\right)^{(-1/\gamma)-1} b_j x_j^{-\gamma-2}\left(\frac{b_j x_j^{-\gamma}}{\sum_{k=1}^{n} b_k x_k^{-\gamma}} - 1\right) \\
&< 0
\end{aligned}$$

since $0 < b_j x_j^{-\gamma} < \sum b_k x_k^{-\gamma}$ and $\gamma + 1 > 0$. So, just as *the Cobb–Douglas functions (7.53)* (with $0 < c_j < 1, j = 1, \ldots, n$), *the CES functions (7.67)* (with $\gamma > -1$) *are strictly concave in each variable (in each production factor quantity).*

Of course, in production theory also production functions are used which are not homogeneous (of any degree) or do not have the property of constant elasticity of substitution. They do not have to be everywhere convex from above (concave) either in their individual variables. For instance, stretches convex from below then from above, then again from below may alternate, compare to the end of Sect. 3.5.

7.5.1 Exercises

1. Give an example of a "production function" $F : \mathbb{R}_+ \to \mathbb{R}_+$ that satisfies both
   (a) strictly decreasing average product for all $x > 0$,
   (b) strictly increasing marginal returns for all $x > x^* > 0$.
2. Give an example of a "production function" $F : \mathbb{R}_+ \to \mathbb{R}_+$ that satisfies both
   (a) strictly increasing average product for all $x > 0$,
   (b) strictly decreasing marginal returns for all $x > \hat{x} > 0$.
3. Let $\varepsilon(\lambda, \mathbf{x})$ be the scale elasticity of the production function $F : \mathbb{R}^2_{++} \to \mathbb{R}_{++}$ at $(\lambda, \mathbf{x}) \in \mathbb{R}^3_{++}$, and $\varepsilon_j(\lambda\mathbf{x})$ the output elasticity of the j-th ($j = 1, 2$) production factor at $\lambda\mathbf{x}$. Show that $\varepsilon(\lambda, \mathbf{x}) = \varepsilon_1(\lambda\mathbf{x}) + \varepsilon_2(\lambda\mathbf{x})$ for the function F given by
   (a) $F(x_1, x_2) = \ln(1 + x_1x_2)$, (b) $F(x_1, x_2) = 1/(1 + e^{1-x_1x_2})$,
   (c) $F(x_1, x_2) = x_1^{1/3} + x_2^{1/2}$, (d) $F(x_1, x_2) = x_1^2x_2^3/(1 + x_1x_2^2)$.
4. Determine the elasticity of substitution $\sigma_{21}(x_1, x_2)$ ($= \sigma_{12}(x_1, x_2)$) for the functions F given in (a), (b), (c) of Exercise 3.
5. Let $F : \mathbb{R}^n_{++} \to D \subset \mathbb{R}_{++}$ be an arbitrary twice differentiable function. Show that $g \circ F$, where $g : D \to \mathbb{R}_{++}$ is an arbitrary function satisfying $g'(u) > 0$ ($u = F(\mathbf{x})$), has the same elasticity of substitution $\sigma_{kj}\mathbf{x}$ ($= \sigma_{kj}(\mathbf{x})$) as F.

7.5.2 Answers

1. $F(x) = x^{1/2}(1/(1 + x^2) + x^{1/2})$, for example, has the properties:
   (a) $F(x)/x = 1/\sqrt{x}(1 + x^2) + 1/\sqrt{x}$ is strictly decreasing for all $x > 0$,
   (b) $F'(x) = 1 + (\frac{1}{2}x^{-1/2} - \frac{3}{2}x^{\frac{3}{2}})/(1 + 2x^2 + x^4)$ is strictly increasing for all $x > \bar{x} = 1.119\ldots$.
2. $F(x) = x^3/(10 + x^2)$, for example, has the properties:
   (a) $F(x)/x = x^2/(10 + x^2)$ is strictly increasing for all $x \geq 0$,
   (b) $F'(x) = (30x^2 + x^4)/(100 + 20x^2 + x^4)$ is strictly decreasing for all $x > x^* = \sqrt{30} = 5.477225575\ldots$.
3. (a) 1, (b) 1, (c) $(2x_1^{1/3} + 3x_2^{1/2})/(x_1^{1/3} + 2x_2^{1/2})$.
4. Hint: Insert $g \circ F$ for F into (7.28), that is, write

$$\frac{\partial g(F(\mathbf{x}))}{\partial x_j} = g'(u)F_j'(\mathbf{x}) \quad \text{for} \quad \frac{\partial F(\mathbf{x})}{\partial x_j} = F_j'(\mathbf{x}) \quad \text{and}$$

$$\frac{\partial^2 g(F(\mathbf{x}))}{\partial x_j \partial x_k} = g''(u)F_j'(\mathbf{x})F_k'(\mathbf{x}) + g'(u)F_{jk}''(\mathbf{x}) \quad \text{for} \quad \frac{\partial^2 F(\mathbf{x})}{\partial x_j \partial x_k} = F_{jk}''(\mathbf{x})$$

in (7.28).

7.6 Nonlinear Vector-Valued Functions, Systems of Equations. Banach's Fixed Point Theorem

In Sect. 7.3 we solved nonlinear equations, such as

$$e^r - 1 = i \quad \text{and} \quad e^{rt} = 2,$$

where $i \cdot 100\%$ is the yearly interest rate, $r \cdot 100\%$ the continuous compounding rate and t is the doubling time. While these and similar equations are easily solved with aid of the logarithm or other inverse functions, in general it is very difficult to solve nonlinear equations or even establish whether solutions exist and, if yes, how many. As we saw in Sect. 7.5, many things can happen. For instance,

$$e^x = x + 2$$

has two solutions,

$$e^x = x + 1$$

has exactly one solution ($X = 0$) and

$$e^x = x$$

has no solution (see Fig. 7.14). The equation

$$\sin x = 1/2$$

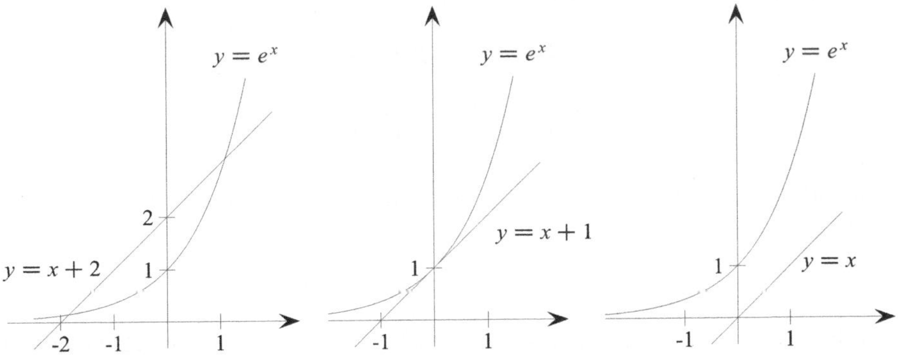

Fig. 7.14 The equations $e^x = x + 2$, $e^x = x + 1$, $e^x = x$ have two solutions, one solution or no solution, respectively

even has infinitely many solutions (check by calculation or drawing):

$$x = \pi/6 + 2kn \quad \text{and} \quad x = 5\pi/6 + 2kn$$

for all $k \in \mathbb{R}$ (that is, see Sect. 1.7.2, $k = 0, 1, -1, 2, -2, \ldots$).

Solving systems of nonlinear equations is, of course, even more difficult. But such systems often arise in economics and other sciences.

For instance, as we saw in Sect. 6.9, the price for which a product can be sold may determine the quantity to be produced next time. We now consider, more realistically, a market on which s products are supplied by each of r competitors, rather than by just one producer with one product. The supposition that each competitor brings the same number of products to the market is only seemingly a restriction: we will allow 0 quantities of products. Also, no producer can bring to the market an unlimited quantity of any product, therefore (or because of differences in quantity, preference or location) not only the cheapest product will sell. So, let the price for which the j-th competitor sells the k-th product be $p_{jk} \in \mathbb{R}_{++}$ ($j = 1, \ldots, r$; $k = 1, \ldots, s$). Based on these prices, the j-th competitor produces or hopes to sell in the next time interval (production or sales "period") the quantity $q_{jk} \in \mathbb{R}_{+}$ of the k-th product. For the sake of brevity we introduce the vectors

$$q_j = (q_{j1}, \ldots, q_{js}) \in \mathbb{R}^s_{+}, \qquad p_j = (p_{j1}, \ldots, p_{js}) \in \mathbb{R}^s_{++} \qquad (j = 1, \ldots, r)$$

(as we see, these are "row-vectors", not "column vectors" as in most places in Chap. 4). Let the functions

$$G_j : \mathbb{R}^{rs}_{++} \longrightarrow \mathbb{R}^r_{+}$$

describe how the quantities q_j ($j = 1, \ldots, r$) in the next time interval depend upon the prices $p_1, \ldots p_r$ previously attained

$$\begin{aligned} G_1(p_1, \ldots, p_r) &= q_1, \\ &\vdots \\ G_r(p_1, \ldots, p_r) &= q_r. \end{aligned} \tag{7.70}$$

Similarly, as in Sect. 6.9, "we"—usually the market researchers—wish, conversely, to determine the prices

$$p_{11}, \ldots, p_{1s}, \ldots, p_{r1}, \ldots, p_{rs}$$

(or the price vectors $p_1, \ldots, p_r$) which would guarantee that the projected (and thus given) quantities

$$q_{11}, \ldots, q_{1s}, \ldots, q_{r1}, \ldots, q_{rs}$$

(or the quantity vectors $q_1, \ldots, q_r$) will indeed be sold in the next time interval.

The system (7.70) consists of rs scalar equations (r vector–equations) which may then serve the market researchers to determine rs unknown numbers (r vectors). If, instead of the quantities "we" the market researchers had projected the *revenues* $q_j \cdot p_j = t_j$ ($j = 1, \ldots, r$; "·" the scalar product, see Sect. 1.5 or (1.3)) then we have to determine rs unknown numbers $p_{11}, \ldots, p_{rs}$ (r vectors $p_1, p_2, \ldots, p_r$) from the r (scalar) equations

$$\begin{aligned} G_1(p_1, \ldots, p_r) \cdot p_1 &= t_1, \\ &\vdots \\ G_r(p_1, \ldots, p_r) \cdot p_r &= t_r. \end{aligned} \tag{7.71}$$

There are further variations on the theme. For instance in knowledge of the "price lists" $p_2 = p_2^0, \ldots, p_r = p_r^0$ of her competitors the first seller wants to know how much she can charge in order to be able to sell the quantities $(q_{11}, \ldots, q_{1s}) = q_1$ or have the revenue t_1 from the sale. Then the rs equations (7.70) or the r equations (7.71) serve to determine the s unknown prices $(p_{11}, \ldots, p_{1s}) = p_1$.

In general we may have m equations

$$\begin{aligned} h_1(x_1, \ldots, x_n) &= 0, \\ &\vdots \\ h_m(x_1, \ldots, x_n) &= 0, \end{aligned}$$

to determine the values of n unknown $x_1, \ldots, x_n$. If at least one of the functions $h_1, \ldots, h_m$ is not linear (compare Sect. 4.8) then we have a *nonlinear system of equations*. If all functions $h_1, \ldots, h_m$ are linear then we are back to a system o linear equations. In Sects. 4.6 and 4.7 we saw quite general methods for deciding whether a system of linear equations has solutions at all and, if yes, to determine them.

For nonlinear systems of equations (as already for nonlinear equations), the solution is much more complicated. Therefore we can give neither necessary and sufficient conditions for the existence or uniqueness of solutions nor formulas which give these solutions. So in this section we will just generalise the method of iteration processes used in Sect. 6.10 for single scalar equations and apply it to determine, under certain conditions, the solution of vector equations, that is of systems of scalar equations. But first we give a few more examples.

As already for some systems of linear equations (Sect. 4.6) *the numbers of equations (m) and of variables (n) does not determine whether there are solutions and, if yes, how many*.

Some *examples*, other than these with $m = n = 1$ at the beginning of this section, of m equations with n real unknowns (whether $m = n$, $m < n$ or $m > n$) with drastically different numbers of solutions are the following.

Example 1 First for $m = n$ we write

$$\begin{aligned} x_1^2 + \ldots + x_n^2 &= 0, \\ x_1^3 + \ldots + x_n^3 &= 1, \\ &\vdots \\ x_1^n + \ldots + x_n^n &= n-2, \\ x_1^{n+1} + \ldots + x_n^{n+1} &= n-1. \end{aligned} \tag{7.72}$$

Since the square of a real number is always nonnegative, the first equation can be satisfied only if all term on the left are 0, so

$$x_1 = 0, \ldots, x_n = 0 \tag{7.73}$$

is the only n-tuple which satisfies the first equation, but it clearly does not satisfy the second (or any further) equation. So this system of nonlinear equations with as many equations as variables (that is $m = n$) has *no solutions*.

If we just take the first m ($1 < m < n$) equations, we also have *no solution*, this time with $m < n$ and if we enlarge the above system (7.72) by several more, completely arbitrary equations, the new system, with $m < n$, still has no solutions.

Example 2 Replace in both the $m = n$ and the $m < n$ cases of Example 1 all the right hand sides in (7.72) by 0. This, of course, does not change the first equation or its only solution (7.73). But this solution $x_1 = 0, \ldots, x_n = 0$ then satisfies also all other equations so, whether $m = n$ or $m < n$, this system of m nonlinear equations has exactly one n-tuple of solutions, namely (7.73).

If we take, similarly, the system

$$\begin{aligned} x_1^2 + \ldots + x_n^2 &= 0, \\ &\vdots \\ x_1^{n+1} + \ldots + x_n^{n+1} &= 0, \\ &\vdots \\ x_1^{m+1} + \ldots + x_n^{m+1} &= 0, \end{aligned}$$

with $m > n$, it has also *exactly the one* n-tuple of solutions (7.73).

Example 3

$$\begin{aligned}(x_1-1)(x_1-2)\cdot\ldots\cdot(x_1-k)(x_2^2+\ldots+x_n^2+1) &= 0,\\ (x_2^2+\ldots+x_n^2)x_1 &= 0,\\ &\vdots\\ (x_2^m+\ldots+x_n^m)x_1 &= 0.\end{aligned}$$

The last factor on the left hand side of the first equation is positive for all real $x_1, x_2, \ldots, x_n$. So the left hand side can be 0 only if

$$\text{either}\quad x_1=1 \quad\text{or}\quad x_1=2 \quad\text{or}\quad \ldots \quad\text{or}\quad x_1=k.$$

For each of these values of x_1, the second equation is satisfied exactly when $x_2=0, \ldots, x_n=0$. So, whether $m<n$, $m=n$ or $m>n$, *the above system of m nonlinear equations in n unknowns has exactly k (n-tuples of) solutions* (and we can make $k \geq 1$ as large or as small as we want to):

$$\begin{aligned}x_1 &= 1, x_2=\ldots=x_n=0,\\ x_1 &= 2, x_2=\ldots=x_n=0,\\ &\vdots\\ x_1 &= k, x_2=\ldots=x_n=0.\end{aligned}$$

Example 4 Let $f_1:\mathbb{R}^n\to\mathbb{R}, \ldots, f_m:\mathbb{R}^n\to\mathbb{R}$ be arbitrary functions. We consider the system

$$\begin{aligned}f_1(x_1,x_2,\ldots,x_n)\sin x_1 &= 0,\\ &\vdots\\ f_m(x_1,x_2,\ldots,x_n)\sin x_1 &= 0.\end{aligned}$$

As we know (Sect. 1.7 **2**; Fig. 6.7) $\sin x_1 = 0$ at the infinitely many places

$$x_1 = 0, \pi, -\pi, 2\pi, -2\pi, 3\pi, -3\pi, \ldots,$$

so, whatever $x_2, \ldots, x_n$ and $f_1, \ldots, f_m$ are (though these may furnish further solutions), with these x_1-values all m equations of the system are solved. Thus, *whether $m<n$, $m=n$ or $m>n$ the above system of m nonlinear equations in n unknown has infinitely many solutions.*

Since it is easier to visualise them, we look now at situations with m equations containing just $n = 2$ variables:

$$\begin{aligned} h_1(x_1, x_2) &= 0, \\ &\vdots, \\ h_m(x_1, x_2) &= 0. \end{aligned} \tag{7.74}$$

While the functions $h_1 : S_1 \to \mathbb{R}, \ldots, h_m : S_m \to \mathbb{R}$ ($S_j \subset \mathbb{R}^2$, $j = 1, \ldots, m$) need not be defined on the same domain, their domains should have at least one point and, preferably, a whole neighbourhood (see Sect. 6.2) in common. We consider

$$S := \bigcap_{j=1}^{n} S_j \subset \mathbb{R}^2,$$

the domain, where *all* our functions $h_1, \ldots, h_m$ are defined. Solving the system (7.74) of equations means to find those points $x = (x_1, x_2) \in S$ where *all* functions $h_1, \ldots, h_m$ are 0, in other words, to find the zeros of the vector-vector function $h : S \to \mathbb{R}^m$ ($h = (h_1, \ldots, h_m)$).

We just said that for $n = 2$, that is in two dimensions, it is easier to visualise the above problem. In Sect. 3.3 we introduced *contour-lines* as set of points where a function of two variables assumes the same value. Here we are, of course interested in the value 0, but for *all* functions $h_1, \ldots, h_m$. So we draw the contour-lines belonging to the value 0 of *all* these functions on the part S of $\mathbb{R}^2$ and look whether there are points which lie on all of them and, if yes, how many.

The following *examples* show again the possibility of no, one, k, or infinitely many solutions.

Example 5 Here $h_1(x_1, x_2) := x_1^2 + x_2^2 - 4$, $h_2(x_1, x_2) := x_1^2 - 3x_2$, $h_3(x_1, x_2) := \sqrt{3}x_1 + x_2 - 4$, $h_4(x_1, x_2) := 2x_1 - 3x_2 + 8$. *The system of equations* (third and fourth equation linear, first and second not)

$$x_1^2 + x_2^2 - 4 = 0, \tag{7.75}$$

$$x_1^2 - 3x_2 = 0, \tag{7.76}$$

$$\sqrt{3}x_1 + x_2 - 4 = 0, \tag{7.77}$$

$$2x_1 - 3x_2 + 8 = 0 \tag{7.78}$$

has, as Fig. 7.15 shows, *no solution*. But the system (7.75), (7.76), (7.77) has exactly one solution, namely $(x_1, x_2) = (\sqrt{3}, 1)$. We see that a system of *three* equations in *two* variables can have a solution. On the other hand,

(continued)

a system of *two* equations in *two* variables can have (see Fig. 7.15 for the first three cases)

- no solution: see (7.75), (7.78),
- exactly one solution: see (7.75), (7.77) and (7.77), (7.78) having the solutions $(x_1, x_2) = (\sqrt{3}, 1)$ and $(x_1, x_2) = ((12\sqrt{3} - 8)/23, (56 + 8\sqrt{3})/23)$, respectively,
- exactly two solutions: those of the system
  - (7.75), (7.76) are $(x_1, x_2) = (\sqrt{3}, 1)$ and $(x_1, x_2) = (-\sqrt{3}, 1)$,
  - (7.76), (7.77) are $(x_1, x_2) = (\sqrt{3}, 1)$ and $(x_1, x_2) = (-4\sqrt{3}, 16)$,
  - (7.76), (7.78) are $(x_1, x_2) = (4, 16/3)$ and $(x_1, x_2) = (-2, 4/3)$,
- three, four, fife, ... solutions,
- infinitely many solutions.

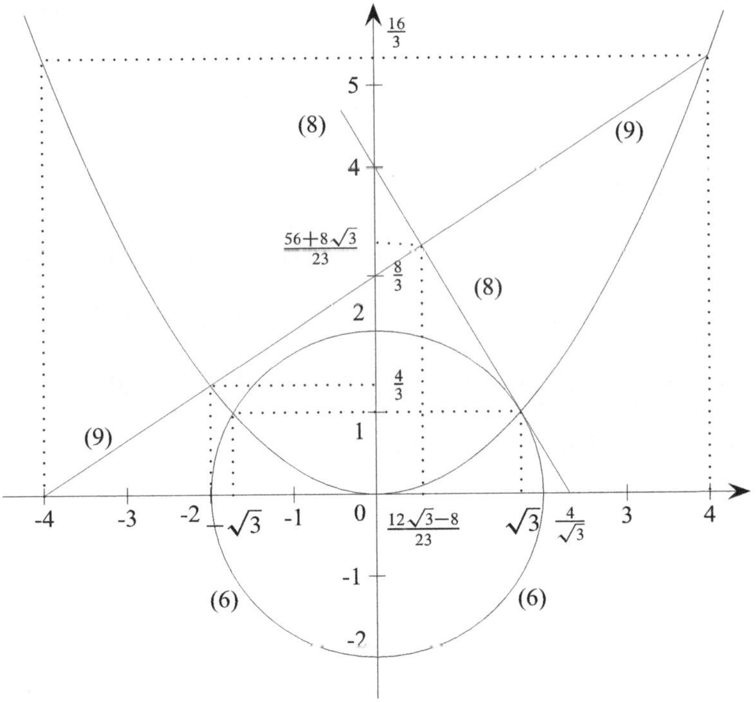

Fig. 7.15 The curves denoted by (7.75), (7.76), (7.77), (7.78) are representations of the solutions to Eqs. (7.75), (7.76), (7.77), (7.78), respectively. If a point (x_1, x_2) lies on two or three of the curves, it is a solution to the corresponding two or three equations in the system of equations (7.75), (7.76), (7.77), (7.78)

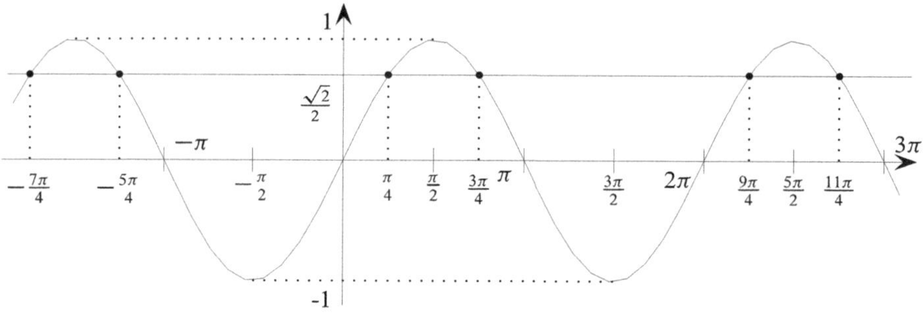

Fig. 7.16 The system of equations (7.79), (7.80) has infinitely many solutions $\ldots$, $(-7\pi/4, \sqrt{2}/2)$, $(-5\pi/4, \sqrt{2}/2)$, $(\pi/4, \sqrt{2}/2)$, $(3\pi/4, \sqrt{2}/2)$, $(9\pi/4, \sqrt{2}/2)$, $(11\pi/4, \sqrt{2}/2)$, $\ldots$

We affirm the last two assertions by

Example 6 Now $h1(x_1, x_2) := \sin x_1 - x_2$, $h_2(x_1, x_2) := 2x_2 - \sqrt{2}$. *The system of equations*

$$\sin x_1 - x_2 = 0, \tag{7.79}$$

$$2x_2 - \sqrt{2} = 0 \tag{7.80}$$

(we see that not all equations have to contain all unknowns—but each unknown should be in at least one equation) *has infinitely many solutions*

$$(x_1, x_2) = (\pi/4 + 2k\pi, \sqrt{2}/2) \quad \text{and} \quad (x_1, x_2) = (-5\pi/4 + 2k\pi, \sqrt{2}/2)$$

($k \in \mathbb{Z}$), see Fig. 7.18. Obviously, if the domain of h_1 is $S_1 = [0, \pi/2 + k^*\pi] \times \mathbb{R}$ for a fixed $k^* \in \{0, 1, 2, \ldots\}$ then the system of equations (7.79), (7.80) has exactly $k^* + 1$ solutions (Fig. 7.16).

Example 7 Here, too, $h_1(x_1, x_2) := x_1^2 + x_2^2 - 1$ but this time $h_2(x_1, x_2) := x_1 + x_2$. The *nonlinear system of equations*

$$x_1^2 + x_2^2 - 1 = 0,$$

$$x_1 + x_2 = 0$$

(continued)

(nonlinear system, because the first equation is not linear, even though the second is linear) has (see Fig. 7.17) *two solutions*: $(-\sqrt{2}/2, \sqrt{2}/2)$ and $(\sqrt{2}/2, -\sqrt{2}/2)$. With $h_3(x_1, x_2) := \sqrt{2}x_1^2 - x_2$ again, the system consisting of the above two equations and of

$$\sqrt{2}x_1^2 - x_2 = 0,$$

however, has only *one solution* $(-\sqrt{2}/2, \sqrt{2}/2)$ (Fig. 7.17).

On the other hand, the system $h_1(x_1, x_2) = 0$, $h_3(x_1, x_2) = 0$ and the system $h_2(x_1, x_2) = 0$, $h_3(x_1, x_2) = 0$ again have two solutions each, $(-\sqrt{2}/2, \sqrt{2}/2)$, $(\sqrt{2}/2, \sqrt{2}/2)$ and $(-\sqrt{2}/2, \sqrt{2})$, $(0, 0)$, respectively. These two pairs of solutions are different from each other and from the solutions of the original system of two equations in this system.

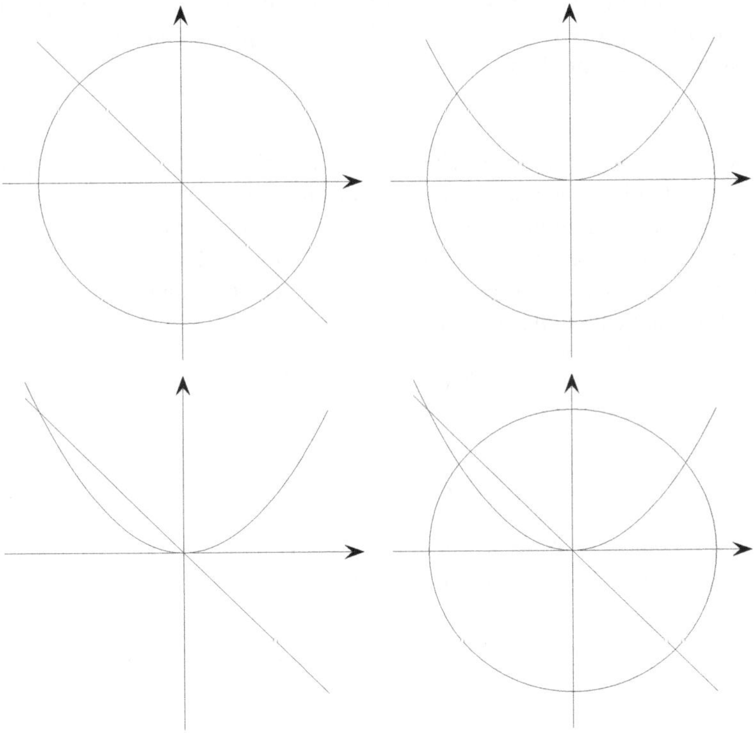

Fig. 7.17 Given are the equations (i) $x_1^2 + x_2^2 - 1 = 0$, (ii) $x_1 + x_2 = 0$, and (iii) $0.5x_1^2 - x_2 = 0$. The system (i), (ii) has two solutions and so have the systems (i), (iii) and (ii), (iii) (but different ones), while the system (i), (ii), (iii) has only one solution

Example 8 Now $h_1(x_1, x_2) := \sin x_1 - x_2$, $h_2(x_1, x_2) := 2x_2 - \sqrt{2}$. The system of equations

$$\sin x_1 - x_2 = 0,$$
$$2x_2 - \sqrt{2} = 0$$

(we see that not all equations have to contain all unknowns—but each unknown should be in at least one equation) has infinitely many solutions

$$(\pi/4 + 2k\pi, \sqrt{2}/2) \quad (k \in \mathbb{Z}) \quad \text{and} \quad (-\pi/4 + 2k\pi, \sqrt{2}/2) \quad (k \in \mathbb{Z})$$

(see Fig. 7.18. With $h_3(x_1, x_2) := \cos x_1 - x_2 = 0$, the system $h_1(x_1, x_2) = 0$, $h_2(x_1, x_2) = 0$, $h_3(x_1, x_2) = 0$ has still infinitely many solutions, but only the first set of the above solutions, which are also the only solutions of $h_1(x_1, x_2) = 0$, $h_3(x_1, x_2) = 0$, while the system $h_2(x_1, x_2) = 0$, $h_3(x_1, x_2) = 0$ has the first set of the above solutions and $(3\pi/4 + 2k\pi, \sqrt{2}/2)$ $(k \in \mathbb{Z})$ as solutions (infinitely many solutions for all these systems). For all systems in this example the domain is $\mathbb{R} \times [-1, 1]$. (In Examples 6 and 7 the domains are the same as in Example 5.)

By now we have probably convinced the reader that having as many equations as unknowns is neither necessary nor sufficient for the system of equations to have solutions, let alone unique solutions. Some *methods of solution*, however, work better in this case and even give (restricted) existence and uniqueness results.

The *iteration process* ("cobweb situation") in Sect. 6.9 can be generalised to such a method. We have now the system

$$h_1(x_1, \dots, x_n) = 0,$$
$$\vdots$$
$$h_n(x_1, \dots, x_n) = 0$$

of n equations in n unknowns, where $h_j : S_j \to \mathbb{R}$ and $S_j \subset \mathbb{R}^n$. We use again the vector notation $\mathbf{x} = (x_1, \dots, x_n)$, also for the functions:

$$\mathbf{h} := (h_1, \dots, h_n) : S \to \mathbb{R}^n, \text{ where } S = \bigcap_{j=1}^{n} \subset \mathbb{R}^n.$$

So our problem is to determine the **zeros** of $\mathbf{h}$, that is, those $\mathbf{x} \in S$ for which $\mathbf{h}(\mathbf{x}) = \mathbf{0}$. As in Sect. 6.10, we define a new function $\mathbf{F}$ by

$$\mathbf{F}(\mathbf{x}) = \mathbf{x} - \mathbf{h}(\mathbf{x}),$$

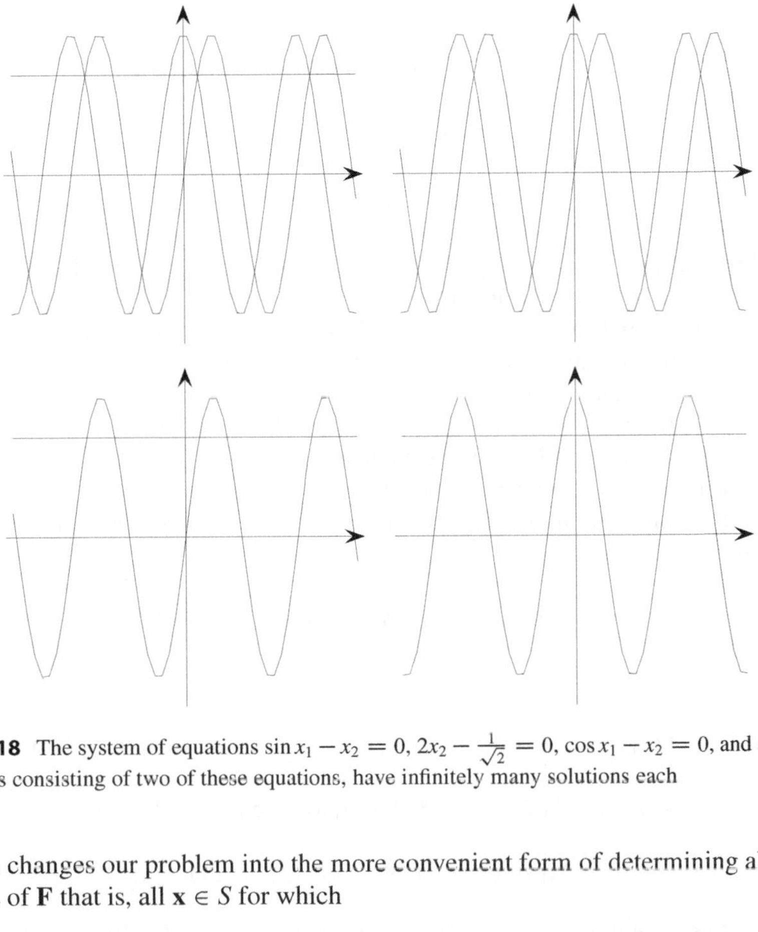

Fig. 7.18 The system of equations $\sin x_1 - x_2 = 0$, $2x_2 - \frac{1}{\sqrt{2}} = 0$, $\cos x_1 - x_2 = 0$, and all three systems consisting of two of these equations, have infinitely many solutions each

which changes our problem into the more convenient form of determining all fixed points of $\mathbf{F}$ that is, all $\mathbf{x} \in S$ for which

$$\mathbf{x} = \mathbf{F}(\mathbf{x}). \tag{7.81}$$

Under certain conditions, analogous to those in Sect. 6.10, such fixed points of $\mathbf{F}$ do exist and can be determined, in generalisation of the method given there, by the following *iteration process*:

$$\mathbf{x}_{n+1} = \mathbf{F}(\mathbf{x}_n) \quad (n = 0, 1, \ldots). \tag{7.82}$$

We are helped by the following result called "Banach's fixed point theorem" (Stefan Banach, 1892–1945): If, for a function $: S \to \mathbb{R}^n$ $(S \subset \mathbb{R}^n)$, there exists a number $c \in [0, 1[$ and a δ-neighbourhood (see Sect. 6.10)

$$N_\delta(\mathbf{x}_0) := \{\mathbf{x} \mid |\mathbf{x} - \mathbf{x}_0| < \delta\} \subset S$$

such that

$$|\mathbf{F}(\mathbf{x}) - \mathbf{F}(\mathbf{y})| \leq c\,|\mathbf{x} - \mathbf{y}| \quad \textit{for all} \quad \mathbf{x}, \mathbf{y} \in N_\delta(\mathbf{x}_0) \tag{7.83}$$

(for norms of vectors see Sect. 1.4) and

$$|\mathbf{F}(\mathbf{x}_0) - \mathbf{x}_0| < (1 - c)\delta \tag{7.84}$$

then the sequence defined by (7.82) converges to a fixed point of $\mathbf{F}$, that is, to a solution of (7.81)—the only one in $N_\delta(\mathbf{x}_0)$.

The proof, which we sketch here, follows the lines of the argument given in Sect. 6.10, with modifications made necessary by the space of more than one dimension (no "squeeze rule"). Actually, (7.83) is the exact analogue of the "Lipschitz condition" (b) there. Furthermore, (7.84) establishes that $\mathbf{x}_1 = \mathbf{F}(\mathbf{x}_0)$ is in $N_\delta(\mathbf{x}_0)$ and use of (7.83) will establish that *all* $\mathbf{x}_n$ $(n = 1, 2, \ldots)$, as defined by (7.82) are in $N_\delta(\mathbf{x}_0)$, which corresponds to the condition (a) in Sect. 6.10.

Indeed we prove for all n (or at least for $n = 1, 2, 3$, which shows already how to proceed),

$$|\mathbf{x}_n - \mathbf{x}_{n-1}| \leq c^{n-1}\,|\mathbf{x}_1 - \mathbf{x}_0| < c^{n-1}(1 - c)\delta \tag{7.85}$$

and

$$|\mathbf{x}_n - \mathbf{x}_0| \leq (1 + c + \ldots + c^{n-1})\,|\mathbf{x}_1 - \mathbf{x}_0| < (1 - c^n)\delta, \text{ so } \mathbf{x}_n \in N_\delta(\mathbf{x}_0) \tag{7.86}$$

(remember $c > 0$). By (7.82) and (7.84) we have already

$$|\mathbf{x}_1 - \mathbf{x}_0| = \left|\mathbf{F}(\mathbf{x}) - \mathbf{x}_0\right| < (1 - c)\delta,$$

that is, (7.85) and (7.86) are true for $n = 1$. Now, using this, (7.85) and (7.82), we have

$$|\mathbf{x}_2 - \mathbf{x}_1| = |\mathbf{F}(\mathbf{x}_1) - \mathbf{F}(\mathbf{x}_0)| \leq c\,|\mathbf{x}_1 - \mathbf{x}_0| < c(1 - c)\delta$$

(since $\mathbf{x}_0 \in N_\delta(\mathbf{x}_0)$, $\mathbf{x}_1 \in N_\delta(\mathbf{x}_0)$) and, from the triangle inequality (see Sect. 1.5),

$$\begin{aligned}|\mathbf{x}_2 - \mathbf{x}_0| &= |(\mathbf{x}_2 - \mathbf{x}_1) + (\mathbf{x}_1 - \mathbf{x}_0)| \\ &\leq |\mathbf{x}_2 - \mathbf{x}_1| + |\mathbf{x}_1 - \mathbf{x}_0| \leq (c + 1)\,|\mathbf{x}_1 - \mathbf{x}_0| \\ &< (1 + c)(1 - c)\delta = (1 - c^2)\delta \leq \delta.\end{aligned}$$

So (7.85) and (7.86) hold also for $n = 2$. We use this, (7.83) and (7.82) again to show

$$|\mathbf{x}_3 - \mathbf{x}_2| = |\mathbf{F}(\mathbf{x}_2) - \mathbf{F}(\mathbf{x}_1)| \leq c\,|\mathbf{x}_2 - \mathbf{x}_1| \leq c^2\,|\mathbf{x}_1 - \mathbf{x}_0| < c^{(}1 - c)\delta$$

(since $\mathbf{x}_1 \in N_\delta(\mathbf{x}_0)$, $\mathbf{x}_2 \in N_\delta(\mathbf{x}_0)$) and, again from the triangle inequality,

$$\begin{aligned}|\mathbf{x}_3 - \mathbf{x}_0| &= |(\mathbf{x}_3 - \mathbf{x}_2) + (\mathbf{x}_2 - \mathbf{x}_0)| \\ &\leq |\mathbf{x}_3 - \mathbf{x}_2| + |\mathbf{x}_2 - \mathbf{x}_0| \leq (c^2 + c + 1)\,|\mathbf{x}_1 - \mathbf{x}_0| \\ &< (c^2 + c + 1)(1 - c)\delta = (1 - c^3)\delta \leq \delta.\end{aligned}$$

Thus we see that (7.85) and (7.86) hold also for $n = 3$. In the same way we get (7.85) and (7.86) for all $n \in \mathbb{N}$ (one can apply induction).

The next step is to get, from (7.85) and from repeated use of the triangle inequality, for all $n > m \geq 1$,

$$\begin{aligned}|\mathbf{x}_n - \mathbf{x}_m| &\leq |\mathbf{x}_n - \mathbf{x}_{n-1}| + |\mathbf{x}_{n-1} - \mathbf{x}_m| \\ &\leq |\mathbf{x}_n - \mathbf{x}_{n-1}| + |\mathbf{x}_{n-1} - \mathbf{x}_{n-2}| + \ldots + |\mathbf{x}_{m+1} - \mathbf{x}_m| \\ &< (c^{n-1} + c^{n-2} + \ldots + c^{m+1} + c^m)(1 - c)\delta \\ &= (c^m - c^{m+1} + c^{m+1} - c^{m+2} + \ldots \\ &\qquad + c^{n+3} - c^{n+2} + c^{n+2} - c^{n+1} - c^{n+1} - c^n)\delta \\ &= (c^m - c^n)\delta < c^m\delta.\end{aligned}$$

In Sect. 7.2 we have proved that $\{c^m\}$ converges to 0 as $m \to \infty$ if $0 \leq c < 1$. So $|\mathbf{x}_n - \mathbf{x}_m|$ converges to 0 when m (and thus also n) tends to ∞. It can be proved (this is called "Cauchy's criterium" after Augustin Louis Cauchy (1789–1857), the founder of modern exact analysis) that this implies

$$\lim_{n\to\infty} \mathbf{x}_n =: \mathbf{x}^* \tag{7.87}$$

(the limit of sequences of vectors being defined the same way as in Sect. 6.2 for scalars and as limits of vector-vector functions in Sect. 6.12.

As in Sect. 6.8, the Lipschitz inequality (7.83) has the continuity of $\mathbf{F}$ as consequence. But, by (7.82) and (7.87) for all $\varepsilon > 0$ there exists an N such that

$$|\mathbf{F}(\mathbf{x}_n) - \mathbf{x}^*| = |\mathbf{x}_{n+1} - \mathbf{x}^*| \quad \text{for} \quad n > N,$$

that is,

$$\mathbf{x}^* = \lim_{n\to\infty} \mathbf{F}(\mathbf{x}_n) = \mathbf{F}(\mathbf{x}^*),$$

(by the continuity of $\mathbf{F}$). So the sequence (7.82) indeed converges to a solution of (7.81). That there is no other solution of (7.81) in $N_\delta(\mathbf{x}_0)$ is proved by contradiction: if there were also a $\mathbf{y}^* \neq \mathbf{x}^*$ with $\mathbf{F}(\mathbf{y}^*) = \mathbf{y}^*$ in $N_\delta(\mathbf{x}_0)$ then, by $0 \le c < 1$ and by (7.83),

$$c\,|\mathbf{x}^* - \mathbf{y}^*| < |\mathbf{x}^* - \mathbf{y}^*| = |\mathbf{F}(\mathbf{x}^*) - \mathbf{F}(\mathbf{y}^*)| \le c\,|\mathbf{x}^* - \mathbf{y}^*|\,,$$

a contradiction indeed. (Actually one has also to prove that $\mathbf{x}^* \in N_\delta(\mathbf{x}_0)$.)

This (with the gaps which we pointed out) concludes the proof of Banach's fixed point theorem.

Returning to our problem of solving

$$\mathbf{h}(\mathbf{x}) = \mathbf{0}, \tag{7.88}$$

this means that, if there exists a $c \in [0, 1[$ such that

$$|\mathbf{x} - \mathbf{y} - (\mathbf{h}(\mathbf{x}) - \mathbf{h}(\mathbf{y}))| \le c\,|\mathbf{x} - \mathbf{y}|$$

for all $\mathbf{x}$, $\mathbf{y}$ in a δ-neighbourhood $N_\delta(\mathbf{x}_0)$ and if

$$|\mathbf{h}(\mathbf{x}_0)| < (1 - c)\delta,$$

then there is exactly one solution (root) of (7.88) in $N_\delta(\mathbf{x}_0)$ and it can be calculated by the iteration process

$$\mathbf{x}_{n+1} = \mathbf{x}_n - \mathbf{h}(\mathbf{x}_n) \qquad (n = 0, 1, 2, \ldots).$$

We recommend that the reader check how these results relate to those in Sects. 6.9 and 6.10.

7.6.1 Exercises

1. Determine by iteration the two solutions $\mathbf{x}^*$, $\mathbf{x}^{**}$ of equation $e^x = x + 2$ up to five decimals. (Hint: One solution is negative, one is positive. Take $x_{j+1} = e^{x_j} - 2$ and $x_{j+1} = \frac{3}{2}x_j - \frac{1}{2}e^{x_j} + 1$, respectively.)
2. Determine by direct calculation all solution points (x_1, x_2) of the nonlinear system of equations $x_1^2 + x_2^2 - 1 = 0$, $\sqrt{2}x_1^2 - x_2 = 0$.
3. Determine by direct calculation all solution points (x_1, x_2) of the nonlinear system of equations $x_1^2 + x_2^2 - 2 = 0$, $x_1^2 - x_2 = 0$, $3x_1^2 + x_2 - 4 = 0$.
4. Add to the equations in Exercise 3 a linear equation such that the new system of four equations has exactly one solution point (x_1, x_2) which is also the only solution point of that linear equation and equation $x_1^2 + x_2^2 - 2 = 0$.
5. Construct a system of three equations in two variables x_1, x_2 which has exactly four solution points (x_1, x_2).

7.6.2 Answers

1. $\mathbf{x}^* = -1.84141, \quad \mathbf{x}^{**} = 1.14619.$
2. $(x_1, x_2) = (-\sqrt{2}/2, \sqrt{2}/2), \quad (x_1, x_2) = (\sqrt{2}/2, \sqrt{2}/2).$
3. $(x_1, x_2) = (-1, 1), \quad (x_1, x_2) = (1, 1).$
4. There are exactly two such linear equations, namely

$$x_1 + x_2 - 2 = 0 \quad \text{and} \quad x_1 - x_2 + 2 = 0.$$

5. For instance, the system $\sin x_1 - x_2 = 0$, $2x_2 - \sqrt{2} = 0$, $\cos(x_1) - x_2 = 0$, where $x_1 \in [0, 7\pi]$, has exactly four solution points, namely $\left(\frac{\pi}{4}, \frac{\sqrt{2}}{2}\right)$, $\left(\frac{9\pi}{4}, \frac{\sqrt{2}}{2}\right)$, $\left(\frac{17\pi}{4}, \frac{\sqrt{2}}{2}\right)$, $\left(\frac{25\pi}{4}, \frac{\sqrt{2}}{2}\right)$.

Nonlinear Optimisation with One or Several Objectives: Kuhn–Tucker Conditions

8

The best is the enemy of the good.

VOLTAIRE (FRANÇOIS–MARIE AROUET, 1694–1778)

8.1 Introduction

As in other introductory sections, we set the stage here too (this time for nonlinear optimisation) by describing a situation from economics. Within the framework of a simple model we are interested in the optimal investment ratio in national economy.

The *investment ratio* x_t in the year t is defined by

$$x_t = \frac{I_t}{Y_t}, \tag{8.1}$$

where I_t is the *gross fixed capital formation (investment)* and Y_t the *gross domestic product* in the year t. Let t run from the year t to the year T. The question is for what choice of the investment ratios $x_1, \ldots, x_T$ is the "*discounted aggregate consumption*" maximal. To get the latter, we have to consider the *capital stock* (that is, the (value of the) aggregate of capital goods in the economy) K_{t-1} at the beginning of the year t. We suppose that the gross domestic product Y_t *depends only upon* K_{t-1}:

$$Y_t = F(K_{t-1}) \tag{8.2}$$

($F : \mathbb{R}_+ \to \mathbb{R}_+$ is the *production function*). By (8.1) and (8.2)

$$I_t = x_t F(K_{t-1}). \tag{8.3}$$

W. Eichhorn, W. Gleißner, *Mathematics and Methodology for Economics*, Springer Texts in Business and Economics, DOI 10.1007/978-3-319-23353-6_8

Now we define the *total consumption in the year t*, C_t, as "gross domestic product less gross capital formation", that is, by

$$C_t = Y_t - I_t = (1 - x_t)F(K_{t-1})$$

(here $(1 - x_t)$ is called the "*average propensity to consume*"). This has to be *discontinued* to year 1, that is, with the *discount factor* $d = (1 + i)^{-1} < 1$ (see Sect. 7.2; i is the interest rate which we here assume to be the *real* rate, that is, the inflation–adjusted rate) we form

$$d^{t-1}C_t = (1 - x_t)F(K_{t-1})d^{t-1}.$$

The *discounted aggregate consumption* C is the sum of these terms for the years $t = 1, t = 2, \ldots, t = T$

$$C = \sum_{t=1}^{T} d^{t-1} = \sum_{t=1}^{T}(1 - x_t)F(K_{t-1})d^{t-1}. \tag{8.4}$$

This is what we have to maximise. The K_t's can be determined *recursively* from K_0 which is given by the following argument. The capital stock K_{t-1} at the beginning of the year t *depreciates* to qK_{t-1} (q the *depreciation factor*, $0 < q < 1$) by the end of the year. At the same time the capital formation (investment) $I_t = x_tF(K_t - q)$ (see (8.3)) is added. so we have the *recursive formula* (compare to the *iteration process* in Sect. 6.9)

$$K_t = qK_{t-1} + x_tF(K_{t-1}) \qquad (t = 1, \ldots, T). \tag{8.5}$$

Our *optimisation problem* is to maximise (8.4) where the function F is usually nonlinear (compare Sect. 7.4). This is a *nonlinear* optimisation problem. We will solve it in Sect. 8.2 under certain assumptions in addition to those which we have already made. This will answer in a particular case a classical question of macroeconomics, which asks *what investment ratio maximises the discounted aggregate consumption.*

One of the strongly restrictive simplifying assumptions is that *the production function is neither depending on the employment level nor on the year (time) t* (which it would, e.g., if the operating rates of the production units would change in the same direction in time). The assumption of independence of the employment level does not seem very restrictive, since unfortunately nowadays the operating rate of the capital stock depends more on the market situation than on the number of people employed. Another restrictive assumption in our model is that the considered economy *is closed*, that is, foreign trade is neglected or the value of exports always equals that of the imports.

We have real life data about investment ratios or, to be exact, about their arithmetic mean for the years 1970–1999. This mean was greater then .32 in Japan, between .24 and .20 in France, Italy and (West)Germany, and less than .17 in Great Britain and the U.S.A.

In Sect. 8.2 we lay the foundations for Sects. 8.3, 8.4, 8.5 and 8.6 which deal with the problem of determining maxima and minima ("extrema") of functions of several variables. These foundations include the notions of convexity and concavity of differentiable functions of several variables, matrix conditions for convexity (concavity), and eigenvalues and eigenvectors of matrices. These in turn will let us find conditions for extrema of such functions (Sect. 8.4) and of functions under constraints (Sect. 8.6) The Kuhn–Tucker conditions (Sect. 8.9) are conditions of this kind. An application of nonlinear optimisation establishes the method of least squares in the theory of linear regression (Sect. 8.5). The concluding Sect. 8.10 deals with optimisation in the case when several functions ("objectives") are to be maximised or minimised at the same time.

8.2 Convexity of Differentiable Functions of Several Variables, Matrix–Conditions for Convexity, Eigenvalues, Eigenvectors

In Sect. 3.5 we introduced convex functions (from below or from above; the latter also called concave functions) of one and of several variables. In Sect. 6.8 we found conditions (some necessary, some sufficient, some both) for *differentiable functions in a single variable* to be convex (from above or below, strictly or otherwise):

In what follows we aim at finding conditions for one or twice differentiable functions of *several* variables to be convex. These will be of use in subsequent sections which will deal with *maxima and minima of functions of several variables.*

We will need the concepts of *interior points* and of *open sets* in $\mathbb{R}^n$ (compare Sect. 3.5. If a *point* $\mathbf{a} \in S \subseteq \mathbb{R}^n$ *has a neighbourhood* (no matter how small) *that* is in S (is a subset of S) then $\mathbf{a}$ is an interior point of S. The set of all interior points of S is the interior of S (see Fig. 8.1. *If all points of S are interior* (are elements of the interior of S) *then S* is an *open set*

Fig. 8.1 Interior point (*left*). Interior of a set (*middle*; the dotted set does not belong to the interior). Open and convex set (*right*): with $\mathbf{x}$ and $\mathbf{x} + \mathbf{h}$ it contains $\mathbf{x} + r\mathbf{h}$ for all $r \in I$

Let now $S \subseteq \mathbb{R}^n$ be a convex open set (for convex sets see Sect. 3.5 and let the function $F : S \to \mathbb{R}$ be convex from below and differentiable. Then, as we have seen in Sect. 6.11, it is also partially differentiable in each variable, has thus a gradient

$$\nabla F(\mathbf{x}) := (\frac{\partial F}{\partial x_1}(\mathbf{x}), \ldots, \frac{\partial F}{\partial x_n}(\mathbf{x})).$$

If both $\mathbf{x}$ and $\mathbf{y} = \mathbf{x} + \mathbf{h}$ are in S then, since S is convex, also $\mathbf{x} + \mathbf{h} \in S$ for $r \in [0, 1]$ but, since S is also open, it contains with each point $\mathbf{x} + \mathbf{h} \in S$ also a neighbourhood, so there exists an open interval I, of which $[0, 1]$ is a subinterval, such that $\mathbf{x} + \mathbf{h} \in S$ for all $r \in I$ (we need the larger interval I to facilitate differentiation). Take now $g : I \to \mathbb{R}$ defined by

$$g(r) = F(\mathbf{x} + r\mathbf{h}) \tag{8.6}$$

The graph of this function is (for $n = 2$) the vertical slice of the graph of F on the ray–segment $\{\mathbf{x} + r\mathbf{h} \mid r \in I\}$ through $\mathbf{x}$. Since F is convex from below, and differentiable, so is g. By the chain rule in Sect. 6.5

$$g'(r) = \frac{\partial F}{\partial x_1}(\mathbf{x} + r\mathbf{h})h_1, \ldots, \frac{\partial F}{\partial x_n}(\mathbf{x} + r\mathbf{h})h_n = \mathbf{h} \cdot \nabla F(\mathbf{x} + r\mathbf{h}). \tag{8.7}$$

By the law of means (Sect. 6.7),

$$F(\mathbf{y}) - F(\mathbf{x}) = F(\mathbf{x} + r\mathbf{h}) - F(\mathbf{x}) = g(1) - g(0) = g'(\Theta) \quad \text{for some } \Theta \in]0, 1[.$$

Since, as we have seen in Sect. 6.8, the derivative of a differentiable convex function is increasing,

$$F(\mathbf{y}) - F(\mathbf{x}) = g'(\Theta) - g'(0) = \mathbf{h} \cdot \nabla F(\mathbf{x}) = (\mathbf{y} - \mathbf{x}) \cdot \nabla F(\mathbf{x}).$$

So, *if F is convex from below and differentiable on an open convex set S and* $\mathbf{x}, \mathbf{y} \in S$ *then*

$$F(\mathbf{y}) - F(\mathbf{x}) \geq (\mathbf{y} - \mathbf{x}) \cdot \nabla F(\mathbf{x}). \tag{8.8}$$

The question arises whether, conversely, (8.8) implies the convexity of F form below. The answer is *yes*. Indeed, take any $\mathbf{y}$ and $\mathbf{z}$ in the open, convex set $S \subseteq \mathbb{R}^n$ and let $\lambda \in [0, 1]$ be arbitrary. Then $\mathbf{x} = \lambda\mathbf{y} + (1 - \lambda)\mathbf{z} \in S$ (since S is a convex set, so (8.8) and the similar inequality

$$F(\mathbf{z}) - F(\mathbf{x}) \geq (\mathbf{z} - \mathbf{x}) \cdot \nabla F(\mathbf{x}).$$

hold. Multiply the latter inequality by $(1 - \lambda)$ and (8.8) by λ and add them:

$$\lambda F(\mathbf{y}) + (1 - \lambda)F(\mathbf{z}) - F(\mathbf{x}) \geq (\lambda\mathbf{z} + (1 - \lambda)\mathbf{z} - \mathbf{x}) \cdot \nabla F(\mathbf{x}) = 0.$$

since $\mathbf{x} = \lambda\mathbf{y} + (1-\lambda)\mathbf{z}$. Thus

$$F(\lambda\mathbf{y} + (1-\lambda)\mathbf{z}) = F(\mathbf{x}) \leq \lambda F(\mathbf{y}) + (1-\lambda)F(\mathbf{z}),$$

which is exactly how we defined, in Sect. 3.5, functions convex from below.

Thus *on an open convex set* $S \subseteq \mathbb{R}^n$ *a differentiable function* $F : S \to \mathbb{R}$ *is convex from below if, and only if, (8.8) holds for all* $x, y \in S$. *A similar statement, with* $\leq$ *in place of* $\geq$ *holds for functions convex from above ("concave").* The geometric interpretation of (8.8) (for $n = 2$) is that the graph of F does not get below any tangent plane and, with $\leq$, that it does not get above any tangent plane (compare Sects. 6.8 and 7.2)

For differentiable functions *strictly* convex from below, by the same argument, (8.8) with $>$ on place of $\geq$, but only for $\mathbf{y} \neq \mathbf{x}$, is necessary and sufficient and (8.8) with $<$ for $\mathbf{y} \neq \mathbf{x}$ "characterises" (is necessary and sufficient for) functions *strictly convex from above* (that is, "*strictly concave*" functions).

While (8.8) and similar conditions involve only first derivatives, we will give now convexity conditions for functions of n variables, involving second derivatives, which are similar to but more complicated than those in Sect. 6.8 for functions of one variable. There we proved that the twice differentiable function $g : I \to \mathbb{R}$ on an open Interval I is convex from below or from above if, and only if

$$g''(r) \geq 0 \quad \text{on } I \quad \text{or} \quad g''(r) \leq 0 \quad \text{on } I,$$

respectively.

As we have seen , F is convex from below on S exactly if the function of one variable g, defined by (8.6), is convex from below on I. If F has continuous second partial derivatives (here it would be enough that F be twice differentiable but we will need later continuous second partial derivatives anyway), then $g : I \to \mathbb{R}$ is twice differentiable and exactly then convex from below on I if

$$0 \geq g''(r) \geq \Sigma_{j=1}^{n} \Sigma_{k=1}^{n} \frac{\partial^2 F}{\partial x_j \partial x_k}(\mathbf{x} + r\mathbf{h})h_j h_k =: \mathbf{h}\mathbf{F}''(\mathbf{x} + r\mathbf{h})\mathbf{h}^T \tag{8.9}$$

for $r \in I$ (cf. Sect. 6.8). Here we applied the chain rule (Sect. 6.5) again on $\partial F(\mathbf{x} + r\mathbf{h})/\partial x_j$ in (8.7). We denote by $\mathbf{h}^T$ the vector $\mathbf{h} = (h_1, \ldots, h_n)$ transformed into a column vector and by $\mathbf{F}''\mathbf{x}$ the *Hessian matrix* (Ludwig Otto Hesse (1811–1874)), "*Hessian*" for short,

$$\mathbf{F}''(\mathbf{x}) := \begin{pmatrix} \frac{\partial^2 F}{\partial x_1^2}(\mathbf{x}) & \ldots & \frac{\partial^2 F}{\partial x_1 \partial x_n}(\mathbf{x}) \\ \vdots & & \vdots \\ \frac{\partial^2 F}{\partial x_n \partial x_1}(\mathbf{x}) & \ldots & \frac{\partial^2 F}{\partial x_n^2}(\mathbf{x}) \end{pmatrix} = \begin{pmatrix} F''_{x_1 x_1}(\mathbf{x}) & \ldots & F''_{x_1 x_n}(\mathbf{x}) \\ \vdots & & \vdots \\ F''_{x_n x_1}(\mathbf{x}) & \ldots & F''_{x_n x_n}(\mathbf{x}) \end{pmatrix},$$

where $F''_{x_jx_k}(\mathbf{x}) := \partial^2 F(\mathbf{x})/\partial x_j \partial x_k$. Since we supposed the second partial derivatives to be continuous, we have (see Sect. 6.11)

$$F''_{x_jx_k}(\mathbf{x}) = F''_{x_kx_j}(\mathbf{x}),$$

therefore *the Hessian matrix is symmetric*.

We proved (8.9) for $r \in I$ ($I \supseteq [0, 1]$), so, in particular, it has to hold for $r = 0$:

$$\mathbf{hF}''(\mathbf{x})\mathbf{h}^T \geq 0 \tag{8.10}$$

and $r = 1$:

$$\mathbf{hF}''(\mathbf{x} + \mathbf{h})\mathbf{h}^T \geq 0$$

for $\mathbf{x}$ and $\mathbf{x} + \mathbf{h}$ in S, respectively. The left hand side is a *quadratic form* (see Sect. 7.4). Multiplication of (8.10) by λ^2 shows that

$$(\lambda \mathbf{h})F''(\mathbf{x})(\lambda \mathbf{h})^T \geq 0$$

that is, *(8.10) holds for all* $\mathbf{h} \in \mathbb{R}^n$ *if* F *is convex in a neighbourhood of* $\mathbf{x}$. Conversely, if F satisfies (8.10) and $\mathbf{F}''$ is continuous (that is, F has continuous second derivatives) then, for sufficiently small r also

$$\mathbf{hF}''(\mathbf{x} + r\mathbf{h})\mathbf{h}^T \geq 0, \qquad \text{for all } \mathbf{h} \in \mathbb{R}^n$$

(values of continuous functions change little when their variables change little). So we have proved that *the function* F*, having continuous second partial derivatives in a neighbourhood of* $\mathbf{x}$*, is convex from below in that neighbourhood if, and only if (8.10) holds for all* $\mathbf{h} \in \mathbb{R}^n$.

A similar result, with $\leq$ *in place of* $\geq$ *in (8.10), holds for functions convex from above (concave)*. As to *strictly* convex functions from below or above, (8.10) with $<$ or $>$, respectively, is sufficient but not necessary as the example of $F(\mathbf{x}) = x_1^4 + \dots x_n^4$ at $\mathbf{0}$ shows (F strictly convex but $\mathbf{F}''(\mathbf{0}) = \mathbf{0}$ so (8.10) holds with $=$, not with $>$). However, *(8.10) with* $\geq$*, but* $=$ *not holding on any open set, is necessary and sufficient for the twice continuously differentiable* F *to be strictly convex from below on a neighbourhood of* $\mathbf{x}$. Again, a *similar result holds for functions strictly convex from above (concave)*.

Quadratic ($n \times n$) $\mathbf{A}$ or the quadratic forms $\mathbf{hAh}^T$ for which in equalities of the form (8.10),

$$\mathbf{hAh}^T \geq 0, \qquad \text{for all } \mathbf{h} \in \mathbb{R}^n, \tag{8.11}$$

hold are called *positive semidefinite*. The same inequality with $\leq$, $>$ or $<$, in place of $\geq$,makes **A** *negative semidefinite, positive definite* or *negative definite*, respectively, So $F : S \to R(S \subseteq \mathbb{R}^n)$ *is convex or strictly convex from below or from above on the neighbourhood of an interior point* **x** *of* S *if the Hessian matrix* $\mathbf{F}''(\mathbf{x})$ *is positive semidefinite or definite or negative semidefinite or definite, respectively*. Notice that we did not suppose that S is an open set, only that is that it has at least one interior point **x**. Notice also that we did not write "and only if" because this is nit true for strict convexity from below or from above (strict concavity).

In what follows, we give conditions and algorithms which help determine whether a matrix (or quadratic form) is positive or negative semidefinite or definite.

Such tools are *eigenvalues* and *eigenvectors*. In (8.11) we have the product $\mathbf{Ah}^T$. Things become considerably simpler of there exists a *scalar* λ such that

$$\mathbf{Ah}^T = \lambda\mathbf{h}^T.$$

If for a quadratic $(n \times n)$-matrix **A** there exist both a scalar λ and a vector $\mathbf{v} = (v_1, \ldots, v_n)$ such that

$$\mathbf{Av}^T = \lambda\mathbf{v}^T. \tag{8.12}$$

then λ is called an *eigenvalue* and **v** an *eigenvector* of **A**. We can write (8.12) as

$$(\mathbf{A} - \lambda\mathbf{I})\mathbf{v}^T = \mathbf{0}^T. \tag{8.13}$$

where **I** is the $n \times n$ unit matrix and $\mathbf{0}^T$ the $n \times I$ zero column vector. This is a vector equation and, as those with which we dealt in Sect. 4.6, it is a compact way to write a system of homogeneous linear equations. If it is supposed to have a nontrivial solution $\mathbf{v} \neq \mathbf{0}$ then (see Sect. 4.7)

$$det(\mathbf{A} - \lambda\mathbf{I}) = det\begin{pmatrix} a_{11} - \lambda & a_{12} & \ldots & a_{1n} \\ a_{21} & a_{22} - \lambda & \ldots & a_{2n} \\ \vdots & \vdots & \ddots & \vdots \\ a_{n1} & a_{n2} & \ldots & a_{nn} - \lambda \end{pmatrix} = 0. \tag{8.14}$$

This equations called the *the characteristic equation of* **A** and $det(\mathbf{a} - \lambda\mathbf{I})$ it's *characteristic polynomial*. Indeed, if we calculate this determinant as in Sect. 4.7, we see that it is a polynomial of n-th degree. But then (8.14) has (with multiplicity) exactly n real or complex zeros. So *a real* $n \times n$ *matrix has, with multiplicity, exactly n real or complex eigenvalues*. So even matrices with only real entries may have complex eigenvalues, since polynomials with real coefficient may have complex zeros (for example $x^2 + 1 = 0$). However, one can show that *all eigenvalues of symmetric real matrices are real*. Since the Hessian matrices, with which we deal here, are, as we had seen, symmetric, we will not have worry about complex eigenvalues.

It will be of importance that *eigenvectors, belonging to different eigenvalues of the same matrix, are orthogonal.* In order to show this, we have to remind the reader that (Sect. 1.6) two vectors are orthogonal, if their product is 0 and that (Sect. 4.4) the dot product of two (say row) vectors $\mathbf{v}_1$ and $\mathbf{v}_2$ with n components each, is the product of an $n \times 1$ matrix (the column vector $\mathbf{v}_1^T$) and of a $1 \times n$ matrix (the row vector $\mathbf{v}_2$). We point also out that *the transpose* $\mathbf{A}^T$ *of a matrix* $\mathbf{A}$ *is obtained by interchanging its rows and columns.* Clearly $(\mathbf{A}^T)^T = \mathbf{A}$ and it is easy to check (do it!) that

$$(\mathbf{AB})^T = \mathbf{A}^T\mathbf{B}^T.$$

Moreover, *for symmetric matrices*, by definition $\mathbf{A}^T = \mathbf{A}$,

$$(\mathbf{A}\mathbf{v}_1^T) \cdot \mathbf{v}_2 = (\mathbf{A}\mathbf{v}_1^T)^T\mathbf{v}_2^T = \mathbf{v}_1\mathbf{A}^T\mathbf{v}_2^T = \mathbf{v}_1\mathbf{A}\mathbf{v}_2^T = \mathbf{v}_1 \cdot (\mathbf{A}\mathbf{v}_2^T)$$

Using this we get, for the eigenvectors $\mathbf{v}_1$, $\mathbf{v}_2$ belonging to two different eigenvalues λ_1, λ_1, that is, satisfying $\mathbf{A}\mathbf{v}_1^T = \lambda_1\mathbf{v}_1^T$, $\mathbf{A}\mathbf{v}_2^T = \lambda_2\mathbf{v}_2^T$, that

$$\begin{aligned}\lambda_1(\mathbf{v}_1 \cdot \mathbf{v}_2) &= (\lambda_1\mathbf{v}_1) \cdot \mathbf{v}_2 = (\mathbf{A}\mathbf{v}_1^T) \cdot \mathbf{v}_2 \\ &= \mathbf{v}_1 \cdot (\mathbf{A}\mathbf{v}_2^T) = \mathbf{v}_1 \cdot (\lambda_2\mathbf{v}_2) = \lambda_2(\mathbf{v}_1 \cdot \mathbf{v}_2)\end{aligned}$$

Since $\lambda_2 \neq \Lambda_2$ this is possible only if $\mathbf{v}_1 \cdot \mathbf{v}_2 = 0$. So eigenvectors belonging into different eigenvalues are indeed orthogonal, as asserted.

Without loss of generality, we may suppose that the eigenvectors $\mathbf{v}$ are unit vectors (by supposition, they are not $\mathbf{0}$ and, it $\mathbf{v}$ is an eigenvector belonging to the eigenvalue λ, then so is the unit vector $(1/|\mathbf{v}|)\mathbf{v})$. A set of orthogonal vectors, each of which has a norm 1, is called *orthonormal.* Ignoring, for the time being eigenvalues with multiplicity >1, we unite the (column) eigenvectors into a matrix $\mathbf{V}$:

$$\mathbf{V} = (\mathbf{v}_1^T, \mathbf{v}_2^T, \ldots, \mathbf{v}_n^T) = \begin{pmatrix} v_{11} & v_{12} & \ldots & v_{1n} \\ v_{21} & v_{22} & \ldots & v_{2n} \\ \vdots & \vdots & \ddots & \vdots \\ v_{n1} & v_{n2} & \ldots & v_{nn} \end{pmatrix}.$$

It follows from the orthonormality of $\mathbf{v}_1^T, \mathbf{v}_2^T, \ldots, \mathbf{v}_n^T$ that *the matrix* $\mathbf{V}$*is of rank* n. Indeed, $\det \mathbf{V} \neq 0$ (compare Sect. 4.7) because

$$\mathbf{V}^T\mathbf{V} = \begin{pmatrix} \mathbf{v}_1 \\ \mathbf{v}_2 \\ \vdots \\ \mathbf{v}_n \end{pmatrix} (\mathbf{v}_1^T, \mathbf{v}_2^T, \ldots, \mathbf{v}_n^T)$$

$$= \begin{pmatrix} v_1 \cdot v_1 & v_1 \cdot v_2 & \dots & v_1 \cdot v_n \\ v_2 \cdot v_1 & v_2 \cdot v_2 & \dots & v_2 \cdot v_n \\ \vdots & \vdots & \ddots & \vdots \\ v_n \cdot v_1 & v_n \cdot v_2 & \dots & v_n \cdot v_n \end{pmatrix} = \begin{pmatrix} 1 & 0 & \dots & 0 \\ 0 & 1 & \dots & 0 \\ \vdots & \vdots & \ddots & \vdots \\ 0 & 0 & \dots & 1 \end{pmatrix} = I,$$

which shows also that $\mathbf{V}^T = \mathbf{V}^{-1}$ for these $\mathbf{V}$.

We transform now the quadratic form

$$\mathbf{hAh}^T = \sum_{j=1}^{n} \sum_{k=1}^{n} a_{jk} h_j h_k \tag{8.15}$$

by the linear transformation (function)

$$\mathbf{h}^T = \mathbf{Vx}^T. \tag{8.16}$$

Transposing the last equation, since, as we have seen, $(\mathbf{W}^T)^T = \mathbf{W}$ and

$$(\mathbf{VW})^T = \mathbf{W}^T\mathbf{V}^T,$$

we get

$$\mathbf{h} = \mathbf{xV}^T.$$

So, since matrix multiplication is *associative* (Sect. 4.4),

$$(\mathbf{hAh})^T = (\mathbf{xV}^T)\mathbf{A}(\mathbf{Vx}^T) = \mathbf{xV}^T\mathbf{AVx}^T.$$

Now

$$\begin{aligned} (\mathbf{AV})^T &= \mathbf{A}(\mathbf{v}_1^T, \dots, \mathbf{v}_n^T) \\ &= (\mathbf{Av}_1^T, \dots, \mathbf{Av}_n^T) \\ &= (\boldsymbol{\lambda}\mathbf{v}_1^T, \dots, \boldsymbol{\lambda}\mathbf{v}_n^T), \end{aligned}$$

since $\mathbf{v}_1^T, \dots, \mathbf{v}_n^T$ are the eigenvectors belonging to the eigenvalues $\lambda_1, \dots, \lambda_n$. Thus, *if all eigenvalues of* $\mathbf{A}$ *are different,*

$$\mathbf{hAh}^T = (\mathbf{xV}^T)(\boldsymbol{\lambda}\mathbf{v}_1^T, \dots, \boldsymbol{\lambda}\mathbf{v}_n^T)\mathbf{x}^T$$

$$= \mathbf{x} \begin{pmatrix} \mathbf{v}_1 \\ \mathbf{v}_2 \\ \vdots \\ \mathbf{v}_n \end{pmatrix} (\lambda \mathbf{v}_1^T, \ldots, \lambda \mathbf{v}_n^T)\mathbf{x}^T = \mathbf{x} \begin{pmatrix} \lambda_1 & 0 & \ldots & 0 \\ 0 & \lambda_2 & \ldots & 0 \\ \vdots & \vdots & \ddots & \vdots \\ 0 & 0 & \ldots & \lambda_n \end{pmatrix} \mathbf{x}^T$$

$$= \lambda_1 x_1^2 + \lambda_2 x_2^2 + \ldots + \lambda_n x_n^2 =: Q(\mathbf{x})$$

which considerably simplifies the quadratic form.

We supposes all eigenvalues to be different. One can show that *a somewhat similar result holds if some eigenvalues are equal.*

Of course, $x_1^2, x_2^2, \ldots, x_n^2$ are nonnegative. So *the quadratic form* $\mathbf{hAh}^T$ *is nonnegative or the matrix* $\mathbf{A}$ *is positive semidefinite if, and only if, all eigenvalues of* $\mathbf{A}$ *are nonnegative* (if even one λ_j would be negative then, for large enough x_j, the value $Q(\mathbf{x})$ could be negative). Similarly, *the quadratic form* $\mathbf{hAh}^T$ *or the matrix* $\mathbf{A}$ *is negative semidefinite if, and only if all eigenvalues of* $\mathbf{A}$ *are nonpositive*. One has to be more careful with positive or negative *definite* quadratic forms, since

$$Q(\mathbf{x}) = \lambda_1 x_1^2 + \lambda_2 x_2^2 + \ldots + \lambda_n x_n^2$$

may be 0 even if all λ_j's are positive (or negative), actually if $x_1 = \ldots = x_n = 0$, that is, $\mathbf{x} = 0$. By (8.15) this is the case exactly if $\mathbf{h} = \mathbf{0}$. So *the quadratic form* $\mathbf{hAh}^T$ *is positive or negative, respectively*. We do not have to worry about $\mathbf{h}$ when we check the positive or negative definiteness of the *matrix* $\mathbf{A}$ rather then that of the quadratic form $\mathbf{hAh}^T$, so *the matrix* $\mathbf{A}$ *is positive or negative definite if, and only if, all its eigenvalues are positive or negative, respectively.*

Quadratic forms $\mathbf{h}^T\mathbf{Ah}$, which are *positive for some* $\mathbf{h}$ *and negative for others* are called *indefinite* and therefore *matrices* which have *both positive and negative eigenvalues* are also called *indefinite*. At points where the Hessian matrix is indefinite the graph of the function may have *line(s) of inflection* (compare Fig. 3.31 at which the function changes from convex from below to convex from above (concave) or from convex from above to convex from below.

Example 1 The characteristic equation of

$$\mathbf{A} = \begin{pmatrix} 6 & -3 \\ -3 & 6 \end{pmatrix} \quad \text{is} \quad det \begin{pmatrix} 6-\lambda & -3 \\ -3 & 6-\lambda \end{pmatrix} = (6-\lambda)^2 - 9 = 0$$

so the eigenvalues are calculated from $6 - \lambda = \pm 3$, that is, $\lambda_1 = 3$, $\lambda_2 = 9$ (we could have used also the usual formula for the solution of an equation of second degree). *Both eigenvalues are positive, so the matrix (and the quadratic form* $6h_1^2 - 6h_1h_2 + 6h_2^2$ *for* $(h_1, h_2) \neq (0, 0)$*) is positive definite.*

(continued)

The eigenvectors belonging to $\lambda_1 = 3$ are the vectors (v_1, v_2) satisfying

$$\begin{pmatrix} 6-\lambda_1 & -3 \\ -3 & 6-\lambda_1 \end{pmatrix}\begin{pmatrix} v_1 \\ v_2 \end{pmatrix} = \begin{pmatrix} 3 & -3 \\ -3 & 3 \end{pmatrix}\begin{pmatrix} v_1 \\ v_2 \end{pmatrix} = \begin{pmatrix} 3v_1 & -3v_2 \\ -3v_1 & 3v_2 \end{pmatrix} = \begin{pmatrix} 0 \\ 0 \end{pmatrix}$$

Obviously, these are the vectors $(v_1, v_2) = (v_1, v_2)$, where $v_1 \in \mathbb{R}$. The eigenvectors belonging to $\lambda_2 = 9$ are the vectors (v_1, v_2) satisfying

$$\begin{pmatrix} 6-\lambda_2 & -3 \\ -3 & 6-\lambda_2 \end{pmatrix}\begin{pmatrix} v_1 \\ v_2 \end{pmatrix} = \begin{pmatrix} -3 & -3 \\ -3 & -3 \end{pmatrix}\begin{pmatrix} v_1 \\ v_2 \end{pmatrix} = \begin{pmatrix} -3v_1 & -3v_2 \\ -3v_1 & -3v_2 \end{pmatrix} = \begin{pmatrix} 0 \\ 0 \end{pmatrix}$$

Obviously, these are the vectors $(v_1, v_2) = (v_1, -v_1)$, where $v_1 \in \mathbb{R}$. Since

$$\left|\begin{pmatrix} v_1 \\ v_2 \end{pmatrix}\right| = \left|\begin{pmatrix} v_1 \\ -v_1 \end{pmatrix}\right| = \sqrt{2v_1^2} = \sqrt{2}v_1$$

these eigenvectors have norm 1 exactly if $v_1 = 1/\sqrt{2} = \sqrt{2}/2$. The eigenvectors $(\sqrt{2}/2, \sqrt{2}/2)$ and $(\sqrt{2}/2, -\sqrt{2}/2)$ are indeed *orthogonal*:

$$\left(\frac{\sqrt{2}}{2}, \frac{\sqrt{2}}{2}\right) \cdot \left(\frac{\sqrt{2}}{2}, -\frac{\sqrt{2}}{2}\right) = \left(\frac{2}{4} - \frac{2}{4}\right) = 0,$$

so *these eigenvectors from an orthonormal set.* Here

$$\mathbf{V} = \begin{pmatrix} \sqrt{1}/2 & \sqrt{1}/2 \\ \sqrt{1}/2 & -\sqrt{1}/2 \end{pmatrix} = \mathbf{V}^T$$

and we have indeed

$$\mathbf{V}^T\mathbf{A}\mathbf{V} = \begin{pmatrix} \sqrt{1}/2 & \sqrt{1}/2 \\ \sqrt{1}/2 & -\sqrt{1}/2 \end{pmatrix}\begin{pmatrix} 6 & -3 \\ -3 & 6 \end{pmatrix}\begin{pmatrix} \sqrt{1}/2 & \sqrt{1}/2 \\ \sqrt{1}/2 & -\sqrt{1}/2 \end{pmatrix}$$
$$= \begin{pmatrix} 3 & 0 \\ 0 & 9 \end{pmatrix} = \begin{pmatrix} \lambda_1 & 0 \\ 0 & \lambda_2 \end{pmatrix}.$$

Example 2 Consider function $F : \mathbb{R}^4 \to \mathbb{R}$, given by

$$F(\mathbf{x}) = F(x_1, x_2, x_3, x_4) = x_1^4 - 2x_1 + x_2^3 - 3x_2^2 + x_3^2 - 4x_x.$$

(continued)

Calculate the second partial derivatives:

$$F''_{x_1x_1}(\mathbf{x}) = 12x_1^2, \quad F''_{x_2x_2}(\mathbf{x}) = 6x_2 - 6, \quad F''_{x_3x_3}(\mathbf{x}) = 2, \quad F''_{x_4x_4}(\mathbf{x}) = 0$$

$$\text{and} \quad F''_{x_jx_k}(\mathbf{x}) = 0 \quad \text{if} \quad j \neq k \quad (j, k = 1, 2, 3, 4).$$

So the Hessian matrix

$$\mathbf{F}''(\mathbf{x}) = \begin{pmatrix} 12x_1^2 & 0 & 0 & 0 \\ 0 & 6x_2 - 6 & 0 & 0 \\ 0 & 0 & 0 & 0 \\ 0 & 0 & 0 & 0 \end{pmatrix}$$

is already of the form

$$\begin{pmatrix} \lambda_1 & 0 & 0 & 0 \\ 0 & \lambda_2 & 0 & 0 \\ 0 & 0 & \lambda_3 & 0 \\ 0 & 0 & 0 & \lambda_4 \end{pmatrix}.$$

Therefore the eigenvalues are

$$\lambda_1 = 12x_1^2, \quad \lambda_2 = 6x_2 - 6, \quad \lambda_3 = 2, \quad \lambda_4 = 0.$$

We see that λ_1, λ_2 and λ_4 are always nonnegative, while $\lambda_2 = 6x_2 - 6 \geq 0$ exactly if $x_2 \geq 1$. *So the Hessian is positive semidefinite and F is convex (but not strictly convex) from below on the domain*

$$\{(x_1, x_2, x_3, x_4) \mid x_1 \in \mathbb{R},\ x_2 \geq 1,\ x_3 \in \mathbb{R},\ x_4 \in \mathbb{R}\}.$$

While $\lambda_4 = 0$ always and $\lambda_1 = 0$ for $x_1 = 0$ and $\lambda_2 = 0$ for $x_2 = 1$, *the Hessian cannot be negative semidefinite and* F *cannot be convex from above (concave) anywhere because* $\lambda_3 = 2 > 0$.

Example 3 The *Cobb–Douglas functions* $F : \mathbb{R}_+^2 \to \mathbb{R}_+$ (compare Sects. 6.12 and 7.5 defined by (we omit the positive constant multiplier)

$$F(\mathbf{x}) = F(x_1, x_2) = x_1^{C1}, x_1^{C2}$$

(continued)

with positive constants c_1, c_2 satisfying $c_1 + c_2 \leq 1$, *are convex from above (concave) on* $\mathbb{R}_+^2$*; in the case* $c_1 + c_2 < 1$ *even strictly concave on* $\mathbb{R}_{++}^2$. Indeed, *the Hessian* matrix is

$$\mathbf{F}''(\mathbf{x}) = \begin{pmatrix} c_1(c_1-1)x_1^{c_1-2}x_2^{c_2} & c_1c_2x_1^{c_1-1}x_2^{c_2-1} \\ c_1c_2x_1^{c_1-1}x_2^{c_2-1} & c_2(c_2-1)x_1^{c_1}x_2^{c_2-2} \end{pmatrix}$$

and the *characteristic equation* is

$$\begin{aligned} \lambda^2 - (c_1(c_1-1)x_1^{c_1-2}x_2^{c_2} + c_2(c_2-1)x_1^{c_1}x_2^{c_2-2})\lambda \\ + c_1c_2((c_1-1)(c_2-1) - c_1c_2)x_1^{2c_1-2}x_2^{2c_2-2} &= 0. \end{aligned}$$

As mentioned before, symmetric matrices with real entries have real eigenvalues and the Hessian matrix of a twice continuously partially differentiable function is symmetric. So the eigenvalues are real. Notice that in the characteristic equation the term not containing λ is, for $x_1 \in \mathbb{R}_{++}, x_1 \in \mathbb{R}_{++}$,

$$c := c_1c_2(1-c_1-c_2)x_1^{2c_1-2}x_2^{2c_2-2} \begin{cases} > 0 & \text{for } c_1+c_2<1 \\ = 0 & \text{for } c_1+c_2=1 \end{cases}$$

while the coefficient of λ is

$$b := -(c_q(c_1-1)x_2^2 + c_2(c_2-1)x_1^2)x_1^{c_1-2}x_1^{c_2-2} > 0$$

for all $x_1,\ x_2 \in \mathbb{R}_{++}$, We can see as follows that b is positive:

$$c_1 + c_2 \leq 1,\ \ c_1 > 0,\ \ c_2 > 0 \quad \text{so} \quad c_1 < 1,\ \ c_2 < 1$$

thus

$$c_1(c_1-1) < 0 \quad \text{and} \quad c_2(c_2-1) < 0.$$

Now we can write the characteristic equation as

$$\lambda^2 + b\lambda + c = 0$$

and its solution are

$$\lambda_1 = \frac{-b+\sqrt{b^2-4c}}{2}, \qquad \lambda_2 = \frac{-b-\sqrt{b^2-4c}}{2}.$$

(continued)

Since $c \geq 0$, we have $b^2 - 4c \leq b^2$ so $b \geq \sqrt{b^2 - 4c}$, $-b + \sqrt{b^2 - 4c} \leq 0$, $-b - \sqrt{b^2 - 4c} \leq 0$. Thus *both eigenvalues are nonpositive, even negative if* $c > 0$, that is *for* $x_1 > 0, x_2 > 0, c_1 \neq c_2 < 1$. Thus, *in the case* $c_1 + c_2 < 0$ *the Cobb–Douglas Function F is strictly convex from above (strictly concave) on* $\mathbb{R}^2_{++}$. *If* $x_1 = 0$ or $x_2 = 0$ *then* b=c=0 *and* $\lambda_1 = \lambda_2 = 0$. Also *if* $c_1 + c_2 = 0$ *then* $c = 0$, *so* $\lambda_1 = 0$ even if $x_1 > 0, x_2 > 0$). Thus, *these cases, the Cobb–Douglas function F is convex from above (concave only in the broader sense on* $\mathbb{R}^2_+$ *(if* $c_1 + c_2 = 1$ *then even on* $\mathbb{R}^2_{++}$), as asserted.

We will need characteristic equations and eigenvalues in Chaps. 11 and 12 in the context of systems of differential and difference equations again.

Of course, we only know how to solve equations of second degree explicitly and there exist such explicit formulas only for equations of up to fourth degree. So the following conditions for positive or negative definiteness or semi-definiteness, which we present without proof, are of importance. For

$$\mathbf{A} = \begin{pmatrix} a_{11} & \dots & a_{1n} \\ \vdots & & \vdots \\ a_{n1} & \dots & a_{nn} \end{pmatrix}$$

the determinants

$$D_j := \det \begin{pmatrix} a_{11} & \dots & a_{1j} \\ \vdots & & \vdots \\ a_{j1} & \dots & a_{jj} \end{pmatrix} \qquad (j = 1, 2, \dots, n)$$

are called *principal minors of* $\det \mathbf{A}$. *A matrix* $\mathbf{A}$ (*or the quadratic form* $\mathbf{h}^T\mathbf{A}\mathbf{h}$) is positive definite if, and only if,

$$D_j > 0 \qquad (j = 1, \dots, n).$$

However $\mathbf{A}$ *is negative definite if, and only if alternatingly*

$$D_1 < 0, \quad D_2 > 0, \quad D_3 < 0, \quad \dots \quad (-1)^j D_j > 0$$

$(j = 1, \dots, n)$. This is easy to see when $\mathbf{A}$ is of the diagonal form

$$\begin{pmatrix} \lambda_1 & 0 & \dots & 0 \\ 0 & \lambda_2 & \dots & 0 \\ \vdots & \vdots & \ddots & \vdots \\ 0 & 0 & \dots & \lambda_n \end{pmatrix}.$$

Indeed, then

$$D_j = \det \begin{pmatrix} \lambda_1 & 0 & \dots & 0 \\ 0 & \lambda_2 & \dots & 0 \\ \vdots & \vdots & \ddots & \vdots \\ 0 & 0 & \dots & \lambda_n \end{pmatrix} = \lambda_1 \lambda_2 \dots \lambda_n$$

is positive for all $j = 1, 2, \dots, n$ if, and only if

$$\lambda_1 > 0, \;\; \lambda_2 > 0, \;\; \dots \;\; \lambda_n > 0.$$

but alternating if, and only if,

$$\lambda_1 < 0, \;\; \lambda_2 < 0, \;\; \dots \;\; \lambda_n < 0.$$

Example 4 The Hessian matrix of the function $F : \mathbb{R}^n \to \mathbb{R}$, defined by

$$F(\mathbf{x}) = F(x_1, x_2, x_3) = 12 + 6x_1 - 3x_2 + 6x_3 + 6x_1x_3 - 3x_1^2 + x_2^2 - 9x_3^2$$

is

$$\mathbf{F}''(\mathbf{x}) = \begin{pmatrix} -6 & 0 & 6 \\ 0 & 6x_2 & 0 \\ 6 & 0 & -18 \end{pmatrix}.$$

On the domain

$$\{(x_1, x_2, x_3) \mid x_1 \in \mathbb{R}, \; x_2 < 0, \; x_3 \in \mathbb{R}\} \tag{8.17}$$

we have

$$D_1 = -6 < 0, \qquad D_2 = -36x_2 > 0$$

$$D_3 = 648x_2 - 216x_2 = 432x_2 < 0.$$

So $\mathbf{F}''(\mathbf{x})$ is negative definite and thus F *is strictly convex from above (concave) on (8.17)*. On the other hand, on the domain

$$\begin{aligned} &\{(x_1, x_2, x_3) \mid x_1 \in \mathbb{R}, \; x_2 \in \mathbb{R}_{++}, \; x_3 \in \mathbb{R}\}, \\ &d_1 = -6 < 0, \; d_2 = -36x_2 < 0, \; d_3 = 432x_3 > 0. \end{aligned} \tag{8.18}$$

So $\mathbf{A}$ is neither positive nor negative definite on (8.18) and thus, F *is not strictly convex either from above or from below on* $\mathbb{R} \times \mathbb{R}_{++} \times \mathbb{R}$.

Actually, this F *is neither convex or concave even broader sense on the domain* (8.18). Indeed, the characteristic equation of $\mathbf{F}''(\mathbf{x})$ is

$$0 = \det \begin{pmatrix} -6-\lambda & 0 & 6 \\ 0 & 6x_2-\lambda & 0 \\ 6 & 0 & 18-\lambda \end{pmatrix}$$

$$= (-6-\lambda)(6x_2-\lambda)(-18-\lambda) - 36(6x_2-\lambda)$$

$$= (6x_2-\lambda)((6+\lambda)(18+\lambda)-36) = (6x_2-\lambda)(\lambda^2+24\lambda+72).$$

Thus

$$\lambda_1 = -12 + \sqrt{12^2-72} = -\ 3.5147\ldots < 0,$$

$$\lambda_2 = -12 - \sqrt{12^2-72} = -20.4852\ldots < 0,$$

$$\lambda_3 = 6x_2,$$

so $\mathbf{F}''(\mathbf{x})$ is *indefinite* in (8.18), from which our statement follows. (Notice that we have had no result connecting *semi*definiteness or *indefiniteness* with $D_1, D_2, D_3, \ldots$).

8.2.1 Exercises

1. (a) Determine the Hessian matrix $\mathbf{H}$ $(= \mathbf{F}'')$ of the function $F : \mathbb{R}^2 \to \mathbb{R}$ given by $F(x, y) = x^2 - 4x + y^2 + 6y$.
   (b) Is F convex (from below or above) on $\mathbb{R}^2$?
2. (a) Determine the Hessian matrix $\mathbf{H}$ $(= \mathbf{G}'')$ of the function $G : \mathbb{R}^2 \to \mathbb{R}$ given by $G(x, y) = x^2 + 5xy^3 + 2y$.
   (b) Is $\mathbf{H}(x, 0)$, that is $\mathbf{H}$ at the points $(x, y) = (x, 0)$, positive or negative definite or neither?
3. Is the function $f : \mathbb{R}^2_{++} \to \mathbb{R}$ given by $f(x, y) = x^2y^2 + 2y + 3$ convex (from below or above) an a neighbourhood of some point $(x*, y*) \in \mathbb{R}^2_{++}$?
4. Determine the Hessian matrix $\mathbf{H}$ for the quadratic form give by $(x_1, \ldots, x_n)\mathbf{A}(x_1, \ldots, x_n)^T$, where $\mathbf{A}$ is a (n,n)-matrix of real constants.
5. On which subset of $\mathbb{R}^2$ is the function $g : \mathbb{R}^2 \to \mathbb{R}$, defined by $g(x, y) = -4xy - y^2$, strictly convex from above?

8.2.2 Answers

1. (a) $\mathbf{H} = \begin{pmatrix} 2 & 0 \\ 0 & 2 \end{pmatrix}$,

   (b) F is convex from below on $\mathbb{R}^2$ since H is positive definite on $\mathbb{R}^2$.

2. (a) $\mathbf{H} = \begin{pmatrix} 2 & 15y^2 \\ 15y^2 & 30xy \end{pmatrix}$,

   (b) $\mathbf{H}(x,0) = \begin{pmatrix} 2 & 0 \\ 0 & 0 \end{pmatrix}$, that is, $\mathbf{H}$ is positive *semide*finite at $(x,0)$.

3. No: The Principal minors of the determinant of the Hessean matrix $\mathbf{H}(x,0) = \begin{pmatrix} 2y^2 & 4xy \\ 4xy & 2y^2 \end{pmatrix}$ are $D_1 = 2y^2$ and $d_2 = 4x^2y^2 - 16x^2y^2$, that is $D_1 > 0$, $D_2 < 0$ at each point of $\mathbb{R}^2_{++}$. The Hessean is indefinite on $\mathbb{R}^2_{++}$, hence there is no neighborhood of a point $(x*, y*) \in \mathbb{R}^2_{++}$ on which f is convex.

4. $\mathbf{H} = 2\mathbf{A}$.

5. The Hessean of g is $\begin{pmatrix} -12x^2 & -4 \\ -4 & -2 \end{pmatrix}$. The principal minor D_1 of the determinant equals $-12x^2 < 0$ for $x \neq 0$. The principal minor $D_2 = 24x_1^2 - 16$. This is <0 for all $|x| < \sqrt{2/3}$ and >0 for all $|x| > \sqrt{2/3}$. Hence g is strictly convex from above on $]-\infty, -\sqrt{2/3}[\times\mathbb{R}$ and on $]\sqrt{2/3}, \infty[\times\mathbb{R}$.

8.3 Quadratic Approximation. Maxima and Minima of Functions of Several Variables

The economic objective of reaching a goal with the least possible effort (expense) or obtain maximal yield by use of given means makes it important to determine maxima and minima of functions, in particular of functions in several variables. If the function is, for instance, the cost function of a firm then the task is to minimise the cost; if it is the utility function of a household or the profit function of a firm then we talk about maximising utility or profit.

We will be able to progress in analogy to and by use of what we learnt in Sects. 5.2, 6.3 and 6.9. We will need just three new concepts: those of *bounded closed (compact) sets*, *orders of magnitude* and *saddle points*.

As we saw in Sect. 6.3 (Properties 1, 2), not even continuous functions of one variable need to be *bounded* on intervals, which are not closed and, even if they are bounded, they need not assume their greatest and/or smallest value (*maximum* or *minimum*) on that interval. Both situations changed, however, when the intervals were *closed*.

In Sect. 3.3 we have defined *closed n-dimensional intervals*. Here we will need more general n-dimensional closed sets. (For $n = 1$ every closed set consists of closed intervals, that is the reason why we did not need this generalisation before). We defined *n-dimensional neighbourhoods* in Sects. 6.11 and 6.12 and we defined limits of one-dimensional sequences in Sect. 6.2. In complete analogy, an *n-dimensional (real) sequence is a function* $f : \mathbb{N} \to \mathbb{R}^n$, denoted by $\{\mathbf{f}(k)\}$ or $\{\mathbf{a}_k\}$ or $\{\mathbf{a}_1, \mathbf{a}_2, \ldots\}$. It *converges to* **a** *or has* **a** *as limit*, in symbols

$$\lim_{k\to\infty} \mathbf{a}_k = \mathbf{a},$$

if, for every neighbourhood of **a** there exists a K such that all $\mathbf{a}_k$ $(k > K)$ will be contained in that neighbourhood. While for one-dimensional sequences we considered also infinity as limit, this is not the case here. Sequences which converge to a (finite) **a** are again called *convergent*. Of course, *all terms* $\mathbf{a}_k$ (function values $\mathbf{f}(k)$ *of a convergent series) may be in a set* $S \subset \mathbb{R}^n$ (that is, *the sequence is in S) but its limit may still not be in S*. For instance, $\mathbf{a}_k = (\frac{1}{k}, \frac{1}{k^2}) \in S = {\mathbb{R}_{++}}^2$ but its limit

$$\lim_{k\to\infty} \left(\frac{1}{k}, \frac{1}{k^2}\right) = (0, 0) \notin \mathbb{R}^2_{++}.$$

If, for every convergent sequence in S, also its limit is in S then S is *closed*.

For instance, as we have just seen, $\mathbb{R}^2_{++}$ is not closed. Neither is the *open ball*

$$\{\mathbf{x} \in \mathbb{R}^n \mid |\mathbf{x}| < 1\}.$$

But the sets

$$\begin{aligned}
S_1 &= \{\mathbf{x} \in \mathbb{R}^n \mid |\mathbf{x} - \mathbf{b}| \le r\}, \qquad \textit{(closed ball)}\\
S_2 &= \left\{(x, y) \in \mathbb{R}^2 \;\middle|\; \tfrac{x^2}{a^2} + \tfrac{y^2}{b^2} \le 1\right\},\\
S_3 &= \mathbb{R}^2_+,\\
S_4 &= \left\{(x, y) \in \mathbb{R}^2_+ \mid xy \ge 1\right\}
\end{aligned}$$

are *closed*. Of course, there are sets which are *neither open* (see Sects. 3.2 and 6.8) *nor closed*, for instance (check this, and also the above; compare Fig. 8.2)

$$\begin{aligned}
S_5 &= \left\{(x, y) \in \mathbb{R}^2 \mid a < x \le b,\ c \le x < d\right\},\\
S_6 &= \left\{(x, y, z) \in \mathbb{R}^3 \mid x \ge 0,\ y > 0,\ z \ge 0\right\}.
\end{aligned}$$

Among the above sets S_1, S_2, S_5 are *bounded*, S_3, S_4, S_6 are *not bounded*. A set $S \subset \mathbb{R}^n$ is *bounded* if there exists an $r \in \mathbb{R}_+$ such that

$$|\mathbf{x}| \le r \qquad \text{for all } \mathbf{x} \in S$$

(compare Fig. 8.3). A set $S \subset \mathbb{R}^n$ which is both closed and bounded is called *compact* (for more general sets there is a more general definition but for $\mathbb{R}^n$ this will do). So, among the above sets S_1 and S_2 are compact.

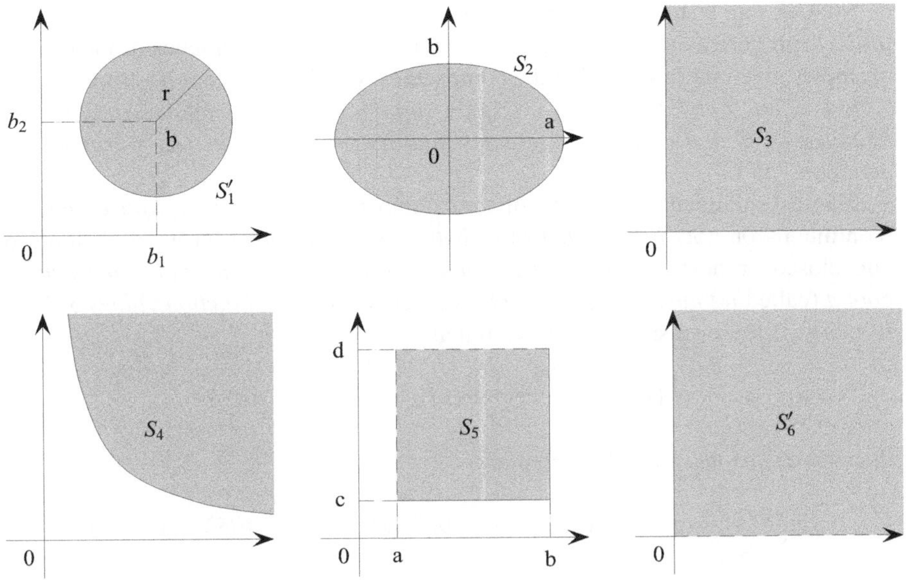

Fig. 8.2 In $\mathbb{R}^2$, S_1', S_2 are compact; S_5 bounded but not closed; S_3, S_4 closed but not bounded; S_6' neither closed nor bounded

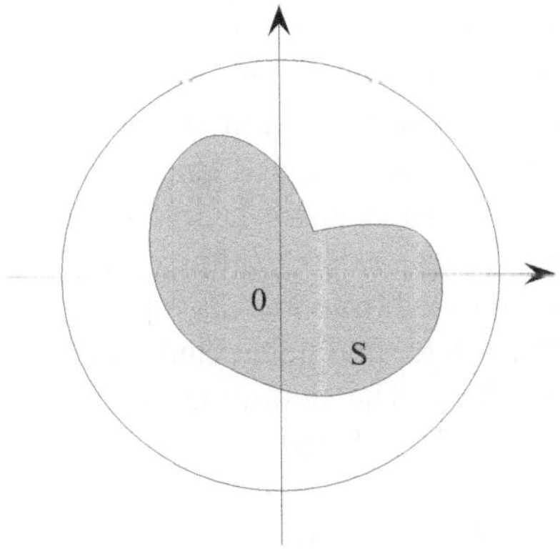

Fig. 8.3 Bounded set S

Of course, here too, a real valued function $F : S \to \mathbb{R}\,(S \subseteq \mathbb{R}^n)$ is called *bounded from above* or *below on* $C \subseteq S$ if there exist $m, M \in \mathbb{R}$ such that

$$F(\mathbf{x}) \geq m \quad \text{or} \quad F(\mathbf{x}) \leq M, \quad \text{respectively, for all } x \in C.$$

If a function is bounded both from above and from below on C the it is *bounded on* C.

Now we can state the analogues of Properties 1 and 2 in Sect. 6.3. (We will not prove them here as we had not proved them there either. Analogues of the counter examples given there show that the compactness condition cannot be dropped).

Every function $F : S \to \mathbb{R}$ $(S \subseteq \mathbb{R}^n)$ *continuous on a compact set* $C \subseteq S$ *is bounded on* C *and assumes its greatest and its smallest value (its maximum and minimum) an* C.

It is a fundamental property of real numbers that, *if a real valued function* (continuous or not) *is bounded from below or from above* on a set S (finite or not, closed or not) *then there exists a greatest lower bound and a least upper bound* (called *infimum* and *supremum*, respectively) *of the function values* S. More formally: If there exist $m, M \in \mathbb{R}$ such that

$$F(\mathbf{x}) \geq m \quad \text{or} \quad F(\mathbf{x}) \leq M \quad \text{for} \quad \mathbf{x} \in S$$

then there exist $m_0, M_0 \in \mathbb{R}$ such that

$$\inf_{\mathbf{x}\in S} F(\mathbf{x}) := m_0 \leq F(\tilde{\mathbf{x}}) \quad \text{or} \quad \sup_{\mathbf{x}\in S} F(\mathbf{x}) := M_o \geq F(\tilde{\mathbf{x}}) \quad (\tilde{\mathbf{x}} \in S),$$

respectively, but for *all* $m' > m_0$ there exists an $\mathbf{x}' \in S$ such that $F(\mathbf{x}') < m'$ or for *all* $M' < M_0$ there exists an $\mathbf{x}'' \in S$ such that $F(\mathbf{x}'') > M'$, respectively. So the above second property can be formulated so that *for continuous functions on a compact set, the infimum is the minimum and the supremum is the maximum.*

These are also nice results but for the above mentioned practical purposes it is of at least as much importance to know *how to calculate the maxima and minima.* (Actually, the above considerations showed also that in some cases there exists no maximum or minimum. For instance, for $(x, y) \mapsto 2x + y$ on $]0, 1[^2$ the infimum is 0, the supremum is 3 but there exists neither a maximum nor a minimum, while $(x, y) \mapsto x/y$ is not even bounded from above on $]0, 2]^2$).

We saw in Sect. 6.8 how to find local and global minima and maxima for functions of several variables. The definition of global and local minima (and maxima) is similar here too: If $\mathbf{a} \in S$ and

$$F(\mathbf{x}) \geq F(\mathbf{a}) \quad \textit{for all } \mathbf{x} \in S$$

then there is *a global minimum at* $\mathbf{a}$ *for* $F : S \to \mathbb{R}$ *(strict global minimum if* $>$ holds for all $\mathbf{x} \neq \mathbf{a}$ in S) and similarly for *(strictly) global maxima*. We did not require here that S is compact. We define *local* minima and maxima at *interior* points $\mathbf{a}$ of S, that is (see Sect. 6.8), if there is a neighbourhood of $\mathbf{a}$ completely in S. *If* there exists a neighbourhood $N(\mathbf{a}) \subset S$ of $\mathbf{a}$ such that

$$F(\mathbf{x}) \geq F(\mathbf{a}) \quad \text{for all} \quad \mathbf{x} \in N(\mathbf{a}) \tag{8.19}$$

then there is a *local minimum at* $\mathbf{a}$ *(strictly local minimum)* if we have $>$ for $\mathbf{x} \neq \mathbf{a}$) and similarly for *(strict) local maxima.*

As we saw in Sect. 6.8, *local maxima and minima need not be global. There are cases, however, when they are.* One is the following. *Let F be convex from below on the open convex set S* (see Sect. 3.4) *and have a local minimum at* $\mathbf{a} \in S$. *Then this will be a global minimum on S.* Indeed, since F is convex from below on the convex set S, we have for all $\mathbf{z} \in S$ and all $\lambda \in]0, 1[$, that $\lambda\mathbf{a} + (1-\lambda)\mathbf{z} \in S$ and

$$F(\lambda\mathbf{a} + (1-\lambda)\mathbf{z}) \leq \lambda F(\mathbf{a}) + (1-\lambda)F(\mathbf{z}).$$

Suppose for contradiction that there exists a $\mathbf{z} \in S$ with

$$F(\mathbf{z}) < F(\mathbf{a}). \tag{8.20}$$

The closer λ is to 1, the closer $\lambda\mathbf{a} + (1-\lambda)\mathbf{z}$ gets to $\mathbf{a}$ so, for λ close enough to 1, we will have $\mathbf{x} = \lambda\mathbf{a} + (1-\lambda)\mathbf{z}$ in that neighbourhood $N(\mathbf{a})$ for which (8.19) holds. Thus

$$\begin{aligned} F(\mathbf{a}) \leq F(\mathbf{x}) &= F(\lambda\mathbf{a} + (1-\lambda)\mathbf{z}) \leq \lambda F(\mathbf{a}) + (1-\lambda)F(\mathbf{z}) \\ &< \lambda F(\mathbf{a}) + (1-\lambda)F(\mathbf{a}) = F(\mathbf{a}), \end{aligned}$$

which is *impossible* (it would mean $F(\mathbf{a}) < F(\mathbf{a})$). So (8.20) *cannot hold for any* $\mathbf{z} \in S$, that is,

$$F(\mathbf{z}) \geq F(\mathbf{a}) \quad \textit{for all} \quad \mathbf{z} \subset S,$$

and so *F has a global minimum at* **a**, as asserted. Similarly, *if F is convex from above (concave) on the open convex set S and has a local maximum at* **a** *then this is also a global maximum on S*. Similar statements hold for *strict* maxima and minima. While we needed here no differentiability of F, we will need it in what follows. Similarly as we did in Sect. 6.8 with convexity, we intend to reduce now the finding of local maxima and minima of functions in several variables to dealings with functions in a single variable.

We first use the *law of the mean for functions of several variables: If* **p** *and* $\mathbf{p}+\mathbf{h}$ *are interior points of the convex set* $S \subseteq \mathbb{R}^n$ *and if* $F : S \to \mathbb{R}$ *is differentiable, then*

$$F(\mathbf{p}+\mathbf{h}) - F(\mathbf{p}) = \mathbf{h} \cdot \nabla F(\mathbf{p} + \theta\mathbf{h}) \quad \text{for some} \quad \theta \in]0, 1[,$$

as we derived it in Sect. 6.8 from the formula (8.20) of that section (we use here **p** rather than **x**). As in Sect. 6.8, we may also write

$$\nabla F = \mathbf{F}' = \left(\frac{\partial F}{\partial x_1}, \ldots, \frac{\partial F}{\partial x_n}\right) \quad \text{and so} \quad F(\mathbf{p}+\mathbf{h}) - F(\mathbf{p}) = \mathbf{F}'(\mathbf{p} + \theta\mathbf{h}) \cdot \mathbf{h}. \tag{8.21}$$

(We wrote ∇ and $\mathbf{F}'$ bold faced to emphasise that they are vectors, even though F is scalar-valued.)

Up to here we needed only the *differentiability* of F.

Suppose now that the partial derivatives $\partial F/\partial x_1, \ldots, \partial F/\partial x_n$ are continuous on S (then it follows also that F is differentiable on S). Since in this case

$$\lim_{\mathbf{x}\to\mathbf{p}} (\mathbf{F}'(\mathbf{x}) - \mathbf{F}'(\mathbf{p})) = \mathbf{0},$$

therefore, writing

$$\mathbf{d}(\mathbf{x}) := \mathbf{F}'(\mathbf{x}) - \mathbf{F}'(\mathbf{p}),$$

we have

$$\mathbf{F}'(\mathbf{x}) = \mathbf{F}'(\mathbf{p}) + \mathbf{d}(\mathbf{x}), \quad \text{where} \quad \lim_{\mathbf{x}\to\mathbf{p}} \mathbf{d}(\mathbf{x}) = \mathbf{0}. \tag{8.22}$$

Thus (8.21) becomes

$$F(\mathbf{p}+\mathbf{h}) - F(\mathbf{p}) = \mathbf{F}'(\mathbf{p}) \cdot \mathbf{h} + \mathbf{d}(\mathbf{p}+\theta\mathbf{h}) \cdot \mathbf{h} = \mathbf{F}'(\mathbf{p}) \cdot \mathbf{h} + R(\mathbf{p}, \mathbf{h}). \tag{8.23}$$

Here $R(\mathbf{p}, \mathbf{h}) := F(\mathbf{p}+\mathbf{h}) - F(\mathbf{p}) - \mathbf{F}'(\mathbf{p}) \cdot \mathbf{h} = \mathbf{d}(\mathbf{p}+\theta\mathbf{h}) \cdot \mathbf{h}$ corresponds to the *remainder* R_1 of the Taylor series in Sect. 6.7. We show that

$$\lim_{|\mathbf{h}|\to 0} \frac{1}{|\mathbf{h}|} R(\mathbf{p}, \mathbf{h}) = \lim_{|\mathbf{h}|\to 0} \mathbf{d}(\mathbf{p}+\theta\mathbf{h}) \cdot \frac{\mathbf{h}}{|\mathbf{h}|} = 0. \tag{8.24}$$

Indeed, $|\mathbf{h}| = (h_1^2 + \cdots h_n^2)^{\frac{1}{2}} \to 0$ implies $h_1 \to 0$, $h_2 \to 0$, $\cdots$, $h_n \to 0$, that is, $\mathbf{h} = (h_1, \cdots, h_n) \to \mathbf{0}$. So, by (8.22),

$$\lim_{|\mathbf{h}|\to 0} \mathbf{d}(\mathbf{p}+\theta\mathbf{h}) = \mathbf{0}, \quad \text{in components} \quad \lim_{|\mathbf{h}|\to 0} d_j(\mathbf{p}+\theta\mathbf{h}) = 0, \quad (j = 1, \ldots, n). \tag{8.25}$$

As a consequence we have (8.24):

$$\lim_{|\mathbf{h}|\to 0} \frac{R(\mathbf{p}, \mathbf{h})}{|\mathbf{h}|} = \lim_{|\mathbf{h}|\to 0} \left(d_1(\mathbf{p}+\theta\mathbf{h}) \cdot \frac{h_1}{|\mathbf{h}|} + \ldots + d_n(\mathbf{p}+\theta\mathbf{h}) \cdot \frac{h_n}{|\mathbf{h}|} \right) = 0, \tag{8.26}$$

since $|h_j| \leq (h_1^2 + \cdots h_n^2)^{\frac{1}{2}} = |\mathbf{h}|$, that is,

$$\left| \frac{h_j}{|\mathbf{h}|} \right| \leq 1 \quad (j = 1, \ldots, n), \quad \text{if} \quad |\mathbf{h}| \neq 0$$

and, by (8.25) and the definition of limit in Sects. 6.2 and 6.10, to every $\varepsilon' > 0$ there exists a $\delta > 0$ such that

$$|d_j(\mathbf{p}+\theta\mathbf{h})| < \varepsilon' = \varepsilon/n \quad \text{if} \quad |\mathbf{h}| < \delta \quad (j = 1, \ldots, n)$$

so, by the triangle inequality (see Sects. 1.6 and 1.7)

$$\left|(d_1(\mathbf{p}+\theta\mathbf{h})\cdot\frac{h_1}{|\mathbf{h}|}+\ldots+d_n(\mathbf{p}+\theta\mathbf{h})\cdot\frac{h_n}{|\mathbf{h}|}\right|$$
$$\le |(d_1(\mathbf{p}+\theta\mathbf{h})+\ldots+d_n(\mathbf{p}+\theta\mathbf{h})|<\varepsilon \quad \text{if} \quad |\mathbf{h}|<\delta$$

which is exactly (8.26) (with ε' also $\varepsilon = n\varepsilon'$ is as small as we want to make it).

The statement (8.24) is verbalised as follows: *R is of order 1 in* $|\mathbf{h}|$ (second order will soon follow). In formula:

$$R(\mathbf{p},\mathbf{h}) = o(|\mathbf{h}|) \quad \text{and} \quad F(\mathbf{p}+\mathbf{h}) - F(\mathbf{p}) = \mathbf{F}'(\mathbf{p})\cdot\mathbf{h} + o(|\mathbf{h}|),$$

in view (8.23); or, equivalently,

$$F(\mathbf{x}) = F(\mathbf{p}) + \mathbf{F}'(\mathbf{p})\cdot(\mathbf{x}-\mathbf{p}) + o(|\mathbf{x}-\mathbf{p}|). \tag{8.27}$$

This is just a more exact formulation of the statement at the end of Sect. 6.10 about *linear approximation and differentials*. The first two terms on the right hand side of (8.27) form an affine linear function of $\mathbf{x}$, the *linear approximation* of F at $\mathbf{p}$ (see Fig. 7.4).

We can often approximate the value of a function better by *quadratic approximation* that is, by a *quadratic polynomial* (see Sect. 7.4 (7.3)) in place of the *linear polynomial* $F(\mathbf{p})+'(\mathbf{p})\cdot(\mathbf{x}-\mathbf{p})$ in (8.27). This will be particularly useful for finding conditions for *maxim and minima of functions of several variables*.

As indicated above, we want to reduce the problem to finding quadratic approximations, maxima and minima for a function of one variable. We define, as in Sect. 8.2 (7.2),

$$g(r) := F(\mathbf{p} + r\mathbf{h}). \tag{8.28}$$

We apply the *Taylor formula with first degree (affine) polynomial part and with the remainder in the Lagrange form* (Sect. 6.7 (6.11)):

$$g(r) = g(0) + g'(0)r + \frac{1}{2}g''(\theta)r^2 \quad (r \in [0, 1]) \tag{8.29}$$

which holds for some $\theta \in]0, 1[$, if g is twice differentiable on an interval containing $[0, r]$ (we have here $a = 0$, so really a "MacLaurin formula", and wrote r in place of x and θ instead of ξ which, of course, does not change its validity). As in Sect. 6.8, the continuity of the second partial derivatives of F in (8.28) implies that g has a continuous second derivative.

By repeated use of the chain rule in Sect. 6.5, we get (compare Sect. 6.8 (7.2) and (7.19):

$$g'(r) = \mathbf{F}'(\mathbf{p} + r\mathbf{h}) \cdot \mathbf{h},$$
$$g''(r) = \mathbf{h}\mathbf{F}''(\mathbf{p} + r\mathbf{h})\mathbf{h}^T.$$

So (8.28) and (8.29) (at $r = 1$) yield

$$F(\mathbf{p} + \mathbf{h}) = F(\mathbf{p}) + \mathbf{F}'(\mathbf{p}) \cdot \mathbf{h} + \frac{1}{2}\mathbf{h}\mathbf{F}''(\mathbf{p} + \theta\mathbf{h})\mathbf{h}^T.$$

(Remember that F is a scalar valued function of a vector, consequently $\mathbf{F}'$ is vector valued and $\mathbf{F}''$ is matrix valued, it is the *Hesseean matrix*). Since $\mathbf{F}''$, (thus also every component of the Hessian matrix $\mathbf{F}''$) is *continuous*, we can write (compare (8.22))

$$\mathbf{F}''(\mathbf{p} + \theta\mathbf{h}) = \mathbf{F}''(\mathbf{p}) + 2\mathbf{D}(\mathbf{p} + \theta^*\mathbf{h}), \qquad \lim_{\mathbf{h}\to 0} \mathbf{D}(\mathbf{p} + \theta^*\mathbf{h}) = \mathbf{0}$$

($\mathbf{D}$ now matrix valued; the factor 2 is harmless but simplifies what follows; $\theta\ \in]0, 1[$, $\theta^*\ \in]0, \theta[$) and get

$$F(\mathbf{p} + \mathbf{h}) = F(\mathbf{p}) + \mathbf{F}'(\mathbf{p}) \cdot \mathbf{h} + \frac{1}{2}\mathbf{h}\mathbf{F}''(\mathbf{p})\mathbf{h}^T + \mathbf{h}\mathbf{D}(\mathbf{p} + \theta^*\mathbf{h})\mathbf{h}^T. \tag{8.30}$$

But exactly as (8.24), one can show that

$$\lim_{\mathbf{h}\to 0} \left(\frac{1}{|\mathbf{h}|^2}\mathbf{h}\mathbf{D}(\mathbf{p} + \theta^*\mathbf{h})\mathbf{h}^T \right) = 0$$

which we express so that $\mathbf{h}\mathbf{D}(\mathbf{p} + \theta^*\mathbf{h})\mathbf{h}^T$ is *of order 2 in* $|\mathbf{h}|$; in formula:

$$\mathbf{h}\mathbf{D}(\mathbf{p} + \theta^*\mathbf{h})\mathbf{h}^T ? o(|\mathbf{h}|^2), \qquad \text{where} \qquad \lim_{\mathbf{h}\to 0} \frac{o(|\mathbf{h}|)^2}{\mathbf{h}^2} = 0.$$

So, in view of (8.30),

$$F(\mathbf{p} + \mathbf{h}) = F(\mathbf{p}) + \mathbf{F}'(\mathbf{p}) \cdot \mathbf{h} + \frac{1}{2}\mathbf{h}\mathbf{F}''(\mathbf{p})\mathbf{h}^T + o(|\mathbf{h}|^2)$$

or, what is the same, we have

$$F(\mathbf{x}) = F(\mathbf{p}) + \mathbf{F}'(\mathbf{p}) \cdot (\mathbf{x} - \mathbf{p}) + \frac{1}{2}(\mathbf{x} - \mathbf{p})\mathbf{F}''(\mathbf{p})(\mathbf{x} - \mathbf{p})^T + o(|\mathbf{x} - \mathbf{p}|^2) \tag{8.31}$$

if F has continuous second partial derivative on a convex set containing $\mathbf{p}$ and bx.

The first three terms on the right hand side of (8.31) form a polynomial of second degree, the *quadratic approximation* of F at $\mathbf{p}$ (see Fig. 8.4).

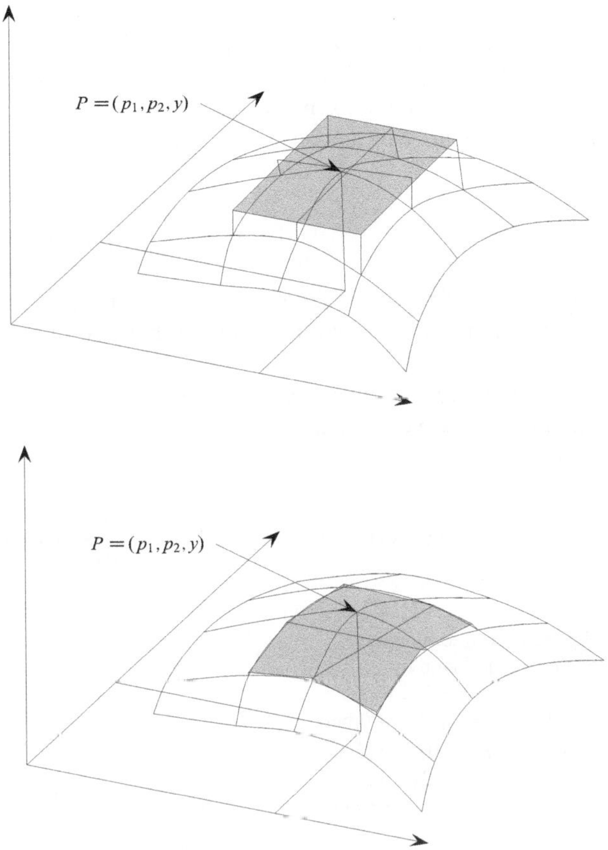

Fig. 8.4 Spatial graphs of linear and quadratic approximations of a function $F : \mathbb{R}^2_+ \to \mathbb{R}$ at $\mathbf{p} = (p_1, p_2)$

Now we get to local maxima and minima of functions of several variables at interior points (as in Sect. 6.7 we will use occasionally "*extremum*" as a common name for maximum or minimum). If $F(\mathbf{p})$ is maximal among *all* values of F on a neighbourhood $N(\mathbf{p})$ then it will be maximal also among the values at

$$(p_1, \ldots, p_{j-1}, x_j, p_{j+1}, \cdots, p_n) \quad \text{in} \quad N(\mathbf{p}).$$

But these values are those of the *partial function*

$$x_j \longmapsto F(p_1, \ldots, p_{j-1}, x_j, p_{j+1}, \cdots, p_n), \tag{8.32}$$

so it too should have a local maximum at $x_j = p_j$. Thus, if the derivative of this function, that is $\partial F/\partial x_j$, exists on a neighbourhood of $\mathbf{p}$ then, by what we learned in Sect. 6.8,

$$\frac{\partial F}{\partial x_j}(\mathbf{p}) = 0 \quad (j = 1, \ldots, n) \tag{8.33}$$

is necessary for F to have a local maximum—or, similarly, a local minimum, generally: *a local extremum*—at $\mathbf{p}$, As in the case $n = 1$ (see (6.19) in Sect. 6.11) we call a point $\mathbf{p}$ satisfying (8.33), a *critical* or *stationary point of F*.

Already for functions in one variable we have seen in Sect. 6.8 that *this condition is not sufficient*. We can construct examples which follow the same pattern as there, for instance $F : \mathbb{R}^2 \to \mathbb{R}$, defined by

$$F(x_1, x_2) = x_1^3 + 2x_2^3,$$

satisfies $\partial F/\partial x_1 = \partial F/\partial x_2 = 0$ (only) at $x_1 = x_2 = 0$ but there is neither local maximum nor local minimum at that point not even for the partial functions (Fig. 8.5).

For functions of several variables we have, however, also a new phenomenon: the *saddle point*. We can see it (Fig. 8.6; compare Fig. 3.26) on the function $F : \mathbb{R}^2 \to \mathbb{R}$ given by

$$F(x_1, x_2) = x_1^2 - x_2^2; \quad \text{here} \quad \frac{\partial F}{\partial x_1}(0, 0) = \frac{\partial F}{\partial x_2}(0, 0) = 0$$

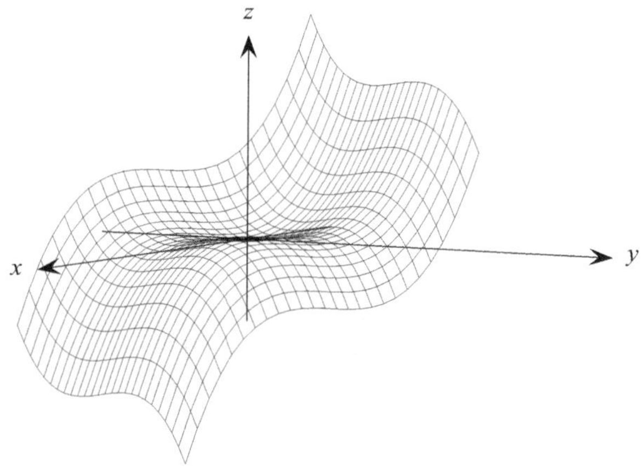

Fig. 8.5 For the function $F : \mathbb{R}^2 \to \mathbb{R}$ given by $F(x_1, x_2) = x_1^3 + 2x_2^3$ we have $\frac{\partial F}{\partial x_1}(0, 0) = \frac{\partial F}{\partial x_2}(0, 0) = 0$ but both $x_1 \mapsto F(x_1, 0) = x_1^3$ and $x_2 \mapsto F(0, x_2) = 2x_2^3$ keep increasing: *no local extremum*

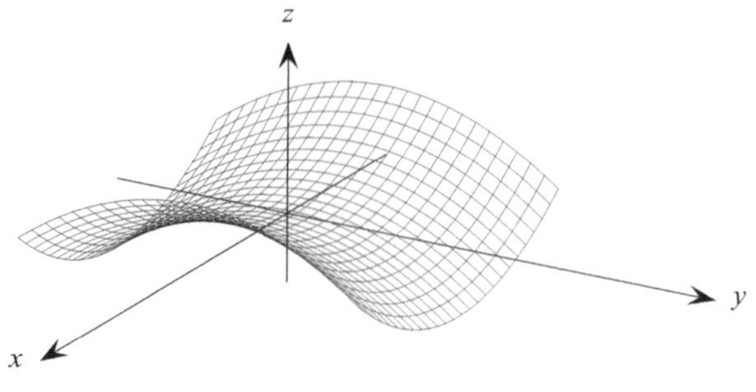

Fig. 8.6 For the function $F : \mathbb{R}^2 \to \mathbb{R}$ given by $F(x_1, x_2) = x_1^2 - x_2^2$ there is $\frac{\partial F}{\partial x_1}(0, 0) = \frac{\partial F}{\partial x_2}(0, 0) = 0$ and the partial functions $x_1 \mapsto F(x_1, 0) = x_1^2$, $x_2 \mapsto F(0, x_2) = -x_2^2$ have (local) extrema at 0 but the first has a minimum, the second a maximum there: F has a *saddle point* at $(0, 0)$

and indeed *both partial functions*

$$x_1 \mapsto F(x_1, 0) = x_1^2 \quad \textit{and} \quad x_2 \mapsto F(0, x_2) = -x_2^2$$

have local extrema at 0, but the first a maximum, the second a minimum. Clearly, F has no local extremum at 0:

$$F(0,0) = 0, \qquad F(0,\varepsilon) = -\varepsilon^2 < 0, \qquad F(\varepsilon, 0) = \varepsilon^2 > 0,$$

no matter how small the positive ε is. Such points are called *saddle points*. In general, *if all partial functions (8.32) have a local extremum at* $x_j = p_j$ $(j = 1, \ldots, n)$ *but at least one partial function has a local maximum and at least one other a local minimum (both strict) then the critical point* $(p_1, \ldots, p_n)$ *is a saddle point of* F. *Saddle points can appear also in other ways.* Take, for instance, the function $F : \mathbb{R}^2 \to \mathbb{R}$ given by

$$F(x_1, x_2) = x_1^2 + x_2^2 - 3x_1x_2$$

(Fig. 8.7). We have

$$\frac{\partial F}{\partial x_1}(0,0) = \frac{\partial F}{\partial x_2}(0,0) = 0,$$

both partial functions $x_1 \mapsto F(x_1, 0) = x_1^2$, $x_2 \mapsto F(0, x_2) = 2x_2^2$ have a minimum at 0 but the vertical 45° "cut" $x \mapsto F(x, x) = -x^2$ has a maximum at 0. Again, F has no local extremum at $\mathbf{0}$: $F(0,0) = 0$, $F(0,\varepsilon) = \varepsilon^2 < 0$, $F(\varepsilon, \varepsilon) = -\varepsilon^2 < 0$, no matter how small the positive ε is. One may stop worrying about the different

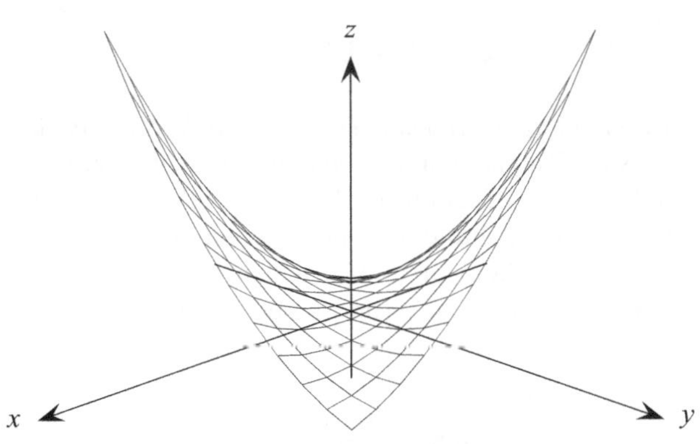

Fig. 8.7 For the function $F : \mathbb{R}^2 \to \mathbb{R}$ given by $F(x_1, x_2) = x_1^2 + x_2^2 - 3x_1x_2$ we have $\frac{\partial F}{\partial x_1}(0,0) = \frac{\partial F}{\partial x_2}(0,0) = 0$ and the partial functions $x_1 \mapsto F(x_1, 0) = x_1^2$, $x_2 \mapsto F(0, x_2) = x_2^2$ have a minimum at 0 but $x \mapsto F(x, x) = -x^2$ has a maximum at 0: F has a *saddle point* at $(0, 0)$

possibilities by defining a *saddle point as a* critical point $\mathbf{p}$ (see (8.33)) at which $F(\mathbf{p})$ is neither a local maximum or a local minimum of F.

So, if (8.33) is not sufficient for F to have a local extremum at $\mathbf{p}$, what is? We go back to (8.31) for that. Since the condition (8.33) is *necessary*, at local extrema (8.31) reduces to

$$F(\mathbf{x}) - F(\mathbf{p}) = \frac{1}{2}(\mathbf{x}-\mathbf{p})\mathbf{F}''(\mathbf{p})(\mathbf{x}-\mathbf{p})^T + o(|\mathbf{x}-\mathbf{p}|^2). \tag{8.34}$$

By definition, *F has a strict (sharp) local minimum at $\mathbf{p}$ if there exists a punctured neighbourhood $N'(\mathbf{p})$ of $\mathbf{p}$ ($\mathbf{p} \notin N'(\mathbf{p})$) such that*

$$F(\mathbf{x}) - F(\mathbf{p}) > 0 \quad \text{for all} \quad \mathbf{x} \in N'(\mathbf{p}).$$

This is certainly the case if, in (8.34), the quadratic form

$$\frac{1}{2}(\mathbf{x}-\mathbf{p})\mathbf{F}''(\mathbf{p})(\mathbf{x}-\mathbf{p})^T$$

is *positive definite*, as defined in Sect. 8.2 (or, what is the same, the Hessian matrix $\mathbf{F}''(\mathbf{p})$ is positive definite). Indeed, then also

$$\frac{1}{2}\frac{\mathbf{x}-\mathbf{p}}{|\mathbf{x}-\mathbf{p}|}\mathbf{F}''(\mathbf{p})\left(\frac{\mathbf{x}-\mathbf{p}}{|\mathbf{x}-\mathbf{p}|}\right)^T \tag{8.35}$$

is positive definite (see Sect. 8.2; as it happens, F *is* also *strictly convex from below* on a neighbourhood of $\mathbf{p}$). This expression depends upon $\mathbf{x}$ only through $\frac{1}{|\mathbf{x}-\mathbf{p}|}(\mathbf{x}-\mathbf{p})$ and $\frac{1}{|\mathbf{x}-\mathbf{p}|}(\mathbf{x}-\mathbf{p})^T$. Both these vectors (really the same vector in row and column form) have norm 1, so they lie on the *n-dimensional unit sphere*

$$C_1 = \{\mathbf{z} \in \mathbb{R}^n \mid |\mathbf{z}| = 1\}$$

which is *a closed and bounded (compact) set on which (8.35) is* clearly *continuous*. So, by what we stated about maxima and minima of continuous functions on compact sets earlier in this section, *(8.35) has a minimum on C_1, and assumes it*. Since (8.35) is positive definite, it is positive everywhere (because $\mathbf{x}-\mathbf{p} = \mathbf{0}$ is excluded). So the minimum of (8.35), which is the value of (8.35) at some $(\mathbf{x}^0-\mathbf{p})/\left|\mathbf{x}^0-\mathbf{p}\right| \in C_1$, is positive and independent of $\mathbf{x}$, say $\mu > 0$. But, by definition,

$$\lim_{\mathbf{x}\to\mathbf{p}} \frac{o(|\mathbf{x}-\mathbf{p}|^2)}{|\mathbf{x}-\mathbf{p}|^2} = 0.$$

Take $\mathbf{x}$ in a punctured neighbourhood $N'(\mathbf{p})$ of the point $\mathbf{p}$ so small, such that $\left|o(|\mathbf{x}-\mathbf{p}|^2)/\,|\mathbf{x}-\mathbf{p}|^2\right| < \mu/2$, then (8.34) gives

$$\frac{F(\mathbf{x})-F(\mathbf{p})}{|\mathbf{x}-\mathbf{p}|^2} = \frac{1}{2}\frac{\mathbf{x}-\mathbf{p}}{|\mathbf{x}-\mathbf{p}|}\mathbf{F}''(\mathbf{p})\left(\frac{\mathbf{x}-\mathbf{p}}{|\mathbf{x}-\mathbf{p}|}\right)^T + \frac{o(|\mathbf{x}-\mathbf{p}|^2)}{|\mathbf{x}-\mathbf{p}|^2}$$

$$> \mu - \frac{\mu}{2} = \frac{\mu}{2} > 0.$$

(The worst effect $\delta = o(|\mathbf{x}-\mathbf{p}|^2)/|\mathbf{x}-\mathbf{p}|^2$ can have on the first term on the right is to diminish it by $|\delta|$. Diminishing it by $\mu/2 > |\delta|$ decreases the right hand side even more). Both then indeed

$$F(\mathbf{x}) - F(\mathbf{p}) > 0 \quad \text{for all} \quad \mathbf{x} \in N'(\mathbf{p}).$$

So we have proved that, *if* $\mathbf{F}'(\mathbf{p}) = \mathbf{0}$ *and* $\mathbf{F}''(\mathbf{p})$ *is positive definite, then* F *has a sharp local minimum at the (interior) point* $\mathbf{p}$. This implies also (just take F' in place of F) that, *if at the interior point* $\mathbf{p}$ *we have* $\mathbf{F}'(\mathbf{p}) = \mathbf{0}$ *and* $\mathbf{F}''(\mathbf{p})$ *is negative definite, then* F *has a sharp local maximum at* $\mathbf{p}$.

One can check also (do it!) that, *if* $\mathbf{F}''(\mathbf{p})$ *is indefinite then* F *cannot have a local extremum at the critical point* $\mathbf{p}$, *not even in the wider sense. Even if* $\mathbf{F}''(\mathbf{p})$ *is (positive or negative) semidefinite and* $\mathbf{F}'(\mathbf{p}) = \mathbf{0}$, *it can happen that* F *has no local extremum at* $\mathbf{p}$.

In Sect. 8.2 we saw how to determine whether a symmetric (Hessian) matrix is positive or negative definite, semidefinite or indefinite. We remind the reader that *global extrema may be assumed*, if at all, not only where there is a local extremum in the interior but *also on the boundary*. Moreover, *even local extrema in the interior may be located at points where* F *is not differentiable.*

Examples

1. $F(\mathbf{x}) = |x_1| + \ldots + |x_n|$ ($\mathbf{x} = (x_1,\ldots,x_n) \in \mathbb{R}^n$). There is a local and global minimum at $\mathbf{x} = 0$ where F is not differentiable (not even partially differentiable with respect to any x_j).
2. $F(\mathbf{x}) = x_1 + x_2^2 + \ldots + x_n^n$. On $[0,1]^n$ there is a global maximum at $\mathbf{1} = (1,1,\ldots,1)$ even though $\mathbf{F}'(\mathbf{1}) = (\frac{\partial F}{\partial x_1}(\mathbf{1}),\ldots,\frac{\partial F}{\partial x_n}(\mathbf{1})) = (1,2,\ldots,n) \neq \mathbf{0}$ (all partial derivatives are taken "from the left").
3. $F(\mathbf{x}) = 1 + 6x_1 - 3x_2 + 6x_3 + 6x_1x_3 - 3x_1^2 + x_2^3 - 9x_3^2$ for $(x_1,x_2,x_3) \in \mathbb{R}^3$ (compare Sect. 7.2, Example **4**). Local extrema can exist only where

$$\frac{\partial F}{\partial x_1} = \quad 6 + 6x_3 - 6x_1 = 0,$$

(continued)

$$\frac{\partial F}{\partial x_2} = -3 + 3x_2^2 = 0,$$

$$\frac{\partial F}{\partial x_3} = \quad 6 + 6x_1 - 18x_3 = 0.$$

We solve this system of equations. From the second equation $x_2 = 1$ or $x_2 = -1$. The first and third equation form a system of two linear equations for the two unknowns x_1 and x_3. Solving it by methods in Sect. 4.6 or directly, we get $x_1 = 2$, $x_3 = 1$. So there *may* be local extrema *only at* $(2, 1, 1)$ *or* $(2, -1, 1)$. In Sect. 8.2, Example **4**, we saw that the Hessian matrix $\mathbf{F}''(\mathbf{x})$ of this function is negative definite for negative x_2 but indefinite for positive x_2 (whatever x_1 and x_3 are). So the above F has *a strict local maximum at* $(2, -1, 1)$ *but no local extremum at* $(2, 1, 1)$.

As we have also seen in Sect. 8.2, this F is strictly convex from above (strictly concave) on

$$\{(x_1, x_2, x_3) \mid x_1 \in \mathbb{R},\ x_2 \in \mathbb{R}_{--},\ x_3 \in \mathbb{R}\}. \tag{8.36}$$

According to the result earlier in this section about the globality of local maxima on domains where F is concave, the above F *has a strict global maximum at* $(2, -1, 1)$ *(8.36). On* $\mathbb{R}^3$, however, *no global maximum exists*—also *no global minimum*—because

$$\lim_{x_2 \to \infty} F(0, x_2, 0) = \lim_{x_2 \to \infty} x_2 \left(\frac{1}{x_2} - 3 + x_2^2 \right) = \infty;$$

$$\lim_{x_2 \to -\infty} F(0, x_2, 0) = -\infty.$$

4. $F(x_1, x_2) = x_1^3 + x_2^3 - 3x_1x_2$ $(x_1,\ x_2 \in \mathbb{R})$. Local extrema can be only where

$$\frac{\partial F}{\partial x_1} = 3x_1^2 - 3x_2 = 0 \quad \text{and} \quad \frac{\partial F}{\partial x_2} = 3x_2^2 - 3x_1 = 0.$$

From the first equation $x_2 = x_1^2$ and from the second $x_1 = x_2^2 = x_1^4$. But $0 = x_1^4 - x_1 = x_1(x_1^3 - 1)$ has only $x_1 = 0$ and $x_1 = 1$ as real solutions. The $x_2 (= x_1^2)$ values belonging to them are 0 and 1. So there can be local extrema only at $(0, 0)$ and $(1, 1)$. Since

$$\frac{\partial^2 F}{\partial x_1^2} = 6x_1, \quad \frac{\partial^2 F}{\partial x_1 \partial x_2} = -3, \quad \frac{\partial^2 F}{\partial x_2^2} = 6x_2,$$

(continued)

the Hessian matrices at $(0, 0)$ and at $(1, 1)$ are

$$\mathbf{F}''(0,0) = \begin{pmatrix} 0 & -3 \\ -3 & 0 \end{pmatrix} \quad \text{and} \quad \mathbf{F}''(1,1) = \begin{pmatrix} 6 & -3 \\ -3 & 6 \end{pmatrix},$$

respectively. The eigenvalues of the first matrix are (see Sect. 8.2 (8.27)) the solution of

$$0 = \det \begin{pmatrix} -\lambda & -3 \\ -3 & -\lambda \end{pmatrix} = \lambda^2 - 9, \quad \text{that is,} \quad \lambda_1 = 3,\ \lambda_2 = -3,$$

one positive, one negative. Therefore, as we saw in Sect. 8.2, $\mathbf{F}''(0, 0)$ is indefinite. So *F has no local extremum at* $(0, 0)$, which is thus a *saddle point*. Indeed, $F(0, 0) = 0$, $F(0, \varepsilon) = \varepsilon^3 > 0$ but $F(-\varepsilon, 0) = -\varepsilon^2 < 0$, no matter how small $\varepsilon > 0$ is. On the other hand, the eigenvalues of the second matrix are the solutions of

$$0 = \det \begin{pmatrix} 6-\lambda & -3 \\ -3 & 6-\lambda \end{pmatrix} = (6-\lambda)^2 - 9,$$

$$\text{that is,} \quad 6 - \lambda = \pm 3, \quad \text{so} \quad \lambda_1 = 3,\ \lambda_2 = 9,$$

both positive (see Sect. 8.2, Example **1**). Thus $\mathbf{F}''(1, 1)$ is positive definite and *F has a strict local minimum at* $(1, 1)$. Again *there exists no global maximum and no global minimum on* $\mathbb{R}^2$ because

$$\lim_{x\to\infty} F(x,x) = \lim_{x\to\infty} (2x^3 - 3x^2) = \lim_{x\to\infty} x^3\left(2 - \frac{3}{x}\right) = \infty$$

$$\text{and} \quad \lim_{x\to-\infty} F(x,x) = -\infty.$$

5. In our concluding example we suppose that a monopolist sells n goods at the prices $p_1, \dots, p_n$, respectively. Let $\mathbf{G} : \mathbb{R}^n_{++} \to \mathbb{R}^n_{++}$ be the *price-demand function* which assigns to the price vector $\mathbf{p} = (p_1, \dots, p_n)$ the vector of quantities $\mathbf{q} = (q_1, \dots, q_n)$ which can be sold at those prices during a fixed time interval:

$$\mathbf{q} = \mathbf{G}(\mathbf{p})$$

So $q_1p_1 + \ldots + q_np_n = \mathbf{q} \cdot \mathbf{p}$ are our monopolists *cash receipts*. If $C(\mathbf{q})$ $(C : \mathbb{R}^n_{++} \to \mathbb{R}_{++})$ is the *cost* connected to the production of the quantities $\mathbf{q}$ *then the gross profit is*

$$F(\mathbf{p}) = \mathbf{q} \cdot \mathbf{p} - C(\mathbf{q}) = \mathbf{G}(\mathbf{p}) \cdot \mathbf{p} - C(\mathbf{G}(\mathbf{p})).$$

(continued)

Let us try to *maximise this profit*. We suppose in this example that $n = 2$ and that both C and $\mathbf{G}$ are *affine*, say, numerically

$$\mathbf{G}(p_1, p_2) = (10 - 3p_1 + 2p_2,\ 15 + p_1 - 4p_2), \qquad C(q_1, q_2) = 5 + q_1 + q_2.$$

So $q_1 = 10 - 3p_1 + 2p_2$, $q_2 = 15 + p_1 - 4p_2$ and

$$\begin{aligned} F(p_1, p_2) &= 10p_1 - 3p_1^2 + 2p_1p_2 + 15p_2 + p_1p_2 - 4p_2^2 \\ &\qquad -5 - 10 + 3p_1 - 2p_2 - 30 - 2p_1 + 8p_2 \\ &= -45 + 11p_1 + 21p_2 - 3p_1^2 + 3p_1p_2 - 4p_2^2. \end{aligned} \tag{8.37}$$

(Notice that, because of the economic context, we here denoted the variables by p_1, p_2 rather than x_1, x_2). Here F can have local extrema where

$$\frac{\partial F}{\partial p_1} = 11 - 6p_1 + 3p_2 = 0, \qquad \frac{\partial F}{\partial p_2} = 21 + 3p_1 - 8p_2 = 0.$$

Solving this system of two linear equations, we get the critical points $p_1 = 151/39, p_2 = 53/13$.

So there can be a local extremum only at $(151/39, 53/13) \sim (3.87, 4.08)$. Since

$$\frac{\partial^2 F}{\partial p_1^2} = -6, \qquad \frac{\partial^2 F}{\partial p_1 \partial p_2} = 3, \qquad \frac{\partial^2 F}{\partial p_2^2} = 8,$$

the Hessian matrix is everywhere constant:

$$\mathbf{F}'' = \begin{pmatrix} -6 & 3 \\ 3 & -8 \end{pmatrix}.$$

By what we learnt near the end of Sect. 8.2, the fact that the sectional determinants (principal minors, see Sect. 8.2)

$$D_1 = -6 < 0, \qquad D_2 = \det \begin{pmatrix} -6 & 3 \\ 3 & -8 \end{pmatrix} = 48 - 9 = 39 > 0$$

are alternating means that the Hessian H is *negative definite everywhere* so, on one hand the above F is *strictly convex from above (strictly concave) everywhere*, on the other hand it *has at* $(151/39,\ 53/13)$ *a strict local*

(continued)

maximum on $\mathbb{R}^2_{++}$. By our result already quoted in Example 3, *this is* also *a strict global maximum on* $\mathbb{R}^2_{++}$.

So the *maximal gross profit* will be

$$F(\frac{151}{39}, \frac{53}{13}) = -45 + 11 \cdot \frac{151}{39} + 21 \cdot \frac{53}{13} - 3 \cdot \frac{151^2}{39^2} + 3 \cdot \frac{151}{39} \times \frac{53}{13}$$
$$- 4 \cdot \frac{53^2}{13^2} \sim 19.10,$$

attained with the prices $p_1 \sim 3.87, p_2 \sim 4.08$ and with the sale quantities

$$q_1 = 10 - 3 \cdot \frac{151}{39} + 2 \cdot \frac{53}{13} = \frac{85}{13} \sim 6.54,$$
$$q_2 = 15 - \frac{151}{39} - 4 \cdot \frac{53}{13} = \frac{100}{13} \sim 2.56.$$

If the quantities q_1, q_2 *are given (constants)*, then the problem is one of a *conditional extremum*, that of determining *the maximum of the function* F of p_1 and p_2 given by (8.37), under the *restriction* (condition)

$$10 - 3p_1 + 2p_2 = q_1, \qquad 15 + p_1 - 4p_2 = q_2$$

with given q_1, q_2. We will deal with *conditional extrema* in Sect. 8.5 but first we give an important application to econometrics of the results and methods which we have just learnt.

8.3.1 Exercises

1. (a) For the function $F : \mathbb{R}^2 \to \mathbb{R}$ given by $F(x, y) = x^2 - 4x + y^2 + 6y$ determine all points (x^*, y^*) at which $\frac{\partial F}{\partial x}(x, y) = 0$ and $\frac{\partial F}{\partial y}(x, y) = 0$.
   (b) At which of these points does F have a local maximum or minimum? Why?
   (c) Determine the values of F at these maximising or minimising points. Which ones are global?
2. Answer similar questions as in Exercise 1 for the function $G : \mathbb{R}^2 \to \mathbb{R}$ given by $G(x, y) = x^2 + 5xy^3 + 2y$.
3. Answer similar questions as in Exercise 1 for the function $g : \mathbb{R}^2 \to \mathbb{R}$ given by $g(x, y) = x - x^4 - 4xy - y^2$.

4. Determine place (x^*, y^*), kind and function value of the local extrema or saddle points of the functions $f : \mathbb{R}^2 \to \mathbb{R}$
   (a) $f(x, y) = (x^2 + 2y^2)e^{-x^2}$,
   (b) $f(x, y) = x^3y^2(1 - x - 2y)$,
   (c) $f(x, y) = x^2(6 - x)y^2e^{-y}$.
5. Consider differentiable functions $F : \mathbb{R}^2 \to \mathbb{R}$ satisfying $\dfrac{\partial F}{\partial x_1}(0, 0) = 0$, $\dfrac{\partial F}{\partial x_2}(0, 0) = 0$.
   (a) Give an example of such a function satisfying
   $x_1 \mapsto F(x_1, 0)$ is strictly increasing,
   $x_2 \mapsto F(0, x_2)$ is strictly decreasing.
   (b) Same problem for the properties
   $x_1 \mapsto F(x_1, 0)$ is strictly increasing,
   $x \mapsto F(x, x)$ is strictly decreasing.
   (c) Same problem for the properties
   $x_1 \mapsto F(x_1, 0)$ is strictly increasing,
   $x_2 \mapsto F(0, x_2)$ has a maximum at $x_2 = 0$,
   $x \mapsto F(x, x)$ has a local maximum at $x = 0$.

8.3.2 Answers

1. (a) $(x^*, y^*) = (2, -3)$.
   (b) At $(x^*, y^*) = (2, -3)$ there is a minimum of F since
$$\mathbf{H} = \mathbf{F}''(x, y) = \begin{pmatrix} 2 & 0 \\ 0 & 2 \end{pmatrix},$$
   the Hessean of F, is positive definit.
   (c) $F(2, -3) = -13$. This minimu is global since $\mathbf{F}''(x, y)$ is positive definit on the whole of $\mathbb{R}^2$.
2. (a) $(x^*, y^*) = (-\frac{15}{2}(\frac{4}{225})^{3/5}, (\frac{4}{225})^{1/5}) = (-0.668325, 0.446658)$.
   (b) This point (x^*, y^*) is a saddle point of G, that is, at (x^*, y^*) there is neither a (local) maximum nor a (local) minimum of G, since the Hessean
$$\mathbf{G}'' = \begin{pmatrix} 2 & 15y^2 \\ 15y^2 & 30xy \end{pmatrix}$$
   with $\det \mathbf{G}'' = 60xy - 225y^4 < 0$ at (x^*, y^*).
3. (a) $(x^*, y^*) = (1.473, -2.946)$.
   (b) At (x^*, y^*) there is a local maximum of g since the principal minor D_1 of the Hessean of g, $\mathbf{g}''(x, y) = \begin{pmatrix} -12x^2 & -4 \\ -4 & -2 \end{pmatrix}$, equals $-12x^2$ (<0 for $x \neq 0$)

and the principal minor D_2 of the Hessean of g equals $24x^2 - 16$ (>0 for $x^* = 1.473$).

(c) $g(x^*, y^*) = 5.444$ is a global maximum on $]+\sqrt{2/3}, \infty[\times\mathbb{R}$, since there the Hessean of g is negative definit. One can show that $g(x^*, y^*) = 5.444$ is also a global maximum on $\mathbb{R}^2$.

4.

	Place	Kind	Function value	Hessean
(a)	$(x_1^*, y^*{}_1) = (0,0)$	Minimum	$f(0,0) = 0$	$\mathbf{f}''(0,0) = \begin{pmatrix} 2 & 0 \\ 0 & -4 \end{pmatrix}$,
	$(x_2^*, y^*{}_2) = (1,0)$	Saddle point	$f(1,0) = 1/e$	$\mathbf{f}''(1,0) = \begin{pmatrix} -4/e & 0 \\ 0 & 4/e \end{pmatrix}$,
	$(x_3^*, y^*{}_3) = (-1,0)$	Saddle point	$f(-1,0) = 1/e$	$\mathbf{f}''(-1,0) = \begin{pmatrix} -4/e & 0 \\ 0 & 4/e \end{pmatrix}$.
(b)	$(x_1^*, y^*{}_1) = (\frac{1}{2}, \frac{1}{6})$	Maximum	$f(\frac{1}{2}, \frac{1}{6}) = \frac{1}{1728}$	$\mathbf{f}''(\frac{1}{2}, \frac{1}{6}) = -\frac{1}{4}\begin{pmatrix} 1/9 & 1/6 \\ 1/6 & 1/2 \end{pmatrix}$,
	$(x_2^*, y^*{}_2) = (x,0)$	Saddle points	$f(x,0) = 0$	$\mathbf{f}''(x,0) = \mathbf{0}$,
	$(x_3^*, y^*{}_3) = (0,y)$	Saddle points	$f(0,y) = 0$	$\mathbf{f}''(0,y) = \mathbf{0}$.
(c)	$(x_1^*, y^*{}_1) = (4,2)$	Maximum	$f(4,2) = 128e^{-2}$	$\mathbf{f}''(4,2) = -e^{-2}\begin{pmatrix} 48 & 0 \\ 0 & 64 \end{pmatrix}$,
	$(x_2^*, y^*{}_2) = (x,0)$	Saddle points	$f(x,0) = 0$	$\mathbf{f}''(x,0) = \mathbf{0}$,
	$(x_3^*, y^*{}_3) = (0,y)$	Saddle points	$f(0,y) = 0$	$\mathbf{f}''(0,y) = \mathbf{0}$.

5. (a) $F(x_1, x_2) = x_1^3 - x_2^3$,
   (b) $F(x_1, x_2) = x_1^3 - 2x_2^3$,
   (c) $F(x_1, x_2) = x_1^3 - x_2^3 - x_1x_2$.

8.4 Bellman's Principle of Dynamic Optimisation; Application to a Maximum Problem

This section presents the solution of the problem formulated in Sect. 8.1. The problem was to maximise

$$C = \sum_{t=1}^{T} d^{t-1}C_t = \sum_{t=1}^{T}(1 - x_t)F(K_{t-1})d^{t-1} \tag{8.38}$$

(see (8.22)), where C is the discounted aggregate consumption in a closed economy during the years $t = 1, \ldots, t = T$, while $d^{t-1}C_t$ is the discounted consumption in the year t, d $(0 < d < 1)$ and $1 - d$ are the discount factor and the discount rate, respectively x_t is the investment ratio in the year t, F is the production function and K_{t-1} is the capital stock of the economy at the beginning of year t. As we showed in Sect. 8.1, the capital stocks $K_1, \dots, K_T$ can be calculated from the given initial capital stock K_0 by the recursive equation

$$K_t = qK_{t-1} + x_t F(K_{t-1}) \quad (t = 1, \dots, T) \tag{8.39}$$

(see (8.23)). Here q is the depreciation factor ($0 < q < 1$). All terms are assumed to be inflation-adjusted. We assume further that the production function is twice differentiable.

In order to maximise (8.38) under the conditions (8.39) we have to calculate values of the investment ratios $x_1, \dots, x_T$ which maximise (8.38), with K_0 given and $K_1, \dots, K_T$ calculated and inserted from (8.39). We insert (8.39) into (8.38) and get

$$C = \sum_{t=1}^{T} (1 - x_t) F(K_{t-1}) d^{t-1}, \tag{8.40}$$

that is,

$$\begin{aligned} C &= F(K_0) + \sum_{t=1}^{T-1} F(K_t) d^t - \sum_{t=1}^{T} x_t F(K_{t-1}) d^{t-1} \\ &= F(K_0) + \sum_{t=1}^{T-1} F(qK_{t-1} + x_t F(K_{t-1})) d^t - \sum_{t=1}^{T} x_t F(K_{t-1}) d^{t-1}. \end{aligned}$$

From this we see that C as function of x_T is greatest when x_T is zero. As we have learned before, in order to get necessary conditions for local maxima of C as function of $x_1, \dots, x_{T-1}$ in the interior of its domain (see (8.19) and assume that $I_t \le Y_t$)

$$\{(x_1, \dots, x_{T-1}) \mid 0 \le x_t \le 1,\ t = 1, \dots, T-1\}, \tag{8.41}$$

we have to set the partial derivatives with respect to $x_1, \dots, x_{T-1}$ equal to zero:

$$\frac{\partial C}{\partial x_t} = 0 \quad (t = 1, \dots, T-1). \tag{8.42}$$

We first differentiate (the last line of the expression for) C with respect to x_{T-1}. Doing this we start applying Bellman's *principle of backward dynamic optimisation* (RICHARD E. BELLMAN (*1920 – †1984)) which says that problems of dynamic optimisation like that considered here can be solved as follows:

(i) *Determine the optimal value of x_T, then that of x_{T-1}, x_{T-2} and so on.*

The optimal value of x_t may depend on the values of $x_{t-1}, \dots, x_1$, that is, $x_t = f_t(x_{t-1}, \dots, x_1)$, $t = 1, \dots, T$. The last step backward is calculating the optimal value

x_t^* of x_1 which is determined by the initial situation of the problem, that is, in our case, by the initial capital stock K_0.

(ii) *Insert* x_1^* *into* $x_2 = f_2(x_1)$ *to get* x_2^*, *then insert* x_1^*, x_2^* *into* $x_3 = f_3(x_2, x_1)$ *to get* x_3^*, *and so on.*

(iii) *For problems like that considered here (see also the assumptions that will follow) the unique (global) solution vector is* $(x_T^*, x_{T-1}^*, \ldots, x_1^*)$.

In what follows we apply (i) and (ii) to our example, that is, we determine the vector of investment ratios $x_T^* = 0$, $x_{T-1}^*, x_{T-2}^*, \ldots, x_1^*$, which is the (unique) solution to our problem (see statement (iii); we omit its proof).

The terms of C which contain x_{T-1} are

$$F(qK_{T-2} + x_{T-1}F(K_{T-2}))d^{T-1} - x_{T-1}F(K_{T-2})d^{T-2}.$$

Differentiating this partially with respect to x_{T-1} and setting the result equal to zero gives, with $u = qK_{T-2} + x_{T-1}F(K_{T-2})$,

$$\frac{dF(u)}{du}\frac{\partial u}{\partial x_{T-1}}d^{T-1} - F(K_{T-2})d^{T-2} = 0.$$

Since

$$\frac{\partial u}{\partial x_{T-1}} = F(K_{T-2}) = Y_{T-1} > 0 \quad \text{(see (8.20))}$$

we get

$$F'(qK_{T-2} + x_{T-1}F(K_{T-2})) = \frac{dF(u)}{du} = \frac{1}{d}. \tag{8.43}$$

We assume that the second derivative of the production function F is smaller than zero for all $u > 0$:

$$\frac{d^2F(u)}{du^2} = F''(u), \tag{8.44}$$

This means that the first derivative of F,

$$\frac{dF(u)}{du} = F'(u),$$

is strictly decreasing for all $u > 0$, which is usually the case for macroeconomic production functions, because the marginal returns are strictly decreasing; see the law of diminishing marginal returns in Sect. 7.5. Then F' has an inverse function

$(F')^{-1}$ (see Sect. 3.2), that is, we have, by (8.43),

$$qK_{T-2} + x_{T-1}F(K_{T-2}) = (F')^{-1}(\frac{1}{d}).$$

We note that we have to assume here that d^{-1} belongs to the domain of $(F')^{-1}$ (which is true, e.g., if F is of Cobb-Douglas form $F(K) = aK^b$, $a > 0$ and $b \in]0, 1[$ being constants). From here we get

$$x_{T-1} = \frac{(F')^{-1}(d^{-1}) - qK_{T-2}}{F(K_{T-2})}. \tag{8.45}$$

Note that this unique x_{T-1}, which depends on the equation $K_{T-2} = qK_{T-3} + x_{T-2}F(K_{T-3})$, that is, on x_{T-2} and (via K_{T-3}) on $x_{T-3}, \ldots,$ maximises C because of (8.44).

Now we have to calculate x_{T-2}. To do this we differentiate C with respect to x_{T-2}. The sum of the terms in (8.40), which contain x_{T-2}, is

$$F(K_{T-2})d^{T-2} - x_{T-1}F(K_{T-2})d^{T-2} - x_{T-2}F(K_{T-3})d^{T-3}. \tag{8.46}$$

In view of (8.39), (8.43) and (8.45), this equals

$$\begin{aligned}
&F(qK_{T-3} + x_{T-2}F(K_{T-3}))d^{T-2} - (F')^{-1}(d^{-1})d^{T-2} \\
&\quad + q(qK_{T-3} + x_{T-2}F(K_{T-3}))d^{T-2}) - x_{T-2}F(K_{T-3})d^{T-3}.
\end{aligned}$$

Derivation with respect to x_{T-2} and setting the result equal to zero gives, with $v = qK_{T-3} + x_{T-2}F(K_{T-3})$,

$$\begin{aligned}
&\frac{dF(v)}{dv}\frac{\partial v}{\partial x_{T-2}}d^{T-2} + qK_{T-3}d^{T-2} - F(K_{T-3})d^{T-3} \\
&\quad = F'(qK_{T-3} + x_{T-2}F(K_{T-3}))F(K_{T-3})d^{T-2} \\
&\quad\quad + qK_{T-3}d^{T-2} - F(K_{T-3})d^{T-3} \quad = 0.
\end{aligned}$$

From here we get analogously as above

$$qK_{T-3} + x_{T-2}F(K_{T-3}) = (F')^{-1}(d^{-1} - q)$$

and

$$x_{T-2} = \frac{(F')^{-1}(d^{-1} - q) - qK_{T-3}}{F(K_{T-3})}. \tag{8.47}$$

From (8.44) it follows that this x_{T-2} maximises C. Notice the difference in the arguments of $(F')^{-1}$ in (8.45) and (8.47). Since both d and q are in $]0, 1[$, we have $d^{-1} - q > 0$. As in the case of the argument d^{-1} in (8.45) we assume that F is such that $d^{-1} - q$ is in the domain of $(F')^{-1}$. Obviously, the same process which resulted in (8.47) gives subsequently

$$\begin{aligned} x_{T-3} &= \frac{(F')^{-1}(d^{-1}-q) - qK_{T-4}}{F(K_{T-4})}, \\ &\vdots \\ x_2 &= \frac{(F')^{-1}(d^{-1}-q) - qK_1}{F(K_1)}, \\ x_1 &= \frac{(F')^{-1}(d^{-1}-q) - qK_0}{F(K_0)}. \end{aligned} \tag{8.48}$$

(no change in the arguments of $(F')^{-1}$ from x_{T-2} on). By definition (see (8.19), (8.20), and (8.23))

$$x_t = \frac{I_t}{Y_t} = \frac{I_t}{F(K_{t-1})},$$

$$I_T = K_t - qK_{t-1},$$

thus

$$x_t = \frac{K_t - qK_{t-1}}{F(K_{t-1})}.$$

But, by (8.47) and (8.48),

$$x_t = \frac{(F')^{-1}(d^{-1}-q) - qK_{t-1}}{F(K_{t-1})}.$$

for $t = 1, \ldots, T-2$. So each of $K_1, K_2, \ldots, K_{T-1}$ has to equal $(F')^{-1}(d^{-1} - q)$ in order to get a (local) maximum of C as function of $x_1, x_2, \ldots, x_{T-2}$. We write

$$K^* = (F')^{-1}(d^{-1} - q)$$

and

$$\tilde{K} = (F')^{-1}(d^{-1}).$$

Thus (remember (i), (ii), (iii) in Bellman's principle) the investment ratios

$$x_T^* = 0,$$

$$x_{T-1}^* = \frac{\tilde{K} - qK^*}{F(K^*)}, \tag{8.49}$$

$$x_{T-2}^* = \frac{K^* - qK^*}{F(K^*)} = x_{T-3}^* = \ldots = x_2^*, \tag{8.50}$$

$$x_1^* = \frac{\tilde{K} - qK_0}{F(K_0)} \tag{8.51}$$

are the solution of our problem; they determine the optional investment ratios.

From now on we assume that the depreciation rate $1 - q$ satisfies

$$0 \leq 1 - q \leq 0.15, \tag{8.52}$$

that is, the depreciation factor q fulfils

$$0.85 \leq q \leq 1.$$

We assume further

$$F(K^*) \geq 0.15K^*, \tag{8.53}$$

that is, the optional gross domestic product $F(K^*)$ is greater than or equal to 0.15 times the optimal capital stock K^*. We note in this connection that the relations (8.52) and (8.53) were satisfied every year during the last 30 years by the actual depreciation rates, capital stocks and gross domestic products in the economies of France, Germany, Italy, Japan, UK, and USA.

Obviously, he relations (8.52) and (8.53) yield (see (8.52))

$$0 \leq x_2^* \leq 1, \ \ldots, \ 0 \leq x_{T-2}^* \leq 1$$

for $x_2^*, \ldots, x_{T-2}^*$ as defined in (8.50), and

$$0 \leq x_1^* \leq 1, \ 0 \leq x_{T-1}^* \leq 1$$

for x_1^* (see (8.51)) and x_{T-1}^* (see (8.50)) if the initial capital stock K_0 and the capital stock

$$K_{T-1} = \tilde{K} = (F')^{-1}(d^{-1})$$

are sufficiently close to K^*.

Notice that in our model the optimal investment policy becomes stationary after the first step and changes only in the last two steps $T-1$ and T. In other words, up to the last two steps the optimal investment ratio and the optimal capital stock remain constant, that is, the growth rate of our model economy is zero from the second step on. We point out here that, *in our model, maximisation of consumption is not only consistent with "zero growth"—it yields zero growth.* (This would not necessarily be the case if the production function F depended on time t.)

Let F be the Cobb-Douglas production function (see Sect. 6.9) in one variable,

$$F(K) = cK^{\gamma} \quad (c > 0,\ \gamma \in]0,1[\text{ constants}).$$

Then

$$F'(K) = c\gamma K^{\gamma-1} =: w,$$

that is,

$$K = (F')^{-1}(w) = \left(\frac{w}{c\gamma}\right)^{1/(\gamma-1)}.$$

Hence,

$$K^* = (F')^{-1}(d^{-1} - q) = \left(\frac{d^{-1} - q}{c\gamma}\right)^{1/(\gamma-1)} = \left(\frac{c\gamma d}{1-qd}\right)^{1/(\gamma-1)},$$

$$x_t^* = \frac{K^* - qK^*}{cK^{*\gamma}} = \frac{(1-q)c\gamma d}{c(1-qd)} = \frac{1-q}{1-qd}\gamma d \quad (t = 2, \ldots, T-2).$$

In the case of the USA we take the somewhat realistic values $\gamma = 0.265$, $q = 0.92$, $d = 0.96$ and get $x_t^* = 0.174$. The investment ratio of the USA in 1997 was 17.4 %. This is a quite encouraging result in a simple (but hopefully not too simple) model.

8.4.1 Exercises

1. Let $F : \mathbb{R}_+ \to \mathbb{R}_+$ be the "CES production function in one variable" given by $F(K) = (\beta K^{-\rho} + \delta)^{-1/\rho}$ ($\beta > 0$, $\delta < 0$, $\rho > -1$, $\rho \neq 0$ real constants). Calculate $(F')^{-1}$, the inverse of the derivative of F.
2. Take the function F in Exercise 1 and determine, in the model formulated in this section, the optimal capital stocks $K_1^*, \ldots, K_{T-2}^*, K_{T-1}^*$.
3. Take the function F in Exercise 1 and determine, in the model formulated in this section, the optimal investment ratios $x_{T-2}^*, \ldots, x_2^*$.
4. Calculate the numerical value of the optimal investment ratio x_3^* determined in Exercise 3 for $\beta = 0.265$, $\rho = -0.2$, $q = 0.92$, $d = 0.96$.
5. Determine the limit of x_3^* in Exercise 4 when β, q and d are the same as in Exercise 4, while $\rho \to 0$.

8.4.2 Answers

1. $(F')^{-1}(w) = \left(\frac{(w/\beta)^{\delta/(1+\delta)}-\beta}{\delta}\right)^{1/\delta}$.
2. $K_1^* = \ldots = K_{T-2}^* = \left(\frac{(d\beta/(1-qd))^{\delta/(1+\delta)}-\beta}{\delta}\right)^{1/\delta} = (F')^{-1}(d^{-1}-q)$,
   $K_{T-1}^* = \left(\frac{(d\beta)^{\delta/(1+\delta)}-\beta}{\delta}\right)^{1/\delta} = (F')^{-1}(d^{-1})$.
3. $x_{T-2}^* = \ldots = x_2^* = (1-q)\left(\frac{d\beta}{1-qd}\right)^{1/(1+\delta)}$.
4. $x_3^* = 0.212$.
5. $x_3^* = 0.174$.

8.5 Linear Regression; the "Method of Least Squares"

The following is a fundamental problem in economics and generally in statistics. Suppose that we have good reasons to presume that *a function of m* ("independent") *variables is* of the following form

$$y = b_0 + b_1x_1 + \ldots + b_mx_m \tag{8.54}$$

with real constants $b_0, b_1, \ldots, b_m$.

(The graphs are, of course, straight lines in the case $m = 1$, planes for $m = 2$ and are called hyperplanes for $m > 2$.) For instance, $x_1, x_2, \ldots, x_m$ may be the advertising expenses of a company in m different categories of advertising efforts and y the amount of sales.

However, by errors of observation or experiment or because of variation of circumstances or (in the above example) because expenses and revenue are measured in consecutive time intervals (of equal length) and external or random occurrences may influence the results of measurement, n observations

$$(x_{11}, \ldots, x_{m1}, y_1), \ldots, (x_{1n}, \ldots, x_{mn}, y_n) \tag{8.55}$$

do not exactly satisfy (8.54). The "cloud of points" representing our observations (for an example in the case $m = 1$, $n = 31$ see Fig. 8.8) rather seems to make an *affine* (in the older terminology "linear") *approximation* possible. Looking for the "best" linear approximation in the sense to be explained below is the object of "linear regression".

Rather than fitting (8.54), the data (8.55) satisfy

$$y_k = b_0 + b_1x_{1k}u_1 + \ldots + b_mx_{mk}u_m \quad (k = 1, \ldots, n), \tag{8.56}$$

where u_k is the deviation from the theoretical (and, eventually, "optimal") value given by (8.54). Equivalently, substituting the observations

$$(x_{1k}, \ldots, x_{mk})$$

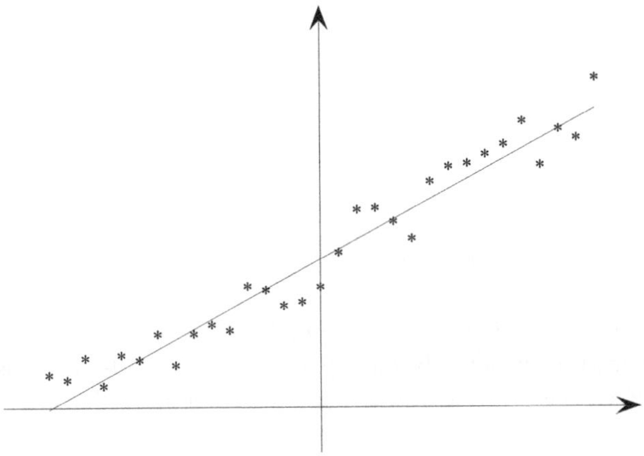

Fig. 8.8 The "cloud" of 31 points (marked by asterisks) are, as can be seen, approximated by the straight line $\{(x, y) \mid y = 2 + \frac{1}{2}x\}$. Linear regression gives the optimal approximating straight line as $\{(x, y) \mid y = 1.66 + 0.56x\}$; see also (8.67)

into the right hand side of (8.54) gives, instead of y_k,

$$y^*{}_k = y_k - u_k \qquad (k = 1, \ldots, n).$$

At least since Carl Friedrich Gauss (1777–1855) the affine function (or the $(m+1)$-tuple $b_0, b_1, \ldots, b_m$) in (8.54) is considered to be the "best" approximation of the data (8.55) if *the* sum of the squares of the deviations

$$\sum_{k=1}^{n} u_k^2 = \sum_{k=1}^{n} (y_k - y^*{}_k)^2 = \sum_{k=1}^{n} (y_k - b_0 - b_1 x_{1k} - \ldots - b_m x_{mk})^2 \tag{8.57}$$

called variance is minimal. This is called the "method of least squares". Gauss certainly gave good reasons for this choice. One is that this gives equal treatment to positive and negative deviations. Another is that in the simplest case $m = 0$, when the approximation is to be done by a *constant* b_0, that is, (8.54) reduces to

$$y = b_0$$

and (8.56) to

$$y_k = b_0 + u_k \qquad (k = 1, \ldots, n)$$

then the method of least squares requires to find that $b_0 = t$ for which

$$\sum_{k=1}^{n} u_k^2 = \sum_{k=1}^{n} (y_k - b_0)^2 = \sum_{k=1}^{n} (y_k - t)^2 \tag{8.58}$$

is minimal. By what we learned in Sect. 8.3 this can be only when

$$0 = \frac{d}{dt}\sum_{k=1}^{n}(y_k - t)^2 = -2\sum_{k=1}^{n}(y_k - t) = 2nt - 2\sum_{k=1}^{n} y_k$$

that is, when

$$b_0 = t = (y_1 + \ldots + y_n)/n,$$

the constant function value is the arithmetic mean of the observed values, very reasonable indeed. (For actually having $t = (y_1 + \ldots + y_n)/n$ as minimum of $\sum_{k=1}^{n}(y_k - t)^2$ we have also to check that the second derivative is positive there:

$$\frac{d^2}{dt^2}\sum_{k=1}^{n}(y_k - t)^2 = \frac{d}{dt}(2nt - 2\sum_{k=1}^{n} y_k) = 2n > 0,$$

which holds trivially since $n \in \mathbb{N}$). One can show that essentially (up to a constant) *only the square has this property*.

In general, we can write (8.56) in the simpler form

$$y_k = \sum_{j=0}^{m} b_j x_{jk} + u_k \qquad (k = 1, \ldots, n)$$

by taking

$$x_{01} = \ldots = x_{0n} = 1.$$

With the vectors and matrix

$$\mathbf{y} = (y_1, \ldots, y_n), \qquad \mathbf{u} = (u_1, \ldots, u_n), \qquad \mathbf{b} = (b_0, b_1, \ldots, b_m),$$

$$\mathbf{X} = \begin{pmatrix} x_{01} & \cdots & x_{0n} \\ x_{11} & \cdots & x_{1n} \\ \vdots & & \vdots \\ x_{m1} & \cdots & x_{mn} \end{pmatrix} = \begin{pmatrix} 1 & \cdots & 1 \\ x_{11} & \cdots & x_{1n} \\ \vdots & & \vdots \\ x_{m1} & \cdots & x_{mn} \end{pmatrix}$$

and, later, $\mathbf{t} = (t_0, t_1, \ldots, t_m)$ we can write the above as

$$\mathbf{y} = \mathbf{bX} + \mathbf{u}.$$

We have to find the minimal value of the "variance"

$$\sum_{k=1}^{n} u_k^2 = \mathbf{u} \cdot \mathbf{u} = \mathbf{u}\mathbf{u}^T$$

(compare Sects. 5.4, 6.8 and 6.9), that is, of

$$\begin{aligned} F(\mathbf{b}) &:= (\mathbf{y} - \mathbf{b}\mathbf{X})(\mathbf{y} - \mathbf{b}\mathbf{X})^T = (\mathbf{y} - \mathbf{b}\mathbf{X})(\mathbf{y}^T - \mathbf{X}^T\mathbf{b}^T) \\ &= \mathbf{y}\mathbf{y}^T - \mathbf{b}\mathbf{X}\mathbf{y}^T - \mathbf{y}\mathbf{X}^T\mathbf{b}^T + \mathbf{b}\mathbf{X}\mathbf{X}^T\mathbf{b}^T. \end{aligned}$$

However, $\mathbf{b}\mathbf{X}\mathbf{y}^T$ *is scalar* (a $1 \times (m+1)$ matrix times a $(m+1) \times n$ matrix times a $n \times 1$ matrix = a 1×1 matrix) and its transpose is $\mathbf{y}\mathbf{X}^T\mathbf{b}^T$; furthermore *the transpose of a scalar is itself*, so $\mathbf{b}\mathbf{X}\mathbf{y}^T = \mathbf{y}\mathbf{X}^T\mathbf{b}^T$. Thus we look for the minimum of

$$F(\mathbf{b}) = \mathbf{y}\mathbf{y}^T - 2\mathbf{b}\mathbf{X}\mathbf{y}^T + \mathbf{b}\mathbf{X}\mathbf{X}^T\mathbf{b}^T,$$

that is, for that $\mathbf{b}$ which, makes the variance minimal.

As in (8.58), we emphasise that $\mathbf{b}$ is the unknown (variable) vector by writing $\mathbf{t} = \mathbf{b}$. So we want to determine the minimum of

$$F(\mathbf{t}) = \mathbf{y}\mathbf{y}^T - 2\mathbf{t}\mathbf{X}\mathbf{y}^T + \mathbf{t}\mathbf{X}\mathbf{X}^T\mathbf{t}^T. \tag{8.59}$$

As we know from Sect. 6.9, there *may* be a local minimum only where

$$\nabla F(\mathbf{t}) = \left(\frac{\partial F}{\partial t_0}(\mathbf{t}), \frac{\partial F}{\partial t_1}(\mathbf{t}), \ldots, \frac{\partial F}{\partial t_m}(\mathbf{t})\right) = \mathbf{0}.$$

Doing the calculations (for instance by determining $F(\mathbf{t})$ as a function of the $m+1$ variables $t_0, t_1, \ldots, t_m$) we get

$$\nabla F(\mathbf{t}) = \mathbf{F}'(\mathbf{t}) = -2\mathbf{X}\mathbf{y}^T + 2\mathbf{X}\mathbf{X}^T\mathbf{t}^T, \tag{8.60}$$

so that F *may have a local minimum where*

$$2\mathbf{X}\mathbf{X}^T\mathbf{t}^T = 2\mathbf{X}\mathbf{y}^T. \tag{8.61}$$

After cancelling 2, we take the transpose of both sides:

$$(\mathbf{X}\mathbf{y}^T)^T = ((\mathbf{X}\mathbf{X}^T)\mathbf{t}^T)^T = (\mathbf{t}^T)^T(\mathbf{X}\mathbf{X}^T)^T,$$

that is,

$$\mathbf{y}\mathbf{X}^T = \mathbf{t}(\mathbf{X}^T)^T\mathbf{X}^T = \mathbf{t}\mathbf{X}\mathbf{X}^T.$$

If the inverse of the matrix $\mathbf{XX}^T$ *exists*, that is, if $\det(\mathbf{XX}^T) \neq 0$ (compare Sect. 4.7), then this equation has a unique solution and F *can have a local minimum only where*

$$\mathbf{t} = \mathbf{yX}^T(\mathbf{XX}^T)^{-1} =: \hat{\mathbf{t}}. \tag{8.62}$$

Exactly *under the condition that the inverse of* $\mathbf{XX}^T$ *exists, the function F given by (8.59) has a strict minimum at* $\hat{\mathbf{t}}$. Indeed, from (8.60), the Hessian matrix is

$$\mathbf{F}''(\mathbf{t}) = 2\mathbf{XX}^T \quad \text{for all} \quad \mathbf{t}. \tag{8.63}$$

This is *positive semidefinite* because, for all $\mathbf{v} = (v_0, v_1, \ldots, v_m) \in \mathbb{R}^{m+1}$,

$$\begin{aligned}\mathbf{vF}''(\mathbf{t})\mathbf{v}^T &= \mathbf{v}(2\mathbf{XX}^T)\mathbf{v}^T = 2(\mathbf{vX})(\mathbf{vX})^T \\ &= 2(\mathbf{vX}) \cdot (\mathbf{vX}) = 2\,|\mathbf{vX}|^2 \geq 0.\end{aligned}$$

As positive semidefinite form it is *positive definite if* 0 *is not an eigenvalue of* $\mathbf{F}''(\mathbf{t})$. *That is* indeed *the case*, since 0 is an eigenvalue of $\mathbf{A}$ exactly when (compare Sect. 6.8)

$$0 = \det \begin{pmatrix} a_{11} - 0 & a_{12} & \cdots & a_{1r} \\ a_{21} & a_{12} - 0 & \cdots & a_{2r} \\ \vdots & \vdots & \ddots & \vdots \\ a_{r1} - 0 & a_{r2} & \cdots & a_{rr} \end{pmatrix} = \det \mathbf{A}$$

(here $r = m + 1$) and, by supposition, $\det \mathbf{F}''(\mathbf{t}) = 2\det(\mathbf{XX}^T) \neq 0$. So $\mathbf{F}''(t)$ *is positive definite* and F *has a strict local minimum at the* $\hat{\mathbf{t}}$ *nm* 0 (compare Sect. 6.9). However, see (8.63), $\mathbf{F}''(\mathbf{t})$ is independent of $\mathbf{t}$ so it is *positive definite for all* $\mathbf{t}$, therefore (Sect. 6.8) *everywhere strictly convex from below* and so its local minimum $\hat{\mathbf{t}}$ *in (8.62) is a strict global minimum.*

We get the *value of the "minimal variance"* by putting (8.62) into (8.59). First we note that, because of $[(\mathbf{XX}^T)^{-1}]^T = [(\mathbf{xX}^T)^T]^{-1} = (\mathbf{XX}^T)^{-1}$, from (8.62) we get

$$\hat{\mathbf{t}}^T = (\mathbf{xX}^T)^{-1}\mathbf{Xy}^T.$$

So the *minimal variance is*

$$\begin{aligned}F(\hat{\mathbf{t}}) &= F(\mathbf{yX}^T(\mathbf{XX}^T)^{-1}) \\ &= \mathbf{yy}^T - 2\mathbf{yX}^T(\mathbf{XX}^T)^{-1}\mathbf{Xy}^T \\ &\quad + \mathbf{yX}^T(\mathbf{XX}^T)^{-1}\mathbf{XX}^T(\mathbf{XX}^T)^{-1}\mathbf{Xy}^T \\ &= \mathbf{yy}^T - \mathbf{yx}^T(\mathbf{XX}^T)^{-1}\mathbf{Xy}^T.\end{aligned} \tag{8.64}$$

Of course, $\mathbf{y}\mathbf{y}^T = \mathbf{y} \circ \mathbf{y} = |\mathbf{y}|^2$ so, if not all observed values $y_1, \ldots, y_n$ are 0—a rather natural supposition—then we may divide by $\mathbf{y}\mathbf{y}^T$:

$$\frac{F(\hat{\mathbf{t}})}{\mathbf{y}\mathbf{y}^T} = \frac{F(\mathbf{y}\mathbf{X}^T(\mathbf{X}\mathbf{X}^T)^{-1})}{\mathbf{y}\mathbf{y}^T} = 1 - \frac{\mathbf{y}\mathbf{X}^T(\mathbf{X}\mathbf{X}^T)^{-1}\mathbf{X}\mathbf{y}^T}{\mathbf{y}\mathbf{y}^T} = 1 - R^2. \tag{8.65}$$

$$R := \left(\frac{\mathbf{y}\mathbf{X}^T(\mathbf{X}\mathbf{X}^T)^{-1}\mathbf{X}\mathbf{y}^T}{\mathbf{y}\mathbf{y}^T}\right)^{\frac{1}{2}} \tag{8.66}$$

is the "sum of square coefficients" which, as can be shown, satisfies $0 \leq R \leq 1$. The closer R is to 1 the less the "dispersion" of the "cloud of points" (8.55) is and the closer is (8.65) to 0.

As an example where the inverse of $\mathbf{X}\mathbf{X}^T$ always exists, we examine the case $m = 1$ in some detail. Here

$$\mathbf{X} = \begin{pmatrix} 1 & \cdots & 1 \\ x_1 & \cdots & x_n \end{pmatrix},$$

where we wrote, for simplicity, $x_1, \ldots, x_n$ in place of $x_{11}, \ldots, x_{1n}$. Thus

$$\mathbf{X}\mathbf{X}^T = \begin{pmatrix} 1 & \cdots & 1 \\ x_1 & \cdots & x_n \end{pmatrix} \begin{pmatrix} 1 & x_1 \\ \vdots & \vdots \\ 1 & x_n \end{pmatrix} = \begin{pmatrix} n & x_1 + \ldots x_n \\ x_1 + \ldots + x_n & x_1^2 + \ldots + x_n^2 \end{pmatrix},$$

so, denoting the arithmetic mean by $\bar{x} = (x_1 + \ldots + x_n)/n$,

$$\begin{aligned} \det(\mathbf{X}\mathbf{X}^T) &= n(x_1^2 + \ldots + x_n^2) - (x_1 + \ldots + x_n)^2 \\ &= n(x_1^2 + \ldots + x_n^2) - 2n^2\bar{x}^2 + n^2\bar{x}^2 \\ &= n(x_1^2 + \ldots x_n^2) - 2n(x_1 + \ldots + x_n)\bar{x} + n^2\bar{x}^2 \\ &= n(x_1^2 + \ldots + x_n^2 - 2x_1\bar{x} - \ldots - 2x_n\bar{x} + n\bar{x}^2) \\ &= n((x_1 - \bar{x})^2 + \ldots + (x_n - \bar{x})^2) > 0 \end{aligned}$$

except if $x_1 = \ldots = x_n = \bar{x}$. The latter case is uninteresting since in practice one wants observations at more than one place for the now single ($m = 1$) independent variables. So $\det(\mathbf{X}\mathbf{X}^T) \neq 0$ and thus (Sect. 4.7) the inverse of $\mathbf{X}\mathbf{X}^T$ indeed exists.

In the case drawn in Fig. 8.8, $m = 1$, $n = 31$, we have

$$\mathbf{X} = \begin{pmatrix} 1&1&1&1&1&1&1&1&1&1&1&1&1&1&1&1&1&1&1&1&1 \\ 2&2&3&3&4&4&5&5&6&6&7&7&7&8&8&8&9&9&10&10 \end{pmatrix},$$

$$\mathbf{y} = (3, 4, 2, 3, 3, 5, 4, 5, 3, 6, 4, 5, 7, 6, 7, 8, 5, 8, 6, 8).$$

Therefore

$$\mathbf{y}\mathbf{X}^T = (102, 697), \qquad \mathbf{X}\mathbf{X}^T = \begin{pmatrix} 20 & 123 \\ 123 & 881 \end{pmatrix},$$

$$(\mathbf{X}\mathbf{X}^T)^{-1} = \frac{1}{2491}\begin{pmatrix} 881 & -123 \\ -123 & 20 \end{pmatrix} \approx \begin{pmatrix} 0.35367 & -0.04938 \\ -0.04938 & 0.00803 \end{pmatrix}.$$

Thus, from (8.62) the function F is minimal at

$$\begin{aligned} \mathbf{b} = \hat{\mathbf{t}} &= (102, 697)\begin{pmatrix} 881 & -123 \\ -123 & 20 \end{pmatrix}\frac{1}{2491} \\ &= \frac{1}{2491}(4131, 1394) \approx (1.65837, 0.55961) \end{aligned} \tag{8.67}$$

and the *minimal variance* (8.64) will be the value of F at the $\mathbf{b} = \hat{\mathbf{t}}$ calculated in (8.67):

$$F(\hat{\mathbf{t}}) = \mathbf{y}\mathbf{y}^T - \hat{\mathbf{t}}\mathbf{X}\mathbf{y}^T = 586 - \frac{1392980}{2491} \approx 26.8$$

while, from (8.66) and (8.62),

$$R = \left(\frac{\hat{\mathbf{t}}\mathbf{X}\mathbf{y}^T}{\mathbf{y}\mathbf{y}^T}\right)^{\frac{1}{2}} \approx 0.9769,$$

is quite close to 1 so the dispersion is small. Indeed

$$\frac{F(\hat{\mathbf{t}})}{\mathbf{y}\mathbf{y}^T} \approx \frac{26.8}{586} \approx 0.0457.$$

8.5.1 Exercises

Consider the case, where in (8.56)

$m = 1, \quad k = 5, \quad (x_{11}, y_1) = (2, 4), \quad (x_{12}, y_2) = 5, 6,$

$(x_{13}, y_3) = (3, 5), \quad (x_{14}, y_4) = (4, 5), \quad (x_{15}, y_5) = (6, 10).$

For $\mathbf{X}$, $\mathbf{y}$, $\mathbf{b} = \mathbf{t}$, $F(\mathbf{t})$ as defined in this section determine:

1. (a) $\mathbf{y}\mathbf{X}^T$, (b) $\mathbf{X}\mathbf{y}^T$,
2. (a) $\mathbf{X}\mathbf{X}^T$, (b) $(\mathbf{X}\mathbf{X}^T)^{-1}$,
3. (a) $\mathbf{b} = \hat{\mathbf{t}}$, (b) $F(\hat{\mathbf{t}})$,
4. (a) $F(\hat{\mathbf{t}})/(\mathbf{y}\mathbf{y}^T)$, (b) $R = \left((\hat{\mathbf{t}}\mathbf{X}\mathbf{y}^T)/(\mathbf{y}\mathbf{y}^T)\right)^{\frac{1}{2}}$.

Consider now the case, where in (8.56) $m = 3$, $k = 8$ and

$(x_{11}, x_{21}, x_{31}, y_1) = (0, 0, 1, 19)$, $(x_{12}, x_{22}, x_{32}, y_2) = (1, 4, 1, 17)$,
$(x_{13}, x_{23}, x_{33}, y_3) = (2, 3, 1, 31)$, $(x_{14}, x_{24}, x_{34}, y_4) = (3, 1, 0, 30)$,
$(x_{15}, x_{25}, x_{35}, y_5) = (4, 0, 0, 16)$, $(x_{16}, x_{26}, x_{36}, y_6) = (0, 0, 0, 16)$,
$(x_{17}, x_{27}, x_{37}, y_7) = (3, 2, 0, 22)$, $(x_{18}, x_{28}, x_{38}, y_8) = (1, 1, 0, 21)$.

For the corresponding $\mathbf{X}$, $\mathbf{y}$,$\mathbf{b} = \hat{\mathbf{t}}$ determine

5. (a) $\mathbf{y}\mathbf{X}^T$, (b) $\mathbf{X}\mathbf{X}^T$,
6. (a) $(\mathbf{X}\mathbf{X}^T)^{-1}$, (b) $\mathbf{b} = \hat{\mathbf{t}}$.

8.5.2 Answers

1. (a) $\mathbf{y}\mathbf{X}^T = (4, 6, 5, 5, 10)\begin{pmatrix} 1 & 2 \\ 1 & 5 \\ 1 & 3 \\ 1 & 4 \\ 1 & 6 \end{pmatrix} = (30, 133)$, (b) $\mathbf{X}\mathbf{y}^T = \begin{pmatrix} 30 \\ 133 \end{pmatrix}$.

2. (a) $\mathbf{X}\mathbf{X}^T - \begin{pmatrix} 5 & 20 \\ 20 & 90 \end{pmatrix}$,

   (b) $(\mathbf{X}\mathbf{X}^T)^{-1} = \frac{1}{50}\begin{pmatrix} 90 & -20 \\ -20 & 5 \end{pmatrix} = \begin{pmatrix} 1.8 & -0.4 \\ -0.4 & 0.1 \end{pmatrix}$.

3. (a) $\mathbf{b} = \hat{\mathbf{t}} = \mathbf{y}\mathbf{X}^T(\mathbf{X}\mathbf{X}^T)^{-1} = (30, 133)\begin{pmatrix} 1.8 & -0.4 \\ -0.4 & 0.1 \end{pmatrix} = (0.8, 1.3)$.

4. (a) $\dfrac{F(\hat{\mathbf{t}})}{\mathbf{y}\mathbf{y}^T} = \dfrac{\mathbf{y}\mathbf{y}^T - \mathbf{y}\mathbf{X}^T(\mathbf{X}\mathbf{X}^T)^{-1}\mathbf{X}\mathbf{y}^T}{\mathbf{y}\mathbf{y}^T} = 1 - R^2$

   $= 1 - \dfrac{(0.8, 1.3)(30, 133)^T}{202} = 1 - \dfrac{24 + 172.9}{202}$

   $= 1 - \dfrac{196.9}{202} \approx 0.025$,

   (b) $R = (\frac{196.9}{202})^{\frac{1}{2}} = \sqrt{0.974752475} \approx 0.987$.

5. (a) $\mathbf{y}\mathbf{X}^T = (72, 320, 256, 67)$,

   (b) $\mathbf{X}\mathbf{X}^T = \begin{pmatrix} 8 & 14 & 11 & 3 \\ 14 & 40 & 20 & 3 \\ 11 & 20 & 31 & 7 \\ 3 & 3 & 7 & 3 \end{pmatrix}$.

6. (a) $(\mathbf{XX}^T)^{-1} = \frac{1}{1937}\begin{pmatrix} 1121 & -328 & -15 & -758 \\ -328 & 172 & -63 & 303 \\ -15 & -63 & 192 & -370 \\ -758 & 303 & -370 & 1964 \end{pmatrix}$,

(b) $\mathbf{b} = \hat{\mathbf{t}} = \mathbf{y}\mathbf{X}^T(\mathbf{XX}^T)^{-1} = \frac{1}{1937}(33226, 2797, 1622, 3452)$.

8.6 Extrema of an Objective Function Under Equality Constraints

In Sect. 8.5 (linear optimisation) we presented methods for determining extrema, that is maxima and minima of linear functions of several variables under linear conditions (constraints). In Sect. 8.3 we considered maxima and minima of nonlinear functions of several variables without constraints other than rather arbitrary limitations of the domain. This was so both when the maxima or minima were local (on neighbourhoods) or global (usually on a predetermined domain). For linear optimisation problems the constraints (equations or inequalities) themselves determined or at least "constrained" the domains. These constraints often came from economic or technical limitations. The function, the maximum or minimum (extremum) of which we are interested in, is called the *objective function*. There and in Sects. 8.7, 8.8 and 8.9 we deal with only one, Sect. 8.10 with several objective functions.

The present section and Sects. 8.7, 8.8 and 8.9 are about *nonlinear optimisation*, that is, at least one of the sets {objective function, constraints} is not linear or affine. Here too, careful analyses of the constraints is needed to determine the *domain*, a task neither easy nor always successful. Note that the domain needs not be a subset of $\mathbb{R}^n$. For instance in Sect. 9.4 (where there will be several objective functions) the objective functions depend on *strategies*, not vectors in $\mathbb{R}^n$. In most optimisation problems in economics or engineering, however, the domains are subsets of $\mathbb{R}^n$ of even of $\mathbb{R}^n_+$.

Example 1 Ms. A intends to spend a budget B of exactly 360$ for two goods 1 and 2. The prices of the goods are $p_1 = 3\$$ and $p_2 = 4\$$, respectively. Let x_1 and x_2 be the quantities of the goods. Then the so-called *budget equation* is

$$3x_1 + 4x_2 = 360.$$

Ms. A's *utility function* $U : \mathbb{R}^2_+ \to \mathbb{R}_+$ is given by

$$U(x_1, x_2) = x_1^{1/4} x_2^{1/2}.$$

(continued)

Which quantities $\widehat{x_1}$, $\widehat{x_2}$ has she to buy in order to maximise her utility? Since U is nonlinear our problem is a *nonlinear optimisation problem*. The constraint $3x_1 + 4x_2 = 360$ is affine. Note that the domain of our problem is the line segment

$$\{(x_1, x_2) \mid x_2 = 90 - \tfrac{3}{4}x_1, 0 \le x_1 \le 120\}$$

rather than $\mathbb{R}^2_+$. We have to find the maximum of U on this segment. To do this we insert $x_2 = 90 - \frac{3}{4}x_1$ into our objective function U:

$$U(x_1, 90 - \frac{3}{4}x_1) = x_1^{1/4}(90 - \frac{3}{4}x_1)^{1/2}.$$

So we get the function $u : [0, 120] \to \mathbb{R}_+$ of only one variable $x_1 \in [0, 120]$ given by

$$u(x_1) = x_1^{1/4}(90 - \frac{3}{4}x_1)^{1/2}.$$

This function is continuous on the closed finite interval $[0, 120]$, that is (see Sect. 6.3), there exist points on $[0, 120]$ at which u is maximal (as well as points at which u is minimal). Since

$$u(0) = u(120) = 0, \qquad u(1) = \sqrt{90 - \frac{3}{4}} > 0,$$

the maximum of u is at some point(s) $\widehat{x_1}$ in the interior of the set of the critical points of u, that is, of the points x_1 satisfying $du(x_1)/dx_1 = 0$. We determine these points:

$$\begin{aligned}\frac{du(x_1)}{dx_1} &= \frac{d(x_1^{1/4}(90 - \frac{3}{4}x_1)^{1/2})}{dx_1}\\ &= \frac{1}{4}x_1^{-3/4}(90 - \frac{3}{4}x_1)^{1/2} + \frac{1}{2}x_1^{1/4}(90 - \frac{3}{4}x_1)^{-1/2}(-\frac{3}{4}) = 0.\end{aligned}$$

Multiplying this by $4x_1^{3/4}(90 - \frac{3}{4}x_1)^{1/2}$ gives

$$(90 - \frac{3}{4}x_1) - \frac{3}{2}x_1 = 0,$$

(continued)

that is, $\widehat{x_1} = 40$ is the only point x_1 that satisfies $du(x_1)/dx_1 = 0$. One can show (prove it!) that

$$\frac{d^2u}{dx_1^2}(40) = -1350 \cdot 40^{-7/4}60^{-3/2} < 0,$$

i.e. (see Sect. 6.3), the (only) maximum of u is at $\widehat{x_1} = 40$. We insert this into the budget equation $3x_1 + 4x_2 = 360$ and get $\widehat{x_2} = 60$. So the (unique) solution point of our problem is $(\widehat{x_1}, \widehat{x_2}) = (40, 60)$ and the maximum utility value is $U(\widehat{x_1}, \widehat{x_2}) = \widehat{x_1}^{1/4}\widehat{x_2}^{1/2} = 40^{1/4} \cdot 60^{1/2} \approx 19.48$.

Example 2 A generalisation of the problem dealt with in Example 1 is, with the (nonlinear) utility function $U : \mathbb{R}^n_+ \to \mathbb{R}, n > 2$, the prices $p1, \ldots, p_n$, the quantities $x_1, \ldots, x_n$ of n goods, and the budget b:

$$\textit{Maximise} \quad U(x_1, \ldots, x_n)$$

under the condition (budget equation)

$$p_1x_1 + \ldots p_nx_n = b.$$

In this case the solution method applied to our problem in Exercise 1, that is, the method of inserting the constraint into the objective function, yields difficulties that generally increase with the number $n > 2$ of goods under consideration.

Example 3 Further difficulties may arise when the number m of the (equality) constraints is greater than one. Forgetting about utility functions and budget equations, more general (and frequently more difficult) problems than those in Examples 1 and 2 are of the form:

$$\textit{Maximise (or minimise)} \quad F(x_1, \ldots, x_n) \tag{8.68}$$

under the conditions (constraints)

$$\begin{aligned} g_1(x_1, \ldots, x_n) &= c_1, \\ &\vdots \\ g_m(x_1, \ldots, x_n) &= c_m, \end{aligned} \tag{8.69}$$

(continued)

where $F : \mathbb{R}^n \to \mathbb{R}$ (objective function), $g_1 : \mathbb{R}^n \to \mathbb{R}$, ..., $g_m : \mathbb{R}^n \to \mathbb{R}$, $x_1, \ldots, x_n$ are real variables, and $c_1, \ldots c_m$ real constants.

Frequently the (common) domain of $F, g_1, \ldots, g_m$ is rather a true subset of $\mathbb{R}^n$ than the $\mathbb{R}^n$ itself.

Unfortunately, there are no general methods for the solution of optimisation problems of this kind when at least one of the functions $F, g_1, \ldots, g_m$ is *nonlinear* and $n > 2$, $m > 1$. Note that this is in contrast to the problems of *linear* optimisation, where the simplex algorithm is a general solution method. We can formulate, however, conditions, either necessary or necessary and sufficient, for a vector $(x_1, \ldots, x_n)$ to maximise (or minimise) $F(x_1, \ldots, x_n)$ under the constraints (8.69).

These conditions can be formulated with aid of the *Lagrange* (Joseph Louis Lagrange (1736–1813)) *multipliers* and the *Lagrange function*. This is a function $L : \mathbb{R}^n \times \mathbb{R}^m \to \mathbb{R}$ or $L : D \times \mathbb{R}^m \to \mathbb{R}$, where D is a domain in $\mathbb{R}^n$, defined by

$$\begin{aligned} &L(x_1, \ldots, x_n, u_1, \ldots, u_m) \\ &\quad = F(x_1, \ldots, x_n) + u_1 \cdot (c_1 - g_1(x_1, \ldots, x_n)) \\ &\quad + \ldots + u_m \cdot (c_m - g_m(x_1, \ldots, x_n)), \end{aligned} \tag{8.70}$$

where F is the objective function and $c_1 - g_1(x_1, \ldots, x_n) = 0$, ..., $c_m - g_m(x_1, \ldots, x_n) = 0$ are the constraints of our nonlinear optimisation problem. The variables $u_1, \ldots, u_m$ are called *Lagrange multipliers*.

We mention without proof: *If the derivatives of the functions $F, g_1, \ldots, g_m$ exist and are continuous, if $\hat{\mathbf{x}} = (\widehat{x_1}, \ldots, \widehat{x_n})$ is a solution of the optimisation problem (8.68), (8.69) and if the rank of the Jacobian matrix*

$$\begin{pmatrix} \frac{\partial g_1}{\partial x_1}(\hat{\mathbf{x}}) & \cdots & \frac{\partial g_1}{\partial x_n}(\hat{\mathbf{x}}) \\ \vdots & & \vdots \\ \frac{\partial g_m}{\partial x_1}(\hat{\mathbf{x}}) & \cdots & \frac{\partial g_m}{\partial x_n}(\hat{\mathbf{x}}) \end{pmatrix}$$

is m ($m < n$) then there exists a vector $\hat{\mathbf{u}} = (\widehat{u_1}, \ldots, \widehat{u_m})$ of Lagrange multipliers such that the vector

$$(\hat{x}, \hat{\mathbf{u}}) = (\widehat{x_1}, \ldots, \widehat{x_n}, \widehat{u_1}, \ldots, \widehat{u_m})$$

(continued)

is a critical point of the Lagrange function (8.70), that is,

$$\frac{\partial L}{\partial x_1}(\hat{\mathbf{x}}, \hat{\mathbf{u}}) = \frac{\partial F}{\partial x_1}(\hat{\mathbf{x}}) + \widehat{u_1}\frac{\partial g_1}{\partial x_1}(\hat{\mathbf{x}}) + \ldots + \widehat{u_m}\frac{\partial g_m}{\partial x_1}(\hat{\mathbf{x}}) = 0,$$

$$\vdots$$

$$\frac{\partial L}{\partial x_n}(\hat{\mathbf{x}}, \hat{\mathbf{u}}) = \frac{\partial F}{\partial x_n}(\hat{\mathbf{x}}) + \widehat{u_1}\frac{\partial g_1}{\partial x_n}(\hat{\mathbf{x}}) + \ldots + \widehat{u_m}\frac{\partial g_m}{\partial x_n}(\hat{\mathbf{x}}) = 0,$$

$$\frac{\partial L}{\partial u_1}(\hat{\mathbf{x}}, \hat{\mathbf{u}}) = c_1 - g_1(\hat{\mathbf{x}}) = 0, \ldots \frac{\partial L}{\partial u_m}(\hat{\mathbf{x}}, \hat{\mathbf{u}}) = c_m - g_m(\hat{\mathbf{x}}) = 0.$$

Note that the last line says that $\hat{\mathbf{x}}$ satisfies the constraints (8.68). (This is not surprising since we *assumed* that $\hat{\mathbf{x}}$ is a solution of our problem). Note further that $\hat{\mathbf{x}}$ maximises F under the constraints (8.69), but that $(\hat{\mathbf{x}}, \hat{\mathbf{u}})$ does *not* maximise L; see, in this connection, Sect. 8.8.

Let us apply this to Example 1. The Lagrange function of the problem

$$\begin{array}{ll} \text{maximise} & U(x_1, x_2) = x_1^{1/4} x_2^{1/2} \\ \text{under the constraint} & 3x_1 + 4x_2 = 360 \end{array} \tag{8.71}$$

is

$$L(x_1, x_2, u) = x_1^{1/4} x_2^{1/2} + u \cdot (360 - 3x_1 + 4x_2). \tag{8.72}$$

The critical points of L are the solutions of

$$\frac{\partial L}{\partial x_1}(x_1, x_2, u) = \frac{1}{4} x_1^{-3/4} x_2^{1/2} - 3u = 0,$$

$$\frac{\partial L}{\partial x_2}(x_1, x_2, u) = \frac{1}{2} x_1^{1/4} x_2^{-1/2} - 4u = 0,$$

$$\frac{\partial L}{\partial u}(x_1, x_2, u) = 360 - 3x_1 - 4x_2 = 0.$$

The (only) solution of this system of nonlinear equations is

$$(\widehat{x_1}, \widehat{x_2}, \hat{u}) = (40, 60, \frac{1}{8} 40^{1/4} 60^{-1/2}). \tag{8.73}$$

(continued)

If we did not know (from Example 1) that $(\widehat{x_1}, \widehat{x_2}) = (40, 60)$ is the unique solution point of our problem, we would need a criterion for deciding whether there is maximum or minimum at this point.

We present such a criterion for the general case of our optimisation problem, that is, for the problem (8.68) and (8.69). Let the Lagrange function L (see (8.70)) of this problem be differentiable. The so-called *bordered Hessian matrix* of L is the $(m+n) \times (m+n)$ matrix

$$\mathcal{H} := \begin{pmatrix} 0 & \dots & 0 & \frac{\partial g_1(\mathbf{x})}{\partial x_1} & \dots & \frac{\partial g_1(\mathbf{x})}{\partial x_n} \\ \vdots & & \vdots & \vdots & & \vdots \\ 0 & \dots & 0 & \frac{\partial g_m(\mathbf{x})}{\partial x_1} & \dots & \frac{\partial g_m(\mathbf{x})}{\partial x_n} \\ \frac{\partial g_1(\mathbf{x})}{\partial x_1} & \dots & \frac{\partial g_m(\mathbf{x})}{\partial x_1} & \frac{\partial^2 L(\mathbf{x})}{\partial x_1 \partial x_1} & \dots & \frac{\partial^2 L(\mathbf{x})}{\partial x_1 \partial x_n} \\ \vdots & & \vdots & \vdots & & \vdots \\ \frac{\partial g_1(\mathbf{x})}{\partial x_n} & \dots & \frac{\partial g_m(\mathbf{x})}{\partial x_n} & \frac{\partial^2 L(\mathbf{x})}{\partial x_n \partial x_1} & \dots & \frac{\partial^2 L(\mathbf{x})}{\partial x_n \partial x_n} \end{pmatrix} \tag{8.74}$$

Compare this to the Hessian matrix in Sect. 8.2. Obviously the bordered Hessian of the Lagrange function (8.72) from Example 1 is

$$\mathcal{H}_3 := \begin{pmatrix} 0 & 3 & 4 \\ 3 & -\frac{3}{16}x_1^{-7/4}x_2^{1/2} & \frac{1}{8}x_1^{-3/4}x_2^{-1/2} \\ 4 & \frac{1}{8}x_1^{-3/4}x_2^{-1/2} & -\frac{1}{4}x_1^{1/4}x_2^{-3/2} \end{pmatrix}. \tag{8.75}$$

Let D_j $(j = 1, \dots, m+n)$ be the *principal minors* (see Sect. 8.2) of $\mathcal{H}$. Without proof we formulate now the announced

Criterion *Let* $(\hat{\mathbf{x}}, \hat{\mathbf{u}})$ *be a critical point of the Lagrange function (8.71). Then at* $\hat{\mathbf{x}}$ *we have a*

(i) local maximum of F under the constraints (8.69) if alternatingly

$$D_{2m+1} > 0,\ D_{2m+2} < 0,\ D_{2m+3} > 0, \dots \quad \text{for } m \text{ odd}, 2m+k \le m+n,$$

$$D_{2m+1} < 0,\ D_{2m+2} > 0,\ D_{2m+3} < 0, \dots \quad \text{for } m \text{ even}, 2m+k \le m+n,$$

(ii) local minimum of F under the constraints (8.69) if

$$D_{2m+k} > 0 \quad \text{for } m \text{ even}, 2m+k \le m+n,$$

$$D_{2m+1} < 0 \quad \text{for } m \text{ odd}, 2m+k \le m+n.$$

(continued)

We apply this criterion to Example 1. Since we had only one constraint there ($m = 1$), we have to determine only $D_{2m+1} = D_3$, that is, the determinant of (8.75):

$$D_3 = \det \mathcal{H}_3 = \frac{9}{4} x_1^{1/4} x_2^{-3/2} + 3x_1^{-3/4} x_2^{-1/2} + 3x_1^{-7/4} x_2^{1/2}. \tag{8.76}$$

Prove (8.76). Obviously D_3 is *positive* for all $(x_1, x_2) \in \mathbb{R}^2_{++}$, in particular for the stationary point (8.73). Our criterion says that at $(\widehat{x_1}, \widehat{x_2}) = (40, 60)$ (see (8.73)) there is a local maximum of the problem (8.71). This maximum is global, since, on one hand, (8.73) is the only critical point *in the interior* of the domain of definition $\mathbb{R}^2_+ \times \mathbb{R}$ of the Lagrange function L to problem (8.71) (see (8.72)) and, on the other hand, at the relevant points of the boundary of $\mathbb{R}^2_+ \times \mathbb{R}$ we have

$$L(0, 90, u) = L(120, 0, u) = 0.$$

Example 4 Determine the extrema of

$$F(x_1, x_2, x_3) = x_1 + x_2 + x_3 \tag{8.77}$$

under the condition

$$(x_1 - 1)^2 + (x_2 - 1)^2 + (x_3 - 1)^2 = 12. \tag{8.78}$$

The Lagrange function of this problem is

$$L(x_1, x_2, x_3, u) = x_1 + x_2 + x_3 + u \cdot (12 - (x_1 - 1)^2 - (x_2 - 1)^2 - (x_3 - 1)^2). \tag{8.79}$$

We determine the critical points of L from

$$\frac{\partial L(x_1, x_2, x_3, u)}{\partial x_1} = 1 - 2u(x_1 - 1) = 0,$$

$$\frac{\partial L(x_1, x_2, x_3, u)}{\partial x_2} = 1 - 2u(x_2 - 1) = 0,$$

$$\frac{\partial L(x_1, x_2, x_3, u)}{\partial x_3} = 1 - 2u(x_3 - 1) = 0,$$

$$\frac{\partial L(x_1, x_2, x_3, u)}{\partial u} = 12 - (x_1 - 1)^2 - (x_2 - 1)^2 - (x_3 - 1)^2 = 0.$$

(continued)

Solving this system of equations by eliminating successively u, x_1, x_2 gives two critical points

$$(x_1^*, x_2^*, x_3^*, u^*) = (3, 3, 3, \frac{1}{4}) \tag{8.80}$$

and

$$(x_1^{**}, x_2^{**}, x_3^{**}, u^{**}) = (-1, -1, -1, -\frac{1}{4}). \tag{8.81}$$

The bordered Hessian (see (8.74)) of (8.79) is

$$\begin{pmatrix} 0 & 2x_1-2 & 2x_2-2 & 2x_3-2 \\ 2x_1-2 & -2u & 0 & 0 \\ 2x_2-2 & 0 & -2u & 0 \\ 2x_3-2 & 0 & 0 & -2u \end{pmatrix},$$

that is, for the critical point (8.80):

$$\begin{pmatrix} 0 & 4 & 4 & 4 \\ 4 & -1/2 & 0 & 0 \\ 4 & 0 & -1/2 & 0 \\ 4 & 0 & 0 & -1/2 \end{pmatrix} \tag{8.82}$$

and for the critical point (8.81):

$$\begin{pmatrix} 0 & -4 & -4 & -4 \\ -4 & 1/2 & 0 & 0 \\ -4 & 0 & 1/2 & 0 \\ -4 & 0 & 0 & 1/2 \end{pmatrix}. \tag{8.83}$$

For (8.82) the principal minors D_3 and D_4 are $D_3 = 16$, $D_4 = -12$, for (8.34) we have $D_3 = -16$, $D_4 = -12$, respectively. Prove this. According to our criterion we have a local maximum of F (see (8.77)) under the condition (8.78) at the critical point (8.80) with $F(3, 3, 3) = 3+3+3 = 9$ (see (i)) and a local minimum at the critical point (8.81) with $F(-1, -1, -1) = -1 - 1 - 1 = -3$ (see (ii)). Since there do not exist any other critical points (or other points at which extrema of (8.77) under (8.78) can exist) both the maximum 9 and the minimum -3 are global.

Until now we were always interested only in certain values and properties of the variables $x_1, \ldots, x_n$ in the vector $(x_1, \ldots, x_n, u_1, \ldots, u_m)$ of the variables

(continued)

of the Lagrange function L (see (8.70)). Now we give an interpretation of the numerical values of the Lagrange multipliers $u_1, \ldots, u_m$ at the critical points

$$(\widehat{x_1}, \ldots, \widehat{x_n}, \widehat{u_1}, \ldots, \widehat{u_m}) = (\hat{\mathbf{x}}, \hat{\mathbf{u}}). \tag{8.84}$$

Let us consider, for the moment, the c_j's in (8.69) as parameters rather than constants. Of course, each critical point $(\hat{\mathbf{x}}, \hat{\mathbf{u}})$ depends upon $\mathbf{c} = (c_1, \ldots, c_m)$. To emphasise that, we will write for now, $(\hat{\mathbf{x}}(\mathbf{c}), \hat{\mathbf{u}}(\mathbf{c}))$ and, from (compare (8.70))

$$\begin{aligned} L(\mathbf{x}, \mathbf{u}, \mathbf{c}) \quad &= \quad F(\mathbf{x}) + \textstyle\sum_{j=1}^m u_j(c_j - g_j(\mathbf{x})), \\ L(\hat{\mathbf{x}}(\mathbf{c}), \hat{\mathbf{u}}(\mathbf{c}), \mathbf{c}) &= F(\hat{\mathbf{x}}(\mathbf{c})) + \textstyle\sum_{j=1}^m \hat{u}_j(\mathbf{c})(c_j - g_j(\hat{\mathbf{x}}(\mathbf{c}))). \end{aligned} \tag{8.85}$$

So L depends both directly and indirectly, through $\hat{\mathbf{x}}$ and $\hat{\mathbf{u}}$ upon $\mathbf{c}$. Thus, partial derivation with respect to c_j gives

$$\begin{aligned} \frac{\partial}{\partial c_j} L(\hat{\mathbf{x}}(\mathbf{c}), \hat{\mathbf{u}}(\mathbf{c}), \mathbf{c}) &= \sum_{k=1}^n \frac{\partial}{\partial x_k} L(\hat{\mathbf{x}}(\mathbf{c}), \hat{\mathbf{u}}(\mathbf{c}), \mathbf{c}) \frac{\partial \hat{x}_k(\mathbf{c})}{\partial c_j} \\ &= + \sum_{k=1}^m \frac{\partial}{\partial u_k} L(\hat{\mathbf{x}}(\mathbf{c}), \hat{\mathbf{u}}(\mathbf{c}), \mathbf{c}) \frac{\partial \hat{u}_k(\mathbf{c})}{\partial c_j} \\ &= + \frac{\partial L(\mathbf{x}, \mathbf{u}, \mathbf{c})}{\partial c_j} \Big|_{\mathbf{x}=\hat{\mathbf{x}}(\mathbf{c}), \mathbf{u}=\hat{\mathbf{u}}(\mathbf{c})} \end{aligned}$$

by the rule of differentiating composite functions (see Sect. 6.12). But by $(*)$,

$$\frac{\partial}{\partial x_k} L(\hat{\mathbf{x}}(\mathbf{c}), \hat{\mathbf{u}}(\mathbf{c}), \mathbf{c}) = \frac{\partial}{\partial u_k} L(\hat{\mathbf{x}}(\mathbf{c}), \hat{\mathbf{u}}(\mathbf{c}), \mathbf{c}) = 0,$$

while, by (8.85), for $j = 1, \ldots, m$,

$$\frac{\partial L(\mathbf{x}, \mathbf{u}, \mathbf{c})}{\partial c_j} = u_j, \quad \text{thus} \quad \frac{\partial L(\mathbf{x}, \mathbf{u}, \mathbf{c})}{\partial c_j} \Big|_{\mathbf{x}=\hat{\mathbf{x}}(\mathbf{c}), \mathbf{u}=\hat{\mathbf{u}}(\mathbf{c})} = \hat{u}_j(\mathbf{c}).$$

On the other hand, $L(\hat{\mathbf{x}}(\mathbf{c}), \hat{\mathbf{u}}(\mathbf{c}), \mathbf{c}) = F(\hat{\mathbf{x}}(\mathbf{c}))$ by $(**)$ and by (8.69). Thus

$$\frac{\partial F(\hat{\mathbf{x}}(\mathbf{c}))}{\partial c_j} = \hat{u}_j(\mathbf{c}) \quad (j = 1, \ldots, m). \tag{8.86}$$

[Notice above the difference between differentiating $L(\hat{\mathbf{x}}(\mathbf{c}), \hat{\mathbf{u}}(\mathbf{c}), \mathbf{c})$ with respect to c_j and differentiating $L(\mathbf{x}, \mathbf{u}, \mathbf{c})$ with respect to c_j, then substituting $\mathbf{x} = \hat{\mathbf{x}}(\mathbf{c})$, $\mathbf{u} = \hat{\mathbf{u}}(\mathbf{c})$].

From (8.86) *we learn that the value of the Lagrange multiplier $\hat{u}_j$ equals the value of the partial derivative (the value of the (partial) slope) with respect to c_j of the objective function F* (see (8.68)) *at the critical point* $\hat{\mathbf{x}}$. To say it in other words: *The multipliers $\hat{u}_j$ measure the sensitivity of the optimal value of F to changes in the right sides c_j of the constraints* (see (8.69)).

To show this in a special case we consider again Example 1, that is, problem (8.71) now becomes the *parameter* $c \in \mathbb{R}_{++}$, and the Lagrange function differs from (8.72) in that 360 is replaced by c:

$$L(x_1, x_2, u) = x_1^{1/4} x_2^{1/2} + u \cdot (c - 3x_1 + 4x_2). \tag{8.87}$$

It is easy to show that for each $c \in \mathbb{R}_{++}$ there exists exactly one critical point, namely

$$(\hat{x}_1, \hat{x}_2, \hat{u}) = (\frac{c}{9}, \frac{c}{6}, \frac{1}{8}(\frac{c}{9})^{1/4}(\frac{c}{9})^{-1/2}). \tag{8.88}$$

(Prove it). Is (8.86) satisfied at this point? The answer is yes, since for the derivative $dU(\hat{x}_1, \hat{x}_2)/dc$ (see (8.71)), the *marginal utility of the money or budget c*, equals

$$\begin{aligned}\frac{d(\frac{c}{9})^{1/4}(\frac{c}{9})^{-1/2}}{dc} &= \frac{1}{4}(\frac{c}{9})^{-3/4}\frac{1}{9}(\frac{c}{6})^{1/2} + (\frac{c}{9})^{1/4}\frac{1}{2}(\frac{c}{6})^{-1/2}\frac{1}{6} \\ &= \frac{1}{8}(\frac{c}{9})^{1/4}(\frac{c}{6})^{-1/2}[\frac{2}{9}(\frac{c}{9})^{-1}\frac{c}{6} + \frac{2}{3}] \\ &= \frac{1}{8}(\frac{c}{9})^{1/4}(\frac{c}{6})^{-1/2} = \hat{u}.\end{aligned}$$

Note that $\frac{2}{9}(\frac{c}{9})^{-1}\frac{c}{6} + \frac{2}{3} = \frac{2}{9}\frac{9}{c}\frac{c}{6} + \frac{2}{3} = \frac{1}{3} + \frac{2}{3} = 1$. For $c = 360$ we have $\hat{u} = \frac{1}{8}40^{1/4}60^{-1/2} \approx 0.04$ (see (8.73)).

8.6.1 Exercises

1. Calculate $d^2u(x)/dx^2$ for the function $u :]0, 120[\to \mathbb{R}_+$ given by $u(x) = x^{1/4} \cdot (90 - \frac{3}{4})^{1/2}$.
2. Determine the determinant of $\mathcal{H}_3$ in (8.75).
3. Determine the principal minors D_3 and D_4 of
   (a) matrix (8.82),
   (b) matrix (8.83).
4. Determine the critical point of the Lagrange function (8.87).
5. Given a differentiable utility function $U : \mathbb{R}_+^2 \to \mathbb{R}$, a budget $b > 0$ to buy two goods 1 and 2 in quantities x_1, x_2 at prices p_1, p_2, respectively, consider the

problem of the classical theory of the consumer household: maximise $U(x_1, x_2)$ under the budget equation $p_1x_1 + p_2x_2 = b$. Let U be so that

- there is exactly one critical point $(\hat{x}_1, \hat{x}_2, \hat{u})$ of the Lagrange function L of this problem,
- at $(\hat{x}_1, \hat{x}_2)$ there is the unique (constrained) maximum of U.

Show for this problem that at $(\hat{x}_1, \hat{x}_2)$
(a) the value $\hat{u}$ of the Lagrange multiplier equals the ratio of the marginal utility and the price of good 1 as well as that of god 2,
(b) the ratio of the marginal utilities of goods 1 and 2 equals the ratio of their prices,
(c) the value $\hat{u}$ equals the marginal utility of the money, that is the derivative, $dU(x_1, x_2)/db$,
(d) the so called budget line, that is, the geometric representation of the budget equation in the coordinate plane, is the tangent at the indifference curve $\{x_1, x_2 \mid U(x_1, x_2) = c,\ c \in \mathbb{R}_+\}$ through (x_1, x_2).

6. For the problem: maximise $F(x_1, x_2) = -3 + \frac{7}{4}x_1 - x_1^2 + \frac{x_1^3}{3} + x_2$ under the condition $x_1 + x_2 = 3$, $x_1 \in \mathbb{R}_+$, $x_2 \in \mathbb{R}_+$ determine
(a) the critical points of the Lagrange function,
(b) the local extrema,
(c) the global extrema.

8.6.2 Answers

1. $\dfrac{d^2u(x)}{dx^2} = -\dfrac{3}{16}x^{-7/4}(90 - \dfrac{3}{4}x)^{1/2}$
$= -\dfrac{3}{16}x^{-3/4}(90 - \dfrac{3}{4}x)^{-1/2} - \dfrac{9}{64}x^{1/4}(90 - \dfrac{3}{4}x)^{-3/2}$. Note that this is <0 for all $x \in]0, 120[$, in particular for $x = 40$.
2. Calculation of the determinant of $\mathcal{H}_3$ in (8.75) by expanding along the first row:

$$\det \mathcal{H}_3 = 0 \cdot \det \begin{pmatrix} -\frac{3}{16}x_1^{-7/4}x_2^{1/2} & \frac{1}{8}x_1^{-3/4}x_2^{-1/2} \\ \frac{1}{8}x_1^{-3/4}x_2^{-1/2} & -\frac{1}{4}x_1^{1/4}x_2^{-3/2} \end{pmatrix}$$

$$+ (-1) \cdot 3 \cdot \det \begin{pmatrix} 3 & \frac{1}{8}x_1^{-3/4}x_2^{-1/2} \\ 4 & -\frac{1}{4}x_1^{1/4}x_2^{-3/2} \end{pmatrix}$$

$$+ 4 \cdot \det \begin{pmatrix} 3 & -\frac{3}{16}x_1^{-7/4}x_2^{1/2} \\ 4 & \frac{1}{8}x_1^{-3/4}x_2^{-1/2} \end{pmatrix}$$

$$= 0 + (-3) \cdot (3 \cdot -\frac{1}{4}x_1^{1/4}x_2^{-3/2}) - (\frac{1}{8}x_1^{-3/4}x_2^{-1/2}) \cdot 4)$$

$$+ 4 \cdot (3 \cdot \frac{1}{8}x_1^{-3/4}x_2^{-1/2} - (-\frac{3}{16}x_1^{-7/4}x_2^{1/2}) \cdot 4)$$

$$= \text{right-hand side of (8.76).}$$

3. (a) Calculation of the principal minors D_3 and D_4 of (8.82) by expanding them along the first column:

$$D_3 = \det\begin{pmatrix} 0 & 4 & 4 \\ 4 & -\frac{1}{2} & 0 \\ 4 & 0 & -\frac{1}{2} \end{pmatrix}$$

$$= 0 \cdot \det\begin{pmatrix} -\frac{1}{2} & 0 \\ 0 & -\frac{1}{2} \end{pmatrix} + (-1) \cdot 4 \cdot \begin{pmatrix} 4 & 4 \\ 0 & -\frac{1}{2} \end{pmatrix} + 4 \cdot \begin{pmatrix} 4 & 4 \\ -\frac{1}{2} & 0 \end{pmatrix}$$

$$= 0 \cdot [(-\tfrac{1}{2})(-\tfrac{1}{2}) - 0 \cdot 0]$$

$$+ (-4) \cdot [4 \cdot (-\tfrac{1}{2}) - 4 \cdot 0] + 4[4 \cdot 0 - 4 \cdot (\tfrac{1}{2})]$$

$$= 0 + 8 + 8 = 16.$$

$$D_4 = \det\begin{pmatrix} 0 & 4 & 4 & 4 \\ 4 & -\frac{1}{2} & 0 & 0 \\ 4 & 0 & -\frac{1}{2} & 0 \\ 4 & 0 & 0 & -\frac{1}{2} \end{pmatrix}$$

$$= 0 \cdot \det\begin{pmatrix} -\frac{1}{2} & 0 & 0 \\ 0 & -\frac{1}{2} & 0 \\ 0 & 0 & -\frac{1}{2} \end{pmatrix} + (-1) \cdot 4 \cdot \det\begin{pmatrix} 4 & 4 & 4 \\ 0 & -\frac{1}{2} & 0 \\ 0 & 0 & -\frac{1}{2} \end{pmatrix}$$

$$+ 4 \cdot \det\begin{pmatrix} 4 & 4 & 4 \\ -\frac{1}{2} & 0 & 0 \\ 0 & 0 & -\frac{1}{2} \end{pmatrix} + (-1) \cdot 4 \cdot \det\begin{pmatrix} 4 & 4 & 4 \\ -\frac{1}{2} & 0 & 0 \\ 0 & -\frac{1}{2} & 0 \end{pmatrix}$$

$$= 0 \cdot (-\tfrac{1}{2})^3 + (-4) \cdot [4 \cdot \det\begin{pmatrix} -\frac{1}{2} & 0 \\ 0 & -\frac{1}{2} \end{pmatrix} + (-1) \cdot 0 \cdot \det\begin{pmatrix} 4 & 4 \\ 0 & -\frac{1}{2} \end{pmatrix}$$

$$+ 0 \cdot \det\begin{pmatrix} 4 & 4 \\ -\frac{1}{2} & 0 \end{pmatrix}] + 4 \cdot [(-1) \cdot 4 \cdot \det\begin{pmatrix} -\frac{1}{2} & 0 \\ 0 & -\frac{1}{2} \end{pmatrix}$$

$$+\,0\cdot\det\begin{pmatrix}4 & 4\\ 0 & -\frac{1}{2}\end{pmatrix} + (-1)\cdot 0\cdot\det\begin{pmatrix}4 & 4\\ -\frac{1}{2} & 0\end{pmatrix}]$$

$$+(-1)\cdot 4\cdot [4\cdot\det\begin{pmatrix}-\frac{1}{2} & 0\\ 0 & -\frac{1}{2}\end{pmatrix} + (-1)\cdot 0\cdot\det\begin{pmatrix}4 & 4\\ 0 & -\frac{1}{2}\end{pmatrix}$$

$$+\,0\cdot\det\begin{pmatrix}4 & 4\\ -\frac{1}{2} & 0\end{pmatrix}]$$

$$= 0 + (-4)\cdot[4\cdot[(-\tfrac{1}{2})\cdot(-\tfrac{1}{2}) - 0\cdot 0] - 0 + 0]$$

$$+4\cdot[(-4)\cdot[(-\tfrac{1}{2})\cdot(-\tfrac{1}{2}) - 0\cdot 0] + 0 - 0]$$

$$(-4)\cdot[4\cdot[(-\tfrac{1}{2})\cdot(-\tfrac{1}{2}) - 0\cdot 0] - 0 + 0]$$

$$= -4 - 4 - 4 = -12.$$

(Note that the determinants of the last two 3×3-matrices are calculated by expanding them along the second and third column, respectively).

(b) Since $\det(-\mathbf{A}) = (-1)^n\det\mathbf{A}$ for any $n\times n$-matrix $\mathbf{A}$, we have $D_3 = -16$, $D_4 = -12$ for (8.83) (matrix (8.83) is (-1) times matrix (8.82).

4. $\dfrac{\partial L(x_1,x_2,u)}{\partial x_1} = \dfrac{1}{4}x_1^{-3/4}x_2^{1/2} - 3u = 0,$

$\dfrac{\partial L(x_1,x_2,u)}{\partial x_2} = \dfrac{1}{2}x_1^{1/4}x_2^{-1/2} - 4u = 0,$

$\dfrac{\partial L(x_1,x_2,u)}{\partial u} = c - 3x_1 - 4x_2 = 0.$

Adding $(-\frac{4}{3})$ times the first equation to the second gives $\frac{1}{2}x_1^{1/4}x_2^{-1/2} - \frac{1}{3}x_1^{-3/4}x_2^{1/2} = 0$ or, multiplying this by $x_1^{3/4}x_2^{1/2}$, $\frac{1}{2}x_1 - \frac{1}{3}x_2 = 0$. Since $x_2 = \frac{c}{4} - \frac{3}{4}x_1$, we get $\frac{1}{2}x_1 - \frac{c}{12} + \frac{1}{4}x_1 = 0$ or $\frac{3}{4}x_1 = \frac{c}{12}$, that is, $\hat{x}_1 = \frac{c}{9}$. Together with $c - 3x_1 - 4x_2 = 0$ (see above) this yields $\hat{x}_2 = \frac{c}{6}$. Inserting $\hat{x}_1 = \frac{c}{9}$ and $\hat{x}_2 = \frac{c}{6}$ into $\frac{1}{2}x_1^{1/4}x_2^{-1/2} - 4u = 0$ gives $\hat{u} = \frac{1}{8}(\frac{c}{9})^{1/4}(\frac{c}{6})^{-1/2}$.

5. (a) $L(x_1,x_2,u) = U(x_1,x_2) + u\cdot(b - p_1x_1 - p_2x_2)$,

$\dfrac{\partial L(x_1,x_2,u)}{\partial x_1} = \dfrac{\partial U(x_1,x_2)}{\partial x_1} - up_1 = 0,$

$\dfrac{\partial L(x_1,x_2,u)}{\partial x_2} = \dfrac{\partial U(x_1,x_2)}{\partial x_2} - up_2 = 0.$

This implies, at $(\hat{x}_1,\hat{x}_2,\hat{u})$,

$\dfrac{\partial U}{\partial x_1}(\hat{x}_1,\hat{x}_2)/p_1 = \dfrac{\partial U}{\partial x_2}(\hat{x}_1,\hat{x}_2)/p_2 = \hat{u}$ and

(b) $\dfrac{\partial U}{\partial x_1}\Big/\dfrac{\partial U}{\partial x_2} = p_1/p_2.$

(c) $\dfrac{\partial L}{\partial b}(\hat{x}_1,\hat{x}_2,\hat{u}) = \hat{u} = \dfrac{\partial U}{\partial b}(\hat{x}_1,\hat{x}_2)$ see (8.85), (8.86).

(d) The slope of an indifference curve at $(\hat{x}_1, \hat{x}_2)$ is
$-\frac{\partial U}{\partial x_1}(\hat{x}_1, \hat{x}_2)/\frac{\partial U}{\partial x_2}(\hat{x}_1, \hat{x}_2)$
(see Sect. 3.4). Because of (b) the budget equation $p_1x_1 + p_2x_2 = b$, that is, $x_2 = -p_1x_1/p_2 + b/p_2$ has the same slope at $(\hat{x}_1, \hat{x}_2)$.

6. (a) $(\frac{3}{2}, \frac{3}{2}, 1)$, $(\frac{1}{2}, \frac{5}{2}, 1)$,
(b) the local (constrained) minima of F are at $(\frac{3}{2}, \frac{3}{2})$ and $(0, 3)$ where $F(\frac{3}{2}, \frac{3}{2}) = 0 = F(0, 3)$, the local (constrained) maxima of F are at $(\frac{1}{2}, \frac{5}{2})$ and $(3, 0)$ where $F(\frac{1}{2}, \frac{5}{2}) = \frac{1}{6}$ and $F(3, 0) = \frac{9}{4}$, respectively,
(c) the global (constrained) minimum of F is $0 = F(\frac{3}{2}, \frac{3}{2}) = F(0, 3)$, the global (constrained) maximum of F is $\frac{9}{4} = F(3, 0)$.

8.7 Extrema of an Objective Function Depending on Parameters. Envelope Theorems. LeChatelier Principle

At the end of the last section we discussed how the solution(s) of constrained optimisation problem (8.68), (8.69) depend upon the constants $c_1, \ldots, c_m$ in (8.90). We considered these constants as *parameters* and found an interesting connection between the partial derivative (with respect to c_j) of the objective function F (see (8.68) in Sect. 8.6) and the Lagrange multiplier u_j at a critical point $(\hat{x}_1, \ldots, \hat{x}_n, \hat{u}_1, \ldots \hat{u}_m)$ of the Lagrange function (see (8.85) in Sect. 8.6, (8.86) in Sect. 8.6).

What we have found there is a special case of a class of theorems called *envelope theorems*. These are theorems on parameterised optimisation problems that describe the impact of a change of one of the parameters on the value of the objective function.

To explain the name "envelope theorem" we start with an example of an unconstrained problem:

Let the function $F : \mathbb{R}^2 \to \mathbb{R}$ be given by

$$F(x, a) = -x^2 + 2ax - a^2 + 4,$$

where $x \in \mathbb{R}$ is a variable and $a \in \mathbb{R}$ a parameter (to be also varied eventually). For each choice of the parameter a

$$\text{maximise} \quad \text{F(x,a)} \quad \text{with respect to } x. \tag{8.89}$$

From

$$\frac{dF}{dx}(x, a) = -2x + 2a = 0, \qquad \frac{d^2F}{dx^2}(x, a) = -2 < 0$$

it follows that at $x = a$ the function F reaches its maximum value. In this context the function $V : \mathbb{R} \to \mathbb{R}$ defined by

$$V(a) = \max_{x} \{F(x, a) \mid x \in \mathbb{R}\} \tag{8.90}$$

is called the *value function* for the maximisation problem (8.89). Obviously, in our example the value function V is given by

$$V(a) = F(a, a) = -a^2 + 2a^2 - a^2 + 4 = 4.$$

As Fig. 8.9 shows the graph of V is a kind of *envelope* of the graphs of the functions $a \mapsto F(x, a)$, that is,

$$a \mapsto -(a - x)^2 + 4.$$

We consider more general situations, where $\mathbf{x} = (x_1, \ldots, x_n) \in M \subset \mathbb{R}^n$, $\mathbf{a} = (a_1, \ldots, a_r) \in A \subset \mathbb{R}^r$, $F : M \times A \to \mathbb{R}$, and M and A are open sets.

Let, for instance, $F(x_1, \ldots, x_n, a_1, \ldots, a_r)$ be the profit of a firm when it sells the quantities $x_1, \ldots, x_n$ of the goods $1, \ldots, n$ in a parameter constellation $a_1, \ldots, a_r$. The numerical values of the parameters $a_1, \ldots, a_r$ may be the prices set for the goods (then $r = n$) and for the prices paid for the inputs necessary to produce the goods or other parameters, for instance macro-economic or political ones that have an impact on the firm's profit. We then call the function $F : M \times A \to \mathbb{R}$ the *profit function* of the firm. We suppose that the firm wants to maximise its profit $F(\mathbf{x}, \mathbf{a})$.

Generalising (8.89), (8.90) for the maximisation problem

$$\text{maximise} \quad \mathrm{F}(\mathbf{x},\mathbf{a}) \quad \text{with respect to } x \in M, \tag{8.91}$$

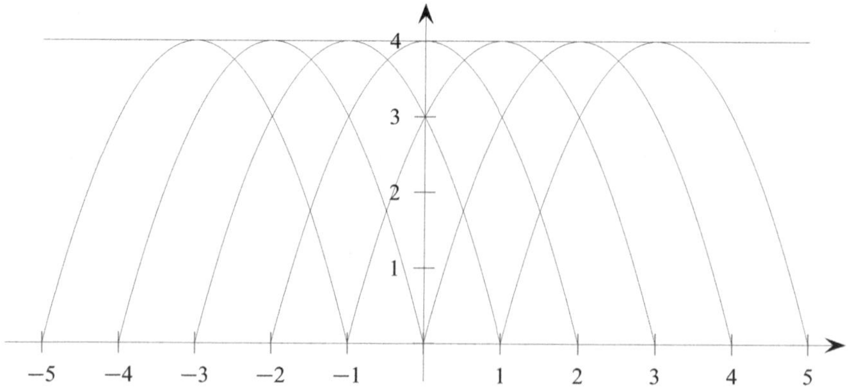

Fig. 8.9 The graph of $a \mapsto V(a) = 4$, that is, the horizontal line through the point $(0, 4)$, is the "envelope" of the graphs of the function $a \mapsto F(x, a) = -(a - x)^2 + 4$, $x \in \mathbb{R}$

we define the *value function* $V : A \to \mathbb{R}$ for each choice of the parameter vector $a \in A$ by

$$V(\mathbf{a}) = \max_{x} \{F(\mathbf{x}, \mathbf{a}) \mid \mathbf{x} \in M \subseteq \mathbb{R}^n\}. \tag{8.92}$$

Notice, that V is defined only if the maximum in (8.92) exists for *each* $\mathbf{a} \in A$. We assume further that the following two conditions are fulfilled:

(i) For each choice of the parameter vector $\mathbf{a} \in A$, F is a continuously differentiable function of $\mathbf{x} \in M$.
(ii) In some neighbourhood $N(\tilde{\mathbf{a}}) \subseteq A$ of $\tilde{\mathbf{a}} \in A$ there is a unique continuously differentiable function $\mathbf{X} : N(\tilde{\mathbf{a}}) \to \mathbb{R}^n$ satisfying

$$V(\mathbf{a}) = F(\mathbf{X}(\mathbf{a}), \mathbf{a}). \tag{8.93}$$

Here $\mathbf{x} = \mathbf{X}(\mathbf{a})$ indicates the dependence of $\mathbf{x}$ upon $\mathbf{a}$, and $\tilde{\mathbf{x}} = \mathbf{X}(\tilde{\mathbf{a}})$ is the maximum point.

Then we get via the chain rule (see Sect. 6.5)

$$\begin{aligned}\frac{\partial V}{\partial a_j}(\tilde{\mathbf{a}}) &= \sum_{k=1}^{n} \frac{\partial F}{\partial x_k}(\mathbf{X}(\tilde{\mathbf{a}}), \tilde{\mathbf{a}}) \frac{\partial x_k}{\partial a_j}(\tilde{\mathbf{a}}) + \frac{\partial F}{\partial x_j}(\mathbf{X}(\tilde{\mathbf{a}}), \tilde{\mathbf{a}}) \\ &= \frac{\partial F}{\partial x_j}(\mathbf{X}(\tilde{\mathbf{a}}), \tilde{\mathbf{a}}) \quad \text{for } j = 1, \ldots, r,\end{aligned}$$

since

$$\frac{\partial F}{\partial x_k}(\mathbf{X}(\tilde{\mathbf{a}}), \tilde{\mathbf{a}}) = 0 \quad \text{for } k = 1, \ldots, n. \tag{8.94}$$

Equations (8.94) hold because F has its maximum at $(\mathbf{X}(\tilde{\mathbf{a}}), \tilde{\mathbf{a}}) = (\tilde{\mathbf{x}}, \tilde{\mathbf{a}})$. In our interpretation of F as a profit function we can say that the conditions (8.94) guarantee that there are no marginal gains from small changes in the quantities of the goods offered when we start from a profit maximising point.

We have proved the so-called *envelope theorem for maximisation of a function depending on parameters:*

Let $F : M \times A \to \mathbb{R}$, where $M \subseteq \mathbb{R}^n$ and $A \subseteq \mathbb{R}^r$ are open sets, be a continuously differentiable function of $x \in M$ for each $\mathbf{a} \in A$. Let $\ddot{\mathbf{x}} \in M$ be a maximiser of F for $\tilde{\mathbf{a}} \in A$ and let for the value function V (see (8.92)) the identity

$$V(\mathbf{a}) = F(\mathbf{X}(\mathbf{a}), \mathbf{a})$$

hold for a unique continuously differentiable function $\mathbf{X}$ defined in a neighbourhood of $\tilde{\mathbf{a}}$ that belongs to A. Then

$$\frac{\partial V}{\partial a_j}(\tilde{\mathbf{a}}) = \frac{\partial F}{\partial a_j}(\mathbf{X}(\tilde{\mathbf{a}}), \tilde{\mathbf{a}}) \quad \text{for } j = 1, \ldots, r. \tag{8.95}$$

Clearly there is a similar envelope theorem for *minimisation* of a function depending on parameters when $\tilde{\mathbf{x}} \in M$ is not a maximiser but a *minimiser* of F for $\tilde{\mathbf{a}} \in A$ and the value function V is defined in (8.92) with min instead of max.

We emphasise that in (8.95) the left-hand side is the partial derivative with respect to a_j of the function V of $\mathbf{a}$ at $\tilde{\mathbf{a}}$, whereas the right-hand side is the partial derivative with respect to a_j of the function

$$(a_1, \ldots, a_r) \mapsto F(x_1, \ldots, x_n, a_1, \ldots a_r)$$

at

$$(\tilde{x}_1, \ldots, \tilde{x}_n, \tilde{a}_1, \ldots, \tilde{a}_r) = (X_1(\tilde{\mathbf{a}}), \ldots, X_n(\tilde{\mathbf{a}}), \tilde{a}_1, \ldots, \tilde{a}_r)$$

(see (8.93)).

The envelope theorem, in particular equation (8.90), is useful, since in many applications the right-hand side of (8.95) or rather the calculations determining it are easier than the way to get the left-hand sides. We show this in Examples 1 and 2:

Example 1 Consider the function $F : \mathbb{R} \times \mathbb{R}^2 \to \mathbb{R}$ given by

$$F(X, a_1, a_2) = -x^2 + 2(a_1 + a_2)x + a_1^2 + a_2^2. \tag{8.96}$$

First we determine the value function V for the problem of maximising F with respect to x. The equation for the maximiser of F is

$$\frac{\partial F}{\partial x}(x, a_1, a_2) = -2x + 2(a_1 + a_2) = 0$$

(notice that the second derivative with respect to x is -2). So,

$$x = a_1 + a_2 = X(a_1, a_2) \qquad (8') \mu rx$$

(see (ii)). Putting this into $F(X(a_1, a_2), a_1, a_2)$ leads to

$$F(X(a_1, a_2), a_1, a_2) = -(a_1 + a_2)^2 + 2(a_1 + a_2)^2 + a_1^1 + a_2^2,$$

(continued)

that is, the value function V is given by

$$V(a_1, a_2) = (a_1 + a_2)^2 + a_1^2 + a_2^2.$$

Now we can calculate the left-hand sides in (8.95) at a (parameter) point $(\tilde{a}_1, \tilde{a}_2)$:

$$\frac{\partial V}{\partial a_1}(\tilde{a}_1, \tilde{a}_2) = 4\tilde{a}_1 + 2\tilde{a}_2, \qquad \frac{\partial V}{\partial a_2}(\tilde{a}_1, \tilde{a}_2) = 2\tilde{a}_1 + 4\tilde{a}_2. \tag{8.97}$$

We show that it is easier to apply the envelope theorem, that is, to start out with the *right-hand sides* in (8.95). This leads (see (8.96)) directly to

$$\frac{\partial F}{\partial a_1}(x, a_1, a_2) = 2a_1 + 2x = 2a_1 + 2a_2, \tag{8.98}$$

$$\frac{\partial F}{\partial a_2}(x, a_1, a_2) = 2a_2 + 2x = 4a_2 + 2a_1, \tag{8.99}$$

10: (8.98) 11: (8.98) since, by (1), $x = X(a_1, a_2) = a_1 + a_2$ (compare to (8.95) and (8.97)).

An interpretation of (8.98) is: When $\tilde{x}$ maximises F in the parameter constellation $(\tilde{a}_1, \tilde{a}_2)$ and $\tilde{a}_1$ is increased then

$$F(\tilde{x}, \tilde{a}_1, \tilde{a}_2) = F(X(\tilde{a}_1, \tilde{a}_2), \tilde{a}_1, \tilde{a}_2) = F(\tilde{a}_1 + \tilde{a}_2, \tilde{a}_1, \tilde{a}_2)$$

will increase at the rate $4\tilde{a}_1 + 2\tilde{a}_2$. Equation (8.99) can be interpreted similarly.

Example 2 We consider the function $F : \mathbb{R}^2_{++} \times \mathbb{R}^3_{++} \to \mathbb{R}$, given by

$$F(x_1, x_2, p_1, p_2, p_3) = x_1^{1/2} x_2^{1/3} p_3 - x_1 p_1 - x_2 p_2, \tag{8.100}$$

the profit function of a firm, where x_1, x_2 are the quantities of the two inputs 1 and 2, $x_1^{1/2} x_2^{1/3}$ is the maximum quantity of an output good that can be produced with the aid of x_1 and x_2, p_3 is the price paid for the output good, and p_1, p_2 are the prices of the inputs, respectively. We consider p_1, p_2 and p_3 as parameters and determine the maximisers of F from the equations

$$\frac{\partial F}{\partial x_1}(x_1, x_2, p_1, p_2, p_3) = \frac{1}{2} x_1^{-1/2} x_2^{1/3} p_3 - p_1 = 0,$$

(continued)

$$\frac{\partial F}{\partial x_2}(x_1, x_2, p_1, p_2, p_3) = \frac{1}{3} x_1^{1/2} x_2^{-2/3} p_3 - p_2 = 0$$

as

$$x_1 = (p_3^3/12p_1^2p_2)^2, \quad x_2 = (p_3^2/6p_1p_2)^3. \tag{8.101}$$

Notice that F takes its maximum at (8.101), since for the Hessian of F,

$$\begin{pmatrix} \frac{\partial^2 F}{\partial x_1^2} & \frac{\partial^2 F}{\partial x_1 \partial x_2} \\ \frac{\partial^2 F}{\partial x_2 \partial x_1} & \frac{\partial^2 F}{\partial x_2^2} \end{pmatrix} = \begin{pmatrix} -\frac{1}{4} x_1^{-3/2} x_2^{1/3} p_3 & \frac{1}{6} x_1^{-1/2} x_2^{-2/3} p_3 \\ \frac{1}{6} x_1^{-1/2} x_2^{-2/3} p_3 & -\frac{2}{9} x_1^{1/2} x_2^{-5/3} p_3 \end{pmatrix} =: \mathbf{H},$$

we have

$$D_1 = -\frac{1}{4} x_1^{-3/2} x_2^{1/3} p_3 < 0,$$

$$D_2 = \det \mathbf{H} = \frac{2}{36} x_1^{-1} x_2^{-4/3} p_3^2 - \frac{1}{36} x_1^{-1} x_2^{-4/3} p_3^2$$

$$= \frac{1}{36} x_1^{-1} x_2^{-4/3} p_3^2 > 0$$

(D_1, D_2 principal minors; see Sect. 8.2).

Putting (8.101) into (8.100) leads to

$$F(x_1, x_2, p_1 p_2, p_3) = \frac{p_3^3}{12p_1^2p_2} \frac{p_3^2}{6p_1p_2} - \left(\frac{p_3^3}{12p_1^2p_2}\right)^2 p_1 - \left(\frac{p_3^2}{6p_1p_2}\right)^3 p_2$$

$$= \frac{p_3^6}{432 p_1^3 p_2^2},$$

that is, the value function V. From this we get

$$\frac{\partial V}{\partial p_1}(p_1, p_2, p_3) = -\frac{p_1^{-4} p_2^{-2} p_3^6}{144}, \tag{8.102}$$

$$\frac{\partial V}{\partial p_2}(p_1, p_2, p_3) = -\frac{p_1^{-3} p_2^{-3} p_3^6}{216}, \tag{8.103}$$

$$\frac{\partial V}{\partial p_3}(p_1, p_2, p_3) = \frac{p_1^{-3} p_2^{-2} p_3^5}{72}. \tag{8.104}$$

(continued)

For instance, equation (8.104) says that when in the price constellation $(p_1, p_2, p_3) = (1, 1, 6)$ the output price $p_3 = 6$ is increased then the maximal profit (gained at $(x_1, x_2) = (324, 216)$; see (8.101)) will increase at a rate of 108.

We point out that equations (8.102), (8.103), (8.104) can be deduced much easier than above with the aid of the envelope theorem, that is, by application of the right-hand side in (8.95): Inserting (8.101) into the partial derivatives of (8.100) with respect to p_1, p_2, p_3 gives (8.102), (8.103), (8.104), respectively.

Our next example applies the envelope theorem to an important problem of production theory.

Example 3 Consider a firm that produces s goods, using n inputs. The amounts (quantities) of the goods are $y_1, \dots, y_s$, those of the inputs $x_1, \dots, x_n$, that is, we have the output vector $\mathbf{y} = (y_1, \dots, y_s)$ and the input vector $\mathbf{x} = (x_1, \dots, x_n)$. Let $\mathbf{y} \in \mathbb{R}^s_{++}$, $\mathbf{x} \in \mathbb{R}^n_{++}$. For the prices p_j of input j $(j = 1, \dots, n)$ we assume $(p_1, \dots, p_n) = \mathbf{p} \in \mathbb{R}^n_{++}$. Given $\mathbf{p}$ the firm wants to minimise the cost

$$\mathbf{x} \cdot \mathbf{p} = x_1 p_1 + \ldots + x_n p_n$$

of production $\mathbf{y}$. Let the firm's input correspondence (see Sect. 2.2), $M : \mathbb{R}^s_{++} \to$ power set of $\mathbb{R}^n_{++}$, be given by

$$M(\mathbf{y}) := \left\{\mathbf{x} \in \mathbb{R}^n_{++} \mid \mathbf{y} \in \mathbb{R}^s_{++} \text{can be produced with the help of } \mathbf{x}\right\}. \tag{8.105}$$

Considering the price vector $\mathbf{p}$ as parameter vector we have to solve the minimisation problem

$$\text{minimise } F(\mathbf{x}, \mathbf{p}) := \mathbf{x} \cdot \mathbf{p} \quad \text{with respect to } \mathbf{x} \in M(\mathbf{y}) \tag{8.106}$$

for each choice of both $\mathbf{p} \in \mathbb{R}^n_{++}$ and $\mathbf{y} \in \mathbb{R}^s_{++}$. Compare this with problem (8.91), where M was a (constant) subset of $\mathbb{R}^n$. In (8.105), $M(\mathbf{y})$ is a subset of $\mathbb{R}^n_{++}$ for *each value* of $\mathbf{y}$, thus a subset of $\mathbb{R}^n_{++}$ *depending upon the parameter(s)* $\mathbf{y} = (y_1, \dots, y_m)$.

Notice that as (8.92) is the value function for (8.91) so the function $C : \mathbb{R}^s_{++} \times \mathbb{R}^n_{++} \to \mathbb{R}_{++}$ given by

$$C(\mathbf{y}, \mathbf{p}) = \min_{\mathbf{y}} \{\mathbf{x} \cdot \mathbf{p} \mid \mathbf{x}\} \in M(\mathbf{y}),\ \mathbf{p} \in \mathbb{R}^n_{++} \tag{8.107}$$

is the value function for (8.106). Here we are interested only in situations, where conditions (a), (b), (c) hold:

1. the minimum in (8.107) exists for *each* choice of the pair of points $\mathbf{y} \in \mathbb{R}^s_{++}$, $\mathbf{p} \in \mathbb{R}^n_{++}$,
2. $\mathbf{p} \mapsto C(\mathbf{y}, \mathbf{p})$ is concave from below and continuously differentiable,
3. there exists a unique function $\mathbf{X} : \mathbb{R}^s_{++} \times \mathbb{R}^n_{++} \to \mathbb{R}^n_{++}$ satisfying

$$C(\mathbf{y}, \mathbf{p}) = \mathbf{X}(\mathbf{y}, \mathbf{p}) \cdot \mathbf{p}. \tag{8.108}$$

3.Obviously, C is the *cost function* of the firm, that is, $C(\mathbf{y}, \mathbf{p})$ is the minimum cost to produce $\mathbf{y}$ when the prices of the inputs are $\mathbf{p}$, and $\mathbf{X}(\mathbf{y}, \mathbf{p})$ is the *cost minimising input vector* for producing $\mathbf{y}$ given the input prices $\mathbf{p}$.

Under these conditions we get from the envelope theorem, in particular from (8.95),

$$\frac{\partial C}{\partial p_j}(\mathbf{y}, \tilde{\mathbf{p}}) = X_j(\mathbf{y}, \tilde{\mathbf{p}}) \quad \mathit{for}\, j = 1, \ldots, n. \tag{8.109}$$

This statement, well known as *Shephard's lemma* (Ronald W. Shephard (1912–1982)), *yields, if* $\mathbf{p} \mapsto C(\mathbf{y}, \mathbf{p})$ *is twice continuously differentiable,*

$$\frac{\partial X_j}{\partial p_j}(\mathbf{y}, \tilde{\mathbf{p}}) = \frac{\partial X_k}{\partial p_j}(\mathbf{y}, \tilde{\mathbf{p}}) \quad \text{for all } j \text{ and } k.$$

Shephard's lemma (8.109) says that the cost minimising quantity of input j in the case of input prices $\tilde{\mathbf{p}}$ and output vector $\mathbf{y}$ equals the marginal cost with respect to p_j at $(\mathbf{y}, \tilde{\mathbf{p}})$.

A direct proof of (8.109) runs as follows. We calculate the derivative of (8.108) with respect to p_j:

$$\begin{aligned}
\frac{\partial C}{\partial p_j}(\mathbf{y}, \mathbf{p}) &= \mathbf{X}_j(\mathbf{y}, \mathbf{p}) + \mathbf{p} \cdot \frac{\partial \mathbf{X}}{\partial p_j}(\mathbf{y}, \mathbf{p}) \\
&= \mathbf{X}_j(\mathbf{y}, \mathbf{p}) + p_1 \cdot \frac{\partial X_1}{\partial p_j}(\mathbf{y}, \mathbf{p}) + \ldots + p_n \cdot \frac{\partial X_n}{\partial p_j}(\mathbf{y}, \mathbf{p}).
\end{aligned}$$

We have to show that

$$\mathbf{p} \cdot \frac{\partial \mathbf{X}}{\partial p_j}(\mathbf{y}, \mathbf{p}) = 0 \quad \text{at} \quad \mathbf{p} = \tilde{\mathbf{p}}. \tag{8.110}$$

Since $\mathbf{X}(\mathbf{y}, \mathbf{p})$ is the cost minimising input vector for producing $\mathbf{y}$ given the input prices $\mathbf{p}$, we have

$$\mathbf{p} \cdot \mathbf{X}(\mathbf{y}, \tilde{\mathbf{p}}) \geq C(\mathbf{y}, \mathbf{p}) = \mathbf{p} \cdot \mathbf{X}(\mathbf{y}, \mathbf{p}).$$

Hence the function $\Phi : \mathbb{R}^s_{++} \times \mathbb{R}^n_{++} \to \mathbb{R}_{++}$ defined by

$$\Phi(\mathbf{y}, \mathbf{p}) := \mathbf{p} \cdot \mathbf{X}(\mathbf{y}, \tilde{\mathbf{p}}) - \mathbf{p} \cdot \mathbf{X}(\mathbf{y}, \tilde{\mathbf{p}}) \tag{8.111}$$

satisfies $\Phi(\mathbf{y}, \mathbf{p}) \geq 0$ and $\Phi(\mathbf{y}, \tilde{\mathbf{p}}) = 0$ for all $\mathbf{p} \in \mathbb{R}^n_{++}$ and since $\mathbf{p} \mapsto \mathbf{p} \cdot \mathbf{X}(\mathbf{y}, \tilde{\mathbf{p}})$ is linear and $\mathbf{p} \mapsto -\mathbf{p} \cdot \mathbf{X}(\mathbf{y}, \mathbf{p})$ is convex by assumption (b). These properties and the differentiability of Φ with respect to p_j imply

$$0 = \frac{\partial \Phi}{\partial p_j}(\mathbf{y}, \mathbf{p}) \quad \text{at } \tilde{\mathbf{p}}$$

that is (see (8.111))

$$0 = X_j(\mathbf{y}, \tilde{\mathbf{p}}) - X_j(\mathbf{y}, \mathbf{p}) - \mathbf{p} \cdot \frac{\partial \mathbf{X}}{\partial p_j}(\mathbf{y}, \mathbf{p}) \quad \text{at } \tilde{\mathbf{p}},$$

which proves (8.110).

We now generalise the above envelope theorem. In our problem (8.91) the variable (vector) $\mathbf{x}$ was constrained by the conditions on the domain in which $\mathbf{x}$ can move.

Let $G_1 : M_1 \times A \to \mathbb{R}, \ldots G_m : M_m \times A \to \mathbb{R}, F : M \times A \to \mathbb{R}$, where $M_1 \subseteq \mathbb{R}^n, \ldots, M_m \subseteq \mathbb{R}^n, M \subseteq \mathbb{R}^n, A \subseteq \mathbb{R}^s$ are open sets and $M_1 \cap M_2 \cap \ldots \cap M_m \cap M \neq \emptyset$. The problem is as follows. For each choice of the parameter vector $\mathbf{a} \in A$

$$\text{maximise} \quad F(\mathbf{x}, \mathbf{a}) \quad \text{with respect to } \mathbf{x}, \tag{8.112}$$

where $\mathbf{x}$ satisfies the conditions

$$\mathbf{x} \in M, \quad G_1(\mathbf{x}, \mathbf{a}) = 0, \ldots, G_m(\mathbf{x}, \mathbf{a}) = 0. \tag{8.113}$$

The value function $V : A \to \mathbb{R}$ for this problem is given by

$$V(\mathbf{a}) = \max_{\mathbf{x}} \{F(\mathbf{x}, \mathbf{a}) \mid \mathbf{x} \in M, G_1(\mathbf{x}, \mathbf{a}) = 0, \ldots, G_m(\mathbf{x}, \mathbf{a}) = 0\}. \tag{8.114}$$

We assume that, for any choice of $\mathbf{a} \in A$, the maximum (8.114) exists. Generalising conditions (i), (ii) at the beginning of this section we assume:

(i') For each choice of $\mathbf{a} \in A$, the functions $F, G_1, \ldots, G_m$ are continuously differentiable functions of $\mathbf{x}$, where

$$\mathbf{x} \in M_1 \cap M_2 \cap \ldots \cap M_m \cap M =: \mathcal{M}$$

(ii') In some neighbourhood $N(\tilde{\mathbf{a}}) \subseteq A$ of $\tilde{\mathbf{a}} \in A$ there is a unique continuously differentiable function $\mathbf{X} : N(\tilde{\mathbf{a}}) \to \mathcal{M}$ satisfying

$$V(\mathbf{a}) = F(\mathbf{X}(\mathbf{a}), \mathbf{a}) \tag{8.115}$$

and

$$G_1(\mathbf{X}(\mathbf{a}), \mathbf{a}) = 0, \ldots, G_m(\mathbf{X}(\mathbf{a}), \mathbf{a}) = 0, \qquad \mathbf{X}(\mathbf{a}) = \mathbf{x}, \mathbf{X}(\tilde{\mathbf{a}}) = \tilde{\mathbf{x}} \tag{8.116}$$

(so $\tilde{\mathbf{x}}$ is a maximiser of F for $\mathbf{a} = \tilde{\mathbf{a}}$).

The Lagrange function $L : \mathcal{M} \times A \times \mathbb{R}^m \to \mathbb{R}$ for problem (8.112), (8.113) is given by

$$L(\mathbf{x}, \mathbf{a}, \mathbf{u}) = F(\mathbf{x}, \mathbf{a}) + u_1 G_1(\mathbf{x}, \mathbf{a}) + \ldots u_m G_m(\mathbf{x}, \mathbf{a}) \tag{8.117}$$

(compare to Sect. 8.6). When $\mathbf{a} = \tilde{\mathbf{a}}$ (see (ii')) we differentiate (8.117) with respect to $\mathbf{x}$ and $\mathbf{u}$ to obtain the conditions for the critical points of the problem:

$$\begin{aligned} &\frac{\partial L}{\partial x_k}(\mathbf{x}, \tilde{\mathbf{y}}, \mathbf{u}) \\ &\quad = \frac{\partial F}{\partial x_k}(\mathbf{x}, \tilde{\mathbf{a}}) + u_1 \frac{\partial G_1}{\partial x_k}(\mathbf{x}, \tilde{\mathbf{a}}) + \ldots \\ &\quad + u_m \frac{\partial G_m}{\partial x_k}(\mathbf{x}, \tilde{\mathbf{a}}) = 0 \qquad (k = 1, \ldots, n), \end{aligned} \tag{8.118}$$

$$\frac{\partial L}{\partial u_l}(\mathbf{x}, \tilde{\mathbf{a}}) = G_l(\mathbf{x}, \tilde{\mathbf{a}}) = 0 \qquad (l = 1, \ldots, m). \tag{8.119}$$

In order to obtain equations easier to handle, we differentiate the value function V given by (8.115) with respect to the parameters. Using the chain rule (see Sect. 6.5) we get

$$\frac{\partial V}{\partial a_j}(\mathbf{a}) = \sum_{k=1}^{n} \frac{\partial F}{\partial x_k}(\mathbf{X}(\mathbf{a}), \mathbf{a}) \frac{\partial X_k}{\partial a_j}(\mathbf{a}) + \frac{\partial F}{\partial a_j}(\mathbf{X}(\mathbf{a}), \mathbf{a}) \qquad (j = 1, \ldots, r).$$

At $\mathbf{a} = \tilde{\mathbf{a}}$, $\mathbf{X}(\tilde{\mathbf{a}}) = \tilde{\mathbf{x}}$ (see (ii')) this becomes, because of (8.118),

$$\begin{aligned} \frac{\partial V}{\partial a_j}(\tilde{\mathbf{a}}) = \; & \frac{\partial F}{\partial a_j}(\mathbf{X}(\mathbf{a}), \mathbf{a}) \\ & - \sum_{k=1}^{n} (u_1 \frac{\partial G_1}{\partial x_k}(\mathbf{X}(\mathbf{a}), \mathbf{a}) + \ldots \\ & + u_m \frac{\partial G_m}{\partial x_k}(\mathbf{X}(\mathbf{a}), \mathbf{a})) \frac{\partial X_k}{\partial a_j}(\tilde{\mathbf{a}}). \end{aligned} \tag{8.120}$$

For $\mathbf{a} \in N(\tilde{\mathbf{a}}) \subseteq A$ and $\mathbf{x} = \mathbf{X}(\mathbf{a}) \in \mathcal{M}$ (see (ii')), in particular (8.116) we differentiate G_l in (8.113) with respect to a_j:

$$\sum_{k=1}^{n} \frac{\partial G_l}{\partial x_k}(\mathbf{X}(\mathbf{a}), \mathbf{a}) \frac{\partial X_k}{\partial a_j}(\mathbf{a}) + \frac{\partial G_l}{\partial a_j}(\mathbf{X}(\mathbf{a}), \mathbf{a}) = 0$$
$$(j = 1, \ldots, r;\ l = 1, \ldots, m).$$

Using the derivative $\frac{\partial G_l}{\partial x_k}(\mathbf{X}(\mathbf{a}), \mathbf{a})$, obtained from these equations at $\mathbf{a} = \tilde{\mathbf{a}}$, the equations (8.120) reduce to

$$\begin{aligned} \frac{\partial V}{\partial a_j}(\mathbf{a}) &= \frac{\partial F}{\partial a_j}(\mathbf{X}(\tilde{\mathbf{a}}), \tilde{\mathbf{a}}) \\ &\quad + u_1 \frac{\partial G_1}{\partial a_j}(\mathbf{X}(\tilde{\mathbf{a}}), \tilde{\mathbf{a}}) + \ldots + u_m \frac{\partial G_m}{\partial a_j}(\mathbf{X}(\tilde{\mathbf{a}}), \tilde{\mathbf{a}}) \\ &\quad (j = 1, \ldots, r). \end{aligned} \tag{8.121}$$

We have proved the *envelope theorem for maximisation of a function under equality constraints, where the function and the constraints depend on parameters:*

In the situation described above consider the problem (8.112), (8.113). Suppose that the above assumptions on the objective function F, on the constraining functions $G_1, \ldots, G_m$ and on the value function V (see (8.114)) hold, in particular assumptions (i'), (ii'). Let, for $\tilde{\mathbf{a}}$ satisfying (ii'), the vector $\tilde{\mathbf{x}} = \mathbf{X}(\tilde{\mathbf{a}})$ be a maximiser of F. Then (8.121) holds. We will not discuss necessary and/or sufficient conditions here under which the Lagrange multipliers $u_1, \ldots, u_m$ in (8.121) can be uniquely determined. Instead, we present an application of this theorem for the particular case described by the problem (8.112), (8.113), where F does *not* depend on parameters, thus

$$\frac{\partial F}{\partial a_j}(\mathbf{X}(\mathbf{a}), \mathbf{a}) = 0 \quad (j = 1, \ldots, m) \quad \text{for all } \mathbf{a} \in A, \tag{8.122}$$

and the conditions are

$$G_1(\mathbf{x}, \mathbf{a}) := a_1 - g_1(\mathbf{x}) = 0, \ldots, G_m(\mathbf{x}, \mathbf{a}) := a_m - g_m(\mathbf{x}) = 0,$$

less general than those in (8.113). In this case the value function V is given by

$$V(a_1, \ldots, a_m) = \max_{\mathbf{x}} \{F(\mathbf{x}) \mid a_1 - g_1(\mathbf{x}) = 0, \ldots, a_m - g_m(\mathbf{x}) = 0\}$$

and (8.121) becomes, for the solution $\tilde{\mathbf{x}} = \mathbf{X}(a_1, \ldots, a_m)$, $\tilde{\mathbf{u}}$ of the Lagrange problem (if it exists),

$$\frac{\partial V}{\partial a_j}(a_1, \ldots, a_m) = \tilde{\mathbf{u}}_j \quad (j = 1, \ldots, m),$$

since, by (8.122),

$$\frac{\partial F}{\partial a_j}(\tilde{\mathbf{x}}) = 0 \qquad (j = 1, \ldots, m).$$

We note that this particular case of the envelope theorem coincides with both our results at the end of Sect. (8.7) (see formula (8.107 there) and our interpretation of the Lagrange multipliers given there.

We conclude this section with some remarks on the extrema of real-valued functions that are not necessarily differentiable with respect to both their variables and their parameters.

Let M and A be two non-void sets and let $x \in M$ and $a \in A$ be a "variable" and a "parameter", respectively. In what follows we deal with functions $F : M \times A \to \mathbb{R}$ and their (real) function values $F(x, a)$. From now on M and A are not necessarily subsets of $\mathbb{R}^n$ or $\mathbb{R}^r$, respectively, that is, x and a are not necessarily n- or r-dimensional real vectors.

We denote by $S(a)$ the set of all $x \in M$ which minimise $F(y, a)$ subject to the condition $y \in M$. Thus

$$S(a) = \{x \in M \mid F(x, a) = \inf_{y \in M} F(y, a)\}.$$

We show the following. *Let* $a^1 \in A$, $a^2 \in A$, $x^1 \in S(a^1)$, $x^2 \in S(a^2)$. *Then*

$$F(x^2, a^2) - F(x^2, a^1) \le F(x^1, a^2) - F(x^1, a^1) \tag{8.123}$$

with equality holding if and only if $x^2 \in S(a^1)$ *and* $x^1 \in S(a^2)$.

The inequality (8.123) follows from

$$F(x^1, a^1) \le F(x^2, a^1) \qquad \text{for } x^1 \in S(a^1) \tag{8.124}$$

and

$$F(x^2, a^2) \le F(x^1, a^2) \qquad \text{for } x^2 \in S(a^2), \tag{8.125}$$

with equality if and only if $x^2 \in S(a^1)$ and $x^1 \in S(a^2)$. Adding (8.124) and (8.125) and then subtracting $F(x^1, a^2) + F(x^2, a^1)$ from both sides at once yields (8.123). Clearly equality in (8.123) is possible if and only if equality holds in both (8.124) and (8.125).

Note that the inequality (8.122) is invariant under a permutation of the indices 1 and 2.

For a maximisation problem we accordingly define

$$S(a) = \{x \in M \mid F(x, a) = \sup_{y \in M} F(y, a)\},$$

and then the inequality (8.123) is reversed.

We can illustrate (8.123) as follows. Suppose that the function F describes a general "system" (economical, chemical, physical). Then x is an element ("point", "intrinsic or endogenous variable") of a "state set" M and a n element ("point", "experimental condition", "control", "extrinsic or exogenous variable") of a "parameter set" A. We define the system as being in "(stable) equilibrium" at (x, a) if $x \in S(a)$, that is, $y \mapsto F(y, a)$ has a minimum at $y = x$. We may furthermore consider the difference $F(x, a^2) - F(x, a^1)$ as the "effect", evaluated at x, of a change in the "conditions" (parameters) from a^1 to a^2.

In this terminology formula (8.123) can be phrased as follows:

Suppose a system which is in equilibrium at the point (x^1, a^1) *is disturbed by a change in the parameter a from* a^1 *to* a^2 *and assumes a new equilibrium state at the point* (x^2, a^2). *Then the difference*

$$F(x^2, a^2) - F(x^1, a^1),$$

which is a measure of he effect of the disturbance (i.e., of the reaction of the system to the change) with respect to the point x^2, *is less than or equal to the difference*

$$F(x^1, a^2) - F(x^1, a^1),$$

which is the corresponding measure with respect to the point x^1.

As an application let the sets M and A both be $\mathbb{R}^n$, and let F have the form

$$F(\mathbf{x}, \mathbf{a}) = H(\mathbf{x}) + \mathbf{a} \cdot \mathbf{x}, \tag{8.126}$$

where $\mathbf{x} = (x_1, \ldots, x_n)$, $\mathbf{a} = (a_1, \ldots, a_n)$, $\mathbf{a} \cdot \mathbf{x} = a_1 x_1 + \ldots a_n x_n$, $H : \mathbb{R}^n \to \mathbb{R}$. Let $\mathbf{a}^1 \in A$, $\mathbf{a}^2 \in A$, $\mathbf{x}^1 \in S(\mathbf{a}^1)$, $\mathbf{x}^2 \in S(\mathbf{a}^2)$. Then $F(\mathbf{x}^2, \mathbf{a}^2) - F(\mathbf{x}^2, \mathbf{a}1) = (\mathbf{a}^2 - \mathbf{a}^1) \cdot \mathbf{x}^2$, $F(\mathbf{x}^1, \mathbf{a}^2) - F(\mathbf{x}^1, \mathbf{a}1) = (\mathbf{a}^2 - \mathbf{a}^1) \cdot \mathbf{x}^1$ and, by (8.123),

$$(\mathbf{a}^2 - \mathbf{a}^1) \cdot (\mathbf{x}^2 - \mathbf{x}^1) \leq 0 \tag{8.127}$$

with equality holding if and only if $\mathbf{x}^2 \in S(\mathbf{a}^1)$ and $\mathbf{x}^1 \in S(\mathbf{a}^2)$.

Paul A. Samuelson (*1915, †2009) relates (8.127) to a *principle of LeChatelier* (Henri Louis LeChatelier (1850–1936)) in physics: "The method employed here is that which underlies LeChatelier's principle in physics." This method of deriving (8.127) together with formula (8.127) itself and a lot of consequences of (8.127) are called, by economists, *LeChatelier–Samuelson principle.*

8.7.1 Exercises

1. Let $x \in \mathbb{R}$, $a \in \mathbb{R}$. For the problem
   maximise $F(x, a) = -x^2 + 4ax + 2a^2$ with respect to x
   determine
   (a) the dependency of the maximiser x on a, that is, $x = X(a)$,

(b) the value function V,
(c) the derivative of V,
(d) the derivative $\dfrac{\partial F}{\partial a}(x, a)$ at $(X(a), a)$.

2. Let $x \in \mathbb{R}$, $(a_1, a_2) \in \mathbb{R}^2_{++}$. For the problem
minimise $F(x, a_1, a_2) = a_1 - a_2 x + x^2$ with respect to x
determine
(a) the dependence of the minimiser x on a_1, a_2, that is, $x = X(a_1, a_2)$,
(b) the value function V,
(c) the derivative of V,
(d) the derivative $\dfrac{\partial F}{\partial a_1}(x, a_1, a_2)$, $\dfrac{\partial F}{\partial a_2}(x, a_1, a_2)$ at $(X(a_1, a_2), a_1, a_2)$.
3. Let $x \in \mathbb{R}$, $a \in \mathbb{R}_{++}$. For the problem maximise $F(x, a) = -a^5 x^4 + 8x^3 - e^a x^2 + 9$ with respect to x, it is difficult to determine both the maximiser $x = X(a)$ and the value function V in a neighbourhood of a certain a, say $a = 1$ (though they exist in such a neighbourhood). Show, with the aid of the envelope theorem, that the maximum of the function $x \mapsto F(x, a)$ decreases as a increases.
4. Let $(x_1, x_2) \in \mathbb{R}^2_{++}$, $c \in \mathbb{R}_{++}$. For the problem
maximise $F(x_1, x_2) = x_1 + x_2$ under the restriction
$g(x_1, x_2) = x_1^2 + x_2^2 + c$
determine
(a) the Lagrange function L,
(b) the critical points of L (they depend on c),
(c) $\partial F / \partial c$ at the critical points $\tilde{x}_1 = x_1(c)$, $\tilde{x}_2 = x_2(c)$.
(d) Show that the critical points $\tilde{x}_1, \tilde{x}_2$ are maximisers.
(e) Let the function $F : \mathbb{R}^n \times \mathbb{R}^n \to \mathbb{R}$ be given by $F(\mathbf{x}, \mathbf{A}) = H(\mathbf{x}) + \mathbf{x}\mathbf{A}\mathbf{x}^T$, where $H : \mathbb{R}^n \to \mathbb{R}$, $\mathbf{x} = (x_1, \ldots, x_n)$, $\mathbf{x}^T$ is the transpose of $\mathbf{x}$, and $\mathbf{A}$ is a quadratic matrix of n^2 real "parameters" a_{jk} ($j = 1, \ldots n; k = 1, \ldots, n$) that is symmetric ($a_{jk} = a_{kj}$ for all j, k), i.e., the transpose $\mathbf{A}^T$ of $\mathbf{A}$ equals $\mathbf{A}$. For the "parameter matrices" $\mathbf{A}^1$ and $\mathbf{A}^2$ let $\mathbf{x}^1$ and $\mathbf{x}^2$ be minimisers of $F(\mathbf{x}, \mathbf{A}^1)$ and $F(\mathbf{x}, \mathbf{A}^2)$, respectively. Show that then $(\mathbf{x}^1 - \mathbf{x}^2)(\mathbf{A}^1 - \mathbf{A}^2)(\mathbf{x}^1 - \mathbf{a}^2)^T \leq 0$.

8.7.2 Answers

1. (a) $x = X(a) = 2a$,
(b) $V(a) = -4a^2 + 8a^2 + 2a^2 = 6a^2$,
(c) $\frac{dV}{da}(a) = 12a$,
(d) $\frac{\partial F}{\partial a}(x, a) = 4x + 4a$, $\frac{\partial F}{\partial a}(X(a), a) = 4X(a) + 4a = 8a + 4a = 12a$.
2. (a) $x = X(a_1, a_2) = a_2/2$,
(b) $V(a_1, a_2) = a_1 - a_2^2/2 + a_2^2/4 = a_1 - a_2^2/4$,
(c) $\frac{\partial V}{\partial a_1}(a_1, a_2) = 1$, $\frac{\partial V}{\partial a_2}(a_1, a_2) = -a_2/2$,
(d) $\frac{\partial F}{\partial a_1}(x, a_1, a_2) = 1$,
$\frac{\partial F}{\partial a_2}(x, a_1, a_2) = -x = -X(a_1, a_2) = -a_2/2$.

3. $\frac{dF}{da}(x, a) = -a^4x^4 - e^a x^2$ is negative at all $x \neq 0$ and all a. Without determining it explicitly we insert $X(a)$ for x. We know that as a increases, $F(X(a), a)$ is a decreasing function of a.
4. (a) $L(x_1, x_2, c, u) = x_1 + x_2 + u \cdot (c - x_1^2 - x_2^2)$,
   (b) $\tilde{x}_1 = \sqrt{2c}/2, \tilde{x}_2 = \sqrt{2c}/2, \tilde{u} = 1/\sqrt{2c}$,
   (c) $\frac{\partial F}{\partial c}(\tilde{x}_1, \tilde{x}_2) = 1/\sqrt{2c}$.
   (d) Since $\frac{\partial^2 L}{\partial x_1^2} = \frac{\partial^2 L}{\partial x_2^2} = -2u$, $u > 0$, $\frac{\partial^2 L}{\partial x_1 \partial x_2} = 0$, we know that $\tilde{x}_1$, $\tilde{x}_2$ are maximizers.
5. From the definition of $\mathbf{x}^1$ and $\mathbf{x}^2$ we have, as in (8.124), (8.125),

$$H(\mathbf{x}^1) + \mathbf{x}^1\mathbf{A}^1\mathbf{x}^{1^T} \leq H(\mathbf{x}^2) + \mathbf{x}^2\mathbf{A}^1\mathbf{x}^{2T}$$

$$H(\mathbf{x}^2) + \mathbf{x}^2\mathbf{A}^2\mathbf{x}^{2^T} \leq H(\mathbf{x}^1) + \mathbf{x}^1\mathbf{A}^2\mathbf{x}^{1^T}.$$

Adding up these inequalities and subtracting $H(\mathbf{x}^1) + H(\mathbf{x}^2)$ on both sides of the resulting inequality gives

$$\mathbf{x}^1\mathbf{A}^1\mathbf{x}^{1^T} + \mathbf{x}^2\mathbf{A}^2\mathbf{x}^{2^T} \leq \mathbf{x}^2\mathbf{A}^1\mathbf{x}^{2^T} + \mathbf{x}^1\mathbf{A}^2\mathbf{x}^{1^T},$$

that is,

$$\mathbf{x}^1\mathbf{A}^1\mathbf{x}^{1^T} - \mathbf{x}^2\mathbf{A}^1\mathbf{x}^{2^T} + \mathbf{x}^2\mathbf{A}^2\mathbf{x}^{2^T} - \mathbf{x}^1\mathbf{A}^2\mathbf{x}^{1^T} \leq 0,$$

(compare to (8.123)). The left-hand side of this inequality is what remains when we calculate $(\mathbf{x}^1 - \mathbf{x}^2)(\mathbf{A}^1 - \mathbf{A}^2)(\mathbf{x}^1 + \mathbf{x}^2)^T$. (Notice that four of the eight terms of this product add up to zero because of $\mathbf{x}^1\mathbf{A}^1\mathbf{x}^{2^T} = (\mathbf{x}^1\mathbf{A}^1\mathbf{x}^{2^T})^T = \mathbf{x}^2\mathbf{A}^{1^T}\mathbf{x}^{1^T} = \mathbf{x}^2\mathbf{A}^1\mathbf{x}^{1^T}$).

8.8 Extrema of an Objective Function Under Inequality Constraints

In Sects. 8.6 and 8.7 we were interested in determining extrema of an objective function of several variables under equality constraints, where at least one of the elements of the set {objective function, constraints} is not linear or affine. In Sect. 8.7 the functions and constraints depended not only on variables but also on parameters.

This and the following section consider the case of an objective function under *inequality* constraints. As in Sects. 8.6 and 8.7 we deal with nonlinear optimisation.

Application 1 Here $\mathbf{x} = (x_1, \ldots, x_n) \in \mathbb{R}_+^n$ is the vector of input quantities (*inputs*) for a company, $\mathbf{p} = (p_1, \ldots, p_n) \in \mathbb{R}_+^n$ the vector of *prices* per unit charged for the n kinds of inputs. Furthermore, $P : \mathbb{R}_+^n \to \mathbb{R}_+$ is the *production function*, the value

$P(\mathbf{x})$ of which is (see Sect. 7.5) the *maximal output* value obtainable from the inputs $x_1, \ldots, x_n$ during a production "period" (fixed time interval). A problem of interest for the company is the following. *Produce with minimal input costs an output worth at least b money units*, that is, *minimise*

$$F(\mathbf{x}) = \mathbf{p} \cdot \mathbf{x} = p_1 x_1 + \ldots p_n x_n \tag{8.128}$$

under the conditions (constraints)

$$P(\mathbf{x}) \geq b, \tag{8.129}$$

$$\mathbf{x} \geq 0. \tag{8.130}$$

Compare this problem to that in Sect. 2.3. Both there and here the objective function is linear or affine but here the constraints (8.129) may be nonlinear. This would be the case, for instance, if P in (8.129) were the Cobb–Douglas function (see Sect. 7.5, (7.29)). As in Sects. 2.4 and 4.8, further constraints may have to be added, reflecting limitation of inputs.

Example 1 We chose in (8.128) $F(\mathbf{x}) = F(x_1, x_2) = 16x_1 + 2x_2$ and, in (8.129), $P(x_1, x_2) = 10x_1^{2/3}x_2^{1/3}$, that is, P is a Cobb–Douglas production function (as in Sect. 6.5) and $b = 20$. Then our optimisation problem is the following.

$$\textit{Minimise} \quad F(x_1, x_2) = 16x_1 + 2x_2 \tag{8.131}$$

under the conditions

$$P(x_1, x_2) = 10x_1^{2/3}x_2^{1/3} \tag{8.132}$$

$$x_1 \geq 0, \quad x_2 \geq 0. \tag{8.133}$$

We will solve this problem by a geometric method and inspection.

We see (Fig. 8.10) that the "*feasible domain*" (the set of admissible solutions, compare Sect. 5.2) of the objective function F in (8.131) is given by (8.132) and (8.133), that is, the part (shaded area) of $\mathbb{R}^2_{++}$ (of the "first quarter plan") above the curve with the equation

$$10x_1^{2/3}x_2^{1/3} = 20, \qquad \text{that is} \quad x_1^{2/3}x_2^{1/3} = 2$$

or explicitly (by taking cubes on both sides and solving with respect to x_2)

$$x_2 = 8x_1^{-1} \tag{8.134}$$

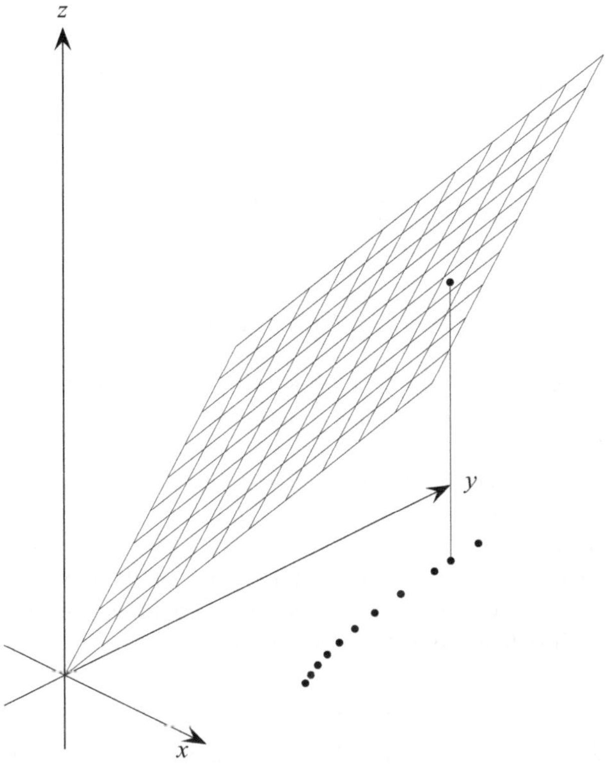

Fig. 8.10 Geometric representation of the optimisation problem: Minimise $16x_1 + 2x_2$ under the constraints $10x_1^{2/3}x_2^{1/3} \geq 20$, $x_1 \geq 0$, $x_2 \geq 0$ (solution point $(\hat{x}_1, \hat{x}_2)$)

This curve (see Fig. 8.10) clearly goes through the points $(1, 8)$ and $(2, 2)$ because $8 = 8 \cdot 1^{-2}$ and $2 = 8 \cdot 2^{-2}$.

The contour lines (see Sect. 3.3) of F in (8.131) are given by

$$16x_1 + 2x_2 = c.$$

These are parallel straight lines with the slope -8 (as seen in form $x_2 = 8x_1 + c/2$ of the equation). Because of $x_1 \geq 0$, $x_2 \geq 0$, we are interested in the segments of these lines in the first quarter plan. The segment with the smallest c, which has points in common with the feasible domain, determines the solution $(\hat{x}_1, \hat{x}_2)$ of the optimisation problem (8.131), (8.132) and (8.133) and the minimal value $16\hat{x}_1 + 2\hat{x}_2$. Since the function $x_1 \mapsto 8x_1^2$ (see (8.134)) is differentiable on $\mathbb{R}++$, the solution will be the point where the derivative of this function (the slope of the curve described by (8.134)) will be -8 (see Fig. 8.10). Since

$$\frac{d(8x_1^{-2})}{dx_1} = -16x_1^{-3},$$

we are looking for the $\hat{x}_1$ with

$$-16\hat{x}_1^{-3} = -8,$$

that is,

$$\hat{x}_1^3 = 2, \quad \hat{x}_1 = \sqrt[3]{2} \approx 1.26,$$

and

$$\hat{x}_2 = 8\hat{x}_1^{-2} \approx 5.04.$$

So *the solution of (8.131), (8.132) and (8.133) is the point* $(1.26, 5.04)$ (up to the third decimal) and *the minimal value is* approximately

$$F(\hat{x}_1, \hat{x}_2) \approx 16 \cdot 1.26 + 2 \cdot 5.04 = 30.24$$

(compare to Fig. 8.10).

If further conditions (constraints) are imposed upon the inputs then the feasible domain *may* be restricted and the previous solution point *may* not be contained in it anymore. If, for instance, the further conditions are

$$x_1 + 3x_2 \leq 9, \tag{8.135}$$

$$x_1 \leq 5, \quad x_2 \leq 2.5 \tag{8.136}$$

then the feasible domain is restricted to the shaded area in Fig. 8.11.

The new solution point will be the "upper left corner" of this area, a point of intersection of the curve segment with equation (8.134) and the straight line segment with equation

$$x_1 + 3x_2 = 9,$$

both with $0 \leq x_1 \leq 5$, $0 \leq x_2 \leq 2.5$. The new solution point $(\bar{x}_1, \bar{x}_2)$ has thus to satisfy

$$\bar{x}_2 = 8\bar{x}_1^{-2},$$

$$\bar{x}_1 + 3\bar{x}_2 = 9 \quad (0 \leq \bar{x}_1 \leq 5,\ 0 \leq \bar{x}_2 \leq 2.5).$$

As Fig. 8.11 shows, $\bar{x}_2 \leq 2.5$ is no genuine restriction. Putting the first equation into the second, we get

$$\bar{x}_1 + 24\bar{x}_2^{-2} = 9$$

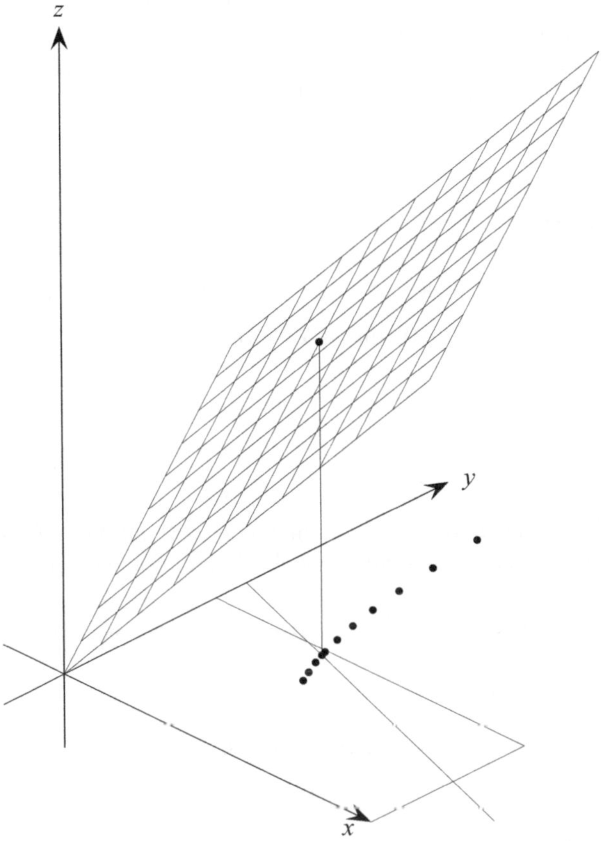

Fig. 8.11 Geometric representation of the optimisation problem dealt with in Fig. 8.10, under the further constraints $x_1 + 3x_2 \leq 9, x_1 \leq 5, x_2 \leq 2.5$ (solution point $(\bar{x}_1, \bar{x}_2)$)

or, multiplied by $\bar{x}_1^2$,

$$\bar{x}_1^3 - 9\bar{x}_2^2 + 24 = 0.$$

This equation of third degree (cubic equation) has three solutions (as Fig. 8.11 shows, one is close to 8.7; another not shown on the figure is close to -1.5) but only one satisfies $0 \leq \bar{x}_1 \leq 5$, it is approximately $\bar{x}_1 \approx 1.83$. Since $\bar{x}_2 = 8\bar{x}_1^{-2} \approx 2.39$, also $0 \leq \bar{x}_2 \leq 2.5$ is satisfied. So the new solution point is (compare Fig. 8.11) $(1.83, 2.39)$ (up to the third decimal) and the new minimal value is $16\bar{x}_1 + 2\bar{x}_2 \approx 34.06$.

Application 2 This time $\mathbf{x} = (x_1, \ldots, x_n) \in \mathbb{R}_+^n$ is the vector of the *quantities* of the n goods and services which a household may use; the vector of their prices per unit is again $\mathbf{p} = (p_1, \ldots, p_n) \in rz^n_+$. Furthermore, let $B \in \mathbb{R}_{++}$ be the *budget* of the household and $U(\mathbf{x}) \in \mathbb{R}$ the *utility* of these quantities of goods and services for the household (notice that the utility may also be negative). According to the *principle of efficiency* in economics, we have to

$$\textit{maximise} \quad U(\mathbf{x}) \tag{8.137}$$

under the conditions

$$\mathbf{p} \cdot \mathbf{x} = p_1 x_1 + \ldots + p_n x_n \leq B, \tag{8.138}$$

$$\mathbf{x} \geq 0, \text{ that is } x_1 \geq 0, \ldots, x_n \geq 0. \tag{8.139}$$

If the utility function $U : \mathbb{R}^n_+ \to \mathbb{R}$ is nonlinear then this too is a nonlinear optimisation problem.

Example 2 We chose in (8.137), budget as $B = 9$ and the prices as $p_1 = 3$, $p_2 = 1$ (in whatever money unit we deal). So the problem is to

$$\textit{maximise} \quad U(x_1, x_2) = 6x_1 - x_1^2 + 4x_2 - x_2^2 \tag{8.140}$$

under the conditions

$$3x_1 + x_2 \leq 9, \tag{8.141}$$

$$x_1 \geq 0, \; x_2 \geq 0. \tag{8.142}$$

The law of diminishing marginal returns in utility from Hermann Heinrich Gossen (1810–1858) requires that the *utility function U* be strictly concave (strictly convex from above). According to Sect. 8.2 this means that *the Hessian matrix* $U''(\mathbf{x})$ be *negative definite*. This is indeed the case:

$$\frac{\partial^2 U}{\partial x_1^2} = -2, \quad \frac{\partial^2 U}{\partial x_1 \partial x_2} = 0, \quad \frac{\partial^2 U}{\partial x_2^2} = -2,$$

$$U''(x_1, x_2) = \begin{pmatrix} -2 & 0 \\ 0 & -2 \end{pmatrix}.$$

The eigenvalues of this matrix are the solution of

$$0 = \det \begin{pmatrix} -2 - \lambda & 0 \\ 0 & -2 - \lambda \end{pmatrix} = (-2 - \lambda)^2.$$

The (double) solution of this equations is *negative* $\lambda_1 = \lambda_2 = -2$. So the Hessian matrix is negative definite and U is indeed strictly convex from above.

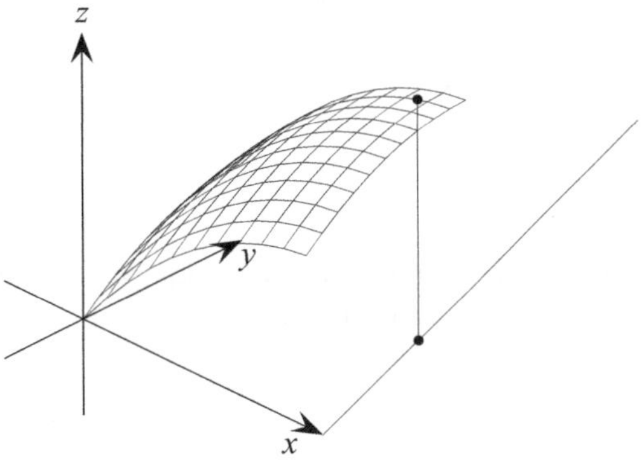

Fig. 8.12 Geometric representation of the problem of maximising $6x_1 - x_1^2 + 4x_2 - x_2^2$ under the conditions $3x_1 + x_2 \leq 9$, $x_1 \geq 0, x_2 \geq 0$ and of its solution $(\hat{\mathbf{x}}_1, \hat{\mathbf{x}}_2)$

We will see later ("convex optimisation") that the convexity (or concavity) of the objective function is very helpful in the solution of optimisation problems. Here, however, we will solve the problem (8.140), (8.141) and (8.142) by intuitive geometric considerations. First a false start: $U(x_1, x_2)$ in (8.140) can be written as

$$U(x_1, x_2) = 13 - (x_1 - 3)^2 - (x_2 - 2)^2. \tag{8.143}$$

This shows immediately that U has at $x_1 = 3$, $x_2 = 2$ a unique global maximum on $\mathbb{R}_+^2$ with the value 13. (It shows also that $U(x_1, x_2)$ can be negative, for instance at $\mathbf{x} = (6, 5)$. This is not absurd: There are households for which consumption of relatively big quantities of nonrenewable goods have negative utility). However, $x_1 = 3$, $x_2 = 2$ do not satisfy (8.141). $(3, 2)$ is *not in the feasible domain*, shaded in Fig. 8.12.

An argument similar to that by which we solved (8.131), (8.132) and (8.133) may be more successful: The contour lines of the function described by (8.143) have the equation

$$13 - (x_1 - 3)^2 - (x_2 - 2)^2 = c, \text{ that is, } (x_1 - 3)^2 + (x_2 - 2)^2 = 13 - c \quad (c \in \mathbb{R}).$$

So *the contour lines are concentric circles around* $(3, 2)$ *with radius* $(13 - c)^{1/2}$. *Accordingly, the smaller the radius, the greater the U-value c*. Thus we are looking for the *circle with the smallest radius which has points in common with the feasible domain* described by (8.141) and (8.142). Since, see Fig. 8.12, this domain is the rectangular triangle (interior and boundaries), bounded by the horizontal and vertical axes and by the segment in the first quarter plane of the straight line with equation

$$3x_1 + x_2 = 9, \tag{8.144}$$

the circle with the smallest radius will be the one whose tangent is this straight line. We could again proceed by differentiating but we know that the tangent of a circle is perpendicular to its radius at that point. Since (8.144) can be written as $x_2 = -3x_1 + 9$, the tangent has the slope -3, so the radius line's slope is $1/3$ and its equation is

$$x_2 = \frac{1}{3}x_1 + b.$$

We can determine b from the knowledge that this straight line goes through the centre $(3, 2)$ of the circle, so

$$2 = \frac{1}{3}3 + b, \qquad b = 1$$

and the equation is

$$x_2 = \frac{1}{3}x_2 + 1.$$

For the point of intersection $(\hat{\mathbf{x}}_1, \hat{\mathbf{x}}_2)$ of this straight line and that described by (8.144) (which is the point where the latter touches the circle)

$$3\hat{\mathbf{x}}_1 + (\frac{1}{3}\hat{\mathbf{x}}_1 + 1) = 9,$$

that is,

$$\hat{\mathbf{x}}_1 = \frac{24}{10} = 2.4, \qquad \hat{\mathbf{x}}_2 = \frac{1}{3}\hat{\mathbf{x}}_1 + 1 = 1.8.$$

So the solution point of the optimisation problem (8.140), (8.141) and (8.142) is $(2.4, 1.8)$ and the conditional maximum value is $6\cdot 2.4 - 2.4^2 + 4\cdot 1.8 + 1.8^2 = 12.6$ (as expected smaller than 13, the unconditional global maximum value of U).

In these examples it was relatively easy to find the solution, because the objective functions and conditions were simple, there were only two variables and the geometric representation was quite intuitive. The methods of solution were ad hoc, suggested by particularities of these problems.

A more systematic and general method but still with geometric motivation, is the *method of steepest ascent* (for maxima) or of steepest descent (for minima) which we encountered in the context of linear optimisation already in Sects. 2.4 and 5.2. The basic idea is to start from a feasible point (which we found somehow) and advance on that straight path on which the values of the objective function show the greatest increase (decrease) till the boundary of the feasible domain (beyond which at least one of the constraints would be violated). The thus obtained point is closer to the solution of the optimisation problem but needs not be the solution point itself.

We will not discuss here in general how to go further from this point but will do so in one particular case, that of the nonlinear optimisation problem (8.140), (8.141) and (8.142).

Let us start, for instance, with the origin $(0, 0)$, which certainly is in the feasible domain, and advance on the straight line leading to the global maximum point (without constraints) which, as we happen to know, is $(3, 2)$. This will be the *path of steepest ascent*. Its equation is

$$x_2 = \frac{2}{3}x_1.$$

We have to stay in the permissible domain, so (8.141) has to be satisfied:

$$3x_1 + x_2 = \frac{1}{3}x_1 \leq 9.$$

The largest (farthest from 0) x_1 satisfying the inequality is $\bar{x}_1 = 27/11$. Then $\bar{x}_2 = 2/3 \cdot 27/11 = 18/11$ and we got the point $(\bar{x}_1, \bar{x}_2) = (27/11, 18/11)$ (see Fig. 8.12). At this feasible point

$$U(\bar{x}_1, \bar{x}_2) = 6 \cdot \frac{27}{11} - \left(\tfrac{27}{11}\right)^2 + 4 \cdot \tfrac{18}{11} - \left(\tfrac{18}{11}\right)^2 \approx 12.57,$$

somewhat short of the actual maximal value 12.60 calculated above:

How should we go further? In what direction? It is easy to see that we have to move on the boundary line

$$3x_1 + x_2 = 9.$$

Indeed, points above it are not feasible (because they do not satisfy (8.32)) and for any point below it we can get one on the boundary with greater U value by moving on the (straight) line of steepest ascent (towards the centre of the concentric circles) till the boundary. Starting with the point $(\bar{x}_1, \bar{x}_2) = (\frac{27}{11}, \frac{18}{11})$ determined above, the points on the boundary can be given in parametric form as

$$(\bar{x}_1, \bar{x}_2) = \left(\tfrac{27}{11}, \tfrac{18}{11}\right) + \lambda(-1, 3), \quad (\lambda \in \mathbb{R}), \tag{8.145}$$

because the boundary is parallel to the vector $(-1, 3)$. The value of the objective function U is on this line

$$U(x_1, x_2) = 6\left(\tfrac{27}{11} - \lambda\right) - \left(\tfrac{27}{11} - \lambda\right)^2 + 4\left(\tfrac{18}{11} + 3\lambda\right) - \left(\tfrac{18}{11} + 3\lambda\right)^2 =: f(\lambda).$$

So we look for the maximum of this function (compare Sect. 5.3). The derivation

$$f'(x) = -6 + 2(\tfrac{27}{11} - \lambda) + 12 - 6(\tfrac{18}{11} + 3\lambda)$$

is 0 at $\hat{\lambda} = \frac{3}{55}$ and

$$f''(\lambda) = -2 - 18 = -20 < 0$$

everywhere. So we have a maximum at $\hat{\lambda} = \frac{3}{55}$. By (8.145), we arrive at the point

$$(\hat{x}_1, \hat{x}_2) = (\tfrac{27}{11}, \tfrac{18}{11}) + \tfrac{3}{55}(-1, 3) = (\tfrac{12}{5}, \tfrac{9}{5}) = (2.4, 1.8),$$

the solution point which we had obtained before.

If the contour lines and constraints are not so simple then we find out, as follows, which way to move from $(\bar{x}_1, \bar{x}_2) = \bar{\mathbf{x}}$. As in Sect. 5.4, we have

$$U(\bar{\mathbf{x}} + \mathbf{h}) \approx U(\bar{\mathbf{x}}) + \frac{\partial U}{\partial x_1}(\bar{\mathbf{x}})h_1 + \frac{\partial U}{\partial x_2}(\bar{\mathbf{x}})h_2 = U(\bar{\mathbf{x}}) + \nabla U(\bar{\mathbf{x}}) \cdot \mathbf{h}$$

approximately. Another way to look at this expression is to notice that

$$\frac{\partial U}{\partial x_1}(\bar{\mathbf{x}})\frac{h_1}{|\mathbf{h}|} + \frac{\partial U}{\partial x_2}(\bar{\mathbf{x}})\frac{h_2}{|\mathbf{h}|} = \frac{\partial U}{\partial x_1}(\bar{\mathbf{x}})\cos\alpha + \frac{\partial U}{\partial x_2}(\bar{\mathbf{x}})\sin\alpha$$

is the *directional derivative* (see Sect. 5.2) of U at $\bar{\mathbf{x}}$ or *measure of ascent* in the direction α (see Fig. 8.13). We *first find out in which direction* $U(\mathbf{x} + \mathbf{h})/|\mathbf{h}|$ *is maximal* in this approximation which is, as we know, the more accurate the smaller $|\mathbf{h}|$ is. So *we have a short step in that direction, staying in the feasible domain.* Let the unit vector of our direction vector $\bar{\mathbf{h}}$ be $\bar{\mathbf{k}}$ and move by $\bar{\mathbf{h}} = \lambda\bar{\mathbf{k}}$ to get to $\bar{\bar{x}} = \bar{\mathbf{x}} + \bar{\mathbf{h}} = \bar{\mathbf{x}} + \lambda\bar{\mathbf{k}}$ with a small $\lambda = |\bar{\mathbf{h}}|$ and the best unit vector $\bar{\mathbf{k}}$ just

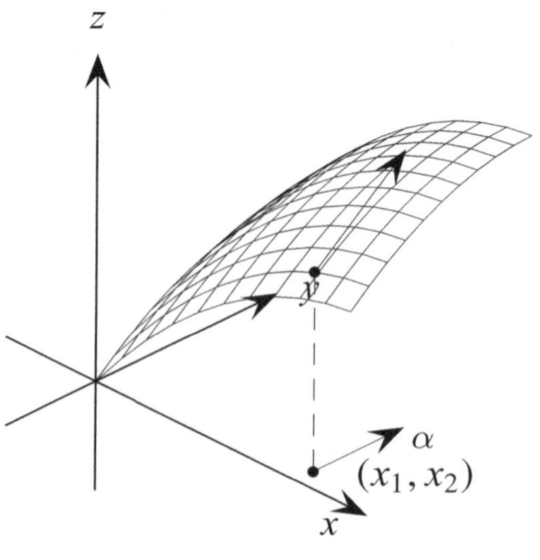

Fig. 8.13 Geometric background concerning the derivative of a function defined by $U : \mathbb{R}^2_+ \to \mathbb{R}$ at $\bar{x} = (\bar{x}_1, \bar{x}_2)$ in the direction α

determined. We will have $U(\bar{\bar{x}})/\left|\bar{\mathbf{h}}\right| \geq U(\bar{\mathbf{x}})/\left|\bar{\mathbf{h}}\right|$, that is, $U(\bar{\bar{x}}) > U(\bar{\mathbf{x}})$ (if λ is small enough). Then we advance from $\bar{\bar{x}}$ by the same method till we reach a still feasible $\hat{\mathbf{x}}$ from which we cannot get to any feasible point with larger (not smaller) U value. This will be at least an *approximation of a local condition maximum point* (not necessarily sharp). This works also for functions of more than two variables.

After these rather intuitive arguments in special cases, we formulate now the general, in particular nonlinear optimisation problem and, mostly without proof, some results about its solution.

In Application 1 we had to minimise a function F while in Application 2 the problem was to maximise U. The latter, however, is equivalent to *minimising* the function $F = -U$. Also, *all conditions* (constraints) (8.20), (8.21), (8.29), (8.30), in particular (8.23), (8.24), (8.32), (8.33), *can be written in the form*

$$G_1(x_1, \ldots, x_n) \leq 0, \ \ldots \ G_m(x_1, \ldots, x_n) \leq 0. \tag{8.146}$$

If we also had equality constraints in our applications, for instance, if we had $g_1(x_1, \ldots, x_n) = 0, \ldots, g_r(x_1, \ldots, x_n) = 0$, they could have been written in the form (8.146): $g_1(x_1, \ldots, x_n) \leq 0, -g_1(x_1, \ldots, x_n) \leq 0, \ldots, g_r(x_1, \ldots, x_n \leq 0, -g_r(x_1, \ldots, x_n) \leq 0$. Notice also that, as with linear programming, the *domain* is described by (part of) the constraints (8.37) (for instance (8.23), (8.24), (8.32)), so it seems that we could leave the functions $F, \ G_1, \ \ldots, \ G_m$ defined on all of $\mathbb{R}^n$ (or, for practical reasons, $\mathbb{R}^n_+$). It may not be possible, however, to define some functions on all of $\mathbb{R}^n$ or $\mathbb{R}^n_+$ (for instance $(x_1, x_2) \mapsto (1 - x_1^2 - x_2^2)^{\frac{1}{2}}$ is not defined if $x_1^2 + x_2^2 > 1$). Therefore we will keep saying "on the whole domain" occasionally, instead of "on $\mathbb{R}^n$" or "on $\mathbb{R}^n_+$".

As mentioned before, *an optimisation problem is nonlinear if at least one of the functions $F, \ G_1, \ \ldots, \ G_m$ is not linear or affine.* An optimisation problem may have infinitely or finitely many solutions or just one (unique) solution or no solution at all.

In the usual vector form with the notations

$$\mathbf{x} = (x_1, \ldots, x_n), \qquad \mathbf{G} = (G_1, \ldots, G_m),$$

the optimisation problem consists of

$$\text{minimising } F(\mathbf{x})$$

under the constraints

$$\mathbf{G}(\mathbf{x}) \leqq \mathbf{0}.$$

Unfortunately, there are no general methods for the solution of nonlinear optimisation problems comparable, for instance, to the simplex algorithm in linear optimisation. We can formulate, however, conditions, neither necessary nor sufficient, for a vector $\mathbf{x}$ to

$$\text{minimise } F(\mathbf{x}) \text{ under the constraints } \mathbf{G}(\mathbf{x}) \leqq \mathbf{0} \tag{8.147}$$

These can be formulated with aid of *Lagrange functions*. These are functions $L : \mathbb{R}^n \times \mathbb{R}^n \to \mathbb{R}$ defined by

$$\begin{aligned} L(\mathbf{x}, \mathbf{u}) &= F(\mathbf{x}) + \mathbf{u} \cdot \mathbf{G}(\mathbf{x}) \\ &= F(\mathbf{x}) + u_1 G_1(\mathbf{x}) + \cdots + u_m G_m(\mathbf{x}), \end{aligned} \tag{8.148}$$

where $\mathbf{x} \in \mathbb{R}^n$, $\mathbf{u} \in \mathbb{R}^m$ (later $\mathbf{u} \in \mathbb{R}^m_+$, $\mathbf{x} \in \mathbb{R}^n_+$). The components $u_1, \ldots, u_m$ of $\mathbf{u}$ are the *Lagrange multipliers*. If a Lagrange multiplier u_k is nonnegative, it can be considered *penalty per unit* by which $G_k(\mathbf{x})$ is above 0. As mentioned, for the time being, we let the domain of F be the entire $\mathbb{R}^n$ or, at least, $\mathbb{R}^n_+$.

It is remarkable that, while in Sect. 8.3 (local) saddle points figured as alternatives to local maxima or minima without constraints, here a certain kind of *global saddle points of the Lagrange function* can help us to solve the global minimum problem (8.147) with constraints. These are points (vectors) $(\hat{\mathbf{x}}, \hat{\mathbf{u}})$ in the $(n+m)$-dimensional space $\mathbb{R}^n \times \mathbb{R}^m$ such that

$$L(\hat{\mathbf{x}}, \mathbf{u}) \le L(\hat{\mathbf{x}}, \hat{\mathbf{u}}) \le L(\mathbf{x}, \hat{\mathbf{u}}) \quad \text{for all } \mathbf{x} \in \mathbb{R}^n;\ \mathbf{u} \in \mathbb{R}^m. \tag{8.149}$$

We called such saddle points *global*, because (8.149) is supposed to hold for *all* $\mathbf{x} \in \mathbb{R}^n$, $\mathbf{u} \in \mathbb{R}^m$,

For $n = m = 1$ such points (x, u) are clearly saddle points of the type in Fig. 8.6 of Sect. 8.3; see also Fig. 8.7. Notice that, because of the special form (8.148) of the Lagrange functions, the functions

$$\mathbf{u} \longmapsto L(\mathbf{x}, \mathbf{u}) = F(\mathbf{x}) + \mathbf{u} \cdot \mathbf{G}(\mathbf{x})$$

are affine for all $\mathbf{x}$. Specially for $\mathbf{x} = \hat{\mathbf{x}}$, by (8.149), $\mathbf{u} \mapsto L(\hat{\mathbf{x}}, \mathbf{u})$ has to have a global maximum at $\hat{\mathbf{u}}$, so, being affine, it has to be constant (compare Fig. 8.14:

$$L(\hat{\mathbf{x}}, \mathbf{u}) = c \quad \text{(constant)}. \tag{8.150}$$

Restricting $\mathbf{u} \geqq \mathbf{0}$, we have, in terms of $L(\mathbf{x}, \mathbf{u})$, the following *sufficient condition for* $\hat{\mathbf{x}}$ *to be a global minimum of* F *under the condition(s)* $\mathbf{G}(\mathbf{x}) \leqq \mathbf{0}$. *If* $(\hat{\mathbf{x}}, \hat{\mathbf{u}})$ *is a saddle point of the Lagrange function* L*, defined by (8.148)* (for $\mathbf{u} \geqq \mathbf{0}$), *that is* (compare (8.149))

$$F(\hat{\mathbf{x}}) + \mathbf{u} \cdot \mathbf{G}(\hat{\mathbf{x}}) \le F(\hat{\mathbf{x}}) + \hat{\mathbf{u}} \cdot \mathbf{G}(\hat{\mathbf{x}}) \le F(\mathbf{x}) + \hat{\mathbf{u}} \cdot \mathbf{G}(\mathbf{x}) \quad \textit{for} \quad \mathbf{u} \geqq \mathbf{0}, \tag{8.151}$$

and for all $\mathbf{x}$ *in the domain then* $\hat{\mathbf{x}}$ *is a solution of the optimisation problem (8.147).*

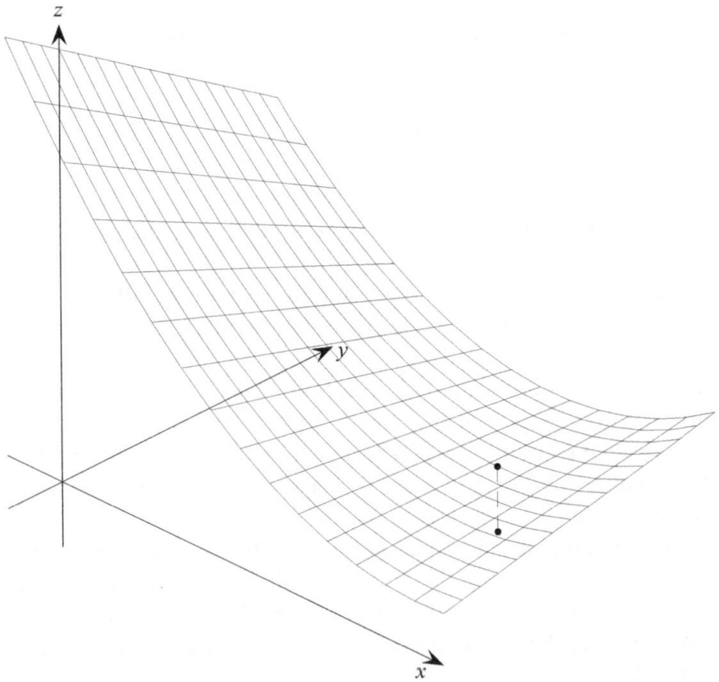

Fig. 8.14 Global saddle point $(\hat{\mathbf{x}}, \hat{\mathbf{u}}) = (5, 3)$ on $\mathbb{R} \times \mathbb{R}$ of the Lagrange function $L : \mathbb{R} \times \mathbb{R} \to \mathbb{R}$, $L(x, u) = x^2 - 13x + 50 + (x - 5)u$, for the problem: Maximise $x^2 - 13x + 50$ under the constraint $x \leq 5$

Indeed, the first inequality in (8.151) can be written as

$$\mathbf{u} \cdot \mathbf{G}(\hat{\mathbf{x}}) \leq \hat{\mathbf{u}} \cdot \mathbf{G}(\hat{\mathbf{x}}). \tag{8.152}$$

For $\mathbf{u} = \mathbf{0}$ this gives $\hat{\mathbf{u}} \cdot \mathbf{G}(\hat{\mathbf{x}}) \geq \mathbf{0}$, while with $\mathbf{u} = 2\hat{\mathbf{u}}$ we get $\hat{\mathbf{u}} \cdot \mathbf{G}(\hat{\mathbf{x}}) \leq \mathbf{0}$. So

$$\hat{\mathbf{u}} \cdot \mathbf{G}(\hat{\mathbf{x}}) = 0. \tag{8.153}$$

But then (8.152) reduces to

$$\mathbf{u} \cdot \mathbf{G}(\hat{\mathbf{x}}) \leq 0, \qquad \text{that is,} \qquad u_1 G_1(\hat{\mathbf{x}}) + \ldots + u_m G_m(\hat{\mathbf{x}}) \leq 0$$

Since $u_k \geq 0$ $(k = 1, \ldots, m)$, this is possible only if $G_k(\hat{\mathbf{x}}) \leq 0$ $(k = 1, \ldots, m)$ (if we had $G_{k_0}(\hat{\mathbf{x}}) > 0$ then, with $u_k = 0$ for $k \neq k_0$ and $u_{k_0} = 1$, we would have $\mathbf{u} \cdot G(\hat{\mathbf{x}}) > 0$, that is,

$$\mathbf{G}(\hat{\mathbf{x}}) \leqq \mathbf{0}. \tag{8.154}$$

So $\hat{\mathbf{x}}$ satisfies the m conditions (one vector condition) $\mathbf{G}(\mathbf{x}) \leqq \mathbf{0}$ in (8.147). With (8.153) the second inequality of (8.151) becomes $F(\hat{\mathbf{x}}) \leq F(\mathbf{x}) + \hat{\mathbf{u}} \cdot G(\mathbf{x})$. For all $\mathbf{x}$ satisfying the constraint $\mathbf{G}(\mathbf{x}) \leqq \mathbf{0}$ in (8.147) we get

$$F(\hat{\mathbf{x}}) \leq F(\mathbf{x}),$$

that is, $\hat{\mathbf{x}}$ indeed minimises $F(\mathbf{x})$ if $\mathbf{G}(\mathbf{x}) \leqq \mathbf{0}$, as asserted.

As we saw (see (8.149), (8.150)), $(\hat{\mathbf{x}}, \hat{\mathbf{u}})$ is a (global) saddle point of L if

(a) the function $\mathbf{u} \mapsto L(\hat{\mathbf{x}}, \mathbf{u})$ is constant (in this case for $\mathbf{u} \geqq \mathbf{0}$) and
(b) $\hat{\mathbf{x}}$ is a global minimum (or maximum) of $\mathbf{x} \mapsto L(\mathbf{x}, \hat{\mathbf{u}})$.

We saw in the previous proof the importance of (8.153) and (of course) of (8.154). We mention without proof that *(a) can be replaced by (8.153) and (8.154) for L given by (8.148)* that is, $(\hat{\mathbf{x}}, \hat{\mathbf{u}})$ *is a saddle point of L belonging through (8.148) to the optimisation problem (8.147) (with* $\mathbf{u} \geqq \mathbf{0}$*) if and only if, (b) and (8.153) and (8.154) are satisfied.*

Notice the remarkable fact that *neither of the two results* just mentioned *supposed any regularity* (continuity, convexity, differentiability) of the functions F, $G_1, \ldots, G_m$. Neither of the conditions (a), (b), (8.153) or (8.154) is, however, easy to check. We will later have conditions which are easier to verify (the Kuhn-Tucker conditions) in the case when these functions are continuously differentiable (that is, their derivatives are continuous functions).

First we mention, however, *convex optimisation*, that is, the case where F, $G_1, \ldots, G_m$ are all *convex from below* on the domain in which we are interested. Multiplying the appropriate functions by -1 shows that *maximising a function F convex from above (concave) and/or changing the constrains into*

$$G_1(\mathbf{x}) \geq 0, \ldots, G_m(\mathbf{x}) \geq 0$$

but then supposing $G_1, \ldots, G_m$ *to be convex from above (concave) are also problems of convex optimisation*, equivalent to the one above. Since affine functions are convex (Sect. 3.6), *linear optimisation is a particular case of convex optimisation.* We speak of *quadratic optimisation* if, in (8.147) or its "maximise" equivalent, F is a quadratic form and $G_1, \ldots, G_m$ are affine functions.

A noteworthy advantage of convex optimisation is that conditional local minima are also conditional global maxima for them. These two concepts are defined as follows. For the optimisation problem (8.147), $\hat{\mathbf{x}}$ is a*conditional local minimum* if $\mathbf{G}(\hat{\mathbf{x}}) \leqq \mathbf{0}$ (that is, (8.154) holds) and there exists an (open) neighbourhood of $\hat{\mathbf{x}}$ such that, for all $\hat{\mathbf{x}}$ satisfying $\mathbf{G}(\mathbf{x}) \leqq \mathbf{0}$ in that neighbourhood,

$$F(\hat{\mathbf{x}}) \leq F(\mathbf{x}). \tag{8.155}$$

On the other hand, $\hat{\mathbf{x}}$ is a *conditional global minimum* for (8.147) if $\mathbf{G}(\hat{\mathbf{x}}) \leqq \mathbf{0}$ and (8.155) holds for all $\mathbf{x}$ satisfying $\mathbf{G}(\mathbf{x}) \leqq \mathbf{0}$ in the domain of F. The equality of local and global conditional minima in *convex* optimisation problems is proved the same way as it was done in Sect. 8.3 for unconditional extrema of convex functions.

As for linear optimisation (Sect. 5.2), also *for convex optimisation problems, the set of solution vectors is a convex set* (Sect. 3.5). Indeed, if $\hat{\mathbf{x}}$ and $\tilde{\mathbf{x}}$ are solution vectors then

$$F(\hat{\mathbf{x}}) \leq F(\mathbf{x}) \quad \text{and} \quad F(\tilde{\mathbf{x}}) \leq F(\mathbf{x})$$

for all $\mathbf{x}$ satisfying $\mathbf{G}(\mathbf{x}) \leqq \mathbf{0}$. If we multiply the second inequality by $\lambda \in]0, 1[$ and the first by $(1 - \lambda)$, we get from the convexity of F from below

$$F((1-\lambda)\hat{\mathbf{x}} + \lambda\tilde{\mathbf{x}}) \leq (1-\lambda)F(\hat{\mathbf{x}}) + \lambda F(\tilde{\mathbf{x}}) \leq F(\mathbf{x})$$

or all $\mathbf{x}$ satisfying $\mathbf{G}(\mathbf{x}) \leqq \mathbf{0}$. So $(1-\lambda)\hat{\mathbf{x}} + \lambda\tilde{\mathbf{x}}$ is also a solution, as asserted.

In the case of convex optimisation, the above *sufficient condition* involving $L(\mathbf{x}, \mathbf{u}) = F(\mathbf{x}) + \mathbf{u} \cdot \mathbf{G}(\mathbf{x})$ is also necessary for F to be *a global minimum*, provided $\mathbf{G}(\mathbf{x}) \leqq \mathbf{0}$, *if* we slightly *strengthen* this last assumption in the following sense: In addition to $\mathbf{G}(\mathbf{x}) \leqq \mathbf{0}$ for all $\mathbf{x}$ in the domain *we suppose that*

$$\text{there exists an } \mathbf{x}' \text{ for which } \mathbf{G}(\mathbf{x}') < \mathbf{0} \text{ (strictly)}. \tag{8.156}$$

This is called the *Slater condition. If, under this condition,* $\hat{\mathbf{x}}$ *minimises the convex function F then there exists a* $\hat{\mathbf{u}} \geqq \mathbf{0}$ *such that* $(\hat{\mathbf{x}}, \hat{\mathbf{u}})$ *is a saddle point of* $L(\mathbf{x}, \mathbf{u}) = F(\mathbf{x}) + \mathbf{u}\mathbf{G}(\mathbf{x})$ $(\mathbf{u} \geqq \mathbf{0})$. We do not prove this but note that, combined with the above sufficient condition, we have now the following result.

If, in the convex optimisation problem (8.147), there exists an $\mathbf{x}'$ *with the strict inequality* $\mathbf{G}(\mathbf{x}') < \mathbf{0}$ *and the Lagrange function defined by (8.148) has a saddle point* $(\hat{\mathbf{x}}, \hat{\mathbf{u}})$ *then F has a conditional global minimum at* $\hat{\mathbf{x}}$. *Conversely, if* $\hat{\mathbf{x}}$ *is a solution of the convex optimisation problem (8.147) under the Slater condition (8.156), then there exists a* $\hat{\mathbf{u}} \geqq \mathbf{0}$ *such that* $(\hat{\mathbf{x}}, \hat{\mathbf{u}})$ *is a saddle point of* $L(x, \mathbf{u}) = F(x) + \mathbf{u} \cdot \mathbf{G}(x)$. *These are the Kuhn-Tucker conditions.*

8.8.1 Exercises

1. (a) Determine, by a graphical method similar to that applied in this section, the solution point $(\widehat{x_1}, \widehat{x_2})$ to the problem:
   Minimise $F(x_1, x_2) = x_1^3 - 3x_1x_2$
   subject to $2x_1 - x_2 + 5 = 0,\ 37 - 5x_1 - 2x_2 \leq 0,\ x_1 \geq 0,\ x_2 \geq 0.$
   (b) Calculate the (constrained) minimum of F.

2. (a) Determine, by graphical method similar to that applied in this section, the solution point $(\widehat{x_1}, \widehat{x_2})$ to the problem:
   Minimise $F(x_1, x_2) = \sqrt{x_1 x_2}$
   subject to $x_1 + 4x_2 \leq 40, x_1 \geq 0, x_2 \geq 0.$
   (b) Calculate the (constrained) maximum of F.
3. (a) Is the problem of quadratic optimisation:
   Minimise $F(x_1, x_2) = x_1^2 + x_2^2$
   subject to $x_1 + 4x_2 \geq 5, 4x_1 + x_2 \geq 5, x_1 \geq 0, x_2 \geq 0$
   a convex optimisation problem?
   (b) Determine the solution of this problem.
4. (a) Is the problem of quadratic optimisation:
   Minimise $F(x_1, x_2) = x_1^2 - x_2^2$
   subject to $x_1 + x_2 \leq 5, x_1 + 5x_2 \leq 10, x_1 \geq 0, x_2 \geq 0$
   a convex optimisation problem?
   (b) Determine the solution of this problem.
5. Let the system of (convex) constraints of a nonlinear optimisation problem be

$$G_1(x_1, x_2, x_3) = x_1^2 + x_2^2 + x_3^2 - 1 \leq 0,$$

$$G_2(x_1, x_2, x_3) = x_1^4 + x_2^4 + x_3^4 - \frac{1}{4} \leq 0,$$

$$G_3(x_1, x_2, x_3) = \quad e^{x_1 + x_2 + x_3} \quad - 5 \ \leq 0.$$

   Is there an (x_1', x_2', x_3') satisfying the Slater condition (8.156)?

8.8.2 Answers

1. (a) $(\hat{\mathbf{x}}_1, \hat{\mathbf{x}}_2) = (\ 5, 15)$, (b) $F(\hat{\mathbf{x}}_1, \hat{\mathbf{x}}_2) = -100.$
2. (a) $(\hat{\mathbf{x}}_1, \hat{\mathbf{x}}_2) = (20,\ 5)$, (b) $F(\hat{\mathbf{x}}_1, \hat{\mathbf{x}}_2) = \quad 10.$
3. (a) Yes, $F(x_1, x_2) = x_1^2 + x_2^2$ is (strictly) convex from below, $5 - x_1 - 4x_2 \leq 0$ and $5 - 4x_1 - x_2 \leq 0$ are convex from below.
   (b) The (unique) solution point is $(\hat{\mathbf{x}}_1, \hat{\mathbf{x}}_2) = (1, 1)$, the (constrained) minimum of F is $F(1, 1) = 2$.
4. (a) No, $F(x_1, x_2) = x_1^2 - x_2^2$ is neither convex from below nor convex from above.
   (b) The (unique) solution point is $(\hat{\mathbf{x}}_1, \hat{\mathbf{x}}_2) = (5, 0)$, the (constrained) minimum of F is $F(5, 0) = 25$.
5. Yes, for instance $(x_1', x_2', x_3') = (\frac{1}{2}, \frac{1}{2}, \frac{1}{2})$:

$$G_1(\tfrac{1}{2}, \tfrac{1}{2}, \tfrac{1}{2}) = \frac{3}{4} - 1 = -\frac{1}{4} < 0,$$

$$G_2(\tfrac{1}{2}, \tfrac{1}{2}, \tfrac{1}{2}) = \frac{3}{16} - \frac{1}{4} = -\frac{1}{16} < 0,$$

$$G_3(\tfrac{1}{2}, \tfrac{1}{2}, \tfrac{1}{2}) = e^{3/2} - 5 \approx 4.481689 - 5 = -0518311 < 0.$$

8.9 The Kuhn–Tucker Conditions

Now we get the promised *Kuhn–Tucker conditions*. They provide a solution method for problem (8.39) (in Sect. 8.8) when the functions $F, G_1, \ldots, G_m$ are *continuously differentiable*. Harold William Kuhn (*1925) and Albert William Tucker (*1905) showed, among others: If $(\hat{\mathbf{x}}, \hat{\mathbf{u}})$ is a saddle point of L (see (8.40) and problem (8.39)) then the conditions (8.157), (8.158), (8.159), (8.160), (8.161), (8.162), (8.163), (8.164), (8.165) and (8.166) below are satisfied. In other words, these conditions are *necessary* for $(\hat{\mathbf{x}}, \hat{\mathbf{u}})$ to be a saddle point of L. *From now on we suppose* $\hat{\mathbf{x}} \geq \mathbf{0}$. We denote the gradient (derivative, compare Sect. 6.11) of L with respect to $\mathbf{x}$ or $\mathbf{u}$ by $\nabla_{\mathbf{x}} L$ and $\nabla_{\mathbf{u}} L$, respectively.

If $(\hat{\mathbf{x}}, \hat{\mathbf{u}}) \in \mathbb{R}^n_+ \times \mathbb{R}^m_+$ *is a saddle point of* $L : \mathbb{R}^n_+ \times \mathbb{R}^m_+ \to \mathbb{R}$ *then*

$$\nabla_{\mathbf{x}} L(\hat{\mathbf{x}}, \hat{\mathbf{u}}) \geq \mathbf{0}, \quad \textit{that is,} \quad \frac{\partial L}{\partial x_j}(\hat{\mathbf{x}}, \hat{\mathbf{u}}) \geq 0 \quad (j = 1, \ldots, n), \tag{8.157}$$

$$\hat{\mathbf{x}} \cdot \nabla_{\mathbf{x}} L(\hat{\mathbf{x}}, \hat{\mathbf{u}}) = \mathbf{0}, \quad \textit{that is,} \quad \sum_{j=1}^{n} \hat{x}_j \frac{\partial L}{\partial x_j}(\hat{\mathbf{x}}, \hat{\mathbf{u}}) = 0, \tag{8.158}$$

$$\nabla_{\mathbf{u}} L(\hat{\mathbf{x}}, \hat{\mathbf{u}}) \leq \mathbf{0}, \quad \textit{that is,} \quad \frac{\partial L}{\partial u_k}(\hat{\mathbf{x}}, \hat{\mathbf{u}}) \leq 0 \quad (k = 1, \ldots, m), \tag{8.159}$$

$$\hat{\mathbf{u}} \cdot \nabla_{\mathbf{u}} L(\hat{\mathbf{x}}, \hat{\mathbf{u}}) = \mathbf{0}, \quad \textit{that is,} \quad \sum_{k=1}^{m} \hat{u}_k \frac{\partial L}{\partial u_k}(\hat{\mathbf{x}}, \hat{\mathbf{u}}) = 0. \tag{8.160}$$

The conditions (8.157), (8.158), (8.159) and (8.160) as well as the following conditions (8.161), (8.162), (8.163), (8.164), (8.165) and (8.166) are the *Kuhn–Tucker conditions*. Because of $\hat{\mathbf{x}} \geq \mathbf{0}$ and (8.157) all terms in (8.158) have to be nonnegative. Also $\hat{\mathbf{u}} \geq \mathbf{0}$ and(8.159) imply that all terms in (8.160) are nonpositive. But they have to add up to 0, so each term has to be zero; therefore *(8.158) and (8.160) can be replaced by*

$$\hat{x}_j \frac{\partial L}{\partial x_j}(\hat{\mathbf{x}}, \hat{\mathbf{u}}) = 0 \quad (j = 1, \ldots, n) \tag{8.161}$$

and

$$\hat{u}_k \frac{\partial L}{\partial u_k}(\hat{\mathbf{x}}, \hat{\mathbf{u}}) = 0 \quad (k = 1, \ldots, m), \tag{8.162}$$

respectively. Since (see (8.40)) $L(\mathbf{x}, \mathbf{u}) = F(\mathbf{x}) + \mathbf{u}G(\mathbf{x})$, (8.161) becomes

$$\frac{\partial F}{\partial x_j}(\hat{\mathbf{x}}) + \sum_{k=1}^{m} \hat{u}_k \frac{\partial G_k}{\partial x_j}(\hat{\mathbf{x}}) \geq 0 \quad (j = 1, \ldots, n), \tag{8.163}$$

$$\hat{x}_j \left(\frac{\partial F}{\partial x_j}(\hat{\mathbf{x}}) + \sum_{k=1}^{m} \hat{u}_k \frac{\partial G_k}{\partial x_j}(\hat{\mathbf{x}}) \right) = 0 \quad (j = 1, \ldots, n), \tag{8.164}$$

and (8.159) and (8.162) can be written as

$$G_k(\hat{\mathbf{x}}) \leq 0 \quad (k = 1, \ldots, m) \tag{8.165}$$

and

$$\hat{u}_k G_k(\hat{\mathbf{x}}) = 0 \quad (k = 1, \ldots, m), \tag{8.166}$$

respectively. Of course, (8.165) is just the condition (8.47). Notice that (8.166) and (8.164) imply that, if $\hat{u}_k > 0$ for a k, then $G_k(\hat{\mathbf{x}}) = 0$ for that k and if, for a j, $\hat{x}_j > 0$ then, for that j,

$$\frac{\partial F}{\partial x_j}(\hat{\mathbf{x}}) + \sum_{k=1}^{m} \hat{u}_k \frac{\partial G_k}{\partial x_j}(\hat{\mathbf{x}}) = 0.$$

In addition to the equation (8.165) and to the continuous differentiability of $F, G_1, \ldots, G_m$ let these functions be also convex and the Slater condition (8.117) be satisfied. Then $\hat{\mathbf{x}} \geq 0$ is a global minimum point of this convex optimisation problem if, and only if, there exists a $\hat{\mathbf{u}} \geq 0$ with which the Kuhn–Tucker conditions (8.163), (8.165) and (8.166) are satisfied. So, in this situation the Kuhn–Tucker conditions are *necessary and sufficient*.

We do not prove these results. We note, however, the following. The conditions (8.165)–(8.166) are in general *not sufficient* for $\hat{\mathbf{x}}$ to be a global minimum in the optimisation problem (8.39) if not all of $F, G_1, \ldots, G_m$ are convex (or we have no proof that they are) or we do not know whether (8.117) is satisfied. However, the Kuhn–Tucker conditions can be *useful* even in these cases. Indeed they then give *all candidates for global minima*: no $\hat{\mathbf{x}} \geq \mathbf{0}$ for which there does not exist a $\hat{\mathbf{u}} \geq \mathbf{0}$ such that (8.165)–(8.166) are satisfied, can be a solution of (8.39). So, if we can solve (8.163), (8.164), (8.165) and (8.166), we substitute the solutions into F; *the solution $\hat{\mathbf{x}}$ which gives the smallest $F(\hat{\mathbf{x}})$ value will give the global minimum* of F under the conditions $\mathbf{G}(\mathbf{x}) \leq \mathbf{0}$.

We *apply* these results *to Example 2* (see Sect. (8.8)) where we can compare the result to what we obtained previously.

In order to bring (8.140), (8.141) and (8.142) to the form (8.39), we define F by

$$F(x_1, x_2) = -U(x_1, x_2) = x_1^2 - 6x_1 + x_2^2 - 4x_2 \tag{8.167}$$

and write (8.141), (8.142) as

$$G_1(x_1, x_2) = 3x_1 + x_2 - 9 \leq 0, \tag{8.168}$$

$$G_2(x_1, x_2) = -x_1 \leq 0, \tag{8.169}$$

$$G_3(x_1, x_2) = -x_2 \leq 0. \tag{8.170}$$

We have shown, when we introduced Example 2 in Sect. 8.8 that U is strictly concave (strictly convex from above); G_1, G_2, G_3 are concave (not strictly), since they are affine. Furthermore the Slater condition (8.117) is satisfied, for instance for $\mathbf{x}' = (1, 1)$: $G_1(\mathbf{x}') = -5 < 0$, $G_2(\mathbf{x}') = G_3(\mathbf{x}') = -1 < 0$. Also, F, G_1, G_2, G_3 are, of course, continuously differentiable, so the Kuhn–Tucker conditions (8.157), (8.161), (8.159), (8.162) are necessary and sufficient for the global conditional minimum of F.

The definition (8.40) of the Lagrange function in Sect. 8.8 is in this example

$$L(\mathbf{x}, \mathbf{u}) = x_1^2 - 6x_1 + x_2^2 - 4x_2 + u_1(3x_1 + x_2 - 9) - u_2x_1 - u_3x_2,$$

so (8.157), (8.161), (8.159), (8.162) give

$$\begin{aligned} \frac{\partial L}{\partial x_1} &= 2x_1 - 6 + 3u_1 - u2 \geq 0, \\ \frac{\partial L}{\partial x_2} &= 2x_2 - 4 + u_1 - u_2 \geq 0, \end{aligned} \tag{8.171}$$

$$\begin{aligned} x_1\frac{\partial L}{\partial x_1} &= x_1(2x_1 - 6 + 3u_1 - u2) = 0, \\ x_2\frac{\partial L}{\partial x_2} &= x_2(2x_2 - 4 + u_1 - u_2) = 0, \end{aligned} \tag{8.172}$$

$$\begin{aligned} \frac{\partial L}{\partial u_1} &= 3x_1 + x_2 - 9 \leq 0, \\ \frac{\partial L}{\partial u_2} &= -x_1 \leq 0, \qquad \frac{\partial L}{\partial u_3} = -x_2 \leq 0, \end{aligned} \tag{8.173}$$

$$\begin{aligned} u_1\frac{\partial L}{\partial u_1} &= u_1(3x_1 + x_2 - 9) = 0, \\ u_2\frac{\partial L}{\partial u_2} &= -u_2x_1 = 0, \qquad u_3\frac{\partial L}{\partial u_3} = -u_3x_2 = 0, \end{aligned} \tag{8.174}$$

respectively. None of the points $\mathbf{x}^1 = (0, t)$, $\mathbf{x}^2 = (s, 0)$ $(s, t \in \mathbb{R}_+)$ are global conditional minimum points, since $F(0, t) = t^2 - 4t > F(1, t) = -5 + t^2 - 4t$, $F(s, 0) = s^2 - 6s > F(s, 1) = s^2 - 6s - 3$. So the conditional minimum points are not on the boundary, and we may suppose $\hat{x}_1 \neq 0$, $\hat{x}_2 \neq 0$ for the solution. Then by (8.174), $\hat{u}_2 = \hat{u}_3 = 0$. If we had also $\hat{u}_1 = 0$, then (8.172) would give $2x_1 - 6 = 0$, $2x_2 - 4 = 0$, that is, $x_1 = 3$, $x_2 = 2$ which does not satisfy (8.168) (or the first inequality of (8.173)). So $\hat{u}_1 \neq 0$ and (8.172), (8.174) give

$$\begin{aligned} 2\hat{x}_1 + \phantom{2\hat{x}_2 +} 3\hat{u}_1 &= 6, \\ 2\hat{x}_2 + \hat{u}_1 &= 4, \\ 3\hat{x}_1 + \hat{x}_2 \phantom{+ \hat{u}_1} &= 9. \end{aligned}$$

The (unique) solution of this system of linear equations is (compare Sect. 4.6) $\hat{x}_1 = 2.4$, $\hat{x}_2 = 1.8$, $\hat{u}_1 = 0.4$. So the (only) vector minimising (8.167) under the conditions (8.168), (8.169) and (8.170) is $(2.4, 1.8)$, in accordance with what we previously found by geometric methods.

8.9.1 Exercises

1. Problem: Minimise $F(x_1, x_2) = x_1^2 + x_2^2 - x_1x_2 - 4x_2 - x_1 + 17$ subject to $x_1 + 2x_2 \leq 6$ and $x_1 \geq 0$, $x_2 \geq 0$.
   (a) Determine the Lagrange function L of this problem.
   (b) Determine all points $(\hat{\mathbf{x}}, \hat{\mathbf{u}})$ that satisfy the Kuhn–Tucker conditions.
   (c) Determine the solution point $\hat{\mathbf{x}}$ and the solution value $F(\hat{\mathbf{x}})$ of the problem.
   (d) Are the conditions satisfied which guarantee that the Kuhn–Tucker conditions are sufficient for $\hat{\mathbf{x}}$ to be a global solution of the problem?
2. Solve the following problems.
   (a) Determine, by a graphical method similar to that applied in Sect. 8.8 the solution point $(\hat{x}_1, \hat{x}_2)$ to the problem:
   Minimise $F(x_1, x_2) = x_1^2 + 4x_2^2 - x_1 - 4x_2$
   subject to $2x_1 + x_2 \leq 1$, $x_1 \geq 0$, $x_2 \geq 0$.
   (b) Calculate the (constrained) minimum of F.
3. Solve the following problems.
   (a) Given the problem in Exercise 2, write out the Kuhn–Tucker conditions for the Lagrange function $L(x_1, x_2, u_1, u_2, u_3) = x_1^2 + 4x_2^2 - x_1 - 4x_2 + u_1(2x_1 + x_2 - 1) + u_2(-x_1) + u_3(-x_2)$.
   (b) Calculate the Lagrange multipliers $\hat{u}_1$, $\hat{u}_2$, $\hat{u}_3$ belonging to $(\hat{x}_1, \hat{x}_2)$ (see Exercise 2).
4. State the Kuhn–Tucker conditions of the Lagrange function
   $L(x_1, x_2, u_1, u_2, u_3) = 16x_1 + 2x_2 + u_1(20 - 10x_1^{2/3}x_2^{1/3}) + u_2(-x_1) + u_3(-x_2)$ for the initial problem in Example 1 of Sect. 8.8. Determine with their aid the Lagrange multipliers $\hat{u}_1, \hat{u}_2, \hat{u}_3$ belonging to the solution point $(\hat{x}_1, \hat{x}_2) = (2^{1/3}, 8 \cdot 2^{-2/3}) \approx (1.26, 5.04)$.
5. Carry through the process described in Exercise 4 with the additional conditions (8.135), (8.136) of Sect. 8.8 taken into consideration.

8.9.2 Answers

1. (a) $L(x_1, x_2, u_1, u_2, u_3)$
   $= F(x_1, x_2) + u_1G_1(x_1, x_2) + u_2G_2(x_1, x_2) + u_3G_3(x_1, x_2)$
   $= x_1^2 + x_2^2 - x_1x_2 - 4x_2 - x_1 + 17$
   $+u_1(x_1 + 2x_2 - 6) - u_2x_1 - u_3x_2$.
   (b) $(\hat{\mathbf{x}}_1, \hat{\mathbf{x}}_2, \hat{\mathbf{u}}_1, \hat{\mathbf{u}}_2, \hat{\mathbf{u}}_3) = (\frac{10}{7}, \frac{16}{7}, \frac{6}{7}, 0, 0)$.
   (c) $(\hat{\mathbf{x}}_1, \hat{\mathbf{x}}_2) = (\frac{10}{7}, \frac{16}{7})$, $F(\hat{\mathbf{x}}_1, \hat{\mathbf{x}}_2) = \frac{73}{7} \approx 10.43$.

(d) Yes: The functions F, G_1, G_2, G_3 given by

$$F(x_1,x_2) = x_1^2 + x_2^2 - x_1x_2 - 4x_2 - x_1 + 17,$$

$$G_1(x_1,x_2) = x_1 + 2x_2 - 6,$$

$$G_2(x_1,x_2) = -x_1, \qquad G_3(x_1,x_2) = -x_2$$

are *convex*, and the *Slater condition (8.117) is satisfied*, for instance by $\mathbf{x}' = (1,1)$:

$$G_1(1,1) = 1 + 2 - 6 < 0,\; G_2(1,1) = -1 < 0,\; G_3(1,1) = -1 < 0.$$

2. (a) $(\hat{\mathbf{x}}_1, \hat{\mathbf{x}}_2) = (\frac{9}{34}, \frac{8}{17})$, (b) $F(\hat{\mathbf{x}}_1, \hat{\mathbf{x}}_2) = -\frac{81}{68}$.
3. (a) We calculate

$$\frac{\partial L}{\partial x_1}(x_1,x_2,u_1,u_2,u_3) = 2x_1 - 1 + 2u_1 - u_2 \geq 0,$$

$$\frac{\partial L}{\partial x_2}(x_1,x_2,u_1,u_2,u_3) = 8x_2 - 4 + u_1 - u_3 \geq 0,$$

$$x_1 \frac{\partial L}{\partial x_1}(x_1,x_2,u_1,u_2,u_3) + x_2\frac{\partial L}{\partial x_2}(x_1,x_2,u_1,u_2,u_3)$$
$$= x_1(2x_1 - 1 + 2u_1 - u_2) + x_2(8x_2 - 4 + u_1 - u_3) = 0,$$

$$\frac{\partial L}{\partial u_1}(x_1,x_2,u_1,u_2,u_3) = 2x_1 + x_2 - 1 \; \geq 0,$$

$$\frac{\partial L}{\partial u_2}(\cdot) = -x_1 \leq 0, \qquad \frac{\partial L}{\partial u_3}(\cdot) = -x_2 \leq 0,$$

$$u_1 \frac{\partial L}{\partial u_1}(\cdot) + u_2\frac{\partial L}{\partial u_2}(\cdot) + u_3\frac{\partial L}{\partial u_3}(\cdot)$$
$$= u_1(2x - 1 + x_2 - 1) + u_2(-x_1) + u_3(-x_2) \; = 0.$$

(b) $(\hat{\mathbf{u}}_1, \hat{\mathbf{u}}_2, \hat{\mathbf{u}}_3) = (\frac{4}{17}, 0, 0)$.

4. $(\hat{\mathbf{u}}_1, \hat{\mathbf{u}}_2, \hat{\mathbf{u}}_3) = (\frac{2^{7/3}+2^{4/3}}{5}, 0, 0) \approx (1.51, 0, 0)$.
5. $(\bar{x}_1, \bar{x}_2) \approx (1.83, 2.39)$, $(\bar{u}_1, \bar{u}_2, \bar{u}_3, \bar{u}_4) \approx (2.41, 1.55, 0, 0)$.

8.10 Optimisation with Several Objective Functions

Up to now we considered optimisation problems (both linear and nonlinear; also unconditional extremum problems) with just one (scalar valued) objective function.

In economics and other practical situations one may have several objectives. For instance a firm may wish to maximise not just its profit but also its market share and the quality of its products (supposed here to be measurable by a numerical measure). If there are several objective functions (or, equivalently, a vector valued objective function) then we have a problem of *multi objective optimisation* (or *vector optimisation*).

So now we have q real valued objective functions

$$F_1 : \mathbb{R}^n \longrightarrow \mathbb{R}, \ \ldots, \ F_q : \mathbb{R}^n \longrightarrow \mathbb{R}$$

or one vector valued objective function

$$\mathbf{F} = (F_1, \ldots, F_q) : \mathbb{R}^n \longrightarrow \mathbb{R}^q.$$

(In practical situations F_l is often not defined on the whole of $\mathbb{R}^n$ but rather on a domain of definition D_l, $l = 1, \ldots, q$, which is a subset of $\mathbb{R}^n$.) In the above example the profit $F_1(\mathbf{x})$, the market share $F_2(\mathbf{x})$, and the measure $F_3(\mathbf{x})$ of a quality gained from the inputs $(x_1, \ldots, x_n) = \mathbf{x} \in \mathbb{R}^n_+$ would have to be maximised or, equivalently $-F_1(\mathbf{x})$, $-F_2(\mathbf{x})$, and $-F_3(\mathbf{x})$ minimised under certain conditions (in what follows we will speak about *minimising*).

When $q \geq 2$, however, then we have problems both with maximising and with minimising since, as we saw in Sects. 1.4 and 2.2, two vectors (with two or more components) need not be comparable with respect to $\leqq$ (or $\leq$ or $<$). In other words, $\leqq$ does *not* induce a *total* order on $\mathbb{R}^q$ if $q \geq 2$ (consider, for instance, which is greater, $(1, 2, 4)$, $(2, 1, 4)$ or $(3, 1, 2)$?)

This gap is partially bridged by the following definition. A vector $\hat{\mathbf{y}}$ is *minimal* of *efficient* or *Pareto-optimal* in a set $S \subseteq \mathbb{R}^q$ if there exists no $\mathbf{y} \in S$ with $\hat{\mathbf{y}} \leq \mathbf{y}$ (which means, as we know, $\mathbf{y} \leqq \hat{\mathbf{y}}$, but $\mathbf{y} \neq \hat{\mathbf{y}}$). *Maximal* vectors are similarly defined. (Note that a set of q-dimensional vectors can have many maximal or minimal elements if $q \geq 2$; for instance, in the above example, all three vectors are minimal, maximal (efficient, Pareto-optimal). But there are also sets of q-dimensional vectors that have neither a maximal nor a minimal element.)

We can now formulate the multi objective optimisation (in this case minimisation) problem: *Determine the vectors* $\hat{\mathbf{x}}$ *for which* $\mathbf{F}(\hat{\mathbf{x}})$ *is minimal in the set*

$$\left\{\mathbf{y} = \mathbf{F}(\mathbf{x}) \;\middle|\; \mathbf{y} \in \mathbb{R}^q,\ \mathbf{x} \in \mathbb{R}^n,\ \mathbf{G}(\mathbf{x}) \leqq \mathbf{0};\ \mathbf{G}(\mathbf{x}) \in \mathbb{R}^m\right\}. \tag{8.175}$$

These vectors $\hat{\mathbf{x}}$ are the *solutions of* this *vector optimisation problem*. Just as for $q = 1$ (without supposing strict convexity), there may be infinitely or finitely many solutions or one (unique) solution or no solution at all (there are no minimum solutions, if the set (8.175) has no minimal element).

It is often not easy to determine all solutions of a multi objective optimisation problem. In such cases one wishes to determine at least a *subset of the set of solutions*. This can happen by *introducing a total order* for the elements of the set (8.175) which, as we have seen, had none. Here are three methods to do this.

(i) *Method of goal priority.* One puts the objectives (goals) in order of priority (importance). Then, for (8.175), one considers first the objective function with first priority goal ("first priority function" for short). If, for an element $\mathbf{y}^{(1)}$ of (8.175), the value of this function is greater than for $\mathbf{y}^{(2)}$ then, by definition, $\mathbf{y}^{(1)}$ is "greater by priority" than $\mathbf{y}^{(2)}$. Only if the values of the first priority function are equal for two elements of (8.175), do we consider the objective function with second priority goal and define that element "greater by priority" for which the value of this "second priority function" is greater, if any, and so on: If the values of the 1-st, ..., k-th priority functions are equal for $\mathbf{y}^{(1)}$ and $\mathbf{y}^{(2)}$, we consider the $(k+1)$-st ($k = 1, \ldots, q-1$). Only if they are equal for all of $F_1, \ldots, F_q$, are $\mathbf{y}^{(1)}$ and $\mathbf{y}^{(2)}$ "equal by priority". In Mathematics, this is called a "*lexicographic order*" (cf. Sect. 1.4).

(ii) *Method of goal weighting.* Weights $a_1, \ldots, a_q$ are assigned to the q goals and thus to the q (scalar valued) objective functions, reflecting, upon their relative importance. We form

$$\Phi = a_1 F_1 + \ldots + a_q F_q \tag{8.176}$$

as a new (scalar valued) objective function and consider that element (of (8.175)) greater for which the value of Φ is larger. Since Φ is scalar valued, this (or, to be exact: $\mathbf{y}^{(1)}$ greater than $\mathbf{y}^{(2)}$, if $a_1 y_1^{(1)} + \ldots + a_q y_q^{(1)} > a_1 y_1^{(2)} + \ldots + a_q y_q^{(2)}$) is a total order. The weights $a_1, \ldots, a_q$ are, of course, positive. It needs to be supposed that they add up to 1, since this can always be attained without changing the order they induce by dividing each of them by their sum $a_1 + \ldots + a_q$. The (conditional) minimum will be at the same point (vector).

So now we have to minimise the *one* scalar valued function Φ (under the constraints $\mathbf{G}(\mathbf{x}) \leqq \mathbf{0}$) and we are back to a single-objective optimisation problem, to which we can apply the methods which we learned in Sects. 8.8 and 8.9.

Example We *keep the conditions (8.141), (8.142)* from Example 2 in Sect. 8.8 but, instead of (8.140) we want to *minimise the vector valued function* $\mathbf{F} : \mathbb{R}_+^2 \to \mathbb{R}^3$, *whose components are given by*

$$F_1(x_1, x_2) = x_1^2 - x_1 x_2 - x_2^2 - \frac{12}{5} x_1 - 12 x_2,$$

(continued)

$$F_2(x_1,x_2) = x_1^2 + \frac{1}{2}x_1x_2 + 4x_2^2 - 9x_1 + 12x_2,$$

$$F_3(x_1,x_2) = x_1^2 + x_1x_2 + \frac{1}{3}x_2^2 - 8x_1 - 12x_2.$$

If we choose the weights as $a_1 = \frac{5}{12}$, $a_2 = \frac{1}{3}$, $a_3 = \frac{1}{4}$ ($a_1 + a_2 + a_3 = 1$) then (conveniently but not typically), we get the optimisation problem (8.140), (8.141) and (8.142) again (for Φ in place of $\mathbf{F}$):

$$\begin{aligned}\Phi(x_1,x_2) &= \frac{5}{12}F_1(x_1,x_2) + \frac{1}{3}F_2(x_1,x_2) + \frac{1}{4}F_3(x_1,x_2)\\ &= \frac{1}{12}(5x_1^2 - 5x_1x_2 - 5x_2^2 - 12x_1 - 60x_2,\\ &\quad +4x_1^2 + 2x_1x_2 + 16x_2^2 - 36x_1 + 48x_2,\\ &\quad 3x_1^2 + 3x_1x_2 + x_2^2 - 24x_1 - 36x_2)\\ &= x_1^2 + x_2^2 - 6x_1 - 4x_2.\end{aligned}$$

The solution of *this* optimisation problem is, as we know, $\hat{\mathbf{x}} = (2.4, 1.8)$. So $\hat{\mathbf{x}} = (2.4, 1.8)$ is *one* solution of the above multi objective optimisation problem. Indeed, and this argument works also for (8.176) in general, if it were *not* a solution, that would mean that there exists an $\mathbf{x}'$ such that $F_k(\mathbf{x}') \leq F_k(\hat{\mathbf{x}})$ for $k = 1, 2, 3$ but $F_{k_0}(\mathbf{x}') < F_{k_0}(\hat{\mathbf{x}})$ for at least one $k_0 \in \{1, 2, 3\}$ (and $\mathbf{G}(\mathbf{x}') \leqq \mathbf{0}$). But then we would have

$$\Phi(x_1,x_2) = \frac{5}{12}F_1(\mathbf{x}')+\frac{1}{3}F_2(\mathbf{x}')+\frac{1}{4}F_3(\mathbf{x}') < \frac{5}{12}F_1(\hat{\mathbf{x}})+\frac{1}{3}F_2(\hat{\mathbf{x}})+\frac{1}{4}F_3(\hat{\mathbf{x}}),$$

that is, $\hat{\mathbf{x}}$ would not be a conditional minimum point of Φ: a contradiction!

Of course, there can be several other solutions of the multi-objective optimisation problem, some of which could be obtained, for instance, by choosing weights different from 5/12, 1/3, 1/4.

(iii) *Method of goal programming.* This method consists of stating a goal, say $\mathbf{c} = (c_1, \ldots, c_q)$, which we would like for $\mathbf{F} = (F_1, \ldots, F_q)$ to reach, then define as new (nonnegative) scalar valued objective function the *distance* (compare Sect. 1.6) *between* $\mathbf{F}(\boldsymbol{x})$ *and* $\mathbf{c}$ or its square, that is

$$\Psi(\mathbf{x}) = (F_1(\mathbf{x}) - c_1)^2 + \ldots + (F_q(\mathbf{x}) - c_q)^2. \tag{8.177}$$

The method would consist in minimising the above expression under the condition $\mathbf{G}(\mathbf{x}) \leqq \mathbf{0}$. If, however, we choose $\mathbf{c}$ as a non-minimal element of (8.175) then we would get the minimum value $\mathbf{0}$ of (8.177) at the non-minimal point $\mathbf{x}'$ for which $\mathbf{F}(\mathbf{x}') = \mathbf{c}$, $\mathbf{G}(\mathbf{x}') \leqq \mathbf{0}$. This would, of course, make the method useless. So we have to choose as $\mathbf{c}$ either a minimal element of (8.175), which we do not know (yet), or an element which is certainly smaller (with respect to the partial order induced by $\leqq$) than each minimal element of (8.175) (if at least one exists). This can be accomplished by *choosing*

$$c_l \leq \min \left\{ y_l = F_l(\mathbf{x}) \mid \mathbf{x} \in \mathbb{R}^n, \ \mathbf{G}(\mathbf{x}) \leqq \mathbf{0} \right\} \qquad (l = 1, \ldots, q)$$

if these minima exist and are all finite (not $-\infty$). If one or more of these minima do not exist but the sets

$$\left\{ y_l = F_l(\mathbf{x}) \mid \mathbf{x} \in \mathbb{R}^n, \ \mathbf{G}(\mathbf{x}) \leqq \mathbf{0} \right\} \qquad (l = 1, \ldots, q)$$

are bounded from below (compare Sect. 7.2), say *by b_l, then choose $c_l \leq b_l$ $(l = 1, \ldots, q)$.* By a similar proof as in our example one can show that a (conditional) minimum point of the function Ψ, defined by (8.177) with such a $\mathbf{c} = (c_1, \ldots, c_q)$, is a solution of our multi-objective optimisation problem.
The set of solutions thus obtained is again in general *not* the set of *all* solutions of the original problem. One way of obtaining further solutions is to consider *other distance definitions*. Such distances different from the "*geometric distance*" which we used here and in Sect. 1.6 do exist. For instance, if a firm wants to choose the best location with regard to suppliers, it may be better advised to minimise the sum of delivery times from suppliers rather than the sum of geometric distances from them. So *the choice of "distance" may depend* also *upon the context of the real-life problem.*

8.10.1 Exercises

1. Determine, by graphical method similar to that applied in Sect. 8.8, the set of the maximal (efficient, Pareto-optimal) points $(\hat{\mathbf{x}}_1, \hat{\mathbf{x}}_2)$ to the problem:

$$\text{Maximise} \quad \begin{pmatrix} F_1(x_1, x_2) \\ F_2(x_1, x_2) \end{pmatrix} = \begin{pmatrix} 4x_1 + 2x_2 \\ 8x_1 + 20x_2 \end{pmatrix}$$

$$\text{subject to} \quad x_1 \leq 60, \ x_2 \leq 40, \ x_1 + x_2 \leq 70$$

$$x_1 + 2x_2 \leq 100, \quad x_1 \geq 0, \ x_2 \geq 0.$$

2. With aid of weights a and $1-a$ $(0 \le a \le 1)$ get *one* objective function $\Phi = aF_1 + (1-a)F_2$ from the two objective functions F_1 and F_2 in Exercise 1. Solve the problem in Exercise 1 for Φ instead of F_1, F_2 *for each value of* a. (We thank Werner Dinkelbach (*1934) for the problems in Exercise 1 and 2.)
3. Determine, by a graphical method, the set of the minimal (efficient, Pareto-optimal) points $(\hat{\mathbf{x}}_1, \hat{\mathbf{x}}_2, \hat{\mathbf{x}}_3)$ to the problem:

$$\text{Maximise} \qquad \begin{pmatrix} F_1(x_1,x_2,x_3) \\ F_2(x_1,x_2,x_3) \end{pmatrix} = \begin{pmatrix} -3x_1 - 2x_2 + 2x_3 \\ 4x_1 + x_2 - x_3 \end{pmatrix}$$

$$\text{subject to} \qquad x_1 + x_2 + x_3 \le 1,\ x_1 \ge 0,\ x_2 \ge 0,\ x_3 \ge 0.$$

4. Let two objective functions F_1 and F_2 be given by

$$F_1(x_1,x_2) = x_1^2 + x_2^2, \qquad F_2(x_1,x_2) = (x_1 + 1)x_2.$$

With the constraints

$$\begin{aligned} x_1 - 4x_2 &\le \ \ 1, & -x_1 - x_2 &\le -2, \\ -3x_1 - x_2 &\le -3, & x_1 \ge 0,\ x_2 &\ge 0 \end{aligned}$$

determine the optimal points $(\hat{\mathbf{x}}_1, \hat{\mathbf{x}}_2)$ and the minima for the problems:
(a) Minimise $F_1(x_1,x_2)$,
(b) Minimise $F_2(x_1,x_2)$,
(c) Minimise $(F_1(x_1,x_2) - c_1)^2 + (F_2(x_1,x_2) - c_2)^2$
for the goal $(c_1,c_2) = (0,0)$.
5. Let $1, 2, 3$ be the order of priority of the objective functions F_1, F_2, F_3. Solve the problem: points $(\hat{\mathbf{x}}_1, \hat{\mathbf{x}}_2)$ for the problem:

$$\text{Maximise by goal priority} \qquad \begin{pmatrix} F_1(x_1,x_2) \\ F_2(x_1,x_2) \\ F_3(x_1,x_2) \end{pmatrix} = \begin{pmatrix} x_1 + x_2 \\ x_1^2 + x_2^2 \\ x_1x_2 + x_1 \end{pmatrix}$$

$$\begin{aligned} \text{subject to} \qquad & 6x_1 + x_2 \le 27,\ x_1 + 6x_2 \le 27 \\ & x_1 + x_2 \le 7, \quad x_1 \ge 0,\ x_2 \ge 0. \end{aligned}$$

8.10.2 Answers

1. $\{(x_1,x_2) \mid x_1 + x_2 = 70,\ 20 \le x_1 \le 49\}$
$\cup \{(x_1,x_2) \mid x_1 + 2x_2 = 100,\ 40 \le x_1 \le 60\}$.

2.

$$(\hat{\mathbf{x}}_1, \hat{\mathbf{x}}_2) = \begin{cases} (20, 40) & \text{for} \quad a \in [0, 2/5[, \\ (20\lambda + 40(1-\lambda), 40\lambda + 30(1-\lambda)) & \text{for} \quad \begin{cases} a = 2/5 \\ (0 \le \lambda \le 1), \end{cases} \\ (40, 30) & \text{for} \quad a \in]2/5, 6/7[, \\ (40\lambda + 60(1-\lambda), 30\lambda + 10(1-\lambda)) & \text{for} \quad \begin{cases} a = 6/7 \\ (0 \le \lambda \le 1), \end{cases} \\ (60, 10) & \text{for} \quad a \in]6/, 1[. \end{cases}$$

4. (a) $(\hat{\mathbf{x}}_1, \hat{\mathbf{x}}_2) = (1, 1), \quad F(1, 1) = 2,$
   (b) $(\hat{\mathbf{x}}_1, \hat{\mathbf{x}}_2) = (1.8, 0.2), \quad F(1.8, 0.2) = 0.56,$
   (c) $(\hat{\mathbf{x}}_1, \hat{\mathbf{x}}_2) \approx (1.359, 0.641), \quad F_1(\hat{\mathbf{x}}_1, \hat{\mathbf{x}}_2)^2 + F_2(\hat{\mathbf{x}}_1, \hat{\mathbf{x}}_2)^2 \approx 7.384,$

$$(F_1(\hat{\mathbf{x}}_1, \hat{\mathbf{x}}_2)^2 + F_2(\hat{\mathbf{x}}_1, \hat{\mathbf{x}}_2)^2)^{\frac{1}{2}} \approx 2.717.$$

5. $(\hat{\mathbf{x}}_1, \hat{\mathbf{x}}_2) = (4, 3), \quad \begin{pmatrix} F_1(\hat{\mathbf{x}}_1, \hat{\mathbf{x}}_2) \\ F_2(\hat{\mathbf{x}}_1, \hat{\mathbf{x}}_2) \\ F_3(\hat{\mathbf{x}}_1, \hat{\mathbf{x}}_2) \end{pmatrix} = \begin{pmatrix} F_1(4, 3) \\ F_2(4, 3) \\ F_3(4, 3) \end{pmatrix} = \begin{pmatrix} 7 \\ 25 \\ 16 \end{pmatrix}.$

   Note that $F_1(3.5, 3.5) = 7$, but $F_2(3.5, 3.5) = 24.5 < 25$, and that $\binom{F_1(3,4)}{F_2(3,4)} = \binom{7}{25}$, but $F_3(3, 4) = 15 < 16$.

Set Valued Functions: Equilibria—Games 9

The development of general equilibrium theory represents one of the greatest advances in economic analysis in the latter half of the twentieth century.

BRYAN ELLICKSON, UCLA

9.1 Introduction

In Chap. 8 we showed how problems of determining maxima and minima (extrema) of scalar valued of (Sect. 8.9 vector valued) function of several variables can be solved. We were interested in *optimal points* in the domains of definition of these functions.

This chapter deals with *set valued* functions and with games. This time the important problem is *not* to find maxima or minima; their definition could be controversial in these situations and their importance would be limited anyway. Accordingly, the main notion in Sects. 9.3 and 9.4 is *not* so much optimal points as *equilibrium points*.

We introduce this concept with a very simple model of a duopoly market. Assume that only two firms (the *duopolists*) supply one good each, and that the two goods are very similar. The profit which each of the two makes in a certain time interval I ("sales period") depends not only on the price she sets during I, but also on the price set by the competitor during I. We assume that the first duopolist takes into consideration only three prices, namely the prices 10, 12 and 14, and the second only two prices, 11 and 13.

This is a simple market model. It can be treated by applying some notions and methods of the *theory of games*—not necessarily of the theory of zero-sum-games which we considered in Sect. 5.4. In the terminology of game theory the two firms are called *players*, the sets $\{10, 12, 14\}$ and $\{11, 13\}$ of prices are called *set of* three or two price *strategies*, respectively, and the profits depending on the six price vectors $(10, 11)$, $(12, 11)$, $(14, 11)$, $(10, 13)$, $(12, 13)$, $(14, 13)$ are called *payoffs*.

W. Eichhorn, W. Gleißner, *Mathematics and Methodology for Economics*,
Springer Texts in Business and Economics, DOI 10.1007/978-3-319-23353-6_9

The set consisting of these price (or strategy) vectors is the domain of definition of the payoff (or profit) function F_1 of the first player (firm, duopolist) and F_2 of the second. We assume that these functions are given by

$$\begin{array}{lll} F_1(10,11)=1000, & F_1(12,11)=\ 950, & F_1(14,11)=980, \\ F_1(10,13)=1100, & F_1(12,13)=1110, & F_1(14,13)=\ 990 \end{array}$$

and

$$\begin{array}{lll} F_2(10,11)=800, & F_2(12,11)=920, & F_2(14,11)=\ 970, \\ F_2(10,13)=750, & F_2(12,13)=910, & F_2(14,13)=1050 \end{array}$$

respectively. For instance, the profit of the first duopolist is 950 if she sets the price 12 and the second one sets the price 11. In our case of only two players and only very few strategies one usually represents the two payoff functions F_1 and F_2 by two matrices, the so-called *payoff matrices* of the players; see Table 9.1

For instance, Table 9.1 shows that $F_1(10, 11) = 1000$, that is, the profit of the first duopolist is 1000 when she sets the price at 10 and the second duopolist sets the price 11. Similarly, $F_2(12, 13) = 910$ means that the second duopolist, who sets the price at 13, gains 910 if the price set by the first is 12. If the second duopolist's price is set at 11, then we see from Table 9.1 that the first duopolist gains most by setting the price at 10. Similarly, if the second duopolist demands the price 13 then 12 is the optimal price for the first. Further we see that, for both the prices 10 and 12 of the first duopolist, the second gains most by setting the price at 11, while for the price setting 14 of the first the price 13 is optimal for the second.

More substantial is the observation that (10, 11) is the one point which is *stable* in the following sense. If the first duopolist adheres to the strategy to keep the price at 10 then keeping the price at 11 is the best strategy for the second and *conversely*, if the second duopolist's strategy is to keep the price at 11 then keeping the price at 10 is the optimal strategy for the first. Such stable points are called *Nash equilibrium points* (see in Sect. 9.4 the definition in a more general situation). A quick check shows that there are no further Nash equilibrium points in Table 9.1.

From now on we suppose that cooperation of the two players (firms, duopolists) is not permitted and that none of them can choose her strategy (price) *after* the other has chosen her strategy.

Finding the Nash equilibrium point is *just one possible way* to act in noncooperative games and *not necessarily the best one*. We see, for instance, in Table 9.1, that in the point (12, 13) which is not stable in the above sense, both the first and the

Table 9.1 The payoff matrices (payoff functions) in a duopoly

F_1	11	13
10	1000	1100
12	950	1110
14	980	990

F_2	11	13
10	800	750
12	920	910
14	970	1050

second duopolist gain more (1110 as compared to 1000 and 910 as compared to 800, respectively). In such cases we say that (10, 11) is *not Pareto-optimal*. Also (10, 11) is *not optimal with respect to the total profit* of the two duopolists: that total profit is 1800 *for* (10, 11) but 2040 for the non stable point (14, 13). While in Table 9.1 the Nash equilibrium point (10, 11) at least represents the *lowest prices*, this is not always so, as we will see in Sect. 9.4.

If one replaces, in Table 9.1, 980 by 1010 (or by any number greater than 1000) then there is *no* Nash equilibrium point. We will give conditions in Sect. 9.4 under which there always exists at least one Nash equilibrium point in an n-person m-goods game. There always exist Nash equilibrium points also in a modified n-person game, where the strategy sets are finite but each strategy is chosen—"played"—with some predetermined probability. Such games are, however, of minor importance in oligopoly theory.

If we replace, in Table 9.1, 910 by 921 then there are *two* Nash equilibrium points, (10, 11) and (12, 13), but in each of them the total profit is smaller than the maximal 2040 in (14, 13).

We get a generalisation of the Nash equilibrium point by considering the following particularly risk-averse behaviour of duopolists or, in the case of more than two, but not too many competitors or players, "oligopolists": When a "match" (a sales period with certain prices set) is over then the j.th player does *not* change strategy if *all* possible changes *could* result in changes of strategy of competitors which lead to less profit for her. Such *equilibrium points* need not be of the Nash type; for instance, in Table 9.1, (12, 13) is such a non-Nash equilibrium point (also the Nash equilibrium point (10, 11) has this property).

Notwithstanding all drawbacks Nash equilibrium points have in achieving profits in noncooperative game theory, they play an important role in some popular *oligopoly* models. In Sect. 9.4 we will give the more general definitions and deal also with k-goods oligopoly models and Nash equilibria in these models.

Equilibria of another (but not completely different) kind will be defined and considered in Sect. 9.3, so called *competitive equilibria*, that is, equilibria in a competitive exchange economy with *many* economic agents. (Note that above we spoke always about two (duopoly) or only few (oligopoly) competitors).

The theory of competitive equilibria rests essentially on the notion of set values functions (correspondences). *Such* functions will be considered in Sect. 9.3.

Another application of correspondences can be made in production theory; see, in this connection, Shephard's axioms in Sect. 9.2.

9.2 Set Valued Functions (Correspondences): Shephard's Axioms

Till now we dealt with functions whose values were scalars (real numbers) or vectors, Only in Sects. 3.2 and 3.3 did we mention, on hand of examples from consumption, preference, utility and production theory, functions which map points (vectors) of $\mathbb{R}_+^n$ into subsets of $\mathbb{R}_+^m$, meaning that a subset is assigned by the mapping to each vector. Such mappings are called *correspondences*. In order to pinpoint those correspondences which are of interest to production theory ("*production correspondences*") the mathematical economist Ronald W. Shephard (*1912–†1982)formulated certain conditions (postulates) which are now called "Shephard's axioms". We remind the reader that we had also described classes of scalar valued functions (for instance price indices in Sect. 3.7) by conditions (suppositions, postulates, "axioms").

Let $E(\mathbb{R}_+^m)$ denote the set of all subsets of $\mathbb{R}_+^m$, the so-called *power set of* $\mathbb{R}_+^m$ (often denoted by $2^{\mathbb{R}_+^m}$). A *production correspondence* (or "*output correspondence*")

$$\mathbf{P} : \mathbb{R}_+^n \longrightarrow E(\mathbb{R}_+^m)$$

is meant to assign to the input vector $\mathbf{x} = (x_1, \dots, x_n)$ the set of all output vectors $\mathbf{u} = (u_1 \dots, u_m)$ which can be produced from $\mathbf{x}$ in a given situation (technological knowledge, etc.) during a production period. In view of this, the following assumptions ("axioms") seem natural.

S1 $\mathbf{P}(\mathbf{0}) = \{\emptyset\}$; $\mathbf{0} \in \mathbf{P}(\mathbf{x})$ *for all* $\mathbf{x} \in \mathbb{R}_+^n$ *but there is at least one* $\mathbf{u} > \mathbf{0}$ *for which there exists an* $\mathbf{x} \geq \mathbf{0}$ *such that* $\mathbf{u} \in \mathbf{P}(\mathbf{x})$.

Interpretation (explanation): No input produces no output ("there ain't no such thing as a free lunch"); it is also possible to produce nothing from any input, but there are also "some things" (positive outputs) which can be produced by some input vector with at least one positive component. At this point we remind the reader that, in the partial ordering of vectors (Sect. 1.4), $\mathbf{x} > \mathbf{y}$ means that *each* component of $\mathbf{x}$ is *greater* than the corresponding component of $\mathbf{y}$, while, if $\mathbf{x} \geqq \mathbf{y}$ then *each* component of $\mathbf{x}$ is just *not smaller* than that of $\mathbf{y}$; finally $\mathbf{x} \geq \mathbf{y}$ is $\mathbf{x} \geqq \mathbf{y}$ but with *at least one* component of $\mathbf{x}$ *greater* than the corresponding component of $\mathbf{y}$.

S2 *For every* $\mathbf{x} \in \mathbb{R}_+^n$, *the set* $\mathbf{P}(\mathbf{x})$ *is bounded*, that is, there exists an m-dimensional interval $[\mathbf{a}, \mathbf{b}]$ depending on $\mathbf{x}$ such that $\mathbf{P}(\mathbf{x}) \subseteq [\mathbf{a}, \mathbf{b}]$ (compare also Sect. 3.2).

This means simply that no finite input can produce arbitrarily large output.

S3 $\mathbf{P}(\lambda\mathbf{x}) \supseteq \mathbf{P}(\mathbf{x})$ *for all* $\mathbf{x} \in \mathbb{R}_+^n$ *and all* $\lambda \in [1, \infty[$.

Explanation: Whatever can be produced with the input $\mathbf{x}$, can also be produced with $\lambda\mathbf{x}$ where $\lambda \geq 1$. This is also described as *free disposal of inputs in case that all input components are proportionally enlarged.* (The producer is free to use the inputs fully or only in part.)

S4 *If* $\mathbf{u} \in \mathbf{P}(\mathbf{x})$ *then also* $\mu\mathbf{u} \in \mathbf{P}(\mathbf{x})$ *for all* $\mu \in [0, 1]$ *and all* $\mathbf{x} \in \mathbb{R}_+$.

Interpretation: If the output $\mathbf{u}$ can be produced from the input $\mathbf{x}$ then any proportionally smaller (not larger) output can also be produced from the same input

$\mathbf{x}$. So there is, in a sense, *free disposal of outputs*. (The producer is free to use the outputs fully or only in part.)

S5 *If, for* $\mathbf{u} \geq \mathbf{0}$, *there exists an* $\mathbf{x} \geq \mathbf{0}$ *with* $\mathbf{u} \in \mathbf{P}(\mathbf{x})$ *then, for these* $\mathbf{x}$ *and* $\mathbf{u}$ *and for all* $\mu \in \mathbb{R}_{++}$, *there exists a* $\lambda \in \mathbb{R}_{++}$ *with* $\mu\mathbf{u} \in \mathbf{P}(\mathbf{x})$.

This means that proportional output changes are possible under appropriate proportional changes of input.

S6 *The graph* $\{(\mathbf{x}, \mathbf{u}) \mid \mathbf{x} \in \mathbb{R}^n_+,\ \mathbf{u} \in \mathbf{P}(\mathbf{x})\}$ *of* $\mathbf{P}$ *is a closed set.*

For the definition of closed sets see Sect. 8.3. From there we also see that this "technical" condition **S6** guarantees, together with **S2**, for instance, that the maximum $\max\{u \mid \mathbf{x} \in \mathbb{R}_+,\ u \in \mathbf{P}(\mathbf{x})\}$ exists (here $m = 1$).

A correspondence $\mathbf{P} : \mathbb{R}^n_+ \to E(\mathbb{R}^m_+)$ is an *output production correspondence* if it satisfies Shephard's axioms **S1–S6**.

For every system of axioms (conditions) the question arises, whether there exist at all objects which satisfy all these conditions. It is easy to find $\mathbf{P} : \mathbb{R}^n_+ \to E(\mathbb{R}^m_+)$ which satisfy **S1–S6**. For example, take functions $F_j : \mathbb{R}^n_+ \to E(\mathbb{R}^m_+)$, continuous and strictly increasing in each variable, homogeneous of degree $r > 0$ ($F_j(\lambda\mathbf{x}) = \lambda^r F_j(\mathbf{x})$ for all $\lambda \in \mathbb{R}_{++}$) and such that $F_j(\mathbf{0}) = 0$ ($j = 1, \ldots, m$), for instance $F_j(\mathbf{x}) = c_{j1}x_1^r + \ldots + c_{jn}x_n^r$ ($c_{jk} \in \mathbb{R}_{++}; j = 1, \ldots, m; k = 1, \ldots, n$) and define

$$\mathbf{P}(\mathbf{x}) = \{\mathbf{u} = (u_1, \ldots, u_m) \mid 0 \leq u_j \leq F_j(\mathbf{x}),\ j = 1, \ldots, m\}.$$

In principle, with *any* output correspondence $\mathbf{P}$, also its "inverse", the "input correspondence" $\mathbf{P}^{-1} : \mathbb{R}^m_+ \to E/\mathbb{R}^n_+$, is defined, by

$$\mathbf{P}^{-1}(\mathbf{u}) = \{\mathbf{x} \mid \mathbf{u} \in \mathbf{P}(\mathbf{x})\}.$$

Notice, that $\mathbf{P}^{-1}$ *assigns to each output vector* $\mathbf{u}$ *the set of all input vectors* $\mathbf{x}$ *from which*, under the given circumstances, $\mathbf{u}$ *can be produced* during a production period. Keep in mind that $\mathbf{P}^{-1}$ *is not the inverse mapping of* $\mathbf{P}$ in the sense of Sect. 3.2 since the values of $\mathbf{P}$ are sets of vectors, while the arguments of $\mathbf{P}^{-1}$ are individual vectors.

Homogeneous correspondences of degree r can be defined, in complete analogy to homogeneous functions (see Sects. 6.11 and 7.4 **3**) by

$$\mathbf{P}(\lambda\mathbf{x}) = \lambda^r\mathbf{P}(\mathbf{x}) := \{\lambda^r\mathbf{u} \mid \mathbf{u} \in \mathbf{P}(\mathbf{x})\} \quad \text{for all} \quad \lambda \in \mathbb{R}_{++},\ \mathbf{x} \in \mathbb{R}^n_+ \tag{9.1}$$

or, spelled out for production correspondences: when the input vector $\mathbf{x}$ can produce the output $\mathbf{u}$, then the proportionally increased (or decreased) input $\lambda\mathbf{x}$ can produce $\lambda^r\mathbf{u}$. Here too, we get *linear homogeneity* in the case $r = 1$. Clearly, (9.1) implies

$$\mathbf{P}^{-1}(\lambda\mathbf{u}) = \lambda^{1/r}\mathbf{P}^{-1}(\mathbf{u}) \quad \text{for all} \quad \lambda \in \mathbb{R}_{++},\ \mathbf{u} \in \mathbb{R}^m_{++}. \tag{9.2}$$

Now *suppose that for the cost* $c(\mathbf{x})$ *of the input vector* $\mathbf{x}$ we have

$$c(\mathbf{x}) \geq c(\mathbf{y}) \Longrightarrow c(\rho\mathbf{x}) \geq c(\rho\mathbf{y}) \quad \text{for all} \quad \rho \in \mathbb{R}_+. \tag{9.3}$$

This is the case, for example, if the prices $(q_1, \ldots, q_n) = \mathbf{q}$ of the inputs are constant, that is, if

$$c(\mathbf{x}) = q_1x_1 + \ldots + q_nx_n = \mathbf{q} \cdot \mathbf{x},$$

but also for more general cost functions. For instance, for any homogeneous function $c : \mathbb{R}^n_+ \to \mathbb{R}_+$ of degree $r > 0$ (not necessarily $r = 1$) we have (9.3):

$$c(\mathbf{x}) \geq c(\mathbf{y}) \Longrightarrow c(\rho\mathbf{x}) = \rho^r c(\mathbf{x}) \geq \rho^r c(\mathbf{y}) = c(\rho\mathbf{y}).$$

The *minimal cost* C of production of the output vector $\mathbf{u} \in \mathbb{R}^n_+$ is defined by

$$C(\mathbf{u}) = \min \left\{ c(\mathbf{x}) \;\middle|\; \mathbf{x} \in \mathbf{P}^{-1}(\mathbf{u}) \right\} + \gamma,$$

where γ represents the fixed cost.

Any input vector $\mathbf{x}$ establishing this minimal cost is a *minimal cost combination for* $\mathbf{u}$ *with respect to* $\mathbf{P}$ *and* c. If c is continuous and $\mathbf{P}^{-1}(\mathbf{u})$ compact, then the minimum defining C exists by what we learned in Sect. 6.8.

In what follows we answer the question how a minimal cost combination will vary if the output is changed from $\mathbf{u}^*$ to $\lambda\mathbf{u}^*$ with $\lambda \in \mathbb{R}_{++}$.

Let $\mathbf{P}$ *be an output production correspondence that is homogeneous of degree* r $(r \in \mathbb{R}_{++})$*. Let the cost* $c(\mathbf{x})$ *of the input vector* $\mathbf{x}$ *satisfy condition (9.3). If* $\mathbf{x}^*$ *is a minimal cost combination for* $\mathbf{u}^*$ *with respect to* $\mathbf{P}$ *and* c *then, for all* $\lambda \in \mathbb{R}_{++}$, $\lambda^{1/r}\mathbf{x}^*$ *is a minimal cost combination for* $\lambda\mathbf{u}^*$ *with respect to* $\mathbf{P}$ *and* c.

To prove this we show first that

$$\lambda\mathbf{u}^* \in \mathbf{P}(\lambda^{1/r}\mathbf{x}^*), \tag{9.4}$$

that is, output $\lambda\mathbf{u}^*$ is obtainable with input $\lambda^{1/r}\mathbf{x}^*$. Indeed, by assumption, $\mathbf{u}^* \in \mathbf{P}(\mathbf{x}^*)$. Hence, by (9.1),

$$\lambda\mathbf{u}^* \in \lambda\mathbf{P}(\mathbf{x}^*) = (\lambda^{1/r})^r\mathbf{P}(\mathbf{x}^*) = \mathbf{P}(\lambda^{1/r}\mathbf{x}^*)$$

and this is (9.4).

It remains to show that

$$c(\mathbf{x}) \geq c(\lambda^{1/r}\mathbf{x}^*) \quad \text{for all} \quad \mathbf{x} \in \mathbf{P}^{-1}(\lambda\mathbf{u}^*) \tag{9.5}$$

that is, no input vector $\mathbf{x}$ yielding at least $\lambda\mathbf{u}^*$ is cheaper than $\lambda^{1/r}\mathbf{x}^*$. Indeed, multiply $\lambda\mathbf{u}^* \in \mathbf{P}(\mathbf{x})$ by λ^{-1} in order to obtain (by (9.1))

$$\mathbf{u}^* \in \lambda^{-1}\mathbf{P}(\mathbf{x}) = (\lambda^{-1/r})^r\mathbf{P}(\mathbf{x}) = \mathbf{P}(\lambda^{-1/r}\mathbf{x}),$$

that is, $\mathbf{u}^*$ is obtainable with $\lambda^{-1/r}\mathbf{x}$. Since $\mathbf{x}$ is assumed to be a minimal cost combination for $\mathbf{u}^*$ with respect to $\mathbf{P}$ and c, we have

$$c(\lambda^{-1/r}) \geq c(\mathbf{x}^*).$$

By (9.3), this inequality implies (9.5). This completes the proof.

We have shown that, for homogeneous $\mathbf{P}$ and for the input cost c satisfying (9.3), the following is true: Given any scalar multiple $\lambda\mathbf{u}^*$ of $\mathbf{u}^*$ ($\lambda \in \mathbb{R}_{++}$) and a minimal cost combination $\mathbf{x}^*$ for $\mathbf{u}^*$, there exists a scalar multiple of $\mathbf{x}^*$, in this case $\lambda^{1/r}\mathbf{x}^*$, which is a minimal cost combination for $\lambda\mathbf{u}^*$. We will say that a homogeneous output production correspondence $\mathbf{P}$ yields *linear expansion paths*.

The (minimal) cost for producing the output vector $\lambda\mathbf{u}^* \neq \mathbf{0}$ is

$$C(\lambda\mathbf{u}^*) = \min\left\{c(\mathbf{x}) \mid \mathbf{x} \in \mathbf{P}^{-1}(\lambda\mathbf{u}^*)\right\} + \gamma.$$

We are interested in establishing explicitly its dependence upon $\lambda \in \mathbb{R}_{++}$. Under the assumptions (9.1) and (9.3), we have, as already shown,

$$C(\lambda\mathbf{u}^*) = c(\lambda^{1/r}\mathbf{x}^*) + \gamma \qquad (\lambda \in \mathbb{R}_{++})$$

or, if c is given by $c(\mathbf{x}) = \mathbf{q} \cdot \mathbf{x}$, where $\mathbf{q}$ is the constant and positive price vector of the inputs,

$$C(\lambda\mathbf{u}^*) = a\lambda^{1/r} + \gamma \qquad (a := \mathbf{q} \cdot \mathbf{x}^* > 0). \tag{9.6}$$

Here, $a := \mathbf{q} \cdot \mathbf{x}^*$ is greater than 0, since $\mathbf{q} > \mathbf{0}$ and, by **S1** and since $\mathbf{u}^* \geq \mathbf{0}$, $\mathbf{x}^* \geq \mathbf{0}$.

Hence, for production systems with homogeneous $\mathbf{P}$ and constant prices of the inputs there exist no "classic cost functions", that is, strictly increasing functions $\lambda \mapsto C(\lambda\mathbf{u}^*)$ which are strictly concave (convex from above) on an interval $[0, b]$ and strictly convex from below to the right of $b \in \mathbb{R}_{++}$; see Fig. 9.1.

9.2.1 Exercises

1. Give an example of a correspondence $\mathbf{Q} : \mathbb{R}^n_+ \to E(\mathbb{R}^m_+)$ satisfying the conditions **S1–S5** and the following further condition:

   The set $\left\{(\tilde{\mathbf{x}}, \mathbf{u}) \mid \tilde{\mathbf{x}} \in \mathbb{R}^n_+ \text{ fixed; } \mathbf{u} \in \mathbf{Q}(\tilde{\mathbf{x}})\right\}$ does not contain any efficient vector.

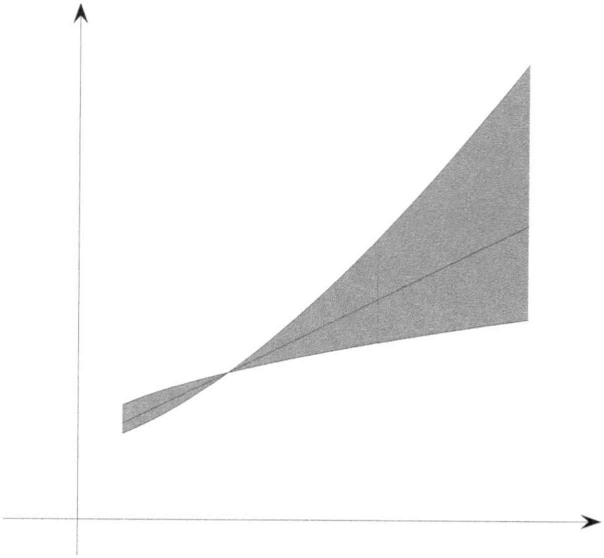

Fig. 9.1 Graphs of a "classic cost function" (*black line*) and of the cost functions (9.6) with three different r's (*shaded area*)

2. Suppose a correspondence $\mathbf{Q} : \mathbb{R}^n_+ \to E(\mathbb{R}^m_+)$ satisfying Shephard's axiom **S2**. Formulate an equivalent axiom for the inverse $\mathbf{Q}^{-1}(\mathbf{u}) = \{\mathbf{x} \mid \mathbf{u} \in \mathbf{Q}(\mathbf{x})\}$ of $\mathbf{Q}$.
3. Same problem as in Exercise 2 for Shephard's axiom **S3**.
4. Same problem as in Exercise 2 for Shephard's axiom **S5**.
5. Take the correspondence $\mathbf{P} : \mathbb{R}^2_+ \to E(\mathbb{R}^2_+)$ given by

$$\mathbf{P}\binom{x_1}{x_2} := \left\{ \binom{u_1}{u_2} \,\middle|\, \binom{u_1}{u_2} = \binom{\lambda_1}{\lambda_2} \in \mathbb{R}^2_+, \begin{pmatrix} \frac{3}{4} & \frac{1}{6} \\ \frac{1}{8} & \frac{5}{6} \end{pmatrix} \binom{\lambda_1}{\lambda_2} \leqq \binom{x_1}{x_2} \right\}$$

("Leontief output correspondence").
(a) Determine $\mathbf{P}^{-1}(\mathbf{u})$.
(b) Show that $\mathbf{P}$ is linearly homogeneous: $\mathbf{P}(t\mathbf{x}) = t\mathbf{P}(\mathbf{x})$, $t > 0$.
(c) Determine the minimal cost combination $\binom{x_1^*}{x_2^*}$ for $\binom{u_1^*}{u_2^*} = \binom{16}{24}$.
(d) Show that the minimal cost combination for $\binom{3u_1^*}{3u_2^*} = \binom{48}{72}$ is three times the minimal cost combination $\binom{x_1^*}{x_2^*}$ determined in (c).

9.2.2 Answers

1. $\mathbf{Q}(\tilde{\mathbf{x}}) = \{\mathbf{u} = (u_1, \ldots, u_m) \mid 0 \leq u_j < F_j(\tilde{\mathbf{x}}),\ j = 1, \ldots, m\}$, where $F_j : \mathbb{R}^n_+ \to \mathbb{R}$ are functions as defined after **S6** in Sect. 9.1. For each $(\tilde{\mathbf{x}}, \mathbf{u})$

satisfying $\mathbf{u} < (F_1(\tilde{\mathbf{x}}), \ldots, F_m(\tilde{\mathbf{x}}))$ there exists a vector $\mathbf{v} < (F_1(\tilde{\mathbf{x}}), \ldots F_m(\tilde{\mathbf{x}}))$ such that $\mathbf{v} > \mathbf{u}$, $(\tilde{\mathbf{x}}, \mathbf{v}) \geq (\tilde{\mathbf{x}}, \mathbf{u})$.

2. An equivalent axiom is:

$$|\mathbf{u}(k) \to \infty| \text{ for } k \to \infty \text{ implies } \bigcap_{k=1}^{\infty} \mathbf{Q}^{-1}(\mathbf{u}(k)) = \emptyset.$$

3. An equivalent axiom is:

$$\mathbf{x} \in \mathbf{Q}^{-1}(\mathbf{u}) \text{ implies } \lambda\mathbf{x} \in \mathbf{Q}^{-1}(\mathbf{u}) \text{ for all } \lambda \in [1, \infty[.$$

4. An equivalent axiom is: If, for $\mathbf{x} \geq \mathbf{0}$, there exists an $\mathbf{u} \geq \mathbf{0}$ such that $\mathbf{x} \in \mathbf{Q}^{-1}(\mathbf{u})$ then, for these $\mathbf{x}$ and $\mathbf{u}$ and for all $\mu \in \mathbb{R}_{++}$, there exists a $\lambda \in \mathbb{R}_{++}$ with $\lambda\mathbf{x} \in \mathbf{Q}(\mu\mathbf{u})$.

5. (a) $\mathbf{P}^{-1}\begin{pmatrix} u_1 \\ u_2 \end{pmatrix} = \left\{ \begin{pmatrix} x_1 \\ x_2 \end{pmatrix} \in \mathbb{R}^2_+ \;\middle|\; \begin{pmatrix} x_1 \\ x_2 \end{pmatrix} \geqq \begin{pmatrix} \frac{3}{4} & \frac{1}{6} \\ \frac{1}{8} & \frac{5}{6} \end{pmatrix} \begin{pmatrix} u_1 \\ u_2 \end{pmatrix} \right\}.$

(b) $\mathbf{P}\begin{pmatrix} tx_1 \\ tx_2 \end{pmatrix} = \left\{ \begin{pmatrix} u_1 \\ u_2 \end{pmatrix} = \begin{pmatrix} \lambda_1 \\ \lambda_2 \end{pmatrix} \in \mathbb{R}^2_+ \;\middle|\; \begin{pmatrix} \frac{3}{4} & \frac{1}{6} \\ \frac{1}{8} & \frac{5}{6} \end{pmatrix} \begin{pmatrix} \lambda_1 \\ \lambda_2 \end{pmatrix} \leqq \begin{pmatrix} tx_1 \\ tx_2 \end{pmatrix} \right\}$

$= \left\{ \begin{pmatrix} u_1 \\ u_2 \end{pmatrix} = \begin{pmatrix} \lambda_1/t \\ \lambda_2/t \end{pmatrix} \in \mathbb{R}^2_+ \;\middle|\; \begin{pmatrix} \frac{3}{4} & \frac{1}{6} \\ \frac{1}{8} & \frac{5}{6} \end{pmatrix} \begin{pmatrix} \lambda_1/t \\ \lambda_2/t \end{pmatrix} \leqq \begin{pmatrix} x_1 \\ x_2 \end{pmatrix} \right\}$

$= t\mathbf{P}(\begin{pmatrix} x_1 \\ x_2 \end{pmatrix}).$

(c) The minimal cost combination for $\binom{\mathbf{u}^*{}_1}{\mathbf{u}^*{}_2} = \binom{16}{24}$ is $\binom{\mathbf{x}^*{}_1}{\mathbf{x}^*{}_2} = \binom{16}{22}$ since

$\mathbf{P}^{-1}\begin{pmatrix} 16 \\ 24 \end{pmatrix} = \left\{ \begin{pmatrix} x_1 \\ x_2 \end{pmatrix} \in \mathbb{R}^2_+ \;\middle|\; \begin{pmatrix} x_1 \\ x_2 \end{pmatrix} \geqq \begin{pmatrix} \frac{3}{4} & \frac{1}{6} \\ \frac{1}{8} & \frac{5}{6} \end{pmatrix} \begin{pmatrix} 16 \\ 24 \end{pmatrix} = \begin{pmatrix} 16 \\ 22 \end{pmatrix} \right\}.$

(d) $\mathbf{P}^{-1}\begin{pmatrix} 3\mathbf{u}^*{}_1 \\ 3\mathbf{u}^*{}_2 \end{pmatrix} = \mathbf{P}^{-1}\begin{pmatrix} 48 \\ 72 \end{pmatrix}$

$= \left\{ \begin{pmatrix} x_1 \\ x_2 \end{pmatrix} \in \mathbb{R}^2_+ \;\middle|\; \begin{pmatrix} x_1 \\ x_2 \end{pmatrix} \geqq \begin{pmatrix} \frac{3}{4} & \frac{1}{6} \\ \frac{1}{8} & \frac{5}{6} \end{pmatrix} \begin{pmatrix} 48 \\ 72 \end{pmatrix} = \begin{pmatrix} 48 \\ 66 \end{pmatrix} \right\}.$

9.3 Competitive Equilibria: Kakutani's Fixed Point Theorem

Set–valued maps (correspondences) play a prominent role not only in production theory, that is, in the theory of *supply*, but also in the theory of *demand* (including consumption preference and utility theory) and in the theory of *economic equilibrium*. The objects of the latter are conditions under which supply and demand balance out.

The model proposed by LEON WALRAS (*1834 – †1910) for an "*exchange economy*" became quite famous. For simplicity it excludes production, still its main point is the balance of supply and demand. Each economic agent has an initial supply (or endowment) of l possible goods in store which it can offer in exchange. If this initial supply (endowment) consists, for the economic agent A, of the quantities e_1, …, e_l, united in the "*vector (or bundle) of commodities*" $\mathbf{e} = (e_1, \ldots, e_l) \in \mathbb{R}^l_+$, and if the prices are $p_1, \ldots, p_l$, respectively, forming the price vector $\mathbf{p} = (p_1, \ldots, p_l) \in \mathbb{R}^l_+$ then the "*wealth*" of A will be

$$\mathbf{p} \cdot \mathbf{e} = p_1 e_1 + \cdots + p_l e_l. \tag{9.7}$$

In the *Walras–model* the supposition is that A can obtain goods only up to this value (no credit). So, A's total *demand* may be represented by any element of the "*budget set*"

$$\mathbf{B}(\mathbf{p}, \mathbf{e}) := \{\mathbf{x} \mid \mathbf{p} \cdot \mathbf{x} \leq \mathbf{p} \cdot \mathbf{e}\} \tag{9.8}$$

(given the prices $\mathbf{p} = (p_1, \cdots, p_l)$ and A's initial supply $\mathbf{e} = (e_1, \ldots, e_l)$). So the mapping $\mathbf{B}$ is a *correspondence* which assigns to $\mathbf{p}$ and $\mathbf{e}$ the *set* of bundles of commodities that can be obtained by using all or part of A's wealth.

Most of the time, however, A is not equally interested in all bundles of commodities which are elements of $\mathbf{B}(\mathbf{p}, \mathbf{e})$. When A is more interested in bundle $\mathbf{y} \in \mathbf{B}(\mathbf{p}, \mathbf{e})$ than in bundle $\mathbf{x} \in \mathbf{B}(\mathbf{p}, \mathbf{e})$ we say that A *prefers* $\mathbf{y}$ *to* $\mathbf{x}$. Note that the *preferences* of A and the other agents in the exchange economy may be different.

The *formal structure of preferences* is often described by the following requirements **P1**, **P2**, **P3**. In what follows, for $\mathbf{x} \in \mathbb{R}^l_+$ and $\mathbf{y} \in \mathbb{R}^l_+$ the "*preference relation*" $\mathbf{x} \preceq \mathbf{y}$ means that $\mathbf{y}$ *is weakly preferred to* $\mathbf{x}$, i.e., $\mathbf{y}$ is preferred to $\mathbf{x}$ or there is indifference between $\mathbf{x}$ and $\mathbf{y}$.

P1 *Reflexivity: for all* $\mathbf{x} \in \mathbb{R}^l_+ : \mathbf{x} \preceq \mathbf{x}$.

In our context this requirement is fulfilled as we will show below.

P2 *Transitivity: for all* $\mathbf{x} \in \mathbb{R}^l_+, \mathbf{y} \in \mathbb{R}^l_+, \mathbf{z} \in \mathbb{R}^l_+$: *the preference relations* $\mathbf{x} \preceq \mathbf{y}$ *and* $\mathbf{y} \preceq \mathbf{z}$ *imply* $\mathbf{x} \preceq \mathbf{z}$.

This *consistency requirement* means the following: If, for A, $\mathbf{y}$ is weakly preferred to $\mathbf{x}$ and $\mathbf{z}$ is weakly preferred to $\mathbf{y}$ then $\mathbf{z}$ is also weakly preferred to $\mathbf{x}$.

P3 *Complete (or total) ordering: for all* $\mathbf{x} \in \mathbb{R}^l_+, \mathbf{y} \in \mathbb{R}^l_+$: *either* $\mathbf{x} \preceq \mathbf{y}$ *or* $\mathbf{y} \preceq \mathbf{x}$ *or both have to hold.*

This means that A is able to choose one ("the preferred one") or, in case of indifference, either bundle from any pair $\mathbf{x}$, $\mathbf{y}$ of bundles of commodities.

A few further definitions and notations follow:

$$\mathbf{y} \succeq \mathbf{x} \quad \text{means the same as} \quad \mathbf{x} \preceq \mathbf{y}$$

(both also verbalised as "$\mathbf{y}$ is at least as desirable as $\mathbf{x}$").

Indifference between **x** and **y** (or *equal desirability*), denoted by $\mathbf{x} \sim \mathbf{y}$, is defined by $\mathbf{x} \preceq \mathbf{y}$ *and* $\mathbf{y} \preceq \mathbf{x}$ holding simultaneously. This can also be written as $\mathbf{y} \sim \mathbf{x}$. Its negation is denoted by $\mathbf{x} \not\sim \mathbf{y}$ (or $\mathbf{y} \not\sim \mathbf{x}$).

Strict preference, denoted by $\mathbf{x} \prec \mathbf{y}$ (or $\mathbf{y} \succ \mathbf{x}$), means $\mathbf{x} \preceq \mathbf{y}$ but $\mathbf{x} \not\sim \mathbf{y}$. In view of **P3** we can have $\mathbf{x} \not\sim \mathbf{y}$ exactly if either $\mathbf{x} \prec \mathbf{y}$ or $\mathbf{x} \succ \mathbf{y}$.

Also by **P3**, $\mathbf{x} \preceq \mathbf{x}$ holds for all $\mathbf{x} \in \mathbb{R}^l_+$ (see **P1**), since either $\mathbf{x} \preceq \mathbf{x}$ or $\mathbf{x} \succeq \mathbf{x}$ or both have to hold but the two are the same. For the same reason $\mathbf{x} \sim \mathbf{x}$ for all $\mathbf{x} \in \mathbb{R}^l_+$, by the definition of $\sim$. So $\mathbf{x} \prec \mathbf{x}$ cannot hold for any $\mathbf{x} \in \mathbb{R}^l_+$.

The requirements **P2**, **P3** should not be taken lightly. We saw, for instance, in Sect. 1.3 that *the inequality* $\leqq$ *between vectors* of $\mathbb{R}^n$, defined by

$$\mathbf{x} \leqq \mathbf{y} \iff x_1 \leq y_1,\ x_2 \leq y_2,\ \ldots, x_n \leq y_n, \tag{9.9}$$

is not a complete order, that is **P3** is not satisfied if $\leqq$ is taken for $\preceq$ (for instance, neither $(2,3) \leqq (4,1)$ nor $(4,1) \leqq (2,3)$ holds). The ordering by (9.9) satisfies both reflexivity and transitivity, **P1** and **P2**. Although transitivity seems normatively compelling, examples are easily devised where it is fair. These are all based on the principle that the source of the ordering is somewhat complex. The most famous example is due to the MARQUIS DE CONDORCET (*1743 – †1794). One speak of *Condorcet's paradox*. He shows that there can be even very simple examples which are *not transitive* (do not satisfy **P2**). Let the economic agent A (for instance a household) consist of three persons N_1, N_2, N_3 with the following individual preferences:

$$\text{for } N_1\text{:}\quad \mathbf{x} \preceq \mathbf{y}, \mathbf{y} \preceq \mathbf{z}, \mathbf{x} \preceq \mathbf{z},$$

$$\text{for } N_2\text{:}\quad \mathbf{x} \preceq \mathbf{y}, \mathbf{z} \preceq \mathbf{y}, \mathbf{z} \preceq \mathbf{x},$$

$$\text{for } N_3\text{:}\quad \mathbf{y} \preceq \mathbf{x}, \mathbf{y} \preceq \mathbf{z}, \mathbf{z} \preceq \mathbf{x},$$

which satisfy **P1**. But the *collective (majority) preference* of the economic agent (household) A, which is decided by democratic vote:

$$\text{for A:}\quad \mathbf{x} \preceq \mathbf{y}, \mathbf{y} \preceq \mathbf{z}, \mathbf{z} \preceq \mathbf{x},$$

does not satisfy **P2**. (One may think that this means that **x**, **y** and **z** are considered "*equally desirable*" for A but *the paradox holds also with strict preference* $\prec$).

Another class of intransitive examples arise when the objects being ordered are multidimensional and the choice pair causes the decision maker to focus on a subset of the dimensions, one subset when judging between **x** and **y**, another when judging between **y** and **z**, and yet another subset when judging between **x** and **z**. The reader is invited to generate a specific example of this character using three dimensions of, say, automobiles.

Nevertheless, **P2** and also **P3** are often *supposed* to hold for preferences, that is, one considers *transitive* and *complete preference orderings*.

Often one supposes that $\mathbf{a} \preceq \mathbf{x}$ and $\mathbf{a} \preceq \mathbf{y}$ imply $\mathbf{a} \preceq \lambda\mathbf{x} + (1-\lambda)\mathbf{y}$ for all $\lambda \in [0,1]$ ($\lambda\mathbf{x}+(1-\lambda)\mathbf{y}$ is the *convex combination of* $\mathbf{x}$ *and* $\mathbf{y}$, compare to Sect. 9.3), that is, that $\{\mathbf{x} \mid \mathbf{x} \succeq \mathbf{a}\}$ is a convex set. The preference ordering $\preceq$ is *convex* if the set $\{\mathbf{x} \in \mathbb{R}^l_+ \mid \mathbf{x} \succeq \mathbf{a}\}$ is convex for all $\mathbf{a} \in \mathbb{R}^l_+$.

Let a preference ordering $\preceq$ for $\mathbf{x} \in \mathbb{R}^l_+$, $\mathbf{y} \in \mathbb{R}^l_+$ satisfy

$$\mathbf{x} \preceq \mathbf{y} \quad \text{if} \quad \mathbf{x} \leqq \mathbf{y}$$

(see (9.9) for the definition of $\mathbf{x} \leqq \mathbf{y}$). It is called *(strictly) monotonic* (*strictly increasing*) if any vectors $\mathbf{x} = (x_1, \ldots, x_l)$, $\mathbf{y} = (y_1, \ldots, y_l)$ satisfying $\mathbf{x} \preceq \mathbf{y}$, that is, $x_1 \le y_1, \ldots, x_l \le y_l$, but $\mathbf{x} \neq \mathbf{y}$, also satisfy $\mathbf{x} \preceq \mathbf{y}$.

The preference ordering $\preceq$ is *continuous* if $\mathbf{x}_n \preceq \mathbf{y}_n$ $(n = 1, 2, \ldots)$ *and*

$$\lim_{n\to\infty} \mathbf{x}_n = \mathbf{x}, \qquad \lim_{n\to\infty} \mathbf{y}_n = \mathbf{y}$$

(in the sense of the convergence of sequences of vectors, see Sect. 6.9 2) imply $\mathbf{x} \preceq \mathbf{y}$.

We mention, without proof, a result by Gerard Debreu (*1921, Nobel Prize in Economics 1983, †2004) which *connects preference orderings with utility functions* (compare Sects. 8.6). *If* $\preceq$ *is a continuous (and, of course, complex and transitive) preference ordering then there exists a utility function, that is, a function* $u : \mathbb{R}^l_+ \to \mathbb{R}$ *which "generates" the preference ordering in the sense that*

$$\mathbf{x} \preceq \mathbf{y} \quad \textit{if and only if} \quad u(\mathbf{x}) \le u(\mathbf{y}). \tag{9.10}$$

Consequently,

$$\mathbf{x} \prec \mathbf{y} \quad \textit{if and only if} \quad u(\mathbf{x}) < u(\mathbf{y}),$$

and

$$\mathbf{x} \sim \mathbf{y} \quad \textit{if and only if} \quad u(\mathbf{x}) = u(\mathbf{y}).$$

We note that the suppositions (transitivity, completeness and continuity of $\preceq$) could have been weakened and the result would still hold, we made the above assumptions of the sake of simplicity.

Obviously, *for any strictly increasing function* $f : \mathbb{R}^l_+ \to \mathbb{R}$, *with* $u : \mathbb{R} \to \mathbb{R}$ *also* $f \circ u$ *satisfies (9.10)*, i.e., *generates the same preference ordering as* u. This means that the utility functions so defined are *ordinal* (as opposed to cardinal) utility functions: The *sign* of the difference $u(\mathbf{x}) - u(\mathbf{y})$ is important because it determines the preferred vector of goods, whereas the *value* of this difference is meaningless since it will change with any transformation f as defined above.

As is easily seen, the set of ordinal utility functions that each represent a particular preference ordering is huge. Much smaller than this set is the set of utility functions $v : \mathbb{R}^l_+ \to \mathbb{R}$ that satisfy: If $v : \mathbb{R}^l_+ \to \mathbb{R}$ represents a preference ordering then only functions $v : \mathbb{R}^l_+ \to \mathbb{R}$ given by

$$v(\mathbf{x}) = a + bu(\mathbf{x}) \qquad (a \in \mathbb{R}, b \in \mathbb{R}_{++} \text{ constants}) \tag{9.11}$$

also represent it. Functions of this kind have the property: Let $\mathbf{x} \in \mathbb{R}^l_+$, $\mathbf{y} \in \mathbb{R}^l_+$, $\mathbf{x}' \in \mathbb{R}^l_+$, $\mathbf{y}' \in \mathbb{R}^l_+$ such that $\mathbf{x}' \prec \mathbf{y}'$ (which implies $u(\mathbf{x}') > u(\mathbf{y}')$, $v(\mathbf{x}') > v(\mathbf{y}')$).

$$\frac{v(\mathbf{x}) - v(\mathbf{y})}{v(\mathbf{x}') - v(\mathbf{y}')} = \frac{a + bu(\mathbf{x}) - (a + bu(\mathbf{y}))}{a + bu(\mathbf{x}') - (a + bu(\mathbf{y}'))} = \frac{u(\mathbf{x}) - u(\mathbf{y})}{u(\mathbf{x}') - u(\mathbf{y}')},$$

that is, the *ratio* of *utility differences is uniquely determined.* (Remember that in the case of ordinal utility functions only the *sign* of the difference $u(\mathbf{x}) - u(\mathbf{y})$ makes sense). A utility representation $u : \mathbb{R}^l_+ \to \mathbb{R}$ that is unique up to an affine transformation (9.11) is called *cordial.*

We move now from the budget set (9.8) to the "*demand set*"

$$\mathbf{D}(\mathbf{p}, \mathbf{e}) := \{\mathbf{x} \in \mathbf{B}(\mathbf{p}, \mathbf{e}) \mid \mathbf{x} \preceq \mathbf{y} \text{ for all } \mathbf{y} \in \mathbf{B}(\mathbf{p}, \mathbf{e})\} \tag{9.12}$$

of the economic agent A. It consists of those bundles of commodities in A's budget set which are most desirable for A, given the prices $\mathbf{p} = (p_1, \ldots, p_l)$ and the initial endowment $\mathbf{e} = (e_1, \ldots, e_l)$. Just as $\mathbf{B}$, also $\mathbf{D}$ is a *correspondence*, that is, may contain several elements $\mathbf{x}$. But note that frequently $\mathbf{D}$ has only one element.

This demand set $\mathbf{D}(\mathbf{p}, \mathbf{e})$ has, in the case of a continuous preference ordering, the following properties, which we enumerate here without proof. It is *nonempty and compact* (see Sect. 8.3) *for all positive price vectors* $\mathbf{p} > \mathbf{0}$ and, *i the preference ordering is also convex, then the set* $\mathbf{D}(\mathbf{p}, \mathbf{e})$ *is convex for all nonnegative price vectors* $\mathbf{p} \geqq \mathbf{0}$. *For monotonic preference orderings Walras's law*

$$\mathbf{p} \cdot \mathbf{x} = \mathbf{p} \cdot \mathbf{e} \quad \text{holds for all } \mathbf{x} \in \mathbf{D}(\mathbf{p}, \mathbf{e}). \tag{9.13}$$

The graph $\{(\mathbf{p}, \mathbf{x}) \mid \mathbf{p} \in \mathbb{R}^l_{++}, \ \mathbf{x} \in \mathbf{D}(\mathbf{p}, \mathbf{e})\}$ *of the correspondence* $\mathbf{p} \mapsto \mathbf{D}(\mathbf{p}, \mathbf{e})$ ($\mathbf{p} > \mathbf{0}$) *is closed for every* $\mathbf{e} \in \mathbb{R}^l_+$ *if the preference ordering is continuous.* Compare this to **S6** in Sect. 9.2, where the closure of the graph of a correspondence is an "axiom".

Now, instead of individual economic agents with initial supplies (endowments) $\mathbf{e}$, consider entire *exchange economies, consisting of the set* $\mathbb{R}^l_+$ *of possible bundles of commodities (goods), of their prices* $(p_1, \ldots, p_l) = \mathbf{p}$, *of m economic agents, of their initial supplies*

$$\mathbf{e}^1 = (e^1_1, \ldots, e^1_l), \ldots, \mathbf{e}^m = (e^m_1, \ldots, e^m_l)$$

and of a continuous preference ordering $\preceq_j$ for each economic agent j ($j = 1, \ldots, m$).

Important vectors and sets of vectors in such exchange economies are the following. *The total supply*

$$\mathbf{e}^1 + \ldots + \mathbf{e}^m = (e_1^1, \ldots, e_l^1) + \ldots + (e_1^m, \ldots, e_l^m)$$

of commodities (goods), the total demand

$$\begin{aligned} &\mathbf{D}_1(\mathbf{p}, \mathbf{e}^1) + \ldots + \mathbf{D}_m(\mathbf{p}, \mathbf{e}^m) \\ &\quad = \left\{\mathbf{x}^1 + \ldots + \mathbf{x}^m \;\middle|\; \mathbf{x}^1 \in \mathbf{D}_1(\mathbf{p}, \mathbf{e}^1) \ldots, \mathbf{x}^m \in \mathbf{D}_m(\mathbf{p}, \mathbf{e}^m)\right\} \end{aligned}$$

(this is how *addition of sets* is done in general, $\mathbf{x}^k = (x_1^k, \ldots, x_l^k)$, $k = 1, \ldots, m$), given the prices $(p_1, \ldots, p_l) = \mathbf{p}$, and *the excess demand*

$$\boldsymbol{\Delta}(\mathbf{p}) := \sum_{k=1}^{m} \mathbf{D}_k(\mathbf{p}, \mathbf{e}^k) - \sum_{k=1}^{m} \mathbf{e}^k := \{\textstyle\sum_{k=1}^m \mathbf{x}^k - \sum_{k=1}^m \mathbf{e}^k \mid \mathbf{x}^{\in}\mathbf{D}_k(\mathbf{p}, \mathbf{e}^k)\} \qquad (9.14)$$

(this shows also how elements are *added to or subtracted from* sets), depending again upon the price vector $\mathbf{p}$. Also $\boldsymbol{\Delta}$ is, of course, a *correspondence.*

The properties of the demand correspondence $\mathbf{D}$ (see (9.12), enumerated above, have for the excess demand correspondence $\mathbf{p} \mapsto \boldsymbol{\Delta}(\mathbf{p})$, in case $\mathbf{p} > \mathbf{0}$, the following consequences. *The graph*

$$\left\{(\mathbf{p}, \mathbf{d}) \;\middle|\; \mathbf{p} \in \mathbb{R}_{++}^l,\; \mathbf{d} \in \boldsymbol{\Delta}(\mathbf{p})\right\}$$

is closed, each $\boldsymbol{\Delta}(\mathbf{p})$ *(*$\mathbf{p} \in \mathbb{R}_{++}^l$*) is compact* (see Sect. 8.3), *each* $\boldsymbol{\Delta}(\mathbf{p})$ *is a convex set if all preference orderings are convex, and Walras's law*

$$\mathbf{p} \cdot \mathbf{d} = 0 \quad \text{holds for all } \mathbf{d} \in \boldsymbol{\Delta}(\mathbf{p}) \qquad (9.15)$$

when all preference orderings are strictly monotonic (compare (9.13)).

A central problem of mathematical economics is *whether in such exchange economies there exists* at least one *(competitive) equilibrium point.* Such a point expresses a balance of exchange in economic competition—competition as far as each agent intends to maximise utility. The balance of exchange *is attained when the excess demand is zero. So the question is whether there exists a* $\hat{\mathbf{p}} \in \mathbb{R}_+^l$ *for which*

$$\mathbf{0} \in \boldsymbol{\Delta}(\hat{\mathbf{p}}). \qquad (9.16)$$

If this is the case then *for this price vector* $\hat{\mathbf{p}}$ *there exists an allocation*

$$\hat{\mathbf{x}}^1 \in \mathbf{D}_1(\hat{\mathbf{p}}, \mathbf{e}^1), \ldots, \hat{\mathbf{x}}^m \in \mathbf{D}_m(\hat{\mathbf{p}}, \mathbf{e}^m)$$

(see (9.12), (9.14)) *which assigns* ("allocates") *to each economic agent a "most desirable" bundle of goods which exactly exhausts* (thus redistributes) *the total initial supply:*

$$\sum_{k=1}^{m} \hat{\mathbf{x}}^k = \sum_{k=1}^{m} \mathbf{e}^k$$

(compare to the definition (9.14) of $\boldsymbol{\Delta}(\mathbf{p})$). *Such an allocation completely satisfies the aspirations of all economic agents.* This is called a Walras *exchange equilibrium* and the price system $\hat{\mathbf{p}} \in \mathbb{R}^l_+$ is a *competitive equilibrium price vector.*

Here we formulate an existence result developed, among others, by Leon Walras and Gerard Debreu: *If,* in the above exchange economy, *all preference orderings are convex, continuous and (strictly) monotonic and if the total initial supply (endowment) is positive* ($\mathbf{e}^1 + \cdots + \mathbf{e}^m > \mathbf{0}$) *then there exists at least one competitive equilibrium price vector and, consequently, a Walras exchange equilibrium.*

Here we only sketch the proof. The vectors $\hat{\mathbf{p}}$ satisfying (9.16) are constructed by a similar algorithm as the solutions $\hat{\mathbf{x}} \in S$ of

$$\mathbf{0} = \mathbf{f}(\mathbf{x})$$

($\mathbf{f} : S \to \mathbb{R}^n$, $S \subseteq \mathbb{R}^n$) in Sect. 6.10, where it was done by successive approximation of a fixed point $\mathbf{x} = \mathbf{F}(\mathbf{x})$ of the function $\mathbf{F} : S \to \mathbb{R}^n$, defined by $\mathbf{F}(\mathbf{x}) := \mathbf{x} - \mathbf{f}(\mathbf{x})$. There we applied the fixed point theorem of Banach, here the following similar fixed point theorem of Shizuo Kakutani (*1911–†2004) for correspondences is useful:

Let $S \neq \emptyset$ be a convex compact subset of $\mathbb{R}^l$ and $\boldsymbol{\Phi} : S \to E(S)$ (the set of all subsets of S) a closed and convex valued correspondence. Then $\boldsymbol{\Phi}$ has at least one fixed point, that is, an $\hat{\mathbf{x}} \in S$ such that $\hat{\mathbf{x}} \in \boldsymbol{\Phi}(\hat{\mathbf{x}})$,

We conclude this section with the following remarks. The existence of equilibria in certain *models* of competitive exchange economies proves that both competition in the above sense and balance of exchange are simultaneously possible. To show this in relatively simple *models* is a first step to find an answer to the following question: Which rules or regulations of competition are, in real market economies, sufficient to yield a competitive equilibrium, i.e., a balance of exchange? In this context another important economic question arises: Are the allocations in such competitive equilibria Pareto-optimal? An allocation

$$\tilde{\mathbf{x}}^1, \ldots, \tilde{\mathbf{x}}^m \quad \text{satisfying} \quad \tilde{\mathbf{x}}^1 + \cdots + \tilde{\mathbf{x}}^m \leqq \mathbf{e}^1 + \cdots + \mathbf{e}^m$$

is *Pareto-optimal* if there is no allocation

$$\mathbf{x}^1, \ldots, \mathbf{x}^m \quad \text{satisfying} \quad \mathbf{x}^1 + \cdots + \mathbf{x}^m \leqq \mathbf{e}^1 + \cdots + \mathbf{e}^m$$

that Pareto-dominates it in the following sense: An allocation $\tilde{\mathbf{x}}^1, \ldots, \tilde{\mathbf{x}}^m$ *Pareto-dominates* another allocation $\mathbf{x}^1, \ldots, \mathbf{x}^m$ if for all $j = 1, \ldots, m$ we have

$$\tilde{\mathbf{x}}^j \succeq \mathbf{x}^j \quad \text{and there is a } k \text{ such that} \quad \tilde{\mathbf{x}}^k \succ \mathbf{x}^k$$

(see, in this context, Sect. 8.9: efficient vectors in sets of vectors). Economically speaking, a Pareto-optimal allocation $\tilde{\mathbf{x}}^1, \ldots, \tilde{\mathbf{x}}^m$ is an allocation of the following kind: The available resources, that is $\mathbf{e}^1 + \ldots + \mathbf{e}^m$, cannot be redistributed from $\tilde{\mathbf{x}}^1, \ldots, \tilde{\mathbf{x}}^m$ to another allocation that would make one or more agents better off without making others worse off.

We asked: Are the allocations in competitive equilibria Pareto-optimal? If they are *not*, that would disappoint those who believe that competition is *the* mean to maximise welfare.

Fortunately, in our *model* of a competitive exchange economy the following holds: *Any competitive equilibrium allocation in an exchange economy with strictly monotonic preferences is Pareto-optimal.*

This theorem is called the *first welfare theorem*. We will not prove it here. Note that this theorem does not require convexity of preferences. This is in contrast to the so called *second welfare theorem*, that is, the above theorem on the existence of (at least one) competitive equilibrium together with the following proposition: *For an exchange economy as defined above, let* $\mathbf{e}^1, \ldots, \mathbf{e}^m$ *be an endowment allocation that is Pareto-optimal for the economy. Let the preferences of the m agents be continuous, convex and strictly monotonic. Then there exists a price vector* $\hat{\mathbf{p}} \in \mathbb{R}^l_+$ *such that* $(\hat{\mathbf{p}}, \mathbf{e}^1, \ldots, \mathbf{e}^m)$ *is a competitive equilibrium for the economy.*

9.3.1 Exercises

1. Let $p_1 = 3$ and $p_2 = 4$ be the prices for the commodities 1 and 2, respectively, that are supplied by an economic agent A. Let the quantities e_1 and e_2 of the commodities, that is the endowment of A, be 2 and 5, respectively. Determine (a) the wealth of A, (b) the budget set $\mathbf{B}(p_1, p_2; e_1, e_2)$ of A.
2. Assume that, in the situation formulated in Exercise 1, the utility function $u_1 : \mathbb{R}^2_+ \to \mathbb{R}_+$ of A is given by $u_1(x_1, x_2) = x_1^{1/3} x_2^{2/3}$. Determine
   (a) A's demand set $\mathbf{D}_1(p_1, p_2; e_1, e_2) = \mathbf{D}_1(3, 4; 2, 5)$ and
   (b) the function value of A's utility function for $(x_1, x_2) = (e_1, e_2) = (2, 5)$ and for $(x_1^*, x_2^*) \in \mathbf{D}_1(3, 4; 2, 5)$.
3. Consider the situation formulated in Exercise 1. Assume that agent A gets a "competitor" (better: partner), agent B, with endowment vector $\mathbf{e}^2 = (3, 4)$ and utility function $u_2 : \mathbb{R}^2_+ \to \mathbb{R}_+$ given by $u_2(x_1, x_2) = x_1^{1/4} x_2^{3/4}$. Determine, for the arbitrary price vector $(p_1, p_2) \in \mathbb{R}^2_{++}$,
   (a) B's demand set $\mathbf{D}_2(p_1, p_2; 3, 4)$ and
   (b) A's demand set $\mathbf{D}_1(p_1, p_2; 2, 5)$.

4. Determine, for the exchange economy defined in Exercise 3,
   (a) the excess demand set $\boldsymbol{\Delta}(p_1, p_2)$ and
   (b) those price vectors (p_1^*, p_2^*) for which $\mathbf{0} \in \boldsymbol{\Delta}(p_1^*, p_2^*)$, that is, for which the excess demand vector is zero.
5. Consider the correspondence $\boldsymbol{\Phi} : \mathbb{R}_+^l \to E(\mathbb{R}_+^l)$ given by

$$\boldsymbol{\Phi}(x_1, \ldots, _l) = \{\mathbf{y} = (y_1, \ldots, y_l) \mid 0 \le y_j \le (x_j/y_j)^(1/2), j = 1, \ldots, l\}.$$

   (a) Which of the points $(0, \ldots, 0)$, $(1, \ldots, 1)$, $(1, 2, \ldots, l) \in \mathbb{R}_+^l$ is a fixed point of $\boldsymbol{\Phi}$?
   (b) Find a fixed point of $\boldsymbol{\Phi}$, different from the points in (a).

9.3.2 Answers

1. (a) $p_1e_1 + p_2e_2 = 3 \cdot 2 + 4 \cdot 5 = 26$
   (b) $\mathbf{B}(p_1, p_2; e_1, e_2) = \mathbf{B}(3, 4; 2, 5) = \{(x_1, x_2) \mid 3x_1 + 4x_2 \le 26\}$.
2. (a) $\mathbf{D}_1(3, 4; 2, 5) = \{(\frac{26}{9}, \frac{13}{3})\}$,
   (b) $u_1(x_1, x_2) = x_1^{1/3} x_2^{2/3} = 2^{1/3} \cdot 5^{2/3} \approx 1.2599 \cdot 2.9240 = 3.6839$,
   $u_1(\frac{26}{9}, \frac{13}{3}) \approx 3.7855$
3. (a) $\mathbf{D}_2(p_1, p_2; 3, 4) = \{(\frac{p_2}{p_1} + \frac{3}{4}, \frac{9}{4}\frac{p_1}{p_2} + 3)\}$,
   (b) $\mathbf{D}_1(p_1, p_2; 2, 5) = \{(\frac{5}{3}\frac{p_2}{p_1} + \frac{2}{3}, \frac{4}{3}\frac{p_1}{p_2} + \frac{10}{3})\}$.
4. (a) $\boldsymbol{\Delta}(p_1, p_2) = \{\boldsymbol{\Delta}_1(p_1, p_2; 2, 5) + \boldsymbol{\Delta}_1(p_1, p_2; 3, 4) - (2, 5) - (3, 4)\}$
   $= \left\{\{(\frac{5}{3}\frac{p_2}{p_1} + \frac{2}{3}, \frac{4}{3}\frac{p_1}{p_2} + \frac{10}{3})\} + \{(\frac{p_2}{p_1} + \frac{3}{4}, \frac{9}{4}\frac{p_1}{p_2} + 3)\} - (5, 9)\right\}$
   $= \left\{(\frac{8}{3}\frac{p_2}{p_1} + \frac{17}{12}, \frac{43}{12}\frac{p_1}{p_2} + \frac{19}{3}) - (5, 9)\right\}$
   $= \left\{(\frac{8}{3}\frac{p_2}{p_1} + \frac{43}{12}, \frac{43}{12}\frac{p_1}{p_2} + \frac{8}{3})\right\}$,
   that is, here the excess demand set contains infinitely many vectors,
   (b) $(\frac{8}{3}\frac{p_2}{p_1} + \frac{43}{12}, \frac{43}{12}\frac{p_1}{p_2} + \frac{8}{3}) = (0, 0)$ exactly for $p_2 = \frac{43}{32}p_1$, that is, the excess demand vector is zero if the ratio p_2/p_1 of the prices p_1 and p_2 is $43/32$.
5. (a) $(0, \ldots, 0) \in \boldsymbol{\Phi}(0, \ldots, 0)$, (b) $(1, \frac{1}{2}, \ldots, \frac{1}{l}) \in \boldsymbol{\Phi}(1, \frac{1}{2}, \ldots, \frac{1}{l})$.

9.4 Applications in the Theory of Games: Nash Equilibrium

In Sect. 8.5 we applied to linear regression what we had learned in Sect. 6.8 about extrema of functions in several variable. They play an important role also in *oligopoly theory* and, in general, in the *theory of games*, more exactly in *interactive decision theory*. The situation here is somewhat more complicated than it is in Sect. 4.8: Each agent/player/oligopolist tries to determine the extrema (usually maxima) of functions or which the values of some variables are determined by her/his opponents.

Notice that this is in contrast to the situation of the agents in the model of an exchange economy considered in the preceding section. There the value of the variables in the utility function of agent A are determined by her-/himself and by nobody else.

We discussed already, in Sect. 5.4, a very special kind of games, the *zero-sum games* for two players. As we saw there, they can be applied to the formalisation of several "parlour games" (bridge, poker, chess, etc.) and to some economic situations, simplified for the purpose of easier analysis, with the assumption that the total gain of one player or coalition of players is the total loss of the other.

On the other hand, the theory of *non-zero-sum games* serves well for the analysis of situations where there are again conflicts of interest among agents or "players" (decision makers such as "natural persons", firms, institutions, organisations, etc.) but the "payoffs" which result from the players' actions do not need to sum up to zero. Such situations often arise in the economy but also in social life, in politics, and in warfare.

Moreover we will deal with *m-person games*, where the payoff (measured in money or utility) resulting from the action of one player depends also on the actions of the other $m - 1$ players. The actions permissible for the players are called *strategies* (we give below a more detailed description). When each player acted upon one of her/his strategies then a *one-move match* has been played. So a one-move match can be considered to be a "vector" whose components are the m strategies which were carried through. The functions assigning the payoffs to each player are called "*payoff functions*" in the theory of games. So there are m payoff functions in an m-player game, one for each player.

We denote by S_j the set of possible strategies of the j-th player. If each player chooses a strategy $s_j \in S_j$ $(j = 1, \ldots, m)$ independently from the other players then the "*strategy vector*" $(s_1, s_2, \ldots, s_m) \in S_1 \times S_2 \times \cdots \times S_m$ (see Sect. 1.4 for the Cartesian product of sets) is fixed for the match. If the functions

$$F_j : S_1 \times S_2 \times \cdots \times S_m \longrightarrow \mathbb{R} \quad (j = 1, \ldots, m)$$

are the *payoff functions*, then the j-th player's payoff in this match is $F_j(s_1, \ldots, s_m)$. Obviously, a game of this kind is completely described by the vector

$$(s_1, s_2, \ldots, s_m; F_1, F_2, \ldots, F_m).$$

It is called an *m-person game in normal form.*

Here each player chooses, independently from the others, just one move, i.e., one strategy and the match is finished, the payoffs are tied up. An *example* is furnished by a market with m suppliers who advertise their prices at the beginning of each sales period (season) and do not change them during the period (season). At the end of the season each player (supplier in this "multiple-good oligopoly") will know the result of the "match".

Extensive games or *games in extensive form* are, on the other hand, games where (as in parlour games) each player has, during a *multi-move match*, at any stage of the game choice of one move (out of many) that precedes, meets, or follows the moves of the other player(s). These games can be reduced to *normal form* if we replace individual behaviour during a series of moves by strategies. A *strategy* of a player is a *complete action plan* which prescribes what the player has to do in any possible situation of the game.

Of course, even simple parlour games permit very large numbers of different strategies. When the number of strategies is finite even for such complicated games as chess, there still does not exist a complete list of all possible strategies of chess. But the advantage of the concept of strategy is, as we hinted above, that with its help *we can reduce extensive games to normal form*. Indeed, these can be considered normal games: each of the players chooses *only one "move"*. As we have learned, these "moves" can be very complicated strategies.

As an example, we consider a *price oligopoly*. This is a market for goods (and/or services) that are offered by m suppliers whose selling strategies depend upon the prices as follows: What oligopolist j ($j = 1, \ldots, m$) sells during a fixed sales period does not only depend upon the prices set by j but also upon the prices set by the $m - 1$ competitors. If $m = 2$ then we have a *duopoly*. Notice that the (supposedly very numerous) households (consumers) on the demand side of the market are *not* regarded as players in the (oligopoly) game, although they are, as a whole, causally related to the demand that meets each of the m suppliers.

For the sake of simplicity we suppose first that there is just one good in the market and that any price $p \in \mathbb{R}_+$ can be demanded. Then the strategy sets of m oligopolists (suppliers, players) are given by $S_j = \mathbb{R}_+$ ($j = 1, \ldots, m$) and the strategies of the j-th oligopolist are given by $s_j = p_j$ ($j = 1, \ldots, m$). Let f_j be the price-demand function during a fixed sales period. This function depends upon the price set by the j-th oligopolist and upon the price $p_1, \ldots, p_{j-1}, p_{j+1}, \ldots, p_m$ set by the competitors. So $f_j(p_1, \ldots, p_m)$ is the quantity of the good which the j-th oligopolist can sell ($j = 1, \ldots, m$). Let us denote the cost function of the j-th supplier by C_j. Then the payoff function, yielding the profit of this supplier is given by

$$F_j(p_1, \ldots, p_m) = p_j f_j(p_1, \ldots, p_m) - C_j(f_j(p_1, \ldots, p_m)) \quad (j = 1, \ldots, m) \tag{9.17}$$

(price times quantity less costs).

It is highly improbable that the total profit

$$F_1(p_1, \ldots, p_m) + \cdots + F_m(p_1, \ldots, p_m)$$

is constant for all price vectors (strategy vectors) $(p_1, \ldots, p_m)$. This was different in parlour games: there the sums of payoff functions are constant:

$$\sum_{j=1}^{m} F_j(s_1, s_2, \ldots, s_m) = c \quad \text{for all} \quad (s_1, s_2, \ldots, s_m) \in S_1 \times S_2 \times \cdots \times S_m.$$

These games are called "*constant-sum games*". If $c = 0$, we are back at zero-sum games which we discussed in the case $m = 2$ (two-person zero-sum games) in Sect. 5.4.

In two-person zero-sum or constant-sum games the gain of one player determines the gain or loss of the other. Therefore there is not much advantage for them in cooperation. In m-person games ($m \geq 3$) or non-constant-sum two-person games, cooperation, if permitted by the rules of the game, may make sense. But such a cooperation in oligopoly or duopoly markets may harm the consumer, therefore it is prohibited in many jurisdictions. Accordingly, in what follows we will deal mostly with *noncooperative games*, even though cooperative games are of interest too (forming of coalitions, their rules of procedure, distribution of profit, etc.).

Of course, for a player in a noncooperative game the vital question is which strategy $s_j \in S_j$ to choose. This would be simplest if the j-th player could wait till all other made their choices of strategies, but this is against the rules. It would also be *non symmetric*: only one player can play a game in this way. We will disregard this in favour of other "winning strategies".

In the introduction of Sect. 9.1 we gave a numerical example and defined Nash equilibrium points in the particular case of one-good duopoly. Before dealing with "multi-good oligopoly models" we give more general definitions. *In a noncooperative m-person game*

$$(S_1, \ldots, S_m; F_1, \ldots, F_m)$$

a strategy vector $(s_1^*, \ldots, s_m^*) \in S_1 \times \cdots \times S_m$ *is a Nash equilibrium point if, for all* $j \in \{1, \ldots, m\}$ *and for all* $s_j \in S_j$, *the inequality*

$$F_j(s_1^*, \ldots, s_{j-1}^*, s_j^*, s_{j+1}^*, \ldots, s_m^*) \geq F_j(s_1^*, \ldots, s_{j-1}^*, s_j, s_{j+1}^*, \ldots, s_m^*)$$

holds. Clearly *in such a point there is no incentive for any player to change strategy if the strategies of the other remain the same.*

One can show that *there is at least one Nash equilibrium point in the m-person game* $(S_1, \ldots, S_m; F_1, \ldots, F_m)$ *if the following three conditions hold for* $j = 1, \ldots, m$*:*

(i) S_j *is a compact convex* (see Sects. 3.3 and 6.7 for definitions) *subset of* $\mathbb{R}_+^{\nu_j}$ ($j = 1, \ldots, m$; the natural numbers $\nu_1, \ldots, \nu_m$ may be different).

(ii) *The partial payoff functions*

$$s_j \longmapsto F_j(s_1, \ldots, s_{j-1}, s_j, s_{j+1}, \ldots, s_m) \quad (j = 1, \ldots, m)$$

are convex from above.

(iii) *The payoff functions*

$$(s_1, \ldots, s_m) \longmapsto F_j(s_1, \ldots, s_m) \qquad (j = 1, \ldots, m)$$

$F_j : S_1 \times \ldots \times S_m \to \mathbb{R}$ *are continuous.*

In what follows we sketch an argument which shows that under certain conditions there exists exactly one Nash equilibrium point—in this case rather *equilibrium vector*—for *noncooperative* games involving m-person, n-good oligopolies, with price settings as strategies, payoffs as profits, and with *affine* (see Sect. 4.2) *cost functions and price-demand functions.* The same argument determines that unique Nash equilibrium vector explicitly. We will compare it to the price vector which maximises the total profit in the case where *all* oligopolists *cooperate*. (This will be our foray into the field of *cooperative games*). Dealing with an n-good model, we will mark the goods for simplicity by $1, 2, \ldots, n$ and denote by N_j the ordered set (in the order of these numbers) of all goods offered by the j-th supplier. It is customary to denote the number of elements in the set N_j by $\#N_j$ (or by $|N_j|$ but we will not use the latter notation to avoid mixup with length of vectors and absolute values of numbers).

For simplicity we will suppose that the m suppliers are *engaging exclusively in price policy*, that is the strategy set S_j consists of the prices $p_r^j \geq 0$ $(r \in N_j)$ which the j-th supplier may set for the goods offered by her/him:

$$S_j := \{p_r^j \geq 0 \mid r \in N_j\} \qquad (j = 1, \ldots, m).$$

We unite the individual prices p_r^j $(r \in N_j)$ into the *price vectors* (column vectors)

$$\mathbf{p}^j = (p_r^j) \qquad (j = 1, \ldots, m).$$

Thus the price of the r-th food can be uniquely found in each price vector which contains it at all.

The essential assumption and restriction in this model *is that the* column *vector of sales* by the j-supplier,

$$\mathbf{x}^j = (x_r^j) \qquad (r \in N_j),$$

is an affine function (see Sect. 4.2) *of all price vectors* (of that supplier and of those of the competitors), that is, *the price-demand function is affine:*

$$\mathbf{x}^j = \sum_{k=1}^{m} \mathbf{A}^{jk}\mathbf{p}^k + \mathbf{c}^j \qquad (j = 1, \ldots, m). \tag{9.18}$$

The so called *saturation quantities* c_r^j are *positive* (since we obviously assume that sales x_r^j are positive, if all prices are zero). We donate the components of the m^2

individual $\#N_j \times \#N_k$ matrices $\mathbf{A}^{jk}$ by

$$a_{rs}^{jk} \quad (j = 1, \ldots, m;\ k = 1, \ldots, m;\ r \in N_j,\ s \in N_k). \tag{9.18'}$$

Not all will be positive. In fact, all *diagonal* elements of the $\mathbf{A}^{jj}$ matrices will be supposed to be negative:

$$a_{rr}^{jj} < 0 \quad (r \in N_j;\ j = 1, \ldots, m), \tag{9.19}$$

that is, *we will suppose that*, everything else being equal, *if the j-th supplier raises the price* p_r^j *of the good r then this supplier's sales of that good will decrease.* However, *for* $j \neq k$ *we will suppose that*

$$a_{rr}^{jk} \geq 0 \quad (r \in N_j \cap N_k; j = 1, \ldots, m; k = 1, \ldots, m), \tag{9.20}$$

that is, *if one other supplier raises the price of good r then*, everything else being equal, *the sales of this good by a supplier who did not raise its price will not decrease.* Clearly, these are quite reasonable suppositions.

The following is not a supposition but enumeration and interpretation of possible case. *For* $r \neq s$ $(r \in N_j, s \in N_k; j = 1, \ldots, m; k = 1, \ldots, m;$ j and k may be different or equal) we have (see (9.18), (9.18'))

(a) $a_{rs}^{jk} > 0$ *if an increase in the price of the good s by supplier k raises the demand for supplier j's good r* (in this case good r is said to be *substitute* for good s),
(b) $a_{rs}^{jk} < 0$ *if an increase in the price of the good s by supplier k lowers the demand for supplier j's god r* (good r is a "*complement*" of good s),
(c) $a_{rs}^{jk} = 0$ *if raising or lowering the price of the good s by supplier k does not influence the demand for supplier j's good r* ("*demand indifference*").

Now we make the further assumption of the *dominance of direct price impacts:* It is possible to choose the quantity units (ounces, kilograms, litres, gallons, etc.) for the n goods so that in (9.18) the resulting components (9.18') of the matrix $\mathbf{A}^{jk}$ satisfy

$$\frac{1}{2} \sum_{\substack{s \in N_j \\ s \neq r}} \left| a_{rs}^{jj} + a_{sr}^{jj} \right| < \left| a_{rr}^{jj} \right| \quad (r \in N_j;\ j = 1, \ldots, m) \tag{9.21}$$

and

$$\sum_{s \in N_j} \left| a_{st}^{jk} \right| < \left| a_{rr}^{jj} \right| \quad (r \in N_j,\ t \in N_k;\ j = 1, \ldots, m;\ k = 1, \ldots, m;\ j \neq k). \tag{9.22}$$

(The summation in the second inequality varies through all elements of N_j while, in the first inequality, it goes through all elements of N_j *except* r). If, as it is reasonable to suppose, a^{ij}_{rs} and a^{ij}_{sr} are either both nonnegative or both nonpositive for any given j, r, s ($r \neq S$) then (9.21) and (9.22) mean, compare (9.18), (9.18'), the following: *When the j-th supplier changes the price of her/his r-th good by h percent then the impact of this price change on the sales of this r-th good is stronger then the effect of an h percent price change of her/his s-th good ($s \neq r$) or of any good supplied by competitor k ($k = 1, \ldots, m$; $k \neq j$).*

If (9.19), (9.20), (9.21) and (9.22) are satisfied then we say that the *direct price impact is negative* (see (9.19)) *and dominant* (see (9.21), (9.22)).

Here the payoff functions in the "game" of the m oligopolists are *profit functions* $F_1, \ldots, F_m$. We determine them *under the supposition that*, similarly as the price-demand function in (9.18), *the cost function C_j is affine*:

$$C_j(\mathbf{x}^j) = \alpha_j + \mathbf{b}^j \cdot \mathbf{x}^j \quad (j = 1, \ldots, m). \tag{9.23}$$

Here $\alpha_j > 0$ are the *fixed costs* of the j-th supplier ($j = 1, \ldots, m$), while the components $b^j_r > 0$ of the vector

$$\mathbf{b}^j = (b^j_r) \quad (r \in N_j;\ j = 1, \ldots, m)$$

are the (partial) marginal costs of the j-th supplier with respect to the r-th good. Of course, $\mathbf{b}^j \cdot \mathbf{x}^j$ is the inner product (Sect. 1.4 **3**) of this vector and of the vector $\mathbf{x}^j$ of quantities of goods offered by the j-th supplier.

As a generalisation of (9.17), now the profit function of the j-th supplier is given by

$$F_j(\mathbf{p}^1, \ldots, \mathbf{p}^m) = \sum_{r \in N_j} p^j_r x^j_r - C_j(\mathbf{x}^j) = \mathbf{p}^j \cdot \mathbf{x}^j - C_j(\mathbf{x}^j) \quad (j = 1, \ldots, m) \tag{9.24}$$

(again sales in money units less costs). Under the assumptions (9.18) and (9.23) this becomes

$$F_j(\mathbf{p}^1, \ldots, \mathbf{p}^m) = (\mathbf{p}^j - \mathbf{b}^j) \cdot \sum_{k=1}^{m} (\mathbf{A}^{jk}\mathbf{p}^k + \mathbf{c}^j) - \alpha_j \quad (j = 1, \ldots, m). \tag{9.25}$$

While the prices are, of course, natural numbers (times 0.01), as usual (see Sects. 3.1 and 5.1), we "interpolate" and let them be any nonnegative real number. While in Sects. 3.1 and 5.1 we emphasised that there are many ways to obtain, by "interpolation", a "smooth" function on an interval from one defined at isolated points, here it is natural to preserve (9.25) as a quadratic function (compare Sect. 6.3 **1**). Of course, quadratic functions can be twice (partially) differentiated. We get

$$\frac{\partial F_j(\mathbf{p}^1,\ldots,\mathbf{p}^m)}{\partial p_r^j} = (0,\ldots,0,1,0,\ldots,0)\cdot\left(\sum_{k=1}^{m}\mathbf{A}^{jk}\mathbf{p}^k + \mathbf{c}^j\right) + (\mathbf{p}^j - \mathbf{b}^j)\cdot\mathbf{A}^{jj}\begin{pmatrix}0\\ \vdots\\ 1\\ \vdots\\ 0\end{pmatrix}. \tag{9.26}$$

As we have seen in Sect. 6.7, F_j can have a *maximum* (that is, *the profit of the j-th supplier can be maximised) only for those positive prices p_r^j for which*

$$\frac{\partial F_j(\mathbf{p}^1,\ldots,\mathbf{p}^m)}{\partial p_r^j} = 0 \quad (r\in N_j;\ j=1,\ldots,m). \tag{9.27}$$

(Since we want the prices p_r^j to be *nonnegative*, therefore we are in the compact set

$$\left\{(\mathbf{p}^1,\ldots,\mathbf{p}^m)\ \middle|\ \mathbf{p}^j=(p_r^j)_{r\in N_j},\quad 0\le p_r^j\le\mu\quad (r\in N_j; j=1,\ldots,m)\right\},$$

where μ is greater than any possible price). We look first at the coefficient of $\mathbf{p}^j$ in (9.26). We see that $\mathbf{p}^j$ figures at two places:

$$(0,\ldots,0,1,0,\ldots,0)\cdot\mathbf{A}^{jj}\mathbf{p}^j \quad\text{and}\quad \mathbf{p}^j\cdot\mathbf{A}^{jj}\begin{pmatrix}0\\ \vdots\\ 1\\ \vdots\\ 0\end{pmatrix}.$$

So *the coefficient of $\mathbf{p}^j$ is the matrix $\mathbf{B}^j$ whose components are $a_{rt}^{jj}+a_{tr}^{jj}$ where both r and t move through N_j*. The vectors $\mathbf{p}^k$ with $k\neq j$ figure only in

$$(0,\ldots,0,1,0,\ldots,0)\cdot\mathbf{A}^{jk}\mathbf{p}^k.$$

So *the coefficients of the $\mathbf{p}^k$ $(k\neq j)$ are the matrices $-\mathbf{C}^{jk}:=\mathbf{A}^{jk}$ whose components are the a_{rs}^{jk}, where r goes through N_j and s through N_k*. We wrote $-\mathbf{C}^{jk}$ because we will carry it over to the right hand side of the equation (9.27) (whose left hand side is determined in (9.26)). Finally, *the terms in (9.26) without* $\mathbf{p}^k$ (which will also go to the right hand side of (9.27)) *from the vector* $-\mathbf{d}^j$, *whose components are*

$$-d_t^j = c_t^j - \sum_{r \in N_j} b_r^j a_{rt}^{jj} \quad (t \in N_j). \tag{9.28}$$

Thus equation (9.27) becomes

$$\mathbf{B}^j \mathbf{p}^j = \sum_{\substack{k=1 \\ k \neq j}}^{m} \mathbf{C}^{jk} \mathbf{p}^k + \mathbf{d}^j. \tag{9.29}$$

We suppose that (9.19), (9.20), (9.21) and (9.22) hold and that the above *case (b) of complementary demand is excluded. By (9.19), the components in the diagonal of* $\mathbf{B}^j$ *are negative and*, by (9.21) and (9.22), *they "dominate" the non diagonal components, which by (a) and (c) are nonnegative.* In (9.29), the components of $\mathbf{p}^k$ are nonnegative and, by (9.20) and (a), (c), those of $\mathbf{C}^{jk} = -\mathbf{A}^{jk}$ $(k \neq j)$ nonpositive. By (9.28), *if the (positive) saturation quantities* c_t^j *are large enough,* $\mathbf{d}^j$ *and thus the right hand side of (9.29) is negative.* At this stage we state without proof the *theorem*:

Let **B** *be a quadratic matrix whose diagonal components are negative and whose non diagonal components are nonnegative. Then for every* $\mathbf{c} \leqq \mathbf{0}$ *there exists a unique* $\mathbf{p} \geqq \mathbf{0}$ *such that* $\mathbf{Bp} = \mathbf{c}$ *if and only if* **B** *has a dominant diagonal. The inverse* $\mathbf{B}^{-1}$ *of such a* **B** *exists and its components are nonpositive.* From this it follows that *the system of linear equations (9.29) has a unique nonnegative solution vector* $\mathbf{p}^j$. This will then be

$$\mathbf{p}^j = (\mathbf{B}^j)^{-1} \left(\sum_{\substack{k=1 \\ j \neq k}}^{m} \mathbf{C}^{jk} \mathbf{p}^k + \mathbf{d}^j \right). \tag{9.30}$$

Of course (see Sect. 6.7), (9.27) is only necessary for F_j to have a maximum at (9.30). However, from (9.26) also

$$\frac{\partial^2 F_j(\mathbf{p}^1, \ldots, \mathbf{p}^m)}{\partial p_r^j \partial p_t^j} = a_{rt}^{jj} + a_{tr}^{jj} \quad (r \in N_j,\ t \in N_j;\ j = 1, \ldots, m)$$

follows and from (9.19) and (9.21) one can conclude that *the matrix with these components is negative definite.* So the function given by *(9.25) has* indeed *a maximum at (9.30).* Obviously it is global, that is the *price vector (9.30) maximises the profit made by the j-th supplier, given the price vectors ("price lists")* $\mathbf{p}^k$ $(k \neq j)$ *of the competitors.*

We arrive now at the result which we have announced. *For a multi-good oligopoly, with the profit functions (9.25), we assume, as in (9.19), negative direct price impacts and also their dominance by requiring (9.21) and (9.22) to hold, furthermore, that the positive saturation quantities* c_r^j *in the affine price-demand function (9.18) are sufficiently large. Then there exists exactly one nonnegative Nash equilibrium vector (*really "vector of vectors"*)*

$$\bar{\mathbf{p}} = \begin{pmatrix} \bar{\mathbf{p}}^1 \\ \vdots \\ \bar{\mathbf{p}}^m \end{pmatrix}$$

and exactly one nonnegative vector of the quantities of goods which sell at the prices $\bar{\mathbf{p}}$ *and which lead to positive profits* $F_j(\bar{\mathbf{p}})$ $(j = 1, \dots, m)$ *for each supplier.*

We do not present the complete proof of the result. We just mention that, writing (9.29) for $j = 1, \dots, m$, we have

$$N := \sum_{j=1}^{m} \# N_j$$

linear equations which we write as

$$(\mathbf{I} - \mathbf{T})\mathbf{p} = \mathbf{a},$$

where $\mathbf{T}$ is the $N \times N$ "hyper matrix" (matrix of matrices)

$$\mathbf{T} = \begin{pmatrix} \mathbf{0} & (\mathbf{B}^1)^{-1}\mathbf{C}^{12} & \cdots & (\mathbf{B}^1)^{-1}\mathbf{C}^{1m} \\ (\mathbf{B}^2)^{-1}\mathbf{C}^{21} & \mathbf{0} & \cdots & (\mathbf{B}^1)^{-1}\mathbf{C}^{1m} \\ \vdots & \vdots & \ddots & \vdots \\ (\mathbf{B}^m)^{-1}\mathbf{C}^{m1} & (\mathbf{B}^m)^{-1}\mathbf{C}^{m2} & \cdots & \mathbf{0} \end{pmatrix},$$

and $\mathbf{a}$ and $\mathbf{p}$ are N-dimensional "vectors of vectors":

$$\mathbf{a} = \begin{pmatrix} (\mathbf{B}^1)^{-1}\mathbf{d}^1 \\ \vdots \\ (\mathbf{B}^m)^{-1}\mathbf{d}^m \end{pmatrix}, \quad \mathbf{p} = \begin{pmatrix} \mathbf{p}^1 \\ \vdots \\ \mathbf{p}^m \end{pmatrix}.$$

Of course, $\mathbf{I}$ is the unit matrix

$$\mathbf{I} = \begin{pmatrix} 1 & 0 & \cdots & 0 \\ 0 & 1 & \cdots & 0 \\ \vdots & \vdots & \ddots & \vdots \\ 0 & 0 & \cdots & 1 \end{pmatrix},$$

consisting of N rows and N columns.

In the above situation ("dominant negative diagonal" and "nonnegative off-diagonal components" in $\mathbf{B}^j$ and "sufficiently large" positive c_t^j), *all components of the vector* ("vector of vectors") $\mathbf{a}$ *are nonnegative* (see the *theorem* mentioned some pages earlier). If in this situation the absolute values of the diagonal components in $\mathbf{B}^j$ are sufficiently large *then the* 1*'s in the diagonal of the matrix* $(\mathbf{I} - \mathbf{T})$ *dominate the non diagonal components, which are nonpositive* since the components of the matrices in $\mathbf{T}$,

$$(\mathbf{B}^j)^{-1}, \quad \mathbf{C}^{jk} \quad (j = 1, \ldots, m;\ k = 1, \ldots, m;\ j \neq k),$$

are nonpositive (note that $-\mathbf{C}^{jk} = (a_{rs}^{jk})$ is nonnegative as supposed in (9.20), (a) and (c)). Thus the *inverse* $(\mathbf{I} - \mathbf{T})^{-1}$ *exists and is nonnegative.* So we get an appropriate *profit-maximising price vector*

$$\bar{\mathbf{p}} = (\mathbf{I} - \mathbf{T})^{-1}\mathbf{a} \geqq \mathbf{0}.$$

Now we come to the question whether the thus obtained Nash equilibrium prices are more favourable for the consumer than those maximising the sum of the profits of *cooperating* oligopolists. One would guess that they are and the next result we quote here gives conditions under which this is indeed so, but read on (also remember the example in Sect. 9.1).

The result, which we again quote without proof, is the following. Here too, we suppose, *for a multi-good oligopoly, that the profit functions are of the form (9.25), that the direct price impacts are negative (see (9.19)) and dominant (see (9.21), (9.22)), that the saturation quantities are large enough and that the demand for any pair of the n goods offered by the m oligopolists is substitutional* (see (a) above). *Then there exists a unique nonnegative price vector* $\hat{\mathbf{p}} = (\hat{\mathbf{p}}^1, \ldots, \hat{\mathbf{p}}^m)$, *which maximises the total profit*

$$F_1(\mathbf{p}^1, \ldots, \mathbf{p}^m) + \cdots + F_m(\mathbf{p}^1, \ldots, \mathbf{p}^m) \tag{9.31}$$

of the m oligopolists. Furthermore, *for this* $\hat{\mathbf{p}}$ *we have* $\hat{\mathbf{p}} \geqq \bar{\mathbf{p}}$, *where* $\bar{\mathbf{p}}$ *is the Nash equilibrium price vector*, calculated above.

If the requirement (a) ("for any pair of the n goods offered by the m oligopolists the demand is substitutional") *is dropped then* $\hat{\mathbf{p}} \geqq \bar{\mathbf{p}}$ *need not hold anymore*, as the following simple example shows.

Two suppliers offer one good each. Let the price-demand functions be given by

$$x = -5p - q + 100,$$
$$y = -p - 5q + 100,$$

where x, y are the sold quantities and p, q, respectively, are the prices of the two goods. As we see, the sales of one supplier for one good go *down* conversely. This is a case (b) of complementary (demand for two) goods (as for coffee and cream: if coffee would be more expensive than it is now, not only the demand for coffee but also for cream would be lower). Let the cost functions be given by

$$C_1(x) = 120 + 2x, \qquad X_2(y) = 120 + 2y.$$

So we get for the profit functions (see (9.24) and (9.27))

$$F_1(p,q) = px - C_1(x) = (p-2)x - 120 = -5p^2 - pq + 110p + 2q - 320,$$

$$F_2(p,q) = qy - C_2(y) = (q-2)y - 120 = -5q^2 - pq + 110p + 2p - 320,$$

$$\frac{\partial F_1(p,q)}{\partial p} = -10p - q + 110, \qquad \frac{\partial F_2(p,q)}{\partial q} = -10q - p + 110.$$

The last two are simultaneously 0 (case of equilibrium) exactly for

$$p = \bar{p} = 10, \qquad q = \bar{q} = 10.$$

Since

$$\frac{\partial^2 F_1(p,q)}{\partial p^2} = -10, \qquad \frac{\partial^2 F_2(p,q)}{\partial q^2} = -10,$$

we have maxima of

$$p \mapsto F_1(p,q), \qquad q \mapsto F_2(p,q)$$

at $\bar{p} = 10$, $\bar{q} = 10$. *so* (10, 10) *is the (only) Nash equilibrium vector.*

Now suppose that the *suppliers cooperate in order to maximise their total profit*

$$F(p,q) = F_1(p,q) + F_2(p,q) = -5p^2 - 5q^2 - 2pq + 112(p+q) - 640.$$

We calculate

$$\frac{\partial F(p,q)}{\partial p} = -10p - 2q + 112 = 0, \qquad \frac{\partial F(p,q)}{\partial q} = -10q - 2p + 112 = 0,$$
$$\frac{\partial^2 F(p,q)}{\partial p^2} = -10, \quad \frac{\partial^2 F(p,q)}{\partial q^2} = -10 \quad \frac{\partial F(p,q)}{\partial p\, \partial q} = \frac{\partial F(p,q)}{\partial q\, \partial p} = -2.$$

The first two equations yield $p = \hat{p} = \frac{28}{3}$, $q = \hat{q} = \frac{28}{3}$. Since (see Sect. 6.8) the matrix

$$\begin{pmatrix} -10 & -2 \\ -2 & -10 \end{pmatrix}$$

is *negative definite*, therefore (see Sect. 6.8) *the total profit F has a maximum at*

$$(\hat{p}, \hat{q}) = \left(\frac{28}{3}, \frac{28}{3}\right).$$

From the form of F it follows that this is the *global maximum* of F. Note that $(\hat{p}, \hat{q}) < (\bar{p}, \bar{q})$. The profits of the suppliers for the two pairs $(\bar{p}, \bar{q})$ and $(\hat{p}, \hat{q})$ of prices will be

$$\begin{aligned} F_1(\bar{p}, \bar{q}) &= F_1(10, 10) = -5 \cdot 10^2 - 10 \cdot 10 + 110 \cdot 10 + 2 \cdot 10 - 320 \\ &= F_2(10, 10), \\ F_1(\hat{p}, \hat{q}) &= F_1(\frac{28}{3}, \frac{28}{3}) = -5 \cdot \frac{28^2}{3^2} - \frac{28}{3} \cdot \frac{28}{3} + 110 \cdot \frac{28}{3} + 2 \cdot \frac{28}{3} - 320 \\ &= 202.67 = F_2(\frac{28}{3}, \frac{28}{3}). \end{aligned}$$

This shows on one hand that *there exist games whose payoffs in certain non equilibrium situations are greater than the payoffs paid in the Nash equilibrium.* On the other hand, *the following is possible in the case (b) of complementary (*demand for two*) goods: A price fixing agreement to maximise the total profit for the suppliers can lead to lower prices for the consumers than competition in prices resulting in equilibrium.*

9.4.1 Exercises

Consider a duopoly, where duopolist A supplies two goods and duopolist B one good. Each of the three goods is a substitute for each of the other two goods. Denote the prices and quantities of A's goods by p, q and x, y, respectively, and the price and quantity of B's good by r and z, respectively. Let the price-demand functions of the market be given by

$$\begin{aligned} x &= -8p + 2q + 4r + 200, \\ y &= 2p - 8q + 4r + 200, \\ z &= 3p + 3q - 9r + 300, \end{aligned}$$

A's cost function C by

$$C(x, y) = 2x + 2y + 40,$$

and B's cost function K by

$$K(z) = z + 30.$$

1. Determine A's profit function F and B's profit function G as functions of the prices p, q and r.
2. Determine the (unique) Nash equilibrium price vector $(\bar{p}, \bar{q}, \bar{r})$ of the market.
3. Determine the (unique) price vector $(\hat{p}, \hat{q}, \hat{r})$ that maximises the total profit $F(p, q, r) + G(p, q, r)$ of the duopolists.
4. Evaluate $F(\bar{p}, \bar{q}, \bar{r})$, $F(\hat{p}, \hat{q}, \hat{r})$, $G(\bar{p}, \bar{q}, \bar{r})$, $G(\hat{p}, \hat{q}, \hat{r})$.
5. In the above set of exercises:
   (a) Is $(\bar{p}, \bar{q}, \bar{r}) < (\hat{p}, \hat{q}, \hat{r})$?
   (b) Is $F(\bar{p}, \bar{q}, \bar{r}) + G(\bar{p}, \bar{q}, \bar{r}) < F(\hat{p}, \hat{q}, \hat{r}) + G(\hat{p}, \hat{q}, \hat{r})$? If yes, why?
6. We consider a set S of matrices

$$\mathbf{A} = \begin{pmatrix} a_{11} & \cdots & a_{1n} \\ \vdots & & \vdots \\ a_{n1} & \cdots & a_{nn} \end{pmatrix} \qquad (a_{jk} \in \mathbb{R})$$

   satisfying $a_{jj} > 0\ (j = 1, \ldots, n)$, $A_{jk} \leq 0, j \neq k$ and $\sum_{k=1}^{n} a_{jk} \geq 0\ (j = 1, \ldots, n)$. We notice that here the diagonals are dominant only in a weak sense.
   (a) Give an example of a matrix $\mathbf{A} \in S$ whose inverse does not exist, that is $\det \mathbf{A} = 0$.
   (b) Give an example of a matrix $\mathbf{A} \in S$ such that $\det \mathbf{A} > 0$, and determine $\mathbf{A}^{-1}$.
   (c) Give an example of a matrix $\mathbf{A} \in S$ that satisfies $a_{j1} + \cdots + a_{jn} = 0$ for at least one $j \in \{1, \ldots, n\}$. Apply the following cancellation method: If

$$\sum_{\substack{k=1 \\ k \neq j}}^{n} |a_{jk}| < a_{jj}$$

   than cancel row j and column j of $\mathbf{A}$. Do the same with the remaining matrix and so forth. If *all* rows and columns of $\mathbf{A}$ can be cancelled with this method then $\det \mathbf{A} > 0$. (We thank WILLI HUMMEL (*(– †*)1931) for this and the following information: If $\det \mathbf{A} > 0$ for $\mathbf{A} \in S$ then *all* rows and columns can be cancelled according to the above method). If $\det \mathbf{A} = 0$ ($\det \mathbf{A} > 0$) in case of your example, give another example $\mathbf{A}^*$ satisfying $\det \mathbf{A}^* > 0$ ($\det \mathbf{A}^* = 0$).

9.4.2 Answers

1. $F(p, q, r) = xp + yq - C(x, y)$
   $= xp + yq - 2x - 2y - 40$
   $= (-8p + 2q + 4r + 200)p + (2p - 8q + 4r + 200)q$
   $\quad -(-8p + 2q + 4r + 200) \cdot 2 - (2p + 8q + 4r + 200) \cdot 2 - 40$
   $= -8p^2 - 8q^2 + 4pq + 4pr + 4qr + 212p + 212q - 16r - 800,$
   $G(p, q, r) = zr - K(z) = zr - z - 30$

$= (3p + 3q - 9r + 300)r - (3p + 3q - 9r + 300) - 30$
$= 3pr + 3qr - 9r^2 - 3p - 3q + 309r - 330.$

2. $(\bar{p}, \bar{q}, \bar{r}) = (\frac{655}{27}, \frac{655}{27}, \frac{178}{9}) \sim (19.78, 19.78, 24.26)$.
3. $(\hat{\mathbf{p}}, \hat{\mathbf{q}}, \hat{\mathbf{r}}) = (\frac{467}{13}, \frac{467}{13}, \frac{479}{13}) \sim (35.92, 35.92, 36.85)$.
4. $F(\bar{p}, \bar{q}, \bar{r}) \approx 5905.695, \quad F(\hat{\mathbf{p}}, \hat{\mathbf{q}}, \hat{\mathbf{r}}) \approx 8905.254,$
   $G(\bar{p}, \bar{q}, \bar{r}) \approx 3522.938, \quad G(\hat{\mathbf{p}}, \hat{\mathbf{q}}, \hat{\mathbf{r}}) \approx 3091.373.$
5. (a) Yes (see 2 and 3),
   (b) Yes (see 4), since $(\hat{\mathbf{p}}, \hat{\mathbf{q}}, \hat{\mathbf{r}})$ maximizes $(p, q, r) + G(p, q, r)$.
6. (a) $\mathbf{A} = \begin{pmatrix} 2 & -2 & 0 \\ -3 & 3 & 0 \\ -2 & -1 & 4 \end{pmatrix}, \quad \det \mathbf{A} = 0.$

   (b) $\mathbf{A} = \begin{pmatrix} 2 & -1 \\ -2 & 3 \end{pmatrix}, \quad \det \mathbf{A} = 4, \quad \mathbf{A}^{-1} = \begin{pmatrix} \frac{3}{4} & \frac{1}{4} \\ \frac{1}{2} & \frac{1}{2} \end{pmatrix}.$

   (c) $\mathbf{A} = \begin{pmatrix} 2 & -1 & -1 \\ -3 & 3 & 0 \\ -2 & -1 & 4 \end{pmatrix}$. Since $-2 - 1 + 4 > 0$, the last row and column can be cancelled, and the matrix $\left(\begin{smallmatrix} 2 & -1 \\ -3 & 3 \end{smallmatrix}\right)$ remains. Here $2 - 1 > 0$, that is, the first row and the first column can be cancelled. So (3) remains which can be cancelled. One knows that then $\det \mathbf{A} > 0$. Indeed, $\det \mathbf{A} = 3$.

   A matrix $\mathbf{A} \in S$ satisfying $\det \mathbf{A} = 0$ is the matrix in (a). Here we can cancel the last row and the last column and get the matrix $\left(\begin{smallmatrix} 2 & -2 \\ -3 & 3 \end{smallmatrix}\right)$ for which further cancelling is impossible.

Integrals 10

Let $f : I \longrightarrow \mathbb{R}$ be given, where $I \subset \mathbb{R}$ is an interval. Try to find all differentiable functions $F : I \longrightarrow \mathbb{R}$ such that $F' = f$. Realize that this is your first step into calculating the areas of certain surfaces.

10.1 Introduction: Definite Integral

In Chaps. 5, 6, 7, 8, and 9 we gave examples of applications to economics of derivation, among others. In this chapter we introduce integration which is, in a sense, the inverse operation of derivation, and which will also prove to be an important tool for describing processes in the social sciences.

Example A money market fund pays, on an investment of \$1,000, the dividend

$$s_4 = 1000\left(\frac{y_{(1)}}{4} + \frac{y_{(2)}}{4} + \frac{y_{(3)}}{4} + \frac{y_{(4)}}{4}\right)$$

where $y_{(j)}$ is the *lowest* of the interest rates of six-month treasury bills during the j-th *quarter* ($j = 1, 2, 3, 4$) of the year. Clearly a dividend

$$S_4 = 1000\left(\frac{Y_{(1)}}{4} + \frac{Y_{(2)}}{4} + \frac{Y_{(3)}}{4} + \frac{Y_{(4)}}{4}\right),$$

where $Y_{(j)}$ is the *highest* treasury bill interest rates during the j-th *quarter*, would be higher than s_4 (or equal to s_4 if by chance the six-month treasury bill interest rates would stay constant during each quarter, though they could still

(continued)

W. Eichhorn, W. Gleißner, *Mathematics and Methodology for Economics*, Springer Texts in Business and Economics, DOI 10.1007/978-3-319-23353-6_10

change from one quarter to the other). In other words, s_4 and S_4 are capital times the arithmetic mean of the lowest or the highest treasury bill interest rates during each of the four quarters, respectively.

In Fig. 10.1 we represented the daily interest rates of the six-month treasury bills, connected by a curve. Clearly s_4 and S_4 are 1000 times the sums of areas of four quadrangles each, that is, the areas of polygons $ABC_1B_2C_2B_3C_3B_4CD$ and of $ABC_1^*B_2^*C_2^*B_3^*C_3^*B_4^*C^*D^*$, respectively. From reflection or inspection of the figure it is clear that, if the dividend were

$$s_{12} = 1000\left(\frac{y_1}{12} + \frac{y_2}{12} + \ldots + \frac{y_{12}}{12}\right)$$

(the capital times the arithmetic mean of lowest six-month treasury bill interest rates y_k during the k-th *month* ($k = 1, \ldots, 12$)), that would be larger (not smaller) than s_4:

$$s_4 \leq s_{12}.$$

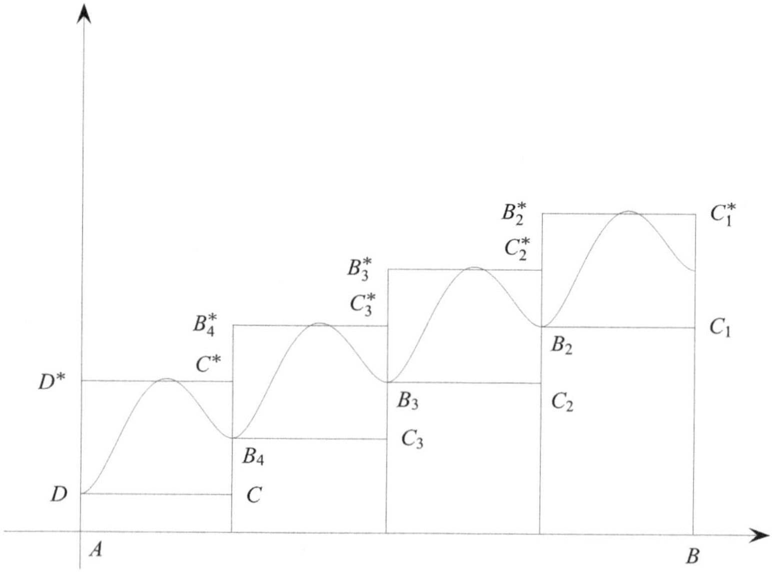

Fig. 10.1 Daily, minimum and maximum monthly, and minimum and maximum quarterly six-month treasury bill interest rates

Similarly

$$S_{12} = 1000\left(\frac{Y_1}{12} + \frac{Y_2}{12} + \ldots + \frac{Y_{12}}{12}\right) \leq S_4$$

where Y_k $(k = 1, \ldots, 12)$ is the highest treasury bill interest rate during the k-th month. Again s_{12} and S_{12} are 1000 times the areas of the corresponding polygons under and above the curve each consisting of twelve rectangles. One sees also that S_{12} is closer to s_{12} than S_4 to s_4:

$$S_{12} - s_{12} \leq S_4 - s_4.$$

If one would move on to lowest and highest interest rates during weeks, s_{52} and S_{52} would be even closer and one gets the impression that 1000 times the *area under the curve BC* would be the fairest dividend. This area is called the *definite integral* of function f whose graph is the curve AB and is written as

$$\int_0^1 f(x)\, dx.$$

Of course, one can similarly define the *definite integral*

$$\int_a^b f(x)\, dx$$

over the (closed) interval $[a, b]$, $a < b$ of the X-axis and under the graph of the function as the area bordered by the X-axis, lines perpendicular to the X-axis at the points a and b and by the graph of the function f.

There is a "catch", though: One would have to define the "area". Why? Because it is not exactly clear what we mean by the area under the graph of a function. Consider, for instance, the function in Example **3** of Sect. 6.4 (for which $f(x) = 1$ if x is irrational and $f(x) = 0$ if x is rational), say between $x = 0$ and $x = 1$. What would be the area under the graph there? (Actually, after appropriate definitions that area would turn out to be 1).

But we will not worry about this: If f is sectionally continuous (see Sect. 6.3) on $[a, b]$ then the area under the graph of f over $[a, b]$, that is, *the definite integral*

$$\int_a^b f(x)\, dx$$

can be defined exactly as the common limit, as $n \to \infty$, of the lower and upper approximating sums s_n and S_n as in Fig. 10.1 but with an arbitrary subdivision of the interval $[a, b]$ (there $[0, 1]$) as long as the length of even the largest subinterval tends to 0 when $n \to \infty$. (One can prove that *for sectionally continuous functions this common limit always exists.*)

Thus

$$\begin{aligned}\int_a^b f(x)\,dx &= \lim_{\substack{n\to\infty \\ \max|x_k-x_{k-1}|\to 0}} \sum_{k=1}^{n}(x_k - x_{k-1})m_k \\ &= \lim_{\substack{n\to\infty \\ \max|x_k-x_{k-1}|\to 0}} \sum_{k=1}^{n}(x_k - x_{k-1})M_k,\end{aligned} \tag{10.1}$$

where

$$m_k = \min_{x_{k-1}\le x\le x_k} f(x),$$

$$M_k = \max_{x_{k-1}\le x\le x_k} f(x),$$

(remember from Sect. 5.3 that functions continuous on a close interval have largest and smallest values, max and min there; the same is true for sectionally continuous functions on closed intervals).

These definitions make sense and the following rules remain true also if $f(x)$ is negative or changes signs.

10.2 Properties of Definite Integrals

From the definition (10.1) of definite integrals and from its interpretation as area under (if f is negative then (-1) times area over) the graph of f and over (under) $[a, b]$, the following properties are obvious:

$$\int_a^c f(x)\,dx = \int_a^b f(x)\,dx + \int_b^c f(x)\,dx \quad (a \le b \le c), \tag{10.2}$$

$$\int_a^a f(x)\,dx = 0, \tag{10.3}$$

$$\int_b^a f(x)\,dx = -\int_a^b f(x)\,dx \tag{10.4}$$

(because then the sign of $x_k - x_{k-1}$ reverses in each term of (10.1)),

$$\int_a^b (Af(x) + Bg(x))\,dx = A\int_a^b f(x)\,dx + B\int_a^b g(x)\,dx \tag{10.5}$$

$(A, B$ constants),

$$m(b-a) \le \int_a^b f(x)\,dx \le M(b-a), \tag{10.6}$$

where $m \le f(x) \le M$ on $[a, b]$, (see Fig. 10.1). In particular, we can have

$$m = \min_{a \le x \le b} f(x), \qquad M = \max_{a \le x \le b} f(x).$$

10.2.1 Exercises

1. Given the definite integrals $\int_a^b f(x)\ dx$, $\int_c^b f(x)\ dx$, $\int_c^d f(x)\ dx$, where $a < c < b < d$, determine $\int_a^d f(x)\ dx$.
2. The inequalities (10.6) give intervals that contain the value of the definite integral $\int_a^b f(x)\ dx$, $a < b$. Determine the smallest of these interval for the case $f(x) = 2 + \sin x$, $a = 0$, $b = 2\pi$.
3. Same problem as in Exercise 2 for $f(x) = \ln x$, $a = e \approx 2.718281828\ldots$, $b = e^2 \approx 7.389056099\ldots$.
4. Same problem as in Exercise 2 for $f(x) = x^3 - x^2 - x + 2$, $a = -1$, $b = 3/2$.
5. Same problem as in Exercise 2 for $f(x) = e^{-x}$, $a = 2$, $b = 5$.

10.2.2 Answers

1. $\displaystyle\int_a^d f(x)\ dx = \int_a^b f(x)\ dx - \int_c^b f(x)\ dx + \int_c^d f(x)\ dx$
$\displaystyle = \int_a^b f(x)\ dx + \int_b^c f(x)\ dx + \int_c^d f(x)\ dx.$
2. $\min(M - m)(b - a) = (3 - 1)(2\pi - 0)$
$= 4\pi \approx 4 \cdot 3.141592654 = 12.566370616.$
3. $\min(M - m)(b - a) = (2 - 1)(e^2 - e)$
$= e^2 - e \approx 7.389056099 - 2.718281828 = 4.670774271.$
4. $\min(M - m)(b - a) = (\frac{59}{27} - 1)(1.5 - (-1))$
$= \frac{32}{27} \cdot 2.5 \approx 1.185185185 \cdot 2.5 = 2.9629629625.$
5. $\min(M - m)(b - a) = (e^{-2} - e^{-5})(5 - 2) = 3 \cdot (e^{-2} - e^{-5})$
$\approx 3 \cdot (0.135335283 - 0.006737947) = 3 \cdot 0.1285973360.385792008.$

10.3 Indefinite Integrals (Antiderivatives)

Suppose now that f is continuous and so the area under the graph of f exists from a_0 till an arbitrary x (at least till an arbitrary x in some interval of positive length). Of course, the area will depend upon x. In accordance with the previous section we write this as

$$F(x) = \int_{a_0}^{x} f(t)\ dt. \tag{10.7}$$

(Just as it makes no difference which letter is used as subscript in a sum, such as

$$\sum_{j=1}^{n} a_j = \sum_{k=1}^{n} a_k, \quad \sum_{k=0}^{\infty} a_k = \sum_{n=0}^{\infty} a_n,$$

it does no make any difference either by which letter the variable inside a definite integral is denoted:

$$\int_{a_0}^{b} f(x)\, dx = \int_{a_0}^{b} f(t)\, dt.$$

However, in (10.7) there is an x outside of the integral (on top of the integral sign), so, in order to avoid confusion, x should *not* be used as inside variable.)

We will prove a highly remarkable relation between F and f:

$$F'(x) = f(x). \tag{10.8}$$

s Indeed, let us form first the difference quotient of F:

$$\begin{aligned} \frac{F(x+h) - F(x)}{h} &= \frac{1}{h}\left(\int_{a_0}^{x+h} f(t)\, dt - \int_{a_0}^{x} f(t)\, dt\right) \\ &= \frac{1}{h}\int_{x}^{x+h} f(t)\, dt, \end{aligned} \tag{10.9}$$

the last equality holding because (see (10.2)) the area from a_0 to $x+h$ equals the area from a to x plus that from x to $x+h$. By (10.6) (compare Fig. 10.2),

$$hm \leq \int_{x}^{x+h} f(t)\, dt \leq hM,$$

Therefore, from (10.9) and from the "squeeze rule" (compare Sect. 6.2),

$$F'(x) = \lim_{h \to 0} \frac{F(x+h) - F(x)}{h} = f(x)$$

which proves (10.8), so $f(x)$ is indeed *the derivative of* $F(x) = \int_{a_0}^{x} f(t)\, dt$ (and that F is *differentiable*) (Fig. 10.3).

The fact alone that, as in (10.8), $F'(x) = f(x)$, makes F, by definition, the *indefinite integral* or antiderivative of f; f is the *integrand*. From (10.2), (10.4) in Sect. 10.2 and from (10.7) we have

$$\int_{a}^{b} f(x)\, dx = \int_{a_0}^{b} f(x)\, dx - \int_{a_0}^{a} f(x)\, dx = F(b) - F(a).$$

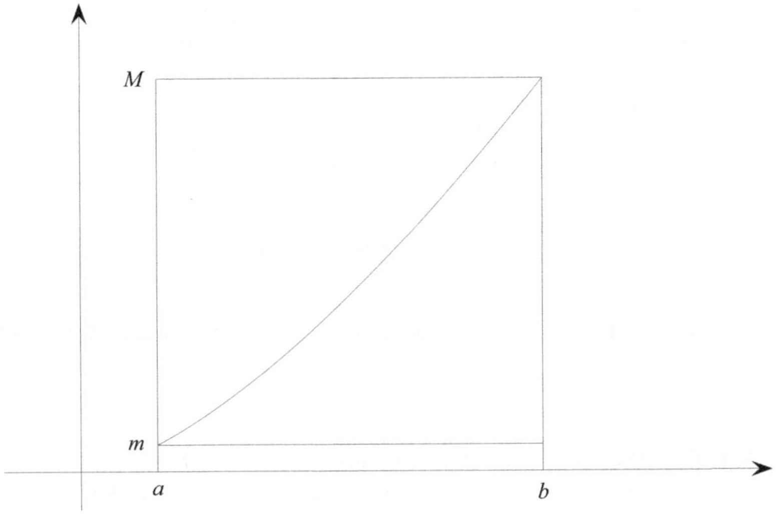

Fig. 10.2 $m(b-a) \leq \int_a^b f(x)\, dx \leq M(b-a)$

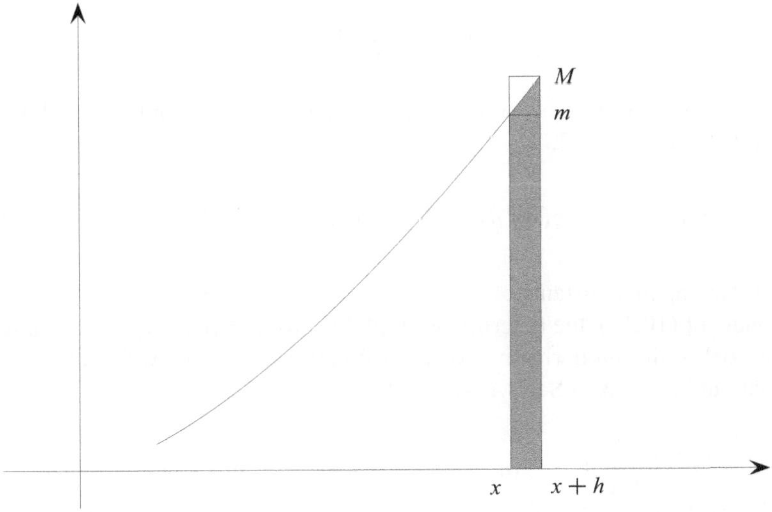

Fig. 10.3 Calculating the difference quotient of $F(x) = \int_a^x f(t)\, dt$

This simple result has the ambitious name "Newton-Leibniz-formula" (Sir ISAAC NEWTON (*1642 – †1727) and Gottfried WILHELM LEIBNIZ (*1646 – †1716) were the founders of differential and integral calculus).

Anyway, we know how to determine the definite integral, that is, the area between the graph of f (of which we know the antiderivative) and the segment $[a, b]$ of the x-axis:

where m and M are the smallest and largest values of $f(x)$ on the interval $[x, x + h]$, respectively. So $m \leq \frac{1}{h}\int_x^{x+h} f(t)\, dt \leq M$, which confines the right hand side of (10.9). Since f is continuous, as $h \to 0$ both m and M tend to $f(x)$.

$$\int_a^b f(x)\, dx = F(b) - F(a), \tag{10.10}$$

where F is any antiderivative of f, that is, as we know, any function for which

$$F'(x) = f(x)$$

holds (see 10.8). It is clear that with $F(X)$ also any $\tilde{F}(x) = F(x) + c$ satisfies (10.8) where c is an arbitrary constant. It is easy to see that *these are the only functions satisfying (10.8).*

We will write the antiderivative as

$$F(x) = \int f(x)\, dx$$

(here we wrote x after the $\int$ sign, since there are no other x on the right hand side). From (10.5) in Sect. 10.2,

$$\int (Af(x) + Bg(x))\, dx = A\int f(x)\, dx + B\int f(x)\, dx. \tag{10.11}$$

(A and B are again constants).

In view of (10.10), the determination of $F(x)$ for a given $f(x)$ is quite important. We give below the antiderivatives of a few functions; they follow from the derivation formulas and rules as in Sects. 6.4, 6.5, 7.2

1. $\int x^n\, dx = \dfrac{x^{n+1}}{n+1} + c \quad (n \neq -1); \qquad \int \dfrac{1}{x}\, dx = \ln x + c \quad (x > 0),$
in particular, taking also (10.11) into consideration, for any constant b.
2. $\int b\, dx = bx + c$ and also $\int 0\, dx = c.$
3. $\int e^x\, dx = e^x + c.$
4. $\int a^x\, dx = \dfrac{a^x}{\ln a} + c \quad (a > 0;\ a \neq 1;$ for $a = 1$ see **2**).
5. $\int \cos x\, dx = \sin x + c, \qquad \int \sin x\, dx = -\cos x + c.$
6. $\int \dfrac{1}{(\cos x)^2}\, dx = \tan x + c, \qquad \int \dfrac{1}{(\sin x)^2}\, dx = -\cot x + c.$

7. $\displaystyle\int \frac{1}{\sqrt{1-x^2}}\,dx = \text{Arc}\sin x + c = -\text{Arc}\cos x + c \qquad (|x| < 1).$
8. $\displaystyle\int \frac{1}{1+x^2}\,dx = \text{Arc}\tan x + c = -\text{Arc}\cot x + c.$

(As we saw in Sect. 6.5 **5** there are several Arc sin, Arc cos, Arc tan, Arc cot functions which differ from Arc sin etc. in constants. These constants can be merged into c.)

The definite integral can then be calculated with the formula (10.10). For instance (notice the use of vertical line to indicate the substitutions),

$$\int_1^4 x^2\,dx = \frac{x^3}{3}\bigg|_{x=1}^{x=4} = \frac{4^3}{3} - \frac{1^3}{3} = \frac{64}{3} - \frac{1}{3} = 21.$$

10.3.1 Exercises

1. Determine the antiderivatives $F : \mathbb{R} \to \mathbb{R}$ of the functions $f : \mathbb{R} \to \mathbb{R}$ given by
   (a) $f(x) = 4x^3 - 3x^2 + 2x - 1$,
   (b) $f(x) = 2^x - 3\cos x + 4\sin x$,
   (c) $f(x) = \frac{5}{1+x^2} - 7x^6$.
2. Determine the antiderivative $F : \mathbb{R}_{++} \to \mathbb{R}$ of the function $f : \mathbb{R}_{++} \to \mathbb{R}$ given by

$$f(x) = \frac{1}{2}x^{-\frac{1}{2}} - x^{-1} - x^{-2} + 2x^{-3} - 3x^{-4}.$$

3. Determine the antiderivative $F :]0, \frac{\pi}{2}[\to \mathbb{R}$ of the function $f :]0, \frac{\pi}{2}[\to \mathbb{R}$ given by

$$f(x) = (\frac{3}{\cos x})^2 - (\frac{2}{\sin x})^2.$$

4. Determine the antiderivative $F :]-1, 1[\to \mathbb{R}$ of the function $f :]-1, 1[\to \mathbb{R}$ given by
   (a) $f(x) = \sqrt{\dfrac{16}{1-x^2} + \dfrac{4-4x^2}{1-x^4}}$,
   (b) $f(x) = \dfrac{\sqrt{1-x^2}}{(1+x)(1-x)}$.
5. Calculate the definite integrals
   (a) $\int_0^2 (4x^3 - 3x^2 + 2x - 1)\,dx$,
   (b) $\int_{\pi/2}^{\pi} (2^x - 3\cos x + 4\sin x)\,dx$,
   (c) $\int_1^2 (5x^{-1} - x^{-2} + 2x^{-3} - 3x^{-4})\,dx$.

10.3.2 Answers

1. (a) $F(x) = x^4 - x^3 + x^2 + c$,
   (b) $F(x) = 2^x/\ln 2 - 3\sin x - 4\cos x + c$,
   (c) $F(x) = 5\arctan x - x^7 + c = -5\,\mathrm{arc\,cot}\,x - x^7 + c$. [-1ex]
2. $F(x) = x^{\frac{1}{2}} - \ln x + x^{-1} - x^{-2} + x^{-3} + c$. [-1ex]
3. $F(x) = 9\tan x + 4\cot x + c$. [-1ex]
4. (a) $F(x) = 4(\arcsin x + \arctan x) + c$,
   (b) $F(x) = \arcsin x + c = -\arccos x + c$.
5. (a) $\int_0^2 (4x^3 - 3x^2 + 2x - 1)\,dx$
   $= (x^4 - x^3 + x^2 - x)\,|_{x=0}^{x=2} = 16 - 8 + 4 - 2 = 10$,
   (b) $\int_{\frac{\pi}{2}}^{\pi} (2^x - 3\cos x + 4\sin x)\,dx$

$$\begin{aligned}
&= (\frac{2^x}{\ln 2} - 3\sin x - 4\cos x)\,|_{x=\frac{\pi}{2}}^{x=\pi} \\
&= \frac{2^\pi}{\ln 2} - 3\sin\pi - 4\cos\pi - \frac{2^{\frac{\pi}{2}}}{\ln 2} - 3\sin\tfrac{\pi}{2} - 4\cos\tfrac{\pi}{2} \\
&= \frac{2^\pi}{\ln 2} - 3\cdot 0 - 4\cdot(-1) - \frac{2^{\frac{\pi}{2}}}{\ln 2} - 3\cdot 1 - 4\cdot 0 \\
&= \frac{2^\pi}{\ln 2} - \frac{2^{\frac{\pi}{2}}}{\ln 2} + 1 \\
&\approx 8.825/0.693 - 2.971/0.693 + 1 \\
&\approx 9.447.
\end{aligned}$$

   (c) $\int_1^2 (5x^{-1} - x^{-2} + 2x^{-3} - 3x^{-4})\,dx$

$$\begin{aligned}
&= (5\ln x + x^{-1} - x^{-2} + x^{-3} + c)\,|_{x=1}^{x=2} \\
&= 5\ln 2 + 1/2 - 1/4 + 1/8 + c - 1 - c \\
&\approx 5\cdot 0.693 - 5/8 = 2.840.
\end{aligned}$$

10.4 Methods to Calculate Integrals

Calculating integrals is by no means so easy or "mechanical" as differentiating. (There are computer programs like Macsyma and Maple which determine indefinite integrals in explicit form; there are many more which calculate definite integrals numerically). Nevertheless we mention a few helpful methods.

1. **Integration by parts or product integration.**

By integrating

$$(u(x)v(x))' = u'(x)v(x) + u(x)v'(x),$$

(see Sect. 6.5 **2**) we get the rules

$$\int u(x)v'(x)\,dx = u(x)v(x) - \int u'(x)v(x)\,dx$$

(often written as

$$\int u\,dv = uv - \int v\,du$$

and

$$\int_a^b u(x)v'(x)\,dx = u(b)v(b) \quad u(a)v(a) - \int_a^b u'(x)v(x)\,dx.$$

Example 1 We assume $x \in \mathbb{R}_{++}$ and $u(x) = \ln x$, $u'(x) = \frac{1}{x}$, $v(x) = x$, $v'(x) = 1$:

$$\int \ln x\,dx = \int \ln x \cdot 1\,dx = x\ln x - \int \frac{1}{x}x\,dx$$
$$= x\ln x - \int 1\,dx = x\ln x - x + c,$$

in view of Sect. 10.3 **2**. [Check: $(x\ln x - x + c)' = \ln x + x\frac{1}{x} - 1 = \ln x$].

Note: It is a good idea to *check* every indefinite integral by derivation.

2. **Substitution.**

From the chain rule of derivation,

$$(f[g(x)])' = f'[g(x)]g'(x)$$

(see Sect. 6.5 **4**) we get by integration

$$\int f'[g(x)]g'(x)\,dx = f[g(x)] + c. \tag{10.12}$$

Example 2 Here $g(x) = \sin x$ (say, $x \in]0, \pi[$), $g'(x) = \cos x$, $f(t) = \ln t$ ($t \in \mathbb{R}_{++}$), $f'(t) = \frac{1}{t}$.

$$\int \cot x \, dx = \int \frac{\cos x}{\sin x} \, dx = \int \frac{1}{\sin x} (\sin x)' \, dx = \ln |\sin x| + c.$$

Similarly for, say, $x \in] - \frac{\pi}{2}, \frac{\pi}{2}[$,

$$\int \tan x \, dx = \int \frac{\sin x}{\cos x} \, dx = - \int \frac{1}{\cos x} (\cos x)' \, dx = - \ln |\cos x| + c.$$

The rule (12) can be written, with $u = g(x)$, as

$$\int f'(u) \, du = f(u) + c$$

but then it has to be used carefully for definite integrals. In the following example $u = \cos x$ and the numbers below and above the $\int$ sign have to be changed accordingly ($\cos 0 = 1$, $\cos \frac{\pi}{4} = \frac{1}{\sqrt{2}}$):

$$\begin{aligned}
\int_0^{\pi/4} \tan x \, dx &= \int_0^{\pi/4} \frac{1}{\cos x} \sin x \, dx \\
&= - \int_0^{\pi/4} \frac{1}{u(x)} u'(x) \, dx = - \int_1^{1/\sqrt{2}} \frac{1}{u} \, dx \\
&= - \ln u \, |_{u=1}^{u=1/\sqrt{2}} = - \ln(1/\sqrt{2}) - (- \ln 1) \\
&= - \ln 1 - (- \ln \sqrt{2}) + \ln 1 = \ln \sqrt{2} = \tfrac{1}{2} \ln 2.
\end{aligned}$$

The following application of rule (12) is much simpler.

Example 3 Here $u = g(x) = x - A$, $u' = g'(x) = 1$. By **1**,

$$\int \frac{1}{x - A} \, dx = \int \frac{1}{u} \, du = \ln |u| + c = \ln |x - A| + c.$$

3. Partial fractions.

As mentioned in Sect. 6.3, every rational function can be broken into primitive partial functions. This can be used to *integrate rational functions*. Since, however, zeros of polynomials with real coefficients can be complex numbers which always appear in *pairs of conjugate complex numbers* $\alpha + \beta$i and $\alpha - \beta$i, we are better off uniting the corresponding primitive partial fractions into real fractions of second degree (called again primitive partial fractions), by

$$\frac{1}{x-\alpha+\beta\mathrm{i}} + \frac{1}{x-\alpha-\beta\mathrm{i}} = \frac{x-\alpha-\beta\mathrm{i}+x-\alpha+\beta\mathrm{i}}{(x-\alpha)^2-(\beta\mathrm{i})^2} = \frac{2(x-\alpha)}{(x-\alpha)^2+\beta^2} \tag{10.13}$$

or

$$\frac{\mathrm{i}}{x-\alpha+\beta\mathrm{i}} - \frac{\mathrm{i}}{x-\alpha-\beta\mathrm{i}} = \frac{2\beta}{(x-\alpha)^2+\beta^2}.$$

Example 4 The denominator of the fraction in the following integral is $x^4 - 4x^3 + 5x^2 = x^2(x^2 - 4x + 5)$. The zeros of $x^2 - 4x + 5$ are

$$x_1 = \frac{4+\sqrt{16-20}}{2} = 2+\frac{\sqrt{-4}}{2} = 2+\mathrm{i} \quad \text{and} \quad x_2 = \frac{4-\sqrt{16-20}}{2} = 2-\mathrm{i}$$

So

$$x^2 - 4x + 5 = (x-2-\mathrm{i})(x-2+\mathrm{i}) = (x-2)^2 + 1.$$

In

$$\int \frac{x^3-3x+10}{x^4-4x^3+5x^2}\,dx = \int \frac{x^3-3x+10}{x^2[(x-2)^2+1]}\,dx \qquad (x>0)$$

we try to expand the integrand as

$$\frac{x^3-3x+10}{x^4-4x^3+5x^2} = \frac{Ax+B}{x^2} + \frac{Cx+D}{x^2-4x+5} \qquad (x>0)$$

(if, in a primitive partial fraction, x is in the denominator with n as greatest exponent, we have to write a polynomial of degree $(n-1)$ in the numerator). Bringing both sides to common denominator, which will, of course, be $x^4 - 4x^3 + 5x^2$, and comparing the two numerators, we get

$$\begin{aligned} x^3 - 3x + 10 &= (Ax+B)(x^2-4x+5) + (Cx+D)x^2 \\ &= (A+C)x^3 + (-4A+B+D)x^2 + (5A-4B)x + 5B. \end{aligned}$$

(continued)

The coefficients of *each* power of x (including $x^0 = 1$) have to be equal, so

$$A + C = 1, \quad -4A + B + D = 0, \quad 5A - 4B = -3, \quad 5B = 10.$$

From the last equation $B = 2$, so the before last becomes

$$5A - 8 = -3, \text{ that is, } 5A = 5, \ A = 1$$

and the first two become

$$1 + C = 1, \text{ that is,; } C = 0$$

and

$$-4 + 2 + D = 0, \text{ that is, } D = 2.$$

Since, as we have seen, $x^2 - 4x + 5 = (x-2)^2 + 1$, we get by using the method **2** of substitution (with $U = x - 2$)

$$\int \frac{Cx + D}{x^2 - 4x + 5}\,dx = \int \frac{2}{(x-2)^2 + 1}\,dx = \int \frac{2}{u^2 + 1}\,du$$
$$= 2\text{Arc}\tan u + c = 2\text{Arc}\tan(x-2) + c$$

and

$$\int \frac{x^3 - 3x + 10}{x^4 - 4x^3 + 5x^2}\,dx = \int \frac{1}{x}\,dx + \int \frac{2}{x^2}\,dx + \int \frac{2}{(x-2)^2 + 1}\,dx$$
$$= \ln x - \frac{2}{x} + 2\text{Arc}\tan(x-2) + c \qquad (x > 0).$$

$$\text{[Check:} \quad (\ln x - \frac{2}{x} + 2\text{Arc}\tan(x-2) + c)' = \frac{1}{x} + \frac{2}{x^2} + \frac{2}{(x-2)^2 + 1}$$
$$= \frac{(x+2)(x^2 - 4x + 5) + 2x^2}{x^2(x^2 - 4x + 5)} = \frac{x^3 - 3x + 10}{x^4 - 4x^3 + 5x^2}].$$

10.4.1 Exercises

1. Apply integration by parts to determine the antiderivatives $F : \mathbb{R} \to \mathbb{R}$ of the functions $f : \mathbb{R} \to \mathbb{R}$ given by

(a) $f(x) = xe^x$, (b) $f(x) = x^2e^x$,

(c) $f(x) = x^3 \sin x$ (d) $f(x) = 4xe^{2x}$.

2. Apply integration by parts to determine the antiderivatives $F : \mathbb{R}_{++} \to \mathbb{R}$ of the functions $f : \mathbb{R}_{++} \to \mathbb{R}$ given by

(a) $f(x) = (x + 2) \ln x$, (b) $f(x) = (3x^2 - 1) \ln x - 2x^2$.

3. Apply integration by substitution to determine the antiderivatives $F : \mathbb{R} \to \mathbb{R}$ of the functions $f : \mathbb{R} \to \mathbb{R}$ given by

(a) $f(x) = \dfrac{2x}{(1 + x^2)^2}$, (b) $f(x) = x^2e^{x^3}$,

(c) $f(x) = e^x \cos e^x$.

4. Apply integration by substitution to determine the antiderivatives $F : \mathbb{R}_{++} \to \mathbb{R}$ of the functions $f : \mathbb{R}_{++} \to \mathbb{R}$ given by

(a) $f(x) = \dfrac{\ln x}{x}$, (b) $f(x) = \dfrac{1}{x \ln c}$,

(c) $f(x) = \dfrac{1 + 2x + 3x^2}{x + x^2 + x^3}$, (d) $f(x) = \dfrac{1}{(1 + x)^2}$.

5. Evaluate

(a) $\int_0^t e^x \cos x \, dx$, (b) $\displaystyle\int_0^3 \frac{2x}{1 + x^2} \, dx$.

6. By applying partial functions determine the antiderivatives F of the rational functions f given by

(a) $f(x) = \dfrac{x + 2}{x^2 + x + 1}$,

(b) $f(x) = \dfrac{3x^4 - 9x^3 + 4x^2 - 34x + 1}{(x - 2)^3(x + 3)^2}$ for $x > 2$,

(c) $f(x) = \dfrac{2x^4 - 5x^3 + 8x^2 + 4x - 20}{x^2(x^2 + 4)(x^2 - 2x + 10)}$ for $x \neq 0$.

10.4.2 Answers

1. (a) $F(x) = xe^x - e^x + c$,
   (b) $F(x) = e^x(x^2 - 2x + 2) + c$,
   (c) $F(x) = -x^3 \cos x + 3x^2 \sin x + 6x \cos x - 6 \sin x + c$,
   (d) $F(x) = 2xe^{2x} - e^{2x} + c$.

2. (a) $F(x) = (\frac{x^2}{2} + 2x)\ln x - \frac{x^2}{4} - 2x + c$,
   (b) $F(x) = (x^3 - x)\ln x - x^3 + x + c$.
3. (a) $F(x) = -\frac{1}{1+x^2} + c$, (b) $F(x) = \frac{1}{3}e^{x^3} + c$,
   (c) $F(x) = \sin e^x + c$.
4. (a) $F(x) = \frac{(\ln x)^2}{2} + c$, (b) $F(x) = \ln(\ln x) + c$,
   (c) $F(x) = \ln(x + x^2 + x^3) + c$, (d) $F(x) = \frac{x}{1+x} + c$.
5. (a) $\int_0^x e^t \cos t\, dt = \int_1^u \cos u\, du = (\sin u + c)\,|_{u=1}^{u=e^x} = \sin e^x - \sin 1$.
   (Here we substituted $x = \ln u$, that is, $u = e^x$. Note that $x = 0$ implies $u = e^0 = 1$).
   (b) $\int_0^3 \frac{2x}{1+x^2}\, dx = \int_0^{10} \frac{du}{u} = (\ln u + c)\,|_{u=1}^{u=10} = \ln 10 + c - \ln 1 - c = \ln 10$.
   (Assuming $u \geq 1$ we substituted here $x = (u-1)^{1/2}$, that is, $u = 1 + x^2$. Note that $x = 0, x = 3$ imply $u = 1, u = 10$, respectively).
6. (a) $F(x) = \frac{1}{2}\ln(x^2 + x + 1) + \sqrt{3}\arctan\frac{2x+1}{\sqrt{3}} + c$,
   (b) $F(x) = \ln(x-2) + \frac{3}{2}\frac{1}{(x-2)^2} + 2\ln(x+3) + \frac{5}{x+3} + c$,
   (c) $F(x) = \frac{1}{2x} - \frac{1}{4}\ln\frac{x^2+4}{x^2-2x+10} + \frac{3}{4}\arctan\frac{x}{2} + \frac{1}{6}\arctan\frac{x-1}{3} + c$.

10.5 An Application: Calculating Present Values

In Sect. 8.3 we saw examples of discrete and continuous compounding and discounting (determining the present value) of *one* amount A. We saw also how one can switch from discrete to continuous compounding ($e^r = 1 + i$ if r is the stated and i the effective yearly interest rate).

Now we first want to determine the present value (e.g., value on January 1, year 1) of a payment of *several* amounts

$$A_1, A_2, \ldots, A_N$$

paid at the end of years $1, 2, \ldots, N$, respectively. If the annual (effective) interest rate is i (that is, $100i\%$) then the present value of the *individual* amounts $A_1, A_2, \ldots, A_N$ is

$$A_1(1+i)^{-1},\ A_2(1+i)^{-2},\ \ldots,\ A_N(1+i)^{-N},$$

respectively

$$\sum_{t=1}^{N} A_t(1+i)^{-t}.$$

(As observed previously, it does not matter, which letter is used as subscript in a sum. Here we write t, because that is what we used in Sect. 7.3 and because it makes this formula more similar to what we will have below in the "continuous case".) For now we stay a little longer *with the "discrete case"*, that is the case, where the payments arrive at discrete points of time.

Suppose payments (including zero amounts) do not only arrive at the end of year $1, 2, \ldots, N$, as above but also at the end of each half year, quarter, month, day or more generally, at the end of every n-th part of each year ($n = 2, 3, 4, \ldots$) with amounts

$$A_{k/n} \quad (k = 1, \ldots, nN).$$

Then we get, with

$$[k/n] := \max\{i \in \mathbb{N}_0 \mid i \leq k/n\}, \qquad n = 2, 3, 4, \ldots,$$

as the present value of all the amounts paid during the N years,

$$\sum_{k=1}^{nN} A_{k/n} \cdot (1 + r)^{-[k/n]}(1 + (k/n - [k/n])r)^{-1},$$

where r is the stated yearly interest rate.

The question is how the present value of payments during N years can be determined when these payments are *not* made at equidistant points of time. Let

$$a_{t_1}, a_{t_2}, \ldots, a_{t_q} \quad (0 \leq t_1 < t_2 < \ldots < t_q \leq N)$$

be the q amounts paid during the N years at arbitrary points in time

$$t_1, t_2, \ldots, t_q \quad (0 \leq t_1 < t_2 < \ldots < t_q \leq N).$$

The present value (value at $t = 0$) of the p payments is

$$\sum_{j=1}^{q} a_{t_j} \cdot (1 + r)^{-[t_j]}(1 + (t_j - [t_j])r)^{-1}. \tag{10.14}$$

Let us now assume that the payments are o frequently and intensely made that they can be satisfactorily described by a *continuous* or *sectionally continuous* function $a : [0, N] \to \mathbb{R}$, a so called *payment density*.

A *payment density* $a : [0, N] \to \mathbb{R}$ is defined by the following property. For each time interval the sum $S(u, u + t)$ of the amounts paid during the interval $[u, u + t] \subseteq [0, N]$, $u \in \mathbb{R}_+$, $t \in \mathbb{R}_{++}$, equals the integral of a form u to $u + t$:

$$S(u, u + t) = \int_u^{u+t} a(s)\, ds.$$

For $u = 0$ this becomes, with $C(t) := S(0, t)$,

$$C(t) = \int_0^t a(s)\, ds. \tag{10.15}$$

As in the *discrete case* we now determine *in the continuous case* the present value of future payments, that is, the value at $t = 0$ of amounts being paid during the time interval $[0, N]$, when the stated yearly interest rate is r.

For this purpose we consider an initial amount that grows by both the receipts (10.15) and the interest paid on its growing cash balance. Let $B(t)$ be this cash balance at the moment t. Then

$$B(t) = \int_0^t a(s)\, ds + \int_0^t B(s) r\, ds, \tag{10.16}$$

with derivative

$$\frac{dB(t)}{dt} = a(t) + B(t)r.$$

This equation is equivalent to

$$\frac{d}{dt}(e^{-rt}B(t)) = e^{-rt}a(t) \tag{10.17}$$

which follows from

$$\begin{aligned} \frac{d}{dt}(e^{-rt}B(t)) &= \frac{de^{-rt}}{dt}B(t) + e^{-rt}\frac{dB(t)}{dt} \\ &= -re^{-rt}B(t) + e^{-rt}\frac{dB(t)}{dt} \end{aligned} \tag{10.18}$$

and from the fact, that we can cancel $e^{-rt} > 0$ after substituting for the left-hand side in (10.17) the right-hand side in (10.18).

Integrating equation (10.17) from 0 to N gives

$$(e^{-rt}B(t))\big|_0^N = \int_0^N a(t)e^{-rt}\, dt,$$

that is, since $B(0) = 0$ (see (10.16)),

$$e^{-rN}B(N) = \int_0^N a(t)e^{-rt}\, dt. \tag{10.19}$$

As we know from Sect. 8.3, e^{-rN} is the *discount factor* in continuous compounding and $e^{-rN}B(N)$ is the *present value* of $B(N)$. Note that $0 < e^{-rN} < 1$. Hence, *in continuous compounding*,

$$\int_0^N a(t)e^{-rt}\, dt \tag{10.20}$$

is the present value (value at $t = 0$) of (future) amounts paid from $t = 0$ to $t = N$, when $a : [0, N] \to \mathbb{R}$ is the (sectionally continuous) payment intensity and r is the stated yearly interest rate.

It is interesting to know that the present values (10.20) and (10.14) in continuous and discrete compounding, respectively, are more similar than it seems at first glance. In (10.14) let $t_1 = 0$ and $t_q = N$. Then the first and the last term of the sum in (10.14) are a_0 and a_N times discount factor, respectively. Similarly, the integrand in (10.20) runs from $a(0)$ to $a(N)$ times discount factor. The discount factors

$$(1+r)^{-[t_j]}(1+(t_j-[t_j])r)^{-1} \quad \text{and} \quad e^{-rt}$$

in (10.14) and (10.20), respectively, are the closer together the smaller > 0 and $t = t_j > 0$ are. Let, for instance, $r = 0.01, t = t_j = 2.25$. Then

$$1.01^{-2}(1+0.25\cdot 0.01)^{-1} \approx 0.97785, \qquad e^{-0.01\cdot 2.25} \approx 0.97775.$$

For $r = 0.03, t = t_j = 5.75$ we get

$$1.03^{-5}(1+0.75\cdot 0.03)^{-1} \approx 0.84363, \qquad e^{-0.03\cdot 5.75} \approx 0.84156.$$

There are situations in which *no bound* can be reasonably set to the duration N (see (10.20)) of the money flow, for instance, when land keeps bringing revenue for a long time (remember that "∞" in mathematics means in practice something like "very long", "very big", "in life span"). In this case the payment intensity $a : [0, N] \to \mathbb{R}$ is defined for *all* nonnegative numbers and its present value is

$$\int_0^\infty a(t)e^{-rt}\,dt := \lim_{N\to\infty}\int_0^N a(t)e^{-rt}\,dt$$

if the limit on the right *exists* (and is *finite*). This is then called an *improper integral*. We will deal with improper integrals in more detail in the next section but we can calculate a simple such improper integral already here, thus showing that this limit may indeed exist and be finite.

Choose $a(t)$ to be constant: $a(t) = b$ ($t \in \mathbb{R}_+$). Then the present value of the money flow during $[0, N]$ is (as we know, $t \mapsto be^{-rt}$ is continuous)

$$\int_0^N be^{-rt}\,dt = \left(b\frac{e^{-rt}}{-r}+c\right)\Bigg|_{t=0}^{t=N} = \frac{b}{r}(1-e^{-rN})$$

(compare to Sect. 10.3 **3**). If $N \to \infty$ then, as shown in Sect. 7.2, $\lim_{N\to\infty} e^{-rN} = \lim_{N\to\infty}(e^{-r})^N = 0$, so that the above expression indeed has a finite limit, the following improper integral exists, is finite and is the *present value of the constant infinite money flow under the stated yearly interest rate*:

$$\int_0^\infty be^{-rt}\,dt = \frac{B}{r}.$$

This is called also the present "asset capitalisation value", under the stated yearly interest rate r, of an asset which has the same yield b every year "from here to eternity", that is, from $t = 0$ to $t = N \to \infty$.

10.5.1 Exercises

1. Let $1, 2, \ldots, 10$ denote the end of years $1, 2, \ldots, 10$, respectively. Determine, for the beginning of year 1, values ("present values") of the individual amounts $A_1 = 1100$, $A_2 = 1200$, …, $A_{10} = 2000$ paid at the end of year $1, 2, \ldots, 10$, respectively,
   (a) when the annual effective interest rate is $i = 0.05$ during each year,
   (b) when the annual effective interest rate during year $1, 2, \ldots, 10$ is
   $i_1 = 0.05$, $i_2 = 0.055$, $i_3 = 0.06$, $i_4 = 0.065$, $i_5 = 0.06$,
   $i_6 = 0.055$, $i_7 = 0.05$, $i_8 = 0.045$, $i_9 = 0.04$, $i_{10} = 0.035$,
   respectively.
   (c) Determine the sum of the present values evaluated in (a).
   (d) Determine the sum of the present values evaluated in (b).
2. Determine the stated (yearly) interest rates $r_1, r_2, \ldots, r_{10}$ belonging to the effective interest rates $t_1, i_2, \ldots, i_{10}$, respectively, that are given in Exercise 1.
3. On the first of January somebody expects a flow of money paid to him during the next four years as follows: At the end of the first, second, third, fourth quarter of the

   - first year: 1.5 thousand dollars (T$),
   - second year: 2.25 T$,
   - third year: 1.75 T$,
   - fourth year: 1.25 T$.

   Consider the function $A : [0, 4] \to \mathbb{R}$ that satisfies $A(0) = 0$, $A(m) =$ sum ("flow") of payments during the year $]m - 1, m[$, $m = 1, \ldots, 4$; $A(t) = A(m)$ for all $t \in]m - 1, m[$.
   (a) Is A uniquely defined everywhere on $[0, 4]$?
   (b) Is A continuous on $[0, 4]$?
   (c) Determine the function values of A at the following values of t:
   $$1/6,\ 2/5,\ 3/4,\ 1,\ 5/4,\ 15/8,\ 2,\ 11/4,\ 3,\ 18/4,\ 4.$$
   (d) The function $A : [0, 4] \to \mathbb{R}$ can be considered to be a payment intensity. Is it *that* payment intensity $a : [0, 4] \to \mathbb{R}$ for which, given any $u \in \mathbb{R}_+$, $t \in \mathbb{R}_{++}$ satisfying $[u, u + t] \subseteq [0, 4]$, the integral $\int_u^{u+t} a(s)\, ds$ is the sum of the amounts paid during $[u, u + t]$?
4. For the function A defined in Exercise 3, evaluate
   (a) $\displaystyle\sum_{t=1}^{4} A(t)e^{-0.04879t} = \sum_{t=1}^{4} A(t)(1 + 0.05)^{-t}$,

(b) $\sum_{t=1}^{16} A(t/4)e^{-0.04879(t/4)}\frac{1}{4}$,

(c) $\int_0^4 A(t)e^{-0.04879t}\,dt$.

5. Determine, for $r > 0$, the integrals

(a) $\int_0^N te^{-rt}\,dt$, (b) $\int_0^N t^2e^{-rt}\,dt$.

10.5.2 Answers

1. (a) 1047.62, 1088.44, 1122.99, 1151.78, 1175.29, 1193.94, 1208.16, 1218.31, 1224.76, 1227.83,
   (b) 1047.62, 1078.14, 1091.51, 1088.25, 1120.89, 1160.39, 1208.16, 1265.73, 1334.91, 1417.84.
   (c) 11659.12, (d) 11813.44.
2. Since $e^r = 1 + i$, this is $r = \ln(1 + i)$, we have

$$r_1 = r_7 = \ln 1.05 \approx 0.04879, \qquad r_2 = r_6 = \ln 1.055 \approx 0.05354,$$

$$r_3 = r_5 = \ln 1.06 \approx 0.05827, \qquad r_4 = \ln 1.065 \approx 0.06297,$$

$$r_8 = \ln 1.045 \approx 0.04402, \qquad r_9 = \ln 1.04 \approx 0.03922,$$

$$r_{10} = \ln 1.035 \approx 0.03440.$$

3. (a) Yes, (b) no (not continuous at $t = 0, \ldots, 4$),
   (c) $A(1/6) = A(2/5) = A(3/4) = A(1) = 6$,
   $A(5/4) = A(15/8) = A(2) = 9$,
   $A(11/4) = A(3) = 7$, $A(18/4) = A(4) = 5$,
   (d) no, since, for instance, $\int_{1.3}^{1.4} a(t)\,dt$, whereas $\int_{1.3}^{1.4} a(t)\,dt = 0.1 \cdot 9 = 0.9$.
4. (a)

$$\int_0^N te^{-rt}\,dt = \left(-e^{-rt}\left(\frac{t}{r} + \frac{1}{r^2}\right) + c\right)\Bigg|_{t=0}^{t=N}$$
$$= \frac{1}{r^2} - e^{-rN}\left(\frac{N}{r} + \frac{1}{r^2}\right),$$

(b)

$$\int_0^N t^2e^{-rt}\,dt = \left(-e^{-rt}\left(\frac{t^2}{r} + \frac{2t}{r^2} + \frac{2}{r^3}\right) + c\right)\Bigg|_{t=0}^{t=N}$$
$$= \frac{2}{r^3} - e^{-rN}\left(\frac{N^2}{r} + \frac{2N}{r^2} + \frac{2}{r^3}\right).$$

10.6 Improper Integrals (Integrals on Infinite Intervals or on Intervals Containing Points Where the Function Tends to Infinity)

In Sect. 10.2 we defined *definite integrals, on finite closed intervals, for functions which are sectionally continuous on these intervals.* Sometimes (not always) the definition can be extended to *infinite intervals* or to intervals on one (finite) end of which *the function tends to infinity.* (If the latter would happen in the *interior of the interval of integration*, then we would prefer to *split the interval and the integral* in the spirit of (10.2) in Sect. 10.2). We saw in Sect. 10.5 an example of the former: an integral on an infinite interval. In this section we give a broader treatment and further examples.

Take (as in (10.7), Sect. 10.3)

$$F(x) = \int_a^x f(t)\,dt.$$

If f is defined on $[a, b]$ exists ($a < b < \infty$) and F *has a limit at* ∞ (as defined in Sect. 6.2) then we define

$$\int_a^{\infty} f(t)\,dt := \lim_{x\to\infty} \int_a^x f(t)\,dt.$$

as the improper integral of f over $[a, \infty[$. Similarly, if f is defined on $]-\infty, b]$ its integral exists over every $[a, b]$ ($-\infty < a < b$), and *the limit* on the right hand side below *exists*, then we define the *improper integral* of f over $]-\infty, b]$ as follows:

$$\int_{-\infty}^{b} f(t)\,dt := \lim_{x\to-\infty} \int_x^b f(t)\,dt.$$

Furthermore, if the function f defined on $]a, b]$, converges (say from the right) to $+\infty$ (or to $-\infty$) at a but the integrals and *the limit* on the right hand side below *exists*, then we define

$$\int_a^{b} f(t)\,dt := \lim_{x\to a+} \int_x^b f(t)\,dt.$$

as the (improper) integral of f over $[a, b]$ even though f is not defined at a and its limit there is $+\infty$ (or $-\infty$). Similarly, if f is defined on $[a, b[$, converges (from the left) to infinity at b but the integrals and the limit below on the right hand exists, then we define the *improper integral* of f over $[a, b]$ as follows:

$$\int_a^{b} f(t)\,dt := \lim_{x\to b-} \int_a^x f(t)\,dt.$$

Examples

1. $\int_1^{\infty} x^{-4}\, dx$. Here x^{-4} is defined on $[1, \infty[$, and by Sect. 10.3 **1**,

$$\int_1^b x^{-4}\, dx = \left.\frac{x^{-3}}{-3}\right|_{x=1}^{x=b} = \frac{1}{3} - \frac{1}{3}\frac{1}{b^3}$$

and, by the definition in Sect. 6.2, noting that $b > 1$, we have

$$\lim_{b\to\infty} \frac{1}{b^3} = 0$$

since, for every $\varepsilon > 0$, there exists an $M(\geq \varepsilon^{-1/3})$ such that $\left|\frac{1}{b^3} - 0\right| = \frac{1}{b^3} < \varepsilon$ if $b > M \geq \varepsilon^{-1/3}$ (in slower steps: $b > 1/\varepsilon^{1/3} \Leftrightarrow b^3 > 1/\varepsilon \Leftrightarrow 1/b^3 < \varepsilon$, the cube being (strictly) increasing and the reciprocal (strictly) decreasing). So, by the rules on the limit (end of Sect. 6.2), the following limit and thus the improper integral, exists and we have

$$\int_1^{\infty} x^{-4}\, dx = \lim_{b\to\infty} \int_1^b x^{-4}\, dx = \lim_{b\to\infty} \left(\frac{1}{3 - \frac{1}{3}\frac{1}{b^3}}\right) = \frac{1}{3}.$$

(In comparison to the first definition above, we wrote here x in place of t and b in place of what was x there: we want the reader to realise that symbols are interchangeable, as long as we use them in a consistent manner.)

2. $\int_2^{\infty} \left(\frac{-1}{t}\right) dt$. Again, $1/t$ is defined on $[2, \infty]$ and even $\lim_{t\to\infty} 1/t = 0$ hold, but by Sect. 10.3 **1**,

$$\int_2^x \left(-\frac{1}{t}\right) dt = (-\ln t)|_{t=2}^{t=x} = \ln 2 - \ln x$$

and, using a logical consequence of the definitions in Sect. 6.2, we have

$$\lim_{x\to\infty} \ln x = +\infty,$$

because, for all (large, positive) M there exists an $M' = e^M$ such that

$$|\ln x| = \ln x > M \quad \text{if} \quad x > M' = e^M.$$

(We used also from Sect. 7.2 that e^M is continuous and $\lim_{x\to\infty} e^x = \infty$). So

(continued)

$$\int_{2}^{\infty}\left(-\frac{1}{t}\right)dt = \lim_{x\to\infty}(\ln 2 - \ln x) = -\infty.$$

Since we accepted infinity as limit we accept it also as value of improper integral, so $\int_2^\infty (-\frac{1}{t})\, dt$ *exists but is* $-\infty$.

3. $\int_{-\infty}^{\pi} \sin t\, dt$. We calculate first $\int_x^{\pi} \sin t\, dt = (-\cos t)|_{t=x}^{t=\pi} = \cos x + 1$. But $\cos x$ (and therefore $\cos x + 1$) has no limit as $x \to -\infty$ (or as $x \to \infty$ for that matter): it keeps oscillating between 1 and -1 (going through all values in between also infinitely often). So $\int_{-\infty}^{\pi} \sin t\, dt$ *does not exist* (and neither does, say $\int_0^\infty \sin t\, dt$).

4. Consider the integral

$$\int_0^1 t^{-2}\, dt = \lim_{x\to 0+}\int_x^1 t^{-2}\, dt = \lim_{x\to 0+}(-t^{-1}+c)\Big|_{t=x}^{t=1}$$
$$= \lim_{x\to 0+}\left(-1+\frac{1}{x}\right) = +\infty,$$

since $\lim_{x\to 0+}\frac{1}{x} = +\infty$ (cf. Sect. 6.2, Example 2). So the improper integral *exists but is* $+\infty$.

5. Consider the integral

$$\int_0^4 (6t^{\frac{1}{2}} + t)\, dt = \lim_{x\to 0+}\int_x^4 (6t^{-\frac{1}{2}} + t)\, dt$$
$$= \lim_{x\to 0+}\left(6\frac{t^{\frac{1}{2}}}{\frac{1}{2}} + \frac{t^2}{2}\right)\Bigg|_{t=x}^{t=4}$$
$$= \lim_{x\to 0+}(12\cdot 4^{\frac{1}{2}} + \frac{16}{2} - 12x^{\frac{1}{2}} - \frac{x^2}{2})$$
$$= 32 - \lim_{x\to 0+}(12x^{\frac{1}{2}} + \frac{x^2}{2}),$$

if the limit exists. Now, $\lim_{t\to 0+} t^{-\frac{1}{2}} = +\infty$, but, by the result in Sect. 6.4, every function which is differentiable at a point is also continuous there and, by formula (6.7) of Sect. 6.5, x^n is differentiable with derivative nx^{n-1} everywhere, where the latter expression exists (make sense) so, for $n = 2$, also at 0. So x^2 is also continuous at 0 and $\lim_{x\to 0+} x^2 = 0$. But also $\lim_{x\to 0+} x^{\frac{1}{2}} = 0$, because for all $\varepsilon > 0$ there exists a δ, for instance $\delta = \varepsilon^2$, such that, $\left|x^{\frac{1}{2}} - 0\right| = x^{\frac{1}{2}} < \varepsilon$ if $0 < x < \delta = \varepsilon^2$. So this improper integral *exists and is finite* (its value is 32):

(continued)

$$\int_0^4 (6t^{-\frac{1}{2}} + t)\, dt = 32 - 12 \lim_{x \to 0+} x^{\frac{1}{2}} - \frac{1}{2} \lim_{x \to 0+} x^2 = 32.$$

10.6.1 Exercises

1. Evaluate

(a) $\int_2^\infty x^{-6}\, dx$ (b) $\int_0^2 x^{-6}\, dx$ (c) $\int_1^\infty \frac{x^3+x^2+1}{x^5+x^3}\, dx.$

2. Calculate

(a) $\int_0^e \frac{1}{x}\, dx,$ (b) $\int_{-\infty}^\infty \frac{1}{1+x^2}\, dx,$ (c) $\int_0^\infty \frac{x}{1+x^2}\, dx.$

3. Compare

(a) $\int_{-\infty}^0 e^x\, dx$ to (b) $\int_{-\infty}^\infty e^x\, dx$ and (c) $\int_{-\infty}^\infty e^{-x}\, dx.$

4. Determine, for $r > 0$, the integrals

(a) $\int_0^\infty te^{-rt}\, dt,$ (b) $\int_0^\infty t^2 e^{-rt}\, dt.$

(c) Calculate the integral in (a) for $r = 0.06$.

(d) Calculate $\int_0^{200} te^{\;0.06t}\, dt$ (see Exercise 10.4.1 5. (a)).

5. Evaluate

(a) $\int_1^\infty x^{-3/2}\, dx,$ (b) $\int_0^\infty \left(\frac{\sin x}{x} + \frac{\cos x}{x^2}\right) dx.$

6. Do the following integrals exist

(a) $\int_0^\infty \frac{\sin x}{x}\, dx,$ (b) $\int_0^\infty (2 \sin x + 3 \cos x)\, dx.$

10.6.2 Answers

1. (a) $5^{-1} \cdot 2^{-5} = 160^{-1} = 0.00625,$ (b) $+\infty,$ (c) $\frac{\pi}{4} + \frac{1}{2} \approx 1.285398.$

2. (a)
$$\int_0^e \frac{1}{x}\,dx = \lim_{a\to 0+}\int_a^e \frac{1}{x}\,dx = \lim_{a\to 0+}(\ln x + c)\Big|_{x=a}^{x=e}$$
$$= \ln e - \lim_{a\to 0+}\ln a = 1-(-\infty) = 1+\infty = \infty.$$

(b)
$$\int_{-\infty}^{\infty}\frac{1}{1+x^2}\,dx$$
$$= \lim_{a\to-\infty}\int_a^b \frac{1}{1+x^2}\,dx + \lim_{r\to\infty}\int_b^r \frac{1}{1+x^2}\,dx \qquad (a<b<r)$$
$$= \lim_{a\to-\infty}(\arctan x + c_1)\Big|_{x=a}^{x=b} + \lim_{r\to\infty}(\arctan x + c_2)\Big|_{x=b}^{x=r}$$
$$= \lim_{a\to-\infty}(\arctan b - \arctan a) + \lim_{r\to\infty}(\arctan r - \arctan b)$$
$$= -(-\frac{\pi}{2}) + \frac{\pi}{2} = \pi.$$

(c)
$$\int_0^{\infty}\frac{x}{1+x^2}\,dx$$
$$= \lim_{r\to\infty}\int_0^r \frac{x}{1+x^2}\,dx = \lim_{r\to\infty}(\frac{1}{2}\ln(1+x^2)+c)\Big|_{x=0}^{x=r}$$
$$= \lim_{r\to\infty}(\frac{1}{2}\ln(1+r^2)) - \frac{1}{2}\ln 1 = \infty - 0 = \infty.$$

3. (a) $\int_{-\infty}^0 e^x\,dx = 1$, (b) $\int_{-\infty}^{\infty} e^x\,dx = \infty$, (c) $\int_{-\infty}^{\infty} e^{-x}\,dx = \infty$.

4. (a) $\int_0^{\infty} te^{-rt}\,dt = 1/r^2$,
   (b) $\int_0^{\infty} t^2e^{-rt}\,dt = 2/r^3$, as follows from Exercise 10.4.1.5,
   (c) $\int_0^{\infty} te^{-0.06t}\,dt = 277.778$,
   (d) $\int_0^{200} te^{-0.06}\,dt = 277.756$.

5. (a) $\displaystyle\int_1^{\infty} x^{-3/2}\,dx = \lim_{r\to\infty}\int_1^r x^{-3/2}\,dx = \lim_{r\to\infty}\left(-2x^{-\frac{1}{2}}+c\right)\Big|_{x=1}^{x=r}$
$$= \lim_{r\to\infty}\left(-\frac{2}{r^{\frac{1}{2}}}\right) - \left(-\frac{2}{1^{\frac{1}{2}}}\right) = 0+2 = 2.$$
   (b) $\displaystyle\int_{\pi}^{\infty}\frac{\sin x}{x}+\frac{\cos x}{x^2}\,dx = \lim_{r\to\infty}\int_{\pi}^{r}\frac{\sin x}{x}+\frac{\cos x}{x^2}\,dx$
$$= \lim_{r\to\infty}\left(\frac{-\cos x}{x}+c\right)\Big|_{x=\pi}^{x=r} = \lim_{r\to\infty}\left(-\frac{\cos r}{r}\right) - \left(-\frac{\cos\pi}{\pi}\right)$$
$$= 0 - \frac{1}{\pi} \approx -0.31831.$$

6. (a) Yes. (b) No.

11 Differential Equations

A (system of) differential equation(s) relates some unknown function(s) with some of its (their) derivatives. In applications, the functions usually represent physical, engineering, biological or economic quantities, and the derivatives represent their rates of change.

11.1 Introduction

Let the amount of money M_1 in an economy (money in circulation and sight deposits by domestic depositors other than banks) be $y(t)$ at a point t in time. Then, h time units later, the amount will be $y(t+h)$, so it increased by $y(t+h)-y(t)$ (it decreased if the difference is negative). Such differences are often denoted by

$$\Delta y(t) \;:=\; y(t+h) - y(t)$$

and accordingly one writes occasionally

$$\Delta t := h, \quad \text{so} \quad y(t) = y(t + \Delta t) - y(t).$$

Now the central bank tries to "smooth" this increase by choosing its policy so that $\Delta y(t)$ is approximately proportional both to the amount $y(t)$ of money at time t and to $h = \Delta t$

$$\Delta y \approx ay(t)\Delta t, \quad a \subset \mathbb{R}_{++}.$$

The sign $\approx$ means

$$\lim_{\Delta t\to 0} \frac{\Delta y}{\Delta t} = ay(t) \quad \text{or} \quad \lim_{h\to 0} \frac{y(t+h) - y(t)}{h} = ay(t).$$

W. Eichhorn, W. Gleißner, *Mathematics and Methodology for Economics*, Springer Texts in Business and Economics, DOI 10.1007/978-3-319-23353-6_11

Here the left hand side is exactly the derivative of the function y at t, so we have

$$y'(t) = ay(t). \tag{11.1}$$

As we see, this is an equation connecting the values of an unknown function $y(t)$ with the values of its derivative at each point t. As we remember, the derivative is sometimes written with the aid of "differentials" as $y'(t) = dy(t)/dt$, so 11.1 can be written as

$$\frac{dy(t)}{dt} = ay(t),$$

and is called a *differential equation*. More generally spoken a differential equation relates some unknown function with some of its derivatives.

Remembering also that the derivative $y'(t)$ is the slope of the tangent of the graph of $y(t)$ at the point t Eq. (11.1) means geometrically that this slope (growth of money) is, in this example, proportional to the amount $y(t)$ of M_1 money at time t.

The question for differential equations, as for other equations is, whether there *exist* "solutions" (in this case functions) which satisfy the equation ("existence problem"), and if yes, *how many* solutions do there exist. Furthermore, can one reduce, possibly by further boundary or initial conditions, the number of solutions to *one* ("uniqueness problem"). In this case substitution shows that the exponential functions of rate a given by

$$y(t) = be^{at} \tag{11.2}$$

satisfy (11.1) whatever the real constant b is. Indeed for this function

$$y'(t) = bae^{at} = ay(t).$$

So there exist solutions of (11.1), namely those given by (11.2). We will show that there are *no other solutions*.

First we have to clarify the domain and range of $y(t)$. While we could permit negative time (time preceeding a starting point agreed upon) and (God forbid!) even 0 or negative amount of money in the whole economy, it seems reasonable to suppose $y : \mathbb{R}_+ \longrightarrow \mathbb{R}_{++}$. Accordingly, we will suppose (11.1) to be valid and (11.2) to be defined for $t \in \mathbb{R}_+$ only, and we shall take $b > 0$ in (11.2).

Now we show that there are no solutions of (11.1) which are not of the form (11.2). For this we write (11.1) as

$$\frac{y'(t)}{y(t)} = a$$

and recognise that the left hand side is the derivative of $\ln y(t)$ with respect to t. while the right hand side is the derivative of at. This means that these two functions can only differ by a constant c:

$$\ln y(t) = at + c, \quad \text{that is,} \quad y(t) = e^{at+c} = be^{at} \quad \text{with} \quad b = e^c.$$

So (11.2) indeed gives all solutions $y : \mathbb{R}_+ \longrightarrow \mathbb{R}_{++}$ of (11.1).

We still have an arbitrary constant ("parameter") in (11.2). This makes it possible to fix an *initial value*, that is, the value y_0 of y at a point t_0 (often, in particular, $t_0 = 0$ is chosen). Indeed, then

$$be^{at_0} = y_0, \quad \text{that is,} \quad b = y_0 e^{-at_0} \quad \text{and} \quad y(t) = y_0 e^{a(t-t_0)},$$

which satisfies both (11.1) and $y(t_0) = y_0$, is *the unique solution of the differential equation* (11.1) with *the initial condition* $y(t_0) = y_0$. (If $t_0 = 0$ then the unique solution is $y(t) = y_0 e^{at}$.) This solves the uniqueness problem.

A geometric representation of the problem is given by direction fields (Fig. 11.1). At every point (t, y) of the plane we draw a little segment of the slope $ay(t)$. This segment may be considered to be a tangent at $(t, y(t))$ to the graph of a still unknown solution of the equation going through that point. One "sees" that every solution "fits" into this direction field. The solution with the initial condition $y(t_0) = y_0$ can be approximated as follows (for the case $t_0 = 0, y_0 = 1, y'(t) = y(t)/2$ as in Fig. 11.1). From $t_0, y(t_0) = (0, y_0) = (0, 1)$ we advance "a little bit" in the direction of the slope $y'(0) = y(0)/2 = 1/2$ to the point $(t_1, y(t_1))$, from where we advance

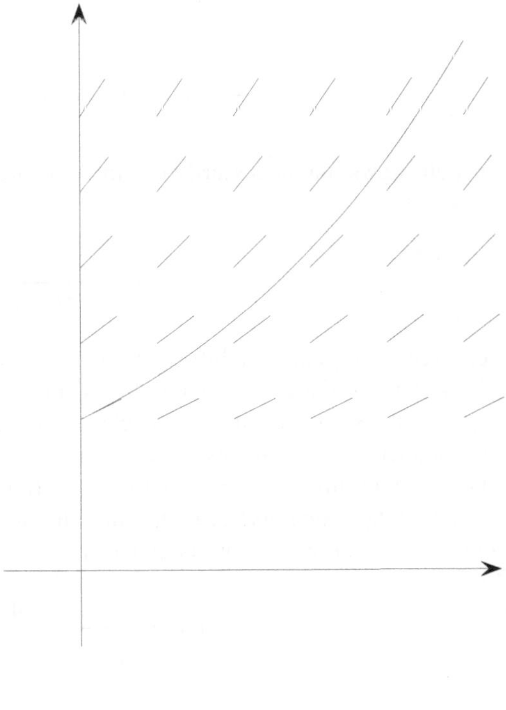

Fig. 11.1 The solution of the differential equation $y'(t) = y(t)/2$ and its vector field

on the segment with slope $y'(t_1) = y(t_1)/2$, again "a little bit", say till $(t_2, y(t_2))$, and so on for ever.

The following is another example leading to an explicit (because $y'(t)$ is on the left hand side) differential equation, actually to a slight modification of the previous one. There the amount of money $y(t) = be^{at}$ remains finite at every finite time t (no hyperinflation). As we remember, this was the solution of the differential equation

$$y'(t) = ay(t)$$

describing constant relative circulation speed of M_1 money. We now ask, what happens, if the amount $y(t)$ of M_1 money supply at time t satisfies the slightly different differential equation

$$y'(t) = ay(t)^{1+\varepsilon} \qquad t \in \mathbb{R}_+, \quad y : \mathbb{R}_+ \longrightarrow \mathbb{R}_{++} \tag{11.3}$$

(slightly, because ε is supposed to be a small positive number). One may be inclined to guess that the situation remains the same (no hyperinflation), if ε is very close to 0 (since (11.3) becomes (11.1) if $\varepsilon \to 0$). This turns out to be false. Indeed , if we write (11.3)

$$y(t)^{-1-\varepsilon} y'(t) = a,$$

we recognise that the left hand side is the derivative of $y(t)^{-\varepsilon}/-\varepsilon$ and the right hand side is the derivative of at. The integrals of both sides can differ at most by a constant c:

$$\frac{y(t)^{-\varepsilon}}{-\varepsilon} + c = at, \quad \text{that is,} \quad y(t) = \frac{1}{\varepsilon(c - at)}$$

(which shows that necessarily $c - at > 0$ since both $y(t)^{\varepsilon}$ and ε are positive). Thus we get that

$$y(t) = \frac{1}{((c - at)\varepsilon)^{1/\varepsilon}}$$

represents the general solution of the differential equation (11.3). (One still has to check that it satisfies (11.3), but this is easy.) This shows that the amount of money y grows to ∞ as the time approaches c/a (hyperinflation!) and this value is even independent of ε as long as ε is a positive constant. We note again that there was no such point in time for $\varepsilon = 0$, that is, the differential equation (11.1).

Incidentally one also sees that the unique solution of (11.3) satisfying also the initial condition $y(t_0) = y_0$, is given by

$$y(t) = \frac{1}{\left(y_0^{-\varepsilon} - a(t - t_0)\varepsilon\right)^{1/\varepsilon}}. \tag{11.4}$$

11.1.1 Exercises

1. Show that for $\varepsilon \to 0$ (11.4) approaches the solution $y(t) = y_0 e^{a(t-t_0)}$ of the differential equation (11.1) with the initial condition $y(t_0) = y_0$. [Hint: take the logarithm of (11.4) and apply the Bernoulli-L'Hospital rule to the derivatives of both sides.]

11.1.2 Answers

1. The answer is obvious.

11.2 Basics

In the differential equations (11.1) and (11.3) the unknown function $y(t)$ was a function of a single (real) variable. Such equations are called *ordinary differential equations*. In the case of unknown functions of several variables *(multiplace functions)*, the equations between them and their partial derivatives are called *partial differential equations (PDEs)*. Here are some examples of the latter:

$$\frac{\partial y(x,t)}{\partial x} + \frac{\partial y(x,t)}{\partial t} = y(x,t)^2 \quad \text{or} \quad \frac{\partial y}{\partial x} + \frac{\partial y}{\partial t} = y^2 \quad \text{for short;}$$

$$\frac{\partial y(x_1,x_2)}{\partial x_1}x_1 + \frac{\partial y(x_1,x_2)}{\partial x_2}x_2 = ry(x_1,x_2) \quad \text{or} \quad \frac{\partial y}{\partial x_1}x_1 + \frac{\partial y}{\partial x_2}x_2 = ry \quad \text{for short;}$$

$$\frac{\partial^2 y(x,t)}{\partial x^2} = a\frac{\partial y(x,t)}{\partial t} \quad \text{or} \quad \frac{\partial^2 y}{\partial x^2} = a\frac{\partial y}{\partial t} \quad \text{for short;}$$

$$\frac{\partial^2 y(x_1,x_2)}{\partial x_1^2} + \frac{\partial^2 y(x_1,x_2)}{\partial x_2^2} = 0 \quad \text{or} \quad \frac{\partial^2 y}{\partial x_1^2} + \frac{\partial^2 y}{\partial x_2^2} = 0 \quad \text{for short.}$$

The second example is Euler's partial differential equation for positively homogeneous function (of two variables) of degree r. We saw in Sect. 7.4 that indeed positively homogeneous functions of degree r, that is, those differentiable functions y which satisfy

$$y(\lambda x_1, \lambda x_2) = \lambda^r y(x_1, x_2)$$

for all $\lambda > 0$ and all $(\lambda x_1, \lambda x_2)$, (x_1, x_2) in the domain of y, are its solutions and only these. The other examples are also important for applications.

The order of the highest order derivative is the *order of the differential equation*. The first and the second equation above are of first oder, the third and the fourth are of second order, while

$$\frac{d^3y(t)}{dt^3} + 2y(t)^4 = \sin t, \quad \text{or}$$

$$\frac{d^3y}{dt^3} + 2y^4 = \sin t \quad \text{or} \quad y''' + 2y^4 = \sin t \quad \text{for short,}$$

is of third order.

If the derivative of the unknown function $y(t)$ only appears in first order, and if there are no products of the unknown function with its derivatives then the differential equation is *linear*, otherwise *nonlinear*. Among the above examples the first and the last equation are nonlinear (They contain the second or the fourth power of the unknown function respectively.), the other three examples are linear.

The general form of an ordinary differential equation of n-th order is

$$F(t, y(t), \frac{dy(t)}{dt}, \cdots, \frac{d^ny(t)}{dt^n}) = 0 \quad \text{or} \quad F(t, y, y', \cdots, y^{(n)}) = 0 \quad \text{for short.}$$

Here F is an $(n+2)$-place function, which is not constant with respect to place $n+2$. Any function $y : I \to \mathbb{R}$ (I a real interval) satisfying the equation for all $t \in I$ is again called a *solution of the differential equation*, or its *integral* (a more old fashioned expression). The graph of this function is accordingly called a *solution-curve* or integral-curve of the differential equation.

The above general form is that of an *implicit ordinary differential equation of n-th order*. If it can be written in the form

$$y^{(n)} = f(t, y, y', \cdots, y^{(n-1)})$$

then it is an *explicit differential equation*.

The set of all solutions of a differential equation is its *general solution*. This set may be empty as for the equation

$$y'^2 + y^2 + 1 = 0$$

or consists of one single solution as for

$$y'^2 + y^2 = 0$$

($y(t) \equiv 0$ is the only solution) but usually the general solution contains as many arbitrary constants ("parameters") as the order of the differential equation. We saw this for the two equations in Sect. 11.1. Often, as in (11.2) the general solution may be represented by one or more formulas. (Note also that the general solution may depend upon the domain and the range of the unknown function. For instance the choice of $\mathbb{R}_{++}$ as range in the examples (11.1) and (11.3) excluded the "trivial solution" $y(t) \equiv 0$. Moreover, if $\mathbb{C}$ is the range of y then the above equation $y'(t)^2 + y(t)^2 = 0$ has not only the "trivial solution" $y(t) \equiv 0$ but also $y(t) = be^{it}$ and $y(t) = be^{-it}$ with arbitrary constant b.) As we also saw, an *initial condition* $y(t_0) = y_0$, may specify the constant and thus give a unique solution of the initial value problem for a differential equation of first order (but not always, as we saw in the example of the real-valued solutions of $y'^2 + y^2 + 1 = 0$ and of $y'^2 + y^2 = 0$).

For an n-th order differential equation

$$F(t, y, y', \cdots, y^{(n)}) = 0,$$

the initial value problem consists in finding a solution y so that $y, y', \cdots, y^{n-1}$ assume given values $y_0, y_0', \cdots, y_0^{n-1}$ at t_0, that is, to find a function y which satisfies

$$F(t, y, y', \cdots, y^{(n)}) = 0, y(t_0) = y_0, y'(t_0) = y_0, \cdots, y^{(n-1)}(t_0) = y_0^{(n-1)}.$$

Again the questions are whether such solutions *exist (existence problem)* and whether there is just one such function *(uniqueness problem).*

11.2.1 Exercises

1. Each differential equation 1–8 listed below belongs to several of the following classes:
   (i) ordinary differential equation,
   (ii) partial differential equation,
   (iii) first order differential equation,
   (iv) second order differential equation,
   (v) third order differential equation,
   (vi) n-th order differential equation (ni>3),
   (vii) linear differential equation,
   (viii) nonlinear differential equation,
   (ix) implicit ordinary differential equation,
   (x) explicit ordinary differential equation.

   State for each differential equation to which classes it belongs.

   1 $\dfrac{dy(t)}{dt} = ay(t) + b$ (a, b real constants).

   2 $\dfrac{dy(t)}{dt} = ay(t)(b - y(t)) + c$ (a, b, c real constants).

   3 $\dfrac{\partial^2 y(x_1, x_2)}{\partial x_1^2} = a\dfrac{\partial y(x_1, x_2)}{\partial x_2}$ (a real constant).

4 $\frac{\partial^3 y(x_1,x_2)}{\partial x_1^2 \partial x_2} + a(x_1,x_2)\frac{\partial^2 y(x_1,x_2)}{\partial x_2^2} + b(x_1,x_2)\frac{\partial y(x_1,x_2)}{\partial x_1} = c(x_1,x_2)$
(a, b, and c real-valued functions).
5 $(y'')^2 + y''y' + (y')^2 + y'y + y^2 + y = 0$
6 $a_1(t) + a_2(t)y(t) + a_3(t)y'(t) + a_4(t)y''(t) + \cdots$
$+a_n(t)y^{(n-2)}(t) + a_{n+1}(t)y^{(n-1)}(t) + a_{n+2}(t)y^{(n)}(t) = 0$
$(a_1, a_2, \cdots, a_{n+2}$ real-valued functions).
7 $(y''')^4 + a(y'')^3 + b(y')^2 + cy + d = 0$
(a, b, c and d real constants).
8 $\frac{\partial^2 y(x_1,x_2,x_3)}{\partial x_1 \partial x_2} \frac{\partial^2 y(x_1,x_2,x_3)}{\partial x_1 \partial x_3} \frac{\partial^2 y(x_1,x_2,x_3)}{\partial x_2 \partial x_3} = a(x_1,x_2,x_3)$
(a real-valued function).

11.2.2 Answers

1. 1 (i), (iii), (vii), (x),
   2 (i), (iii), (viii), (x),
   3 (ii), (iv), (vii),
   4 (ii), (v), (vii),
   5 (i), (iv), (viii), (ix),
   6 (i), (vi), (viii), (ix);
   (x) instead of (ix) whenever $y^{(n)}(t)$ can be brought to the right-hand side of the equation. This is possible, whenever $a_{n+1}(t)$ has no zeroes.
   7 (i), (v), (viii), (ix);
   8 (ii), (iv), (viii).

11.3 Linear Differential Equations of First Order

In a way, the simplest ordinary differential equations are those which are both linear and of first order:

$$a_1(t)y'(t) + a_2(t)y(t) + a_3(t) = 0.$$

Clearly (11.1) is a very special case of this. We suppose this equation to be valid on an interval I, where a_1, a_2, a_3 are defined and where a_1 has no zero and look for solutions $y : I \to \mathbb{R}$. Because of the condition on a_1, we can divide by a_1 and get

$$y'(t) = a(t)y(t) + b(t) \quad (t \in I) \tag{11.5}$$

$(a(t) := -a_2(t)/a_1(t), b(t) := -a_3(t)/a_1(t))$, an *explicit linear differential equation of first order*. If $b(t) \equiv 0$ on I then the equation is *homogeneous* (because, rearranged and written as $y' - ay = 0$, the left hand side is a *linearly homogeneous function of y and* y'), otherwise *inhomogeneous*. The term $b(t)$, if not identically 0, is sometimes called a *perturbation*.

Remembering the motivation for (11.1) we can explain first the *homogeneous explicit linear differential equation*

$$y'(t) = a(t)y(t) \quad (t \in I) \tag{11.6}$$

by saying that, in view of changing circumstances (for instance rate of growth—or decline—of the net national product), the central bank "smoothes" the increase of the amount M_1 of money supply by a factor depending on time. By the same argument as in Sect. 11.1, this leads to (11.6).

We will suppose that the function $a : I \to \mathbb{R}$ is sectionally continuous (see Sect. 6.3) and solve the differential equation (11.6) (that is, determine its general solution). First we show that the function $\tilde{y}$ given by

$$\tilde{y}(t) = e^{\int_\alpha^t a(x)dx} \tag{11.7}$$

satisfies (11.6). Here α is a constant (for instance the left end of the interval I if it is bounded and closed from below) and the integral in the exponent exists, as mentioned in Sect. 10.5, because a is sectionally continuous.

$$\tilde{y}'(t) = a(t)e^{\int_\alpha^t a(x)dx}.$$

Clearly, with $\tilde{y}$, also every $\beta\tilde{y}$ (β an arbitrary constant) satisfies (11.6). We prove that there are no other solutions, that is, the general solution of (11.6) is given by

$$y(t) = \beta e^{\int_\alpha^t a(x)dx}. \tag{11.8}$$

To see this take an arbitrary solution y of (11.6) and calculate

$$\left(\frac{y(t)}{\tilde{y}(t)}\right)' = \frac{\tilde{y}(t)y'(t) - \tilde{y}'(t)y(t)}{\tilde{y}(t)^2} = \frac{\tilde{y}(t)a(t)y(t) - a(t)y(t)\tilde{y}(t)}{\tilde{y}(t)^2} = 0,$$

because both y and $\tilde{y}$ satisfy (11.6). Note that, by (11.7), $\tilde{y}(t) > 0$ for every $t \in I$. So $y(t)/\tilde{y}(t)$ is constant on I:

$$\frac{y(t)}{\tilde{y}(t)} = \beta, \quad y(t) = \beta e^{\int_\alpha^t a(x)dx}$$

as asserted. If we also want the initial condition $y(t_0) = y_0$ to be satisfied then (see (11.8))

$$y_0 = y(t_0) = \beta e^{\int_\alpha^{t_0} a(x)dx}, \quad \beta = y_0 e^{-\int_\alpha^{t_0} a(x)dx},$$

and

$$y(t) = y_0 e^{\int_{t_0}^{t} a(x)dx} \quad (t \in I)$$

is the unique solution of the differential equation (11.6) which also satisfies the initial condition $y(t_0) = y_0$.

Example 1 $y'(t) = 2ty(t)$ $(t \in \mathbb{R}, y(0) = 3$, the function $t \longmapsto 2t$ is continuous, $\int_{\alpha}^{t} 2xdx = t^2 - \alpha^2$, so the general solution of the differential equation $y' = 2ty$ is given by

$$y(t) = \beta e^{t^2-\alpha^2} = \gamma e^{t^2} \quad (t \in \mathbb{R}),$$

where $\gamma(:= \beta e^{-\alpha^2})$ is an arbitrary constant and the solution of the initial value problem is

$$y(t) = 3e^{t^2}.$$

We now consider the *inhomogeneous equation*

$$y'(t) = a(t)y(t) + b(t) \quad (t \in I,\ b(t) \not\equiv 0), \tag{11.9}$$

that is equation (11.5), where $b(t) \not\equiv 0$. It can again be motivated by the central bank regulating the M_1 money supply, taking this time external influences (perturbation) into consideration, too, for instance the demand abroad for the currency. We give two ways to find the general solution of (11.9), again under the supposition that the function $a : I \to \mathbb{R}$ is sectionally continuous.

In the first procedure we suppose that we know one ("particular") solution y_p of (11.9):

$$y_p'(t) = a(t)y_p(t) + b(t) \quad (t \in I).$$

Then, for any solution y

$$\begin{aligned}(y(t) - y_p(t))' &= y'(t) - y_p'(t) = a(t)y(t) + b(t) - a(t)y_p(t) - b(t)\\ &= a(t)(y(t) - y_p(t)),\end{aligned}$$

that is, the function $t \longmapsto y(t) - y_p(t)$ satisfies the corresponding homogeneous equation (11.6). So, by (11.8),

$$y(t) = y_p(t) + \beta e^{\int_\alpha^t a(x)dx} \tag{11.10}$$

gives the general solution of (11.9). (It is easy to check that this function indeed satisfies (11.9).) In other words: The general solution of the inhomogeneous equation equals a particular solution plus the general solution of the corresponding homogeneous equation. So finding one solution of (11.9) is essentially sufficient to determine all its solutions.

Our second method is called (*variation of constants*). We suppose now that b also is sectionally continuous. We look for solutions of (11.9) of the form

$$y(t) = \beta(t)e^{\int_\alpha^t a(x)dx}, \tag{11.11}$$

that is, we replace in the general solution (11.8) of (11.6) the constant β by a function of t. (Therefore "variation of constants".) Note that not only every solution of (11.9), but every function y can be written in this form, one only has to define

$$\beta(t) := y(t)e^{-\int_\alpha^t a(x)dx}.$$

Now let us try to satisfy (11.9) by a function of the form (11.11).

$$\left(e^{\int_\alpha^t a(x)dx}\right)' = a(t)e^{\int_\alpha^t a(x)dx},$$

we want to have

$$0 = y'(t) - a(t)y(t) - b(t)$$

$$= \beta(t)a(t)e^{\int_\alpha^t a(x)dx} + \beta'(t)e^{\int_\alpha^t a(x)dx} - a(t)\beta(t)e^{\int_\alpha^t a(x)dx} - b(t),$$

that is,

$$\beta'(t) = b(t)e^{-\int_\alpha^t a(x)dx}.$$

Since b is sectionally continuous and $\int_\alpha^t a(x)dx$ is continuous, it remains continuous when composed with the exponential function and the product remains sectionally continuous:

$$\beta(t) = \int b(t)e^{-\int_\alpha^t a(x)dx} + \gamma.$$

Notice that this second exterior integral is an indefinite integral. So it contains an arbitrary constant γ. (We also could have replaced $\int_\alpha^t a(x)dx$ by an indefinite integral, only the resulting formula would look messy.) So, from (11.11), the general solution of (11.9) has to be of the form

$$y(t) = \left(\int b(t)e^{-\int_\alpha^t a(x)dx}dt\right)\left(e^{\int_\alpha^t a(x)dx}\right) + \gamma e^{\int_\alpha^t a(x)dx}.$$

Here, too, the second term is the general solution of the homogeneous equation (11.6), while the first one is a solution of (11.9), as in our first method. But we now have a formula to calculate it. Of course, sometimes it may be easier to calculate a particular solution y_p by trial and error.

Example 2 $y' = 3y + 2 + t^2$. We look for a particular solution of the form

$$y_p(t) = c_0 + c_1 t + c_2 t^2.$$

Comparing the coefficients of t^2, t, and the constant terms, we get the equations

$$0 = 3c_2 + 1, \quad 2c_2 = 3c_1, \quad \text{and} \quad c_1 = 3c_0 + 2.$$

This yields

$$c_2 = -\frac{1}{3}, \quad c_1 = -\frac{2}{9}, \quad \text{and} \quad c_0 = -\frac{20}{27}$$

and

$$y_p(t) = -\frac{20}{27} - \frac{2}{9}t - \frac{1}{3}t^2,$$

which indeed satisfies $y' = 3y + 2 + t^2$. By (11.10) (choosing $\alpha = 0$) *the general solution of*

$$y' = 3y + 2 + t^2$$

is given by

$$y(t) = \frac{-20 - 6t - 9t^2}{27} + \beta e^{3t}.$$

(continued)

If we want to solve initial value problems, it is easier, in cases such as this, to make a straight forward substitution. If the initial condition is $y(0) = 2$, then

$$2 = -\frac{20}{27} + \beta, \quad \text{so} \quad \beta = \frac{74}{27}$$

and the solution of the initial value problem is

$$y(t) = \frac{74e^{3t} - 20 - 6t - 9t^2}{27}.$$

There are some important differential equations which are nonlinear but can be reduced to linear differential equations. We give two examples.

1. The Bernoulli equation, which is named after Jakob Bernoulli(1654–1705) (not the same as Johann Bernoulli of the Bernoulli-L'Hospital rule; there was a whole dynasty of mathematicians in the Swiss Bernoulli family) is a nonlinear explicit first order equation

$$y'(t) = a(t)y(t) + b(t)y(t)^r \quad (t \in I), \tag{11.12}$$

where r is a real constant. If $r = 0$ or $r = 1$, we get, of course, linear differential equations. But we can reduce the Bernoulli equation (11.12) to linear equations for $r \notin \{0, 1\}$. We suppose a to be sectionally continuous and look for solutions $y : I \to \mathbb{R}_{++}$ of (11.12). (I an interval, we want the values of y to be positive in order that y^r makes sense for any real r.) For this we define $z : I \to \mathbb{R}_{++}$ by

$$z(t) = y(t)^{1-r}.$$

Then, by the usual rules for derivations and, since y satisfies (11.12),

$$z'(t) = (1-r)y(t)^{-r}y'(t) = (1-r)a(t)y(t)^{1-r} + (1-r)b(t),$$

that is,

$$z'(t) = (1-r)a(t)z(t) + (1-r)b(t), \tag{11.13}$$

which is indeed a linear first order differential equation which we solve as above and then the general solution of the Bernoulli equation (11.12) is given by

$$y(t) = z(t)^{1/(1-r)}.$$

Clearly one can also determine the solution of (11.12) with the initial condition $y(t_0) = y_0$ by choosing that solution of (11.13) which satisfies $z(t_0) = y_0^{1-r}$. As we have seen, there is exactly one such solution for the linear equation (11.13); its $1/(1-r)$-th power will be the solution of the initial value problem for the Bernoulli equation (11.12).

2. The Riccati equation (due to Jacopo Francesco Riccati (1676–1754))

$$y'(t) = f(t)y(t)^2 + g(t)y(t) + h(t) \tag{11.14}$$

is clearly a generalisation both of the case $r = 2$ of the Bernoulli equation (for $h(t) = 0$) and of the general explicit inhomogeneous linear equation (for $f(t) = 0$). We can reduce it to a Bernoulli equation with $r = 2$ if we know a particular solution y_p of (11.14). Indeed, then we subtract

$$y_p'(t) = f(t)y_p^2(t) + g(t)y_p(t) + h(t)$$

from the equation (11.14) which contains its general solution y. We simplify by defining

$$u(t) := y(t) - y_p(t)$$

and thus get

$$y(t) + y_p(t) = u(t) + 2y_p(t).$$

So we obtain (by use of $y^2 - y_p^2 = (y - y_p)(y + y_p)$)

$$y'(t) - y_p'(t) = u'(t) = f(t)u(t)(u(t) + 2y_p(t)) + g(t)u(t),$$

that is,

$$u'(t) = (2f(t)y_p(t) + g(t))u(t) + f(t)u(t)^2,$$

a Bernoulli equation of the form (11.12) with $r = 2$. As we saw

$$u(t) = z(t)^{-1}$$

where $z(t)$ is the general solution of the linear differential equation

$$z'(t) = -(2f(t)y_p(t) + g(t))z(t) - f(t). \tag{11.15}$$

Thus the general solution of the Riccati equation (11.14) is given by

$$y(t) = y_p(t) + z(t)^{-1},$$

where y_p is a particular solution of (11.14) and z is the general solution of the linear differential equation (11.15). Again it is preferable to solve initial value problems for Riccati equations by straight forward substitution.

11.3.1 Exercises

Determine the general solutions of the following differential equations.

1. $y'(t) = y(t)\cos t$,
2. $y'(t) = y(t)\cos t + 1 - \cos t$,
3. $y'(t) = ay(t)(1 - y(t))$, ($a \neq 0$ a real constant),
4. $y'(t) = 3t^2y(t) - 3t^2y(t)^2$,
5. $y'(t) = y(t) - y(t)^3$.
6. Determine the solution of the first equation which satisfies
   (a) $y(0) = 3$,
   (b) $y(\frac{\pi}{2}) = e^2$.
7. Determine the solution of the third equation which satisfies
   (a) $y(0) = 1/2$,
   (b) $y(0) = 3/4$.

11.3.2 Answers

1. $y(t) = Ae^{\sin t}$ (A here and in 1 to 5 is an arbitrary real constant.),
2. $y(t) = t + Ae^{\sin t}$,
3. $y(t) = \dfrac{Ae^{at}}{1 + Ae^{at}}$, $y(t) = 1$,
4. $y(t) = \dfrac{Ae^{t^3}}{1 + Ae^{t^3}}$, $y(t) \equiv 1$,
5. $y(t) = \dfrac{Ae^t}{\sqrt{1 + A^2e^{2t}}}$, $y(t) \equiv 1$, $y(t) \equiv -1$.
6. (a) $y(t) = 3e^{\sin t}$,
   (b) $y(t) = ee^{\sin t}$.

   (a) $y(t) = \dfrac{e^{at}}{1 + e^{at}}$,

   (b) $y(t) = \dfrac{3e^{at}}{1 + 3e^{at}}$.

11.4 An Application: Saturation of Markets: "Logistic Growth"

Suppose that a product has the market share $y(t)$ at time t and let $\tilde{y}$ be the "saturation share", the greatest achievable market share. This clearly cannot be greater than 1, that is 100 % (total saturation). Suppose further that, at an initial time t_0, the market share $y(t_0)$ is small but positive. Neglecting external influences, it is reasonable to suppose that the growth of the market share at any moment depends upon the market share $y(t)$ already achieved, and upon the "market potential", the "room for improvement" $\tilde{y} - y(t)$:

$$y'(t) = F(y(t), \tilde{y} - y(t)).$$

A particularly attractive (and simple) supposition is

$$y'(t) = cy(t)(\tilde{y} - y(t)), \quad (c \in \mathbb{R}_{++}) \tag{11.16}$$

An interpretation of this differential equation could be that, at the initial time t_0, the market share is small and grows "almost proportionally" with its size since $cy\tilde{y} - cy^2 \approx cy\tilde{y}$, if y is small $\left(\lim\limits_{y\to 0} \frac{cy\tilde{y} - cy^2}{cy\tilde{y}} = 1\right)$. When the market share comes closer to $\tilde{y}$ then $u(t) = \tilde{y} - y(t)$ gets "small" and $cy(\tilde{y} - y) = cyu$ as $\lim\limits_{y\to 0} \frac{cy\tilde{y} - cy^2}{cy\tilde{y}} = 1$: the small factor $u = \tilde{y} - y$ "dampens the growth".

Now we choose $t_0 = 0$ and $y(t_0) = y_0$ "positive but small", certainly under the saturation share $\tilde{y}$ $(0 < y_0 < \tilde{y})$. Equation (11.16) is a Bernoulli equation, see (11.12) in the previous section. Here

$$y'(t) = c\tilde{y}y(t) - cy(t)^2,$$

that is, $r = 2$, $a(t)c\tilde{y}$, $b(t) = -c$. So $z(t) = 1/y(t)$ satisfies the linear differential equation

$$z'(t) = -c\tilde{y}z(t) + c, \tag{11.17}$$

corresponding to (11.13). A particular solution clearly is the constant

$$z_p(t) = \frac{1}{\tilde{y}}.$$

So the general solution of (11.17) is, according to Sect. 11.3

$$z(t) = \frac{1}{\tilde{y}} + \gamma e^{-c\tilde{y}t}, \quad (\gamma := \beta e^{c\tilde{y}\alpha} \text{ with some constant } \alpha).$$

Therefore

$$y(t) = \frac{1}{z(t)} = \frac{\tilde{y}}{1 + \gamma\tilde{y}e^{-c\tilde{y}t}} = \frac{\tilde{y}}{1 + Ce^{-c\tilde{y}t}}$$

is the general solution of the differential equation (11.16), where C is an arbitrary constant, in this context positive. If we also take the initial condition $y(0) = y_0$ into consideration, we get

$$y_0 = \frac{\tilde{y}}{1+C}, \quad C = \frac{\tilde{y} - y_0}{y_0},$$

which shows that C is indeed positive since $\tilde{y} > y_0 > 0$, and

$$y(t) = \frac{\tilde{y}}{1 + \frac{\tilde{y}-y_0}{y_0} e^{-c\tilde{y}t}} \tag{11.18}$$

is the solution of the initial value problem. This function is often called a logistic function and its graph the logistic curve (see Fig. 11.2)

Exercise Show that $\left(\frac{\ln(\tilde{y} - y_0) - \ln y_0}{c\tilde{y}}, \frac{\tilde{y}}{2}\right)$ is the point of inflection of the logistic curve.

In nature the so called "organic growth" (growth of plants, or of a species of animals) can, in absence of external disturbing influence, often be well approximated by logistic functions with well chosen "parameters" y_0, $\tilde{y}$, c.

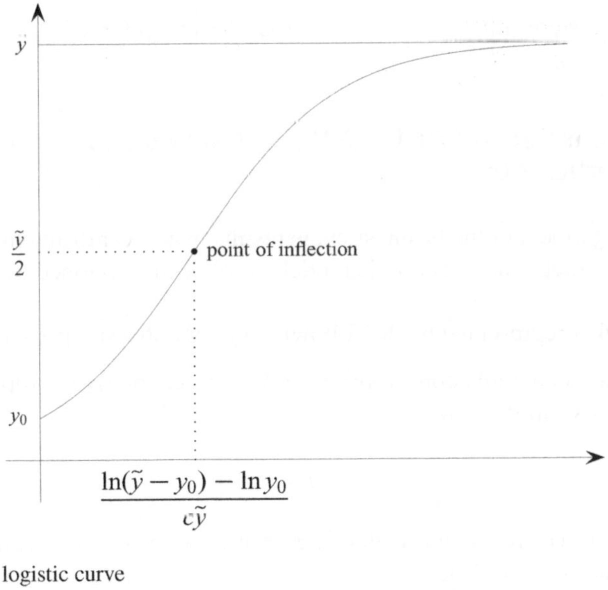

Fig. 11.2 The logistic curve

If one chooses $\tilde{y} = 1$ ("norming to 1", really just a choice of units), then (11.15) becomes the "market share function"

$$y(t) = \frac{1}{1 + (\frac{1}{y_0} - 1)e^{-ct}}.$$

As $t \to \infty$ this converges to 1 (because $c > 0$). Since now $y_0 < 1$, this market share function increases (because $t \mapsto \left(\frac{1}{y_0} - 1\right) e^{-ct}$ decreases and thus is less than 1 for every finite t.

We can also write the last equation in the equivalent forms

$$\frac{1 - y(t)}{y(t)} = \left(\frac{1}{y_0} - 1\right) e^{-ct} \quad \text{or} \quad \ln \frac{y(t)}{1 - y(t)} = ct + \ln \frac{y_0}{1 - y_0}.$$

Since the right hand side of the last equation is affine, it is particularly easy to draw a figure, even from approximate data, smoothing them by Gausses method of least squares (see Sect. 8.5).

The above method has been quite successful in researching the "conquest of markets" of different sources of energy (wood, coal, oil, gas, hydro or nuclear energy). If a new source of energy prevailed over the existing ones then the market share was well approximated by (11.18), that is, for the latter the observed points $(t, \ln \frac{y(t)}{1 - y(t)})$ were situated "in a cloud around a straight line".

11.5 Linear Second Order Differential Equations with Constant Coefficients

The following model of the business cycle results in a linear differential equation of second order. Such macroeconomic models have been developed by A.W. Phillips (1914–1975).

Our model is represented by the following system of assumptions.

A1: The macroeconomic consumption $C(t)$ in the economy is proportional to the national gross product $y(t)$:

$$C(t) = cy(t). \tag{11.19}$$

Here $c \in]0, 1[$. We suppose that the consumption $C_h(t)$ during $[t, t + h]$ is approximately $hC(t)$. Thus

$$C(t) = \lim_{h \to 0} \frac{C_h(t)}{h}.$$

A2: If for the point t in time a capital stock (rent capital which can be used for production) $\tilde{K}(t)$ is projected and, say, $K(t)$ is in fact realised, then also $\tilde{K}(t)$ is proportional to the national product $y(t)$:

$$\tilde{K}(t) = \gamma y(t),$$

where γ is a positive constant.

A3: The investment is the total investment less depreciation. Let $\tilde{I}(t)$ and $I(t)$ be projected or realised net investments, respectively, during the year following t. As in **A1**, we suppose that the projected or realised net investments in the time interval $[t, t+h]$ will approximately be $h\tilde{I}(t)$ and $hI(t)$ respectively. Net investments equal the change in the projected or realised capital stock, respectively. Therefore

$$\tilde{K}(t+h) + \tilde{K}(t) = h\tilde{I}(t) \quad \text{and} \quad K(t+h) - K(t) = hI(t),$$

which yield

$$\tilde{I}(t) = \lim_{h \to 0} \frac{\tilde{K}(t+h) - \tilde{K}(t)}{h} = \tilde{K}'(t),$$

and

$$I(t) = \lim_{h \to 0} \frac{K(t+h) - K(t)}{h} = K'(t).$$

$K(t)$ and $\tilde{K}(t)$ are differentiable by **A1** and **A2**, and so are $C(t)$ and $y(t)$. We suppose moreover that $I(t)$ and $\tilde{I}(t)$ are differentiable, thus $K(t)$, $\tilde{K}(t)$, $C(t)$, $y(t)$, and $A(t)$ (see **A5**) are twice differentiable.

A4: The projected net investment $\tilde{I}(t)$ for the time $[t, t+1]$ is proportional to the difference between the projected and the realised capital stock at that time:

$$\tilde{I}(t) = \beta(\tilde{K}(t) - K(t)).$$

Here β is a positive constant.

A5: One has to add the "exogenous, autonomous" demand $A(t)$ to the real or projected intrinsic demand (consumption or investment) $C(t)+I(t)$ or $C(t)+\tilde{I}(t)$, respectively. So the supply in the economy, which we identify with the national gross product, will be

$$y(t) = C(t) + I(t) + A(t). \tag{11.20}$$

The excess demand will be $C(t) + \tilde{I}(t) + A(t) - y(t) = \tilde{I}(t) - I(t)$ during the year $[t, t+1]$, so $h(\tilde{I}(t) - I(t))$ during $[t, t+h]$.

A6: The change $y(t+h)-y(t)$ in the supply is proportional to the demand during $[t, t+h]$. So

$$y(t+h)-y(t) = \alpha h\left(\tilde{I}(t)-I(t)\right),$$

and as in **A3**

$$\alpha\left(\tilde{I}(t)-I(t)\right) = \lim_{h\to 0}\frac{y(t+h)-y(t)}{h} = y'(t). \tag{11.21}$$

Here α is a positive constant.

We now differentiate the equation in **A4** (possible because of **A3**)

$$\tilde{I}'(t) = \beta\left(\tilde{K}'(t)-K'(t)\right) = \beta\Big(\gamma y'(t)-I(t)\Big).$$

The assumption in **A3** similarly permits us to differentiate (11.21) and the equation in **A1**. So we get

$$y''(t) = \alpha\left(\tilde{I}'(t)-I'(t)\right) = \alpha\Big(\beta\gamma y'(t)-\beta I(t)-I'(t)\Big).$$

Now

$$I(t) = y(t)-C(t)-A(t) = y(t)-cy(t)-A(t),$$

follows from (11.18) and from **A1**, and we get

$$y''(t) = \alpha\Big(\beta\gamma y'(t)-\beta(1-c)y(t)+\beta A(t)-(1-c)y'(t)+A'(t)\Big),$$

that is,

$$y''(t)+\alpha(1-c+\beta\gamma)y'(t)+\alpha\beta(1-c)y(t) = \alpha\beta A(t)+\alpha A'(t).$$

This is a linear differential equation of second order with constant coefficients

$$a = \alpha(1-c+\beta\gamma) \quad \text{and} \quad b = \alpha\beta(1-c).$$

The right hand side

$$f(t) = \alpha\beta A(t)+\alpha A'(t)$$

is the so-called perturbation, and the whole equation is the general form of a linear differential equation of second order with constant coefficients:

$$y''(t) + ay'(t) + by(t) = f(t). \tag{11.22}$$

Notice that the perturbation needs not be constant.—These differential equations are called homogeneous if $f(t) \equiv 0$, otherwise they are called inhomogeneous. The homogeneous differential equations

$$y''(t) + a(t)y'(t) + b(t)y(t) = 0 \tag{11.23}$$

$$y''(t) + ay'(t) + by(t) = 0 \tag{11.24}$$

corresponding to the inhomogeneous equation (11.22) with a function $f(t)$ on the right hand side are of importance.

One proves exactly as in Sect. 11.3 that the general solution of the inhomogeneous equation (11.22) is the sum of a particular solution of the inhomogeneous equation and the general solution of the corresponding homogeneous equation. The same holds, if a and b are functions of t. There is also an analogue of the "variation of constants" (compare Sect. 11.3) for these equations with a slight twist. In order to explain what we mean by this, we give the details for Eq. (11.22).

First we find the general solution of the homogeneous equation (11.24). Of course $y(t) = 0$ gives a *trivial solution*. Using an idea going back to Leonhard Euler, we look for particular solutions of the form

$$y(t) = e^{\lambda t}, \tag{11.25}$$

where λ is a constant which we want to determine such that (11.24) is satisfied. Putting (11.25) into (11.24) we get

$$\lambda^2 e^{\lambda t} + a\lambda e^{\lambda t} + be^{\lambda t} = 0.$$

Since the exponential function is nowhere zero, this equation can hold if and only if the "characteristic equation"

$$\lambda^2 + a\lambda + b = 0 \tag{11.26}$$

of the differential equation (11.24) is satisfied.

As we know from high school, the solutions of the algebraic equation (11.26) are given by

$$\lambda = \frac{-a \pm \sqrt{a^2 - 4b}}{2}.$$

Here are three cases, according to whether the "discriminant"

$$D = a^2 - 4b$$

is positive, zero or negative.

Case 1: $D = a^2 - 4b > 0$. In this case (11.26) has two distinct real solutions

$$\lambda_1 = \frac{-a + \sqrt{a^2 - 4b}}{2} \quad \text{and} \quad \lambda_2 = \frac{-a - \sqrt{a^2 - 4b}}{2}$$

The differential equation (11.24) has two distinct solutions of the form

$$y_1(t) = e^{\lambda_1 t} \quad \text{and} \quad y_2(t) = e^{\lambda_2 t}.$$

A general principle for homogeneous linear differential equations is the principle of linear combinations: if y_1 and y_2 are solutions of such an equation then every function $y = c_1 y_1 + c_2 y_2$ with constants $c_1, c_2 \in \mathbb{R}$ is also a solution (compare linearity in Sects. 4.2 and 4.3). One sees this immediately by substitution. So

$$y(t) = c_1 y_1(t) + c_2 y_2(t) = c_1 e^{\lambda_1 t} + c_2 e^{\lambda_2 t} \tag{11.27}$$

with arbitrary constants c_1 and c_2 is also a solution of (11.24).

In Sect. 11.3 we were able to find a solution of the homogeneous linear differential equation of first order, which satisfies an arbitrary initial condition $y(t_0) = y_0$. Here our differential equation is of second order, so we need two initial conditions

$$y(t_0) = y_0, \quad y'(t_0) = y'_0 \tag{11.28}$$

to be satisfied by a solution of the form (11.27). Indeed this requires that

$$c_1 e^{\lambda_1 t_0} + c_2 e^{\lambda_2 t_0} = y_0 \quad \text{and} \quad c_1 \lambda_1 e^{\lambda_1 t_0} + c_2 \lambda_2 e^{\lambda_2 t_0} = y'_0.$$

We obtained the second equation by differentiating (11.27) for some c_1, c_2. We solve this system of linear equation and get

$$c_1 = \frac{\lambda_2 y_0 - y'_0}{\lambda_2 - \lambda_1} e^{\lambda_1 t_0} \quad \text{and} \quad c_2 = \frac{y'_0 - \lambda_1 y_0}{\lambda_2 - \lambda_1} e^{\lambda_2 t_0}$$

as unique solutions of the initial value problem (11.24) and (11.28). We see that such a solution indeed always exists, since the denominator $\lambda_2 - \lambda_1$ is not zero, because the characteristic equation has two distinct real solutions.

Case 2: $D = a^2 - 4b = 0$. In this case the characteristic equation (11.26) has just one solution of multiplicity 2

$$\lambda_1 = \lambda_2 = \lambda = -\frac{a}{2}.$$

Setting

$$y(t) = c_1 e^{\lambda t} + c_2 t e^{\lambda t} \quad \text{with} \quad c_1, c_2 \in \mathbb{R},$$

one sees that this function satisfies (11.24). So we have a linear combination of two essentially different solutions $e^{\lambda t}$ and $te^{\lambda t}$. There is exactly one choice of c_1 and c_2 for which the two initial conditions $y(t_0) = y_0$ and $y'(t_0) = y_0'$ are satisfied. They give the following system of linear equations:

$$e^{\lambda t_0} c_1 + t_0 e^{\lambda t_0} c_2 = y_0 \quad \text{and} \quad e^{\lambda t_0} c_1 + e^{\lambda t_0} c_2 + t_0 e^{\lambda t_0} c_2 = y_0'.$$

It has the unique solution

$$c_1 = (y_0 + (\lambda y_0 - y_0')t_0)e^{-\lambda t_0} \quad \text{and} \quad c_2 = (y_0' - \lambda y_0)e^{-\lambda t_0}.$$

Case 3: $D = a^2 - 4b < 0$. The two distinct complex solutions λ_1 and λ_2 are

$$\lambda_1 = \alpha + \beta i, \quad \lambda_2 = \alpha - \beta i \quad \text{with}$$

$$\alpha = \frac{a}{2} \quad \text{and} \quad \beta = \sqrt{4b - a^2}.$$

One complex general solution of (11.24) is as in case 1

$$\hat{y}_1(t) = \tilde{c}_1 e^{\lambda_1 t} + \tilde{c}_2 e^{\lambda_2 t} = (\tilde{c}_1 + \tilde{c}_2)e^{\alpha t} \cos \beta t,$$

which is in fact real. Another general solution is

$$\hat{y}_2(t) = \tilde{c}_1 e^{\lambda_1 t} - \tilde{c}_2 e^{\lambda_2 t} = (\tilde{c}_1 - \tilde{c}_2)e^{\alpha t} \sin \beta t,$$

which is purely imaginary. Dividing the latter by the complex unit i, we get two real-valued solutions of (11.24) which are linearly independent. The differential equation is satisfied if and only if the real and the imaginary part are satisfied separately as follows from Sect. 1.7. Hence the general real solution is

$$y(t) = c_1 e^{\alpha t} \cos \beta t \; + \; c_2 e^{\alpha t} \sin \beta t$$

with

$$c_1 = \tilde{c}_1 + \tilde{c}_2, \quad c_2 = \frac{\tilde{c}_1 - \tilde{c}_2}{i}.$$

We now determine the constants c_1 and c_2 such that the initial conditions $y(t_0) = y_0$ and $y'(t_0) = y_0'$ are satisfied. This yields the following system of linear equations

$$c_1 e^{\alpha t_0} \cos \beta t_0 \ + \ c_2 e^{\alpha t_0} \sin \beta t_0 = y_0,$$

$$c_1 \alpha e^{\alpha t_0} \cos \beta t_0 \ - \ c_1 e^{\alpha t_0} \sin \beta t_0 \ + \ c_2 \alpha e^{\alpha t_0} \cos \beta t_0 = y_0'.$$

The reader should calculate or at least verify that its solutions are

$$c_1 = \frac{e^{-\alpha t_0}(\beta y_0 \cos \beta t_0 + \alpha y_0 \sin \beta t_0 - y_0' \sin \beta t_0)}{\beta},$$

$$c_2 = \frac{e^{-\alpha t_0}(\beta y_0 \sin \beta t_0 + y_0' \cos \beta t_0 - \alpha y_0 \cos \beta t_0)}{\beta}.$$

Since this solution is unique the differential equation (11.23) with the initial values (11.27) only has one solution.

Exactly as for first order equations one verifies that the general solution of the inhomogeneous equation is the sum of a particular solution of the inhomogeneous equation and the general solution of the homogeneous equation. Thus the problem is to find a particular solution of the inhomogeneous equation. One could do it by a modified version of the variation of constants known from first order equations. But in most cases it is better to make sophisticated guesses as we shall demonstrate in two examples.

Example 1 $y''(t) - 2y'(t) + y(t) = 3$
A particular solution is $y_p(t) = 3$. The characteristic equation of the homogeneous equation is $\lambda^2 - 2\lambda + 1 = 0$. Its double zero is $\lambda = 1$. Therefore the general solution of the inhomogeneous equation is

$$y(t) = c_1 e^t + c_2 t e^t + 3.$$

Example 2 $y''(t) - y'(t) - 2y(t) = 4t$
Its characteristic equation $\lambda^2 - \lambda - 2 = 0$ has the zeros $\lambda_1 = -1, \lambda_2 = 2$. Thus the general solution of the homogeneous equation is

$$\tilde{y}(t) = \lambda_1 e^{-t} + \lambda_2 e^{2t}.$$

To find a particular solution $y^*(t)$ of the inhomogeneous equation one looks for a solution of the form

$$y^*(t) = \alpha t + \beta.$$

Comparing coefficients one finds $\alpha = -2$ and $\beta = 1$. Thus the general solution is

$$y(t) = \lambda_1 e^{-t} + \lambda_2 e^{2t} - 2t + 1.$$

11.5.1 Exercises

Solve the following second order equations:

1. $y''(t) - 2y'(t) + y(t) = t^2$
2. $y''(t) + y(t) = 3t$
3. $y''(t) + 4y(t) = t^3$

11.5.2 Answers

1. $y(t) = c_1 e^t + c_2 t e^t + 4t + 6$
2. $y(t) = c_1 \sin t + c_2 \cos t + 3t$
3. $y(t) = c_1 \sin 2t + c_2 \cos 2t + \frac{2t^3 - 3t}{8}$

11.6 The Predator-Prey Model

A famous example for a system of two nonlinear differential equations is the predator-prey model of Alfred James Lotka (1880–1949) and Vito Volterra (1860–1940), which has both economic and ecological aspects. In this model we have at any given time t a population of $x(t)$ animals (the preys) with inexhaustible natural resources (food etc.) and a single natural enemy, the predators, a population of $y(t)$ animals of another species, which feed exclusively on these preys. One supposes, which is indeed a good approximation, that the original prey population would, in the absence of predators, grow exponentially with time, that is, it would satisfy the differential equation

$$x'(t) = ax(t), \quad a > 0.$$

With the predators, however, the instantaneous increase $x'(t)$ is reduced by a magnitude proportional to both $x(t)$, the number of preys, and $y(t)$, the number of predators:

$$x'(t) = ax(t) - bx(t)y(t), \quad a > 0,\ b > 0. \tag{11.29}$$

Similarly, in absence of preys ("food"), the predator population would decrease exponentially, that is, satisfy

$$y'(t) = -cy(t), \quad c > 0.$$

But with preys the instantaneous decrease from above will be reduced again by a magnitude proportional to both the number of predators and the number of preys, that is,

$$y'(t) = -cy(t) + kx(t)y(t), \quad c > 0,\ k > 0. \tag{11.30}$$

The "Lotka-Volterra equations" (11.29) and (11.30) form a nonlinear dynamical system. First we look for equilibrium points, that is, stationary (constant) solutions:

$$x(t) = \alpha, \quad y(t) = \beta.$$

Clearly these satisfy (11.29) and (11.30) if and only if

$$0 = \alpha(a - b\beta) \quad \text{and} \quad 0 = \beta(-c + k\alpha).$$

From these equations we get the trivial solution

$$x(t) = y(t) = 0,$$

and the less trivial one

$$x(t) = \frac{c}{k}, \quad y(t) = \frac{a}{b}.$$

We now proceed to determine the trajectories, at least in $\mathbb{R}^2_{++}$. If $y = f(x)$ $(x > 0,\ y > 0)$ is the equation of a trajectory, then, as we have seen, the differential equation (11.29) has to be satisfied, which in this case is

$$f'(x) = \frac{-cf(x) + kxf(x)}{ax - bxf(x)} = \frac{-c + kx}{x}\frac{f(x)}{a - bf(x)} \tag{11.31}$$

that is,

$$\left(\frac{a}{f(x)} - b\right) f'(x) = -\frac{c}{x} + k.$$

From Sect. 10.2 we know that the indefinite integral of both sides can differ only by a constant C:

$$a \ln f(x) - bf(x) = -c \ln x + kx + C. \tag{11.32}$$

While this produces $f(x)$ only implicitly as a solution of an equation, we can write it in a more agreeable form by taking the exponential function on both sides:

$$\frac{f(x)^a}{e^{bf(x)}}\frac{x^c}{e^{kx}} = D, \tag{11.33}$$

where $D = e^C$ is a constant.

The trajectory is an implicit function. It is remarkable that it is either a closed curve, or that it remains in a bounded part of the plane. Independent of our calculations such trajectories show up in several situations. One is covered by the following result of Henri Poincaré (1854–1912) and Ivar Bendixson (1861–1935). To formulate it we need to know what closed and connected point sets are in the plane.

A closed set S is one which contains all its accumulation points. A point P, which does not necessarily belong to S is an accumulation point or cluster point, if every neighbourhood of P contains infinitely many points of S.

A set S in the plane is connected (path-connected to be more exact), if for any two points $A, B \in S$ there exists a continuous curve connecting them, that is, a pair of real numbers a, b and a pair of continuous functions $x : [a, b] \to \mathbb{R}$, $y : [a, b] \to \mathbb{R}$ such that $(x(a), y(a)) = A$, $(x(b), y(b)) = B$. (Note that these definitions correspond to the intuitive notion of accumulation points, closed, and connected sets.)

Now we can formulate (but will not prove) the Poincaré-Bendixson theorem. If the closed, connected, and bounded set $M \subset \mathbb{R}^2$ contains an equilibrium point of the system

$$x'(t) = F(x(t), y(t)), \quad y'(t) = G(x(t), y(t)), \tag{11.34}$$

where the functions $F : \mathbb{R}^2 \longrightarrow \mathbb{R}$ and $G : \mathbb{R}^2 \longrightarrow \mathbb{R}$ have continuous partial derivatives with respect to x and y (Note that the equations (11.29) and (11.30) are special cases of (11.34).), then every trajectory $\{(x(t), y(t)); \quad t \geq t_0\}$ is either a cycle or a spiral converging to a cycle from one side as $t \to \infty$.

The Poincaré-Bendixson theorem shows that all solution curves (trajectories), which stay in the closed, connected, and bounded set M, are "stable" in the following sense (not to be confused with "asymptotically stable" as defined earlier in this section): they are either cycles or converge to cycles as $t \to \infty$.

For the Lotka-Volterra equations (11.29), and (11.30) the cycle trajectories mean that neither the predators nor the preys become extinct (quite reassuring), as long as these equations and their cyclic solutions describe adequately what is happening.

The cycle-solutions are also called "periodic", because the pair of functions (x, y) describing them is periodic:

$$x(0) = x(T), \quad y(0) = y(T)$$

for the time $t = T$, at which the point $(x(t), y(t))$ on the trajectory returns to the point where it was at the initial time 0 and from there one has

$$x(t) = x(t + T), \quad y(t) = y(t + T).$$

We now calculate for such a periodic solution (cycle trajectory) of the above predator-prey process the average sizes of the predator and the prey populations during this time T, that is,

$$\bar{x} = \frac{1}{T} \int_0^T x(t)dt, \quad \bar{y} = \frac{1}{T} \int_0^T y(t)dt.$$

We obtain

$$\bar{x} = \frac{c}{k}, \quad \bar{y} = \frac{a}{b}.$$

The reader is invited to verify these results. So, remarkably, the average sizes of the predator and the prey populations on any cycle-trajectory of the Lotka-Volterra model equal the respective population at the nontrivial equilibrium point $\left(\frac{c}{k}, \frac{a}{b}\right)$.

11.6.1 Exercise

Let the parameters in the Lotka-Volterra equations (11.29) and (11.30) be $a = 1.0$, $b = 0.001$, $c = 0.1$, $k = 0.00001$, respectively. With the aid of formula (11.33), determine D and the points
$(551, y_0)$, $(2000, y_1)$, $(2000, y_2)$, $(12000, y_3)$, $(12000, y_4)$,
$(30000, y_5)$, $(30000, y_6)$, $(40000, y_7)$, $(40000, y_8)$, $(44470, y_9)$

of the trajectory running through the point $(x, y) = (8000, 500)$. Make a sketch of the trajectory.

11.6.2 Answer

$D = 687.68$
$y_0 = 1000, y_1 = 594.5, y_2 = 1558, y_3 = 499.4, y_4 = 1758,$
$y_5 = 608.5, y_6 = 1532, y_7 = 761, y_8 = 1284, y_9 = 1000.$

Difference Equations 12

Difference equations relate to differential equations as discrete mathematics does to continuous mathematics.

12.1 Introduction

In most of our previous deliberations we considered the variables to move "continuously" (regardless whether the functions were continuous or not) to assume all values in an interval of the number line. This was an abstraction: In reality arbitrarily small weights, lengths, amounts of money, etc., even time either do not exist or cannot be measured. There are smallest units appropriate to the problem, even if they are as small as milligrams, cents (or one 100-th of the smallest unit of the most inflated currency), nanoseconds, etc., and in these cases the variables assume only discrete values. Moreover, even if we could locate the present value of either of these variables and of others anywhere on the (say real) number line their increases or decreases may be measured only in such units. Indeed we stressed that observing h tends to 0 stretches or shrinks the imagination. This was for us just a convenient device to describe processes with the help of powerful (differential, integral) calculi.

Often, however, we can take only these discrete values and/or increases (decreases) of the variables (and functions) into consideration. While for the description of the "continuous" processes in the above sense (and even for approximating the "discrete" ones) the differential equations method described in the last Chap. 11 was particularly appropriate, one can go quite far also with their "discrete analogues", the difference equations, in describing discrete processes, where variables change by "finite increments" (one milligram, millicron, cent, nanosecond, or an integer multiple thereof). This will be the subject in this chapter. If only one unit and its multiples figure for one variable, then this unit may always be described by the number 1 and differences of functions (independent of whether they are defined on real intervals or only on a set of consecutive integers, and of what the function values are) as long as subtraction makes sense in the range.

W. Eichhorn, W. Gleißner, *Mathematics and Methodology for Economics*,
Springer Texts in Business and Economics, DOI 10.1007/978-3-319-23353-6_12

One defines the operator Δ as follows

$$\Delta f(t) = f(t+1) - f(t). \tag{12.1}$$

It has the following properties

$$\Delta^2 f(t) = \Delta(\Delta f(t)) = \Delta(f(t+1) - f(t)) = f(t+2) - 2f(t+1) + f(t),$$

or in general

$$\Delta^n f(t) = \Delta(\Delta^{n-1} f(t)) = f(t+n) - \binom{n}{n-1} f(t+n-1)$$
$$+\binom{n}{n-2} f(t+n-2) \cdots + (-1)^{n-2}\binom{n}{2} f(t+2)$$
$$+(-1)^{n-1}\binom{n}{1} f(t+1) + (-1)^n f(t), \quad n = 1, 2, \cdots;$$

for the definition of the binomial coefficients $\binom{n}{k}$ see Sect. 7.2. The formula for $\Delta^n f(t)$ should really be proved by induction from that of $\Delta^{n-1} f(t)$. But a few trials $\Delta^2 f(t)$, $\Delta^3 f(t)$, $\Delta^4 f(t)$ may be convincing enough. Equations containing such differences are called difference equations.

Accordingly the difference analogue, for instance, of the second order linear differential equation with constant coefficients (Sect. 11.5) will be the second order homogeneous difference equation with constant coefficients

$$\Delta^2 Y(t) + a\Delta Y(t) + bY(t) = 0. \tag{12.2}$$

There is, however, also another way of writing difference equations. Take, for instance (12.2) and insert (12.1):

$$Y(t+2) - 2Y(t+1) + Y(t) + aY(t+1) - aY(t) + bY(t) = 0.$$

So, with the new constants $\alpha := a - 2$, $\gamma := b - a + 1$, we can write

$$Y(t+2) + \alpha Y(t+1) + \gamma Y(t) = 0. \tag{12.3}$$

Sometimes the form (12.2), at other times (12.3) is more advantageous for this and other difference equations. The advantage of (12.2) is its similarity to the corresponding differential equation (in this case equation (11.24) in Sect. 11.5). Also the methods of solution are similar, the characteristic equation (11.25) plays a similar role, too. The advantage of the form (12.3) is that, if Y is defined on $\mathbb{N}$ only, and if the values of Y are known say at 1 and 2 (or at some other t_0, $t_0 + 1 \in \mathbb{N}$), then (12.3) and similar more general solutions completely determine Y on $\mathbb{N}$ (or on $\{t_0, t_0 + 1, t_0 + 2, \cdots\}$, respectively). Indeed if, say $Y(1) = Y_1$, $Y(2) = Y_2$ then

$$Y(3) = -\alpha Y_2 - \gamma Y_1, \quad Y(4) = -\alpha Y(3) - \gamma Y_2 = (\alpha^2 - \gamma)Y_2 + \alpha\gamma Y_1,$$

and so on. In (slight) analogy to differential equations, Y_1 and Y_2 are called "initial values". Other ways of writing (12.3) are

$$\begin{aligned} Y(t) &= -\alpha Y(t-1) - \gamma Y(t-2), \\ Y_t &= -\alpha Y_{t-1} - \gamma Y_{t-2}, \quad \text{or} \\ Y_n &= \alpha Y_{n-1} - \gamma Y_{n-2}, \end{aligned} \tag{12.4}$$

in which case, however, $t, n \geq 3$ (or $\geq t_0 + 2$) should be supposed. In the latter form the sequence Y_n is given recursively from Y_1 and Y_2 by (12.4). With a slight abuse of language we speak in such cases about recursive sequences (or in the case (12.4) of "two-step recursions").

However, if Y is also defined for negative integers, say for all of $\mathbb{Z}$, and if (12.4) is valid on $\mathbb{Z}$ then Y is determined by (12.4) on all of $\mathbb{Z}$ as long as $\gamma \neq 0$ (if we had $\gamma = 0$, then Y_2 would not be needed and $\{Y_n\}$ would be determined by the "one-step reduction" $Y_n = -\alpha Y_{n-1}$). Indeed then

$$Y_{t-2} = -\frac{\alpha}{\gamma} Y_{t-1} - \frac{1}{\gamma} Y_t.$$

So

$$Y_0 = -\frac{\alpha}{\gamma} Y_1 - \frac{1}{\gamma} Y_2, \quad Y_{-1} = -\frac{\alpha}{\gamma} Y_0 - \frac{1}{\gamma} Y_1 = \frac{\alpha^2 - \gamma}{\gamma^2} Y_1 + \frac{\alpha}{\gamma^2} Y_2,$$

and so on.

If t means time, then we speak again about dynamical models, this time about "discrete dynamical models". One really needs them because data are always collected during finite time intervals (a year, a quarter, a month, a day), and in real life one cannot (except by "abstraction" or "guessing") consider arbitrarily small time intervals. The price is the loss of some convenience of the "smooth" methods of differential calculus in favour of somewhat "rougher" tools.

As an example we deal with a situation similar to that at the beginning of Sect. 11.5 constructing this time a simplified discrete dynamical model for the business cycle in a closed economy following Paul A. Samuelson (1915–2009; Nobel laureate in 1970). A closed economy is one without foreign trade. The following three assumptions are made.

S1: The consumption $C(t)$ during the time interval $[t, t+1]$ is an affine function of the national income (or national product) $Y(t-1)$ during $[t-1, t]$:

$$C(t) = c_0 + cY(t-1)$$

(compare to **A1** in Sect. 11.5). Here $c_0 \in \mathbb{R}_+$, $c \in]0,1[$ are constants.

S2: The amount of investments $I(t)$ during $[t, t+1[$ is proportional to the difference between the consumptions during $[t, t+1[$ and that during $[t-1, t[$ ($C(t)$ and $C(t-1)$ respectively):

$$I(t) = \beta(C(t) - C(t-1)).$$

This is the "investment originating from the growth of consumption". $\beta \in \mathbb{R}_+$ is again a constant.

S3: The national income $Y(t)$ is the sum of the consumption $C(t)$, the investment $I(t)$ "originating from the growth of consumption", and the autonomous investment $A(t)$ (independent from the other quantities in the model) during $[t, t+1[$:

$$Y(t) = C(t) + I(t) + A(t)$$

(compare to **A5** in Sect. 11.5).

This model is in a way simpler and more restricted than that in Sect. 11.5, for instance, the capital stock is left out of consideration, no distinction is made between projected and realised investments, and $A(t)$ is restricted to the investment part of the autonomous demand. But, on the other hand, no "asymptotic equations", limits, and derivatives are needed, in accordance with the "discrete" nature of economic life.

Economists call the "delay" between $t-1$ and t in **S1** the "Robertson delay" and the equations in **S2** and **S3** the "principle of acceleration" and the "equilibrium equation", respectively. Indeed, the former says that the increase in consumption, $C(t) - C(t-1)$, if positive, "accelerates" the investment $I(t)$, while the latter finds equilibrium in the closed economy, when the national income is the sum of consumption, investment and autonomous investment. The constants β and c are called "accelerator" and "marginal consumption rate", respectively. The latter reflects the vague notion that, in the simplified, because time-independent equation $C = c_0 + cY$ we would have $\frac{dC}{dY} = c$. Anyway, even c is only "approximately" (this time not in the sense of any limit value) constant in time and may be and usually is different for different countries.

Substituting the equations in **S1** and **S2** into **S3** we get

$$Y(t) = c_0 + cY(t-1) + \beta\,(c_0 + cY(t-1) - c_0 - cY(t-2)) + A(t),$$

that is,

$$Y(t) = c(1+\beta)Y(t-1) + \beta c Y(t-2) = c_0 + A(t) \tag{12.5}$$

an inhomogeneous linear difference equation of second order with constant coefficients or, if $c_o + A(t) = 0$, a homogeneous one of the form (12.3) (put $\tilde{t} = t-2$ into (12.5)).

As mentioned above, if $Y(1)$ and $Y(2)$ or $Y(0)$ and $Y(1)$ or, more generally,

$$Y^0 = Y(t_0), \quad Y^1 = Y(t_1) \tag{12.6}$$

are given then, by equation (12.5), so is

$$Y(t_0 + 2) = c(1 + \beta)Y^1 - \beta c Y^0 = c_0 + A(t_0 + 2)$$

and similarly $Y(t_0 + 3)$, $Y(t_0 + 4)$ and so on: "In principle", Y is determined on the domain $t_0, t_0 + 1, t_0 + 2, \cdots$ (on $\mathbb{N}$ if $t_0 = 1$). Thus the "the initial value problem" of the difference equation (12.5) with (12.6) is solved. Here we point out that initial value problems for differential equations cannot be solved in such a "step by step" way, no matter how small these "steps" are: we saw in Sect. 11.1 that even by advancing by very small steps in the "direction field" one can deviate from the exact solution curve.

As an example of solving a second order linear difference equation with constant coefficients, in this case the difference equation (12.5) for the national income $Y(t)$, with the initial condition (12.6), we specify

$$\beta = 1,\ c = \frac{3}{4},\ c_0 = 20,\ A(t) = 80,\ t_0 = 0,\ Y^0 = Y_0 = 320,\ Y^1 = Y_1 = 340$$

that is, we have the initial value problem

$$Y(t) = \tfrac{3}{2}Y(t-1) - \tfrac{3}{4}Y(t-2) + 100 \tag{12.7}$$

$$Y(0) = 320,\ Y(1) = 340. \tag{12.8}$$

We get as above

$$Y(2) = 370, ,\ Y(3) = 400,\ Y(4) = 423,\ Y(5) = 434,\ Y(5) = 434,$$

$$Y(6) = 434, ,\ Y(7) = 425,\ Y(8) = 413,\ Y(9) = 400,\ Y(10) = 391,$$

$$Y(11) = 386, ,\ Y(12) = 386,\ Y(13) = 389,\ Y(14) = 395,\ Y(15) = 400,$$

$$Y(16) = 404, ,\ Y(17) = 406,\ Y(18) = 406,\ Y(19) = 405,\ Y(20) = 402,$$

$$Y(21) = 400, ,\ Y(22) = 398,\ Y(23) = 397,\ Y(24) = 397,\ Y(25) = 398,$$

$$Y(26) = 399, ,\ Y(27) = 400,\ Y(28) = 401,\ Y(29) = 401,\ Y(30) = 401,$$

While these values are rounded up or down to the next integer, it is clearly "visible"—and even more so from Fig. 12.1—that the graph of Y shows a "damped oscillation" around a value y which seems to be close to 400. Actually for $t \to \infty$ y tends to 400 as we can see in (12.7). This equation shows that $Y(t) = 400$ is itself a solution, a constant solution giving the equilibrium value of the national income.

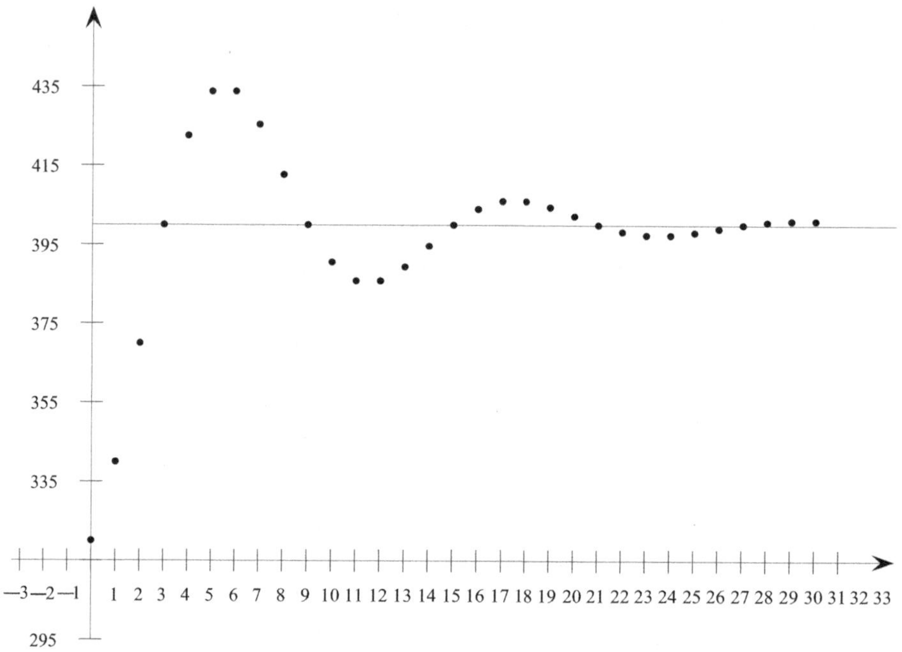

Fig. 12.1 Difference equation for the national income

If, as above, $Y(t)$ is the national income in the time interval $[t, t+1[$, and the consumption $C(t)$ in the same time interval satisfies **S1**, then C has to show the same kind of damped oscillation with 3/4 times as large an amplitude (amount of oscillation) and "retarded" by the shift 1.

As we see, difference equations are extremely easy to solve numerically step by step. However, general formulas would be useful and, independently, fundamental questions about the qualitative behaviour of the solution, whether we know it quantitatively or not, are important. For instance, for which β, c, c_0, and linear A has the difference equation (12.5) uniform, exploding or damped (stabilising) or oscillating solutions, strictly monotonic solutions, solutions converging, as $y \to \infty$ to the value of a constant "equilibrium" solution. We will discuss such questions in the next section.

12.1.1 Exercises

1. Solve the first order difference equation $x_n = 2x_{n-1} + 1$ with the initial condition $x_1 = 1$. (Remark: This equation determines either the number of moves for the towers of Hanoi or the number of knots in a maximal balanced binary tree. Look up the backgrounds in the internet!)

12.1.2 Answers

1. $x_n = 2^n - 1$

12.2 Linear Difference Equations

We summarise some definitions.

Let n be any natural number. An equation is called a linear difference equation, if it is of the form

$$b_n(t)Y(t+n) + b_{n-1}(t)Y(t+n-1) + \cdots + b_1Y(t+1) + b_0(t)Y(t) = g(t), \tag{12.9}$$

where $b_0, b_1, \cdots, b_n$, and the perturbation g are given real-valued functions on

$$D := \{t_0, t_0+1, t_0+2, \cdots\},$$

where t_0 is a fixed number (often $t_0 = 0$) and Y is an unknown real-valued function on D. Equation (12.9) is exactly of n-th order if $b_n(t) \not\equiv 0$ and if $b_0(t) \not\equiv 0$ then, as we see by writing $\tau := t + 1$, the equation is really of (n − 1)-th order. If $b_n(t) \neq 0$ for all $t \in D$, we can divide (12.9) by $b_n(t)$ and get, by defining

$$a_{n-1}(t) := \frac{b_{n-1}(t)}{b_n(t)}, \cdots, a_0(t) := \frac{b_0(t)}{b_n(t)}, f(t) := \frac{g(t)}{b_n(t)},$$

the linear difference equation of n-th order ($a_0(t) \neq 0$) in explicit form.

$$Y(t+n) + a_{n-1}(t)Y(t+n-1) + \cdots + a_1(t)Y(t+1) + a_0(t)Y(t) = f(t) \tag{12.10}$$

An example of a linear difference equation of second order is

$$tY(t+2) + 2Y(t+1) - (t+1)^2Y(t) = \sin t. \tag{12.11}$$

The difference equation of third order

$$Y(t+3)Y(t+1) - Y(t+1) + Y(t)^2 = t$$

is not linear because of the terms $Y(t+3)Y(t+1)$ and $Y(t)^2$.

It may happen that a linear difference equation has no solution for some initial conditions and many solutions for others. For instance the equation

$$tY(t+1) + Y(t) = 0 \tag{12.12}$$

has no solution that satisfies the initial condition $Y(0) = 1$. However, for the initial condition $Y(0) = 0$ there are infinitely many solutions as $Y(1)$ can be chosen arbitrarily.

The linear difference equations (12.9) and (12.10) are called homogeneous if $g(t) \equiv 0$ and $f(t) \equiv 0$, respectively. Compare similar notions in the last chapter, Sect. 11.5. For example,

$$tY(t+2) + 2Y(t+1) - (t+1)^2 Y(t) = 0$$

is a linear homogeneous equation of second order. As it is the same equation as (12.11) except for the term $\sin t$, it is called the homogeneous difference equation corresponding to (12.11). In general, if the coefficients $a_0(t), \cdots, a_{n-1}(t)$ and the perturbation $f(t)$ in (12.10) are defined for all $t \in D$, and if the initial values

$$Y_0 := Y(t_0), Y_1 := Y(t_0+1), \cdots, Y_{n-1} := Y(t_0+n-1) \tag{12.13}$$

are given one calculates from (12.10)

$$Y_n := f(t_0) - a_0(t_0)Y_0 - a_1(t_0)Y_1 - \cdots - a_{n-1}(t_0)Y_{n-1},$$

$$Y_{n+1} := f(t_0+1) - a_0(t_0+1)Y_1 - a_1(t_0+1)Y_2 - \cdots - a_{n-1}(t_0+1)Y_n,$$

and so on. So one gets, step by step, the unique solution of equation (12.10) that satisfies the initial conditions (12.13).

By the same argument as in the case of linear differential equations in the last chapter one can show the following. The general solution of (12.10) on D is obtained by adding one particular solution of (12.10) to the general solution of the corresponding homogeneous equation.

If Y_1 and Y_2 are solutions of the equation (12.10) with $f(t) \equiv 0$ on D then any linear combination

$$Y = c_1 Y_1 + c_2 Y_2, \tag{12.14}$$

where $c_1, c_2 \in \mathbb{R}$ are arbitrary constants, is a solution of (12.10) with $f(t) \equiv 0$ on D.

From now on we assume that the coefficients of equation (12.10) are real constants. Let us first consider equation (12.10) in the case

$$n = 1,\ f(t) \equiv 0,\ a_0(t) = -a \quad (a \in \mathbb{R},\ \text{constant}),$$

that is, we consider the homogeneous linear difference equation of first order

$$Y(t+1) - aY(t) = 0 \quad \text{or} \quad Y(t+1) = aY(t), \tag{12.15}$$

or equivalently

$$Y(t+1) - Y(t) + (1-a)Y(t) = \Delta Y(t) + (1-a)Y(t) = 0 \qquad (12.16)$$

that is Y grows, if $a > 1$, Y remains constant, if $a = 1$, and Y decreases, if $a \in]0, 1[$.

Now let Y have the value Y_0 at $t = 0$: $Y(0) = Y_0$. Then, from (12.15)

$$Y(1) = aY(0) = aY_0,$$

$$Y(2) = aY(1) = a(aY_0) = a^2 Y_0,$$

$$\vdots \qquad \vdots$$

$$Y(t) = aY(t-1) = a(a^{t-1}Y_0 = a^t Y_0.$$

Hence the unique solution of (12.15) that satisfies the initial condition $Y(0) = Y_0$ is $Y(t) = Y_0 a^t$.

Now let us compare this unique solution of the initial value problem for the difference equation or the equivalent initial value problem

$$\Delta Y(t+1) = (a-1)Y(t), \quad Y(0) = Y_0 \qquad (12.17)$$

to the solution of the initial value problem of the following differential equation

$$\frac{dy(t)}{dt} = (a-1)y(t), \quad Y(0) = Y_0. \qquad (12.18)$$

Setting

$$\Delta_h Y(t) = \frac{Y(t+h) - Y(t)}{h}, \qquad (12.19)$$

one obtains the difference equation $\Delta_1 Y(t) = Y(t+1) - Y(t) = (a-1)Y(t)$, and for $h \to 0$ one gets

$$\lim_{h\to 0} \Delta_h Y(t) = \lim_{h\to 0} \frac{Y(t+h) - Y(t)}{h} = \frac{Y(t)}{h} = (a-1)Y(t).$$

The difference equation has the solution

$$Y(t) = Y_0 a^t = Y_0 e^{t \ln a},$$

and the differential equation has the solution

$$y(t) = Y_0 e^{(a-1)t}.$$

Using the Bernoulli-L'Hospital rule one sees that the exponents are asymptotically equal for $a \to 1$:

$$\lim_{a\to 1} \frac{\ln a}{a-1} = \lim_{a\to 1} \frac{1/a}{1} = 1.$$

The domains of y and Y may be different: it is enough if Y is defined on $\{0, 1, 2, \cdots\}$.

In the case of the amount of M_1 money in Sect. 11.1 a is indeed close to 1 under normal economic situations. However, the solutions of (12.17) and (12.18) are close to each other, if we are more careful changing the time span from 1 to h. Indeed, if we replace $\Delta Y(t)$ in (12.17) by $\Delta_h Y(T)$ as defined in (12.19), we get

$$Y(t+h) - Y(t) = h(a-1)Y(t), \quad Y(0) = Y_0, \tag{12.20}$$

that is,

$$Y(t+h) = (ha - h + 1)Y(t) = \alpha_h Y(t), \quad Y(0) = Y_0,$$

where

$$\alpha_h = h(a-1) + 1.$$

As before we get

$$Y(h) = \alpha_h Y_0, \; Y(2h) = \alpha_h^2 Y_0, \; \cdots \; Y(nh) = \alpha_h^n Y_0,$$

that is,

$$Y(t) = \alpha_h^{t/h} Y_0 = (h(a-1)+1)^{t/h} Y_0, \quad \text{for} \quad t \in \{0, h, 2h, \cdots\}. \tag{12.21}$$

Taking logarithms we get

$$\ln Y(t) = \frac{t \ln(ha - h + 1)}{h} + \ln Y_0.$$

For $h \to 0$ one can apply the Bernoulli-L'Hospital rule and gets

$$\lim_{h\to 0} \frac{t \ln(h(a-1)+1)}{h} = \lim_{h\to 0} \frac{t(a-1)}{h(a-1)+1} = t(a-1).$$

Thus the solution (12.21) of (12.20) converges, as $h \to 0$, to the solution of (12.18), that is, to

$$Y(t) = Y_0 e^{(a-1)t}.$$

Next we look at the second order linear difference equations with constant coefficients:

$$Y(t+2) + aY(t+1) + bY(t) = f(t), \quad a \in \mathbb{R},\ b \in \mathbb{R},\ b \neq 0.$$

In particular, we look at the homogeneous version of it:

$$Y(t+2) + aY(t+1) + bY(t) = 0 \quad (t \in \{t_0, t_0+1, \cdots\},\ t_0 \in \mathbb{N} \cup \{0\}). \tag{12.22}$$

Similarly as we experimented with $y(t) = e^{\lambda t}$ as solution of the homogeneous linear differential equation of second order with constant coefficients, here we try

$$Y(t) = \lambda^t.$$

Trial and error shows that $e^{\lambda t}$ does not work in this case, but λ^t does. We supposed in (12.22) that t is a nonnegative integer. λ^t also makes sense, if λ is negative or even complex. ($\lambda = 0$ is excluded because 0^0 is not defined.) We substitute this into (12.22) and get

$$\lambda^{t+2} + a\lambda^{t+1} + b\lambda^t = 0.$$

We divide by λ^t (again $\lambda \neq 0$ is important) in order to obtain

$$\lambda^2 + a\lambda + b = 0. \tag{12.23}$$

This is the same equation as for the linear differential equation in Sect. 11.5, and here, too, it is called the characteristic equation—and $\lambda^2 + a\lambda + b$ the characteristic polynomial—of the difference equation (12.22). (We see again that $\lambda = 0$ can safely be excluded, since $Y(t+2) + bY(t+1) = 0$ is not really a second order difference equation, see our remark after equation (12.9).)

We continue as in Sect. 11.5: The characteristic equation (12.23) has either two distinct real solutions

$$\lambda_1 = \frac{-a + \sqrt{a^2 - 4b}}{2}, \quad \lambda_2 = \frac{-a - \sqrt{a^2 - 4b}}{2}, \tag{12.24}$$

one real solution

$$\lambda_1 = \lambda_2 = -\frac{a}{2},$$

or two distinct conjugate complex solutions

$$\lambda_1 = \frac{-a + i\sqrt{4b - a^2}}{2}, \quad \lambda_2 = \frac{-a - i\sqrt{4b - a^2}}{2}, \tag{12.25}$$

according to whether the discriminant $D := a^2 - 4b$ is positive, zero or negative. The solutions of the characteristic equation are also called zeros or roots or eigenvalues of the characteristic polynomial $\lambda^2 + a\lambda + b$.

Since as mentioned about the more general homogeneous linear difference equation (12.14), if Y_1 and Y_2 are solutions of (12.22), so is

$$Y = c_1 Y_1 + c_2 Y_2$$

for arbitrary real constants c_1 and c_2. In the first case, where $D > 0$, we have

$$Y(t) = c_1 \lambda_1^t + c_2 \lambda_2^t \tag{12.26}$$

as a solution, where λ_1 and λ_2 are given by (12.24).

In the second case, $D = 0$, one solution is $Y_1(t) = \lambda^t$. Another solution is, in analogy to the same situation in Sect. 11.5 $Y_2(t) = t\lambda^t$. The reader is asked to verify this contention. So

$$Y(t) = c_1 \lambda^t + c_2 t \lambda^t \tag{12.27}$$

is a solution of (12.22), where $\lambda = -\dfrac{a}{2}$.

In the third case, where $D < 0$ and $\lambda_1 \neq \lambda_2$ are conjugate complex numbers, one can immediately give the general complex solution

$$Y(t) = \tilde{c}_1 \lambda_1^t + \tilde{c}_2 \lambda_2^t.$$

By the same way of arguments as in Sect. 11.5 one can give the general real solution. The difference equation must be fulfilled by the real and the imaginary part separately. From Sect. 1.7 we know that λ_1 and λ_2 can be written in the form

$$\lambda_1 = r(\cos\phi + i \sin\phi), \quad \text{and} \quad \lambda_2 = r(\cos\phi - i \sin\phi).$$

Therefore the general real solution of (12.22) is

$$Y(t) = r^t (c_1 \cos t\phi + c_2 \sin t\phi), \tag{12.28}$$

where c_1 and c_2 are arbitrary real constants.

Next we show that in all three cases the initial conditions

$$Y(t_0) = Y_0, \quad Y(t_0 + 1) = Y_1 \tag{12.29}$$

with arbitrary $Y_0, Y_1 \in \mathbb{R}$ can be uniquely satisfied by choosing the constants $c_1,\ c_2$ appropriately in (12.26), (12.27), and (12.28). As in Sect. 11.5, we write these solutions from now on as

$$Y(t) = c_1 Y_1(t) + c_2 Y_2(t).$$

We are looking for real numbers c_1, c_2 such that

$$c_1 Y_1(t) + c_2 Y_2(t) = Y_0,$$

$$c_1 Y_1(t+1) + c_2 Y_2(t+1) = Y_1.$$

This again is a system if inhomogeneous linear algebraic equations, which, as we saw in Sect. 4.7, has a unique pair of solutions if and only if

$$\begin{vmatrix} Y_1(t_0) & Y_2(t_0) \\ Y_1(t_0+1) & Y_2(t_0+1) \end{vmatrix} \neq 0.$$

If $\lambda_1 \neq \lambda_2$ (i.e. cases 1 and 3), the determinant equals:

$$\begin{vmatrix} \lambda_1^{t_0} & \lambda_1^{t_0+1} \\ \lambda_2^{t_0} & \lambda_2^{t_0+1} \end{vmatrix} = \lambda_1^{t_0}\lambda_2^{t_0+1} - \lambda_1^{t_0+1}\lambda_2^{t_0} = (\lambda_1\lambda_2)^{t_0}(\lambda_2 - \lambda_1) \neq 0,$$

as $\lambda_1\lambda_2 \neq 0$ and $\lambda_1 \neq \lambda_2$. If $\lambda_1 = \lambda_2 = \lambda \neq 0$ (i.e. case 2), the determinant is

$$\begin{vmatrix} \lambda^{t_0} & t_0\lambda^{t_0} \\ \lambda^{t_0+1} & (t_0+1)\lambda^{t_0+1} \end{vmatrix} = \lambda^{2t_0+1} \neq 0.$$

So in every case the initial value problem (12.29) for the linear second order difference equation (12.22) can be solved uniquely. For this difference equation it is even easier to prove that (12.26), (12.27), and (12.28) (with arbitrary c_1, c_2) give the general solution of (12.22) alone (without initial conditions) than the similar statement in Sect. 11.5. Indeed, for any solution Y of (12.22) on $\{t_0, t_0+1, t_0+2, \cdots\}$ ($t_0 \in \mathbb{N} \cup \{0\}$) we now denote the values assumed by Y at t_0 and $t_0 + 1$ by

$$Y_0 := Y(t_0), \quad Y_1 := Y(t_0+1)$$

respectively. By our step-by-step algorithm in Sect. 12.1 we showed that these values, that is the initial conditions (12.29) uniquely determine Y on $\{t_0, t_0+1, \cdots\}$, On the other hand, we just proved that all solutions of (12.22) satisfying (12.29) are of one of the forms (12.26), (12.27) or (12.28). So (12.22) has no other solution than these.

As mentioned before, while the difference equation itself yields step by step the values of the unknown function at all places $t_0, t_0+1, \cdots$, we are still interested in the qualitative behaviour of the solutions, in particular as $t \to \infty$. In the case of equation (12.22) the solution formulas (12.26), (12.27), (12.28) are of great help. We distinguish the three cases whether the discriminant $D = a^2 - 4b$ is positive, zero or negative. The constants c_1 and c_2 are supposed not to be both zero. This would yield the trivial solution $Y(t) \equiv 0$, which is of no particular interest. If one of c_1

and c_2 is zero the discussion for this case can easily be derived from the arguments below. Hence we assume $c_1c_2 \neq 0$.

Case 1: $D = a^2 - 4b > 0$. We assume $|\lambda_1| > |\lambda_2|$ and distinguish three subcases:

Case 1.1: $|\lambda_1| > 1$. This implies $|Y(t)| \to \infty$ for $t \to \infty$. This solution again is of no particular interest.

Case 1.2: $|\lambda_1| = 1$ implies $|Y(t)| \to |c_1|$. The solution is convergent if $\lambda_1 = 1$ and oscillating, if $\lambda_1 = -1$.

Case 1.3: $|\lambda_1| < 1$. Hence $|Y(t)| \to 0$, but this case is not of great interest.

Case 2: $D = a^2 - 4b = 0$. We set $\lambda_1 = \lambda_2 = \lambda$ and distinguish three subcases again.

Case 2.1: $|\lambda| > 1$. See case 1.1 above.

Case 2.2: $|\lambda| = 1$. $Y(t) = c_1\lambda^t + c_2t\lambda^t$ and $|Y(t)| \to \infty$, if $c_2 \neq 0$. If $c_2 = 0$ then $Y(t) \equiv c_1$, if $\lambda = 1$ or $Y(t) = \pm c_1$, if $\lambda = -1$.

Case 2.3: $|\lambda| < 1$. In this case $|Y(t)| \to 0$ even if $c_2 \neq 0$ as $\lim_{t\to\infty} |t\lambda^t| = 0$.

Case 3: $D = a^2 - 4b < 0$. In this case the solution can be written in the form

$$\lambda_1 = r(\cos\phi + i\sin\phi), \quad \lambda_2 = r(\cos\phi - i\sin\phi), \quad \phi \in [0, 2\pi),$$

and the general real solution is of the form

$$Y(t) = r^t(c_1\cos t\phi + c_2\sin t\phi).$$

Again we distinguish three cases.

Case 3.1: $r > 1$. This implies $|Y(t)| \to \infty$ for $t \to \infty$

Case 3.2: $r = 1$. This is the most interesting case, as the solution is an equilibrium solution. If $\phi \in \mathbb{Q}$, the solution only assumes finitely many discrete values. If $\phi \in \mathbb{R}\backslash\mathbb{Q}$, the values $Y(t)$ are uniformly distributed on the unit circle. This is stated without proof. It is interesting to investigate, how (small) changes in the parameters of the corresponding economic model can change r such that the solution fluctuates around 1 such that the solution neither diverges nor collapses to zero.

Case 3.3: $r < 1$. This implies $|Y(t)| \to 0$ for $t \to \infty$.

This explains why the initial value problem (12.7), (12.8) in Sect. 12.1 had damped oscillations around 400 as solutions. Setting

$$\tilde{Y} = Y - 400, \tag{12.30}$$

one obtains the homogeneous equation

$$\tilde{Y}(t+2) - \frac{3}{2}\tilde{Y}(t+1) + \frac{3}{4} = 0.$$

Its characteristic equation has the solutions

$$\lambda_1 = \frac{3}{4} + i\frac{\sqrt{3}}{4}, \quad \lambda_2 = \frac{3}{4} - i\frac{\sqrt{3}}{4}.$$

Their absolute value is $\frac{\sqrt{3}}{2} < 1$, and by the above arguments, we have a damped solution around $\tilde{Y}(t) = 0$, i.e. $Y(t) = 400$.

A substitution similar to (12.30), actually

$$\tilde{Y}(t) = Y(t) - \frac{c}{1 + a + b},$$

reduces every second order inhomogeneous linear difference equation with constant coefficients a, b and constant perturbation c,

$$Y(t + 2) + aY(t + 1) + bY(t) = c, \tag{12.31}$$

to a homogeneous equation

$$\tilde{Y}(t + 2) + a\tilde{Y}(t + 1) + b\tilde{Y}(t) = 0,$$

if $1 + a + b \neq 0$, no matter what the initial conditions are. The "secret" of how to choose $\tilde{Y}$ is to substitute $Y = \tilde{Y} - K$ into equation (12.31):

$$\tilde{Y}(t + 2) - K + a\tilde{Y}(t + 1) - aK + b\tilde{Y}(t) - bK = c$$

and then choose K such that the perturbation becomes 0, that is, $c + K + aK + bK = 0$. Notice that the constant function given by $Y(t) = c/(1 + a + b)$ is also a solution of (12.31) (if $1 + a + b \neq 0$), the equilibrium solution (just as $Y(t) = 400$ was for (12.7)). This equilibrium solution is stable, if for all initial value conditions

$$Y(t_0) = Y_0, \quad Y(t_1) = Y_1,$$

the solutions of (12.31) tend to K as $t \to \infty$. By the above this happens exactly when the solutions of the characteristic equation (12.23) either both have absolute value smaller than 1 or, trivially, if one of λ_1, λ_2 has absolute value smaller than 1 and the coefficient c_j of the other is zero.

Since the equilibrium solution $Y^*(t) = c/(1 + a + b)$ is a particular solution of the inhomogeneous linear difference equation (12.31) (in particular $Y^*(t) = 400$ is a solution of (12.7)), if $1 + a + b \neq 0$ and since, as mentioned early in this section, the general solution of an inhomogeneous linear difference equation is the sum of a particular solution and of the general solution of the corresponding homogeneous equation, we obtain the general solution of (12.31) as the sum of $Y^*(t)$ and of one of the functions given by (12.26), (12.27) or (12.28) depending on the value of the discriminant $D = a^2 - 4b$, except if $a + b = -1$. In the hitherto excluded case $a + b = -1$, we look for a particular solution of the form $Y(t) = kt$. Substitution into (12.31) gives

$$k(t + 2) + ak(t + 1) + bkt = c, \quad \text{that is,} \quad kt(1 + a + b) + k(2 + a) = c.$$

Since we here deal with the case $1 + a + b = 0$, we get $k = c/(2 + a)$ and

$$Y^*(t) = \frac{c}{2 + a}t,$$

if $2 + a \neq 0$. If both $1 + a + b = 0$ and $2 + a = 0$, that is $a = -2$, $b = 1$, then we try $Y^*(t) = \gamma t^2$ and get

$$\gamma t^2 + 4\gamma t + 4\gamma - 2(\gamma t^2 + 2\gamma t + \gamma) + \gamma t^2 = c,$$

that is $2\gamma = c$ and

$$Y^*(t) = \frac{c}{2}t^2$$

is a particular solution of (12.31) in this last case. Again adding to this Y^* the functions given by (12.26), (12.27) or (12.28) gives the general solution of (12.31) depending on the value of the discriminant $D = a^2 - 4b$, which we now have for all values of a, b, and c. We can exclude $c = 0$, since (12.31) is homogeneous. We had already excluded $b = 0$ previously.

So quadratic polynomials were the right kind of particular solutions to experiment with: we could have substituted right away $Y^*(t) = \gamma t^2 + kt + K$. For the somewhat more general equation

$$Y(t + 2) + aY(t + 1) + bY(t) = c\alpha^t, \tag{12.32}$$

($\alpha \neq 0, c \neq 0, b \neq 0$, a arbitrary real constants) the exponential polynomial

$$Y^*(t) = \left(\gamma t^2 + kt + K\right)\alpha^t$$

will be an adequate candidate for a particular solution. Substitution and comparing coefficients gives for t^2, t and the constants respectively the following equations:

$$\begin{aligned} \gamma(\alpha^2 + a\alpha + b) &= 0, \\ k\left(\alpha^2 + a\alpha + b\right) + 2\gamma(2\alpha + a) &= 0, \\ K\left(\alpha^2 + a\alpha + b\right) + k\alpha(2\alpha + a) + \gamma\alpha(4\alpha + a) &= c. \end{aligned} \tag{12.33}$$

Let us proceed the discussion in analogy to the simpler case above and assume at first $\alpha^2 + a\alpha + b \neq 0$. Then we infer $\gamma = 0$ and $k = 0$ from the first two equations, hence $K = \dfrac{c}{\alpha^2 + a\alpha + b}$. If

$$\alpha^2 + a\alpha + b = 0, \tag{12.34}$$

then there are two subcases:

Case 1: $\gamma = 0$ then $k = \dfrac{c}{\alpha(2\alpha + a)}$.

Case 2: $\gamma \neq 0$ then

$$2\alpha + a = 0 \tag{12.35}$$

and $\gamma = \dfrac{c}{2\alpha^2}$. So there is a nonzero term $\gamma t^2 \alpha^2$ in the particular solution of (12.32) if and only if α is a double solution of the characteristic polynomial of the difference equation.

For the inhomogeneous equation

$$Y(t+2) + aY(t+1) + bY(t) = f(t)$$

we now give without proof the following table of perturbations and forms of particular solutions Y^*

$$\begin{aligned} f(t) = ct^m, &\quad Y^*(t) = k_m t^m + \cdots + k_1 t + k_0, \\ f(t) = c\alpha^t t^m, &\quad Y^*(t) = (k_m t^m + \cdots + k_1 t + k_0)\alpha^t, \\ f(t) = A\cos\phi t + B\sin\phi t, &\quad Y^*(t) = k_1 \cos\phi t + k_2 \sin\phi t, \\ f(t) = \alpha^t (A\cos\phi t + B\sin\phi t), &\quad Y^*(t) = \alpha^t (k_1 \cos\phi t + k_2 \sin\phi t). \end{aligned}$$

Similar results also hold for n-th order linear difference equations with constant coefficients

$$Y(t+n) + a_{n-1}Y(t+n-1) + \cdots + a_1 Y(t+1) + a_0 Y(t) = f(t).$$

But it is beyond the scope of this book to deal with them in detail.

12.2.1 Exercises

1. Solve the second order homogeneous difference equation $x_n = x_{n-1} + x_{n-2}$ with the initial condition $x_1 = 1, x_2 = 1$. (Remark: This problem determines the Fibonacci numbers. Look up the background in the internet.)
2. Solve the second order inhomogeneous difference equation $x_n = x_{n-1} + x_{n-2} + 1$ with the initial conditions $x_1 = 1, x_2 = 2$. (Remark: This equation determines the number of knots in a minimal balanced binary tree. Look up the background in the internet!)

3. For the second order difference equation $x_n + x_{n-1} + x_{n-2} = 0$ find the complex and the real solutions and show that they are periodic of length 3 and calculate their values.

12.2.2 Answers

1. The roots of the characteristic polynomial are $x_p = \frac{1+\sqrt{5}}{2}$ and $x_m = \frac{1-\sqrt{5}}{2}$. The initial conditions yield the constants $c_p = \frac{5+\sqrt{5}}{10}$, and $c_m = \frac{5-\sqrt{5}}{10}$. This gives the solution $x_n = c_p x_p^n + c_m x_m^n$.
2. The roots of the characteristic polynomial are $x_p = \frac{1+\sqrt{5}}{2}$ and $x_m = \frac{1-\sqrt{5}}{2}$ again. A particular solution of the inhomogeneous equation is the constant -1. The initial conditions yield the constants $c_p = \frac{5+3\sqrt{5}}{10}$, and $c_m = \frac{5-3\sqrt{5}}{10}$. This gives the solution $x_n = c_p x_p^n + c_m x_m^n - 1$.
3. The zeroes of the characteristic equation are $x_p = \frac{-1+i\sqrt{3}}{2}$ and $x_m = \frac{-1-i\sqrt{3}}{2}$. The complex general solution is $x_n = c_1 x_p^n + c_2 x_m^n$, and the real solution is $x_n = r_1 \cos\frac{2\pi}{3}n + r_2 \sin\frac{2\pi}{3}n$. From this is t follow that the length of the period is 3. $x_0 = 1, x_1 = \frac{-c_1+\sqrt{3}}{2}, x_2 = \frac{c_1-\sqrt{3}}{2}$.

12.3 Some Applications of Linear Difference Equations

12.3.1 The Growth Model of Roy Forbes Harrod (1900–1978)

We can describe the assumptions as follows.

H1: The total savings $S(t)$ in the time interval $[t, t+1[$ are proportional to the national income $Y(t)$ in the same interval

$$S(t) = sY(t),$$

where the positive constant $s < 1$ is the "savings rate".

H2: The net investment $I(t)$ projected for $[t, t+1[$ is proportional to the increase of national income for $[t, t+1[$ compared to $[t-1, t[$:

$$I(t) = a(Y(t) - Y(t-1)),$$

where the positive constant a is called "accelerator".

H3: There is an equilibrium in the economy in the sense that the total savings equal the projected net investment:

$$S(t) = I(t).$$

It follows from **H1**, **H2**, and **H3** that

$$sY(t) = aY(t) - aY(t-1)$$

or replacing t by $t + 1$, as we had done before,

$$Y(t+1) = \frac{a}{a-s}Y(t),$$

gives an explicit first order homogeneous linear difference equation with constant coefficients, if $a - s \neq 0$. If $a - s = 0$, the equation cannot be made explicit and reduces to $aY(t) = 0$, that is, since we had supposed $a > 0$, we only get the equilibrium solution $Y(t) = 0$. It would be a very poor economy indeed, where the national income would be 0 in every year. So we may exclude this trivial solution. Furthermore, in experience, the accelerator a is greater than the savings rate s. If the national income during the starting year is

$$Y(0) =: Y_0$$

then, as we had calculated in Sect. 12.1 the solution of this initial value problem is

$$Y(t) = Y_0\left(\frac{a}{a-s}\right)^t = Y_0\left(1 + \frac{s}{a-s}\right)^t,$$

that is, the national income grows in this model by the constant growth rate $s/(a-s)$, if the accelerator is greater than the savings rate.

12.3.2 Settlement of Bond Issues

A debt Y_0 is repaid by constant payments R at the end of each year (or at other agreed regular time intervals). The debt $Y(t)$ at the point t in time (start of the year $[t, t+1[$ grows by a yearly interest rate i which remains constant. So, the debt at time $t + 1$ will be

$$Y(t+1) = (1+i)Y(t) - R, \quad (t = 0, 1, 2, \ldots). \tag{12.36}$$

This clearly is a first order inhomogeneous linear difference equation with constant coefficients (the perturbation is $-R$). A particular solution is the constant equilibrium solution

$$Y^*(t) = \frac{R}{i},$$

which we obtain by setting $Y^*(t) = K$ (a constant to be determined) in (12.36), while the general solution of the corresponding homogeneous equation, just as in

(12.1) and (12.15) above is

$$Y(t) = C(1+i)^t$$

with an arbitrary constant C. Therefore the general solution of (12.36) is

$$Y(t) = C(1+i)^t + \frac{R}{i}.$$

The initial condition $Y(0) = Y_0$ is satisfied if and only if

$$Y_0 = C + \frac{R}{i}, \quad \text{that is,} \quad C = Y_0 - \frac{R}{i}.$$

So the solution of this initial value problem is

$$Y(t) = \left(Y_0 - \frac{R}{i}\right)(1+i)^t + \frac{R}{i} = Y_0(1+i)^t - R\frac{(1+i)^t - 1}{i}.$$

It is easy to interpret this formula: the first term on the right hand side is the amount to which the original debt Y_0 grew in t years, while the second subtracted term equals

$$R + R(1+i) + R(1+i)^2 + \cdots + R(1+i)^{t-1},$$

the accrued value of the yearly payments R (annuities).

The same model also serves for calculating the capital K_n decreased from K_0 by yearly payments R (annuities). In this case n is written for t and K_n is written for $Y(n)$:

$$K_n = K_0(1+i)^n - R\frac{(1+i)^n - 1}{i}.$$

This is how the "constant year-end payments formula" is usually written.

If the "year-end" (or end of another time interval) payments are not constant but equal to $R(t)$, which depends on t, then (12.36) is replaced by the difference equation

$$Y(t+1) = (1+i)Y(t) - R(t).$$

If $Y(0) = Y_0$ is the initial debt then

$$Y(1) = (1+i)Y_0 - R(0),$$

$$Y(2) = (1+i)Y(1) - R(1) = (1+i)^2Y_0 - (1+i)R(0) - R(1),$$

$$Y(3) = (1+i)^3Y_0 - (1+i)^2R(0) - (1+i)R(1) - R(2),$$

and so on. So the solution of this initial value problem is

$$Y(t) = (1+i)^t Y_0 - \sum_{j=1}^{t} (1+i)^{t-j} R(j-1),$$

and the capital K_n decreased from K_0 in the beginning after n years of year-end payments $R(0), R(1), \cdots, R(n-1)$ is

$$K_n = (1+i)^n K_0 - \sum_{j=1}^{t} (1+i)^{t-j} R(j-1).$$

12.3.3 Distribution of Wealth

David Gawen Champernowne (1912–2000) suggested the following simple model for the distribution of households into distinct income classes $C_0, C_1, C_2, \cdots$. These contain $Y(0), Y(1), Y(2), \cdots$ households respectively. The probability (chance) of "descending" into the previous income class is everywhere $10\,\% = 0.1$ (say). The chance to "ascend" from C_0 into C_1 is $30\,\%$ (say), from C_1 into C_2 is $15\,\%$, from C_2 into C_3 is $10\,\%$, in general the chance from C_t into C_{t+1} is $30/(t+1)\,\% = 0.3/(t+1)$.

This time $Y(0) - Y_0$ and $Y(1) = 3Y_0$ are given. According to Champernowne there is an equilibrium in this model, if (here t does not denote time)

$$Y(t) = \frac{0.3}{t-1} Y(t-1) + \left(1 - 0.1 - \frac{0.3}{t+1}\right) Y(t) + 0.1 Y(t+1).$$

The number $Y(t)$ of households in class C_t consists of those descended from C_{t+1}, whose number is $0.1Y(t+1)$, plus those ascended from C_{t-1}, whose number is $(0.3/t)Y(t-1)$, and those which remained in C_t after $10\,\%$ of them descended into C_{t-1} and $30/(t+1)\,\%$ ascended into C_{t+1}, which gives additional

$$Y(t) - 0.1Y(t) + \frac{0.3}{t+1} Y(t)$$

households. Multiplying by 10 and replacing t by $t+1$ gives

$$Y(t+2) = \left(1 - \frac{3}{t+2}\right) Y(t+1) - \frac{3}{t+1} Y(t),$$

which is a second order homogeneous difference equation with non constant coefficients. (The reader should verify this equation along the above given hints.)

We want to draw attention to the fact that in the above model the time aspect was neglected.

12.3.4 The Multi-sector Multiplier Model

To prepare the next section, systems of linear difference equations, we introduce an elementary version of the multi-sector multiplier model developed by Richard M. Goodwin (1913–1996) and John S. Chipman (1926–). Here the economy is divided into two sectors 1 and 2. We assume that sector 1 purchases some of its own output and some of the output of sector 2, and vice versa. Let $Y_{11}(t)$ represent the purchase of sector 1 of its own output during the time interval $[t, t+1[$, and $Y_{21}(t)$ represent its purchase of the output of sector 2. Analogously we define $Y_{12}(t)$ and $Y_{22}(t)$. If $Y_1(t)$ and $Y_2(t)$ are the respective total incomes of sectors 1 and 2 during $[t, t+1[$ we have summing up the incomes of each sector

$$Y_1(t) = Y_{11}(t) + Y_{12}(t), \quad Y_2(t) = Y_{21}(t) + Y_{22}(t). \tag{12.37}$$

Suppose now that the purchases of sector 1 of its own output and of the output of sector 2 during the time interval $[t, t+1[$ are affine functions of the current income of sector 1 and that the same assumption applies to sector 2. Then

$$\begin{aligned}
Y_{11}(t+1) &= c_{11} + a_{11}Y_1(t),\\
Y_{21}(t+1) &= c_{21} + a_{21}Y_1(t),\\
Y_{12}(t+1) &= c_{12} + a_{12}Y_2(t),\\
Y_{22}(t+1) &= c_{22} + a_{22}Y_2(t),
\end{aligned}$$

where the c's and a's are real constants. Substituting these relationships into the above equations (12.37) we obtain

$$\begin{aligned}
Y_1(t+1) &= (c_{11} + c_{12}) + a_{11}Y_1(t) + a_{12}Y_2(t),\\
Y_2(t+1) &= (c_{21} + c_{22}) + a_{21}Y_1(t) + a_{22}Y_2(t),
\end{aligned}$$

a system of two linear difference equations for the unknown functions Y_1 and Y_2. In the next section we shall discuss this kind of systems of difference equations.

12.4 Systems of Linear Difference Equations

It will be useful to use vectors (in particular vector-valued functions) in place of n scalar functions and matrices. Since Chap. 4 we are accustomed to using lower case and bold face letters for vectors and upper case and bold face letters for matrices. So we revert to denoting the unknown functions in a difference equation by $\mathbf{y}$ or

$$\mathbf{y} = \begin{pmatrix} y_1 \\ y_2 \\ \vdots \\ y_n \end{pmatrix}$$

in a system of difference equations (or in a vector difference equation) rather than by Y as we did in this chapter till now. The following is a system of n explicit linear first order difference equations with constant coefficients and constant perturbation

$$\begin{aligned} y_1(t+1) &= a_{11}y_1(t) + a_{12}y_2(t) + \cdots + a_{1n}y_n(t) + b_1, \\ y_2(t+1) &= a_{21}y_1(t) + a_{22}y_2(t) + \cdots + a_{2n}y_n(t) + b_2, \\ \vdots\ \vdots\ \vdots \\ y_1(t+1) &= a_{n1}y_1(t) + a_{n2}y_2(t) + \cdots + a_{nn}y_n(t) + b_n \end{aligned} \tag{12.38}$$

or in vector notation

$$\mathbf{y}(t+1) = \mathbf{A}\mathbf{y}(t) + \mathbf{b}, \tag{12.39}$$

where $\mathbf{y}$ is as above, while the constant matrix $\mathbf{A}$ and the constant vector $\mathbf{b}$ are given by

$$\mathbf{A} = \begin{pmatrix} a_{11} & a_{12} & \cdots & a_{1n} \\ a_{21} & a_{22} & \cdots & a_{2n} \\ \vdots & \vdots & & \vdots \\ a_{11} & a_{12} & \cdots & a_{1n} \end{pmatrix}, \quad \mathbf{b} = \begin{pmatrix} b_1 \\ b_2 \\ \vdots \\ b_n \end{pmatrix}.$$

Each vector-valued function $\mathbf{y} : D \longrightarrow \mathbb{R}$ ($D := \{t_0, t_0 + 1, \cdots\}$, $t_0 \in \mathbb{N} \cup \{0\}$ which satisfies (12.39) is called a solution of (12.39) and the components of $\mathbf{y}$ are solutions of the system (12.38) on D. We now impose on (12.39) the initial conditions

$$\mathbf{y}(t_0) = \mathbf{y}_0.$$

With this initial value we get from equation (12.39) for $t = t_0$

$$\mathbf{y}(t_0+1) = \mathbf{A}\mathbf{y}(t_0) + \mathbf{b} = \mathbf{A}\mathbf{y}_0 + \mathbf{b},$$

and

$$\mathbf{y}(t_0+2) = \mathbf{A}\mathbf{y}(t_0+1) + \mathbf{b} = \mathbf{A}(\mathbf{A}\mathbf{y}(t_0) + \mathbf{b}) + \mathbf{b} = \mathbf{A}^2\mathbf{y}_0 + (\mathbf{A} + I)\mathbf{b},$$

where $\mathbf{I}$ is the unit matrix. The same process gives

$$\mathbf{y}(t+3) = \mathbf{A}^3\mathbf{y}_0 + \left(\mathbf{A}^2 + \mathbf{A} + \mathbf{I}\right)\mathbf{b},$$

and in general for each natural number k,

$$\mathbf{y}(t_0+k) = \mathbf{A}^k\mathbf{y}_0 + \sum_{j=0}^{k-1} \mathbf{A}^j\mathbf{b} \quad (\mathbf{A}^0 := \mathbf{I}).$$

Setting $k = t - t_0$, we get

$$\mathbf{y}(t) = \mathbf{A}^{t-t_0}\mathbf{y}_0 + \sum_{j=t_0}^{t-1} \mathbf{A}^{j-t_0}\mathbf{b}, \quad (t = t_0 + 1, t_0 + 2, \cdots). \tag{12.40}$$

Obviously this is the only solution of (12.39) and its components are the only solutions of the system (12.38) that satisfy the initial condition $\mathbf{y}(t_0) = \mathbf{y}_0$.

If the matrix $(\mathbf{I} - \mathbf{A})$ has an inverse, that is, if

$$\det(\mathbf{I} - \mathbf{A}) \neq 0, \tag{12.41}$$

then it follows from

$$\left(\mathbf{I} + \mathbf{A} + \mathbf{A}^2 + \cdots + \mathbf{A}^{t-t_0-1}\right)(\mathbf{I} - \mathbf{A}) = \mathbf{I} - \mathbf{A}^{t-t_0}$$

that

$$\mathbf{I} + \mathbf{A} + \mathbf{A}^2 + \cdots + \mathbf{A}^{t-t_0-1} = (\mathbf{I} - \mathbf{A}^{t-t_0})(\mathbf{I} - \mathbf{A})^{-1}.$$

In this case one can write the solution (12.40) in the form

$$\mathbf{y}(t) = \mathbf{A}^{t-t_0}\mathbf{y}_0 + (\mathbf{I} - \mathbf{A}^{t-t_0})(\mathbf{I} - \mathbf{A})^{-1}\mathbf{b} \quad (t = t_0 + 1, t_0 + 2, \cdots). \tag{12.42}$$

Again it is of interest to examine the behaviour of the solution of (12.39) for large t, that is, to ask what happens to (12.42) when $t \to \infty$. For this purpose we need some facts about series and limits of matrices, which we list here without proof.

The convergence of a matrix series means the convergence of all components. It can be proved that the matrix series

$$\mathbf{I} + \mathbf{A} + \mathbf{A}^2 + \cdots \tag{12.43}$$

converges, if all eigenvalues of $\mathbf{A}$ have absolute values smaller than 1, that is, if for all solution λ of $\det(\mathbf{A} - \lambda\mathbf{I}) = 0$ we have $|\lambda| < 1$. In this case we can define

$$(\mathbf{I} - \mathbf{A})^{-1} = \mathbf{I} + \mathbf{A} + \mathbf{A}^2 + \cdots,$$

in analogy to the series in Sect. 6.7

$$(1 - a)^{-1} = 1 + a + a^2 + \cdots, \quad \text{if } |a| < 1$$

(there $t = -a$). What is important for us is that $\mathbf{I} - \mathbf{A}$ has an inverse if all eigenvalues of $\mathbf{A}$ have absolute values smaller than 1, an alternative condition for (12.41).

Just as we defined the convergence of a series of matrices component wise, the limit of a matrix-valued function of a real variable is the matrix consisting of all limits of the components of that matrix-valued function, if they exist. One handles such limits similarly to those of real-valued functions in Sect. 6.7. As it happens, also

$$\lim_{t\to\infty} \mathbf{A}^t = 0$$

holds, if all eigenvalues of **A** have absolute values smaller than 1. Actually, exactly this guarantees the convergence of the series (12.43). Thus in this case

$$\begin{aligned}\lim_{t\to\infty} \mathbf{y}(t) &= \lim_{t\to\infty} \left(\mathbf{A}^{t-t_0}\mathbf{y}_0 + (\mathbf{I}-\mathbf{A}^{t-t_0})(\mathbf{I}-\mathbf{A})^{-1}\mathbf{b}\right)\\ &= \lim_{t\to\infty} \mathbf{A}^{t-t_0}\mathbf{y}_0 + \lim_{t\to\infty}(\mathbf{I}-\mathbf{A}^{t-t_0})(\mathbf{I}-\mathbf{A})^{-1}\mathbf{b}\\ &= 0\mathbf{y}_0 + (\mathbf{I}-0)(\mathbf{I}-\mathbf{A})^{-1}\mathbf{b} = (\mathbf{I}-\mathbf{A})^{-1}\mathbf{b}.\end{aligned}$$

Notice that this limit is independent of the initial value $\mathbf{y}_0$ and that the limit is 0 if and only if $\mathbf{b} = 0$, that is, in the case of systems of explicit homogeneous linear difference equations. Notice also that the only constant solution of (12.39) is

$$\mathbf{y}(t) = (\mathbf{I}-\mathbf{A})^{-1}\mathbf{b}.$$

Here, too, this is called the equilibrium solution, and it and the vector difference equation (12.39) are called stable, if every solution of (12.39) converges to this equilibrium solution as $t \to \infty$. As we have seen this stability occurs if all absolute values of the eigenvalues of **A** are smaller than 1 and, as can be shown, in no other case.

Example Take the system

$$\begin{aligned}\mathbf{y}_1(t+1) &= \tfrac{1}{2}\mathbf{y}_1(t) + \tfrac{1}{4}\mathbf{y}_2(t) + 1,\\ \mathbf{y}_2(t+1) &= \tfrac{1}{4}\mathbf{y}_1(t) + \tfrac{1}{2}\mathbf{y}_2(t) + 2.\end{aligned} \tag{12.44}$$

We look for a solution $\mathbf{y} = \begin{pmatrix} y_1 \\ y_2 \end{pmatrix}$ which has at $t_0 = 0$ the initial value

$$\mathbf{y}(0) = \begin{pmatrix} y_1(0) \\ y_2(0) \end{pmatrix} = \begin{pmatrix} 0 \\ 1 \end{pmatrix}.$$

(continued)

With the notations

$$\mathbf{A} = \begin{pmatrix} \frac{1}{2} & \frac{1}{4} \\ \frac{1}{4} & \frac{1}{2} \end{pmatrix}, \quad \mathbf{b} = \begin{pmatrix} 1 \\ 2 \end{pmatrix}, \quad \mathbf{y}(t) = \begin{pmatrix} y_1(t) \\ y_2(t) \end{pmatrix}$$

the problem to be solved is

$$\mathbf{y}(t+1) = \mathbf{A}\mathbf{y}(t) + \mathbf{b}, \quad \mathbf{y}(0) = \begin{pmatrix} 0 \\ 1 \end{pmatrix}. \tag{12.45}$$

Since

$$\det(\mathbf{I} - \mathbf{A}) = \det\begin{pmatrix} \frac{1}{2} & -\frac{1}{4} \\ -\frac{1}{4} & \frac{1}{2} \end{pmatrix} = \frac{3}{16} \neq 0,$$

the formula (12.42) can be used, this time with $t_0 = 0$:

$$\mathbf{y}(t) = \mathbf{A}^t\mathbf{y}_0 + \left(\mathbf{I} - \mathbf{A}^t\right)(\mathbf{I} - \mathbf{A})^{-1}\mathbf{b}.$$

Now (check $((\mathbf{I} - \mathbf{A})^{-1}(\mathbf{I} - \mathbf{A}) = \mathbf{I})$,

$$(\mathbf{I} - \mathbf{A})^{-1} = \begin{pmatrix} \frac{8}{3} & \frac{4}{3} \\ \frac{4}{3} & \frac{8}{3} \end{pmatrix}, \quad (\mathbf{I} - \mathbf{A})^{-1}\mathbf{b} = \begin{pmatrix} \frac{8}{3} & \frac{4}{3} \\ \frac{4}{3} & \frac{8}{3} \end{pmatrix}\begin{pmatrix} 1 \\ 2 \end{pmatrix} = \begin{pmatrix} \frac{16}{3} \\ \frac{20}{3} \end{pmatrix},$$

and thus the solution of (12.45) can be written as

$$\mathbf{y}(t) = \mathbf{A}^t\begin{pmatrix} 0 \\ 1 \end{pmatrix} + (\mathbf{I} - \mathbf{A}^t)\begin{pmatrix} \frac{16}{3} \\ \frac{20}{3} \end{pmatrix}. \tag{12.46}$$

To examine the behaviour of this solution (12.46) for large values of t, the eigenvalues of $\mathbf{A}$ must be calculated. These are the roots of

$$\det\left(\mathbf{A} - \lambda\mathbf{I}\right) = \det\begin{pmatrix} \frac{1}{2} - \lambda & \frac{1}{4} \\ \frac{1}{4} & \frac{1}{2} - \lambda \end{pmatrix} = \lambda^2 - \lambda + \frac{3}{16},$$

that is,

$$\lambda_1 = \frac{3}{4}, \quad \lambda_1 = \frac{1}{4}.$$

Thus the eigenvalues of $\mathbf{A}$ are both positive and less than 1, that is, the system is stable. Therefore $\mathbf{A}^t$ converges to the zero matrix as $t \to \infty$ and from

(continued)

(12.46) it follows that

$$\lim_{t\to\infty} \mathbf{y}(t) = \begin{pmatrix} \frac{16}{3} \\ \frac{20}{3} \end{pmatrix}.$$

We note that $\mathbf{y}(t) = \begin{pmatrix} \frac{16}{3} \\ \frac{20}{3} \end{pmatrix}$ is the equilibrium solution of the system (12.44) as it should be.

Again some of our results can be generalised to systems of n (explicit) inhomogeneous linear difference equations of order m (we use m since n already denotes the number of equations in the system). In vector notation this is written as

$$\mathbf{y}(t+m) + \mathbf{A}_m(t)\mathbf{y}(t+m-1) + \cdots + \mathbf{A}_1(t)\mathbf{y}(t+1) + \mathbf{A}_0(t)\mathbf{y}(t) = \mathbf{f}(t), \tag{12.47}$$

where $t \in D := \{t_0, t_0+1, \ldots\}$ ($t_0 \in \mathbb{N} \cup \{0\}$), the $\mathbf{A}_0(t) \not\equiv 0, \mathbf{A}_1(t), \cdots, \mathbf{A}_{m-1}(t)$ are $n \times n$-matrix-valued coefficients depending on $t \in D$, $\mathbf{f} : D \longrightarrow \mathbb{R}^n$ is the perturbation, and $\mathbf{y} : D \longrightarrow \mathbb{R}^n$ is the unknown function. Compare this to the scalar difference equations in Sect. 12.2, which reads in our present notation

$$\mathbf{y}(t+m) + \mathbf{A}_m(t)\mathbf{y}(t+m-1) + \cdots + \mathbf{A}_1(t)\mathbf{y}(t) + \mathbf{A}_0(t)\mathbf{y}(t) = \mathbf{f}(t).$$

In both situations the initial conditions

$$\mathbf{y}(t_0) = \mathbf{y}_0, \mathbf{y}(t_0+1) = \mathbf{y}_1, \ldots, \mathbf{y}(t_0+m-1) = \mathbf{y}_{m-1}$$

determine step by step the solution of (12.47), so also this initial value problem has a unique solution. Also in complete analogy to the scalar case (12.47) is called homogeneous if $f(t) \equiv 0$ (actually it is the homogeneous equation corresponding to (12.47)) and the general solution of (12.47) is the sum of one of its particular solutions and of the general solution of the corresponding homogeneous equation. The latter is a linear combination $c_1\mathbf{y}_1 + \cdots + c_n\mathbf{y}_n$ ($c_1, \cdots, c_n$ arbitrary real constants) of n linearly independent solutions $\mathbf{y}_1, \cdots, \mathbf{y}_n$ (that is, for which $\alpha_1\mathbf{y}_1(t) + \cdots + \alpha_n\mathbf{y}_n(t) \equiv 0$ can hold if and only if $\alpha_1 = \cdots = \alpha_n = 0$). In particular, for any two solutions $\mathbf{y}_1, \mathbf{y}_2$ of the homogeneous linear difference equation and any two real constants c_1, c_2, also $c_1\mathbf{y}_1 + c_2\mathbf{y}_2$ is a solution.

However, even the system (12.47) of n linear difference equations of order m can be reduced to a system of mn first order linear difference equations as follows: Define

$$\mathbf{y}_1(t) := \mathbf{y}(t), \mathbf{y}_2(t) := \mathbf{y}(t+1), \cdots, \mathbf{y}_m(t) := \mathbf{y}(t+m-1).$$

Then, for these "unknown functions", (12.47) is equivalent to the system

$$\mathbf{y}_1(t+1) = \mathbf{y}_2(t)$$

$$\mathbf{y}_2(t+1) = \mathbf{y}_3(t)$$

$$\vdots$$

$$\mathbf{y}_{m-1}(t+1) = \mathbf{y}_m(t)$$

$$\mathbf{y}_m(t+1) + \mathbf{A}_{m-1}(t)\mathbf{y}_m(t) + \cdots + \mathbf{A}_1(t)\mathbf{y}_2(t) + \mathbf{A}_0(t)\mathbf{y}_1(t) = \mathbf{f}(t)$$

of m first order n-component vector difference equations, that is, of mn first order scalar difference equations. A similar statement also holds for nonlinear vector difference equations. But, if in (12.47) the matrix-valued coefficient functions $\mathbf{A}_0, \mathbf{A}_1, \cdots, \mathbf{A}_{m-1}$, and the vector-valued perturbation function are constants then it can be reduced to a system of mn (rather than n) scalar first order linear difference equations with constant coefficients and constant perturbation of the form (12.38), with which we started this section.

12.5 Nonlinear Difference Equations, Chaos

An explicit nonlinear difference equation is of the form

$$Y(t+n) = G(t, Y(t+n-1), \cdots, Y(t+1), Y(t)),$$

where the function G is not affine, in particular not linear, in its last n variables. As with linear difference equations it is of n-th order, if G is not constant in its last variable $Y(t)$.

With initial conditions

$$Y(t_0) = Y_0,\ Y(t_0+1) = Y_1,\ \cdots,\ Y(t_0+n-1) = Y_{n-1}$$

it can be solved step-by-step, just as we solved linear difference equations in Sects. 12.1 and 12.2:

$$\begin{aligned} Y(t_0+n) &= G(t_0, Y(n-1), \cdots, Y(1), Y(0)) \\ Y(t_0+n+1) &= G(t_0+1, Y(t_0+n), Y(t_0+n-1), \cdots, Y(t_0+1)) \\ \vdots\ &\ \vdots\ \vdots \end{aligned}$$

Of course, here too, this algorithm does not yield the qualitative properties of the solution, such as the behaviour of $Y(t)$ as $t \to \infty$. Now we will deal with such questions.

Nonlinear difference equations show a peculiarity which we did not encounter in the case of linear ones, and which makes their application to exact medium or long range forecasts (for example in economics or meteorology) difficult, if not impossible: Small errors in the initial conditions (resulting, for instance, from small inaccuracies in measurement) may lead to very large deviations from the solution of the initial value problem with exact initial conditions already after a few steps.

We give an example often used in the so-called theory of chaos: the "logistic difference equation"

$$y(t+1) = b_1 y(t) - b_2 y(t)^2 \tag{12.48}$$

(b_1, b_2 positive constants). Its name comes from its connection to the logistic differential equation (defining "logistic functions")

$$y'(t) = \beta y(t) - \gamma y(t)^2$$

(β, γ positive constants in Sect. 11.4. If we replace, as we had done in other occasions, see Sect. 12.1, the derivative $y'(t)$ by the difference quotient $(y(t+h) - y(t))/h$, we then get

$$\frac{y(t+h) - y(t)}{h} = \beta y(t) - \gamma y(t)^2.$$

If, in particular, $h = 1$ and if we write $b_1 = \beta + 1$, $b_2 = \gamma$ then we indeed obtain

$$y(t+1) = b_1 y(t) - b_2 y(t)^2.$$

While, as we have seen in Sect. 11.4 the solutions of the logistic differential equation are continuous, strictly monotonic and bounded, the solutions of the logistic difference equation show for certain values of the constant parameters a peculiar behaviour, which is called "chaotic".

First we simplify (12.48) by introducing

$$Y := \frac{b_1}{b_2} y$$

and b for b_2. We get the explicit difference equation of order 1

$$Y(t+1) = bY(t)(1 - Y(t)). \tag{12.49}$$

As in Sect. 12.4 we are interested in nonnegative solutions of (12.49). The left hand side, that is Y, is nonnegative if and only if the right hand side is nonnegative, which happens if and only if

$$Y(t) \in [0, 1] \quad \text{for all} \quad t. \tag{12.50}$$

The right hand side of (12.49) can be written (omitting t) as

$$bY - bY^2 = \frac{b}{4} - b\left(Y - \frac{1}{2}\right)^2.$$

Since $\left(Y - \frac{1}{2}\right)^2$ is nonnegative, this difference is not greater than $\frac{b}{4}$ and equal to $\frac{b}{4}$ exactly if $Y = \frac{1}{2}$.

The sequence $\{Y(t)\}$ defined by (12.49) converges to a finite

$$Y^* := \lim_{t\to\infty} Y(t),$$

if $Y_0 := Y(0)$ is in a neighbourhood of Y^*, in which

$$|F'(Y)| \leq c < 1 \tag{12.51}$$

(c independent of Y) and if, with any Y, also $F(Y)$ is in that neighbourhood. Here F is defined by

$$F(Y) = bY(1 - Y)$$

and Y^* is a fixed point of F, that is, the only fixed points in this case are

$$Y_1^* = 0 \quad \text{and} \quad Y_2^* = \frac{b-1}{b}.$$

If $b \in]0, 1[$ then the second fixed point Y_2 is negative, which contradicts (12.50) (The limit of a sequence with nonnegative terms cannot be negative.), so we can neglect it. Also $0 < F(Y) = bY(1 - Y) < 1$, if $0 < Y < 1$. Moreover

$$|F'(Y) = b|1 - 2Y| \leq b < 1 \text{for all} \quad Y \in [0, 1],$$

so (12.51) is satisfied. Therefore starting with any $Y(0) \in [0, 1[$, even with $Y(0) = 1$, as one easily calculates, one will have

$$\lim_{t\to\infty} Y(t) = 0.$$

One can prove that this is true for $b = 1$, too. Note that in this case the two fixed points are equal.

If $b \in]1, 3[$ then the second nonzero fixed point $Y^* = (b-1)/b$ will be in $]0, 1[$. Moreover, from $F(Y) = bY(1 - Y)$ we get

$$F'(Y) = b|1 - 2Y| \leq c < 1,$$

if

$$|1 - 2Y| \leq \frac{c}{b}, \quad \text{that is,} \quad -\frac{c}{b} \leq 1 - 2Y \leq \frac{c}{b}$$

or, what is the same

$$Y \in \left[\frac{1}{2}(1 - \frac{c}{b}), \frac{1}{2}(1 + \frac{c}{b})\right] =: I.$$

If c is close enough to 1 then the fixed point $Y^* = (b-1)/b$ is in the same interval I because

$$\frac{1}{2}(1 - \frac{1}{b}) = \frac{b-1}{2b} < \frac{b-1}{b} < \frac{b+1}{2b} = \frac{b-1}{b} + \frac{3-b}{2b}.$$

Furthermore, at least for $b < \sqrt{3}$, and if $Y \in I$ then also $F(Y) \in I$ as required. Without going into details we state that one can also prove that starting with any $Y(0) \in]0, 1[$ one has

$$\lim_{t \to \infty} Y(t) = \frac{b-1}{b},$$

where $\{Y(t)\}$ satisfies the "logistic difference equation" and $b \in]1, 3]$ ($b = 3$ included). Such a fixed point is called an attractor or an attractive fixed point, as the sequence converges to it, if it starts in its neighbourhood.

The reader is invited to verify this contention for two examples. Setting $b = 2$ and $Y_0 = 0.8$ the solution converges to 0.5, and setting $b = 2.8$ and $Y_0 = 0.2$ the solution converges to 0.643. In addition the convergence is rather fast in these two cases.

The situation changes dramatically when $b > 3$. Then

$$\begin{aligned} |F'(Y_2^*| &= b|1 - 2\tfrac{b-1}{b}| = |b - 2b + 2| = |2 - b| > 1, \\ |F'(Y_1^*| &= b|1 - 0| = b > 1, \\ \text{and} \quad |F'(Y_1)| &= b|1 - 0| = b > 1. \end{aligned}$$

Numerical calculations with some computer algebra program such as Maple, Mathematica, Matlab or Sage show that $Y(t)$ is repulsed as it approaches Y_2^*. Playing with different initial values, which are very close, shows that after few iterations already the solutions differ greatly. They jump around in the interval [0, 1] erratically. This behaviour is called chaotic.

So one has to be careful when problems of forecasting say in economics or meteorology lead to models with nonlinear difference equations.

12.5.1 Exercises

1. Let the difference equation $x(n+1) = ax(n)^2$, $a \in \mathbb{R}_{++}$ be given. Find the general solution for the initial value $x(0) = x_0 \in \mathbb{R}_{++}$.
2. Find a constant nontrivial solution of the difference equation in 1. choosing a and x_0 appropriately.
3. For the difference equation in 1. choose a and the initial values x_0 and $\hat{x}_0$ with $|x_0 - \hat{x}_0| < 0.01$ such that for the respective solutions $\{x(n)\}_{n\in\mathbf{N}_0}$ and $\{\hat{x}(n)\}_{n\in\mathbf{N}_0}$ the following holds $\lim_{n\to\infty} |x(n) - \hat{x}(n)| = \infty$.
4. For the difference equation in 1. choose a and the initial values x_0 and $\hat{x}_0$ such that for the respective solutions $\{x(n)\}_{n\in\mathbf{N}_0}$ and $\{\hat{x}(n)\}_{n\in\mathbf{N}_0}$ the following holds $\lim_{n\to\infty} |x(n) - \hat{x}(n)| = 0$.

12.5.2 Answers

1. $x(n) = a^{2^n-1} x_0^{2^n}$
2. $a = 1, x_0 = 1$
3. Take for example $a = 1, x_0 = 1$, and $\hat{x}_0 = 1 + \varepsilon$ with $0 < \varepsilon < 0.01$.
4. Take for example $a = \frac{1}{2}, x_0 = 1$, and $\hat{x}_0 = 1 + \varepsilon$ with $0 < \varepsilon < 1$.

Methodology: Models and Theories in Economics

13

A model shall be as simple as possible, but not simpler.
ALBERT EINSTEIN (1879–1955)

One of the tragedies of life is the murder of a beautiful theory by a gang of brutal facts.
BENJAMIN FRANKLIN (1706–1790)

13.1 Introduction

According to Wikipedia, the free encyclopedia, "methodology is the systematic, theoretical analysis of the methods applied to a field of study. It comprises the theoretical analysis of the body of methods and principles associated with a branch of knowledge". In this book the "field of study" and the "branch of knowledge" is economics.

In what follows, we present tentative steps into the methodology of economics. For this purpose we concentrate ourselves on the terms "model" and "theory" since they play an important role in the empirical sciences. Here the focus of our attention is economics, that is, most of the examples given are from economics.

According to ERICH SCHNEIDER (1900–1970) "the task of economics is to uncover and explain the *interdependencies* within the economy". These *"interdependencies"* are connections, regularities and laws, regularly observed in practice, concerning events, facts, processes, trends etc. which are to be encompassed or covered by a theory. A *theory* is a collection of axioms, hypotheses, assumptions, and theorems followed by a chain of consequences, which are obtained by logical (mostly mathematical) deduction, leading eventually to the mentioned *interdependencies* (laws, regularities, connections). A good theory does more: it *predicts* new events, rules and connections not yet observed, sometimes "with certainty", *deterministically* (within the theory), at other times only *stochastically*, "with certain probability".

W. Eichhorn, W. Gleißner, *Mathematics and Methodology for Economics*, Springer Texts in Business and Economics, DOI 10.1007/978-3-319-23353-6_13

The famous philosopher and economist JOHN STUART MILL (1806–1873) and, later, CARL GUSTAV HEMPEL (1905–1997) and others defined scientific explanation as uniting facts under one or more "laws of nature" (or of economy) which serve as starting points for logical deductions which eventually, under certain "initial", "boundary" and other conditions lead to a description of the observed facts. This "deductive method", applied first in the natural but then also in the social sciences, became the standard method to *explain* already observed facts and regularities, and *forecast* new ones. In spite of that, it is usual in the scientific community to speak of *explanations, forecasts* and *theories* also if there are *no laws* involved in the scientific systems of statements which are built up.

What we just said, may give the impression that the theory is primary to the practice and observations. Really, the situation is quite different: In many cases one needs lots of observations to subsume them under a theory, the assumptions of which are chosen exactly so that statements about the observed facts and connections can be deduced from them. The more "facts of life" (or of science) a theory can explain the better it is (and it is even better if it can forecast future events). So, before going from theory to practice, science (both natural and social) goes from experience to assumptions. Clearly, very often this transition is (pre) theoretically impregnated by concepts and conceptions. Such concepts and conceptions lead to so-called "models".

In mathematics, natural sciences, engineering and in the social sciences, in particular economics, the notion of a "model" usually means somewhat different things. We will deal with them in Sects. 13.2 and 13.3. We will see that models in economics are simplified images, reflections, reconstructions of (parts of) economic reality. They can be presented for instance verbally, graphically, analytically, and abstractly as a system of assumptions. These assumptions (Sect. 13.4) become "kernels" of economic theory (Sect. 13.5). The purpose of Sect. 13.6 is to clarify the role of models and theory in economics and list at the same time, some of the most important types of models and theories. In economics theories serve not only for explanation and forecasting but also as a basis for decision making. Theories everywhere have to be *checked* and *rechecked* by confrontation with empirical facts and *corrected*, if necessary. This will be the subject of Sect. 13.7.

13.2 Models in Engineering, Natural Sciences and Mathematics

We give short descriptions of what is meant with "models" in engineering, natural sciences and mathematics so that we can point out, in the next section, the similarities and differences between these models and the models in the social sciences (in particular in economics).

1. *Models in engineering.* In engineering, models are reduced, real size or enlarged spatial renditions of a technical project or product serving for teaching purposes (for example a cross section of a machine), as a toy (for instance model railway), for experiments (e.g. wind channel for testing airplane shapes), as production tool

(for example a wooden model for producing a cast to mold a metallic object), etc. They are simplified material concretisations of the respective machines or transformation systems to visualise and make available transformation functions, e.g. for manipulation, tuning of variables, etc.

2. *Models in the natural sciences.* Here not a man-made product or project is represented but an object or process of nature. As distinguished from the models mentioned above, models in science usually reflect only those aspects of the original object or process which are considered important for the actual research, while neglecting others ("abstraction"). This procedure serves to explain processes, to forecast future events or to plan experiments. Examples: models of stars and galaxies in astronomy and cosmology; models of "ideal gas", "incompressible fluids", of atoms and nuclei in physics; models of cells and genes in biology.
3. *Models in mathematics.* In mathematics and mathematical logics one describes a research domain by a system of axioms. The *axioms* are statements which sound sufficiently obvious to be accepted by most people but which in their entirety (that is, all axioms of the system) imply as many theorems (descriptions of facts) as possible in the axiomatised field. (Concerning "implication" or "deduction" see Sect. 13.5 **a**).

The famous logician ALFRED TARSKI (1902–1983) gave the following simple *example*. The field here is the part of geometry which deals with straight line segments and their congruence. Intuitively, two segments are *congruent* if they are of equal length but the point is that we use about the objects and their relations only what is contained in the axioms. For easier writing we call S the set of all segments; we denote its elements (that is, the segments) by x, y, z etc. and the congruence relation by $\sim$, so that $x \in S$, $y \in S$, $x \sim y$ means that the segments x and y are congruent. Our system consists of two axioms:

A1 *Reflexivity.* If $x \in S$ then $x \sim x$ (every segment is congruent to itself).
A2 *Skew-transitivity.* If $x, y, z \in S$, $x \sim z$ and $y \sim z$ then $x \sim y$ (if each of the two segments is congruent to a third then the two are congruent)

Already from this skimpy system of axioms one can prove simple theorems. For instance,

T. If $y, z \in S$ and $y \sim z$ then $z \sim y$ (congruence is *symmetric*).

Proof Take the particular case of **A1** where x and z are the same: If $z \sim z$ and $y \sim z$ then $z \sim y$. But, by **A1**, $z \sim z$ always holds, so this says "if $y \sim z$ then $z \sim y$" as asserted.

A consequence (called "corollary" in mathematics) is:

C. If $x, y, z \in S$, $x \sim z$ *and* $z \sim y$ *then* $x \sim y$ (congruence is *transitive*).

This follows since, by **T**, one can replace in **A2** the supposition $y \sim z$ by $z \sim y$.

Notice that nowhere in the proofs did we use the geometric meaning of segments and of congruences. We used about them only the formal properties codified in the axioms. That is how systems of axioms and theory work in mathematics. Therefore, they are *abstractly formulated*. By *model in mathematics* we mean any set of *objects and relations* (mostly within mathematics) *which satisfy the axioms* and thus also their consequences. So the straight line segments and their congruences (each of two congruent segments can be moved so as to completely cover the other) form a model for the system of axioms **A1, A2** (and all its consequences).

It is noteworthy that a *system of axioms can have several models* (actually every meaningful system of axioms has more than one model). For instance, the above system **A1, A2** of axioms allows also the following model in arithmetic. Now $S = \mathbb{R}$ (the set of all real numbers) and two real numbers x, y are congruent if their difference is an integer (positive, negative or 0), for instance $1.223 \sim 4.223$ and $4.223 \not\sim 1.777$. *These* objects, interpreted as segments, and *this* relation, considered to be the congruence, clearly satisfy the axioms **A1** and **A2** and thus also its consequences, for example **T** and **C** which therefore give valid theorems of arithmetics.

In addition to models of systems of axioms in "pure" mathematics, an important task of applied mathematics is to construct models which are mathematical descriptions of fundamental properties of objects (systems) and their interdependencies (relations) in practice. These are simplifications by necessity but, with the advent of computers, much more complicated relations and systems can be so described. In any case, one important purpose of such models is to predict the future behaviour of the system. And this connects them to *models in the social sciences, in particular economics*.

13.3 Models in Economics

To quote Erich Schneider again: "The economist's job necessarily entrains considering models. The evolution theory, from its beginnings to the present time is a continuing search for productive and successful models. The history of economic theory is the history of thinking in models which were constructed successively to deal with different problems and situations". Recent literature in economics and business administration shows that nowadays the significance of "thinking in models" (modelling) is generally accepted and appreciated. So what is a model in economics? We have more than one answer to this question:

1. *Models in economics as simplified images of parts of economic reality.* As in science (and in applied mathematics), also in economics, models are artificially constructed representations of real objects. But while in natural sciences the objects and interdependencies, which are to be reflected, are parts of nature of the world (more or less) independent of humanity, in economics they come from the "world of economy" which is overwhelmingly dependent on human decisions and actions. The choice of objects and interdependencies and the degree of simplification needed and/or permissible to construct a model depends on human purposes etc. (compare Sect. 13.6). So the builders of model in economics have to choose those aspects, properties and relations of (mostly social) reality, which are essential for the object of their research. In writings

about models in economics the relation between reality and model is often described as isomorphy. This is neither quite correct mathematically speaking nor (conceptually because it indicated a "one-to-one" (bijective; compare Sect. 3.2) mapping of the set of objects and of the interdependencies of the (economic) reality onto the model. But the idea in modelling is exactly to simplify in order to see the interdependencies more clearly so that one can make predictions and decisions which could not be based on the complicated, messy, detail-ridden, full reality. The word homomorphism corresponding to injection (compare Sect. 3.2) would be mathematically and conceptually more correct.

Note that in the above we were *describing* the concept of model ("paraphrasing" it, just as we did "set" in Sect. 1.2) *not defining* it in any strict logical or mathematical sense. In **2-5** below we will deal with the question, how models can be *represented*. In research activity, *construction* of a model often comes after such *representations*. Both serve to facilitate solutions of practical or theoretical problems but, in final analysis, they are highly individual and subjective activities of researchers (individuals or teams of them).

2. *Verbal representation (description) of more or less realistic models*. This is quite frequent in the literature of economics. In order for such a representation to be scientific, it has to be more exact than everyday colloquial discourse or writing, In particular, the terminology should be *unambiguous*.

Example 1 Based on a model constructed for this purpose, a researcher predicts that the price level in a country will go down in the following year. If the concept of price level (compare Sect. 3.1 **13**) is not defined exactly then this researcher can claim success even if only one price went down (slightly) during that year, while all other prices went up (considerably). All that is needed to achieve this is to choose skillfully the weights determining the price level with the "wisdom" of hindsight.

Sometimes, even when the "simplifications" in its construction, and verbal description are quite unreal, a model can be rather useful.

Example 2 JOHANN HEINRICH VON THÜNEN (1783–1850) described his famous model of an isolated state as follows: Imagine a big city in the centre of a big, fertile plane containing, near the city, mines as needed. The plane in turn is surrounded by unpenetrable wildness. There are no other villages, towns or cities in the plane and no navigable waterways through it. (Notice that there were no airplanes at Thünen's times). So the plane has to supply the central city with all food and other raw materials while the population of the city has to furnish all necessary tolls for production and living.—Thünen asks the question how the intensity of agricultural production at different parts of

(continued)

the plane depends on their distance from the city.—Unrealistic as this model may sound, it turned out to be very useful and, to quote Erich Schneider again, "it became quite fundamental for modern economic and agricultural theory".

We give one more example of a model in economics which proved to be useful even though it approximates reality only from afar.

Example 3 A *perfect market* is one which satisfies the following conditions:

(i) objective homogeneity of goods (no differences in quality);
(ii) no subjective (personal) preferences of buyers for certain sellers or the other way round;
(iii) no time—or place—dependent differences among sellers or among buyers;
(iv) complete market transparency.

Again, the model of a perfect market, while impossible to realise completely, is a basic tool in economics.

It is worth repeating that such "oversimplified" models proved to be useful also in physics (and other sciences). For instance, there exists no "ideal gas" (by definition, its molecules, which are considered mathematical points without spatial dimensions, move freely on straight lines and no forces connect them) but this concept helped develop the (statistical) kinetic theory of gases, approximately satisfied by "real" gases, the thinner and of the higher temperature they are, the better.

3. *Graphic representation of models*. Geographic maps can be considered models of the part of country they describe. Similarly, but not so frequently models in economics are also occasionally represented by drawings, in particular *"circular-flow models"*. In these models "currents" (for instance of goods, payments, etc.) flow through "directed edges" (or segments) between "vertices" (or poles; for instance households, enterprises, governments, countries), forming the "circular flow" (mathematically: "directed graph", an object of the mathematical discipline of combinatorics; those graphs are *not* the graphs of functions defined in Sect. 3.2; combinatorics has also important applications in other branches of economics and of operations research). One can also indicate the intensity of the currents by the width of the edges (herein the circular-flow models differ from mathematical "directed graphs" whose edges have no width; however, there is also a branch of combinatorics which deals with weighted graphs: these correspond exactly to this situation).—Models and graphs may be represented

graphically even in three dimensions with aid of mechanical models and holograms, by the way.

At the Sorbonne university of Paris, in 1958 the bicentennial of an early and important circular-flow model was celebrated, that of the "tableau de Quesnay". FRANÇOIS QUESNAY (1694–1774) was the founder of the "physiocratic school" of thought in economics and also the private physician of Madame Pompadour, the mistress of King Louis XV. In analogy to blood circulation in medicine, his endeavour was to construct a simple and intuitive model of circulation in economics. In this "tableau" (figure) he represented three "sectors" (land owners, producers of raw materials, and industry, including commerce) by vertices, which he connected by edges representing the transactions between sectors. The physiocratic theory impressed and influenced KARL MARX (1818–1883) to construct a geometrically similar but economically different graphic model. His classification was "functional" rather then "institutional", his sectors were "departments" of capital goods (i.e., means of production), consumption goods and households (embodying, among others, the labour force). Moreover, while Quesnay's "tableau" described just stationary processes, Marx analysed also economies with increasing capital stock. A graphic representation of Marx's model for simple reproduction of an economy is reproduced in Fig. 13.1. In Marx's terminology the reproduction is *simple* if in the process of production the means of production and the labour force are just renewed (without expansion). In order to renew the labour force, the necessary amount of durable (houses, appliances, etc.) and perishable (food, clothing,etc.) consumer

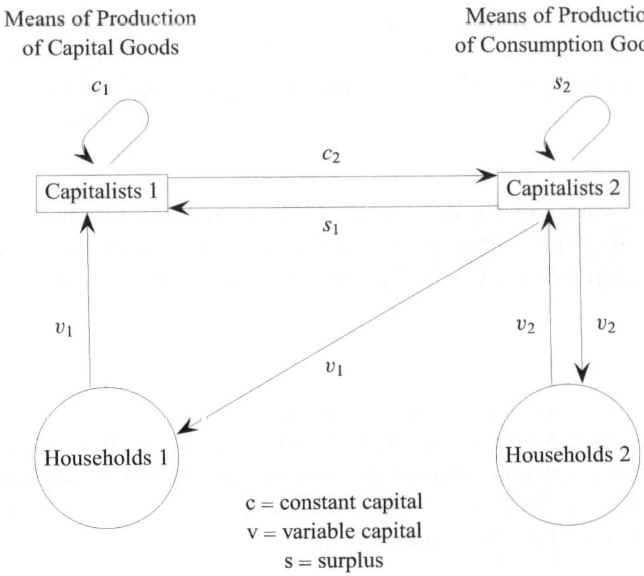

Fig. 13.1 A graphic representation of the model for simple reproduction of an economy

goods have to be available. In order to produce them, again means of production are needed. The goods pass through several stages of the process till they get as products, established by the means of production, into the sector of consumer goods and from there, after passing through several stages of consumer goods production, into households. There they are consumed and thus reproduce the labour force, which in turn produces both the means of production and the consumer goods.—The boxes and arrows in Fig. 13.1 may be replaced by vertices and directed, even weighted edges of the weighted directed graphs mentioned above. The *quantitative* aspects of this model, in particular the question, when an equilibrium is achieved, could be analysed by use of its analytic representation (compare to **4** below)

The relation between a model and its graphic representation can be reversed. For instance, the boxes (vertices) and arrows (directed edges) in Fig. 13.1 can be considered to be an abstract pattern (directed graph) and then the above economic interpretation is a model for this pattern in the same mathematical-logical sense as, in Sect. 13.2 **3**, the "geometry of segments" and the "arithmetic of reals with integer differences" were models of the axioms **A1, A2**. Here too, as there, the same pattern may have several (economic, business administration) models. For instance, the following is another model for the pattern in Fig. 13.1.

Replace "capital goods" by "raw material processing", "consumption goods" by "production goods", and "households" by "warehouses and centres of other activities, such as purchasing, sale" on the vertices in Fig. 13.1. If we now reappraise the arrows from the warehouses as flows of material, those from "raw material processing" to "consumption goods" as flows of "finished" raw material, the arrows between two "raw material processing" boxes as flows of raw materials in different stages of processing, those between two "production goods" boxes as flows of (partially finished) products and, finally the arrows to "warehouses, sales" as flows of finished products, then we get a simple model for some sort of *enterprise*.—We give two more examples of directed graphs serving as graphical representations of models.

Example 4 We represent the model of the information structure, say in an enterprise, by a directed graph in which an edge is directed from a vertex A to a vertex B if office B can be directly informed by office (or workplace) A.

Example 5 Now the vertices correspond to enterprises and an edge is directed from the vertex A to the vertex B if enterprise A supplies enterprises B. This directed graph represents a model of the network of delivery and receipt relations in a branch of the economy. If the graph is weighted, it can represent also the *value* of the delivered goods.

(continued)

Models for planning of projects can be represented by special graphs, networks, which reflect the logical structure of the project and can be converted into the practical logistics of realising the project.

4. *Analytic representation of models.* In order to make quantitative statements, problem solving and forecasting possible, analytic representation of models became prevalent in economics, that is, their description by mathematical relations between economic variables in the sense of models in applied mathematics, as mentioned at the end of Sect. 13.2 and as we have shown by several examples in this book. Such relations may be equations or inequalities or functions deterministically (with certainty) or stochastically (depending on probabilities) connecting economic variables or defining them or stating conditions. Here is an example:

Example 6 The variables are the national income Y and the sums C and I of the expenses for consumption and investment, respectively, all during a fixed period in the economy. The following *conditions* are again simplifications (compare Sects. 13.2 **3** and 13.3 **2**).

C1 *The economy is "closed"* (no commercial or other economic exchange with other economies), and there is no economic activity of the government.

C2 *The planned consumption C of all households depends "linearly"* (really, *"affinely"*, see Sect. 3.1) *upon the national income Y*:

$$C = cY + d \quad (c \in]0, 1[, \ d \in \mathbb{R}_+ \text{ constants}). \tag{13.1}$$

C3 *The groups of all producers plans to invest, during the period under consideration, the fixed amount A*:

$$I = A \quad (A \in \mathbb{R}_{++}, \text{ constant}). \tag{13.2}$$

C4 *Equilibrium condition.* The national income Y should equal the sum of all expenses for consumption and investment:

$$Y = C + I. \tag{13.3}$$

The three variables C, I and Y in the model are often called "endogenous", meaning that they have to be determined within the model, while the constants or "parameters" c, d, a are "exogenous" that is, they are imposed (given) from outside the model. In the present context, (13.1) and (13.2) describe C and I as *functions* of Y (affine or constant, respectively), while Eq. (13.3) connects them as a further "equilibrium" condition. We called the constants c, d and A "parameters" because C, I and Y can be determined from (13.1), (13.2) and (13.3) uniquely, depending only upon c, d and A: We obtain $Y = cY + d + A$, so (since, by supposition, $0 < c < b$) we get the following *equilibrium values* of the model:

$$Y = \frac{d + A}{1 - c}, \quad C = cY + d = \frac{cA + d}{1 - c}, \quad I = A. \tag{13.4}$$

The *existence* of such a solution guarantees that, *no matter how the parameters* $c \in]0, 1[$, $d \in \mathbb{R}_{+}$, $A \in \mathbb{R}_{++}$ *are given*, both the investment (I) and the consumption (C) expenditure can be planned so that they be constant (A) resp. affine functions (with coefficients c, d) of the national income Y and satisfy the equilibrium condition (13.3), while the *uniqueness* of the solution shows that, for any given c, d, A, there is *just one* such triple Y, C, I.

This example gives an indication of the advantage of models described quantitatively, in the language of mathematics, as compared to models represented verbally, in everyday language: Mathematics makes a concise, lucid representation of the model possible. Furthermore, the exact, "syntactic" rules (which concern the logical, formal structure) of the mathematical language make it possible to reach unambiguous conclusions, independently of "semantic" aspects (which depend on the meaning of words and expressions) through formal manipulations ("calculation"). Then the semantic analysis of the result permits pragmatic, practical applications of the model. We have just seen, for instance, how formal (syntactic) manipulation of the conditions (13.1), (13.2) and (13.3) led "automatically" to the result (13.4). In these calculations the semantic meaning of C, I and Y was irrelevant. But then the semantic analysis of the result (13.4), applied to the present problem, showed that there exist unique equilibrium values of expenditure for consumption and investment and of the value of national income in our model.

Just as this Example 6, also the Examples 2, 3, 4 and 5 could be given a mathematical representation in the form of exactly formulated assumptions and, as a further refinement, even of quantitative relationships.

The great physicist GALILEO GALILEI (1564–1642) wrote that "the book of nature is written in the languages of mathematics". Nowadays practically everybody agrees in the importance of the language of mathematics also in the representation of models in economics.

5. *Representation of models by systems of assumptions.* If the concept of assumptions is taken broadly enough, including axioms, principles, suppositions, premises, postulates, hypotheses, initial-, boundary- and other conditions then clearly every model, as defined at the beginning of this section, can be represented by systems of assumptions. While we want now our assumptions to be exactly formulated in the mathematical sense, we will keep talking about *systems of assumptions* rather than *system of axioms* (though we sometimes used

the word axiom parenthetically). The reason is that, historically, by an "axiom" in mathematics and in mathematical logic (compare Sect. 13.2 **3**) something like "generally accepted fundamental principle" has been meant and the assumptions concerning models in economics are, as a rule, not quite so fundamental.

As mentioned above, every model in economics can be represented by a system of assumption. One can ask, what requirements, conversely, a system of assumptions should satisfy in order to serve as representation of a model in economics. In other words, what kind of assumptions can characterise a simplified image of (a part of) reality.

We will examine this in the next section.

13.4 Systems of Assumptions

A "subjective" theory in philosophy doubts whether an objective reality exists at all and, if it does, whether it can be explored. If it did not or could not, then models could not be considered anymore "simplified images of reality" *represented* by a system of assumptions. Then these assumptions would rather *define* the model.

Be as it may, the question, what requirements a system of assumption has to satisfy in order to represent a model or to be a model itself, makes sense. Again, opinions differ not surprisingly, since, for instance, it is to a certain degree a matter of taste what one accepts as "simplified image of reality". For the same reason, it is not our aim here to put together a rigid list of such requirements. Actually, most such lists would exclude some classic models in economics. In view of the purpose of creating models one could require, however, that the assumptions for a model in economics should at least be *realistic and informative* (see **2** below), *corroborated* by previous experience and checking (see **3** below) and *consistent*, that is, it should not lead to contradiction (see Sect. 13.5 **3**).

1. *Formulation of assumptions; inductive reasoning.* The formulation of assumptions is often aided by psychological factors, beyond rational reasoning. Often "creative intuition" is mentioned in this context, without spelling out what this is supposed to mean. The so called "factor analysis" originating in experimental psychology deals, among others, with the subject of systematising intuition by means of observed data ("factors").

One of the theoretical explanations of how assumptions are formed is *inductive reasoning*, as formulated by KARL POPPER (1902–1994). From particular observations, experiments, etc. one intuitively reaches general conclusions or hypotheses, which have then to be checked by confronting them with reality (further observations, experiments, and so on). For instance, in economics one observes several times that full employment leads to rising prices. We advance therefrom by "inductive reasoning" to the general assumption that rising levels of employment always (or usually) lead to higher price levels. If further observation confirms this (in most cases) then this assumption can be accepted as part of a model. (Somewhat analogous to this process is the often unformulated presumption that all or most people will act as we ourselves act and do as those whom we know.)

The expression "inductive reasoning" is often abbreviated to "induction" which, of course, has nothing to do with "mathematical induction". (That says in its simplest form the following about a statement S_n concerning $n \in \mathbb{N}$: If S_1 is true and S_n implies S_{n+1} for all $n \in \mathbb{N}$ then S_n is true for all $n \in \mathbb{N}$.)

Of course, assumptions may have also other sources. In addition to the *purpose* of a system of assumptions (see Sect. 13.6) *logical* (Sect. 13.5 **1, 2, 3**), *historical* and *practical* considerations may play a role. For instance, the individual researcher's experience in and knowledge of economics, history, sociology, of political situations, etc. can have an impact, and so can his or her subjective political-ideological-religious persuasion, personal experiences, ways of thinking, etc. Even personal likes or dislikes of mathematics and logic make a difference. Those mathematical-logically inclined will be careful to deal with well defined concepts and may try to formulate their assumptions on relations and conditions in a mathematical form. A researcher even more advanced in mathematics and logic may also be interested, whether the assumptions in the system are *independent* (no assumptions follows from another, compare Sect. 13.5 **2**), *complete* (all those results following from it, for the explanation of which has been created) and, necessarily, whether it is *consistent*, that is, *does not lead to contradictions* (no assumption should contradict any consequence of the others).

Of course, for a system of assumptions in economics to be useful, it has to be at least approximately *correct* in its forecast about a *wide* range of *future* events and observations (see **2** and **3** below).

2. *Information content of assumptions and their link to reality.* Since the purpose of creating a model in economics is presumably to obtain information about one or more aspects of reality and to reach optimal decisions on them, it is natural to require that the assumptions be *realistic* and *informative* (contain essential information). The two requirements are not the same. For instance, the statement that if nothing changes in the US economy then its growth rate will continue to develop as before is quite close to reality but contains little if any information. (There is also the joke about a parachutist landing on top of a tree in a part of the country which she does not know. But she sees somebody walking by and yells down to him "where am I?". He stops, thinks, then answers "you are on the top of that tree". She says "so you are a pure mathematician". He: "yes, how did you know?". She: "you carefully deliberated before you answered, your answer was perfectly exact and completely useless"—or, as we would say, conformed with reality but gave no new information whatsoever.)

For a system of assumptions in economics to be realistic and informative, it is important that, in addition to economic data, it should also pay attention to legal, technical and social conditions.

3. *Corroboration of an assumption.* One way to ascertain that an assumption contains much information is to check *what possibilities it does exclude*. For this,

Karl Popper (whom we already quoted before) argues essentially as follows. If an assumption contains more information then, at least in principle, it is easier to check its correctness because then there are *more things which could contradict it.* If an assumption that stands out against attempts to *falsify* it proves correct in many practical situations then it is well *corroborated.* Theories containing empirical content are characterised by being *falsifiable.* Of course, one can never confirm a (falsifiable) hypothesis (assumption) absolutely but, if we confront it with a great variety of situations where it could fail (be falsified) and it does not, then we have a good reason to accept it, at least for the time being.

For instance, in Example 6 of Sect. 13.3 we made the assumption that the consumption C depends upon the national income Y linearly ("in an affine way", to be exact):

$$C = cY + d \tag{13.5}$$

but we assumed about the constants ("parameters") only $0 < c < 1$ and $d > 0$. If our assumption contained, more narrowly, $1/2 < c < 3/4$ (while about d we still know only that it is positive) then its information content increases. Indeed, if numerical checking of the data gave (13.5) with $c = d = 1/4$ that does not contradict the first assumption but it does contradict the second. If the assumption is further restricted by specifying, say $c = 3/4$, $d = 0$ (in which case (13.5) reduces to $C = 3Y/4$, a truly "linear" relation), then the information content increases dramatically and so does the ease of rejection of this final hypothesis: one or, in practice preferably several, observations with $C \neq 3Y/4$ are sufficient to reject it. However, even the original assumption (13.5) with the lower information content is vulnerable: Even if *up till now* all relations between Y and C prove to be affine, *future* observations may turn out not to be. That is why it is still an *assumption* (although well corroborated) and not an unconditionally true *law*.

13.5 Theories in the Sciences, in Particular in Economics

We pointed out already in Sect. 13.1 that, in general, a theory T is a system of statements, that is, assumptions (axioms, postulates, hypotheses, with initial, boundary and other conditions) and theorems (propositions, lemmata, corollaries) concerning a field of research. We will deal with particularities of theories in economics below, in **4**. First, we will clarify basic notations concerning theories in general. Independently of its origins, a theory T is presumed to contain all present and future consequences (theorems) which *follow* in a purely logical manner from the assumptions (axioms) of T. We start, in **1** with the description of this "purely logical deduction".

1. *The method of deduction.* A theory T is developed in the following way from a system S of statements on the basic objects in a field of research. A statement

t belongs to the theory T if it is a *logical consequence* of one or more (or all) statements s of S. One says in this case that S *implies* t (in formula: $S \Rightarrow t$) or that t can be *deduced* (follows) from S or that some (or all) s of S form *sufficient conditions* for t. This means also that not all s of S are valid if t is not valid (because the validity of S would imply that of T). So, conversely, t is a *necessary condition* for S. Clearly, $S \subset T$.

Having S would, in principle, yield the knowledge of *all* consequences of S, that is, of the *whole theory* T; but *only in principle*. An analogue of the superhuman "demon" of the famous mathematician and physicist PIERRE SIMON MARQUIS DE LAPLACE (1749–1827; the demon would be able, in knowledge of the laws of physics and the initial state, to determine the present and future state and behaviour of everything) could deduce from S all of T. But researchers who are only human, have to be content with deducing a part of the set T of all consequences of S. The amount and importance of consequences deduced from a system S of assumptions may be quite impressive but may still only be a minuscule part of T ("our knowledge is small, our ignorance is immense" to paraphrase again Laplace). It is particular annoying when statements, which can be formulated in a quite simple way in the terminology of a theory T, resist for centuries a logical deduction from the system of axioms (and their consequences) in T. For instance we know (from high school or from Sect. 1.4, Fig. 1.4) that the lengths of the longest (r) and the two shorter (x_1, x_2) sides of a rectilinear triangle are connected by Pythagoras' equation $r^2 = x_1^2 + x_2^2$. There are (infinitely) many triples (x_1, x_2, r) of positive integers satisfying this equation, for instance (3, 4, 5). The great French mathematician PIERRE FERMAT (1601–1665) thought to have proved (he wrote it on the margin of a book with the comment "I have discovered a truly remarkable proof of this theorem but this margin is too small to write it here") that *there are no positive integers p, q, r satisfying $p^n + q^n = r^n$ for any integer $n > 2$*. It took more than 300 years till a long and pretentious proof of this theorem (that is, a deduction of this statement from the axioms of number theory and their consequences) was presented that satisfied a great number of experts. The following example is easier. We deduce, from a system of three simple assumptions in economics, both related and (seemingly) unrelated results.

Example 1 Experience shows that the *production function* $F : \mathbb{R}^n_+ \to \mathbb{R}_+$ for a certain one-product-enterprise or even for some country's economy satisfies the following assumptions:

E1. Output resp. national income *strictly increases* with increasing input, that is (compare Sect. 3.4),

$$\mathbf{x} \geq \mathbf{v} \Rightarrow F(\mathbf{x}) > F(\mathbf{v}). \quad (\mathbf{x} \in \mathbb{R}^n_+, \mathbf{v} \in \mathbb{R}^n_+)$$

(continued)

E2. The production function is positively *linearly homogeneous* (compare Sect. 3.4):

$$F(\lambda\mathbf{x}) = \lambda F(\mathbf{x}) \quad \text{for all} \quad \lambda \in \mathbb{R}_{++}, \mathbf{x} \in \mathbb{R}^n_+$$

or writing F as function of the n (scalar) inputs:

$$F(\lambda x_1, \ldots, \lambda x_n) = \lambda F(x_1, \ldots, x_n) \quad \text{for all} \quad \lambda > 0, x_1 > 0, \ldots, x_n > 0.$$

E3. The production function $F : \mathbb{R}^n_+ \to \mathbb{R}_+$ *as function of* $(n-1)$ *of its variables (inputs) is* positively *homogeneous of some degree* (compare Sect. 6.12): for some $r_j \in \mathbb{R}_{++}$,

$$F(\lambda x_1, \cdots, \lambda x_{j-1}, x_j, \lambda x_{j+1}, \ldots, \lambda x_n) = \lambda^{r_j} F(x_1, \ldots, x_n) \quad (j = 1, \ldots, n)$$
$$\text{for all} \quad \lambda > 0, x_1 \geq 0, \ldots, x_n \geq 0.$$

We will *prove* that *this implies that F has the Cobb-Douglas form* (see Sects. 6.12, and 8.4.)

$$F(x_1, x_2, \ldots, x_n) = C x_1^{1-r_1} x_2^{1-r_2} \ldots x_n^{1-r_n} \tag{13.6}$$

with a positive constant C, furthermore $0 < r_j < 1$ $(j = 1, 2, \ldots, n)$ *and* $r_1 + r_2 + \ldots + r_n = n - 1$. Conversely, every function F so given satisfies **E1, E2** and **E3**.—The second, converse part of the statement is easily checked by putting (13.6) into **E1, E2, E3**.

In order to prove the first part, that (13.6) and the restrictions on $C, r_1, r_2, \ldots, r_n$ follow from **E1, E2, E3**, we use first **E2** then **E3**:

$$F(x_1, x_2, \ldots, x_n) = x_1 x_2 \ldots x_n F\left(\frac{1}{x_2 x_3 \ldots x_n}, \frac{1}{x_1 x_3 \ldots x_n}, \ldots, \frac{1}{x_1 x_2 \ldots x_{n-1}}\right)$$

$$= \left(\frac{1}{x_1}\right)^{r_1} x_1 x_2 \ldots x_n F\left(\frac{1}{x_2 x_3 \ldots x_n}, \frac{1}{x_3 x_4 \ldots x_n}, \ldots, \frac{1}{x_2 x_3 \ldots x_{n-1}}\right) = \cdots$$
$$= x_1^{1-r_1} x_2^{1-r_2} \ldots x_n^{1-r_n} F(1, 1, \ldots, 1)$$

which, with $C = F(1, 1, \ldots, 1)$, is exactly (13.6). Since the values of F are nonnegative ($F : \mathbb{R}^n_+ \to \mathbb{R}_+$), we have ($\geq 0$) and; by **E1** ($F$ strictly increasing in *each* variable, compare Sect. 3.2), $1 - r_1 > 0, 1 - r_2 > 0, \ldots, 1 - r_n > 0, C > 0$. Finally, putting (13.6) into **E2** gives $1 - r_1 + 1 - r_2 + \ldots + 1 - r_n = 1$, which concludes the proof.

Since $x \mapsto x^{1-r}$ is strictly convex from above if $0 < r < 1$ (see Sect. 3.5) we get, as an added bonus, the *strict law of diminishing returns* for each input variable, that is,

$$x_j \mapsto F(x_1, \dots, x_{j-1}, x_j, x_{j+1}, \dots, x_n) \tag{13.7}$$

is *strictly convex from above on all of* $\mathbb{R}_+$ for all positive n-dimensional vectors $x_1, \dots, x_{j-1}, x_{j+1}, \dots, x_n$ $(j = 1, 2, \dots, n)$. In other words, *the output (income) increases keep strictly decreasing from* 0 *on all the way* (see Fig. 13.1). This is remarkable, since nothing in the assumptions seems to have to do with convexity or diminishing returns.

Often there is not or not yet sufficient evidence to justify assumptions like **E1, E2, E3** above, or assumptions are such that they cannot be directly verified. Then "the proof of the pudding is in its eating": the assumptions are first just conjectures and it are their consequences (that is, the theory T which follows from them) which have to be tested in order to justify (or reject) the assumptions.

2. *Independence of assumptions.* Whether the assumptions come from experience or are just guesses (lucky or otherwise, as shown by testing their consequences), it is of some advantage to *keep the number of assumptions*, on which the theory is built, *as small as possible*, for otherwise the sheer size of the system S of assumptions may cause confusion. In particular it is redundant to include in S a statement which *follows* from the other assumptions in S. The assumptions in S are *independent* if none of them follows from the others. This requirement, however, is often more a matter of logical beauty and conciseness than of practicability: sometimes the deduction of a very simple and obviously sounding statement from a minimal system of assumptions may be quite complicated. In such cases (and if S cannot be replaced by a more convenient system of independent assumptions), one may sacrifice independence for simplicity. Be it as it may, one can *prove independence by giving for each assumption a "counter-example" which does not satisfy that assumption but satisfies all others.*

Example 2 The assumptions **A1, A2, A3, A4** *for price indices* in Sect. 3.7 *are independent.* To see this, check that among

$$P_1(\mathbf{q}^0, \mathbf{p}^0, \mathbf{q}, \mathbf{p}) := (p_1/p_1^0)^{b_1}(p_2/p_2^0)^{b_2} \dots (p_n/p_n^0)^{b_n}$$

(continued)

with constant $b_1 \in \mathbb{R}_-$, $b_k \in \mathbb{R}_{++}$ $(k = 2, \ldots, n)$ such that $b_1 + b_2 + \ldots + b_n = 1$,

$$P_2(\mathbf{q}^0, \mathbf{p}^0, \mathbf{q}, \mathbf{p}) := \frac{p_1}{p_1^0} + \frac{p_2}{p_2^0} + \cdots + \frac{p_n}{p_n^0},$$

$$P_3(\mathbf{q}^0, \mathbf{p}^0, \mathbf{q}, \mathbf{p}) := \frac{1}{\mathbf{q}^0 \cdot \mathbf{p}^0 + 1} \left(\frac{\mathbf{q}^0 \cdot \mathbf{p}^0}{n} \sum_{k=1}^{n} \frac{p_k}{p_k^0} + \max \left\{ \frac{p_1}{p_1^0}, \cdots, \frac{p_n}{p_n^0} \right\} \right),$$

and

$$P_4(\mathbf{q}^0, \mathbf{p}^0, \mathbf{q}, \mathbf{p}) := \frac{p_1 + p_2 + \ldots + p_n}{p_1^0 + p_2^0 + \ldots + p_n^0},$$

P_j *does not satisfy* $\mathbf{A}_j$ $(j = 1, 2, 3, 4)$ but satisfies all others ($P_j : \mathbb{R}^n_{++} \times \mathbb{R}^n_{++} \times \mathbb{R}^n_{++} \times \mathbb{R}^n_{++} \to \mathbb{R}_{++}; j = 1, 2, 3, 4; P_1, P_2, P_4$ are independent of $\mathbf{q}^0$ and of $\mathbf{q}$, and P_3 is independent of $\mathbf{q}$, but that does not matter).

Example 3 Shephard's six "axioms" **P1-P6** for production correspondences in Sect. 8.7 are independent. Again the proof consists of giving examples C_j of correspondences which do not satisfy P_j $(j = 1, \ldots, 6)$ but satisfy the other five "axioms":

3. *Consistency of a system of assumptions or of a theory.* As we just saw, it is quite easy to prove the *independence* of the assumptions of a system (by constructing appropriate counter-examples, as in Examples 2 and 3), but it is not a very important property of a system of assumptions. The situation is quite the opposite for the *consistency* of a system S of statements (assumptions and theorems; see Sect. 13.5 **1**) or, equivalently, of the theory T consisting of all their consequences: it is vitally *important* (at least theoretically) but *very difficult to prove*. Indeed this *consistency* means, as indicated at the beginning of Sect. 13.4, that the theory T *should not contain two contradictory statements*. If it would, then the theory would clearly be *useless* (so much the more because it can be shown that it would then lead to infinitely many contradictions). The trouble is, that such contradictions may show up only very late in the development and elaboration of the theory. So nothing guarantees that even centuries-old systems of assumptions will not eventually lead to a contradiction. This is a problem even in pure mathematics and logic: Let T be a (formalised) theory formulated in the language of logic and mathematics. It is well

(continued)

known that the consistency of such a T cannot be proved within T itself, only *for* T. If one can construct for such a T, with the language of T, a "well-formed" formula (i.e., a formula fitting into T according to the formation rules of the language of T) which can *not* be derived within T then T is consistent.

We certainly do not want to go into this here in more detail, we offer instead comfort in three ways: (i) *If we can find objects* in practice *which satisfy all assumptions* of a system *S then S and the theory based upon it is consistent.* (ii) Also, if a reasonably developed theory T *did not lead to contradiction during a considerable time* of intensive research, we may *use* T *for the time being*. (iii) On the other hand, it is often (but not always, see above) quite easy to *prove, by an example, that a theory or a system of assumptions is not consistent.*

Take, for instance, the assumptions **E1, E2, E3** of Example 1 in **1** above and add this fourth assumption:

E4 There exists a $j \in \{1, 2, \ldots, n\}$ and a $b \in \mathbb{R}_{++}$ such that the partial function in Fig. 13.2 be *convex from below on* $[0, b]$. The system **E1, E2, E3, E4** of assumptions is *not consistent* (compare(iii)), since we proved in **1** that **E1, E2, E3** imply that in Fig. 13.2 is strictly convex from above, which contradicts **E4**.— But, as we have seen, (13.6) does satisfy **E1, E2, E3**, so the system **E1, E2, E3** of assumptions is *consistent*, by (i).

4. *Theories in the natural and social sciences, in particular in economics.* As we have seen in Sects. 13.2 and 13.3, in particular in Sects. 13.2 **2** and 13.3 **5**, *models* in the natural and social sciences *are simplified images of* (different parts of) *reality* and *can be described by systems of assumptions*. By what we learnt in the present section, such a system of assumptions and the consequences deduced from it form a theory. The *theory describes the model in more detail* and helps us bring order and transparency into the often confusing appearance of nature or of say, the economy, *permits us to understand it for the present time* and even *make forecasts for the future*. The more important aspects of reality we captured in the model and in the theory, the better our understanding and forecasts will

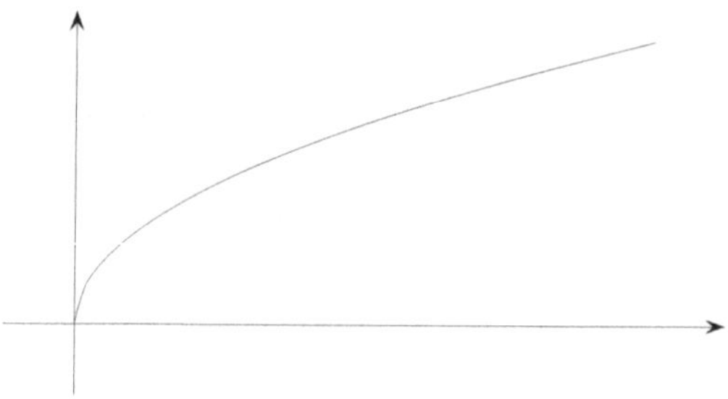

Fig. 13.2 The strict law of diminishing returns

be. In Sect. 13.2 **3** we saw that the situation is similar in applied mathematics but that it is, in a way, reversed in pure mathematics and in logic: there a model is a *realisation* of a theory, a bunch of objects and relations which satisfy the axioms (assumptions) and thus the whole theory. Similarly as in Sect. 13.3 **3** with graphic representations, also the relation between economic model and systems of assumptions (axioms) or theory can be reversed to conform with the pure mathematical—logical usage. Then the system of axioms and thus the whole theory can be considered in an abstract—formal way (as in pure mathematics or logic) and *the* (say, economic) *model is a realisation of the theory*. As in Sects. 13.2 **3** and 13.3 **3**, again several models (realisations) may be attached to the same theory or system of assumptions.

For instance, the assumptions in Example 6 (Sect. 13.3 **4**, (13.1), (13.2), (13.3)), may be recognised as pure mathematical *equations* (with constants $A \in \mathbb{R}_{++}$, $c \in]0, 1[$, $d \in \mathbb{R}_+$),

$$C = cY + d,$$

$$I = A,$$

$$Y = C + I.$$

As model (meaning now "realisation") we had there national income as Y, expenditure for consumption (C) and investment (I). *Another model* could be *income* Y, *consumption* C and *savings* I *of private household*, with the values (13.4) assumed again.

We state some further requirements for economic theories to be specifically "*empirical*" (similar specifications apply to theories in other social, behavioural and even natural sciences). *In order for a consistent theory to be relevant to economics, its system of assumptions should represent a model of a branch of economics.* Furthermore, *a theory is empirical if it contains statements* ("theorems") *which have been checked in practice and are not (yet) falsified.* It is also desirable that *the theory should contain statements about processes of significant extent* ("dimension") *in space and time.*

The above shows the importance of the method of deduction also for economics. It allows to *condense* a branch of economics *into* consistent *assumptions* from which by logical *deduction* a *theory* is built containing statements ("theorems") which can be *confronted with the economic reality* and thus *justified or falsified*. Nowadays much of economic theory is created in this way.

However, there exist useful theories in economics and in other sciences, no assumptions or theorems of which can be *exactly* verified by experience. We mentioned in Sect. 13.3 **2** the theory of "ideal gases" in physics and that of "perfect markets" (Sect. 13.3, Example 3) in economics. Since their basic assumptions are abstractions (idealisations) they cannot be exactly verified by experience. Nevertheless, the *consequences* of these assumptions (the theory built upon them) are *approximately* correct and help to explain many phenomena of the physical or economic reality, respectively. In particular, logical consequences of the assumptions describing a perfect market (Sect. 13.3, Example 3) and of assumptions about human behaviour, which are also abstractions (such as "maximisation of profit" and "maximisation of utility") show, among others, that

(i) *equilibria* exist, that is, supply and demand can be balanced and that
(ii) such equilibria are *efficient*, that is, one "economic agent" (e.g., individual enterprise, etc.) can do better than in the equilibrium situation only if one or more others do worse (see Sect. 13.3).

The purpose is *not* so much to find *quantitatively (numerically)* the points of equilibrium, say. The results mentioned are rather important *qualitative (structural)* results. The closer *real* markets get to fulfilling the perfect market conditions (Sect. 13.3, Example 3) as is increasingly the case with stock markets the better the results of this theory approximate the real situation in them.

13.6 Why Construct Models and Theories? Types of Models and Theories

Depending upon the *purpose* of constructing models and theories, different types of them evolved. In this section we shall touch, by means of examples, on some important types of models and theories, as determined by their purposes, such as description (**1**), working hypothesis (**2**), explanation (**3**), forecasting (**4**), decision making (**5**) and political justification (**6**).

1. *Description.* If one wants to *describe* a complex situation in the economy one uses (knowingly or intuitively) *models*. For instance, if we wish to describe the flow of goods, services, work and money in the economy of a country, we can use a circular-flow scheme as in Sect. 13.3 **3**, the vertices being the households, enterprises, the state, foreign countries, national savings and investment. If we wish to obtain more detailed information we have to disaggregate the national data into those for sectors of the economy, regions of the country, etc.

   In general, whenever we use a system A, that is neither directly nor indirectly interacting with a system B, to obtain information about the system B, we are using A as a model for B.

2. *Working hypothesis.* The purpose of some simple models is more modest than explaining processes, they just present "working hypotheses", the logical consequences of which we can then compare to observations.

   For instance the affine relation $C = cY + d$ (with constant $c \in]0, 1[$, $d \in \mathbb{R}_{++}$) between consumption and national income in Example 6 (Sect. 13.3), which we have quoted repeatedly, can be considered as such a working hypothesis. On the one hand, if observations do not corroborate it, we can replace it by another working hypothesis, say that C is a function of Y strictly convex from above. On the other hand, if observations confirm it (which is the case in the long run in most economies) then we can use them to determine or estimate "econometrically" the values of the constants ("parameters") c and d. This gives the model a *quantitative* character.

Example 1 We spoke of "econometric" determination or estimation because such relations are rarely "deterministic", they are more often "stochastic" (depending on chance, on minor or rare accidental oscillations). It is often supposed that two variable quantities (say x and y which may also be vectors, with x uniting all specific influences upon y) are connected by a single-valued deterministic function f, that is, $y = f(x)$, but only "in average". More exactly, this means that there exists a *"random variable"* u (depending again on chance) so that $y = f(x) + u$. Of course the models differ depending on f and u. If f is affine, that is, $y = ax+b+u$, *by* an abuse of language one often still speaks about a *linear* (really: *affine*) *model*.

3. *Explanation.* Models which serve for explanation of observable processes (*"explanatory models"*) usually contain "laws" (such as "Newton's law of gravity" in physics), as mentioned in Sect. 13.1. For our purposes, a *law* is a truly universal statement (without temporal and spatial as well as any other, e.g. cultural, social etc., limitation of applicability) of interdependencies
   (i) which stand till now all tests, no matter where or when they were carried out, and
   (ii) which, based on past experiences, are expected by an overwhelming majority of experts to be valid also in the future.

Notice that this definition of a law contains not only objective but also subjective criteria ("the overwhelming majority of experts")—this is the case also with the notation of explanatory model itself, in particular in the social sciences.

Whether a model contains a law in the above sense or not, it is not enough that one should be able to deduce from it one or a *few* observed phenomena. As already the famous economist VILFREDO PARETO (1848–1923) noted, to every statement or observation B there exists a system of assumptions $A_1, \ldots, A_n$ from which B follows by logical deduction. The point is that the consequences of the assumptions in an explanatory model should conform with *many* observations. On the other hand there should be enough latitude in the model to explain future or hypothetical situations such as: what would happen if
(a) the enterpreneur or the consumers behaved differently (in a certain way), or
(b) taxes and/or custom went up or down.

We have often compared models and theories of economics (or, more generally, the social sciences) to those of physics (or the natural sciences). However, there are differences: The existence of deterministic and stochastic laws in physics permit logical deductive explanations of observed definitive or statistical interdependencies, respectively. In economics the "laws" are not so general and categorical, and therefore there are proposals to speak only of "quasi laws", trends like regularities, etc. Here the models and theories are based upon *suppositions* (hypotheses) about human economic behaviour. This is in particular true about microeconomic models and theories which deal with the behaviour and plans of individual households and enterprises. This behaviour and these plans can change rather abruptly and unpredictably. It is doubtful whether laws will ever be found from which these changes can be deduced. For statistical reasons macroeconomic models and theories, which deal with aggregate quantities (of national income, consumption, investment, etc.) rather than individual ones, can be based on interdependencies more stable for a longer period. An example of such stable macroeconomics interdependency is the affine relation (13.1) in Sect. 13.3 ($C = cY + d$) between national income and consumption. The existence of such (relatively) stable macroeconomic laws makes *forecasting* possible.

4. *Forecasting*. While theories in the social sciences which can predict future events with anything even close to the exactness of natural sciences (and even there, predictions are often not very exact: think of weather forecasts), they are still better than lucky (or unlucky) guesses and prophecies: "*forecasting models*" should contain "laws" as described in **3** and be *dynamic*, that is, at least one assumption should contain the *time* (as process, not just one point in time). Models, which are not dynamic, are called *static*. A further classification is into *total* and *partial* models. The first reflects the *entire* economy of a country or of a group of countries, while the second is more restricted, for instance, to a sector or to a market.

   We should not forget that in economics (and in other social sciences) the *forecast may influence the future*. That is why one talks about "*self-fulfilling*" (or "*self-destroying*") *prophecies*. A really good model may take also this influence into account. But economic forecasting is anyway difficult enough. Even short- and medium-term forecasts for the entire economy make the solution of several hundred equations and inequalities necessary, in particular if *decision making* is to be based on them. The solution is nowadays done with computers. The following model can be handled without the aid of computers.

> *Example 2* Samuelson's dynamic macroeconomic total model for the *"boom and bust" business cycle* (Sect. 8.7) shows that simple assumptions about the (linear or affine) interdependencies between the national income, consumption and investment may imply, if they take also the time delay into consideration, cyclic oscillations of these quantities around equilibrium values.

5. *Decision making*. Mainly in business administration but also in (all of) economics, the importance of models for rational decision making is growing in particular in *operations research* and *game theory*. Examples are models

for linear, nonlinear and dynamic optimisation, inventory control, replacement, queueing, games (against nature or other opponents; zero-sum and non-zero-sum-games). As in Sect. 13.3 **3**, also for these decision making models, *graphs* (and *graph theory*) are often used.

As to models for (static, deterministic) models for optimisation, the problem is usually to find the *maximum* or *minimum* of a function $f : \mathbb{R}^n_+ \to \mathbb{R}$ under m *conditions (restrictions)* of the form

$$g_j(x_1, \ldots, x_n) \le c_j \quad (g_j; \mathbb{R}^n_+ \to \mathbb{R};\ j = 1, 2, \ldots, m)$$

(the c_j could be immerged into the g_j and thus replaced by 0). For instance, the variables could be the quantities of goods produced by an enterprise, the value of the function f, the gain, expected from the production of the goods in these quantities and the restrictions could express bounds of capacity in the enterprise.

If all the above functions $f, g_1, g_2, \ldots, g_m$ are linear then we have a *linear optimisation* problem (see Sect. 4.8), otherwise one of *nonlinear optimisation* (see Chap. 8) If the values of $f, g_1, \ldots, g_m$ may in part depend on chance, then this new, different problem belongs to *stochastic optimisation*. Nowadays all these theories are sufficiently advanced to make easier and better decisions possible than "trial and error".

More microeconomic models (dealing, say, with individual enterprises) exist and have been used for decision making than macroeconomic ones (dealing with sectors of the national or international economy) but it can be expected that increasingly also the latter will be used in making decisions on questions of national and even global economics.

6. *Justification of politics.* In addition to the purposes mentioned till now, construction of models and theories may also be *politically* or *ideologically motivated*, and the more so, the less they can be tested by experience. The goal of creating a theory conforming to and explaining one's ideological bias can be done even under semblance of objectivity, for instance by excluding (or including) non-economic influences under the catch phrase "all other things being equal" (compare Sect. 13.7 **1**), by separating production and distribution, or by overidealized assumptions such as unlimited flexibility of all "factors", complete information or foresight, rigidity of market-evolution, etc.

13.7 Control, Correction and Applicability of Models and Theories

Of course, models and theories should not remain unchecked, without feedback from (further) observations of the real world. This can be done in several ways, considering also what kind of models are involved.

1. *Control and correction of explanatory models and theories in economics.* If a theory in economics claims to be able to explain essential phenomena of economic reality and thus to assist decision making (compare to **4** and **5** in Sect. 13.6) then it should be possible to *test* (check, *control*) its assumptions and their consequences empirically (by observation of the real world). But even if they turn out to be correct several times, this does *not prove* or even guarantee the truth of the theory, since several further observations may disprove (falsify) it. Karl Popper (whom we quoted in Sect. 13.4 **1** on inductive reasoning) observed that progress in empirical sciences is mostly made not by those who try to justify or "save" a theory, but by those who *try to disprove* (falsify) it. If they succeed, then the theory has to be *corrected* or a *new theory* has to be created; if not then the original theory is greatly strengthened (corroborated). This is the basis of "*critical rationalism*".

Figure 13.3, by HANS KARL SCHNEIDER (1920–2011), is itself a graphic representation of a model (compare Sect. 13.3 **3**): it shows how theories in empirical sciences should be created, controlled and corrected. It fits natural sciences somewhat better than social sciences, understandable, since the models of social scientists, in particular of economists, have to take into account also a rather difficult and at times irregular subject: human behaviour.

Unfortunately, even assumptions soundly rejected by experience do survive in the social sciences. For instance, microeconomic theories are often based on the assumption that entrepreneurs are always moved by the desire to maximise profit. While *in this generality* the assumption has been falsified by counter-examples, it stubbornly keeps reappearing. Note that the profit-maximising assumption is legitimate if *restricted* to carefully outlined occasional activities (see **2** below).

It is quite another story that *some theories* in the social sciences, in particular in economics, *cannot be confronted with reality at all*. This is often achieved by the "*all other things being equal*" stipulation. As mentioned in Sect. 13.6 **6**, in total models (see Sect. 13.6 **4**) this achieves the exclusion of non-economic influences. In partial models it can be used to exclude influences from other parts of the economy. Whenever observations disprove a tenet of such a theory then its advocates deflect the blame to "not all other things were equal" (they seldom are). For instance, Marxist theory postulated that accumulation of capital causes dramatic increase of poverty, misery, exploitation and degradation of the working masses. When this did not materialise then, for instance ROSA LUXEMBURG (1870–1919) explained it by increased exploitation of colonies and other formerly non-capitalistic markets. Even after decolonialization it lasted some forty years till the theory collapsed.

2. *Control and application of models of limited validity.* When, as it often happens in economics and in other social (and even natural) sciences, no general theories have (yet) been formulated which contain laws explaining certain processes, sometimes "*ad hoc models*" (models of restricted validity and with less scientific foundation; compare also Sect. 13.3 **2**) can serve well. The assumptions in the representation of such models may (a) be of limited validity and/or (b) not be realistic, but their consequences (or some of them), as far as they go, may conform with and be applicable to economic reality. Thünen's "isolated state"

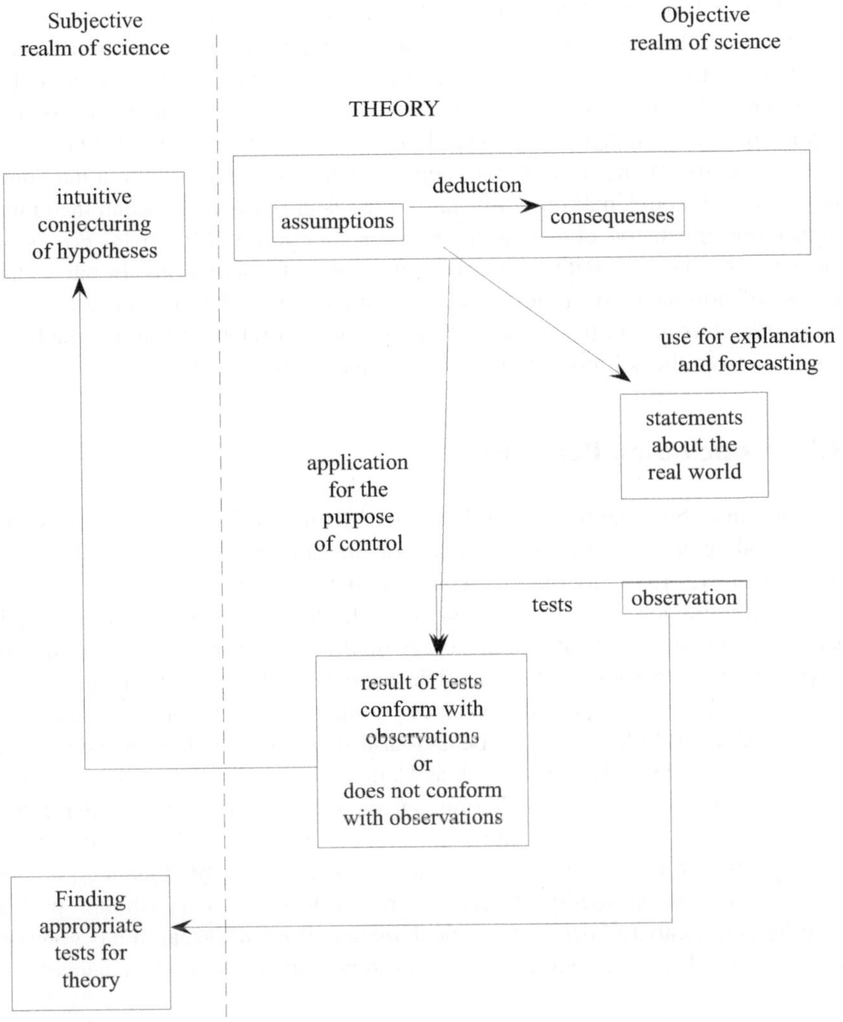

Fig. 13.3 H. K. Schneider's graph on creation, control and correction of theories in empirical sciences

and the "perfect market" (Examples 2 and 3) in Sect. 13.3 **2** are rather of type (b), while the "affine equilibrium" of national income, consumption and investment (Example 6 in Sect. 13.3 **4**) and production functions satisfying **E1, E2, E3** in Example 1 (Sect. 13.5 **1**) are rather of type (a). Of course, before one uses such models and their consequences, it has to be *checked* (empirically, at least "econometrically", compare Sects. 13.5 **4** and 13.6 **2** *whether the conditions of their validity are satisfied*, at least approximately.

Models of limited validity can be *applied to short or medium term* (say, one to twelve months) *forecasting* in the realistic expectation that in time spans of such brevity the economic circumstances do not change significantly and that "disturbances" (interdependencies neglected in the model) will not become significant. But such changes may make long term forecasts incorrect. On the other hand, taking disturbances (more interdependencies) into consideration may unduly *complicate* the model, making for instance the system of equations and inequalities, which represent it, too cumbersome even for computers. Moreover, *errors* often propagate and increase with the number of steps and calculations. In particular, in the case of nonlinear dynamical systems (compare Sect. 12.5) small errors in the data (parameters and initial conditions) may lead in a relatively short time to huge deviations from the solutions which would result from correct data.

13.8 Concluding Remarks

Opinions vary about the role of models, theories and methods in economics and in other social sciences. While the reaction of the famous economist ROY FORBES HARROD (1900–1978) to the advocates of methodology was "stop talking and get on with the job", in the opinion of Vilfredo Pareto (whom we quoted in Sect. 13.6 **3** about the role of models) every method is fine, whether using historical analysis or mathematics, as long as it is expedient. This points in the direction of *interdisciplinary* research, for instance by integration of studies in economics, sociology and psychology (such as behavioural analysis), which is indeed gaining in importance nowadays. This may lead, as Hans Karl Schneider (also quoted before, in Sect. 13.7 **1**, see Fig. 13.3 on theories in sciences) observed, to a general theory connecting social sciences by integrating economic, sociologic, psychologic and other aspects of human behaviour. In order to advance towards this goal, no doubt further research in the theory of scientific research is necessary (though probably not sufficient, compare Sect. 13.5 **1**). Such *methodological* research has gone quite far in the natural sciences but much remains to be done in the social sciences.

13.9 Exercises

Note. Many of the following exercises require longer answers (almost "essays") than most others in this book. Also, there is more than one "correct answer" (the answer we give to some of these exercises at the end of this book is "one" of several correct answers). This situation is more common in economics and even mathematics than one may think at this level.

1. Is a model in economics
   (a) a system of assumptions containing economic notations,
   (b) a system of equations containing economic variables?

2. Give argument for and against models based upon idealistic assumptions (unrealistic but abstracted from reality).
3. What is the role of inductive reasoning in the social sciences?
4. Compare the notions of model in
   (a) logic and pure mathematics on one hand and in
   (b) applied mathematics and in the sciences on the other.
   (c) Can they merge?
5. Formulate statements in economics which have
   (a) all three,
   (b) exactly two (each couple)
   of the following properties: realistic, informative, true.
6. In what sense can a theory T, which "helps us find our way in the vast and confusing economic reality", be considered a representation of a model?
7. Show that the assumptions **E1, E2, E3** in Example 1 of Sect. 13.5 **1** are
   (a) consistent and
   (b) for all $n \geq 3$, independent.
8. What is the difference between
   (a) explanatory models and
   (b) models based on working hypotheses?
9. Is the model in Example
   (a) 6 in Sect. 13.3 **4**,
   (b) 1 in Sect. 13.3 **1**,
   (c) 1 in Sect. 13.3 **2**,
   (d) 2 in Sect. 13.3 **4** (Samuelson's model of the business cycle; see Sect. 8.7)
   an affine or nonlinear, deterministic or stochastic, micro- or macroeconomic, total or partial, static or dynamic one?
10. May one see that the practical applicability of a model in economics is better the more realistic its assumptions are?

13.10 Answers

1. (a) A model in economics can be represented by a system of assumptions containing economic notions, but not every system of such assumptions is a model in economics in the sense of a "simplified image of a part of economic reality". The assumptions can, for instance,
      (i) contradict experience,
      (ii) be logically inconsistent,
      (iii) give no information.
   (b) There are models in economics which are *not* easily representable by a system of equations containing economic variables (see, e.g., Thünen's model of the "isolated state" in Sect. 13.7 **2**). On the other hand, such a system can be logically inconsistent (i.e., can have no solution) or its solution(s) may contradict experience.

2. Frequently idealistic assumptions make logical deductions and clear insight possible. The logical consequences often can be utilized as useful informations (compare Example 3 in Sect. 13.7 **2** and, following there, the "ideal gas" model of physics). That is not necessarily so: Sometimes both the idealistic assumptions and the consequences deduced from them are so idealized that confrontaton with reality makes no sense or is impossible. Nevertheless, models based on such assumptions are used for justification of political measures.
3. Inductive reasoning supports the finding of hypotheses. Since only a finite number of observations or experiments can be made, such hypotheses are preliminary starting points rather than reliable knowledge.
4. A model is in
   (a) logic and pure mathematics any set of objects and relations which satisfy the axioms of an axiom system,
   (b) in applied mathematics and in the sciences a simplified image of a part of reality.
   (c) The two notions of a model can be merged if for any simplified image A of a part of reality there can be abstractly formulated a system S of assumptions of the following kind. There exists a useful covering of the sysmbols in the assumptions with meaning such that S with this realisation of the symbols represents A.
5. The following statement is
   (a) realistic, informative, true:
   DM 1.- -=US$ .6627 on Tuesday, September 10, 1996, at 11:28 a.m. in New York,
   (b) realistic, informative, wrong:
   DM 1.- -=US$ .6543 on Tuesday, September 10, 1996, at 11:28 a.m. in New York,
   realistic, not informative, true:
   if DM 1.- -=US$ .6627 then 1US$=DM 1.5090
   (not informative for anybody who knows that 1/.6627=1.5090),
   not realistic, informative, true:
   if DM 1.- -had been US$ .7000 on Tuesday, September 10, 1996, at 11.28 a.m. in New York then, in view of the transaction costs, all people who had bought DM 1000.- - for less than US$650.- - some time ago had been winners.
6. The system of statements, which constitute a theory T, can be considered to be a system of assumptions. If this represents a simplified image of (a part of) economic reality, it is a model in economics.
7. (a) The assumptions **E1, E2, E3** in Example 1 of Sect. 12.4 **1** are consistent, since the Cobb-Douglas function $F : \mathbb{R}^n_+ \to \mathbb{R}_+$ given by

$$F(x_1, x_2, \ldots, x_n) = Cx_1^{1-r_1}x_2^{1-r_2} \ldots x_n^{1-r_n} \tag{13.8}$$

(with a positive constant C; $0 < r_j < 1; j = 1, 2, \ldots, n, r_1+r_2+\ldots+r_n = n-1$) satisfies them all.

(b) The assumptions **E1, E2, E3** are independent, since (13.8) with $C > 0$, $r_1 > 1, r_2 > 0, \ldots, r_n > 0, r_1+r_2+\ldots+r_n = n-1$ ($n \geq 3$) satisfies **E2, E3**, but not **E1**; since (13.8) with $C > 0, 0 < r_j < 1, r_1 + r_2 + \ldots + r_n \neq n-1$ satisfies **E1,E3**, but not **E2**; since the function $F : \mathbb{R}_+^n \to \mathbb{R}_+$ given by $F(x_1, \ldots, x_n) = x_1 + \ldots + x_n$ satisfies **E1, E2**, but not **E3**.

8. The difference lies in the purpose of the models.
   (a) Explanatory models aim at the logical deduction of events or laws from hypotheses and already known laws.
   (b) Models based on working hypotheses have the purposes of finding laws or (at least) "interdependencies".
9. (a) affine, deterministic, macroeconomic, total, static,
   (b) nonlinear, deterministic, microecnomic (if F is the production function of an enterprise), macroeconomic (if F is the production function of a country), partial, static,
   (c) affine or nonlinear (if F is affine or nonlinear), stochastic (further properties depend on the meaning of F),
   (d) affine, deterministic, macroeconomic, total, dynamic.
10. In many cases the answer is no. The more details are taken into consideration in the assumptions of a model in economics, the more extrusive becomes the basic structure of the model, that is, in many applications, the system of equations describing this structure. The system may be too complicated to be solved. But even if such a system of equations or such a structure still can be mastered logically or numerically, practical applicability may be limited because of the following reasons:
    (i) The inevitable inaccuracy with the determination of data yields propagation of error such that the solutions generally become the more inexact the more extrusive the system of equations is.
    (ii) Propagation of error of a very disappointing kind can also emerge, if one tries to solve initial value problems of certain nonlinear difference equations (which may be not at all complicated or extrusive; see Sect. 12.4).

Index

W. Eichhorn, W. Gleißner, *Mathematics and Methodology for Economics*,
Springer Texts in Business and Economics, DOI 10.1007/978-3-319-23353-6

9783319233529